Lial/Miller/Greenwell

Student's Solution Manual
to accompany
Finite Mathematics and
Calculus with Applications,
Fourth Edition

Prepared with the assistance of

Jane Brandsma
Suffolk County Community College

Brian Hayes
Triton College

Kathleen Pellissier

August Zarcone
College of DuPage

Richard Zylstra

HarperCollins*CollegePublishers*

Student's Solution Manual to accompany *Lial /Miller/Greenwell: FINITE MATHEMATICS AND CALCULUS WITH APPLICATIONS,* **FOURTH EDITION**

Copyright © 1993 by HarperCollins College Publishers

ISBN 0-673-46764-3

93 94 95 9 8 7 6 5 4 3 2

PREFACE

This book provides complete solutions for many of the exercises in <u>Finite Mathematics</u> <u>and</u> <u>Calculus</u> <u>with</u> <u>Applications</u>, fourth edition, by Margaret L. Lial, Charles D. Miller, and Raymond N. Greenwell. Solutions are included for odd–numbered exercises. Solutions are not provided for exercises with open–response answers or for those that require the use of a computer. Sample tests are provided to help you determine if you have mastered the concepts in a given chapter.

This book should be used as an aid as you work to master your course work. Try to solve the exercises that your instructor assigns before you refer to the solutions in this book. Then, if you have difficulty, read these solutions to guide you in solving the exercises. The solutions have been written so that they are consistent with the methods used in the textbook.

You may find that some of the solutions are presented in greater detail than others. Thus, if you cannot find an explanation for a difficulty that you encountered in one exercise, you may find the explanation in the solution for a similar exercise elsewhere in the exercise set.

Solutions that require graphs will refer you to the answer section of the textbook. These graphs are not included in this book.

In addition to solutions, you will find a list of suggestions on how to be successful in mathematics. A careful reading will be helpful for many students.

The following people have made valuable contributions to the production of this <u>Student's</u> <u>Solution</u> <u>Manual</u>: Brian Hayes and Marjorie Seachrist, editors; Judy Martinez and Sheri Minkner, typists; Therese Brown and Charles Sullivan, artists; and Carmen Eldersveld, proofreader.

We also want to thank Tommy Thompson of Seminole Community College for his suggestions for the essay "To the Student: Success in Mathematics" that follows this preface.

TO THE STUDENT: SUCCESS IN MATHEMATICS

The main reason students have difficulty with mathematics is that they don't know how to study it. Studying mathematics *is* different from studying subjects like English or history. The key to success is regular practice.

This should not be surprising. After all, can you learn to play the piano or to ski well without a lot of regular practice? The same thing is true for learning mathematics. Working problems nearly every day is the key to becoming successful. Here is a list of things you can do to help you succeed in studying mathematics.

1. *Attend class regularly.* Pay attention in class to what your instructor says and does, and make careful notes. In particular, note the problems the instructor works on the board and copy the complete solutions. Keep these notes separate from your homework to avoid confusion when you read them over later.

2. Don't hesitate to ask questions in class. It is not a sign of weakness, but of strength. There are always other students with the same question who are too shy to ask.

3. *Read your text carefully.* Many students read only enough to get by, usually only the examples. Reading the complete section will help you to be successful with the homework problems. Most exercises are keyed to specific examples or objectives that will explain the procedures for working them.

4. Before you start on your homework assignment, rework the problems the instructor worked in class. This will reinforce what you have learned. Many students say, "I understand it perfectly when you do it, but I get stuck when I try to work the problem myself."

5. Do your homework assignment only *after* reading the text and reviewing your notes from class. Check your work with the answers in the back of the book. If you get a problem wrong and are unable to see why, mark that problem and ask your instructor about it. Then practice working additional problems of the same type to reinforce what you have learned.

6. Work as neatly as you can. Write your symbols clearly, and make sure the problems are clearly separated from each other. Working neatly will help you to think clearly and also make it easier to review the homework before a test.

7. After you have completed a homework assignment, look over the text again. Try to decide what the main ideas are in the lesson. Often they are clearly highlighted or boxed in the text.

8. Use the chapter review exercises at the end of each chapter as a practice test. Work through the problems under test conditions, without referring to the text or the answers until you are finished. You may want to time yourself to see how long it takes you. When you have finished, check your answers against those in the back of the book and study those problems that you missed. Answers are referenced to the appropriate sections of the text. As an additional review, use the sample test with answers at the end of each chapter of solutions in this <u>Student's</u> <u>Solution</u> <u>Manual</u>.

9. Keep any quizzes and tests that are returned to you and use them when you study for future tests and the final exam. These quizzes and tests indicate what your instructor considers most important. Be sure to correct any problems on these tests that you missed, so you will have the corrected work to study.

10. Don't worry if you do not understand a new topic right away. As you read more about it and work through the problems, you will gain understanding. Each time you look back at a topic you will understand it a little better. No one understands each topic completely right from the start.

CONTENTS

12 APPLICATIONS OF THE DERIVATIVE

13 EXPONENTIAL AND LOGARITHMIC FUNCTIONS

14 INTEGRATION

SOLUTIONS
TO
ODD–NUMBERED EXERCISES

ALGEBRA REFERENCE

Section R.1

1. $(2x^2 - 6x + 11) + (-3x^2 + 7x - 2)$

$= 2x^2 - 6x + 11 - 3x^2 + 7x - 2$

$= (2 - 3)x^2 + (7 - 6)x + (11 - 2)$

$= -x^2 + x + 9$

3. $-3(4q^2 - 3q + 2) + 2(-q^2 + q - 4)$

$= -12q^2 + 9q - 6 - 2q^2 + 2q - 8$

$= -14q^2 + 11q - 14$

5. $(.613x^2 - 4.215x + .892)$

$- .47(2x^2 - 3x + 5)$

$= .613x^2 - 4.215x + .892$

$- .94x^2 + 1.41x - 2.35$

$= -.327x^2 - 2.805x - 1.458$

7. $-9m(2m^2 + 3m - 1)$

$= -9m(2m^2) - 9m(3m) - 9m(-1)$

$= -18m^3 - 27m^2 + 9m$

9. Use the FOIL method to find

$(5r - 3s)(5r + 4s)$

$= (5r)(5r) + (5r)(4s) + (-3s)(5r)$

$+ (-3s)(4s)$

$= 25r^2 + 20rs - 15rs - 12s^2$

$= 25r^2 + 5rs - 12s^2.$

11. $\left(\frac{2}{5}y + \frac{1}{8}z\right)\left(\frac{3}{5}y + \frac{1}{2}z\right)$

$= \left(\frac{2}{5}y\right)\left(\frac{3}{5}y\right) + \left(\frac{2}{5}y\right)\left(\frac{1}{2}z\right) + \left(\frac{1}{8}z\right)\left(\frac{3}{5}y\right)$

$+ \left(\frac{1}{8}z\right)\left(\frac{1}{2}z\right)$

$= \frac{6}{25}y^2 + \frac{1}{5}yz + \frac{3}{40}yz + \frac{1}{16}z^2$

$= \frac{6}{25}y^2 + \left(\frac{8}{40} + \frac{3}{40}\right)yz + \frac{1}{16}z^2$

$= \frac{6}{25}y^2 + \frac{11}{40}yz + \frac{1}{16}z^2$

13. $(12x - 1)(12x + 1)$

$= (12x)(12x) + (12x)(1)$

$- 1(12x) - 1(1)$

$= 144x^2 + 12x - 12x - 1$

$= 144x^2 - 1$

15. $(3p - 1)(9p^2 + 3p + 1)$

$= (3p - 1)(9p^2) + (3p - 1)(3p)$

$+ (3p - 1)(1)$

$= 3p(9p^2) - 1(9p^2) + 3p(3p)$

$- 1(3p) + 3p(1) - 1(1)$

$= 27p^3 - 9p^2 + 9p^2 - 3p + 3p - 1$

$= 27p^3 - 1$

17. $(2m + 1)(4m^2 - 2m + 1)$

$= 2m(4m^2 - 2m + 1) + 1(4m^2 - 2m + 1)$

$= 8m^3 - 4m^2 + 2m + 4m^2 - 2m + 1$

$= 8m^3 + 1$

19. $(m - n + k)(m + 2n - 3k)$

$= m(m + 2n - 3k) - n(m + 2n - 3k)$

$+ k(m + 2n - 3k)$

$= m^2 + 2mn - 3km - mn - 2n^2 + 3kn$

$+ km + 2kn - 3k^2$

$= m^2 + mn - 2km - 2n^2 + 5kn - 3k^2$

Section R.2

1. $8a^3 - 16a^2 + 24a$

$= 8a \cdot a^2 - 8a \cdot 2a + 8a \cdot 3$

$= 8a(a^2 - 2a + 3)$

3. $25p^4 - 20p^3q + 100p^2q^2$

$= 5p^2 \cdot 5p^2 - 5p^2 \cdot 4pq + 5p^2 \cdot 20q^2$

$= 5p^2(5p^2 - 4pq + 20q^2)$

5. $m^2 + 9m + 14 = (m + 2)(m + 7)$

since $2 \cdot 7 = 14$ and $2 + 7 = 9$.

7. $z^2 + 9z + 20 = (z + 4)(z + 5)$

since $4 \cdot 5 = 20$ and $4 + 5 = 9$.

9. $a^2 - 6ab + 5b^2 = (a - b)(a - 5b)$

since $(-b)(-5b) = 5b^2$ and
$-b + (-5b) = -6b$.

11. $y^2 - 4yz - 21z^2$

$= (y + 3z)(y - 7z)$

since $(3z)(-7z) = -21z^2$ and
$3z + (-7z) = -4z$.

13. $3m^3 + 12m^2 + 9m$

$= 3m(m^2 + 4m + 3)$

$= 3m(m + 1)(m + 3)$

15. $3a^2 + 10a + 7$

The possible factors of $3a^2$ are $3a$
and a and the possible factors of 7
are 7 and 1. Try various combinations until one works.

$3a^2 + 10a + 7 = (a + 1)(3a + 7)$

17. $15y^2 + y - 2 = (5y + 2)(3y - 1)$

19. $24a^4 + 10a^3b - 4a^2b^2$

$= 2a^2(12a^2 + 5ab - 2b^2)$

$= 2a^2(4a - b)(3a + 2b)$

21. $x^2 - 64 = x^2 - 8^2$

$= (x + 8)(x - 8)$

23. $121a^2 - 100$

$= (11a)^2 - 10^2$

$= (11a - 10)(11a + 10)$

25. $z^2 + 14zy + 49y^2$

$= z^2 + 2 \cdot 7zy + 7^2y^2$

$= (z + 7y)^2$

27. $9p^2 - 24p + 16$

$= (3p)^2 - 2 \cdot 3p \cdot 4 + 4^2$

$= (3p - 4)^2$

29. $8r^3 - 27s^3$

$= (2r)^3 - (3s)^3$

$= (2r - 3s)(4r^2 + 6rs + 9s^2)$

Section R.3

When writing rational expressions in lowest terms in the following exercises,
first write numerators and denominators
in factored form. Then use the fundamental property of rational expressions.

1. $\dfrac{7z^2}{14z} = \dfrac{7z \cdot z}{2(7)z} = \dfrac{z}{2}$

3. $\dfrac{8k + 16}{9k + 18} = \dfrac{8(k + 2)}{9(k + 2)} = \dfrac{8}{9}$

5. $\dfrac{8x^2 + 16x}{4x^2} = \dfrac{8x(x + 2)}{4x^2}$

$= \dfrac{2(x + 2)}{x}$ or $\dfrac{2x + 4}{x}$

7. $\dfrac{m^2 - 4m + 4}{m^2 + m - 6}$

$= \dfrac{(m - 2)(m - 2)}{(m - 2)(m + 3)}$

$= \dfrac{m - 2}{m + 3}$

9. $\dfrac{x^2 + 3x - 4}{x^2 - 1}$

$= \dfrac{(x - 1)(x + 4)}{(x - 1)(x + 1)}$

$= \dfrac{x + 4}{x + 1}$

11. $\dfrac{8m^2 + 6m - 9}{16m^2 - 9}$

$= \dfrac{(4m - 3)(2m + 3)}{(4m - 3)(4m + 3)}$

$= \dfrac{2m + 3}{4m + 3}$

13. $\dfrac{9k^2}{25} \cdot \dfrac{5}{3k} = \dfrac{3 \cdot 3 \cdot 5k^2}{5 \cdot 5 \cdot 3k} = \dfrac{3k^2}{5k} = \dfrac{3k}{5}$

15. $\dfrac{a + b}{2p} \cdot \dfrac{12}{5(a + b)}$

$= \dfrac{(a + b)12}{2p(5)(a + b)} = \dfrac{12}{10p}$

$= \dfrac{6}{5p}$

17. $\dfrac{2k + 8}{6} \div \dfrac{3k + 12}{2}$

$= \dfrac{2(k + 4)}{6} \cdot \dfrac{2}{3(k + 4)}$

$= \dfrac{2(k + 4)(2)}{2 \cdot 3 \cdot 3(k + 4)}$

$= \dfrac{2}{9}$

19. $\dfrac{4a + 12}{2a - 10} \div \dfrac{a^2 - 9}{a^2 - a - 20}$

$= \dfrac{4(a + 3)}{2(a - 5)} \cdot \dfrac{(a - 5)(a + 4)}{(a - 3)(a + 3)}$

$= \dfrac{2(a + 4)}{a - 3}$

21. $\dfrac{k^2 - k - 6}{k^2 + k - 12} \cdot \dfrac{k^2 + 3k - 4}{k^2 + 2k - 3}$

$= \dfrac{(k - 3)(k + 2)(k - 1)(k + 4)}{(k + 4)(k - 3)(k + 3)(k - 1)}$

$= \dfrac{k + 2}{k + 3}$

23. $\dfrac{2m^2 - 5m - 12}{m^2 - 10m + 24} \div \dfrac{4m^2 - 9}{m^2 - 9m + 18}$

$= \dfrac{2m^2 - 5m - 12}{m^2 - 10m + 24} \cdot \dfrac{m^2 - 9m + 18}{4m^2 - 9}$

$= \dfrac{(2m + 3)(m - 4)(m - 6)(m - 3)}{(m - 6)(m - 4)(2m - 3)(2m + 3)}$

$= \dfrac{m - 3}{2m - 3}$

25. $\dfrac{a + 1}{2} - \dfrac{a - 1}{2}$

$= \dfrac{(a + 1) - (a - 1)}{2}$

$= \dfrac{a + 1 - a + 1}{2}$

$= \dfrac{2}{2}$

$= 1$

27. $\dfrac{2}{y} - \dfrac{1}{4} = \left(\dfrac{4}{4}\right)\dfrac{2}{y} - \left(\dfrac{y}{y}\right)\dfrac{1}{4}$

$= \dfrac{8 - y}{4y}$

29. $\dfrac{1}{m - 1} + \dfrac{2}{m}$

$= \dfrac{m}{m}\left(\dfrac{1}{m - 1}\right) + \dfrac{m - 1}{m - 1}\left(\dfrac{2}{m}\right)$

$= \dfrac{m + 2m - 2}{m(m - 1)}$

$= \dfrac{3m - 2}{m(m - 1)}$

31. $\dfrac{8}{3(a - 1)} + \dfrac{2}{a - 1}$

$= \dfrac{8}{3(a - 1)} + \dfrac{3}{3}\left(\dfrac{2}{a + 1}\right)$

$= \dfrac{8 + 6}{3(a - 1)}$

$= \dfrac{14}{3(a - 1)}$

33. $\dfrac{2}{x^2 - 2x - 3} + \dfrac{5}{x^2 - x - 6}$

$= \dfrac{2}{(x - 3)(x + 1)} + \dfrac{5}{(x - 3)(x + 2)}$

$= \left(\dfrac{x + 2}{x + 2}\right)\dfrac{2}{(x - 3)(x + 1)}$

$\quad + \left(\dfrac{x + 1}{x + 1}\right)\dfrac{5}{(x - 3)(x + 2)}$

$= \dfrac{2x + 4 + 5x + 5}{(x + 2)(x - 3)(x + 1)}$

$= \dfrac{7x + 9}{(x + 2)(x - 3)(x + 1)}$

35. $\dfrac{3k}{2k^2 + 3k - 2} - \dfrac{2k}{2k^2 - 7k + 3}$

$= \dfrac{3k}{(2k - 1)(k + 2)} - \dfrac{2k}{(2k - 1)(k - 3)}$

$= \left(\dfrac{k - 3}{k - 3}\right)\dfrac{3k}{(2k - 1)(k + 2)}$

$\quad - \left(\dfrac{k + 2}{k + 2}\right)\dfrac{2k}{(2k - 1)(k - 3)}$

$= \dfrac{(3k^2 - 9k) - (2k^2 + 4k)}{(2k - 1)(k + 2)(k - 3)}$

$= \dfrac{k^2 - 13k}{(2k - 1)(k + 2)(k - 3)}$

$= \dfrac{k(k - 13)}{(2k - 1)(k + 2)(k - 3)}$

Section R.4

1. $.2m - .5 = .1m + .7$

$\quad 10(.2m - .5) = 10(.1m + .7)$

$\quad\quad 2m - 5 = m + 7$

$\quad\quad\quad m - 5 = 7$

$\quad\quad\quad\quad m = 12$

The solution is 12.

3. $3r + 2 - 5(r + 1) = 6r + 4$

$\quad 3r + 2 - 5r - 5 = 6r + 4$

$\quad\quad\quad -3 - 2r = 6r + 4$

$\quad\quad\quad\quad -3 = 8r + 4$

$\quad\quad\quad\quad -7 = 8r$

$\quad\quad\quad\quad -\dfrac{7}{8} = r$

The solution is $-7/8$.

5. $|3x + 2| = 9$

$3x + 2 = 9 \quad$ or $\quad -(3x + 2) = 9$

$\quad 3x = 7 \quad\quad\quad\quad -3x - 2 = 9$

$\quad\quad\quad\quad\quad\quad\quad\quad\quad -3x = 11$

$\quad x = \dfrac{7}{3} \quad\quad\quad\quad\quad x = -\dfrac{11}{3}$

Substituting each solution in the original equation shows that both $7/3$ and $-11/3$ are solutions.

7. $|2x + 8| = |x - 4|$

$2x + 8 = x - 4$

$\quad\quad x = -12$

or

$-(2x + 8) = x - 4$

$\quad -2x - 8 = x - 4$

$\quad\quad -3x = 4$

$\quad\quad x = -\dfrac{4}{3}$

The solutions are -12 and $-4/3$.

9. $x^2 + 5x + 6 = 0$

$\quad (x + 3)(x + 2) = 0$

$\quad\quad x + 3 = 0$

$\quad x + 3 - 3 = 0 - 3$

$\quad\quad\quad x = -3$

or

$\quad\quad x + 2 = 0$

$\quad x + 2 - 2 = 0 - 2$

$\quad\quad\quad x = -2$

The solutions are -3 and -2.

11.
$$m^2 + 16 = 8m$$
$$m^2 + 16 - 8m = 8m - 8m$$
$$m^2 + 16 - 8m = 0$$
$$m^2 - 8m + 16 = 0$$
$$(m)^2 - 2(4m) + (4)^2 = 0$$
$$(m - 4)^2 = 0$$
$$m - 4 = 0$$
$$m - 4 + 4 = 0 + 4$$
$$m = 4$$

The solution is 4.

13.
$$6x^2 - 5x = 4$$
$$6x^2 - 5x - 4 = 0$$
$$(3x - 4)(2x + 1) = 0$$
$$3x - 4 = 0$$
$$3x - 4 + 4 = 4$$
$$3x = 4$$
$$\frac{1}{3}(3x) = \frac{1}{3}(4)$$
$$x = \frac{4}{3}$$

or

$$2x + 1 = 0$$
$$2x + 1 - 1 = 0 - 1$$
$$2x = -1$$
$$\frac{1}{2}(2x) = \frac{1}{2}(-1)$$
$$x = -\frac{1}{2}$$

The solutions are 4/3 and -1/2.

15.
$$9x^2 - 16 = 0$$
$$(3x)^2 - (4)^2 = 0$$
$$(3x + 4)(3x - 4) = 0$$

$$3x + 4 = 0$$
$$3x + 4 - 4 = 0 - 4$$
$$3x = -4$$
$$\frac{1}{3}(3x) = \frac{1}{3}(-4)$$
$$x = -\frac{4}{3}$$

or

$$3x - 4 = 0$$
$$3x - 4 + 4 = 0 + 4$$
$$3x = 4$$
$$\frac{1}{3}(3x) = \frac{1}{3}(4)$$
$$x = \frac{4}{3}$$

The solutions are -4/3 and 4/3.

17.
$$12y^2 - 48y = 0$$
$$12y(y) - 12y(4) = 0$$
$$12y(y - 4) = 0$$
$$12y = 0 \quad \text{or} \quad y - 4 = 0$$
$$\frac{1}{12}(12y) = \frac{1}{12}(0) \quad y - 4 + 4 = 0 + 4$$
$$y = 0 \quad\quad\quad y = 4$$

The solutions are 0 and 4.

19.
$$2m^2 = m + 4$$
$$2m^2 - m - 4 = 0$$

Use the quadratic formula.

$$x = \frac{-(-1) \pm \sqrt{(-1)^2 - 4(2)(-4)}}{2(2)}$$
$$x = \frac{1 \pm \sqrt{1 + 32}}{4}$$
$$x = \frac{1 \pm \sqrt{33}}{4}$$

The solutions are $(1 + \sqrt{33})/4 \approx 1.686$ and $(1 - \sqrt{33})/4 \approx -1.186$.

21.
$$k^2 - 10k = -20$$
$$k^2 - 10k + 20 = 0$$

$$k = \frac{-(-10) \pm \sqrt{(-10)^2 - 4(1)(20)}}{2(1)}$$

$$k = \frac{10 \pm \sqrt{100 - 80}}{2}$$

$$k = \frac{10 \pm \sqrt{20}}{2}$$

$$k = \frac{10 \pm \sqrt{4}\sqrt{5}}{2}$$

$$k = \frac{10 \pm 2\sqrt{5}}{2}$$

$$k = \frac{2(5 \pm \sqrt{5})}{2}$$

$$k = 5 \pm \sqrt{5}$$

The solutions are $5 + \sqrt{5} \approx 7.236$ and $5 - \sqrt{5} \approx 2.764$.

23.
$$2r^2 - 7r + 5 = 0$$
$$(2r - 5)(r - 1) = 0$$
$$2r - 5 = 0 \quad \text{or} \quad r - 1 = 0$$
$$2r = 5 \qquad\qquad r = 1$$
$$r = \frac{5}{2}$$

The solutions are 5/2 and 1.

25.
$$3k^2 + k = 6$$
$$3k^2 + k - 6 = 0$$

$$k = \frac{-1 \pm \sqrt{1 - 4(3)(-6)}}{2(3)}$$

$$= \frac{-1 \pm \sqrt{73}}{6}$$

The solutions are $(-1 + \sqrt{73})/6 \approx 1.257$ and $(-1 - \sqrt{73})/6 \approx -1.591$.

27.
$$\frac{3x - 2}{7} = \frac{x + 2}{5}$$

$$35\left(\frac{3x - 2}{7}\right) = 35\left(\frac{x + 2}{5}\right)$$

$$5(3x - 2) = 7(x + 2)$$
$$15x - 10 = 7x + 14$$
$$8x = 24$$
$$x = 3$$

29.
$$\frac{4}{x - 3} - \frac{8}{2x + 5} + \frac{3}{x - 3} = 0$$

$$\frac{4}{x - 3} + \frac{3}{x - 3} - \frac{8}{2x + 5} = 0$$

$$\frac{7}{x - 3} - \frac{8}{2x + 5} = 0$$

Multiply both sides by $(x - 3)(2x + 5)$. Note that $x \neq 3$ and $x \neq -5/2$.

$$(x - 3)(2x + 5)\left(\frac{7}{x - 3} - \frac{8}{2x + 5}\right)$$
$$= (x - 3)(2x + 5)(0)$$

$$7(2x + 5) - 8(x - 3) = 0$$
$$14x + 35 - 8x + 24 = 0$$
$$6x + 59 = 0$$
$$6x = -59$$
$$x = -\frac{59}{6}$$

Note: It is especially important to check solutions of equations that involve rational expressions. Here, a check shows that $-59/6$ is a solution.

31.
$$\frac{2}{m} + \frac{m}{m + 3} = \frac{3m}{m^2 + 3m}$$

$$\frac{2}{m} + \frac{m}{m + 3} = \frac{3m}{m(m + 3)}$$

Multiply both sides by $m(m + 3)$. Note that $m \approx 0$, and $m \neq -3$.

$$m(m + 3)\left[\frac{2}{m} + \frac{m}{m + 3}\right] = m(m + 3)\left[\frac{3m}{m^2 + 3m}\right]$$

$$2(m + 3) + m(m) = 3m$$

$$2m + 6 + m^2 = 3m$$

$$m^2 - m + 6 = 0$$

$$m = \frac{-(-1) \pm \sqrt{(-1)^2 - 4(1)(6)}}{2(1)}$$

$$= \frac{1 \pm \sqrt{1 - 24}}{2}$$

$$= \frac{1 \pm \sqrt{-23}}{2}$$

There are no real-number solutions.

33. $\dfrac{1}{x - 2} - \dfrac{3x}{x - 1} = \dfrac{2x + 1}{x^2 - 3x + 2}$

$$\frac{1}{x - 2} - \frac{3x}{x - 1} = \frac{2x + 1}{(x - 2)(x - 1)}$$

Multiply both sides by $(x - 2)(x - 1)$.

Note that $x \neq 2$ and $x \neq 1$.

$$(x - 2)(x - 1)\left[\frac{1}{x - 2} - \frac{3x}{x - 1}\right]$$

$$= (x - 2)(x - 1)$$
$$\cdot \left[\frac{2x + 1}{(x - 2)(x - 1)}\right]$$

$$(x - 2)(x - 1)\left(\frac{1}{x - 2}\right) - (x - 2)(x - 1) \cdot \left(\frac{3x}{x - 1}\right)$$

$$= \frac{(x - 2)(x - 2)(2x + 1)}{(x - 2)(x - 1)}$$

$$(x - 1) - (x - 2)(3x) = (2x + 1)$$

$$x - 1 - 3x^2 + 6x = 2x + 1$$

$$-3x^2 + 7x - 1 = 2x + 1$$

$$-3x^2 + 5x - 2 = 0$$

$$3x^2 - 5x + 2 = 0$$

$$(3x - 2)(x - 1) = 0$$

$$3x - 2 = 0 \quad \text{or} \quad x - 1 = 0$$

$$x = \frac{2}{3} \qquad\qquad x = 1$$

1 is not a solution since $x \neq 1$.

The solution is 2/3.

35. $\dfrac{2b^2 + 5b - 8}{b^2 + 2b} + \dfrac{5}{b + 2} = \dfrac{-3}{b}$

$$\frac{2b^2 + 5b - 8}{b(b + 2)} + \frac{5}{(b + 2)} = \frac{-3}{b}$$

Multiply both sides by $b(b + 2)$.

Note that $b \neq 0$ and $b \neq -2$.

$$b(b + 2)\left(\frac{2b^2 + 5b - 8}{b^2 + 2b}\right) + b(b + 2)\left(\frac{5}{b + 2}\right)$$

$$= b(b + 2)\left(-\frac{3}{b}\right)$$

$$2b^2 + 5b - 8 + 5b = (b + 2)(-3)$$

$$2b^2 + 10b - 8 = -3b - 6$$

$$2b^2 + 13b - 2 = 0$$

$$b = \frac{-(13) \pm \sqrt{(13)^2 - 4(2)(-2)}}{2(2)}$$

$$= \frac{-13 \pm \sqrt{169 + 16}}{4}$$

$$b = \frac{-13 \pm \sqrt{185}}{4}$$

The solutions are $(-13 + \sqrt{185})/4 \approx$ .150 and $(-13 - \sqrt{185})/4 \approx -6.650$.

Section R.5

For Exercises 1–23, see the answer graphs in the back of the textbook.

1. $\qquad -3p - 2 \geq 1$

$$-3p \geq 3$$

$$\left(-\frac{1}{3}\right)(-3p) \leq \left(-\frac{1}{3}\right)(3)$$

$$p \leq -1$$

The solution in interval notation is $(-\infty, -1]$.

3. $m - (4 + 2m) + 3 < 2m + 2$

$m - 4 - 2m + 3 < 2m + 2$

$-m - 1 < 2m + 2$

$-3m - 1 < 2$

$-3m < 3$

$-\frac{1}{3}(-3m) > -\frac{1}{3}(3)$

$m > -1$

The solution is $(-1, \infty)$.

5. $3p - 1 < 6p + 2(p - 1)$

$3p - 1 < 6p + 2p - 2$

$3p - 1 < 8p - 2$

$-5p - 1 < -2$

$-5p < -1$

$-\frac{1}{5}(-5p) > -\frac{1}{5}(-1)$

$p > \frac{1}{5}$

The solution is $(1/5, \infty)$.

7. $-7 < y - 2 < 4$

$-7 + 2 < y - 2 + 2 < 4 + 2$

$-5 < y < 6$

The solution is $(-5, 6)$.

9. $-4 \leq \frac{2k - 1}{3} \leq 2$

$3(-4) \leq 3\left(\frac{2k - 1}{3}\right) \leq 3(2)$

$-12 \leq 2k - 1 \leq 6$

$-12 + 1 \leq 2k - 1 + 1 \leq 6 + 1$

$-11 \leq 2k \leq 7$

$\frac{1}{2}(-11) \leq \frac{1}{2}(2k) \leq \frac{1}{2}(7)$

$-\frac{11}{2} \leq k \leq \frac{7}{2}$

The solution is $[-11/2, 7/2]$.

11. $\frac{3}{5}(2p + 3) \geq \frac{1}{10}(5p + 1)$

$10\left(\frac{3}{5}\right)(2p + 3) \geq 10\left(\frac{1}{10}\right)(5p + 1)$

$6(2p + 3) \geq 5p + 1$

$12p + 18 \geq 5p + 1$

$7p \geq -17$

$p \geq -\frac{17}{7}$

The solution is $[-17/7, \infty)$.

13. $(m + 2)(m - 4) < 0$

Solve $(m + 2)(m - 4) = 0$.

$m = -2$ or $m = 4$

Intervals: $(-\infty, -2)$, $(-2, 4)$, $(4, \infty)$

For $(-\infty, -2)$, choose -3 to test   fo

m.

$(-3 + 2)(-3 - 4) = -1(-7) = 8 \not< 0$

For $(-2, 4)$, choose 0.

$(0 + 2)(0 - 4) = 2(-4) = -8 < 0$

For $(4, \infty)$, choose 5.

$(5 + 2)(5 - 4) = 7(1) = 7 \not< 0$

The solution is $(-2, 4)$.

15. $y^2 - 3y + 2 < 0$

$(y - 2)(y - 1) < 0$

Solve $(y - 2)(y - 1) = 0$.

$y = 2$ or $y = 1$

Intervals: $(-\infty, 1)$, $(1, 2)$, $(2, \infty)$

For $(-\infty, 1)$, choose $y = 0$.

$0^2 - 3(0) + 2 = 2 \not< 0$

For (1, 2), choose y = 3/2.

$$\left(\frac{3}{2}\right)^2 - 3\left(\frac{3}{2}\right) + 2 = \frac{9}{4} - \frac{9}{2} + 2$$

$$= \frac{9 - 18 + 8}{4}$$

$$= -\frac{1}{4} < 0$$

For (2, ∞), choose 3.

$$3^2 - 3(3) + 2 = 2 \nless 0$$

The solution is (1, 2).

17. $q^2 - 7q + 6 \leq 0$

Solve $q^2 - 7q + 6 = 0$.

(q − 1)(q − 6) = 0
q = 1   or   q = 6

These solutions are also solutions of the given inequality because the symb ≤ indicates that the end- points are included.

Intervals (−∞, 1), (1, 6), (6, ∞)

For (−∞, 1), choose 0.

$$0^2 - 7(0) + 6 = 6 \nless 0$$

For (1, 6), choose 2.

$$2^2 - 7(2) + 6 = -4 \leq 0$$

For (6, ∞), choose 7.

$$7^2 - 7(7) + 6 = 6 \nless 0$$

Since 1 and 6 are solutions, the solution is [1, 6].

19. $6m^2 + m > 1$

Solve $6m^2 + m = 1$.

$$6m^2 + m - 1 = 0$$
(2m + 1)(3m − 1) = 0

$$m = -\frac{1}{2} \quad \text{or} \quad m = \frac{1}{3}$$

Intervals: (−∞, −1/2), (−1/2, 1/3), (1/3, ∞)

For (−∞, −1/2), choose −1.

$$6(-1)^2 + (-1) = 5 > 1$$

For (−1/2, 1/3), choose 0.

$$6(0)^2 + 0 = 0 \ngtr 1$$

For (1/3, ∞), choose 1.

$$6(1)^2 + 1 = 7 > 1$$

The solution is
(−∞, −1/2) ∪ (1/3, ∞).

21. $2y^2 + 5y \leq 3$

Solve $2y^2 + 5y = 3$.

$$2y^2 + 5y - 3 = 0$$
(y + 3)(2y − 1) = 0

$$y = -3 \quad \text{or} \quad y = \frac{1}{2}$$

Intervals: (−∞, −3), (−3, 1/2), (1/2, ∞)

For (−∞, −3), choose −4.

$$2(-4)^2 + 5(-4) = 12 \nleq 3$$

For (−3, 1/2), choose 0.

$$2(0)^2 + 5(0) = 0 \leq 3$$

For (1/2, ∞), choose 1.

$$2(1)^2 + 5(1) = 7 \nleq 3$$

The solution is [−3, 1/2].

23. $x^2 \leq 25$

Solve $x^2 = 25$.

x = −5   or   x = 5

Intervals: (−∞, −5), (−5, 5), (5, ∞)

For $(-\infty, -5)$, choose -6.

$$(-6)^2 = 36 \not\le 25$$

For $(-5, 5)$, choose 0.

$$0^2 = 0 \le 25$$

For $(5, \infty)$, choose 6.

$$6^2 = 36 \not\le 25$$

The solution is $[-5, 5]$.

25. $\dfrac{m-3}{m+5} \le 0$

Solve $\dfrac{m-3}{m+5} = 0$.

$$(m+5)\dfrac{m-3}{m+5} = (m+5)(0)$$

$$m - 3 = 0$$

$$m = 3$$

Set the denominator equal to 0 and solve.

$$m + 5 = 0$$

$$m = -5$$

Intervals: $(-\infty, -5)$, $(-5, 3)$, $(3, \infty)$

For $(-\infty, -5)$, choose -6.

$$\dfrac{-6-3}{-6+5} = 9 \not\le 0$$

For $(-5, 3)$, choose 0.

$$\dfrac{0-3}{0+5} = -\dfrac{3}{5} \le 0$$

For $(3, \infty)$, choose 4.

$$\dfrac{4-3}{4+5} = \dfrac{1}{9} \not\le 0$$

Although the $\le$ symbol is used, in‑
cluding -5 in the solution would caus
the denominator to be zero.
The solution is $(-5, 3]$.

27. $\dfrac{k-1}{k+2} > 1$

Solve $\dfrac{k-1}{k+2} = 1$.

$$k - 1 = k + 2$$

$$-1 \ne 2$$

The equation has no solution.
Solve $k + 2 = 0$.

$$k = -2$$

Intervals: $(-\infty, -2)$, $(-2, \infty)$
For $(-\infty, -2)$, choose -3.

$$\dfrac{-3-1}{-3+2} = 4 > 1$$

For $(-2, \infty)$, choose 0.

$$\dfrac{0-1}{0+2} = -\dfrac{1}{2} \not> 1$$

The solution is $(-\infty, -2)$.

29. $\dfrac{2y+3}{y-5} \le 1$

Solve $\dfrac{2y+3}{y-5} = 1$.

$$2y + 3 = y - 5$$

$$y = -8$$

Solve $y - 5 = 0$.

$$y = 5$$

Intervals: $(-\infty, -8)$, $(-8, 5)$, $(5, \infty)$

For $(-\infty, -8)$, choose $y = -10$.

$$\dfrac{2(-10)+3}{-10-5} = \dfrac{17}{15} \not\le 1$$

For $(-8, 5)$, choose $y = 0$.

$$\dfrac{2(0)+3}{0-5} = -\dfrac{3}{5} \le 1$$

For $(5, \infty)$, choose $y = 6$.

$$\frac{2(6) + 3}{6 - 5} = \frac{15}{1} \not< 1$$

The solution is $[-8, 5)$.

31. $\dfrac{7}{k + 2} \geq \dfrac{1}{k + 2}$

Solve $\dfrac{7}{k + 2} = \dfrac{1}{k + 2}$

$$\frac{7}{k + 2} - \frac{1}{k + 2} = 0$$

$$\frac{6}{k + 2} = 0$$

The equation has no solution.

Solve $k + 2 = 0$.

$$k = -2$$

Intervals: $(-\infty, -2)$, $(-2, \infty)$

For $(-\infty, -2)$, choose $k = -3$.

$$\frac{6}{-3 + 2} = -6 \not\geq 0$$

For $(-2, \infty)$, choose $k = 0$.

$$\frac{6}{0 + 2} = 3 \geq 0$$

The solution is $(-2, \infty)$.

33. $\dfrac{3x}{x^2 - 1} < 2$

Solve $\dfrac{3x}{x^2 - 1} = 2$.

$$3x = 2x^2 - 2$$

$$-2x^2 + 3x + 2 = 0$$

$$(2x + 1)(-x + 2) = 0$$

$$x = -\frac{1}{2} \quad \text{or} \quad x = 2$$

Set $x^2 - 1 = 0$.

$$x = 1 \quad \text{or} \quad x = -1$$

Intervals: $(-\infty, -1)$, $(-1, -1/2)$, $(-1/2, 1)$, $(1, 2)$, $(2, \infty)$

For $(-\infty, -1)$, choose $x = -2$.

$$\frac{3(-2)}{(-2)^2 - 1} = -\frac{6}{3} = -2 < 2$$

For $(-1, -1/2)$, choose $x = -3/4$.

$$\frac{3(-3/4)}{(-3/4)^2 - 1} = \frac{-9/4}{9/16 - 1} = \frac{36}{7} \not< 2$$

For $(-1/2, 1)$, choose $x = 0$.

$$\frac{3(0)}{0^2 - 1} = 0 < 2$$

For $(1, 2)$, choose $x = 3/2$.

$$\frac{3(3/2)}{(3/2)^2 - 1} = \frac{9/2}{5/4} = \frac{18}{5} \not< 2$$

For $(2, \infty)$, choose $x = 3$.

$$\frac{3(3)}{3^2 - 1} = \frac{9}{8} < 2$$

The solution is
$(-\infty, -1) \cup (-1/2, 1) \cup (2, \infty)$.

35. $\dfrac{z^2 + z}{z^2 - 1} \geq 3$

Solve $\dfrac{z^2 + z}{z^2 - 1} = 3$.

$$z^2 + z = 3z^2 - 3$$

$$-2z^2 + z + 3 = 0$$

$$(-z - 1)(2z - 3) = 0$$

$$z = -1 \quad \text{or} \quad z = \frac{3}{2}$$

Set $z^2 - 1 = 0$.

$$z^2 = 1$$

$$z = -1 \quad \text{or} \quad z = 1$$

Intervals: $(-\infty, -1)$, $(-1, 1)$, $(1, 3/2)$, $(3/2, \infty)$

For $(-\infty, -1)$, choose $x = -2$.

$$\frac{(-2)^2 + 3}{(-2)^2 - 1} = \frac{7}{3} \not\geq 3$$

For $(-1, 1)$, choose $x = 0$.

$$\frac{0^2 + 3}{0^2 - 1} = -3 \not\geq 3$$

For $(1, 3/2)$, choose $x = 3/2$.

$$\frac{(3/2)^2 + 3}{(3/2)^2 - 1} = \frac{21}{5} \geq 3$$

For $(3/2, \infty)$, choose $x = 2$.

$$\frac{2^2 + 3}{2^2 - 1} = \frac{7}{3} \not\geq 3$$

The solution is $(1, 3/2]$.

Section R.6

1. $8^{-2} = \frac{1}{8^2} = \frac{1}{64}$

3. $6^{-3} = \frac{1}{6^3} = \frac{1}{216}$

5. $(-12)^0 = 1$, by definition.

7. $-2^{-4} = -\frac{1}{2^4} = -\frac{1}{16}$

9. $-(-3)^{-2} = -\frac{1}{(-3)^2} = -\frac{1}{9}$

11. $\left(\frac{5}{8}\right)^2 = \frac{5^2}{8^2} = \frac{25}{64}$

13. $\left(\frac{1}{2}\right)^{-3} = \frac{1^{-3}}{2^{-3}} = \frac{\frac{1}{1^3}}{\frac{1}{2^3}} = \frac{1}{\frac{1}{8}}$

$$= 1 \div \frac{1}{8} = 1 \cdot \frac{8}{1} = 8$$

15. $\left(\frac{2}{7}\right)^{-2} = \frac{2^{-2}}{7^{-2}} = \frac{\frac{1}{2^2}}{\frac{1}{7^2}} = \frac{\frac{1}{4}}{\frac{1}{49}}$

$$= \frac{1}{4} \div \frac{1}{49} = \frac{1}{4} \cdot \frac{49}{1} = \frac{49}{4}$$

17. $\frac{7^5}{7^9} = 7^{5-9} = 7^{-4} = \frac{1}{7^4}$

19. $\frac{2^{-5}}{2^{-2}} = 2^{-5-(-2)} = 2^{-5+2} = 2^{-3} = \frac{1}{2^3}$

21. $4^{-3} \cdot 4^6 = 4^{-3+6} = 4^3$

23. $\frac{10^8 \cdot 10^{-10}}{10^4 \cdot 10^2}$

$$= \frac{10^{8+(-10)}}{10^{4+2}} = \frac{10^{-2}}{10^6}$$

$$= 10^{-2-6} = 10^{-8}$$

$$= \frac{1}{10^8}$$

25. $\frac{x^4 \cdot x^3}{x^5} = \frac{x^{4+3}}{x^5} = \frac{x^7}{x^5} = x^{7-5} = x^2$

27. $\frac{(4k^{-1})^2}{2k^{-5}} = \frac{4^2 k^{-2}}{2k^{-5}} = \frac{16k^{-2-(-5)}}{2}$

$$= 8k^{-2+5} = 8k^3$$

$$= 2^3 k^3$$

29. $\frac{2^{-1}x^3 y^{-3}}{xy^{-2}} = 2^{-1}x^{3-1}y^{-3-(-2)}$

$$= 2^{-1}x^2 y^{-3+2} = 2^{-1}x^2 y^{-1}$$

$$= \frac{1}{2}x^2 \cdot \frac{1}{y} = \frac{x^2}{2y}$$

31. $\left(\frac{a^{-1}}{b^2}\right)^{-3} = \frac{(a^{-1})^{-3}}{(b^2)^{-3}} = \frac{a^{(-1)(-3)}}{b^{2(-3)}}$

$$= \frac{a^3}{b^{-6}} = \frac{a^3}{\frac{1}{b^6}} = a^3 \cdot \frac{b^6}{1}$$

$$= a^3 b^6$$

For Exercises 33–37, a = 2, b = –3.

33. $a^{-1} + b^{-1} = 2^{-1} + (-3)^{-1}$

$$= \frac{1}{2} + \frac{1}{-3}$$

$$= \frac{1}{2}\left(\frac{3}{3}\right) + \frac{1}{-3}\left(\frac{2}{2}\right)$$

$$= \frac{3}{6} + \left(-\frac{2}{6}\right)$$

$$= \frac{1}{6}$$

35. $\dfrac{2b^{-1} - 3a^{-1}}{a + b^2} = \dfrac{2(-3)^{-1} - 3(2)^{-1}}{2 + (-3)^2}$

$$= \dfrac{\dfrac{2}{-3} - 3\left(\dfrac{1}{2}\right)}{2 + (-3)^2}$$

$$= \dfrac{-\dfrac{2}{3}\left(\dfrac{2}{2}\right) - \dfrac{3}{2}\left(\dfrac{3}{3}\right)}{2 + 9}$$

$$= \dfrac{-\dfrac{4}{6} - \dfrac{9}{6}}{11} = \dfrac{-\dfrac{13}{6}}{\dfrac{11}{1}}$$

$$= -\frac{13}{6} \cdot \frac{1}{11} = -\frac{13}{66}$$

37. $\left(\dfrac{a}{3}\right)^{-1} + \left(\dfrac{b}{2}\right)^{-2} = \left(\dfrac{2}{3}\right)^{-1} + \left(-\dfrac{3}{2}\right)^{-2}$

$$= \frac{2^{-1}}{3^{-1}} + \frac{(-3)^{-2}}{2^{-2}}$$

$$= \dfrac{\dfrac{1}{2}}{\dfrac{1}{3}} + \dfrac{\dfrac{1}{(-3)^2}}{\dfrac{1}{2^2}}$$

$$= \frac{1}{2} \div \frac{1}{3} + \frac{1}{9} \div \frac{1}{4}$$

$$= \frac{1}{2} \cdot \frac{3}{1} + \frac{1}{9} \cdot \frac{4}{1}$$

$$= \frac{3}{2} + \frac{4}{9}$$

$$= \frac{9}{9} \cdot \frac{3}{2} + \frac{2}{2} \cdot \frac{4}{9}$$

$$= \frac{27}{18} + \frac{8}{18} = \frac{35}{18}$$

39. $81^{1/2} = (9^2)^{1/2} = 9^{2(1/2)} = 9^1 = 9$

41. $8^{2/3} = (8^{1/3})^2 = (2)^2 = 4$

We could also write $8^{2/3} = (8^2)^{1/3}$, but this procedure involves taking a cube root of a larger number, 64. Whenever possible, take the root first.

43. $32^{2/5} = (32^{1/5})^2 = 2^2 = 4$

45. $\left(\dfrac{4}{9}\right)^{1/2} = \dfrac{4^{1/2}}{9^{1/2}} = \dfrac{2}{3}$

47. $16^{-5/4} = (16^{1/4})^{-5} = 2^{-5}$

$$= \frac{1}{2^5} \quad \text{or} \quad \frac{1}{32}$$

49. $\left(\dfrac{27}{64}\right)^{-1/3} = \dfrac{27^{-1/3}}{64^{-1/3}} = \dfrac{\dfrac{1}{27^{1/3}}}{\dfrac{1}{64^{1/3}}} = \dfrac{\dfrac{1}{3}}{\dfrac{1}{4}}$

$$= \frac{4}{3}$$

51. $2^{1/2} \cdot 2^{3/2} = 2^{1/2 + 3/2} = 2^{4/2}$

$$= 2^2 \quad \text{or} \quad 4$$

53. $\dfrac{4^{2/3} \cdot 4^{5/3}}{4^{1/3}} = \dfrac{4^{2/3 + 5/3}}{4^{1/3}}$

$$= 4^{7/3 - 1/3} = 4^{6/3}$$

$$= 4^2 \quad \text{or} \quad 16$$

55. $\dfrac{7^{-1/3} \cdot 7r^{-3}}{7^{2/3} \cdot (r^{-2})^2}$

$$= \frac{7^{-1/3 + 1}r^{-3}}{7^{2/3} \cdot r^{-4}}$$

$$= 7^{-1/3 + 3/3 - 2/3}r^{-3 - (-4)}$$

$$= 7^0 r^{-3 + 4} = 1 \cdot r^1 = r$$

57. $\dfrac{6k^{-4} \cdot (3k^{-1})^{-2}}{2^3 \cdot k^{1/2}}$

$= \dfrac{2 \cdot 3k^{-4}(3^{-2})(k^2)}{2^3 k^{1/2}}$

$= 2^{1-3}3^{1+(-2)}k^{-4+2-1/2}$

$= 2^{-2}3^{-1}k^{-5/2}$

$= \dfrac{1}{2^2} \cdot \dfrac{1}{3} \cdot \dfrac{1}{k^{5/2}} = \dfrac{1}{2^2 3k^{5/2}}$

$= \dfrac{1}{12k^{5/2}}$

59. $\dfrac{a^{4/3}}{a^{2/3}} \cdot \dfrac{b^{1/2}}{b^{-3/2}} = a^{4/3-2/3}b^{1/2-(-3/2)}$

$= a^{2/3}b^2$

61. $\dfrac{k^{-3/5} \cdot h^{-1/3} \cdot t^{2/5}}{k^{-1/5} \cdot h^{-2/3} \cdot t^{1/5}}$

$= k^{-3/5-(-1/5)}h^{-1/3-(-2/3)}t^{2/5-1/5}$

$= k^{-3/5+1/5}h^{-1/3+2/3}t^{2/5-1/5}$

$= k^{-2/5}h^{1/3}t^{1/5}$

$= \dfrac{h^{1/3}t^{1/5}}{k^{2/5}}$

63. $(x^2 + 2)(x^2 - 1)^{-1/2}(x) + (x^2 - 1)^{1/2}(2x)$

$= (x^2 + 2)(x^2 - 1)^{-1/2}(x)$

$\quad + (x^2 - 1)^1(x^2 - 1)^{-1/2}(2x)$

$= x(x^2 - 1)^{-1/2}[(x^2 + 2) + (x^2 - 1)(2)]$

$= x(x^2 - 1)^{-1/2}(x^2 + 2 + 2x^2 - 2)$

$= x(x^2 - 1)^{-1/2}(3x^2)$

$= 3x^3(x^2 - 1)^{-1/2}$

65. $(2x + 5)^2\left(\dfrac{1}{2}\right)(x^2 - 4)^{-1/2}(2x)$

$\quad + (x^2 - 4)^{1/2}(2)(2x + 5)$

$= (2x + 5)^2(x^2 - 4)^{-1/2}(x)$

$\quad + (x^2 - 4)^{1/2}(2)(2x + 5)$

$= (2x + 5)^2(x^2 - 4)^{-1/2}(x)$

$\quad + (x^2 - 4)^1(x^2 - 4)^{-1/2}(2)(2x + 5)$

$= (2x + 5)(x^2 - 4)^{-1/2}$

$\quad \cdot [(2x + 5)(x) + (x^2 - 4)(2)]$

$= (2x + 5)(x^2 - 4)^{-1/2}$

$\quad \cdot (2x^2 + 5x + 2x^2 - 8)$

$= (2x + 5)(x^2 - 4)^{-1/2}(4x^2 + 5x - 8)$

Section R.7

1. $\sqrt[3]{125} = 5$ because $5^3 = 125$.

3. $\sqrt[5]{-3125} = -5$ because $(-5)^5 = -3125$.

5. $\sqrt{2000} = \sqrt{4 \cdot 100 \cdot 5} = 2 \cdot 10\sqrt{5}$

$= 20\sqrt{5}$

7. $7\sqrt{2} - 8\sqrt{18} + 4\sqrt{72}$

$= 7\sqrt{2} - 8\sqrt{9 \cdot 2} + 4\sqrt{36 \cdot 2}$

$= 7\sqrt{2} - 8(3)\sqrt{2} + 4(6)\sqrt{2}$

$= 7\sqrt{2} - 24\sqrt{2} + 24\sqrt{2}$

$= 7\sqrt{2}$

9. $2\sqrt{5} - 3\sqrt{20} + 2\sqrt{45}$

$= 2\sqrt{5} - 3\sqrt{4 \cdot 5} + 2\sqrt{9 \cdot 5}$

$= 2\sqrt{5} - 3(2)\sqrt{5} + 2(3)\sqrt{5}$

$= 2\sqrt{5} - 6\sqrt{5} + 6\sqrt{5}$

$= 2\sqrt{5}$

11. $\sqrt[3]{2} - \sqrt[3]{16} + 2\sqrt[3]{54}$

$= \sqrt[3]{2} - (\sqrt[3]{8 \cdot 2}) + 2(\sqrt[3]{27 \cdot 2})$

$= \sqrt[3]{2} - \sqrt[3]{8}\sqrt[3]{2} + 2(\sqrt[3]{27}\sqrt[3]{2})$

$= \sqrt[3]{2} - 2\sqrt[3]{2} + 2(3\sqrt[3]{2})$

$= \sqrt[3]{2} - 2\sqrt[3]{2} + 6\sqrt[3]{2}$

$= 5\sqrt[3]{2}$

13. $\sqrt[3]{32} - 5\sqrt[3]{4} + 2\sqrt[3]{108}$

$\quad = \sqrt[3]{8 \cdot 4} - 5\sqrt[3]{4} + 2\sqrt[3]{27 \cdot 4}$

$\quad = \sqrt[3]{8}\sqrt[3]{4} - 5\sqrt[3]{4} + 2\sqrt[3]{27}\sqrt[3]{4}$

$\quad = 2\sqrt[3]{4} - 5\sqrt[3]{4} + 2(3\sqrt[3]{4})$

$\quad = 2\sqrt[3]{4} - 5\sqrt[3]{4} + 6\sqrt[3]{4}$

$\quad = 3\sqrt[3]{4}$

15. $\sqrt{98r^3s^4t^{10}}$

$\quad = \sqrt{(49 \cdot 2)(r^2 \cdot r)(s^4)(t^{10})}$

$\quad = \sqrt{(49r^2s^4t^{10})(2r)}$

$\quad = \sqrt{49r^2s^4t^{10}}\sqrt{2r}$

$\quad = 7rs^2t^5\sqrt{2r}$

17. $\sqrt[4]{x^8y^7z^{11}} = \sqrt[4]{(x^8)(y^4 \cdot y^3)(z^8z^3)}$

$\quad = \sqrt[4]{(x^8y^4z^8)(y^3z^3)}$

$\quad = \sqrt[4]{x^8y^4z^8}\,\sqrt[4]{y^3z^3}$

$\quad = x^2yz^2\,\sqrt[4]{y^3z^3}$

19. $\sqrt{p^7q^3} - \sqrt{p^5q^9} + \sqrt{p^9q}$

$\quad = \sqrt{(p^6p)(q^2q)} - \sqrt{(p^4p)(q^8q)}$
$\qquad + \sqrt{(p^8p)q}$

$\quad = \sqrt{(p^6q^2)(pq)} - \sqrt{(p^4q^8)(pq)}$
$\qquad + \sqrt{(p^8)(pq)}$

$\quad = \sqrt{p^6q^2}\sqrt{pq} - \sqrt{p^4q^8}\sqrt{pq} + \sqrt{p^8}\sqrt{pq}$

$\quad = p^3q\sqrt{pq} - p^2q^4\sqrt{pq} + p^4\sqrt{pq}$

$\quad = p^2pq\sqrt{pq} - p^2q^4\sqrt{pq} + p^2p^2\sqrt{pq}$

$\quad = p^2\sqrt{pq}(pq - q^4 + p^2)$

21. $\dfrac{-2}{\sqrt{3}} = \dfrac{-2}{\sqrt{3}} \cdot \dfrac{\sqrt{3}}{\sqrt{3}} = \dfrac{-2\sqrt{3}}{\sqrt{9}} = -\dfrac{2\sqrt{3}}{3}$

23. $\dfrac{4}{\sqrt{8}} = \dfrac{4}{\sqrt{8}} \cdot \dfrac{\sqrt{8}}{\sqrt{8}} = \dfrac{4\sqrt{8}}{\sqrt{64}} = \dfrac{4\sqrt{8}}{8} = \dfrac{4(\sqrt{4}\sqrt{2})}{8}$

$\quad = \dfrac{4(2\sqrt{2})}{8} = \dfrac{8\sqrt{2}}{8} = \sqrt{2}$

25. $\dfrac{5}{2 - \sqrt{6}} = \dfrac{5}{2 - \sqrt{6}} \cdot \dfrac{2 + \sqrt{6}}{2 + \sqrt{6}}$

$\quad = \dfrac{5(2 + \sqrt{6})}{4 + 2\sqrt{6} - 2\sqrt{6} - \sqrt{36}}$

$\quad = \dfrac{5(2 + \sqrt{6})}{4 - \sqrt{36}} = \dfrac{5(2 + \sqrt{6})}{4 - 6}$

$\quad = \dfrac{5(2 + \sqrt{6})}{-2}$

$\quad = -\dfrac{5(2 + \sqrt{6})}{2}$

27. $\dfrac{1}{\sqrt{10} + \sqrt{3}} = \dfrac{1}{\sqrt{10} + \sqrt{3}} \cdot \dfrac{\sqrt{10} - \sqrt{3}}{\sqrt{10} - \sqrt{3}}$

$\quad = \dfrac{\sqrt{10} - \sqrt{3}}{\sqrt{100} - \sqrt{30} + \sqrt{30} - \sqrt{9}}$

$\quad = \dfrac{\sqrt{10} - \sqrt{3}}{\sqrt{100} - \sqrt{9}} = \dfrac{\sqrt{10} - \sqrt{3}}{10 - 3}$

$\quad = \dfrac{\sqrt{10} - \sqrt{3}}{7}$

29. $\dfrac{5}{\sqrt{m} - \sqrt{5}} = \dfrac{5}{\sqrt{m} - \sqrt{5}} \cdot \dfrac{\sqrt{m} + \sqrt{5}}{\sqrt{m} + \sqrt{5}}$

$\quad = \dfrac{5(\sqrt{m} + \sqrt{5})}{\sqrt{m^2} + \sqrt{5m} - \sqrt{5m} - \sqrt{25}}$

$\quad = \dfrac{5(\sqrt{m} + \sqrt{5})}{\sqrt{m^2} - \sqrt{25}} = \dfrac{5(\sqrt{m} + \sqrt{5})}{m - 5}$

31. $\dfrac{z - 11}{\sqrt{z} - \sqrt{11}} = \dfrac{z - 11}{\sqrt{z} - \sqrt{11}} \cdot \dfrac{\sqrt{z} + \sqrt{11}}{\sqrt{z} + \sqrt{11}}$

$\quad = \dfrac{(z - 11)(\sqrt{z} + \sqrt{11})}{\sqrt{z^2} + \sqrt{11z} - \sqrt{11z} - \sqrt{121}}$

$\quad = \dfrac{(z - 11)(\sqrt{z} + \sqrt{11})}{\sqrt{z^2} - \sqrt{121}}$

$\quad = \dfrac{(z - 11)(\sqrt{z} + \sqrt{11})}{(z - 11)}$

$\quad = \sqrt{z} + \sqrt{11}$

33. $\dfrac{\sqrt{p} + \sqrt{p^2 - 1}}{\sqrt{p} - \sqrt{p^2 - 1}}$

$= \dfrac{\sqrt{p} + \sqrt{p^2 - 1}}{\sqrt{p} - \sqrt{p^2 - 1}} \cdot \dfrac{\sqrt{p} + \sqrt{p^2 - 1}}{\sqrt{p} + \sqrt{p^2 - 1}}$

$= \dfrac{\sqrt{p^2} + \sqrt{p}\sqrt{p^2 - 1} + \sqrt{p}\sqrt{p^2 - 1} + (\sqrt{p^2 - 1})^2}{\sqrt{p^2} + \sqrt{p}\sqrt{p^2 - 1} - \sqrt{p}\sqrt{p^2 - 1} - (\sqrt{p^2 - 1})^2}$

$= \dfrac{p + 2\sqrt{p}\sqrt{p^2 - 1} + (p^2 - 1)}{p - (p^2 - 1)}$

$= \dfrac{p^2 + p + 2\sqrt{p(p^2 - 1)} - 1}{-p^2 + p + 1}$

35. $\dfrac{1 - \sqrt{3}}{3} = \dfrac{1 - \sqrt{3}}{3} \cdot \dfrac{1 + \sqrt{3}}{1 + \sqrt{3}}$

$= \dfrac{1 + \sqrt{3} - \sqrt{3} - \sqrt{9}}{3(1 + \sqrt{3})}$

$= \dfrac{1 - \sqrt{9}}{3(1 + \sqrt{3})} = \dfrac{1 - 3}{3(1 + \sqrt{3})}$

$= -\dfrac{2}{3(1 + \sqrt{3})}$

37. $\dfrac{\sqrt{p} + \sqrt{p^2 - 1}}{\sqrt{p} - \sqrt{p^2 - 1}}$

$= \dfrac{\sqrt{p} + \sqrt{p^2 - 1}}{\sqrt{p} - \sqrt{p^2 - 1}} \cdot \dfrac{\sqrt{p} - \sqrt{p^2 - 1}}{\sqrt{p} - \sqrt{p^2 - 1}}$

$= \dfrac{\sqrt{p^2} - \sqrt{p}\sqrt{p^2 - 1} + \sqrt{p}\sqrt{p^2 - 1} - (\sqrt{p^2 - 1})^2}{\sqrt{p^2} - \sqrt{p}\sqrt{p^2 - 1} - \sqrt{p}\sqrt{p^2 - 1} + (\sqrt{p^2 - 1})^2}$

$= \dfrac{p - (p^2 - 1)}{p - 2\sqrt{p(p^2 - 1)} + p^2 - 1}$

$= \dfrac{p - p^2 + 1}{p - 2\sqrt{p(p^2 - 1)} + p^2 - 1}$

$= \dfrac{-p^2 + p + 1}{p^2 + p - 2\sqrt{p(p^2 - 1)} - 1}$

39. $\sqrt{4y^2 + 4y + 1}$

$= \sqrt{(2y + 1)(2y + 1)}$

$= \sqrt{(2y + 1)^2}$

$= |2y + 1|$

Since $\sqrt{}$ denotes the nonnegative root, we must have $2y + 1 \geq 0$.

41. $\sqrt{9k^2 + h^2}$

The expression $9k^2 + h^2$ is the sum o two squares and cannot be fac– tored Therefore, $\sqrt{9k^2 + h^2}$ cannot be simplified.

FUNCTIONS AND GRAPHS

Section 1.1

1. The x-value of 82 corresponds to two y-values, 93 and 14. In a function, each x must correspond to exactly one y.
   The rule is not a function.

3. Each x-value corresponds to exactly one y-value.
   The rule is a function.

5. $y = x^3$
   Each x-value corresponds to exactly one y-value.
   The rule is a function.

7. $x = |y|$
   Each value of x (except 0) corresponds to two y-values.
   The rule is not a function.

In Exercises 9-23, the domain is $\{-2, -1, 0, 1, 2, 3\}$. See the answer graphs in the back of your textbook.

9. $y = x - 1$

x	-2	-1	0	1	2	3
y	-3	-2	-1	0	1	2

   Pairs: (-2, -3), (-1, -2), (0, -1), (1, 0), (2, 1), (3, 2)
   Range: $\{-3, -2, -1, 0, 1, 2\}$

11. $y = -4x + 9$

x	-2	-1	0	1	2	3
y	17	13	9	5	1	-3

    Pairs: (-2, 17), (-1, 13), (0, 9), (1, 5), (2, 1), (3, -3)
    Range: $\{-3, 1, 5, 9, 13, 17\}$

13. $2x + y = 9$
    $y = -2x + 9$

x	-2	-1	0	1	2	3
y	13	11	9	7	5	3

    Pairs: (-2, 13), (-1, 11), (0, 9), (1, 7), (2, 5), (3, 3)
    Range: $\{3, 5, 7, 9, 11, 13\}$

15. $2y - x = 5$
    $2y = 5 + x$
    $y = \frac{1}{2}x + \frac{5}{2}$

x	-2	-1	0	1	2	3
y	3/2	2	5/2	3	7/2	4

    Pairs: (-2, 3/2), (-1, 2), (0, 5/2), (1, 3), (2, 7/2), (3, 4)
    Range: $\{3/2, 2, 5/2, 3, 7/2, 4\}$

17. $y = x(x + 1)$
    $y = x^2 + x$

x	-2	-1	0	1	2	3
y	2	0	0	2	6	12

    Pairs: (-2, 2), (-1, 0), (0, 0), (1, 2), (2, 6), (3, 12)
    Range: $\{0, 2, 6, 12\}$

19. $y = x^2$

x	-2	-1	0	1	2	3
y	4	1	0	1	4	9

Pairs: (-2, 4), (-1, 1), (0, 0),
      (1, 1), (2, 4), (3, 9)

Range: $\{0, 1, 4, 9\}$

21. $y = \dfrac{1}{x + 3}$

x	-2	-1	0	1	2	3
y	1	1/2	1/3	1/4	1/5	1/6

Pairs: (-2, 1), (-1, 1/2), (0, 1/3),
      (1, 1/4), (2, 1/5), (3, 1/6)

Range: $\{1, 1/2, 1/3, 1/4, 1/5, 1/6\}$

23. $y = \dfrac{3x - 3}{x + 5}$

x	-2	-1	0	1	2	3
y	-3	-3/2	-3/5	0	3/7	6/8

Pairs: (-2, -3), (-1, -3/2),
      (0, -3/5), (1, 0),
      (2, 3/7), (3, 3/4)

Range: $\{-3, -3/2, -3/5, 0, 3/7, 3/4\}$

For Exercises 25–29, see the answer
graphs in the back of the textbook.

25. $x < 0$

This expression represents all numbers less than 0, or (-∞, 0), The
number 0 is not included in the
interval.

27. $1 \le x < 2$

This expression represents all numbers greater than or equal to 1 and
less than 2, or [1, 2). The number
1 is included in the interval, but
the number 2 is not included.

29. $-9 > x$

This expression represents all numbers less than -9, or (-∞, -9). The
number -9 is not included in the
interval.

31. $(-4, 3)$

This interval represents all numbers greater than -4 and less than
3. Neither -4 nor 3 is included.
As an inequality, this is represented as

$$-4 < x < 3.$$

33. $(-\infty, -1]$

This interval represents all numbers
less than or equal to -1. -1 is
included in the interval. As an inequality, this is represented as

$$x \le -1.$$

35. This interval represents all numbers
greater than or equal to -2 and less
than 6. -2 is included but 6 is
not. As an inequality, this is represented by

$$-2 \le x < 6.$$

37. This interval represents all numbers less than or equal to -4 or greater than or equal to 4. As an inequality, this is represented as

$$x \le -4 \quad \text{or} \quad x \ge 4.$$

39. $f(x) = 2x$

x can take on any value, so the domain is the set of real numbers, $(-\infty, \infty)$.

41. $f(x) = x^4$

x can take on any value, so the domain is the set of real numbers, $(-\infty, \infty)$.

43. $f(x) = \sqrt{16 - x^2}$

For $f(x)$ to be a real number, $16 - x^2 \ge 0$.
Solve $16 - x^2 = 0$.

$$(4 - x)(4 + x) = 0$$
$$x = 4 \quad \text{or} \quad x = -4$$

The numbers form the intervals $(-\infty, -4)$, $(-4, 4)$ and $(4, \infty)$. Only values in the interval $(-4, 4)$ satisfy the inequality. The domain is $[-4, 4]$.

45. $f(x) = (x - 3)^{1/2} = \sqrt{x - 3}$

For $f(x)$ to be a real number,

$$x - 3 \ge 0$$
$$x \ge 3.$$

The domain is $[3, \infty)$.

47. $f(x) = \dfrac{2}{x^2 - 4}$

$$= \dfrac{2}{(x - 2)(x + 2)}$$

Since division by zero is not defined, $(x - 2)(x + 2) \ne 0$.
When $(x - 2)(x + 2) = 0$,

$$x - 2 = 0 \quad \text{or} \quad x + 2 = 0$$
$$x = 2 \quad \text{or} \quad x = -2$$

Thus, x can be any real number except ± 2.
The domain is

$$(-\infty, -2) \cup (-2, 2) \cup (2, \infty).$$

49. $f(x) = -\sqrt{\dfrac{2}{x^2 + 9}}$

x can take on any value. No choice for x will cause the denominator to be zero. Also, no choice for x will produce a negative number under the radical. The domain is $(-\infty, \infty)$.

51. $f(x) = \sqrt{x^2 - 4x - 5}$
$\qquad = \sqrt{(x - 5)(x + 1)}$

See the method used in Exercise 43.

$$(x - 5)(x + 1) \ge 0$$

when $x \ge 5$ and when $x \le -1$.
The domain is $(-\infty, -1] \cup [5, \infty)$.

53. $f(x) = \dfrac{1}{\sqrt{x^2 - 6x + 8}}$

$$= \dfrac{1}{\sqrt{(x - 4)(x - 2)}}$$

$(x - 4)(x - 2) > 0$, since the radicand cannot be negative and the denominator of the function cannot be zero.

Solve $(x - 4)(x - 2) = 0$

$x - 4 = 0$ or $x - 2 = 0$

$x = 4$ or $x = 2$

Use the values 2 and 4 to divide the number line into 3 intervals, $(-\infty, 2)$, $(2, 4)$ and $(4, \infty)$. Only the values in the intervals $(-\infty, -2)$ and $(4, \infty)$ satisfy the inequality.

The domain is $(-\infty, -2) \cup (4, \infty)$.

55. By reading the graph, the domain is all numbers greater than or equal to −5 and less than or equal to 4. The range is all numbers greater than or equal to −2 and less than or equal to 6.

Domain: $[-5, 4]$; range: $[-2, 6]$

57. By reading the graph, x can take on any value, but y is less than or equal to 12.

Domain: $(-\infty, \infty)$; range: $(-\infty, 12]$

59. $f(x) = 3x + 2$

(a) $f(4) = 3(4) + 2 = 12 + 2 = 14$

(b) $f(-3) = 3(-3) + 2 = -9 + 2$
$= -7$

(c) $f\left(-\frac{1}{2}\right) = 3\left(-\frac{1}{2}\right) + 2$

$= -\frac{3}{2} + \frac{4}{2}$

$= \frac{1}{2}$

(d) $f(a) = 3(a) + 2 = 3a + 2$

(e) $f\left(\frac{2}{m}\right) = 3\left(\frac{2}{m}\right) + 2$

$= \frac{6}{m} + 2$

61. $f(x) = -x^2 + 5x + 1$

(a) $f(4) = -(4)^2 + 5(4) + 1$
$= -16 + 20 + 1$
$= 5$

(b) $f(-3) = -(-3)^2 + 5(-3) + 1$
$= -9 - 15 + 1$
$= -23$

(c) $f\left(-\frac{1}{2}\right) = -\left(-\frac{1}{2}\right)^2 + 5\left(-\frac{1}{2}\right) + 1$

$= -\frac{1}{4} - \frac{5}{2} + 1$

$= -\frac{7}{4}$

(d) $f(a) = -(a)^2 + 5(a) + 1$
$= -a^2 + 5a + 1$

(e) $f\left(\frac{2}{m}\right) = -\left(\frac{2}{m}\right)^2 + 5\left(\frac{2}{m}\right) + 1$

$= -\frac{4}{m^2} + \frac{10}{m} + 1$

or $\dfrac{-4 + 10m + m^2}{m^2}$

63. $f(x) = \dfrac{2x + 1}{x - 2}$

(a) $f(4) = \dfrac{2(4) + 1}{4 - 2} = \dfrac{9}{2}$

(b) $f(-3) = \dfrac{2(-3) + 1}{-3 - 2} = \dfrac{-6 + 1}{-5}$

$= \dfrac{-5}{-5} = 1$

(c) $f\left(-\frac{1}{2}\right) = \dfrac{2(-1/2) + 1}{-1/2 - 2}$

$= \dfrac{-1 + 1}{5/2}$

$= \dfrac{0}{5/2} = 0$

(d) $f(a) = \dfrac{2(a) + 1}{(a) - 2} = \dfrac{2a + 1}{a - 1}$

(e) $f\left(\dfrac{2}{m}\right) = \dfrac{2\left(\dfrac{2}{m}\right) + 1}{\dfrac{2}{m} - 2}$

$= \dfrac{\dfrac{4}{m} + \dfrac{m}{m}}{\dfrac{2}{m} - \dfrac{2m}{m}}$

$= \dfrac{\dfrac{4 + m}{m}}{\dfrac{2 - 2m}{m}}$

$= \dfrac{4 + m}{m} \cdot \dfrac{m}{2 - 2m}$

$= \dfrac{4 + m}{2 - 2m}$

In Exercises 65–67, count squares on the graph paper. On the horizontal axis, note that two squares correspond to one unit.

65. **(a)** $f(-2) = 0$

   (b) $f(0) = 4$

   (c) $f\left(\dfrac{1}{2}\right) = 3$

   (d) $f(4) = 4$

67. **(a)** $f(-2) = -3$

   (b) $f(0) = -2$

   (c) $f\left(\dfrac{1}{2}\right) = -1$

   (d) $f(4) = 2$

In Exercises 69–73, $f(x) = 6x - 2$ and $g(x) = x^2 - 2x + 5$.

69. $f(m - 3) = 6(m - 3) - 2$

$= 6m - 18 - 2$

$= 6m - 20$

71. $g(r + h)$

$= (r + h)^2 - 2(r + h) + 5$

$= r^2 + 2hr + h^2 - 2r - 2h + 5$

73. $g\left(\dfrac{3}{q}\right) = \left(\dfrac{3}{q}\right)^2 - 2\left(\dfrac{3}{q}\right) + 5$

$= \dfrac{9}{q^2} - \dfrac{6}{q} + 5$

or $\dfrac{9 - 6q + 5q^2}{q^2}$

75. A vertical line drawn anywhere through the graph will intersect the graph in only one place. The graph represents a function.

77. A vertical line drawn through the graph may intersect the graph in two places. The graph does not represent a function.

79. A vertical line drawn anywhere through the graph will intersect the graph in only one place. The graph represent a function.

81. $f(x) = x^2 - 4$

   (a) $f(x + h) = (x + h)^2 - 4$

$= x^2 + 2hx + h^2 - 4$

   (b) $f(x + h) - f(x)$

$= [(x + h)^2 - 4] - (x^2 - 4)$

$= x^2 + 2hx + h^2 - 4 - x^2 + 4$

$= 2hx + h^2$

(c) $\dfrac{f(x + h) - f(x)}{h}$

$= \dfrac{[(x + h)^2 - 4] - (x^2 - 4)}{h}$

$= \dfrac{x^2 + 2hx + h^2 - 4 - x^2 + 4}{h}$

$= \dfrac{2hx + h^2}{h}$

$= 2x + h$

83. $f(x) = 6x + 2$

(a) $f(x + h) = 6(x + h) + 2$

$\qquad\qquad\quad = 6x + 6h + 2$

(b) $f(x + h) - f(x)$

$\qquad = [6(x + h) + 2] - (6x + 2)$

$\qquad = 6x + 6h + 2 - 6x - 2$

$\qquad = 6h$

(c) $\dfrac{f(x + h) - f(x)}{h}$

$= \dfrac{[6(x + h) + 2] - (6x + 2)}{h}$

$= \dfrac{6x + 6h + 2 - 6x - 2}{h} = \dfrac{6h}{h}$

$= 6$

85. $f(x) = \dfrac{1}{x}$

(a) $f(x + h) = \dfrac{1}{x + h}$

(b) $f(x + h) - f(x)$

$= \dfrac{1}{x + h} - \dfrac{1}{x}$

$= \left(\dfrac{x}{x}\right)\dfrac{1}{x + h} - \dfrac{1}{x}\left(\dfrac{x + h}{x + h}\right)$

$= \dfrac{x - (x + h)}{x(x + h)}$

$= \dfrac{-h}{x(x + h)}$

(c) $\dfrac{f(x + h) - f(x)}{h}$

$= \dfrac{\dfrac{1}{x + h} - \dfrac{1}{x}}{h}$

$= \dfrac{\dfrac{1}{x + h}\left(\dfrac{x}{x}\right) - \dfrac{1}{x}\left(\dfrac{x + h}{x + h}\right)}{h}$

$= \dfrac{\dfrac{x - (x + h)}{(x + h)x}}{h}$

$= \dfrac{1}{h}\left[\dfrac{x - x - h}{(x + h)x}\right]$

$= \dfrac{1}{h}\left[\dfrac{-h}{(x + h)x}\right]$

$= \dfrac{-1}{(x + h)x}$

87.   **(a)** To determine I(1976), locate 1976 on the horizontal axis, find the corresponding point on the energy intensity graph and read across (to the left) to 26,300 BTU per dollar.

(b) To determine S(1984), locate 1984 on the horizontal axis, find the corresponding point on the savings graph and read across (to the right) to $120 billion.

(c) When the two graphs cross, I(t) = 23,500 BTU per dollar and S(t) = $85 billion.   These values occur in 1980.

(d) There is no significance to the point at which the two graphs cross. They each measure different, independent outcomes.

89. If x is a whole ounce, the cost of mailing a letter in cents is $C(x) = 29 + 23(x - 1)$. For x in whole ounces and a fraction of an ounce, substitute the next whole number for x in $C(x) = 29 + 23(x - 1)$ because a fraction of an ounce is billed as a whole ounce.

(a) $C\left(\frac{2}{3}\right) = C(1) = 29 + 23(1 - 1)$
$$= 29 + 23(0) = 29\text{¢}$$

(b) $C\left(1\frac{1}{3}\right) = C(2) = 29 + 23(2 - 1)$
$$= 29 + 23 = 52\text{¢}$$

(c) $C(2) = 29 + 23(2 - 1) = 29 + 23$
$$= 52\text{¢}$$

(d) $C\left(3\frac{1}{8}\right) = 29 + 23(4 - 1)$
$$= 29 + 23(3) = 98\text{¢}$$

(e) To determine the postage on a letter that weighs 2 5/8 ounces, calculate $C(2\ 5/8)$.

$$C\left(2\frac{5}{8}\right) = 29 + 23(3 - 1)$$
$$= 29 + 23(2) = 75\text{¢}$$

It costs 75¢ to send a letter that weighs 2 5/8 ounces.

(g) See the answer graph in the back of the textbook.

(h) The independent variable is x, the weight of the letter.

(i) The dependent variable is $C(x)$, the cost to mail the letter.

Section 1.2

1. Find the slope of the line through (4, 5) and (−1, 2).

$$m = \frac{5 - 2}{4 - (-1)}$$
$$= \frac{3}{5}$$

3. Find the slope of the line through (8, 4) and (8, −7).

$$m = \frac{4 - (-7)}{8 - 8}$$
$$= \frac{11}{0}$$

The slope is not defined; the line is vertical.

5. $y = 2x$

Using the slope–intercept form, $y = mx + b$, we see that the slope is 2.

7. $5x - 9y = 11$

Rewrite the equation in slope–intercept form.

$$9y = 5x - 11$$
$$y = \frac{5}{9}x - \frac{11}{9}$$

The slope is 5/9.

9. $x = -6$

This is a vertical line and the slope is not defined.

11. Find the slope of the line parallel to $2y - 4x = 7$. Rewrite the equation in slope-intercept form.

$$2y = 4x + 7$$
$$y = 2x + \frac{7}{2}$$

This slope is 2, so a parallel line will also have slope 2.

13. Find the slope of the line through $(-1.978, 4.806)$ and $(3.759, 8.125)$. (Note that there are four digits in each number.)

$$m = \frac{8.125 - 4.806}{3.759 - (-1.978)}$$
$$= \frac{3.319}{5.737}$$
$$= .5785$$

(We give the answer correct to four significant digits.)

15. As shown on the graph, the line goes through the points $(0, 2)$ and $(-5, 0)$.

$$m = \frac{0 - 2}{-5 - 0} = \frac{2}{5}$$

17. As shown on the graph, the line goes through the points $(4, 0)$ and $(0, 1)$.

$$m = \frac{1 - 0}{0 - 4} = -\frac{1}{4}$$

19. The line goes through $(1, 3)$, with slope $m = -2$.
Use point-slope form.

$$y - 3 = -2(x - 1)$$
$$y = -2x + 2 + 3$$
$$2x + y = 5$$

21. The line goes through $(6, 1)$, with slope $m = 0$.
Use point-slope form.

$$y - 1 = 0(x - 6)$$
$$y - 1 = 0$$
$$y = 1$$

23. The line goes through $(4, 2)$ and $(1, 3)$.
Find the slope, then use point-slope form with either of the two given points.

$$m = \frac{3 - 2}{1 - 4}$$
$$= -\frac{1}{3}$$

$$y - 3 = -\frac{1}{3}(x - 1)$$
$$y = -\frac{1}{3}x + \frac{1}{3} + 3$$
$$y = -\frac{1}{3}x + \frac{10}{3}$$
$$3y = -x + 10$$
$$x + 3y = 10$$

25. The line goes through $(1/2, 5/3)$ and $(3, 1/6)$.

$$m = \frac{\frac{1}{6} - \frac{5}{3}}{3 - \frac{1}{2}} = \frac{\frac{1}{6} - \frac{10}{6}}{\frac{6}{2} - \frac{1}{2}}$$
$$= \frac{-\frac{9}{6}}{\frac{5}{2}} = -\frac{18}{30} = -\frac{3}{5}$$

$$y - \frac{5}{3} = -\frac{3}{5}\left(x - \frac{1}{2}\right)$$

$$y - \frac{5}{3} = -\frac{3}{5}x + \frac{3}{10}$$

$$30\left(y - \frac{5}{3}\right) = 30\left(-\frac{3}{5}x + \frac{3}{10}\right)$$

$$30y - 50 = -18x + 9$$

$$18x + 30y = 59$$

27. The line has x-intercept 3 and y-intercept -2.

Two points on the line are (3, 0) and (0, -2). Find the slope, then use slope–intercept form.

$$m = \frac{0 - (-2)}{3 - 0}$$

$$= \frac{2}{3}$$

$$b = -2$$

$$y = \frac{2}{3}x - 2$$

$$3y = 2x - 6$$

$$2x - 3y = 6$$

29. The vertical line through (-6, 5) goes through the point (-6, 0), so the equation is

$$x = -6.$$

31. The line goes through (-1.76, 4.25) and has slope -5.081.

Use point–slope form.

$$y - 4.25 = -5.081[x - (-1.76)]$$

$$y = -5.081x - 8.94256$$

$$+ 4.25$$

$$y = -5.081x - 4.69256$$

$$5.081x + y = -4.69$$

For Exercises 33–45, see the answer graphs in the back of your textbook.

33. Graph the line through (-1, 3), $m = 3/2$. The slope is $\frac{3}{2}$, so

$m = \frac{\Delta y}{\Delta x} = \frac{3}{2}$. Thus, $\Delta y = 3$ and

$\Delta x = 2$. Another point on the line has coordinates

$$[(-1 + 2), (3 + 3)] = (1, 6).$$

Graph the line through (-1, 3) and (1, 6).

35. Graph the line through (3, -4), $m = -1/3$.

Since $m = -1/3$, $\Delta y = -1$ and $\Delta x = 3$. Another point on the line has coordinates

$$[(3 + 3), (-4 - 1)] = (6, -5).$$

Graph the line through (3, -4) and (6, -5).

37. Graph the line $3x + 5y = 15$.

Let $x = 0$.

$$3(0) + 5y = 15$$

$$y = 3$$

One point is (0, 3).

Let $y = 0$.

$$3x + 5(0) = 15$$

$$x = 5$$

A second point is (5, 0).

Draw the graph through (0, 3) and (5, 0).

39. Graph line $4x - y = 8$.

Let $x = 0$.

$$4(0) - y = 8$$
$$y = -8$$

One point on the line is $(0, -8)$.
Let $y = 0$.

$$4x - 0 = 8$$
$$4x = 8$$
$$x = 2$$

Another point on the line is $(2, 0)$.
Draw the graph through $(0, -8)$ and $(2, 0)$.

41. Graph the line $x + 2y = 0$.

Let $x = 0$.

$$0 + 2y = 0$$
$$y = 0$$

One point on the line is $(0, 0)$.
Let $x = -2$.

$$-2 + 2y = 0$$
$$2y = 2$$
$$y = 1$$

Another point on the line is $(-2, 1)$.
Draw the graph through $(0, 0)$ and $(-2, 1)$.

43. Graph the line $x = -1$. Since this line has the form $x = a$, it is a vertical line through the point $(-1, 0)$.

45. Graph the line $y = -3$.
Since $y = -3$ for all values of x, the line is horizontal and passes through the point $(0, -3)$.

47. Write the equation of the line through $(-1, 4)$, parallel to $x + 3y = 5$. Rewrite the equation in slope-intercept form.

$$x + 3y = 5$$
$$3y = -x + 5$$
$$y = -\frac{1}{3}x + \frac{5}{3}$$

The slope is $-1/3$.
Use $m = -1/3$ and the point $(-1, 4)$ in the point-slope form.

$$y - 4 = -\frac{1}{3}[x - (-1)]$$
$$y = -\frac{1}{3}(x + 1) + 4$$
$$y = -\frac{1}{3}x - \frac{1}{3} + 4$$
$$= -\frac{1}{3}x + \frac{11}{3}$$
$$x + 3y = 11$$

49. Write the equation of the line through $(3, -4)$, perpendicular to $x + y = 4$.
Rewrite the equation as

$$y = -x + 4.$$

The slope of this line is -1. To find the slope of a perpendicular line, solve

$$-1m = -1.$$
$$m = 1$$

Use $m = 1$ and $(3, -4)$ in the point-slope form.

$$y - (-4) = 1(x - 3)$$
$$y = x - 3 - 4$$
$$y = x - 7$$
$$x - y = 7$$

51. Write the equation of the line with x-intercept -2, parallel to y = 2x.

The given line has slope 2. A parallel line will also have m = 2. Since the x-intercept is -2, the point (-2, 0) is on the required line. Using point-slope form, we have

$$y - 0 = 2[x - (-2)]$$
$$y = 2x + 4$$
$$-2x + y = 4.$$

53. Write the equation of the line with y-intercept 2, perpendicular to 3x + 2y = 6.

Find slope of given line.

$$3x + 2y = 6$$
$$2y = -3x + 6$$
$$y = -\frac{3}{2}x + 3$$

The slope is -3/2, so the slope of the perpendicular line will be 2/3. If the y-intercept is 2, then the point (0, 2) lies on the line. Using point-slope form, we have

$$y - 2 = \frac{2}{3}(x - 0)$$
$$y = \frac{2}{3}x + 2$$
$$2x - 3y = -6.$$

55. Do (4, 3), (2, 0), (-18, -12) lie on the same line?

The line containing (4, 3) and (2, 0) has slope

$$m = \frac{3 - 0}{4 - 2}$$
$$= \frac{3}{2}.$$

The line containing (2, 0) and (-18, -12) has slope

$$m = \frac{0 - (-12)}{2 - (-18)}$$
$$= \frac{12}{20}$$
$$= \frac{3}{5}.$$

The two lines do not have the same slope. Therefore, the three points do not lie on the same line.

57. Show that the quadrilateral with vertices (1, 3), (-5/2, 2), (-7/2, 4), (2, 1) is a parallelogram. A parallelogram is a four-sided plane figure which has opposite sides parallel. A sketch will show which pairs of sides are opposite.

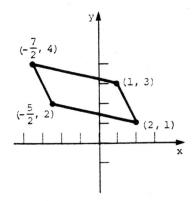

Show that the line containing (-7/2, 4) and (-5/2, 2) is parallel to the line containing (1, 3) and (2, 1) and that the line containing (-7/2, 4) and (1, 3) is parallel to the line containing (-5/2, 2) and (2, 1).

For $(-7/2, 4)$ and $(-5/2, 2)$, the slope is

$$m = \frac{2 - 4}{\left(-\frac{5}{2}\right) - \left(-\frac{7}{2}\right)}$$

$$= \frac{-2}{1}$$

$$= -2$$

For $(1, 3)$ and $(2, 1)$, the slope is

$$m = \frac{1 - 3}{2 - 1} = \frac{-2}{1} = -2.$$

Since the slopes are equal, these lines are parallel.

For $(-7/2, 4)$ and $(1, 3)$, the slope is

$$m = \frac{3 - 4}{1 - \left(-\frac{7}{2}\right)}$$

$$= \frac{-1}{\frac{9}{2}}$$

$$= -\frac{2}{9}.$$

For $(-5/2, 2)$ and $(2, 1)$, the slope is

$$m = \frac{1 - 2}{2 - \left(-\frac{5}{2}\right)}$$

$$= \frac{-1}{\frac{9}{2}}$$

$$= -\frac{2}{9}.$$

Since the slopes are equal, these lines are parallel.

Both pairs of opposite sides are parallel so the quadrilateral is a parallelogram.

59. Let P, P_1, and P_2 be points on a nonvertical line L.
Draw a line N through point P parallel to the x-axis. Draw vertical lines through P_1 and P_2 that intersect N at points Q_1 and Q_2. Angle P is common to triangles PP_1Q_1 and PP_2Q_2.

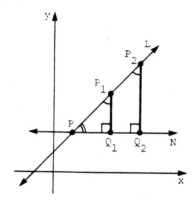

Angle Q_1 = angle Q_2 because they are right angles.

Angle P_1 = angle P_2 because they are corresponding angles formed by parallel lines intersected by a transversal. Since corresponding angles of the triangles are equal, the triangles are similar.

Let m_1 be the slope determined by P and P_1 and m_2 be the slope determined by P and P_2. In geometry, recall that P_1Q_1 is the symbol for the length of segment P_1Q_1.

$$m_1 = \frac{P_1Q_1}{PQ_1} \quad \text{and} \quad m_2 = \frac{P_2Q_2}{PQ_2}$$

Since the triangles are similar, $P_2Q_2 = k(P_1Q_1)$ and $PQ_2 = k(PQ_1)$

so $m_2 = \dfrac{k(P_1Q_1)}{k(PQ_1)} = \dfrac{P_1Q_1}{PQ_1}$

$$= m_1.$$

61. **(a)** See the figure in your textbook. Segment MN is drawn perpendicular to segment PQ. Recall that MQ is the length of segment MQ.

$$m_1 = \frac{\Delta y}{\Delta x} = \frac{MQ}{PQ}$$

From the diagram, we know that PQ = 1. Thus, $m_1 = \frac{MQ}{1}$ so MQ has length m_1.

(b) $m_2 = \frac{\Delta y}{\Delta x} = \frac{-QN}{PQ} = \frac{-QN}{1}$

$QN = -m_2$

(c) Triangles MPQ, PNQ, and MNP are right triangles by construction. In triangles MPQ and MNP

angle M = angle M

and in right triangles PNQ and MNP

angle N = angle N.

Since all right angles are equal and since triangles with two equal angles are similar,

triangle MPQ is similar to triangle MNP

triangle PNQ is similar to t iangle MNP.

Therefore, triangles MPQ and PNQ are similar to each other.

(d) Since corresponding sides in similar triangles are proportional,

MQ = k • PQ   and   PQ = k • QN.

$$\frac{MQ}{PQ} = \frac{k \cdot PQ}{k \cdot QN}$$

$$\frac{MQ}{PQ} = \frac{PQ}{QN}$$

From the diagram, we know that PQ = 1.

$$MQ = \frac{1}{QN}$$

From (a) and (b), m_1 = MQ and $-m_2$ = QN.

Substituting, we get

$$m_1 = \frac{1}{-m_2}.$$

Multiplying both sides by m_2, we have

$$m_1 m_2 = -1.$$

63. **(a)** Graph $S(q) = \frac{2}{5}q$ and

$$D(q) = 100 - \frac{2}{5}q.$$

See the answer graph in the back your textbook.

(b) The equilibrium quantity is the number of units, q, of wool sweaters for which the price on the demand curve equals the price on the supply curve.

Set $\frac{2}{5}q = 100 - \frac{2}{5}q$ and solve for q.

$$\frac{2}{5}q + \frac{2}{5}q = 100$$

$$\frac{4}{5}q = 100$$

$$q = \frac{5}{4}(100)$$

$$= 125 \text{ units}$$

(c) The equilibrium price is the price corresponding to q = 125 units.

Use $S(q) = \frac{2}{5}q$ and $q = 125$.

$$S(125) = \frac{2}{5}(125) = \$50$$

Or use

$$D(q) = 100 - \frac{2}{5}q \text{ and } q = 125.$$

$$D(q) = 100 - \frac{2(125)}{5}$$

$$= \$50$$

65. $C(x) = 27(x) + 25$, C in dollars, where x is a whole number of days. If x includes a part of a day, then the next whole number is used to find $C(x)$ in the formula.

(a) $C\left(\frac{3}{4}\right) = C(1)$

$$= 27(1) + 25$$
$$= \$52$$

(b) $C\left(\frac{9}{10}\right) = C(1)$

$$= 27(1) + 25$$
$$= \$52$$

(c) $C(1) = 27(1) + 25 = \$52$

(d) $C\left(1\frac{5}{8}\right) = C(2)$

$$= 27(2) + 25$$
$$= \$79$$

(e) $C(2.4) = C(3)$

$$= 27(3) + 25$$
$$= \$106$$

(f) Graph $y = C(x)$.

See the answer graph in the back of your textbook.

(g) Yes.
C is function because a vertical line cuts the graph in no more than one point.

(h) No.
C is not a linear function because its graph is not a straight line.

67. **(a)** To find the equation, find the slope using the points (40, 177) and (43, 185).

$$m = \frac{185 - 177}{43 - 40} = \frac{8}{3}$$

Then use the point-slope form with (40, 177).

$$h - 177 = \frac{8}{3}(t - 40)$$

$$h = \frac{8}{3}t - \frac{8}{3}(40) + 177$$

$$h = \frac{8}{3}t + \frac{211}{3}$$

(b) $t = 38$ cm

$$h = \frac{8}{3}t + \frac{211}{3}$$

$$= \frac{8}{3}(38) + \frac{211}{3}$$

$$= \frac{304 + 211}{3}$$

$$\approx 172$$

$t = 45$ cm

$$h = \frac{8}{3}(45) + \frac{211}{3}$$

$$= \frac{360 + 211}{3}$$

$$\approx 190$$

Males with tibias having lengths of 38 cm to 45 cm would have been about 172 cm to 190 cm tall.

(c)   h = 190 cm

$$190 = \frac{8}{3}t + \frac{211}{3}$$

$$\frac{8}{3}t = 190 - \frac{211}{3}$$

$$\frac{8}{3}t = \frac{570 - 211}{3}$$

$$t = \frac{359}{8}$$

$$\approx 45$$

A male whose height is 190 cm would have a tibia bone about 45 cm long.

69. **(a)** Find the slope using (45, 42.5) and (55, 67.5).

$$m = \frac{67.5 - 42.5}{55 - 45} = \frac{25}{10} = 2.5 \text{ or } \frac{5}{2}$$

Use point–slope form with (45, 42.5).

$$(y - 42.5) = \frac{5}{2}(x - 45)$$

$$y = \frac{5}{2}x - \frac{5}{2}(45) + 42.5$$

$$y = \frac{5}{2}x - 70$$

$$= 2.5x - 70$$

(b) Let y = 60.

$$60 = 2.5x - 70$$
$$130 = 2.5x$$
$$52 = x$$

The Democrats would need 52% of the vote.

71. **(a)** If the temperature rises .3°C per decade, and a decade is 10 years, then the temperature rises $\frac{.3}{10} = .03$°C per year. This is the slope of the line, the change in

temperature per year. When t = 0, in 1970, the temperature was 15°C. This is the intercept. The equation can be written in slope–intercept form as

$$T = .03t + 15.$$

(b) Let T = 19.

$$19 = .03t + 15$$
$$4 = .03t$$
$$133 \approx t$$

Since t = 0 corresponds to 1970, t = 133 corresponds to 1970 + 133 = 2103. The sea level will increase by 65 cm in about the year 2103.

Section 1.3

1. $12 is the fixed cost and $1 is the cost per hour.
    Let x = number of hours
    and
    C(x) = cost of renting a saw for x hours.

    So,

    C(x) = fixed cost + (cost per hour)
    • (number of hours)

    C(x) = 12 + 1x
    = 12 + x.

3. 50¢ is the fixed cost and 35¢ is the cost per half–hour.
    Let x = the number of half–hours
    and
    C(x) = the cost of parking a car for x half–hours.

So,

$$C(x) = 50 + 35x$$
$$= 35x + 50.$$

5. Fixed cost, $100; 50 items cost $1600 to produce.

$C(x)$ = cost of producing x items.

$C(x) = mx + b$ where b is the fixed cost.

$$C(x) = mx + 100$$

Now,

$C(x) = 1600$ when $x = 50$.

$$1600 = m(50) + 100$$
$$1500 = 50m$$
$$30 = m$$

Thus, $C(x) = 30x + 100$.

7. Fixed cost, $1000; 40 items cost $2000 to produce.

$$C(x) = mx + 1000$$

Now, $C(x) = 2000$ when $x = 40$.

$$2000 = m(40) + 1000$$
$$1000 = 40m$$
$$25 = m$$

Thus, $C(x) = 25x + 1000$.

9. Marginal cost, $50; 80 items cost $4500 to produce.

Let x = number of items produced and

$C(x)$ = cost of producing x items.

$C(x) = mx + b$, where m is the marginal cost.

$$C(x) = 50x + b$$

Now, $C(x) = 4500$ when $x = 80$.

$$4500 = 50(80) + b$$
$$4500 = 4000 + b$$
$$500 = b$$

Thus, $C(x) = 50x + 500$.

11. Marginal cost, $90; 150 items cost $16,000 to produce.

$$C(x) = 90x + b$$

Now $C(x) = 16,000$ when $x = 150$.

$$16,000 = 90(150) + b$$
$$16,000 = 13,500 + b$$
$$2500 = b$$

Thus, $C(x) = 90x + 2500$.

13. $S(x) = 300x + 2000$; 1987 is year 0, so $x = 0$ for 1987.

(a) 1987: $x = 0$

$$S(0) = 300(0) + 200$$
$$= 2000$$

(b) 1990: $x = 3$

$$S(3) = 300(3) + 2000$$
$$= 900 + 2000$$
$$= 2900$$

(c) 1991: $x = 4$

$$S(4) = 300(4) + 2000$$
$$= 1200 + 2000$$
$$= 3200$$

(d) 1996: $x = 9$

$$S(9) = 300(9) + 2000$$
$$= 2700 + 2000$$
$$= 4700$$

Yes, they will reach their goal of 4000.

(e) Annual rate of change of sales is the slope of the linear function, or 300.

15. **(a)** Let $S(x)$ represent sales. $x = 0$ represents 1981, so

$$S(0) = 200,000.$$

$x = 7$ represents 1988, so

$$S(7) = 1,000,000.$$

$$m = \frac{1,000,000 - 200,000}{7 - 0}$$

$$= \frac{800,000}{7}$$

The value of $S(x)$ when $x = 0$ is the y-intercept of the line, b.
So,

$$b = 200,000.$$

Thus,

$$S(x) = \frac{800,000}{7}x + 200,000.$$

(b) 1992: $x = 11$

$$S(11) = \frac{800,000}{7}(11) + 200,000$$

$$= 1,257,142.8 + 200,000$$

$$= 1,457,142.8 \text{ or about}$$

$$\$1,457,000$$

(c) When sales reach \$2,000,000,

$$2,000,000 = \frac{800,000}{7}x + 200,000$$

$$1,800,000 = \frac{800,000}{7}x$$

$$12,600,000 = 800,000x$$

$$15.75 = x.$$

They will want to negotiate a contract in about 16 years, or in 1997.

17. $C(x) = .097x$

(a) $C(1000) = .097(1000)$

$$= \$97$$

(b) $C(1001) = .097(1001)$

$$= \$97.097$$

(c) Marginal cost of 1001st cup

$$= C(1001) - C(1000)$$

Use parts (a) and (c).

$$= 97.097 - 97$$

$$= .097 \quad \text{or} \quad 9.7¢$$

(d) This value is the slope so the marginal cost of any cup is .097 or 9.7¢.

19. $\bar{C}(x) = \dfrac{C(x)}{x} = \dfrac{800 + 20x}{x}$

$$= \frac{800}{x} + \frac{20x}{x}$$

$$= \frac{800}{x} + 20$$

(a) $\bar{C}(10) = \dfrac{800}{10} + 20$

$$= 80 + 20$$

$$\bar{C}(10) = \$100$$

(b) $\bar{C}(50) = \dfrac{800}{50} + 20$

$$= 16 + 20$$

$$\bar{C}(50) = \$36$$

(c) $\bar{C}(200) = \dfrac{800}{200} + 20$

$$= 4 + 20$$

$$\bar{C}(200) = \$24$$

21. $C(x) = 50x + 5000$

$R(x) = 60x$

Let $C(x) = R(x)$ to find the number of units at the break-even point.

$$50x + 5000 = 60x$$
$$5000 = 10x$$
$$x = 500 \text{ units}$$

Revenue for 500 units:

$$R(500) = 60(500)$$
$$= \$30,000$$

23. $C(x) = 85x + 900$

$R(x) = 105x$

Set $C(x) = R(x)$ to find the number of units at the break-even point.

$$85x + 900 = 105x$$
$$900 = 20x$$
$$45 = x$$

The break-even point would be 45 units, so you decide not to produce since not more than 38 units can be sold.

25. $C(x) = 70x + 500$

$R(x) = 60x$

$$70x + 500 = 60x$$
$$10x = -500$$
$$x = -50$$

It is impossible to make a profit when the break-even point is negative. Cost will always be greater than revenue.

27. From the graph, the break-even point was about $140 billion, which occurred in the middle of 1981.

29. Let $x = 0$ correspond to 1900. Then the "life expectancy from birth" line contains the points (0, 46) and (75, 75).

$$m = \frac{75 - 46}{75 - 0} = \frac{29}{75}$$

Use point slope form.

$$y - 46 = \frac{29}{75}(x - 0)$$
$$y = \frac{29}{75}x + 46$$

The "life expectancy from age 65" line contains the points (0, 76) and (75, 80).

$$m = \frac{80 - 76}{75 - 0} = \frac{4}{75}$$

$$y - 76 = \frac{4}{75}(x - 0)$$
$$y = \frac{4}{75}x + 76$$

Set the two equations equal to determine where the lines intersect. At this point life expectancy should increase no further.

$$\frac{29}{75}x + 46 = \frac{4}{75}x + 76$$
$$29x + 3450 = 4x + 5700$$
$$25x = 2250$$
$$x = 90$$

Determine the y-value when x = 90.

$$y = \frac{4}{75}(90) + 76$$

$$y = 4.8 + 76$$

$$= 80.8$$

Thus, the maximum life expectancy for humans is about 81 years.

31. Let x represent the force and y represent the speed.
The linear function contains the points (.75, 2) and (.93, 3).

$$m = \frac{3 - 2}{.93 - .75} = \frac{1}{.18} = \frac{1}{\frac{18}{100}}$$

$$= \frac{100}{18} = \frac{50}{9}$$

Use point-slope form to write the equation.

$$y - 2 = \frac{50}{9}(x - .75)$$

$$y - 2 = \frac{50}{9}x - \frac{50}{9}(.75)$$

$$y = \frac{50}{9}x - \frac{75}{18} + 2$$

$$= \frac{50}{9}x - \frac{39}{18}$$

Now determine y, the speed, when x, the force, is 1.16.

$$y = \frac{50}{9}(1.16) - \frac{39}{18}$$

$$= \frac{58}{9} - \frac{39}{18}$$

$$= \frac{77}{18} \approx 4.3$$

The pony switches from a trot to a gallop at approximately 4.3 m/sec.

33. y = mx + b
y = 1.25x − 5

(a) y = 1.25(30) − 5
y = 37.5 − 5
y = 32.5 min

(b) y = 1.25(60) − 5
y = 75 − 5
y = 70 min

(c) y = 1.25(120) − 5
y = 150 − 5
y = 145 min

(d) y = 1.25(180) − 5
y = 225 − 5
y = 220 min

35. (a) $C = \frac{5}{9}(F - 32)$

$$C = \frac{5}{9}(58 - 32)$$

$$C = \frac{5}{9}(26)$$

$$C = 14.4°$$

The temperature is 14.4° C.

(b) $F = \frac{9}{5}C + 32$

$$F = \frac{9}{5}(50) + 32$$

$$F = 90 + 32$$

$$F = 122°$$

The temperature is 122°F.

37. Water freezes at 32° F and 0° C. It boils at 212° F and 100° C. The ordered pairs (32, 0) and (212, 100) lie on the graph giving the Celsius temperature (C) as a linear function of the Fahrenheit temperature.

$$m = \frac{100 - 0}{212 - 32} = \frac{100}{180} = \frac{5}{9}$$

$$C - 0 = \frac{5}{9}(F - 32)$$

$$C = \frac{5}{9}(F - 32)$$

39. $R(x) = -8x + 240$

(a) $R(0) = -8(0) + 240$

$= 0 + 240$

$= 240$

(b) $R(5) = -8(5) + 240$

$= -40 + 240$

$= 200$

(c) $R(10) = -8(10) + 240$

$= -80 + 240$

$= 160$

(d) The rate of change is the slope of the line, or -8 students per hour of study. The **negative sign** indicates that student enrollment decreases if the hours of required study increases.

(e) $R(x) = 8x + 240$

$16 = -8x + 240$

$16 - 240 = -8x$

$-224 = -8x$

$28 = x$

28 hr of study are required.

Section 1.4

For Exercises 1–21, see the answer graphs in the back of the textbook.

1. $f(x) = kx^2$ (k is a constant)

(e) All the curves have the same vertex: $(0, 0)$. As the coefficient k increases in absolute value, the parabola becomes narrower.

3. $f(x) = x^2 + k$ (k is a constant)

(e) All the curves have the same shape. The coordinates of the vertices are $(0, k)$ so the graphs are shifted upward or downward.

5. $y = x^2 + 6x + 5$

$y = (x + 5)(x + 1)$

Set $y = 0$ to find the x–intercepts.

$0 = (x + 5)(x + 1)$

$x = -5, \; x = -1$

The x–intercepts are -5 and -1.
Set $x = 0$ to find the y–intercept.

$y = 0^2 + 6(0) + 5$

$y = 5$

The y–intercept is 5.

The x–coordinate of the vertex is

$x = \frac{-b}{2a} = \frac{-6}{2} = -3$.

Substitute to find the y–coordinate.

$y = (-3)^2 + 6(-3) + 5$

$= 9 - 18 + 5$

$= -4$

The vertex is $(-3, -4)$.
The axis is $x = -3$, the vertical
line through the vertex.

7. $y = x^2 - 4x + 4$
    $= (x - 2)(x - 2)$

Let $y = 0$.

$0 = (x - 2)(x - 2)$

2 is the x-intercept.

Let $x = 0$.

$y = 0^2 - 4(0) + 4$
4 is the y-intercept.

Vertex: $x = \dfrac{-b}{2a} = \dfrac{4}{2} = 2$

$y = 2^2 - 4(2) + 4$
$= 4 - 8 + 4 = 0$

The vertex is $(2, 0)$.
The axis is $x = 2$, the vertical line
through the vertex.

9. $y = -2x^2 - 12x - 16$
    $= -2(x^2 + 6x + 8)$
    $= -2(x + 4)(x + 2)$

Let $y = 0$.

$0 = -2(x + 4)(x + 2)$

$x = -4, \; x = -2$

-4 and -2 are the x-intercepts.
Let $x = 0$.

$y = -2(0)^2 + 12(0) - 16$
-16 is the y-intercept.

Vertex: $x = \dfrac{-b}{2a} = \dfrac{12}{-4} = -3$

$y = -2(-3)^2 - 12(-3) - 16$
$= -18 + 36 - 16$
$= 2$

The vertex is $(-3, 2)$.
The axis is $x = -3$, the vertical
line through the vertex.

11. $y = 2x^2 + 12x - 16$
    $= 2(x^2 + 6x - 8)$

Let $y = 0$.

$0 = 2(x^2 + 6x - 8)$

Divide by 2.

$0 = x^2 + 6x - 8$

Use the quadratic formula.

$x = \dfrac{-6 \pm \sqrt{6^2 - 4(1)(-8)}}{2(1)}$

$= \dfrac{-6 \pm \sqrt{36 - (-32)}}{2}$

$= \dfrac{-6 \pm \sqrt{68}}{2}$

$= \dfrac{-6 \pm 2\sqrt{17}}{2}$

$= -3 \pm \sqrt{17}$

The x-intercepts are $-3 + \sqrt{17} \approx 1.12$
and $-3 - \sqrt{17} \approx -7.12$.
Let $x = 0$.

$y = 2(0)^2 + 12(0) - 16$
-16 is the y-intercept.

Vertex: $x = \dfrac{-b}{2a} = \dfrac{-12}{2(2)} = -\dfrac{12}{4} = -3$

$y = 2(-3)^2 + 12(-3) - 16$
$= 18 - 36 - 16$
$= -34$

The vertex is $(-3, -34)$.
The axis is $x = -3$, the vertical
line through the vertex.

13. $y = 3x^2 + 6x + 12$

Let $y = 0$.

$0 = 3x^2 + 6x + 2$

$x = \dfrac{-6 \pm \sqrt{6^2 - 4(3)(2)}}{2(3)}$

$= \dfrac{-6 \pm \sqrt{36 - 24}}{6} = \dfrac{-6 \pm \sqrt{12}}{6}$

$= \dfrac{-6 \pm 2\sqrt{3}}{6} = -1 \pm \dfrac{\sqrt{3}}{3}$

The x-intercepts are $-1 + \dfrac{\sqrt{3}}{3} \approx -.42$

and $-1 - \dfrac{\sqrt{3}}{3} \approx -1.58$.

Let $x = 0$.

$y = 3(0)^2 + 6(0) + 2$

2 is the y-intercept.

Vertex: $x = \dfrac{-b}{2a} = \dfrac{-6}{2(3)} = -\dfrac{6}{6} = -1$

$y = 3(-1)^2 + 6(-1) + 2$

$= 3 - 6 + 2$

$= -1$

The vertex is $(-1, -1)$.
The axis is $x = -1$.

15. $y = -x^2 + 6x - 6$

Let $y = 0$.

$0 = -x^2 + 6x - 6$

$x = \dfrac{-6 \pm \sqrt{6^2 - 4(-1)(-6)}}{2(-1)}$

$= \dfrac{-6 \pm \sqrt{36 - 24}}{-2} = \dfrac{-6 \pm \sqrt{12}}{-2}$

$= \dfrac{-6 \pm 2\sqrt{3}}{-2}$

$= 3 \pm \sqrt{3}$

The x-intercepts are $3 + \sqrt{3} \approx 4.73$
and $3 - \sqrt{3} \approx 1.27$.

Let $x = 0$.

$y = -0^2 + 6(0) - 6$

-6 is the y-intercept.

Vertex: $x = \dfrac{-b}{2a} = \dfrac{-6}{2(-1)} = \dfrac{-6}{-2} = 3$

$y = -3^2 + 6(3) - 6$

$= -9 + 18 - 6$

$= 3$

The vertex is $(3, 3)$.
The axis is $x = 3$.

17. $y = -3x^2 + 24x - 36$

$= -3(x^2 - 8x + 12)$

$= -3(x - 6)(x - 2)$

Let $y = 0$.

$0 = -3(x - 6)(x - 2)$

Divide by -3.

$0 = (x - 6)(x - 2)$

$x = 6, \ x = 2$

6 and 2 are the x-intercepts.

Let $x = 0$.

$y = -3(0)^2 + 24(0) - 36$

-36 is the y-intercept.

Vertex: $x = \dfrac{-b}{2a} = \dfrac{-24}{2(-3)} = \dfrac{-24}{-6} = 4$

$y = -3(4)^2 + 24(4) - 36$

$= -48 + 96 - 36$

$= 12$

The vertex is $(4, 12)$.
The axis is $x = 4$.

19. $y = \frac{5}{2}x^2 + 10x + 8$

Let $y = 0$.

$0 = \frac{5}{2}x^2 + 10x + 8$

$x = \dfrac{-10 \pm \sqrt{10^2 - 4\left(\frac{5}{2}\right)(8)}}{2\left(\frac{5}{2}\right)}$

$ = \dfrac{-10 \pm \sqrt{100 - 80}}{5}$

$ = \dfrac{-10 \pm \sqrt{20}}{5}$

$ = -2 \pm \dfrac{2\sqrt{5}}{5}$

The x–intercepts $-2 + 2\sqrt{5}/5 \approx -1.11$ and $-2 - 2\sqrt{5}/5 \approx -2.89$.
Let $x = 0$.

$y = \frac{5}{2}(0)^2 + 10(0) + 8$

8 is the y–intercept.

Vertex: $x = \dfrac{-b}{2a} = \dfrac{-10}{2\left(\frac{5}{2}\right)} = \dfrac{-10}{5} = -2$

$y = \frac{5}{2}(-2)^2 + 10(-2) + 8$

$ = \frac{5}{2}(4) - 20 + 8$

$ = 10 - 20 + 8$

$ = -2$

The vertex is $(-2, -2)$.
The axis is $x = -2$.

21. $y = \frac{2}{3}x^2 - \frac{8}{3}x + \frac{5}{3}$

Let $y = 0$.

$0 = \frac{2}{3}x^2 - \frac{8}{3}x + \frac{5}{3}$

Multiply by 3.

$0 = 2x^2 - 8x + 5$

$x = \dfrac{-(-8) \pm \sqrt{(-8)^2 - 4(2)(5)}}{2(2)}$

$ = \dfrac{8 \pm \sqrt{64 - 40}}{4} = \dfrac{8 \pm \sqrt{24}}{4}$

$ = \dfrac{8 \pm 2\sqrt{6}}{4} = 2 \pm \dfrac{\sqrt{6}}{2}$

The x–intercepts are $2 + \sqrt{6}/2 \approx 3.22$ and $2 - \sqrt{6}/2 \approx .78$.
Let $x = 0$.

$y = \frac{2}{3}(0)^2 - \frac{8}{3}(0) + \frac{5}{3}$

5/3 is the y–intercept.

Vertex: $x = \dfrac{-b}{2a} = \dfrac{-\left(-\frac{8}{3}\right)}{2\left(\frac{2}{3}\right)} = \dfrac{\frac{8}{3}}{\frac{4}{3}} = 2$

$y = \frac{2}{3}(2)^2 - \frac{8}{3}(2) + \frac{5}{3}$

$ = \frac{8}{3} - \frac{16}{3} + \frac{5}{3}$

$ = -\frac{3}{3} = -1$

The vertex is $(2, -1)$.
The axis is $x = 2$.

23. Let $y = x(1 - x)$.

$y = x - x^2$

$ = -x^2 + x$

Let $y = 0$.

$0 = x(1 - x)$

$x = 0 \text{ or } x = 1$

0 and 1 are the x–intercepts.
Let $x = 0$.

$y = 0(1 - 0)$

0 is the y–intercept.

Vertex: $x = \dfrac{-b}{2a} = \dfrac{-1}{2(-1)} = \dfrac{-1}{-2} = \dfrac{1}{2}$

$$y = -\left(\frac{1}{2}\right)^2 + \frac{1}{2}$$

$$= -\frac{1}{4} + \frac{1}{2}$$

$$= \frac{1}{4}$$

The vertex is at $\left(\frac{1}{2}, \frac{1}{4}\right)$.

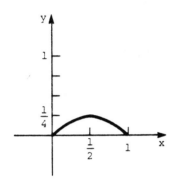

This parabola never reaches a height greater than $\frac{1}{4}$, thus the product $x(1 - x)$ is always less than or equal to $\frac{1}{4}$. The product equals $\frac{1}{4}$ at the vertex, when $x = \frac{1}{2}$.

25. $R(x) = 5000 + 50x - x^2$

$$= -x^2 + 50x + 5000$$

The maximum revenue occurs at the vertex.

$$x = \frac{-b}{2a} = \frac{-50}{2(-1)} = 25$$

$$y = 5000 + 50(25) - (25)^2$$

$$= 5000 + 1250 - 625$$

$$= 5625$$

The vertex is (25, 5625).
The maximum revenue of $5625 is realized when 25 seats are left unsold.

27. $p = 500 - x$

(a) Revenue is $R(x) = px$

$$= (500 - x)(x)$$

$$= 500x - x^2$$

(b) See the answer graph in your textbook.

(c) From the graph, the vertex is halfway between $x = 0$ and $x = 500$ so $x = 250$ units corresponds to maximum revenue. Then the price is

$$p = 500 - x$$

$$= 500 - 250 = \$250.$$

Note that price, p, cannot be read directly from the graph of

$$R(x) = 500x - x^2.$$

(d) $R(x) = 500x - x^2$

$$= -x^2 + 500x$$

Find the vertex.

$$x = \frac{-b}{2a} = \frac{-500}{-2} = 250$$

$$y = -(250)^2 + 500(250)$$

$$= 62,500$$

The vertex is (250, 62,500).
The maximum revenue is $62,500.

29. Let x = the number of $20 increases.

(a) Rent per apartment: $200 + 20x$

(b) Number of apartments rented:
$80 - x$

(c) Revenue:

$$R(x) = \text{(number of apartments rented)(rent per apart-ment)}$$
$$= (80 - x)(200 + 20x)$$
$$= 16,000 + 1400x - 20x^2$$

(d) Find the vertex.

$$x = \frac{-b}{2a} = \frac{-1400}{2(-20)} = \frac{-1400}{-40} = 35$$

$$y = 16,000 + 1400(35) - 20(35)^2$$
$$= 40,500$$

The vertex is (35, 40,500).
The maximum revenue occurs when x = 35.

(e) The maximum revenue is the y-coordinate of the vertex or $40,500.

31. $C(x) = 10x + 50$ (June–September)
$C(x) = -20(x - 5)^2 + 100$
            (October–December)

(a) June: x = 1

$$C(x) = 10 + 50$$
$$= 60$$

(b) July: x = 2

$$C(x) = 10(2) + 50$$
$$= 70$$

(c) September: x = 4

$$C(x) = 10(4) + 50$$
$$= 90$$

(d) October: x = 5

$$C(x) = -20(5 - 5)^2 + 100$$
$$= 100$$

(e) November: x = 6

$$C(x) = -20(6 - 5)^2 + 100$$
$$= -20 + 100$$
$$= 80$$

(f) December: x = 7

$$C(x) = -20(7 - 5)^2 + 100$$
$$= -80 + 100$$
$$= 20$$

33. $h = 32t - 16t^2$
$$= -16t^2 + 32t$$

Find the vertex.

$$x = \frac{-b}{2a} = \frac{-32}{-32} = 1$$

$$y = -16(1)^2 + 32(1)$$
$$= 16$$

The vertex is (1, 16), so the maximum height is 16 ft. When the object hits the ground, h = 0, so

$$32t - 16t^2 = 0$$
$$16t(2 - t) = 0$$
$$t = 0 \quad \text{or} \quad t = 2$$

When t = 0, the object is about to be thrown. When t = 2, the object hits the ground; that is, after 2 sec.

35. Let x = the width.
Then 320 - 2x = the length.

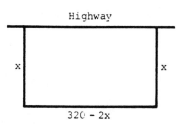

Area = x(320 - 2x)

    = -2x² + 320x

Find the vertex:

$x = \dfrac{-b}{2a} = \dfrac{-320}{-4} = 80$

y = -2(80)² + 320(80)

  = 12,800

The graph of the area function is a parabola with vertex (80, 12,800). The maximum area of 12,800 sq ft occurs when the width is 80 ft and the length is

$$320 - 2x = 320 - 2(80)$$
$$= 160 \text{ ft.}$$

37. Let x = the one number.
Then 20 - x = the other number.
Product = x(20 - x)

       = 20x - x²

       = -x² + 20x

Find the vertex.

$x = \dfrac{-b}{2a} = \dfrac{-20}{-2} = 10$

y = -10² + 20(10)

  = 100

The vertex is (10, 100), so the product is a maximum when

$$x = 10.$$

The other number is

$$20 - 10 = 10.$$

39. Draw a sketch of the arch with the base on the x-axis and the vertex at (0, 15).

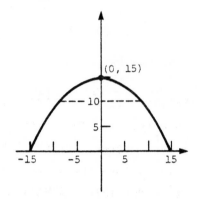

The equation of the parabola is in the form

$$y = -ax² + 15.$$

The points (-15, 0) and (15, 0) are on the parabola. Use (15, 0) as one point on the parabola.

$$0 = -a(15)² + 15$$
$$-15 = -a(225)$$
$$a = \frac{1}{15}$$

The equation is

$$y = -\frac{1}{15}x² + 15.$$

When y = 10,

$$10 = -\frac{x²}{15} + 15$$
$$-5 = -\frac{x²}{15}$$
$$x² = 75$$
$$x = \pm\sqrt{75}$$
$$= \pm5\sqrt{3}.$$

The width of the arch is

$5\sqrt{3} + \left|-5\sqrt{3}\right|$

$= 10\sqrt{3} \text{ m} \approx 17.32 \text{ m}.$

Section 1.5

1. The range is $[-.5, \infty)$, which is of the form $[k, \infty)$. Thus, the function is of even degree. Since there are 3 turning points the degree is 4 or greater. Possible values for the degree are 4, 6, and so on. The shape is similar to the first graph of Figure 34 so the sign of the x^n term is +.

3. Range: $(-\infty, \infty)$

   The function is of odd degree.
   Turning points: 4
   The function has degree 5 or greater. Possible values for degree: 5, 7, and so on. Shape: similar to the first graph of Figure 33. The sign of the x^n term is +.

5. Range: $(-\infty, \infty)$

   The function is of odd degree.
   Turning points: 4
   The function is of degree 5 or greater. Possible values for degree: 5, 7, and so on. Shape: similar to the first graph of Figure 33. The sign of the x^n term is +.

7. Range: $(-\infty, \infty)$

   The function is of odd degree.
   Turning points: 6
   The function is of degree 7 or greater. Possible values for degree: 7, 9, and so on. Shape: similar to the second graph of Figure 33. The sign of the x^n term is -.

For Exercises 9–55, see the answer graphs in the back of your textbook.

9. $y = \dfrac{-4}{x - 3}$

   The function is undefined for $x = 3$, so the line $x = 3$ is a vertical asymptote.

x	-97	-7	-2	0	2	4	13	103
x - 3	-100	-10	-5	-3	-1	1	10	100
y	.04	.4	.8	1.3	4	-4	-.4	-.04

   The graph approaches $y = 0$, so the line $y = 0$ (the x-axis) is a horizontal asymptote.
   Asymptotes: $y = 0$, $x = 3$
   x-intercept:
   none, because the x-axis is an asymptote

   y-intercept:
   4/3, the value when $x = 0$

11. $y = \dfrac{2}{3 + 2x}$

    $3 + 2x = 0$ when $2x = -3$ or $x = -\dfrac{3}{2}$, so the line $x = -3/2$ is a vertical asymptote.

x	-51.5	-6.5	-2	-1	3.5	48.5
3 + 2x	-100	-10	-1	1	10	100
y	-.02	-.2	-2	2	.2	.02

The graph approaches y = 0, so the line y = 0 (the x-axis) is a horizontal asymptote.

Asymptote: y = 0, x = -3/2

x-intercept:
none, since the x-axis is an asymptote

y-intercept:
2/3, the value when x = 0

13. $y = \dfrac{3x}{x - 1}$

x - 1 = 0 when x = 1, so the line x = 1 is a vertical asymptote.

x	-99	-9	-1
3x	-297	-27	-3
x - 1	-100	-10	-2
y	2.97	2.7	1.5

0	.5	1.5	12	11	101
0	1.5	4.5	6	33	303
-1	-.5	.5	1	10	100
0	-3	9	6	3.3	3.03

As x gets larger, $\dfrac{3x}{x - 1} \approx \dfrac{3x}{x} = 3$.

Thus, y = 3 is a horizontal asymptote.

Asymptotes: y = 3, x = 1

x-intercept:
0, the value when y = 0

y-intercept:
0, the value when x = 0

15. $y = \dfrac{x + 1}{x - 4}$

x - 4 = 0 when x = 4, so x = 4 is a vertical asymptote.

x	-96	-6	-1	0	3	3.5	4.5
x + 1	-95	-5	0	1	4	4.5	5.5
x - 4	-100	-10	-5	-4	-1	-.5	.5
y	.95	.5	0	-.25	-4	-9	11

5	14	104
6	15	105
1	10	100
6	1.5	1.05

As x gets larger, $\dfrac{x + 1}{x - 4} \approx \dfrac{x}{x} = 1$.

Thus y = 1 is a horizontal asymptote.

Asymptotes: y = 1, x = 4

x-intercept:
-1, the value when y = 0

y-intercept:
-.25 = -1/4, the value when x = 0

17. $y = \dfrac{1 - 2x}{5x + 20}$

5x + 20 = 0 when 5x = -20 or x = -4, so the line x = -4 is a vertical asymptote.

x	-7	-6	-5	-3	-2	-1	0
$1 - 2x$	15	13	11	7	5	3	1
$5x + 20$	-15	-10	-5	5	10	15	20
y	-1	-1.3	-2.2	1.4	.5	.2	.05

As x gets larger, $\dfrac{1 - 2x}{5x + 20} \approx \dfrac{-2x}{5x} = \dfrac{-2}{5}$.

Thus, the line $y = -2/5$ is a horizontal asymptote.

Asymptotes: $x = -4$, $y = -2/5$

x-intercept:

1/2, the value when $y = 0$

y-intercept:

.05 = 1/20, the value when $x = 0$

19. $y = \dfrac{-x - 4}{3x + 6}$

$3x + 6 = 0$ when $3x = -6$ or $x = -2$, so the line $x = -2$ is a vertical asymptote.

x	-5	-4	-3	-1	0	1
$-x - 4$	1	0	-1	-3	-4	-5
$3x + 6$	-9	-6	-3	3	6	9
y	-.11	0	.33	-1	-.67	-.56

As x gets larger, $\dfrac{-x - 4}{3x + 6} \approx \dfrac{-x}{3x} = -\dfrac{1}{3}$.

The line $y = -1/3$ is a horizontal asymptote.

Asymptotes: $y = -1/3$, $x = -2$

x-intercept:

-4, the value when $y = 0$

y-intercept:

-.67 = -2/3, the value when $x = 0$

21. $C(x) = \dfrac{500}{x + 30}$

(a) $C(10) = \dfrac{500}{10 + 30}$

$= \dfrac{500}{40}$

$= \$12.50$

$C(20) = \dfrac{500}{20 + 30}$

$= \dfrac{500}{50}$

$= \$10$

$C(50) = \dfrac{500}{50 + 30}$

$= \dfrac{500}{80}$

$= \$6.25$

$C(75) = \dfrac{500}{75 + 30}$

$= \dfrac{500}{105}$

$\approx \$4.76$

$C(100) = \dfrac{500}{100 + 30}$

$= \dfrac{500}{130}$

$= 3.8461538$

$\approx \$3.85$

(b) $(0, \infty)$ would be a more reasonable domain for average cost than $[0, \infty)$. If zero were included in the domain, there would be no units produced. It is not reasonable to discuss the average cost per unit of zero units.

(c) See the answer graph in the back of your textbook.

23. $y = x(100 - x)(x^2 + 500)$,

y = tax revenue in hundreds of thousands of dollars

x = tax rate.

(a) $x = 10$

$$y = 10(100 - 10)(10^2 + 500)$$
$$= 10(90)(600)$$
$$= \$54 \text{ billion}$$

(b) $x = 40$

$$y = 40(100 - 40)(40^2 + 500)$$
$$= 40(60)(2100)$$
$$= \$504 \text{ billion}$$

(c) $x = 50$

$$y = 50(100 - 50)(50^2 + 500)$$
$$= 50(50)(3000)$$
$$= \$750 \text{ billion}$$

(d) $x = 80$

$$y = 80(100 - 80)(80^2 + 500)$$
$$= 80(20)(6900)$$
$$= \$1104 \text{ billion}$$

(e) See the answer graph in the back of the textbook.

25. $y = \dfrac{60x - 6000}{x - 120}$,

y = revenue in millions

x = tax rate.

(a) $x = 50$

$$y = \frac{60(50) - 6000}{50 - 120}$$
$$= \frac{-3000}{-70}$$
$$= \$42.9 \text{ million}$$

(b) $x = 60$

$$y = \frac{60(60) - 6000}{60 - 120}$$
$$= \frac{-2400}{-60}$$
$$= \$40 \text{ million}$$

(c) $x = 80$

$$y = \frac{60(80) - 6000}{80 - 120}$$
$$= \frac{-1200}{-40}$$
$$= \$30 \text{ million}$$

(d) $x = \dfrac{60(100) - 6000}{100 - 120}$

$$= \$0$$

(e) See the answer graph in the back of the textbook.

27. $y = \dfrac{6.7x}{100 - x}$,

y = cost in thousands

x = percent of pollutant.

(a) $x = 50$

$$y = \frac{6.7(50)}{100 - 50}$$
$$= 6.7$$

The cost is $6700.

(b) $x = 70$

$$y = \frac{6.7(70)}{100 - 70}$$
$$\approx 15.6$$

The cost is $15,600.

(c) $x = 80$

$$y = \frac{6.7(80)}{100 - 80}$$
$$= 26.8$$

The cost is $26,800.

(d) x = 90

$$y = \frac{6.7(90)}{100 - 90}$$

$$= 60.3$$

The cost is $60,300.

(e) x = 95

$$y = \frac{6.7(95)}{100 - 95}$$

The cost is $127,300.

(f) x = 98

$$y = \frac{6.7(98)}{100 - 98}$$

$$= 328.3$$

The cost is $328,300.

(g) x = 99

$$y = \frac{6.7(99)}{100 - 99}$$

$$= 663.3$$

The cost is $663,300.

(h) No, because x = 100% makes the denominator zero so x = 100% is a vertical asymptote.

(i) See the answer graph in the back of your textbook.

29. **(a)** See the answer graph in the back of your textbook.

(b) The function defined by

g(x) = -.006x⁴ + .140x³ - .053x² + 1.79x,

a fourth-degree polynomial, will have a shape similar to the graphs in Figure 34. The negative coefficient on the x⁴ term indicates that both "ends" of the graph will

eventually head toward -∞, as shown in the second graph in Figure 34. So, the function will not keep increasing but will turn and begin to decrease.

31. A(x) = -.015x² + 1.058x

(a) See the answer graph at the back of your textbook.

(b) Reading the graph, we find that concentration is maximum between x = 4 hr and x = 5 hr, but closer to 5 hr.

(c) Concentration exceeds.08% from less than 1 hr to about 8.4 hr.

33. $f(x) = \dfrac{\lambda x}{1 + (ax)^b}$

(a) A reasonable domain for the function is [0, ∞). Populations are not measured using negative numbers and they may get extremely large.

(b) See the answer graph in the back of the textbook.

(c) See the answer graph in the back of the textbook.

(d) As seen from the graphs, when b increases, the population of the next generation, f(x), gets smaller when the current generation, x, is larger.

35. $G(R) = \dfrac{r}{r + R}$

R is resistance; r = 1000 ohms

(a) $G(R) = \dfrac{R}{1000 + R}$

G is undefined if R = −1000 ohms, so
R = −1000 is a vertical asymptote.

(b) $G(R) = \dfrac{R}{1000 + R}$

As R gets larger,

$$\frac{R}{1000 + R} \approx \frac{R}{R} = 1.$$

Thus G(R) = 1 is a horizontal
asymptote.

(c) See the answer graph at the back
of your textbook.

37. $f_t(u) = u^3 + tu$

(a) t = 1

$f_1(u) = u^3 + u$

u	−2	−1	0	1	2
f(u)	−10	−2	0	2	10

See the answer graph in the back of
the textbook.

(b) t = −1

$f_{-1}(u) = u^3 + (-1)u$

$= u^3 - u$

u	−2	−1	−.5	0	.5	1	2
f(u)	−6	0	.375	0	−.375	0	6

See the answer graph in the back of
the textbook.

39. $C(x) = \dfrac{10x}{49(101 - x)}$

C(x) is in thousands of dollars; x
represents points.

(a) $C(99) = \dfrac{10(99)}{49(101 - 99)}$

$= \dfrac{990}{49(2)}$

$= \dfrac{990}{98}$

$= 10.102040$

The cost is about $10,000.

(c) $C(100) = \dfrac{10(100)}{49(101 - 100)}$

$= \dfrac{1000}{49(1)}$

$= \dfrac{1000}{49}$

$= 20.408163$

The cost is about $20,000.

41. $f(x) = -x^3 + 4x^2 + 3x - 8$

x	−3	−2	−1	0	1	2	3	4	5
f(x)	46	10	−6	−8	−2	6	10	4	−18

See the answer graph in the textbook.

43. $f(x) = 2x^3 + 4x - 1$

x	−2	−1	0	1	2
f(x)	−25	−7	−1	5	23

See the answer graph in the textbook.

45. $f(x) = x^4 + x^3 - 2$

x	-2	-1	0	1	2
f(x)	6	-2	-2	0	22

See the answer graph in the textbook.

47. $f(x) = -x^4 - 2x^3 + 3x^2 + 3x + 5$

x	-3	-2	-1	0	1	2
f(x)	-4	11	6	5	8	-9

See the answer graph in the textbook.

49. $f(x) = -x^5 + 6x^4 - 11x^3 + 6x^2 + 5$

x	-2	-1	-.5	0	.5
f(x)	245	29	5.1	5	5.2

x	1	1.5	2	2.5	3
f(x)	5	4.2	5	7.3	5

See the answer graph in the textbook.

51. $f(x) = \dfrac{-2x^2 + x - 1}{2x + 3}$

Vertical asymptote:

$$2x + 3 = 0$$
$$2x = -3$$
$$x = -\frac{3}{2}$$

Horizontal asymptote:
None, because the degree of the numerator is greater than the degree of the denominator.

x	-5	-4	-3	-2	-1	0	1	2
f(x)	8	7.4	7.3	12	-4	-.3	-.4	-1

See the answer graph in the textbook.

53. $f(x) = \dfrac{2x^2 - 5}{x^2 - 1}$

Vertical asymptotes:

$$x^2 - 1 = 0$$
$$(x + 1)(x - 1) = 0$$
$$x = -1, \ x = 1$$

Horizontal asymptote:
y = 2, since f(x) approaches 2 as x gets larger.

x	-4	-3	-2	0	2	3	4
f(x)	1.8	1.6	3	5	3	1.6	1.8

See the answer graph in the textbook.

55. $f(x) = \dfrac{-2x^2}{x^2 - 10}$

Vertical asymptotes:

$$x^2 - 10 = 0$$
$$x^2 = 10$$
$$x = \pm\sqrt{10} \approx \pm 3.16$$

Horizontal asymptote:
y = -2, since f(x) approaches -2 as x gets larger.

x	-5	-4	-3	-2	-1	0
f(x)	-3.3	-5.3	18	1.3	.22	0

x	1	2	3	4	5
f(x)	.22	1.3	18	5.3	3.3

See the answer graph in the textbook.

Section 1.6

For Exercises 1–15, see the answer
graphs in the back of the textbook.

1. $f(x) = (x + 2)^3$

    Translate the graph of $f(x) = x^2$
    2 units left.

3. $f(x) = (4 - x)^3 + 1$
    $= -(x - 4)^3 + 1$

    Translate the graph of $f(x) = x^2$
    4 units right and 1 unit up.
    Reflect it vertically.

5. $f(x) = -2(x + 1)^3 + 5$

    Translate the graph of $f(x) = x^3$
    1 unit left and 5 units up. Reflect
    the graph vertically and stretch it
    vertically by a factor of 2.

7. $f(x) = -2x^5 + 5$

    Translate the graph of $f(x) = x^5$
    5 units up. Reflect it vertically.
    Then stretch it vertically by a
    factor of 2.

9. $f(x) = \dfrac{2}{x + 3} - 2$

    Translate the graph of $f(x) = \dfrac{1}{x}$

    3 units left and 2 units down.
    Stretch it vertically by a factor
    of 2.

11. $f(x) = \dfrac{-1}{x - 1} + 2$

    Translate the graph of $f(x) = \dfrac{1}{x}$

    1 unit right and 2 units up.
    Reflect it vertically.

13. $f(x) = \sqrt{x - 1} + 3$

    Translate the graph of $f(x) = \sqrt{x}$
    1 unit right and 3 units up.

15. $f(x) = -\sqrt{-4 - x} - 2$
    $= -\sqrt{-(x + 4)} - 2$

    Translate the graph of $f(x) = \sqrt{x}$
    4 units left and 2 units down.
    Reflect vertically and horizontally.

For Exercises 17–23, see the answer
graphs in the textbook.

17. $f(x) = x^2 - 2x - 2$
    $= x^2 - 2x + \left(\dfrac{-2}{2}\right)^2 - \left(\dfrac{-2}{2}\right)^2 - 2$
    $= (x^2 - 2x + 1) - 1 - 2$
    $= (x - 1)^2 - 3$

    Translate: 1 unit right and 3 units
    down

19. $f(x) = x^2 - 4x - 2$
    $= x^2 - 4x + \left(\dfrac{-4}{2}\right)^2 - \left(\dfrac{-4}{2}\right)^2 - 2$
    $= (x^2 - 4x + 4) - 4 - 2$
    $= (x - 2)^2 - 6$

    Translate: 2 units right and 6 units
    down

21. $f(x) = 2x^2 + 4x - 1$

$\qquad = 2(x^2 + 2x) - 1$

$\qquad = 2\left[x^2 + 2x + \left(\dfrac{2}{2}\right)^2 - \left(\dfrac{2}{2}\right)^2\right] - 1$

$\qquad = 2(x^2 + 2x + 1) - 2(1) - 1$

$\qquad = (x + 2)^2 - 3$

Translate: 2 units left and 3 units down

Stretch vertically by a factor of 2.

23. $f(x)$

$\qquad = -3x^2 + 24x - 55$

$\qquad = -3(x^2 - 8x) - 55$

$\qquad = -3\left[x^2 - 8x + \left(-\dfrac{8}{2}\right)^2 - \left(-\dfrac{8}{2}\right)^2\right] - 55$

$\qquad = -3(x^2 - 8x + 16) - 3(-16) - 55$

$\qquad = -3(x - 4)^2 - 7$

Translate: 4 units right and 7 units down

Reflect: vertically

Stretch vertically by a factor of 3.

For Exercises 25–33, see the answer graphs in the textbook.

25. If $0 < a < 1$, the graph of $f(ax)$ will be flatter and wider than the graph of $f(x)$.
Multiplying x by a fraction makes the y-values less than the original y-values.

27. If $-1 < a < 0$, the graph of $f(ax)$ will be reflected horizontally, since a is negative. It will be flatter because multiplying x by a fraction decreases the corresponding y-values.

29. If $0 < a < 1$, the graph of $af(x)$ will be flatter and wider than the graph of $f(x)$. Each y-value is only a fraction of the height of the original y-values.

31. If $-1 < a < 0$, the graph will be reflected vertically, since a will be negative. Also, because a is a fraction, the graph will be flatter because each y-value will only be a fraction of its original height.

33. **(a)** If $f(x) = \dfrac{c}{x}$, then

$$af(x - h) + k = a\left[\dfrac{c}{x - h}\right] + k$$

$$= \dfrac{ac}{x - h} + k$$

$$= \dfrac{ac + kx - hk}{x - h}$$

(c) and

(d) See the answer graphs in the textbook.

35. $S(x) = \dfrac{1}{25}(x - 10)^3 + 40$

(a) $S(0) = \dfrac{1}{25}(0 - 10)^3 + 40$

$\qquad = \dfrac{1}{25}(-1000) + 40$

$\qquad = -40 + 40$

$\qquad = \$0$

(b) $S(10) = \dfrac{1}{25}(10 - 10)^3 + 40$

$\qquad = \dfrac{1}{25} \cdot 0 + 40$

$\qquad = 40 \quad\text{or}\quad \$40,000$

(c) See the answer graph in the textbook.

(d) $A(x) = \bar{S}(x)$

$$= \frac{S(x)}{x} = \frac{\frac{1}{25}(x-10)^3 + 40}{x}$$

(e) $A(10) = \dfrac{\frac{1}{25}(10-10)^3 + 40}{x}$

$$= \frac{0 + 40}{10} = \$4$$

(f) $A(50) = \dfrac{\frac{1}{25}(50-10)^3 + 40}{50}$

$$= \frac{2560 + 40}{50} = \$52$$

Chapter 1 Review Exercises

For Exercises 5–13, the domain is $\{-3, -2, -1, 0, 1, 2, 3\}$. See the graphs in the answer section of your textbook.

5. $2x - 5y = 10$

$\qquad -5y = -2x + 10$

$\qquad\quad y = \frac{2}{5}x - 2$

x	−3	−2	−1	0
y	−16/5	−14/5	−12/5	−2

x	1	2	3
y	−8/5	−6/5	−4/5

Pairs: (−3, −16/5), (−2, −14/5), (−1, −12/5), (0, −2), (1, −8/5), (2, −6/5), (3, −4/5)

Range: $\{-16/5, -14/5, -12/5, -2,$ $-8/5, -6/5, -4/5\}$

7. $y = (2x + 1)(x - 1)$

$\qquad = 2x^2 - x - 1$

x	−3	−2	−1	0	1	2	3
y	20	9	2	−1	0	5	14

Pairs: (−3, 20), (−2, 9), (−1, 2), (0, −1), (1, 0), (2, 5), (3, 14)

Range: $\{-1, 0, 2, 5, 9, 14, 20\}$

9. $y = -2 + x^2$

x	−3	−2	−1	0	1	2	3
y	7	2	−1	−2	−1	2	7

Pairs: (−3, 7), (−2, 2), (−1, −1), (0, −2), (1, −1), (2, 2), (3, 7)

Range: $\{-2, -1, 2, 7\}$

11. $y = \dfrac{2}{x^2 + 1}$

x	−3	−2	−1	0	1	2	3
y	1/5	2/5	1	2	1	2/5	1/5

Pairs: (−3, 1/5), (−2, 2/5), (−1, 1), (0, 2), (1, 1), (2, 2/5), (3, 1/5)

Range: $\{1/5, 2/5, 1, 2\}$

13. $y + 1 = 0$

$\qquad\quad y = -1$

x	−3	−2	−1	0	1	2	3
y	−1	−1	−1	−1	−1	−1	−1

Pairs: $(-3, -1)$, $(-2, -1)$, $(-1, -1)$,
$(0, -1)$, $(1, -1)$, $(2, -1)$,
$(3, -1)$

Range: $\{-1\}$

15. $f(x) = 4x - 1$

   (a) $f(6) = 4(6) - 1$
          $= 24 - 1$
          $= 23$

   (b) $f(-2) = 4(-2) - 1$
           $= -8 - 1$
           $= -9$

   (c) $f(-4) = 4(-4) - 1$
           $= -16 - 1$
           $= -17$

   (d) $f(r + 1) = 4(r + 1) - 1$
            $= 4r + 4 - 1$
            $= 4r + 3$

17. $f(x) = -x^2 + 2x - 4$

   (a) $f(6) = -(6)^2 + 2(6) - 4$
          $= -36 + 12 - 4$
          $= -28$

   (b) $f(-2) = -(-2)^2 + 2(-2) - 4$
           $= -4 - 4 - 4$
           $= -12$

   (c) $f(-4) = -(-4)^2 + 2(-4) - 4$
           $= -16 - 8 - 4$
           $= -28$

   (d) $f(r + 1)$
       $= -(r + 1)^2 + 2(r + 1) - 4$
       $= -(r^2 + 2r + 1) + 2r + 2 - 4$
       $= -r^2 - 3$

19. $f(x) = 5x - 3$
    $g(x) = -x^2 + 4x$

   (a) $f(-2) = 5(-2) - 3$
           $= -10 - 3$
           $= -13$

   (b) $g(3) = -(3)^2 + 4(3)$
          $= -9 + 12$
          $= 3$

   (c) $g(-4) = -(-4)^2 + 4(-4)$
          $= -16 - 16$
          $= -32$

   (d) $f(5) = 5(5) - 3$
          $= 25 - 3$
          $= 22$

   (e) $g(-k) = -(-k)^2 + 4(-k)$
           $= -k^2 - 4k$

   (f) $g(3m) = -(3m)^2 + 4(3m)$
           $= -9m^2 + 12m$

   (g) $g(k - 5)$
      $= -(k - 5)^2 + 4(k - 5)$
      $= -(k^2 - 10k + 25) + 4k - 20$
      $= -k^2 + 10k - 25 + 4k - 20$
      $= -k^2 + 14k - 45$

   (h) $f(3 - p) = 5(3 - p) - 3$
            $= 15 - 5p - 3$
            $= 12 - 5p$

For Exercises 21–27, see the answer graphs in your textbook.

21. $y = 4x + 3$

   Let $x = 0$.

        $y = 4(0) + 3$
        $y = 3$

Let $y = 0$.

$$0 = 4x + 3$$
$$-3 = 4x$$
$$-\frac{3}{4} = x$$

Draw the line through $(0, 3)$ and $(-3/4, 0)$.

23. $3x - 5y = 15$
$$-5y = -3x + 15$$
$$y = \frac{3}{5}x + 3$$

When $x = 0$, $y = -3$.
When $y = 0$, $x = 5$.
Draw the line through $(0, -3)$ and $(5, 0)$.

25. $x + 2 = 0$
$$x = -2$$

This is the vertical line through $(-2, 0)$.

27. $y = 2x$

When $x = 0$, $y = 0$.
When $x = 1$, $y = 2$.
Draw the line through $(0, 0)$ and $(1, 2)$.

29. Through $(-2, 5)$ and $(4, 7)$
$$m = \frac{5 - 7}{-2 - 4}$$
$$= \frac{-2}{-6}$$
$$= \frac{1}{3}$$

31. Through the origin and $(11, -2)$
$$m = \frac{-2 - 0}{11 - 0} = -\frac{2}{11}$$

33. $2x + 3y = 15$
$$3y = -2x + 15$$
$$y = -\frac{2}{3}x + 15$$
$$m = -\frac{2}{3}$$

35. $x + 4 = 9$
$$x = 9 - 4$$
$$x = 5$$

The line is vertical. The slope is not defined.

37. Through $(5, -1)$, with slope $\frac{2}{3}$.
Use point-slope form.

$$y - (-1) = \frac{2}{3}(x - 5)$$
$$y + 1 = \frac{2}{3}(x - 5)$$
$$y = \frac{2}{3}x - \frac{10}{3} - 1$$
$$3y = 2x - 10 - 3$$
$$2x - 3y = 13$$

39. Through $(5, -2)$ and $(1, 3)$
$$m = \frac{-2 - 3}{5 - 1} = \frac{-5}{4}$$

Use point-slope form.

$$y - (-2) = \frac{-5}{4}(x - 5)$$
$$y + 2 = \frac{-5}{4}x + \frac{25}{4}$$
$$4y + 8 = -5x + 25$$
$$5x + 4y = 17$$

41. Undefined slope, through $(-1, 4)$

Undefined slope means the line is vertical. The vertical line through $(-1, 4)$ is $x = -1$.

43. Through $(0, 5)$ and perpendicular to $8x + 5y = 3$ Find the slope of the given line first.

$$8x + 5y = 3$$
$$5y = -8x + 3$$
$$y = \frac{-8}{5}x + \frac{3}{5}$$
$$m = -\frac{8}{5}$$

The perpendicular line has $m = 5/8$. Use point-slope form.

$$y - 5 = \frac{5}{8}(x - 0)$$
$$y = \frac{5}{8}x + 5$$
$$5x - 8y = -40$$

45. Through $(3, -5)$, parallel to $y = 4$

Find the slope of the given line. $y = 0x + 4$, so $m = 0$, and the required line will also have slope 0.
Use the point-slope form.

$$y - (-5) = 0(x - 3)$$
$$y + 5 = 0$$
$$y = -5$$

Another method: Note $y = 4$ is a horizontal line. The required line then will be horizontal, and have an equation of the form $y = b$, that is, all points on the line have the same y-coordinate. Since $(3, -5)$ is on the line, the y-coordinate is -5, and the equation of the line is $y = -5$.

For Exercises 47–69, see the answer graphs in the back of your textbook.

47. Through $(2, -4)$, $m = \frac{3}{4}$

One point on the line is $(2, -4)$

Slope $= \frac{3}{4} = \frac{\Delta y}{\Delta x}$

$x_2 = x_1 + \Delta x = 2 + 4 = 6$
$y_2 = y_1 + \Delta y = -4 + 3 = -1$

A second point on the line is $(6, -1)$.
Draw the line through these points.

49. Through $(-4, -1)$, $m = 3$
One point on the line is $(-4, 1)$.

Slope $= \frac{3}{1} = \frac{\Delta y}{\Delta x}$

$x_2 = x_1 + \Delta x = -4 + 1 = -3$
$y_2 = y_1 + \Delta y = 1 + 3 = 4$

A second point is $(-3, 4)$.
Draw the line through these points.

51. $y = x^2 - 4$
     $= (x + 2)(x - 2)$

Set $y = 0$ to find the x-intercepts.

$0 = (x + 2)(x - 2)$
$x = -2$ or $x = 2$

Set x = 0 to find the y-intercept.

$y = 0^2 - 4$

$y = -4$

The graph is a parabola.

Vertex: $x = \dfrac{-b}{2a} = \dfrac{-0}{2} = 0$

$y = 0^2 - 4 = -4$

The vertex is $(0, -4)$.

53. $y = 3x^2 + 6x - 2$

The graph is a parabola.
Let $y = 0$.

$$0 = 3x^2 + 6x - 2$$

$x = \dfrac{-6 \pm \sqrt{6^2 - 4(3)(-2)}}{2(3)}$

$= \dfrac{-6 \pm \sqrt{36 + 24}}{6} = \dfrac{-6 \pm \sqrt{60}}{6}$

$= \dfrac{-6 \pm 2\sqrt{15}}{6} = -1 \pm \dfrac{\sqrt{15}}{3}$

The x-intercepts are $-1 + \sqrt{15}/3 \approx$
.29 and $-1 - \sqrt{15}/3 \approx -2.29$.
Let $x = 0$.

$y = 3(0)^2 + 6(0) - 2$

-2 is the y-intercept.

Vertex: $x = \dfrac{-b}{2a} = \dfrac{-6}{2(3)} = -1$

$y = 3(-1)^2 + 6(-1) - 2$

$= 3 - 6 - 2$

$= -5$

The vertex is $(-1, -5)$.

55. $y = x^2 - 4x + 2$

Let $y = 0$.

$$0 = x^2 - 4x + 2$$

$x = \dfrac{-(-4) \pm \sqrt{(-4)^2 - 4(1)(2)}}{2(1)}$

$= \dfrac{4 \pm \sqrt{8}}{2}$

$= 2 \pm \sqrt{2}$

The x-intercepts are $2 + \sqrt{2} \approx 3.41$
and $2 - \sqrt{2} \approx -1.41$.

Let $x = 0$.

$y = 0^2 - 4(0) + 2$

2 is the y-intercept.

Vertex: $x = \dfrac{-b}{2a} = \dfrac{-(-4)}{2(1)} = \dfrac{4}{2} = 2$

$y = 2^2 - 4(2) + 2 = -2$

The vertex is $(2, -2)$.

57. $f(x) = x^3 + 5$

Translate the graph of $f(x) = x^3$
5 units up.

59. $y = -(x - 1)^3 + 4$

Translate the graph of $y = x^3$
1 unit right and 4 units up.
Reflect vertically.

61. $y = 2\sqrt{x + 3} + 1$

Translate the graph of $y = \sqrt{x}$
3 units left and 1 unit up. Stretch
the graph vertically by a factor
of 2.

63. $f(x) = \dfrac{8}{x}$

Vertical asymptote:

$x = 0$

Horizontal asymptote:

$\dfrac{8}{x}$ approaches zero as x gets larger.

$y = 0$ is an asymptote.

x	−4	−3	−2	−1	1	2	3	4
y	−2	−2.7	−4	−8	8	4	2.7	2

65. $f(x) = \dfrac{4x - 2}{3x + 1}$

Vertical asymptote:

$$3x + 1 = 0$$

$$x = -\frac{1}{3}$$

Horizontal asymptote:

As x get larger,

$$\frac{4x - 2}{3x - 1} \approx \frac{4x}{3x} = \frac{4}{3}.$$

$y = \dfrac{4}{3}$ is an asymptote.

x	−3	−2	−1	0	1	2	3
y	1.75	2	3	−2	.5	.86	1

67. Exercise 53 is

$$y = 3x^2 + 6x - 2.$$

Completing the square produces

$$y = 3(x + 1)^2 - 5.$$

Translate $y = x^2$ 1 unit left and 5 units down. Stretch vertically by a factor of 3.

69. Exercise 55 is

$$y = x^2 - 4x + 3.$$

Completing the square produces

$$y = (x - 2)^2 - 2.$$

Translate $y = x^2$ 2 units right and 2 units down.

71. Supply: $p = q + 3$
Demand: $p = 19 - 2q$

(a)
$10 = 6q + 3$	$10 = 19 - 2q$
$7 = 6q$	$2q = 19 - 10$
	$2q = 9$
$\dfrac{7}{6} = q$	$q = \dfrac{9}{2}$

Supply: 7/6   Demand: 9/2

(b)
$15 = 6q + 3$	$15 = 19 - 2q$
$12 = 6q$	$2q = 19 - 15$
$2 = q$	$2q = 4$
	$q = 2$

Supply: 2   Demand: 2

(c)
$18 = 6q + 3$	$18 = 19 - 2q$
$15 = 6q$	$2q = 19 - 18$
$\dfrac{15}{6} = q$	$2q = 1$
$\dfrac{5}{2} = q$	$q = \dfrac{1}{2}$

Supply: 5/2   Demand: 1/2

(d) See the answer graphs in the back of your textbook.

(e) The equilibrium price is 15 (refer to (b) above or read graph).

(f) The equilibrium quantity is 2 (refer to (b) above or read graph).

73. Eight units cost $300; fixed cost is $60.

The fixed cost is the cost if zero units are made. (8, 300) and (0, 60) are points on the line.

$$m = \frac{60 - 300}{0 - 8} = 30$$

Use slope–intercept form.

$$y = 30x + 60$$

$$C(x) = 30x + 60$$

$$A(x) = \frac{C(x)}{x} = \frac{30x + 60}{x} = 30 + \frac{60}{x}$$

Thirty units cost $1500; 120 units cost $5640.

75. Thirty units cost $1500; 120 units cost $5640.

Points on the line are (30, 1500), (120, 5640).

$$m = \frac{5640 - 1500}{120 - 30}$$

$$= \frac{4140}{90}$$

$$= 46$$

Use point–slope form.

$$y - 1500 = 46(x - 30)$$

$$y = 46x - 1380 + 1500$$

$$y = 46x + 120$$

$$C(x) = 46x + 120$$

$$A(x) = \frac{C(x)}{x} = \frac{46x + 120}{x}$$

$$= 46 + \frac{120}{x}$$

77. **(a)** For x in the interval $0 < x \le 1$, the renter is charged the fixed cost of $40 and 1 days rent of $40 so

$$C\left(\frac{3}{4}\right) = \$40 + \$40(1)$$

$$= \$40 + \$40$$

$$= \$80.$$

(b) $C\left(\frac{9}{10}\right) = \$40 + \$40(1)$

$$= \$40 + \$40$$

$$= \$80$$

(c) $C(1) = \$40 + \$40(1)$

$$= \$40 + \$40$$

$$= \$80$$

(d) For x in the interval $1 < x \le 2$, the renter is charged the fixed cost of $40 and 2 days rent of $80. So

$$C\left(1\frac{5}{8}\right) = \$40 + \$40(2)$$

$$= \$40 + \$80$$

$$= \$120.$$

(e) For x in the interval $2 < x \le 3$ the renter is charged the fixed cost of $40 and 3 days rent of $120. So

$$C\left(2\frac{1}{9}\right) = \$40 + \$40(3)$$

$$= \$40 + \$120$$

$$= \$160.$$

(f) See the answer graph in text-book.

(g) The independent variable is the number of days, or x.

(h) The dependent variable is the cost, or $C(x)$.

79. **(a)** As seen on the graph, the rate was the highest halfway through 1984.

(b) As seen on the graph, the maximum rate was 160,000.

(c) The range of the function is the interval of possible y–values or $[0, 160,000]$.

81. **(a)** The dosages are equal when

$$\frac{12}{A + 12} = \frac{A + 1}{24}$$

$$24A = (A + 12)(A + 1)$$

$$= A^2 + 13A + 12$$

$$0 = A^2 - 11A + 12$$

$$A = \frac{11 \pm \sqrt{121 - 48}}{2}$$

$$A = \frac{11 - \sqrt{72}}{2} \quad \text{or} \quad A = \frac{11 + \sqrt{72}}{2}$$

$$\approx 1.2 \qquad\qquad \approx 9.8.$$

The two dosages are at approximately 1.2 years and 9.8 years.

(b) See the answer graph in the back of your textbook.

$f(A) > g(A)$ for $1.2 < A < 9.8$

$g(A) > f(A)$ for $A < 1.2$ or $A > 9.8$.

In the interval shown in the graph,

$f(A) > g(A)$ for $2 < A < 9.8$

$g(A) > f(A)$ for $9.8 < A < 13$.

(c) The functions f and g seem to differ most at 5 yr and 13 yr.

83. **(a)** From the graph, the maximum horsepower occurs at 5750 rpm.

(b) From the graph, the maximum horsepower is about 310.

(c) When the horsepower is at its maximum, the graph shows the corresponding torque as 280 ft–lb.

85. $C(x) = \dfrac{5x + 3}{x + 1}$

(a) See the answer graph in the textbook.

(b)

$C(x + 1) - C(x)$

$$= \frac{5(x + 1) + 3}{(x + 1) + 1} - \frac{5x + 3}{x + 1}$$

$$= \frac{5x + 8}{x + 2} - \frac{5x + 3}{x + 1}$$

$$= \frac{(5x + 8)(x + 1) - (5x + 3)(x + 2)}{(x + 2)(x + 1)}$$

$$= \frac{5x^2 + 13x + 8 - 5x^2 - 13x - 6}{(x + 2)(x + 1)}$$

$$= \frac{2}{(x + 2)(x + 1)}$$

(c) $A(x) = \dfrac{C(x)}{x} = \dfrac{\dfrac{5x + 3}{x + 1}}{x}$

$$= \frac{5x + 3}{x(x + 1)}$$

(d)

$A(x + 1) - A(x)$

$$= \frac{5(x + 1) + 3}{(x + 1)[(x + 1) + 1]} - \frac{5x + 3}{x(x + 1)}$$

$$= \frac{5x + 8}{(x + 1)(x + 2)} - \frac{5x + 3}{x(x + 1)}$$

$$= \frac{x(5x + 8) - (5x + 3)(x + 2)}{x(x + 1)(x + 2)}$$

$$= \frac{5x^2 + 8x - 5x^2 - 13x - 6}{x(x + 1)(x + 2)}$$

$$= \frac{-5x - 6}{x(x + 1)(x + 2)}$$

Extended Application

1. $y = .133x + 10.09$, x in millions of units. Selling price is $10.73. Then

$$10.73 = .133x + 10.09$$
$$.64 = .133x$$
$$4.8 \approx x.$$

So the company has marginal revenue = marginal cost when it makes 4.8 million units.

2. For product B,

$$y = .0667x + 10.29.$$

When $x = 3.1$, $y = 10.50$.
When $x = 5.7$, $y = 10.67$.
The graph is the line segment connecting the points

$$(3.1, 10.5) \quad \text{and} \quad (5.7, 10.67).$$

See the graph in the answer section at the back of your textbook.

3. For product B,

$$y = .0667x + 10.29,$$

which is always ≥ 10.29 for $x \geq 0$. The marginal revenue is only 9.65. Therefore, there is no production level possible where the marginal cost = marginal revenue.

4. For product C, the marginal cost y is

$$y = .133x + 9.46.$$

(a) At 3.1 million units,

$$y = .133(3.1) + 9.46$$
$$= \$9.87.$$

At 5.7 million units,

$$y = .133(5.7) + 9.46$$
$$= \$10.22.$$

(b) The marginal cost graph is the portion of a line connecting the points (3.1, 9.87) and (57, 10.22).

See the graph in the answer section at the back of your textbook.

(c) At a selling price of $9.57, the cost equals the selling price when

$$9.57 = .133x + 9.46$$
$$x = .83 \quad \text{million units,}$$

which is not in the interval under discussion.

CHAPTER 1 TEST

1. List the ordered pairs obtained from each of the following if the domain of x for each exercise is $\{-3, \ 2, -1, 0, 1, 2, 3\}$. Graph each set of ordered pairs. Give the range.

   (a) $2x + 3y = 6$ **(b)** $y = \dfrac{1}{x^2 - 2}$

2. Let $f(x) = 3x - 4$ and $g(x) = -x^2 = 5x$. Find each of the following.

   (a) $f(-3)$ **(b)** $g(-2)$ **(c)** $f(2m)$ **(d)** $g(k - 1)$

   (e) $f(x + h)$

Graph each of the following.

3. $3x + 5y = 15$ 4. $2x + y = 0$ 5. $x - 3 = 0$

6. The line through $(-2, 3)$, $m = -3$

In Exercises 7–9, find the slope of each line for which the slope is defined.

7. Through $(2, -5)$ and $(-1, 7)$ 8. $3x - 7y = 9$

9. Through $(9, 5)$ and $(9, 2)$

In Exercises 10–14, find an equation in the form ax + by = c for each line.

10. Through the origin and $(5, -3)$

11. Undefined slope through $(1, -2)$

12. Through $(0, 3)$, parallel to $2x - 4y = 1$

13. Through $(1, -4)$, perpendicular to $3x + y = 1$

14. Through $(2, 5)$, perpendicular to $x = 5$

15. Let the supply and demand functions for a certain product be given by the following equations.

   Supply: $p = .20q - 5$ Demand: $p = 100 - .15q$

   where p represents the price at a supply or demand, respectively, of q units.

   (a) Graph these equations on the same axes.

   (b) Find the equilibrium price.

   (c) Find the equilibrium quantity.

16. (a) Find the appropriate linear cost function if eight units cost $450, while forty units cost $770.

   (b) What is the fixed cost?

   (c) What is the marginal cost per item?

17. The manufacturer of a certain product has determined that his profit in dollars for making x units of this product is given by the equation

   $$p = -2x^2 + 120x + 3000.$$

   (a) Find the number of units that will maximize the profit.

   (b) What is the maximum profit for making this product?

Graph each of the following.

18. $y = 2x^2 - 4x - 2$ 19. $f(x) = x^3 - 2x^2 - x + 2$

20. $f(x) = \dfrac{3}{2x - 1}$ 21. $f(x) = \dfrac{x - 1}{3x + 6}$

22. $f(x) = \dfrac{2x}{(x + 1)(x - 2)}$

23. Suppose a cost-benefit model is given by the equation

   $$y = \frac{20x}{110 - x}$$

   where y is the cost in thousands of dollars of removing x percent of a certain pollutant. Find the cost in thousands of dollars of removing each of the following percents of pollution.

   (a)   70%   (b)   90%   (c)   100%

Use translations and reflections to graph the following.

24. $y = 4 - x^2$

25. $f(x) = \frac{1}{2}x^3 + 1$

CHAPTER 1   TEST ANSWERS

1. **(a)** (−3, 4), (−2, 10/3), (−1, 8/3),
    (0, 2), (1, 4/3), (2, 2/3), (3, 0);
    range: $\{0, 2/3, 4/3, 2, 8/3, 10/3, 4\}$

    (b) (−3, 1/7), (−2, 1/2), (−1, −1),
    (0, −1/2), (1, −1), (2, 1/2), (3, 1/7);
    range: $\{−1, −1/2, 1/7, 1/2\}$

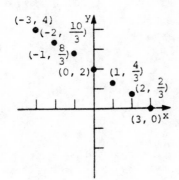

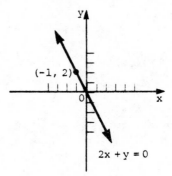

2. **(a)** −13   **(b)** −14   **(c)** 6m − 4   **(d)** $−k^2 + 7k − 6$ **(e)** 3x + 3h − 4

3.

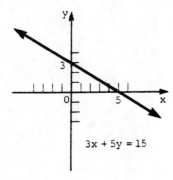

3x + 5y = 15

4.

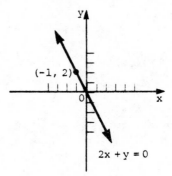

2x + y = 0

5.

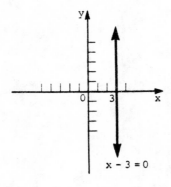

x − 3 = 0

6.

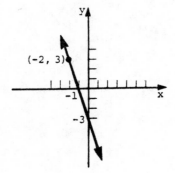

7. −4   **8.** 3/7   **9.** Undefined   **10.** $3x + 5y = 0$ **11.** $x = 1$

12. $x - 2y = -6$ **13.** $x - 3y = 13$ **14.** $y = 5$

15. (a)   **(b)** 55   **(c)** 300

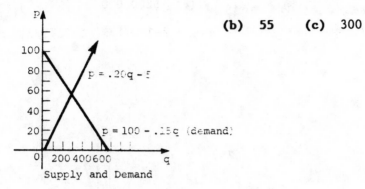

Supply and Demand

16. (a) $y = 10x + 370$ **(b)** \$370   **(c)** \$10

17. (a) 30   **(b)** \$4800

18.

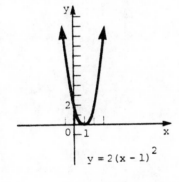

$y = 2(x - 1)^2$

19.

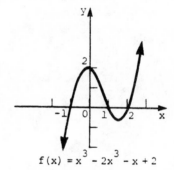

$f(x) = x^3 - 2x^3 - x + 2$

20.

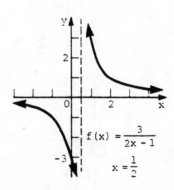

$f(x) = \dfrac{3}{2x - 1}$

$x = \dfrac{1}{2}$

21.

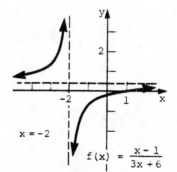

$x = -2$

$f(x) = \dfrac{x - 1}{3x + 6}$

22.

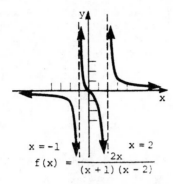

$$x = -1 \qquad x = 2$$
$$f(x) = \frac{2x}{(x + 1)(x - 2)}$$

23. **(a)** $35,000

(b) $90,000

(c) $200,000

24.

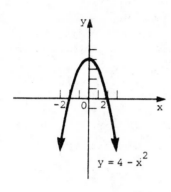

$$y = 4 - x^2$$

25.

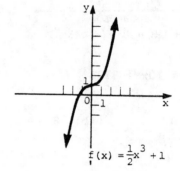

$$f(x) = \frac{1}{2}x^3 + 1$$

CHAPTER 2   SYSTEMS OF LINEAR EQUATIONS AND MATRICES

Section 2.1

1. $x + y = 9$ *(1)*
    $2x - y = 0$ *(2)*

    Multiply equation (1) by -2 and add the result to equation (2). The new system is

    $$x + y = 9 \qquad (1)$$
    $$(-2)R_1 + 1R_2 \to R_2 \qquad -3y = -18. \qquad (3)$$

    Multiply equation (3) by $-\frac{1}{3}$ to get

    $$x + y = 9 \qquad (1)$$
    $$\left(-\frac{1}{3}\right)R_2 \to R_2 \qquad y = 6. \qquad (4)$$

    Substitute 6 for y in equation (1) to get $x = 3$. The solution of the system is $(3, 6)$.

3. $5x + 3y = 7$ *(1)*
    $7x - 3y = -19$ *(2)*

    Multiply equation (1) by 7 and equation (2) by -5 and add the results together. The new system is

    $$5x + 3y = 7 \qquad (1)$$
    $$7R_1 + (-5)R_2 \to R_2 \qquad 36y = 144. \qquad (3)$$

    Now make the coefficient of the first term in each row equal to 1. To accomplish this, multiply equation (1) by $\frac{1}{5}$ and equation (3) by $\frac{1}{36}$. We get the system

    $$\tfrac{1}{5}R_1 \to R_1 \qquad x + \tfrac{3}{5}y = \tfrac{7}{5} \qquad (4)$$
    $$\tfrac{1}{36}R_2 \to R_2 \qquad y = 4. \qquad (5)$$

    Back-substitution gives

    $$x + \tfrac{3}{5}(4) = \tfrac{7}{5}$$
    $$x + \tfrac{12}{5} = \tfrac{7}{5}$$
    $$x = -1.$$

    The solution of the system is $(-1, 4)$.

5. $3x + 2y = -6$ *(1)*
    $5x - 2y = -10$ *(2)*

    Eliminate x in equation (2) to get the system

    $$3x + 2y = -6 \qquad (1)$$
    $$5R_1 + (-3)R_2 \to R_2 \qquad 16y = 0. \qquad (3)$$

    Make the coefficient of the first term in each equation equal 1.

    $$\tfrac{1}{3}R_1 \to R_1 \qquad x + \tfrac{2}{3}y = -2 \qquad (4)$$
    $$\tfrac{1}{16}R_2 \to R_2 \qquad y = 0 \qquad (5)$$

    Substitute 0 for y in equation (4) to get $x = -2$. The solution of the system is $(-2, 0)$.

7. $2x - 3y = -7$ *(1)*
    $5x + 4y = 17$ *(2)*

    Eliminate x in equation (2).

    $$2x - 3y = -7 \qquad (1)$$
    $$5R_1 + (-2)R_2 \to R_2 \qquad -23y = -69 \qquad (3)$$

    Make the coefficient of the first term in each equation equal 1.

    $$\tfrac{1}{2}R_1 \to R_1 \qquad x - \tfrac{3}{2}y = -\tfrac{7}{2} \qquad (4)$$
    $$\left(-\tfrac{1}{23}\right)R_2 \to R_2 \qquad y = 3 \qquad (5)$$

    Substitute 3 for y in equation (4) to get $x = 1$. The solution of the system is $(1, 3)$.

9. $5p + 7q = 6$ *(1)*

$10p - 3q = 46$ *(2)*

Eliminate p in equation (2).

$$5p + 7q = 6 \quad (1)$$

$(-2)R_1 + 1R_2 \rightarrow R_2 \qquad -17q = 34 \quad (3)$

Make the coefficient of the first term in each equation equal 1.

$\frac{1}{5}R_1 \rightarrow R_1 \quad p + \frac{7}{5}q = \frac{6}{5} \quad (4)$

$\left(-\frac{1}{17}\right)R_2 \rightarrow R_2 \qquad q = -2 \quad (5)$

Substitute −2 for q in equation (4) to get p = 4. The solution of the system is (4, −2).

11. $6x + 7y = -2$ *(1)*

$7x - 6y = 26$ *(2)*

Eliminate x in equation (2).

$$6x + 7y = -2 \quad (1)$$

$7R_1 + (-6)R_2 \rightarrow R_2 \qquad 85y = -170 \quad (3)$

Make the coefficient of the first term in each equation equal 1.

$\frac{1}{6}R_1 \rightarrow R_1 \quad x + \frac{7}{6}y = -\frac{1}{3} \quad (4)$

$\frac{1}{85}R_2 \rightarrow R_2 \qquad y = -2 \quad (5)$

Substitute −2 for y in equation (4) to get x = 2. The solution for the system is (2, −2).

13. $3x + 2y = 5$ *(1)*

$6x + 4y = 8$ *(2)*

Eliminate x in equation (2).

$$3x + 2y = 5 \quad (1)$$

$(-2)R_1 + 1R_2 \rightarrow R_2 \qquad 0 = -2 \quad (3)$

Equation (3) is a false statement. The system is inconsistent and has no solution.

15. $4x - y = 9$ *(1)*

$-8x + 2y = -18$ *(2)*

Eliminate x in equation (2).

$$4x - y = 9 \quad (1)$$

$2R_1 + 1R_2 \rightarrow R_2 \qquad 0 = 0 \quad (3)$

The true statement in equation (3) indicates that there are an infinite number of solutions for the system. Solve equation (1) for x.

$$4x - y = 9 \quad (1)$$

$$4x = y + 9$$

$$x = \frac{y + 9}{4} \quad (4)$$

For each value of y, equation (4) indicates that $x = \frac{y + 9}{4}$, and all ordered pairs of the form $\left(\frac{y + 9}{4}, y\right)$ are solutions.

17. The solution of an inconsistent system is no solution.

19. $\frac{x}{2} + \frac{y}{3} = 8$ *(1)*

$\frac{2x}{3} + \frac{3y}{2} = 17$ *(2)*

Rewrite the equations without fractions.

$6R_1 \rightarrow R_1 \quad 3x + 2y = 48 \quad (3)$

$6R_2 \rightarrow R_2 \quad 4x + 9y = 102 \quad (4)$

Eliminate x in equation (4).

$$3x + 2y = 48 \quad (5)$$

$(-4)R_1 + 3R_2 \rightarrow R_2 \qquad 19y = 114 \quad (6)$

Make each leading coefficient equal 1.

$$\frac{1}{3}R_1 \rightarrow R_1 \qquad x + \frac{2}{3}y = 16 \qquad (7)$$

$$\frac{1}{19}R_2 \rightarrow R_2 \qquad \qquad y = 6 \qquad (8)$$

Substitute 6 for y in equation (7) to get x = 12. The solution of the system is (12, 6).

21. $\dfrac{x}{2} + y = \dfrac{3}{2}$ (1)

$\dfrac{x}{3} + y = \dfrac{1}{3}$ (2)

Rewrite the equations without fractions.

$$2R_1 \rightarrow R_1 \qquad x + 2y = 3 \qquad (3)$$
$$3R_2 \rightarrow R_2 \qquad x + 3y = 1 \qquad (4)$$

Eliminate x in equation (4).

$$x + 2y = 3 \qquad (3)$$
$$(-1)R_1 + 1R_2 \rightarrow R_2 \qquad y = -2 \qquad (4)$$

Substitute -2 for y in equation (3) to get x = 7. The solution of the system is (7, -2).

23. $x + y + z = 2$ (1)

$2x + y - z = 5$ (2)

$x - y + z = -2$ (3)

Eliminate x in equations (2) and (3).

$$x + y + z = 2 \qquad (1)$$
$$(-2)R_1 + 1R_2 \rightarrow R_2 \qquad -y - 3z = 1 \qquad (4)$$
$$(-1)R_1 + 1R_3 \rightarrow R_3 \qquad -2y \quad\;\; = -4 \qquad (5)$$

Eliminate y in equation (5).

$$x + y + z = 2 \qquad (1)$$
$$-y - 3z = 1 \qquad (4)$$
$$(-2)R_2 + 1R_3 \rightarrow R_3 \qquad 6z = -6 \qquad (6)$$

Make each leading coefficient equal 1.

$$x + y + z = 2 \qquad (1)$$
$$(-1)R_2 \rightarrow R_2 \qquad y + 3z = -1 \qquad (7)$$
$$\frac{1}{6}R_3 \rightarrow R_3 \qquad \qquad z = -1 \qquad (8)$$

Substitute -1 for z in equation (7) to get y = 2. Finally, substitute -1 for z and 2 for y in equation (1) to get x = 1. The solution of the system is (1, 2, -1).

25. $x + 3y + 4z = 14$ (1)

$2x - 3y + 2z = 10$ (2)

$3x - y + z = 9$ (3)

Eliminate x in equations (2) and (3).

$$x + 3y + 4z = 14 \qquad (1)$$
$$(-2)R_1 + 1R_2 \rightarrow R_2 \qquad -9y - 6z = -18 \qquad (4)$$
$$(-3)R_1 + 1R_3 \rightarrow R_3 \qquad -10y - 11z = -33 \qquad (5)$$

Eliminate y in equation (5).

$$x + 3y + 4z = 14 \qquad (1)$$
$$-9y - 6z = -18 \qquad (4)$$
$$10R_2 + (-9)R_3 \rightarrow R_3 \qquad 39z = 117 \qquad (6)$$

Make each leading coefficient equal 1.

$$x + 3y + 4z = 14 \qquad (1)$$
$$\left(-\frac{1}{9}\right)R_2 \rightarrow R_2 \qquad y + \frac{2}{3}z = 2 \qquad (7)$$
$$\frac{1}{39}R_3 \rightarrow R_3 \qquad \qquad z = 3 \qquad (8)$$

Substitute 3 for z in equation (2) to get y = 0. Finally, substitute 3 for z and 0 for y in equation (1) to get x = 2. The solution of the system is (2, 0, 3).

27. $x + 2y + 3z = 8$ (1)

$3x - y + 2z = 5$ (2)

$-2x - 4y - 6z = 5$ (3)

Eliminate x in equations (2) and (3).

$$x + 2y + 3z = 8 \quad (1)$$
$$(-3)R_1 + 1R_2 \rightarrow R_2 \qquad -7y - 7z = -19 \quad (4)$$
$$2R_1 + 1R_3 \rightarrow R_3 \qquad 0 = 21 \quad (5)$$

Equation (5) is a false statement. The system is inconsistent and has no solution.

29. $\quad 2x - 4y + z = -4 \quad (1)$
$\qquad x + 2y - z = 0 \quad (2)$
$\qquad -x + y + z = 6 \quad (3)$

Eliminate x in equations (2) and (3).

$$2x - 4y + z = -4 \quad (1)$$
$$1R_1 + (-2)R_2 \rightarrow R_2 \qquad -8y + 3z = -4 \quad (4)$$
$$1R_1 + 2R_3 \rightarrow R_3 \qquad -2y + 3z = 8 \quad (5)$$

Eliminate y in equation (5).

$$2x - 4y + z = -4 \quad (1)$$
$$-8y + 3z = -4 \quad (4)$$
$$1R_2 + (-4)R_3 \rightarrow R_3 \qquad -9z = -36 \quad (6)$$

Make each leading coefficient equal 1.

$$\tfrac{1}{2}R_1 \rightarrow R_1 \quad x - 2y + \tfrac{1}{2}z = -2 \quad (7)$$
$$\left(-\tfrac{1}{8}\right)R_2 \rightarrow R_2 \qquad y - \tfrac{3}{8}z = \tfrac{1}{2} \quad (8)$$
$$\left(-\tfrac{1}{9}\right)R_3 \rightarrow R_3 \qquad z = 4 \quad (9)$$

Substitute 4 for z in equation (8) to get y = 2. Finally, substitute 4 for z and 2 for y in equation (7) to get x = 0. The solution of the system is (0, 2, 4).

31. $\quad x + 4y - z = 6 \quad (1)$
$\qquad 2x - y + z = 3 \quad (2)$
$\qquad 3x + 2y + 3z = 16 \quad (3)$

Eliminate x in equations (2) and (3).

$$x + 4y - z = 6 \quad (1)$$
$$(-2)R_1 + 1R_2 \rightarrow R_2 \qquad -9y + 3z = -9 \quad (4)$$
$$(-3)R_1 + 1R_3 \rightarrow R_3 \qquad -10y + 6z = -2 \quad (5)$$

Eliminate y in equation (5).

$$x + 4y - z = 6 \quad (1)$$
$$-9y + 3z = -9 \quad (4)$$
$$(-10)R_2 + 9R_3 \rightarrow R_3 \qquad 24z = 72 \quad (6)$$

Make each leading coefficient equal 1.

$$x + 4y - z = 6 \quad (1)$$
$$\left(-\tfrac{1}{9}\right)R_2 \rightarrow R_2 \qquad y - \tfrac{1}{3}z = 1 \quad (7)$$
$$\tfrac{1}{24}R_3 \rightarrow R_3 \qquad z = 3 \quad (8)$$

Substitute 3 for z in equation (7) to get y = 2. Finally, substitute 3 for z and 2 for y in equation (1) to get x = 1. The solution of the system is (1, 2, 3).

33. $\quad 5m + n - 3p = -6 \quad (1)$
$\qquad 2m + 3n + p = 5 \quad (2)$
$\qquad -3m - 2n + 4p = 3 \quad (3)$

Eliminate m in equations (2) and (3).

$$5m + n - 3p = -6 \quad (1)$$
$$2R_1 + (-5)R_2 \rightarrow R_2 \qquad -13n - 11p = -37 \quad (4)$$
$$3R_1 + 5R_3 \rightarrow R_3 \qquad -7n + 11p = -3 \quad (5)$$

Eliminate n in equation (5).

$$5m + n - 3p = -6 \quad (1)$$
$$-13n - 11p = -37 \quad (4)$$
$$7R_2 + (-13)R_3 \rightarrow R_3 \qquad -220p = -220 \quad (6)$$

Make each leading coefficient equal 1.

$$\tfrac{1}{5}R_1 \rightarrow R_1 \quad m + \tfrac{1}{5}n - \tfrac{3}{5}p = -\tfrac{6}{5} \quad (7)$$
$$\left(-\tfrac{1}{13}\right)R_2 \rightarrow R_2 \qquad n + \tfrac{11}{13}p = \tfrac{37}{13} \quad (8)$$
$$\left(-\tfrac{1}{220}\right)R_3 \rightarrow R_3 \qquad p = 1 \quad (9)$$

Substitute 1 for p in equation (8) to get n = 2. Finally, substitute 1 for p and 2 for n in equation (7) to get m = −1. The solution of the system is (−1, 2, 1).

35.

$$a - 3b - 2c = -3 \quad (1)$$
$$3a + 2b - c = 12 \quad (2)$$
$$-a - b + 4c = 3 \quad (3)$$

Eliminate a in equations (2) and (3).

$$a - 3b - 2c = -3 \quad (1)$$
$$(-3)R_1 + 1R_2 \rightarrow R_2 \quad 11b + 5c = 21 \quad (4)$$
$$1R_1 + 1R_3 \rightarrow R_3 \quad -4b + 2c = 0 \quad (5)$$

Eliminate b in equation (5).

$$a - 3b - 2c = -3 \quad (1)$$
$$11b + 5c = 21 \quad (4)$$
$$4R_2 + 11R_3 \rightarrow R_3 \quad 42c = 84 \quad (6)$$

Make each leading coefficient equal 1.

$$a - 3b - 2c = -3 \quad (1)$$
$$\frac{1}{11}R_2 \rightarrow R_2 \quad b + \frac{5}{11}c = \frac{21}{11} \quad (7)$$
$$\frac{1}{42}R_3 \rightarrow R_3 \quad c = 2 \quad (8)$$

Substitute 2 for c in equation (7) to get b = 1. Finally, substitute 2 for c and 1 for b in equation (1) to get a = 4. The solution of the system is (4, 1, 2).

39.

$$3x + y - z = 0 \quad (1)$$
$$2x - y + 3z = -7 \quad (2)$$

Eliminate x in equation (2).

$$3x + y - z = 0 \quad (1)$$
$$2R_1 + (-3)R_2 \rightarrow R_2 \quad 5y - 11z = 21 \quad (3)$$

Make each leading coefficient equal 1.

$$\frac{1}{3}R_1 \rightarrow R_1 \quad x + \frac{1}{3}y - \frac{1}{3}z = 0 \quad (4)$$

$$\frac{1}{5}R_2 \rightarrow R_2 \quad y - \frac{11}{5}z = \frac{21}{5} \quad (5)$$

Solve equation (5) for y in terms of z.

$$y = \frac{11}{5}z + \frac{21}{5}$$

Substitute this expression for y in equation (4) and solve the equation for x.

$$x + \frac{1}{3}\left(\frac{11}{5}z + \frac{21}{5}\right) - \frac{1}{3}z = 0$$

$$x + \frac{11}{15}z + \frac{7}{5} - \frac{1}{3}z = 0$$

$$x + \frac{2}{5}z = -\frac{7}{5}$$

$$x = -\frac{2}{5}z - \frac{7}{5}$$

The solution of the system is

$$\left(-\frac{2}{5}z - \frac{7}{5}, \frac{11}{5}z + \frac{21}{5}, z\right) \text{ or}$$

$$\left(\frac{-2z - 7}{5}, \frac{11z + 21}{5}, z\right).$$

41.

$$-x + y - z = -7 \quad (1)$$
$$2x + 3y + z = 7 \quad (2)$$

Eliminate x in equation (2).

$$-x + y - z = -7 \quad (1)$$
$$2R_1 + 1R_2 \rightarrow R_2 \quad 5y - z = -7 \quad (3)$$

Make each leading coefficient equal 1.

$$(-1)R_1 \rightarrow R_1 \quad x - y + z = 7 \quad (4)$$
$$\frac{1}{5}R_2 \rightarrow R_2 \quad y - \frac{1}{5}z = -\frac{7}{5} \quad (5)$$

Solve equation (5) for y in terms of z.

$$y = \frac{1}{5}z - \frac{7}{5}$$

Substitute this expression for y in equation (4) and solve the equation for x.

$$x - \left(\frac{1}{5}z - \frac{7}{5}\right) + z = 7$$

$$x - \frac{1}{5}z + \frac{7}{5} + z = 7$$

$$x + \frac{4}{5}z = \frac{28}{5}$$

$$x = -\frac{4}{5}z + \frac{28}{5}$$

The solution of the system is

$$\left(-\frac{4}{5}z + \frac{28}{5}, \ \frac{1}{5}z - \frac{7}{5}, \ z\right) \text{ or}$$

$$\left(\frac{-4z + 28}{5}, \ \frac{z - 7}{5}, \ z\right).$$

43.
$$\begin{aligned} 4x - 3y + z + w &= 21 \quad (1) \\ -2x - y + 2z + 7w &= 2 \quad (2) \\ x + 4y + 2z - w &= -12 \quad (3) \end{aligned}$$

Eliminate x in equations (2) and (3).

$$\begin{aligned} & 4x - 3y + z + w = 21 \quad (1) \\ 1R_1 + 2R_2 \to R_2 \quad & -5y + 5z + 15w = 25 \quad (4) \\ 1R_1 + (-4)R_3 \to R_3 \quad & -19y - 7z + 5w = 69 \quad (5) \end{aligned}$$

Eliminate y in equation (5).

$$\begin{aligned} & 4x - 3y + z + w = 21 \quad (1) \\ & -5y + 5z + 15w = 25 \quad (4) \\ 19R_2 + (-5)R_3 \to R_3 \quad & 130z + 260w = 130 \quad (6) \end{aligned}$$

Make each leading coefficient equal 1.

$$\begin{aligned} \tfrac{1}{4}R_1 \to R_1 \quad & x - \tfrac{3}{4}y + \tfrac{1}{4}z + \tfrac{1}{4}w = \tfrac{21}{4} \quad (7) \\ \left(-\tfrac{1}{5}\right)R_2 \to R_2 \quad & y - z - 3w = -5 \quad (8) \\ \tfrac{1}{130}R_3 \to R_3 \quad & z + 2w = 1 \quad (9) \end{aligned}$$

Solve equation (9) for z in terms of w.

$$z = -2w + 1 \quad (10)$$

Substitute this expression for z in equation (8) and solve for y.

$$\begin{aligned} y - (-2w + 1) - 3w &= -5 \\ y + 2w - 1 - 3w &= -5 \\ y - w &= -4 \\ y &= w - 4 \quad (11) \end{aligned}$$

Substitute the results of equations (10) and (11) in equation (7) and solve the equation for x.

$$x - \frac{3}{4}(w - 4) + \frac{1}{4}(-2w + 1) + \frac{1}{4}w = \frac{21}{4}$$

$$x - \frac{3}{4}w + 3 - \frac{1}{2}w + \frac{1}{4} + \frac{1}{4}w = \frac{21}{4}$$

$$x - w = 2$$

$$x = w + 2$$

The solution of the system is

$$(w + 2, \ w - 4, \ -2w + 1, \ w).$$

45. Let x = the number of units of ROM chips; and y = the number of units of RAM chips.

Use a chart to organize the information.

	ROM	RAM	Totals
Hours on Line A	1	3	15
Hours on Line B	2	1	15

The system to be solved is

$$\begin{aligned} x + 3y &= 15 \quad (1) \\ 2x + y &= 15. \quad (2) \end{aligned}$$

Eliminate x in equation (2).

$$\begin{aligned} & x + 3y = 15 \quad (1) \\ (-2)R_1 + 1R_2 \to R_2 \quad & -5y = -15 \quad (3) \end{aligned}$$

Make each leading coefficient equal 1.

$$\begin{aligned} & x + 3y = 15 \quad (1) \\ \left(-\tfrac{1}{5}\right)R_2 \to R_2 \quad & y = 3 \quad (4) \end{aligned}$$

Substitute 3 for y in equation (1) to get x = 6. 6 units of ROM chips and 3 units of RAM chips can be produced in a day.

47.   Let x = the number of fives;

y = the number of tens; and

z = the number of twenties.

Since the number of fives is three times the number of tens, x = 3y.

The system to be solved is

$$x + y + z = 70 \quad (1)$$
$$x - 3y \quad\quad = 0 \quad (2)$$
$$5x + 10y + 20z = 960. \quad (3)$$

Eliminate x in equations (2) and (3).

$$x + y + z = 70 \quad (1)$$

$R_1 + (-1)R_2 \to R_2 \quad\quad 4y + z = 70 \quad (4)$

$(-5)R_1 + 1R_3 \to R_3 \quad\quad 5y + 15z = 610 \quad (5)$

Eliminate y in equation (5).

$$x + y + z = 70 \quad (1)$$
$$4y + z = 70 \quad (4)$$

$(-5)R_2 + 4R_3 \to R_3 \quad\quad 55z = 2090 \quad (6)$

Make each leading coefficient equal 1.

$$x + y + z = 70 \quad (1)$$

$\frac{1}{4}R_2 \to R_2 \quad\quad y + \frac{1}{4}z = \frac{35}{2} \quad (7)$

$\frac{1}{55}R_3 \to R_3 \quad\quad z = 38 \quad (8)$

Substitute 38 for z in equation (7) to get y = 8. Finally, substitute 38 for z and 8 for y in equation (1) to get x = 24.

There are 24 fives, 8 tens, and 38 twenties.

49.   (a) Let x = the amount borrowed at 13%; y = the amount borrowed at 14%; and z = the amount borrowed at 12%.

Note that y = $\frac{1}{2}$x + 2000, and use the formula for simple interest, I = Prt, to obtain .13x + .14y + .12z = 3240.

The system to be solved is

$$x + y + z = 25{,}000 \quad (1)$$
$$-\frac{1}{2}x + y + \quad = 2000 \quad (2)$$
$$.13x + .14y + .12z = 3240. \quad (3)$$

Rewrite the system without decimals or fractions.

$$x + y + z = 25{,}000 \quad (1)$$

$2R_2 \to R_2 \quad\quad -x + 2y \quad = 4000 \quad (4)$

$100R_3 \to R_3 \quad\quad 13x + 14y + 12z = 324{,}000 \quad (5)$

Eliminate x in equations (4) and (5).

$$x + y + z = 25{,}000 \quad (1)$$

$1R_1 + 1R_2 \to R_2 \quad\quad 3y + z = 29{,}000 \quad (6)$

$(-13)R_1 + 1R_3 \to R_3 \quad\quad y - z = -1000 \quad (7)$

Eliminate y in equation (7).

$$x + y + z = 25{,}000 \quad (1)$$
$$3y + z = 29{,}000 \quad (6)$$

$1R_2 + (-3)R_3 \to R_3 \quad\quad 4z = 32{,}000 \quad (8)$

Make each leading coefficient equal 1.

$$x + y + z = 25{,}000 \quad (1)$$

$\frac{1}{3}R_2 \to R_2 \quad\quad y + \frac{1}{3}z = \frac{29{,}000}{3} \quad (9)$

$\frac{1}{4}R_3 \to R_3 \quad\quad z = 8000 \quad (10)$

Substitute 8000 for z in equation (9) to get y = 7000. Finally, substitute 8000 for z and 7000 for y in equation (1) to get x = 10,000.

The company borrowed $10,000 at 13%, $7000 at 14%, and $8000 at 12%.

(b) Let x, y, z remain as they were defined in part (a). This time, the system to be solved is

$$x + y + z = 25{,}000 \quad (1)$$
$$.13x + .14y + .12z = 3240. \quad (2)$$

Rewrite the system without decimals.

$$x + y + z = 25,000 \quad (1)$$
$$100R_2 \rightarrow R_2 \quad 13x + 14y + 12z = 324,000 \quad (3)$$

Eliminate x in equation (3).

$$x + y + z = 25,000 \quad (1)$$
$$(-13)R_1 + 1R_2 \rightarrow R_2 \quad y - z = -1000 \quad (4)$$

Solve equation (4) for y in terms of z.

$$y = z - 1000$$

Substitute this expression for y in equation (1) and solve the equation for x.

$$x + (z - 1000) + z = 25,000$$
$$x + 2z = 26,000$$
$$x = 26,000 - 2z$$

We have solved a dependent system, and the parameter is z, the amount borrowed at 12%.

For any (positive) amount borrowed at 12%, the amount borrowed at 13% must be twice that amount less than $26,000, and the amount borrowed at 14% must be $1000 less than the amount at 12%.

(c) No, the conditions given in part (a) cannot be satisfied if the company can borrow only $6000 at 12%, since the solution to part (a) requires the company to borrow $8000 at 12%. (However, the conditions of part (b) would still be satisfied if z = 6000.)

Section 2.2

1. $2x + 3y = 11$
    $x + 2y = 8$

    The equations are already in proper form. The augmented matrix obtained from the coefficients and the constants is

    $$\begin{bmatrix} 2 & 3 & 11 \\ 1 & 2 & 8 \end{bmatrix}.$$

3. $2x + y + z = 3$
    $3x - 4y + 2z = -7$
    $x + y + z = 2$

    leads to an augmented matrix

    $$\begin{bmatrix} 2 & 1 & 1 & 3 \\ 3 & -4 & 2 & -7 \\ 1 & 1 & 1 & 2 \end{bmatrix}.$$

5. We are given the augmented matrix

    $$\begin{bmatrix} 1 & 0 & 2 \\ 0 & 1 & 3 \end{bmatrix}.$$

    This is equivalent to the system of equations

    $$x \quad = 2$$
    $$y = 3, \text{ or}$$
    $$x = 2, \ y = 3.$$

7. $$\begin{bmatrix} 1 & 0 & 0 & 2 \\ 0 & 1 & 0 & 3 \\ 0 & 0 & 1 & -2 \end{bmatrix}$$

    The system associated with this matrix is

    $$x \quad = 2$$
    $$y \quad = 3$$
    $$z = -2, \text{ or}$$
    $$x = 2, \ y = 3, \ z = -2.$$

9. Row operations on a matrix correspond to transformations of a system of equations.

11. $\begin{bmatrix} 2 & 3 & 8 & | & 20 \\ 1 & 4 & 6 & | & 12 \\ 0 & 3 & 5 & | & 10 \end{bmatrix}$

$R_1 + (-2)R_2 \rightarrow R_2$ $\begin{bmatrix} 2 & 3 & 8 & | & 20 \\ 2 + (-2)(1) & 3 + (-2)(4) & 8 + (-2)(6) & | & 20 + (-2)(12) \\ 0 & 3 & 5 & | & 10 \end{bmatrix} = \begin{bmatrix} 2 & 3 & 8 & | & 20 \\ 0 & -5 & -4 & | & -4 \\ 0 & 3 & 5 & | & 10 \end{bmatrix}$

13. $\begin{bmatrix} 1 & 4 & 2 & | & 9 \\ 0 & 1 & 5 & | & 14 \\ 0 & 3 & 8 & | & 16 \end{bmatrix}$

$-4R_2 + R_1 \rightarrow R_1$ $\begin{bmatrix} -4(0) + 1 & -4(1) + 4 & -4(5) + 2 & | & -4(14) + 9 \\ 0 & 1 & 5 & | & 14 \\ 0 & 3 & 8 & | & 16 \end{bmatrix} = \begin{bmatrix} 1 & 0 & -18 & | & -47 \\ 0 & 1 & 5 & | & 14 \\ 0 & 3 & 8 & | & 16 \end{bmatrix}$

15. $\begin{bmatrix} 3 & 0 & 0 & | & 18 \\ 0 & 5 & 0 & | & 9 \\ 0 & 0 & 4 & | & 8 \end{bmatrix}$

$\frac{1}{3}R_1 \rightarrow R_1$ $\begin{bmatrix} \frac{1}{3}(3) & \frac{1}{3}(0) & \frac{1}{3}(0) & | & \frac{1}{3}(18) \\ 0 & 5 & 0 & | & 9 \\ 0 & 0 & 4 & | & 8 \end{bmatrix} = \begin{bmatrix} 1 & 0 & 0 & | & 6 \\ 0 & 5 & 0 & | & 9 \\ 0 & 0 & 4 & | & 8 \end{bmatrix}$

17. $x + y = 5$
$x - y = -1$

has augmented matrix

$$\begin{bmatrix} 1 & 1 & | & 5 \\ 1 & -1 & | & -1 \end{bmatrix}.$$

Use row operations as follows.

$-1R + R_2 \rightarrow R_2$ $\begin{bmatrix} 1 & 1 & | & 5 \\ 0 & -2 & | & -6 \end{bmatrix}$

$-\frac{1}{2}R_2 \rightarrow R_2$ $\begin{bmatrix} 1 & 1 & | & 5 \\ 0 & 1 & | & 3 \end{bmatrix}$

$-1R_2 + R_1 \rightarrow R_1$ $\begin{bmatrix} 1 & 0 & | & 2 \\ 0 & 1 & | & 3 \end{bmatrix}$

Read the solution from the last column of the matrix. The solution of the system is (2, 3).

19. $3x + 3y = -9$
$2x - 5y = -6$

The augmented matrix is

$$\begin{bmatrix} 3 & 3 & | & -9 \\ 2 & -5 & | & -6 \end{bmatrix}.$$

$2R_1 + (-3)R_2 \rightarrow R_2$ $\begin{bmatrix} 3 & 3 & | & -9 \\ 0 & 21 & | & 0 \end{bmatrix}$

$-7R_1 + R_2 \rightarrow R_1$ $\begin{bmatrix} -21 & 0 & | & 63 \\ 0 & 21 & | & 0 \end{bmatrix}$

$-\frac{1}{21}R_1 \rightarrow R_1$ $\begin{bmatrix} 1 & 0 & | & -3 \\ 0 & 1 & | & 0 \end{bmatrix}$
$\frac{1}{21}R_2 \rightarrow R_2$

Read the solution from the last column of the matrix. The solution of the system is (-3, 0).

21. $2x = 10 + 3y$

$2y = 5 - 2x$

First rewrite the system in proper form as follows.

$$2x - 3y = 10$$
$$2x + 2y = 5$$

Now, write the augmented matrix of the system and use row operations.

$$\begin{bmatrix} 2 & -3 & | & 10 \\ 2 & 2 & | & 5 \end{bmatrix}$$

$$-R_1 + R_2 \rightarrow R_2 \begin{bmatrix} 2 & -3 & | & 10 \\ 0 & 5 & | & -5 \end{bmatrix}$$

$$5R_1 + 3R_2 \rightarrow R_1 \begin{bmatrix} 10 & 0 & | & 35 \\ 0 & 5 & | & -5 \end{bmatrix}$$

$$\frac{1}{10}R_1 \rightarrow R_1 \begin{bmatrix} 1 & 0 & | & \frac{7}{2} \\ 0 & 1 & | & -1 \end{bmatrix}$$
$$\frac{1}{5}R_2 \rightarrow R_2$$

The solution of the system is $\left(\frac{7}{2}, -1\right)$.

23. $2x - 5y = 10$

$4x - 5y = 15$

Write the augmented matrix and use row operations.

$$\begin{bmatrix} 2 & -5 & | & 10 \\ 4 & -5 & | & 15 \end{bmatrix}$$

$$-2R_1 + R_2 \rightarrow R_2 \begin{bmatrix} 2 & -5 & | & 10 \\ 0 & 5 & | & -5 \end{bmatrix}$$

$$R_1 + R_2 \rightarrow R_1 \begin{bmatrix} 2 & 0 & | & 5 \\ 0 & 5 & | & -5 \end{bmatrix}$$

$$\frac{1}{2}R_1 \rightarrow R_1 \begin{bmatrix} 1 & 0 & | & \frac{5}{2} \\ 0 & 1 & | & -1 \end{bmatrix}$$
$$\frac{1}{5}R_2 \rightarrow R_2$$

The solution of the system is $\left(\frac{5}{2}, -1\right)$.

25. $2x - 3y = 2$

$4x - 6y = 1$

Write the augmented matrix and use row operations.

$$\begin{bmatrix} 2 & -3 & | & 2 \\ 4 & -6 & | & 1 \end{bmatrix}$$

$$-2R_1 + R_2 \rightarrow R_2 \begin{bmatrix} 2 & -3 & | & 2 \\ 0 & 0 & | & -3 \end{bmatrix}$$

The system associated with the last matrix is

$$2x - 3y = 2$$
$$0x + 0y = -3.$$

Since the second equation is $0 = -3$, the system is inconsistent and therefore has no solution.

27. $6x - 3y = 1$

$-12x + 6y = -2$

Write the augmented matrix of the system and use row operations.

$$\begin{bmatrix} 6 & -3 & | & 1 \\ -12 & 6 & | & -2 \end{bmatrix}$$

$$2R_1 + R_2 \rightarrow R_2 \begin{bmatrix} 6 & -3 & | & 1 \\ 0 & 0 & | & 0 \end{bmatrix}$$

$$\frac{1}{6}R_1 \rightarrow R_1 \begin{bmatrix} 1 & -\frac{1}{2} & | & \frac{1}{6} \\ 0 & 0 & | & 0 \end{bmatrix}$$

This is as far as we can go with the Gauss-Jordan method. To complete the solution, write the equation that corresponds to the first row of the matrix.

$$x - \frac{1}{2}y = \frac{1}{6}$$

Solve this equation for x in terms of y.

$$x = \frac{1}{2}y + \frac{1}{6}$$

The solution of the system is

$$\left(\frac{1}{2}y + \frac{1}{6}, y\right).$$

29. $2x - 2y \qquad = -2$

$\qquad y + z = 4$

$x \qquad + z = 1$

Write the augmented matrix and use row operations.

$$\begin{bmatrix} 2 & -2 & 0 & | & -2 \\ 0 & 1 & 1 & | & 4 \\ 1 & 0 & 1 & | & 1 \end{bmatrix}$$

$$R_1 + (-2)R_3 \to R_3 \begin{bmatrix} 2 & -2 & 0 & | & -2 \\ 0 & 1 & 1 & | & 4 \\ 0 & -2 & -2 & | & -4 \end{bmatrix}$$

$$\begin{aligned} 2R_2 + R_1 \to R_1 \\ 2R_2 + R_3 \to R_3 \end{aligned} \begin{bmatrix} 2 & 0 & 2 & | & 6 \\ 0 & 1 & 1 & | & 4 \\ 0 & 0 & 0 & | & 4 \end{bmatrix}$$

This matrix corresponds to the system

$$2x + 2z = 6$$

$$y + z = 4$$

$$0 = 4.$$

The false statement $0 = 4$ indicates that the system is inconsistent and therefore has no solution.

31. $4x + 4y - 4z = 24$

$2x - y + z = -9$

$x - 2y + 3z = 1$

Write the augmented matrix and use row operations.

$$\begin{bmatrix} 4 & 4 & -4 & | & 24 \\ 2 & -1 & 1 & | & -9 \\ 1 & -2 & 3 & | & 1 \end{bmatrix}$$

$$\begin{aligned} R_1 + (-2)R_2 \to R_2 \\ R_1 + (-4)R_3 \to R_3 \end{aligned} \begin{bmatrix} 4 & 4 & -4 & | & 24 \\ 0 & 6 & -6 & | & 42 \\ 0 & 12 & -16 & | & 20 \end{bmatrix}$$

$$\begin{aligned} -3R_1 + 2R_2 \to R_1 \\ -2R_2 + R_3 \to R_3 \end{aligned} \begin{bmatrix} -12 & 0 & 0 & | & 12 \\ 0 & 6 & -6 & | & 42 \\ 0 & 0 & -4 & | & -64 \end{bmatrix}$$

$$2R_2 - 3R_3 \to R_2 \begin{bmatrix} -12 & 0 & 0 & | & 12 \\ 0 & 12 & 0 & | & 276 \\ 0 & 0 & -4 & | & -64 \end{bmatrix}$$

$$\begin{aligned} -\tfrac{1}{12}R_1 \to R_1 \\ \tfrac{1}{12}R_2 \to R_2 \\ -\tfrac{1}{4}R_3 \to R_3 \end{aligned} \begin{bmatrix} 1 & 0 & 0 & | & -1 \\ 0 & 1 & 0 & | & 23 \\ 0 & 0 & 1 & | & 16 \end{bmatrix}$$

The solution of the system is $(-1, 23, 16)$.

33. $y = x - 1$

$y = 6 + z$

$z = -1 - x$

First write the system in proper form.

$$-x + y \qquad = -1$$

$$y - z = 6$$

$$x \qquad + z = -1$$

Write the augmented matrix and use row operations.

$$\begin{bmatrix} -1 & 1 & 0 & | & -1 \\ 0 & 1 & -1 & | & 6 \\ 1 & 0 & 1 & | & -1 \end{bmatrix}$$

$$R_1 + R_3 \to R_3 \begin{bmatrix} -1 & 1 & 0 & | & -1 \\ 0 & 1 & -1 & | & 6 \\ 0 & 1 & 1 & | & -2 \end{bmatrix}$$

$$\begin{aligned} -1R_2 + R_1 \to R_1 \\ -1R_2 + R_3 \to R_3 \end{aligned} \begin{bmatrix} -1 & 0 & 1 & | & -7 \\ 0 & 1 & -1 & | & 6 \\ 0 & 0 & 2 & | & -8 \end{bmatrix}$$

$$\begin{aligned} -2R_1 + R_3 \to R_1 \\ 2R_2 + R_3 \to R_2 \end{aligned} \begin{bmatrix} 2 & 0 & 0 & | & 6 \\ 0 & 2 & 0 & | & 4 \\ 0 & 0 & 2 & | & -8 \end{bmatrix}$$

$$\begin{aligned} \tfrac{1}{2}R_1 \to R_1 \\ \tfrac{1}{2}R_2 \to R_2 \\ \tfrac{1}{2}R_3 \to R_3 \end{aligned} \begin{bmatrix} 1 & 0 & 0 & | & 3 \\ 0 & 1 & 0 & | & 2 \\ 0 & 0 & 1 & | & -4 \end{bmatrix}$$

The solution of the system is $(3, 2, -4)$.

35. $3x + 5y - z = 0$

$4x - y + 2z = 1$

$-6x - 10y + 2z = 0$

Write the augmented matrix and use row operations.

$$\begin{bmatrix} 3 & 5 & -1 & | & 0 \\ 4 & -1 & 2 & | & 1 \\ -6 & -10 & 2 & | & 0 \end{bmatrix}$$

$4R_1 + (-3)R_2 \to R_2$
$2R_1 + R_3 \to R_3$
$$\begin{bmatrix} 3 & 5 & -1 & | & 0 \\ 0 & 23 & -10 & | & -3 \\ 0 & 0 & 0 & | & 0 \end{bmatrix}$$

$23R_1 + (-5)R_2 \to R_1$
$$\begin{bmatrix} 69 & 0 & 27 & | & 15 \\ 0 & 23 & -10 & | & -3 \\ 0 & 0 & 0 & | & 0 \end{bmatrix}$$

$\frac{1}{69}R_1 \to R_1$
$\frac{1}{23}R_2 \to R_2$
$$\begin{bmatrix} 1 & 0 & \frac{9}{23} & | & \frac{5}{23} \\ 0 & 1 & -\frac{10}{23} & | & -\frac{3}{23} \\ 0 & 0 & 0 & | & 0 \end{bmatrix}$$

The row of zeros indicates dependent equations. Solve the first two equations respectively for x and y in terms of z to obtain

$$x = -\frac{9}{23}z + \frac{5}{23}$$

and

$$y = \frac{10}{23}z - \frac{3}{23}.$$

The solution of the system is

$\left(-\frac{9}{23}z + \frac{5}{23}, \frac{10}{23}z - \frac{3}{23}, z\right).$

37. $2x + 3y + z = 9$

$4x - y + 3z = -1$

$6x + 2y - 4z = -8$

Write the augmented matrix and use row operations.

$$\begin{bmatrix} 2 & 3 & 1 & | & 9 \\ 4 & -1 & 3 & | & -1 \\ 6 & 2 & -4 & | & -8 \end{bmatrix}$$

$-2R_1 + R_2 \to R_2$
$-3R_1 + R_3 \to R_3$
$$\begin{bmatrix} 2 & 3 & 1 & | & 9 \\ 0 & -7 & 1 & | & -19 \\ 0 & -7 & -7 & | & -35 \end{bmatrix}$$

$7R_1 + 3R_2 \to R_1$

$R_2 + (-1)R_3 \to R_3$
$$\begin{bmatrix} 14 & 0 & 10 & | & 6 \\ 0 & -7 & 1 & | & -19 \\ 0 & 0 & 8 & | & 16 \end{bmatrix}$$

$8R_1 + (-10)R_3 \to R_1$
$-8R_2 + R_3 \to R_2$
$$\begin{bmatrix} 112 & 0 & 0 & | & -112 \\ 0 & 56 & 0 & | & 168 \\ 0 & 0 & 8 & | & 16 \end{bmatrix}$$

$\frac{1}{112}R_1 \to R_1$
$\frac{1}{56}R_2 \to R_2$
$\frac{1}{8}R_3 \to R_3$
$$\begin{bmatrix} 1 & 0 & 0 & | & -1 \\ 0 & 1 & 0 & | & 3 \\ 0 & 0 & 1 & | & 2 \end{bmatrix}$$

The solution of the system is $(-1, 3, 2)$.

39. $5x - 4y + 2z = 4$

$5x + 3y - z = 17$

$15x - 5y + 3z = 25$

Write the augmented matrix and use row operations.

$$\begin{bmatrix} 5 & -4 & 2 & | & 4 \\ 5 & 3 & -1 & | & 17 \\ 15 & -5 & 3 & | & 25 \end{bmatrix}$$

$-1R_1 + R_2 \to R_2$
$-3R_1 + R_3 \to R_3$
$$\begin{bmatrix} 5 & -4 & 2 & | & 4 \\ 0 & 7 & -3 & | & 13 \\ 0 & 7 & -3 & | & 13 \end{bmatrix}$$

$7R_1 + 4R_2 \to R_1$

$-1R_2 + R_3 \to R_3$
$$\begin{bmatrix} 35 & 0 & 2 & | & 80 \\ 0 & 7 & -3 & | & 13 \\ 0 & 0 & 0 & | & 0 \end{bmatrix}$$

$\frac{1}{35}R_1 \to R_1$
$\frac{1}{7}R_2 \to R_2$
$$\begin{bmatrix} 1 & 0 & \frac{2}{35} & | & \frac{16}{7} \\ 0 & 1 & -\frac{3}{7} & | & \frac{13}{7} \\ 0 & 0 & 0 & | & 0 \end{bmatrix}$$

The row of zeros indicates dependent equations. Solve the first two equations respectively for x and y in terms of z to obtain

$x = -\frac{2}{35}z + \frac{16}{7}$ and $y = \frac{3}{7}z + \frac{13}{7}$.

The solution of the system is

$\left(-\frac{2}{35}z + \frac{16}{7}, \frac{3}{7}z + \frac{13}{7}, z\right)$.

41.
$$
\begin{array}{rcl}
x + 2y \quad\;\; - \;\; w &=& 3 \\
2x \quad\;\; + 4z + 2w &=& -6 \\
x + 2y - \;\; z \quad\quad\;\; &=& 6 \\
2x - \;\; y + \;\; z + \;\; w &=& -3
\end{array}
$$

Write the augmented matrix and use row operations.

$$
\left[\begin{array}{cccc|c}
1 & 2 & 0 & -1 & 3 \\
2 & 0 & 4 & 2 & -6 \\
1 & 2 & -1 & 0 & 6 \\
2 & -1 & 1 & 1 & -3
\end{array}\right]
$$

$$
\begin{array}{l}
-2R_1 + R_2 \rightarrow R_2 \\
-1R_1 + R_3 \rightarrow R_3 \\
-2R_1 + R_4 \rightarrow R_4
\end{array}
\left[\begin{array}{cccc|c}
1 & 2 & 0 & -1 & 3 \\
0 & -4 & 4 & 4 & -12 \\
0 & 0 & -1 & 1 & 3 \\
0 & -5 & 1 & 3 & -9
\end{array}\right]
$$

$$
\begin{array}{l}
2R_1 + R_2 \rightarrow R_1 \\
\\
\\
-5R_2 + 4R_4 \rightarrow R_4
\end{array}
\left[\begin{array}{cccc|c}
2 & 0 & 4 & 2 & -6 \\
0 & -4 & 4 & 4 & -12 \\
0 & 0 & -1 & 1 & 3 \\
0 & 0 & -16 & -8 & 24
\end{array}\right]
$$

$$
\begin{array}{l}
4R_3 + R_1 \rightarrow R_1 \\
4R_3 + R_2 \rightarrow R_2 \\
\\
16R_3 + (-1)R_4 \rightarrow R_4
\end{array}
\left[\begin{array}{cccc|c}
2 & 0 & 0 & 6 & 6 \\
0 & -4 & 0 & 8 & 0 \\
0 & 0 & -1 & 1 & 3 \\
0 & 0 & 0 & 24 & 24
\end{array}\right]
$$

$$
\begin{array}{l}
-4R_1 + R_4 \rightarrow R_1 \\
-3R_2 + R_4 \rightarrow R_2 \\
-24R_3 + R_4 \rightarrow R_3
\end{array}
\left[\begin{array}{cccc|c}
-8 & 0 & 0 & 0 & 0 \\
0 & 12 & 0 & 0 & 24 \\
0 & 0 & 24 & 0 & -48 \\
0 & 0 & 0 & 24 & 24
\end{array}\right]
$$

$$
\begin{array}{l}
-\frac{1}{8}R_1 \rightarrow R_1 \\[4pt]
\frac{1}{12}R_2 \rightarrow R_2 \\[4pt]
\frac{1}{24}R_3 \rightarrow R_3 \\[4pt]
\frac{1}{24}R_4 \rightarrow R_4
\end{array}
\left[\begin{array}{cccc|c}
1 & 0 & 0 & 0 & 0 \\
0 & 1 & 0 & 0 & 2 \\
0 & 0 & 1 & 0 & -2 \\
0 & 0 & 0 & 1 & 1
\end{array}\right]
$$

The solution of the system is $x = 0$, $y = 2$, $z = -2$, $w = 1$, or $(0, 2, -2, 1)$.

43. Let x_1 = the number of cars sent from I to A;

x_2 = the number of cars sent from II to A;

x_3 = the number of cars sent from I to B; and

x_4 = the number of cars sent from II to B.

	A	B
I	x_1	x_3
II	x_2	x_4

Plant I has 28 cars, so

$$x_1 + x_3 = 28. \quad (1)$$

Plant II has 8 cars, so

$$x_2 + x_4 = 8. \quad (2)$$

Dealer A needs 20 cars, so

$$x_1 + x_2 = 20. \quad (3)$$

Dealer B needs 16 cars, so

$$x_3 + x_4 = 16. \quad (4)$$

The total transportation cost is $10,640, so

$$220x_1 + 400x_2 + 300x_3 + 180x_4$$
$$= 10,640. \quad (5)$$

The system to be solved is

$$
\begin{array}{rcl}
x_1 \quad\quad + \quad x_3 \quad\quad &=& 28 \\
x_2 \quad\quad + \quad x_4 &=& 8 \\
x_1 + \quad x_2 \quad\quad\quad &=& 20 \\
x_3 + \quad x_4 &=& 16 \\
220x_1 + 400x_2 + 300x_3 + 180x_4 &=& 10,640.
\end{array}
$$

Write the augmented matrix and use row operations.

$$
\left[\begin{array}{cccc|c}
1 & 0 & 1 & 0 & 28 \\
0 & 1 & 0 & 1 & 8 \\
1 & 1 & 0 & 0 & 20 \\
0 & 0 & 1 & 1 & 16 \\
220 & 400 & 300 & 180 & 10,640
\end{array}\right]
$$

$-1R_1 + R_3 \rightarrow R_3$
$-220R_1 + R_5 \rightarrow R_5$
$$\begin{bmatrix} 1 & 0 & 1 & 0 & 28 \\ 0 & 1 & 0 & 1 & 8 \\ 0 & 1 & -1 & 0 & -8 \\ 0 & 0 & 1 & 1 & 16 \\ 0 & 400 & 80 & 180 & 4480 \end{bmatrix}$$

$-1R_2 + R_3 \rightarrow R_3$
$-400R_2 + R_5 \rightarrow R_5$
$$\begin{bmatrix} 1 & 0 & 1 & 0 & 28 \\ 0 & 1 & 0 & 1 & 8 \\ 0 & 0 & -1 & -1 & -16 \\ 0 & 0 & 1 & 1 & 16 \\ 0 & 0 & 80 & -220 & 1280 \end{bmatrix}$$

$R_1 + R_3 \rightarrow R_1$
$R_3 + R_4 \rightarrow R_4$
$80R_3 + R_5 \rightarrow R_5$
$$\begin{bmatrix} 1 & 0 & 0 & -1 & 12 \\ 0 & 1 & 0 & 1 & 8 \\ 0 & 0 & -1 & -1 & -16 \\ 0 & 0 & 0 & 0 & 0 \\ 0 & 0 & 0 & -300 & 0 \end{bmatrix}$$

There is a 0 now in row 4, column 4, where we would like to get a 1. To proceed, interchange the fourth and fifth rows.

$$\begin{bmatrix} 1 & 0 & 0 & -1 & 12 \\ 0 & 1 & 0 & 1 & 8 \\ 0 & 0 & -1 & -1 & -16 \\ 0 & 0 & 0 & -300 & 0 \\ 0 & 0 & 0 & 0 & 0 \end{bmatrix}$$

$300R_1 + (-1)R_4 \rightarrow R_1$
$300R_2 + (-1)R_4 \rightarrow R_2$
$300R_3 + (-1)R_4 \rightarrow R_3$
$$\begin{bmatrix} 300 & 0 & 0 & 0 & 3600 \\ 0 & 300 & 0 & 0 & 2400 \\ 0 & 0 & -300 & 0 & -4800 \\ 0 & 0 & 0 & -300 & 0 \\ 0 & 0 & 0 & 0 & 0 \end{bmatrix}$$

$\frac{1}{300}R_1 \rightarrow R_1$
$\frac{1}{300}R_2 \rightarrow R_2$
$-\frac{1}{300}R_3 \rightarrow R_3$
$-\frac{1}{300}R_4 \rightarrow R_4$
$$\begin{bmatrix} 1 & 0 & 0 & 0 & 12 \\ 0 & 1 & 0 & 0 & 8 \\ 0 & 0 & 1 & 0 & 16 \\ 0 & 0 & 0 & 1 & 0 \\ 0 & 0 & 0 & 0 & 0 \end{bmatrix}$$

Each of the original variables has a value, so the last row of all zeros may be ignored.

The solution of the system is

$x_1 = 12$, $x_2 = 8$, $x_3 = 16$, $x_4 = 0$.

12 cars should be sent from I to A, 8 cars from II to A, 16 cars from I to B, and no cars from II to B.

45. (a) Let x_1 = the number of units from first supplier for Roseville;

x_2 = the number of units from first supplier for Akron;

x_3 = the number of units from second supplier for Roseville; and

x_4 = the number of units from second supplier for Akron.

(b)
$$\begin{aligned} x_1 + x_2 &= 75 \\ x_3 + x_4 &= 40 \\ x_1 + x_3 &= 40 \\ x_2 + x_4 &= 75 \\ 70x_1 + 90x_2 + 80x_3 + 120x_4 &= 10{,}750 \end{aligned}$$

(c) Write the augmented matrix and use row operations.

$$\begin{bmatrix} 1 & 1 & 0 & 0 & 75 \\ 0 & 0 & 1 & 1 & 40 \\ 1 & 0 & 1 & 0 & 40 \\ 0 & 1 & 0 & 1 & 75 \\ 70 & 90 & 80 & 120 & 10{,}750 \end{bmatrix}$$

$-1R_1 + R_3 \rightarrow R_3$
$-70R_1 + R_5 \rightarrow R_5$
$$\begin{bmatrix} 1 & 1 & 0 & 0 & 75 \\ 0 & 0 & 1 & 1 & 40 \\ 0 & -1 & 1 & 0 & -35 \\ 0 & 1 & 0 & 1 & 75 \\ 0 & 20 & 80 & 120 & 5500 \end{bmatrix}$$

Interchange rows 2 and 4.

$$\begin{bmatrix} 1 & 1 & 0 & 0 & 75 \\ 0 & 1 & 0 & 1 & 75 \\ 0 & -1 & 1 & 0 & -35 \\ 0 & 0 & 1 & 1 & 40 \\ 0 & 20 & 80 & 120 & 5500 \end{bmatrix}$$

$-1R_2 + R_1 \rightarrow R_1$
$R_2 + R_3 \rightarrow R_3$
$-20R_2 + R_5 \rightarrow R_5$
$$\begin{bmatrix} 1 & 0 & 0 & -1 & 0 \\ 0 & 1 & 0 & 1 & 75 \\ 0 & 0 & 1 & 1 & 40 \\ 0 & 0 & 1 & 1 & 40 \\ 0 & 0 & 80 & 100 & 4000 \end{bmatrix}$$

$-1R_3 + R_4 \rightarrow R_4$
$-80R_3 + R_5 \rightarrow R_5$
$$\begin{bmatrix} 1 & 0 & 0 & -1 & 0 \\ 0 & 1 & 0 & 1 & 75 \\ 0 & 0 & 1 & 1 & 40 \\ 0 & 0 & 0 & 0 & 0 \\ 0 & 0 & 0 & 20 & 800 \end{bmatrix}$$

Interchange rows 4 and 5.

$$\begin{bmatrix} 1 & 0 & 0 & -1 & 0 \\ 0 & 1 & 0 & 1 & 75 \\ 0 & 0 & 1 & 1 & 40 \\ 0 & 0 & 0 & 20 & 800 \\ 0 & 0 & 0 & 0 & 0 \end{bmatrix}$$

$$\frac{1}{20}R_4 \to R_4 \begin{bmatrix} 1 & 0 & 0 & -1 & 0 \\ 0 & 1 & 0 & 1 & 75 \\ 0 & 0 & 1 & 1 & 40 \\ 0 & 0 & 0 & 1 & 40 \\ 0 & 0 & 0 & 0 & 0 \end{bmatrix}$$

$$\begin{matrix} R_1 + R_4 \to R_1 \\ R_2 + (-1)R_4 \to R_2 \\ R_3 + (-1)R_4 \to R_3 \end{matrix} \begin{bmatrix} 1 & 0 & 0 & 0 & 40 \\ 0 & 1 & 0 & 0 & 35 \\ 0 & 0 & 1 & 0 & 0 \\ 0 & 0 & 0 & 1 & 40 \\ 0 & 0 & 0 & 0 & 0 \end{bmatrix}$$

Each of the original variables has a value, so the last row of all zeros may be ignored. The solution of the system is $x_1 = 40$, $x_2 = 35$, $x_3 = 0$, $x_4 = 40$, or $(40, 35, 0, 40)$.
The manufacturer should purchase 40 units for Roseville from the first supplier, 35 units for Akron from the first supplier, 0 units for Roseville from the second supplier, and 40 units for Akron from the second supplier.

47. This exercise should be solved by computer methods. The solution may vary based on the computer and software that are used.
The solution is $x \approx 30.7209$, $y \approx 39.6513$, $z \approx 31.386$, $w \approx 50.3966$.

49. Let x = the number of kilograms of the first chemical;

    y = the number of kilograms of the second chemical; and

    z = the number of kilograms of the third chemical.

The system to be solved is

$$x + y + z = 750$$
$$x = .108(750)$$
$$\frac{y}{z} = \frac{4}{3}.$$

Rewrite this system as

$$x + y + z = 750$$
$$x \qquad\qquad = 81$$
$$3y - 4z = 0.$$

Use computer methods to solve this system. The solution is that 81 kg of the first chemical, about 382.286 kg of the second chemical, and about 286.714 kg of the third chemical should be used.

51. Let x = the number of species A;

    y = the number of species B; and

    z = the number of species C.

Use a chart to organize the information.

		Species			
		A	B	C	Totals
	I	1.32	2.1	.86	490
Food	II	2.9	.95	1.52	897
	III	1.75	.6	2.01	653

The system to be solved is

$$1.32x + 2.1y + .86z = 490$$
$$2.9x + .95y + 1.52z = 897$$
$$1.75x + .6y + 2.01z = 653.$$

Use computer methods to solve this system. The solution is that about 243 fish of species A, 38 fish of species B, and 101 fish of species C should be stocked in the lake.

Section 2.3

1. $\begin{bmatrix} 1 & 3 \\ 5 & 7 \end{bmatrix} = \begin{bmatrix} 1 & 5 \\ 3 & 7 \end{bmatrix}$

This statement is false, since not all corresponding elements are equal.

3. $\begin{bmatrix} x \\ y \end{bmatrix} = \begin{bmatrix} 3 \\ 5 \end{bmatrix}$ if x = 3 and y = 5.

This statement is true. The matrices are the same size and corresponding elements are equal.

5. $\begin{bmatrix} 1 & 9 & -4 \\ 3 & 7 & 2 \\ -1 & 1 & 0 \end{bmatrix}$ is a square matrix.

This statement is true. The matrix has 3 rows and 3 columns.

7. $\begin{bmatrix} -4 & 8 \\ 2 & 3 \end{bmatrix}$ is a 2 × 2 square matrix.

Its additive inverse is $\begin{bmatrix} 4 & -8 \\ -2 & -3 \end{bmatrix}$.

9. $\begin{bmatrix} -6 & 8 & 0 & 0 \\ 4 & 1 & 9 & 2 \\ 3 & -5 & 7 & 1 \end{bmatrix}$ is a 3 × 4 matrix.

Its additive inverse is

$\begin{bmatrix} 6 & -8 & 0 & 0 \\ -4 & -1 & -9 & -2 \\ -3 & 5 & -7 & -1 \end{bmatrix}$.

11. $\begin{bmatrix} 2 \\ 4 \end{bmatrix}$ is a 2 × 1 column matrix.

Its additive inverse is $\begin{bmatrix} -2 \\ -4 \end{bmatrix}$.

13. The sum of an n × m matrix and its additive inverse is the n × m zero matrix.

15. $\begin{bmatrix} 2 & 1 \\ 4 & 8 \end{bmatrix} = \begin{bmatrix} x & 1 \\ y & z \end{bmatrix}$

Corresponding elements must be equal for the matrices to be equal.
Therefore, x = 2, y = 4, and z = 8.

17. $\begin{bmatrix} x + 6 & y + 2 \\ 8 & 3 \end{bmatrix} = \begin{bmatrix} -9 & 7 \\ 8 & k \end{bmatrix}$

Corresponding elements must be equal.

$\begin{array}{lll} x + 6 = & -9 & y + 2 = 7 \qquad k = 3 \\ x = & -15 & y = 5 \end{array}$

19. $\begin{bmatrix} -7 + z & 4r & 8s \\ 6p & 2 & 5 \end{bmatrix} + \begin{bmatrix} -9 & 8r & 3 \\ 2 & 5 & 4 \end{bmatrix}$

$\quad = \begin{bmatrix} 2 & 36 & 27 \\ 20 & 7 & 12a \end{bmatrix}$

Add the two matrices on the left side of this equation to obtain

$\begin{bmatrix} -7 + z & 4r & 8s \\ 6p & 2 & 5 \end{bmatrix} + \begin{bmatrix} -9 & 8r & 3 \\ 2 & 5 & 4 \end{bmatrix}$

$\quad = \begin{bmatrix} (-7 + z) + (-9) & 4r + 8r & 8s + 3 \\ 6p + 2 & 7 & 9 \end{bmatrix}$

$\quad = \begin{bmatrix} -16 + z & 12r & 8s + 3 \\ 6p + 2 & 7 & 9 \end{bmatrix}$.

Corresponding elements of this matrix and the matrix on the right side of the original equation must be equal.

$\begin{array}{lll} -16 + z = & 2 & 12r = 36 \qquad 8s + 3 = 27 \\ z = 18 & r = 3 \qquad\quad s = 3 \end{array}$

$\begin{array}{ll} 6p + 2 = 30 & 9 = 12a \\ p = 3 & a = \dfrac{3}{4} \end{array}$

21. $\begin{bmatrix} 1 & 2 & 5 & -1 \\ 3 & 0 & 2 & -4 \end{bmatrix} + \begin{bmatrix} 8 & 10 & -5 & 3 \\ -2 & -1 & 0 & 0 \end{bmatrix}$

$\quad = \begin{bmatrix} 1 + 8 & 2 + 10 & 5 + (-5) & -1 + 3 \\ 3 + (-2) & 0 + (-1) & 2 + 0 & -4 + 0 \end{bmatrix}$

$\quad = \begin{bmatrix} 9 & 12 & 0 & 2 \\ 1 & -1 & 2 & -4 \end{bmatrix}$

23. $\begin{bmatrix} 1 & 3 & -2 \\ 4 & 7 & 1 \end{bmatrix} + \begin{bmatrix} 3 & 0 \\ 6 & 4 \\ -5 & 2 \end{bmatrix}$

These matrices cannot be added since the first matrix has size 2×3, while the second has size 3×2. Only matrices that are the same size can be added.

25. The matrices have the same size, so the subtraction can be done. Let A and B represent the given matrices. Using the definition of subtraction, we have

$A - B = A + (-B)$

$= \begin{bmatrix} 2 & 8 & 12 & 0 \\ 7 & 4 & -1 & 5 \\ 1 & 2 & 0 & 10 \end{bmatrix} + \begin{bmatrix} -1 & -3 & -6 & -9 \\ -2 & 3 & 3 & -4 \\ -8 & 0 & 2 & -17 \end{bmatrix}$

$= \begin{bmatrix} 1 & 5 & 6 & -9 \\ 5 & 7 & 2 & 1 \\ -7 & 2 & 2 & -7 \end{bmatrix}.$

27. $\begin{bmatrix} 2 & 3 \\ -2 & 4 \end{bmatrix} + \begin{bmatrix} 4 & 3 \\ 7 & 8 \end{bmatrix} - \begin{bmatrix} 3 & 2 \\ 1 & 4 \end{bmatrix}$

$= \begin{bmatrix} 2+4-3 & 3+3-2 \\ -2+7-1 & 4+8-4 \end{bmatrix} = \begin{bmatrix} 3 & 4 \\ 4 & 8 \end{bmatrix}$

29. $\begin{bmatrix} 1 & 5 \\ -3 & 7 \end{bmatrix} - \begin{bmatrix} 6 & 3 \\ 2 & 4 \end{bmatrix} + \begin{bmatrix} 8 & 10 \\ -1 & 0 \end{bmatrix}$

$= \begin{bmatrix} 1-6+8 & 5-3+10 \\ -3-2+(-1) & 7-4+0 \end{bmatrix}$

$= \begin{bmatrix} 3 & 12 \\ -6 & 3 \end{bmatrix}$

31. $\begin{bmatrix} -4x+2y & -3y+y \\ 6x-3y & 2x-5y \end{bmatrix} + \begin{bmatrix} -8x+6y & 2x \\ 3y-5x & 6x+4y \end{bmatrix}$

$= \begin{bmatrix} (-4x+2y)+(-8x+6y) & (-3x+y)+2x \\ (6x-3y)+(3y-5x) & (2x-5y)+(6x+4y) \end{bmatrix}$

$= \begin{bmatrix} -12x+8y & -x+y \\ x & 8x-y \end{bmatrix}$

33. The additive inverse of

$X = \begin{bmatrix} x & y \\ z & w \end{bmatrix}$

is

$-X = \begin{bmatrix} -x & -y \\ -z & -w \end{bmatrix}.$

35. Show that $X + (T + P) = (X + T) + P$.

On the left side, the sum $T + P$ is obtained first, and then

$X + (T + P).$

This gives the matrix

$\begin{bmatrix} x+(r+m) & y+(s+n) \\ z+(t+p) & w+(u+q) \end{bmatrix}.$

For the right side, first the sum $X + T$ is obtained, and then

$(X + T) + P.$

This gives the matrix

$\begin{bmatrix} (x+r)+m & (y+s)+n \\ (z+t)+p & (w+u)+q \end{bmatrix}.$

Comparing corresponding elements, we see that they are equal by the associative property of addition of real numbers. Thus,

$X + (T + P) = (X + T) + P.$

37. Show that $P + O = P$.

$P + O = \begin{bmatrix} m & n \\ p & q \end{bmatrix} + \begin{bmatrix} 0 & 0 \\ 0 & 0 \end{bmatrix}$

$= \begin{bmatrix} m+0 & n+0 \\ p+0 & q+0 \end{bmatrix}$

$= \begin{bmatrix} m & n \\ p & q \end{bmatrix}$

$= P$

Thus, $P + O = P.$

39. **(a)** The production cost matrix for Chicago is

$$\begin{array}{c}\\ \text{Material} \\ \text{Labor}\end{array}\begin{array}{cc}\text{Phones} & \text{Calculators}\end{array} \\ \left[\begin{array}{cc} 4.05 & 7.01 \\ 3.27 & 3.51 \end{array}\right].$$

The production cost matrix for Seattle is

$$\begin{array}{c}\\ \text{Material} \\ \text{Labor}\end{array}\begin{array}{cc}\text{Phones} & \text{Calculators}\end{array} \\ \left[\begin{array}{cc} 4.40 & 6.90 \\ 3.54 & 3.76 \end{array}\right].$$

(b) $\dfrac{4.27 + 4.05 + 4.40}{3} = 4.24$

$\dfrac{3.45 + 3.27 + 3.54}{3} = 3.42$

$\dfrac{6.94 + 7.01 + 6.90}{3} = 6.95$

$\dfrac{3.65 + 3.51 + 3.76}{3} = 3.64$

The average production cost matrix is

$$\begin{array}{c}\\ \text{Material} \\ \text{Labor}\end{array}\begin{array}{cc}\text{Phones} & \text{Calculators}\end{array} \\ \left[\begin{array}{cc} 4.24 & 6.95 \\ 3.42 & 3.64 \end{array}\right].$$

(c) The new production cost matrix for Chicago is

$$\begin{array}{c}\\ \text{Material} \\ \text{Labor}\end{array}\begin{array}{cc}\text{Phones} & \text{Calculators}\end{array} \\ \left[\begin{array}{cc} 4.05 + .37 & 7.01 + .42 \\ 3.27 + .11 & 3.51 + .11 \end{array}\right]$$

or

$$\left[\begin{array}{cc} 4.42 & 7.43 \\ 3.38 & 3.62 \end{array}\right].$$

(d) $\dfrac{4.42 + 4.40}{2} = 4.41$

$\dfrac{3.38 + 3.54}{2} = 3.46$

$\dfrac{7.43 + 6.90}{2} \approx 7.17$

$\dfrac{3.62 + 3.76}{2} = 3.69$

The new average production cost matrix is

$$\begin{array}{c}\\ \text{Material} \\ \text{Labor}\end{array}\begin{array}{cc}\text{Phones} & \text{Calculators}\end{array} \\ \left[\begin{array}{cc} 4.41 & 7.17 \\ 3.46 & 3.69 \end{array}\right].$$

41. **(a)** There are four food groups and three meals. To represent the data by a 3×4 matrix, we must use the rows to correspond to the meals: breakfast, lunch, and dinner, and the columns to correspond to the four food groups. Thus, we obtain the matrix

$$\left[\begin{array}{cccc} 2 & 1 & 2 & 1 \\ 3 & 2 & 2 & 1 \\ 4 & 3 & 2 & 1 \end{array}\right].$$

(b) There are four food groups. These will correspond to the four rows. There are three components in each food group: fat, carbohydrates, and protein. These will correspond to the three columns. The matrix is

$$\left[\begin{array}{ccc} 5 & 0 & 7 \\ 0 & 10 & 1 \\ 0 & 15 & 2 \\ 10 & 12 & 8 \end{array}\right].$$

(c) The matrix is

$$\left[\begin{array}{c} 8 \\ 4 \\ 5 \end{array}\right].$$

43.

$$\begin{array}{c}\\ \text{Patient Took Painfree} \\ \text{Patient Took Placebo}\end{array}\begin{array}{c}\textit{Obtained Pain Relief} \\ \begin{array}{cc}\text{Yes} & \text{No}\end{array}\end{array} \\ \left[\begin{array}{cc} 22 & 3 \\ 8 & 17 \end{array}\right]$$

(a) Of the 25 patients who took the placebo, 8 got relief.

(b) Of the 25 patients who took Painfree, 3 got no relief.

(c) $\left[\begin{array}{cc} 22 & 3 \\ 8 & 17 \end{array}\right] + \left[\begin{array}{cc} 21 & 4 \\ 6 & 19 \end{array}\right] + \left[\begin{array}{cc} 19 & 6 \\ 10 & 15 \end{array}\right]$

$+ \left[\begin{array}{cc} 23 & 2 \\ 3 & 22 \end{array}\right]$

$= \left[\begin{array}{cc} 85 & 15 \\ 27 & 73 \end{array}\right]$

(d) Yes, it appears that Painfree is effective. Of the 100 patients who took the medication, 85% got relief.

Section 2.4

1. Since A is 2 × 2 and B is 2 × 2, we have the following diagram:

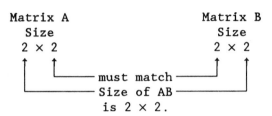

The product AB exists because A has two columns and B has two rows. The size of the product AB is 2 × 2. A similar diagram with B and A interchanged shows that the product BA exists and its size is also 2 × 2.

3.

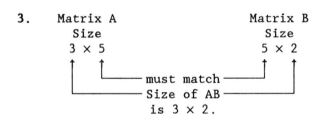

5.

Matrix A   Matrix B
Size   Size
4 × 2   3 × 4

Do not match — AB does not exist.

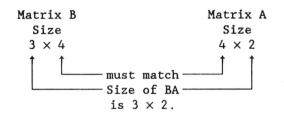

7. $2A = 2 \begin{bmatrix} -2 & 4 \\ 0 & 3 \end{bmatrix} = \begin{bmatrix} -4 & 8 \\ 0 & 6 \end{bmatrix}$

9. $-4B = -4 \begin{bmatrix} -6 & 2 \\ 4 & 0 \end{bmatrix} = \begin{bmatrix} 24 & -8 \\ -16 & 0 \end{bmatrix}$

11. $-4A + 5B = -4 \begin{bmatrix} -2 & 4 \\ 0 & 3 \end{bmatrix} + 5 \begin{bmatrix} -6 & 2 \\ 4 & 0 \end{bmatrix}$

$= \begin{bmatrix} 8 & -16 \\ 0 & -12 \end{bmatrix} + \begin{bmatrix} -30 & 10 \\ 20 & 0 \end{bmatrix}$

$= \begin{bmatrix} -22 & -6 \\ 20 & -12 \end{bmatrix}$

13. To find the product matrix AB, the number of columns of A must be the same as the number of rows of B.

15. Call the first matrix A and the second matrix B. The product matrix AB will have size 2 × 1.

Step 1: Multiply the elements of the first row of A by the corresponding elements of the column of B and add.

$\begin{bmatrix} 1 & 2 \\ 3 & 4 \end{bmatrix} \begin{bmatrix} -1 \\ 7 \end{bmatrix}$ $1(-1) + 2(7) = 13$

Therefore, 13 is the first row entry of the product matrix AB.

Step 2: Multiply the elements of the second row of A by the corresponding elements of the column of B and add.

$$\begin{bmatrix} 1 & 2 \\ \boxed{3} & \boxed{4} \end{bmatrix}\begin{bmatrix} \boxed{-1} \\ \boxed{7} \end{bmatrix} \qquad 3(-1) + 4(7) = 25$$

The second row entry of the product is 25.

Step 3: Write the product using the two entries found above.

$$AB = \begin{bmatrix} 1 & 2 \\ 3 & 4 \end{bmatrix}\begin{bmatrix} -1 \\ 7 \end{bmatrix} = \begin{bmatrix} 13 \\ 25 \end{bmatrix}$$

17. $\begin{bmatrix} 1 & 5 & 3 \\ -1 & 2 & 7 \end{bmatrix}\begin{bmatrix} 4 \\ 2 \\ -3 \end{bmatrix}$

$$= \begin{bmatrix} 1 \cdot 4 + 5 \cdot 2 + 3(-3) \\ -1(4) + 2 \cdot 2 + 7(-3) \end{bmatrix}$$

$$= \begin{bmatrix} 5 \\ -21 \end{bmatrix}$$

19. $\begin{bmatrix} 5 & 1 \\ 2 & 3 \end{bmatrix}\begin{bmatrix} 3 & -1 & 0 \\ 1 & 0 & 2 \end{bmatrix}$

$$= \begin{bmatrix} 5 \cdot 3 + 1 \cdot 1 & 5(-1) + 1 \cdot 0 & 5 \cdot 0 + 1 \cdot 2 \\ 2 \cdot 3 + 3 \cdot 1 & 2(-1) + 3 \cdot 0 & 2 \cdot 0 + 3 \cdot 2 \end{bmatrix}$$

$$= \begin{bmatrix} 16 & -5 & 2 \\ 9 & -2 & 6 \end{bmatrix}$$

21. $\begin{bmatrix} 2 & 2 & -1 \\ 3 & 0 & 1 \end{bmatrix}\begin{bmatrix} 0 & 2 \\ -1 & 4 \\ 0 & 2 \end{bmatrix}$

$$= \begin{bmatrix} 2 \cdot 0 + 2(-1) + (-1)(0) & 2 \cdot 2 + 2 \cdot 4 + (-1)(2) \\ 3 \cdot 0 + 0(-1) + 1(0) & 3 \cdot 2 + 0 \cdot 4 + 1 \cdot 2 \end{bmatrix}$$

$$= \begin{bmatrix} -2 & 10 \\ 0 & 8 \end{bmatrix}$$

23. $\begin{bmatrix} 1 & 2 \\ 3 & 4 \end{bmatrix}\begin{bmatrix} -1 & 5 \\ 7 & 0 \end{bmatrix}$

$$= \begin{bmatrix} 1(-1) + 2 \cdot 7 & 1 \cdot 5 + 2 \cdot 0 \\ 3(-1) + 4 \cdot 7 & 3 \cdot 5 + 4 \cdot 0 \end{bmatrix}$$

$$= \begin{bmatrix} 13 & 5 \\ 25 & 15 \end{bmatrix}$$

25. $\begin{bmatrix} -2 & -3 & 7 \\ 1 & 5 & 6 \end{bmatrix}\begin{bmatrix} 1 \\ 2 \\ 3 \end{bmatrix}$

$$= \begin{bmatrix} -2(1) + (-3)(2) + 7 \cdot 3 \\ 1 \cdot 1 + 5 \cdot 2 + 6 \cdot 3 \end{bmatrix}$$

$$= \begin{bmatrix} 13 \\ 29 \end{bmatrix}$$

27. $\left(\begin{bmatrix} 4 & 3 \\ 0 & 2 \\ 0 & -5 \end{bmatrix}\begin{bmatrix} 2 & -2 \\ 1 & -1 \end{bmatrix}\right)\begin{bmatrix} 10 \\ 0 \end{bmatrix}$

$$= \begin{bmatrix} 11 & -11 \\ 4 & -4 \\ -5 & 5 \end{bmatrix}\begin{bmatrix} 10 \\ 0 \end{bmatrix} = \begin{bmatrix} 110 \\ 40 \\ -50 \end{bmatrix}$$

29. $\begin{bmatrix} 2 & -2 \\ 1 & -1 \end{bmatrix}\left(\begin{bmatrix} 4 & 3 \\ 1 & 2 \end{bmatrix} + \begin{bmatrix} 7 & 0 \\ -1 & 5 \end{bmatrix}\right)$

$$= \begin{bmatrix} 2 & -2 \\ 1 & -1 \end{bmatrix}\begin{bmatrix} 11 & 3 \\ 0 & 7 \end{bmatrix}$$

$$= \begin{bmatrix} 22 & -8 \\ 11 & -4 \end{bmatrix}$$

31. **(a)** $AB = \begin{bmatrix} -2 & 4 \\ 1 & 3 \end{bmatrix}\begin{bmatrix} -2 & 1 \\ 3 & 6 \end{bmatrix} = \begin{bmatrix} 16 & 22 \\ 7 & 19 \end{bmatrix}$

(b) $BA = \begin{bmatrix} -2 & 1 \\ 3 & 6 \end{bmatrix}\begin{bmatrix} -2 & 4 \\ 1 & 3 \end{bmatrix} = \begin{bmatrix} 5 & -5 \\ 0 & 30 \end{bmatrix}$

(c) No, AB and BA are not equal here.

(d) No, AB does not always equal BA.

33. Verify that P(X + T) = PX + PT. Find P(X + T) and PX + PT separately and compare their values to see if they are the same.

P(X + T)

$$= \begin{bmatrix} m & n \\ p & q \end{bmatrix}\left(\begin{bmatrix} x & y \\ z & w \end{bmatrix} + \begin{bmatrix} r & s \\ t & u \end{bmatrix}\right)$$

$$= \begin{bmatrix} m & n \\ p & q \end{bmatrix}\left(\begin{bmatrix} x + r & y + s \\ z + t & w + u \end{bmatrix}\right)$$

$$= \begin{bmatrix} m(x + r) + n(z + t) & m(y + s) + n(w + u) \\ p(x + r) + q(z + t) & p(y + s) + q(w + u) \end{bmatrix}$$

$$= \begin{bmatrix} mx + mr + nz + nt & my + ms + nw + nu \\ px + pr + qz + qt & py + ps + qw + qu \end{bmatrix}$$

PX + PT

$$= \begin{bmatrix} m & n \\ p & q \end{bmatrix} \begin{bmatrix} x & y \\ z & w \end{bmatrix} + \begin{bmatrix} m & n \\ p & q \end{bmatrix} \begin{bmatrix} r & s \\ t & u \end{bmatrix}$$

$$= \begin{bmatrix} mx + nz & my + nw \\ px + qz & py + qw \end{bmatrix} + \begin{bmatrix} mr + nt & ms + nu \\ pr + qt & ps + qu \end{bmatrix}$$

$$= \begin{bmatrix} (mx+nz)+(mr+nt) & (my+nw)+(ms+nu) \\ (px+qz)+(pr+qt) & (py+qw)+(ps+qu) \end{bmatrix}$$

$$= \begin{bmatrix} mx + nz + mr + nt & my + nw + ms + nu \\ px + qz + pr + qt & py + qw + ps + qu \end{bmatrix}$$

$$= \begin{bmatrix} mx + mr + nz + nt & my + ms + nw + nu \\ px + pr + qz + qt & py + ps + qw + qu \end{bmatrix}$$

Observe that the two results are identical. Thus, P(X + T) = PX + PT.

35. Verify that (k + h)P = kP + hP for any real numbers k and h.

$$(k + h)P = (k + h) \begin{bmatrix} m & n \\ p & q \end{bmatrix}$$

$$= \begin{bmatrix} (k + h)m & (k + h)n \\ (k + h)p & (k + h)q \end{bmatrix}$$

$$= \begin{bmatrix} km + hm & kn + hn \\ kp + hp & kq + hq \end{bmatrix}$$

$$= \begin{bmatrix} km & kn \\ kp & kq \end{bmatrix} + \begin{bmatrix} hm & hn \\ hp & hq \end{bmatrix}$$

$$= k \begin{bmatrix} m & n \\ p & q \end{bmatrix} + h \begin{bmatrix} m & n \\ p & q \end{bmatrix}$$

$$= kP + hP$$

Thus, (k + h)P = kP + hP for any real numbers k and h.

37. $\begin{bmatrix} 2 & 3 & 1 \\ 1 & -4 & 5 \end{bmatrix} \begin{bmatrix} x_1 \\ x_2 \\ x_3 \end{bmatrix} = \begin{bmatrix} 2x_1 + 3x_2 + x_3 \\ x_1 - 4x_2 + 5x_3 \end{bmatrix}$,

and $\begin{bmatrix} 2x_1 + 3x_2 + x_3 \\ x_1 - 4x_2 + 5x_3 \end{bmatrix} = \begin{bmatrix} 5 \\ 8 \end{bmatrix}$.

This is equivalent to

$$2x_1 + 3x_2 + x_3 = 5$$
$$x_1 - 4x_2 + 5x_3 = 8$$

since corresponding elements of equal matrices must be equal. Reversing this, observe that the given system of linear equations can be written as the matrix equation

$$\begin{bmatrix} 2 & 3 & 1 \\ 1 & -4 & 5 \end{bmatrix} \begin{bmatrix} x_1 \\ x_2 \\ x_3 \end{bmatrix} = \begin{bmatrix} 5 \\ 8 \end{bmatrix}.$$

39. (a) $\begin{bmatrix} 10 & 4 & 3 & 5 & 6 \\ 7 & 2 & 2 & 3 & 8 \\ 4 & 5 & 1 & 0 & 10 \\ 0 & 3 & 4 & 5 & 5 \end{bmatrix} \begin{bmatrix} 2 & 3 \\ 1 & 1 \\ 4 & 3 \\ 3 & 3 \\ 1 & 2 \end{bmatrix}$

$$= \begin{matrix} & A & B \\ \text{Dept. 1} & \left[\begin{matrix} 57 & 70 \\ 41 & 54 \\ 27 & 40 \\ 39 & 40 \end{matrix} \right] \\ \text{Dept. 2} \\ \text{Dept. 3} \\ \text{Dept. 4} \end{matrix}$$

(b) The total cost to buy from supplier A is 57 + 41 + 27 + 39 = $164, and the total cost to buy from supplier B is 70 + 54 + 40 + 40 = $204. The company should make the purchase from supplier A, since $164 is a lower total cost than $204.

41. (a) The matrices are

$$S = \begin{bmatrix} .027 & .009 \\ .030 & .007 \\ .015 & .009 \\ .013 & .011 \\ .019 & .011 \end{bmatrix} \text{ and}$$

$$P = \begin{bmatrix} 1596 & 218 & 199 & 425 & 214 \\ 1996 & 286 & 226 & 460 & 243 \\ 2440 & 365 & 252 & 484 & 266 \\ 2906 & 455 & 277 & 499 & 291 \end{bmatrix}.$$

(b) $PS =$
$$\begin{array}{c} \\ 1960 \\ 1970 \\ 1980 \\ 1990 \end{array}\begin{array}{cc} \text{Births} & \text{Deaths} \\ \left[\begin{array}{cc} 62.208 & 24.710 \\ 76.459 & 29.733 \\ 91.956 & 35.033 \\ 108.28 & 40.522 \end{array}\right] \end{array}$$

This product matrix gives the total
number of births and deaths in each
year (in millions).

43. **(a)** Use a computer to find the
product matrix. The answer is

$$AC = \begin{bmatrix} 6 & 106 & 158 & 222 & 28 \\ 120 & 139 & 64 & 75 & 115 \\ -146 & -2 & 184 & 144 & -129 \\ 106 & 94 & 24 & 116 & 110 \end{bmatrix}$$

(b)   CA does not exist.

(c)   AC and CA are clearly not equal,
since CA does not even exist.

45. Use a computer to find the matrix
products and sums.
The answers are as follows.

(a) $C + D =$
$$\begin{bmatrix} -1 & 5 & 9 & 13 & -1 \\ 7 & 17 & 2 & -10 & 6 \\ 18 & 9 & -12 & 12 & 22 \\ 9 & 4 & 18 & 10 & -3 \\ 1 & 6 & 10 & 28 & 5 \end{bmatrix}$$

(b) $(C + D)B =$
$$\begin{bmatrix} -2 & -9 & 90 & 77 \\ -42 & -63 & 127 & 62 \\ 413 & 76 & 180 & -56 \\ -29 & -44 & 198 & 85 \\ 137 & 20 & 162 & 103 \end{bmatrix}$$

(c) $CB =$
$$\begin{bmatrix} -56 & -1 & 1 & 45 \\ -156 & -119 & 76 & 122 \\ 315 & 86 & 118 & -91 \\ -17 & -17 & 116 & 51 \\ 118 & 19 & 125 & 77 \end{bmatrix}$$

(d) $DB =$
$$\begin{bmatrix} 54 & -8 & 89 & 32 \\ 114 & 56 & 51 & -60 \\ 98 & -10 & 62 & 35 \\ -12 & -27 & 82 & 34 \\ 19 & 1 & 37 & 26 \end{bmatrix}$$

(e) $CB + DB =$
$$\begin{bmatrix} -2 & -9 & 90 & 77 \\ -42 & -63 & 127 & 62 \\ 413 & 76 & 180 & -56 \\ -29 & -44 & 198 & 85 \\ 137 & 20 & 162 & 103 \end{bmatrix}$$

(f)   Yes, (C + D)B and CB + DB are
equal, as can be seen by observing
that the answers to parts (b) and
(e) are identical.

Section 2.5

1. $\begin{bmatrix} 2 & 3 \\ 1 & 1 \end{bmatrix}\begin{bmatrix} -1 & 3 \\ 1 & -2 \end{bmatrix} = \begin{bmatrix} 1 & 0 \\ 0 & 1 \end{bmatrix} = I$

$\begin{bmatrix} -1 & 3 \\ 1 & -2 \end{bmatrix}\begin{bmatrix} 2 & 3 \\ 1 & 1 \end{bmatrix} = \begin{bmatrix} 1 & 0 \\ 0 & 1 \end{bmatrix} = I$

Yes, these matrices are inverses of
each other since their product matrix
(both ways) is I.

3. $\begin{bmatrix} 2 & 1 \\ 3 & 2 \end{bmatrix}\begin{bmatrix} 2 & 1 \\ -3 & 2 \end{bmatrix} = \begin{bmatrix} 1 & 4 \\ 0 & 7 \end{bmatrix} \neq I$

No, these matrices are not inverses
of each other since their product
matrix is not I.

5. $\begin{bmatrix} 1 & 2 & 0 \\ 0 & 1 & 0 \\ 0 & 1 & 0 \end{bmatrix}\begin{bmatrix} 1 & -2 & 0 \\ 0 & 1 & 0 \\ 0 & -1 & 1 \end{bmatrix}$

$= \begin{bmatrix} 1 & 0 & 0 \\ 0 & 1 & 0 \\ 0 & 1 & 0 \end{bmatrix} \neq I$

No, these matrices are not inverses
of each other.

7. $\begin{bmatrix} 1 & 3 & 3 \\ 1 & 4 & 3 \\ 1 & 3 & 4 \end{bmatrix} \begin{bmatrix} 7 & -3 & -3 \\ -1 & 1 & 0 \\ -1 & 0 & 1 \end{bmatrix}$

$= \begin{bmatrix} 1 & 0 & 0 \\ 0 & 1 & 0 \\ 0 & 0 & 1 \end{bmatrix} = I$

$\begin{bmatrix} 7 & -3 & -3 \\ -1 & 1 & 0 \\ -1 & 0 & 1 \end{bmatrix} \begin{bmatrix} 1 & 3 & 3 \\ 1 & 4 & 3 \\ 1 & 3 & 4 \end{bmatrix}$

$= \begin{bmatrix} 1 & 0 & 0 \\ 0 & 1 & 0 \\ 0 & 0 & 1 \end{bmatrix} = I$

Yes, these matrices are inverses of each other.

9. No, a matrix with a row of all zeros does not have an inverse; the row of all zeros makes it impossible to get all the 1's in the main diagonal of the identity matrix.

11. Let $A = \begin{bmatrix} 1 & -1 \\ 2 & 0 \end{bmatrix}$.

Form the augmented matrix $[A \mid I]$.

$[A \mid I] = \begin{bmatrix} 1 & -1 & 1 & 0 \\ 2 & 0 & 0 & 1 \end{bmatrix}$

Perform row operations on $[A \mid I]$ to get a matrix of the form $[I \mid B]$.

$\begin{bmatrix} 1 & -1 & 1 & 0 \\ 2 & 0 & 0 & 1 \end{bmatrix}$

$-2R_1 + R_2 \rightarrow R_2 \begin{bmatrix} 1 & -1 & 1 & 0 \\ 0 & 2 & -2 & 1 \end{bmatrix}$

$2R_1 + R_2 \rightarrow R_1 \begin{bmatrix} 2 & 0 & 0 & 1 \\ 0 & 2 & -2 & 1 \end{bmatrix}$

$\begin{matrix} \frac{1}{2}R_1 \rightarrow R_1 \\ \frac{1}{2}R_2 \rightarrow R_2 \end{matrix} \begin{bmatrix} 1 & 0 & 0 & \frac{1}{2} \\ 0 & 1 & -1 & \frac{1}{2} \end{bmatrix} = [I \mid B]$

The matrix B in the last transformation is the desired multiplicative inverse.

$$A^{-1} = \begin{bmatrix} 0 & \frac{1}{2} \\ -1 & \frac{1}{2} \end{bmatrix}$$

This answer may be checked by showing that $AA^{-1} = I$ and $A^{-1}A = I$.

13. Let $A = \begin{bmatrix} 3 & -1 \\ -5 & 2 \end{bmatrix}$.

$[A \mid I] = \begin{bmatrix} 3 & -1 & 1 & 0 \\ -5 & 2 & 0 & 1 \end{bmatrix}$

$5R_1 + 3R_2 \rightarrow R_2 \begin{bmatrix} 3 & -1 & 1 & 0 \\ 0 & 1 & 5 & 3 \end{bmatrix}$

$R_1 + R_2 \rightarrow R_1 \begin{bmatrix} 3 & 0 & 6 & 3 \\ 0 & 1 & 5 & 3 \end{bmatrix}$

$\frac{1}{3}R_1 \rightarrow R_1 \begin{bmatrix} 1 & 0 & 2 & 1 \\ 0 & 1 & 5 & 3 \end{bmatrix} = [I \mid B]$

The desired inverse is

$$A^{-1} = \begin{bmatrix} 2 & 1 \\ 5 & 3 \end{bmatrix}.$$

15. Let $A = \begin{bmatrix} -6 & 4 \\ -3 & 2 \end{bmatrix}$.

$[A \mid I] = \begin{bmatrix} -6 & 4 & 1 & 0 \\ -3 & 2 & 0 & 1 \end{bmatrix}$

$R_1 + (-2)R_2 \rightarrow R_2 \begin{bmatrix} -6 & 4 & 1 & 0 \\ 0 & 0 & 1 & -2 \end{bmatrix}$

Because the last row has all zeros to the left of the vertical bar, there is no way to complete the desired transformation. A has no inverse.

17. Let $A = \begin{bmatrix} 1 & 0 & 0 \\ 0 & -1 & 0 \\ 1 & 0 & 1 \end{bmatrix}$.

$[A \mid I] = \begin{bmatrix} 1 & 0 & 0 & 1 & 0 & 0 \\ 0 & -1 & 0 & 0 & 1 & 0 \\ 1 & 0 & 1 & 0 & 0 & 1 \end{bmatrix}$

$-1R_1 + R_3 \rightarrow R_3$ $\begin{bmatrix} 1 & 0 & 0 & | & 1 & 0 & 0 \\ 0 & -1 & 0 & | & 0 & 1 & 0 \\ 0 & 0 & 1 & | & -1 & 0 & 1 \end{bmatrix}$

$-1R_2 \rightarrow R_2$ $\begin{bmatrix} 1 & 0 & 0 & | & 1 & 0 & 0 \\ 0 & 1 & 0 & | & 0 & -1 & 0 \\ 0 & 0 & 1 & | & -1 & 0 & 1 \end{bmatrix}$

$A^{-1} = \begin{bmatrix} 1 & 0 & 0 \\ 0 & -1 & 0 \\ -1 & 0 & 1 \end{bmatrix}$

Because the last row has all zeros to the left of the vertical bar, there is no way to complete the desired transformation. A has no inverse.

19. Let $A = \begin{bmatrix} -1 & -1 & -1 \\ 4 & 5 & 0 \\ 0 & 1 & -3 \end{bmatrix}$.

$[A \mid I] = \begin{bmatrix} -1 & -1 & -1 & | & 1 & 0 & 0 \\ 4 & 5 & 0 & | & 0 & 1 & 0 \\ 0 & 1 & -3 & | & 0 & 0 & 1 \end{bmatrix}$

$4R_1 + R_2 \rightarrow R_2$ $\begin{bmatrix} -1 & -1 & -1 & | & 1 & 0 & 0 \\ 0 & 1 & -4 & | & 4 & 1 & 0 \\ 0 & 1 & -3 & | & 0 & 0 & 1 \end{bmatrix}$

$R_1 + R_2 \rightarrow R_1$ $\begin{bmatrix} -1 & 0 & -5 & | & 5 & 1 & 0 \\ 0 & 1 & -4 & | & 4 & 1 & 0 \\ 0 & 0 & 1 & | & -4 & -1 & 1 \end{bmatrix}$
$-1R_2 + R_3 \rightarrow R_3$

$5R_3 + R_1 \rightarrow R_1$ $\begin{bmatrix} -1 & 0 & 0 & | & -15 & -4 & 5 \\ 0 & 1 & 0 & | & -12 & -3 & 4 \\ 0 & 0 & 1 & | & -4 & -1 & 1 \end{bmatrix}$
$4R_3 + R_2 \rightarrow R_2$

$-1R_1 \rightarrow R_1$ $\begin{bmatrix} -1 & 0 & 0 & | & -15 & 4 & -5 \\ 0 & 1 & 0 & | & -12 & -3 & 4 \\ 0 & 0 & 1 & | & -4 & -1 & 1 \end{bmatrix}$

$A^{-1} = \begin{bmatrix} 15 & 4 & -5 \\ -12 & -3 & 4 \\ -4 & -1 & 1 \end{bmatrix}$

21. Let $A = \begin{bmatrix} 1 & 2 & 3 \\ -3 & -2 & -1 \\ -1 & 0 & 1 \end{bmatrix}$.

$[A \mid I] = \begin{bmatrix} 1 & 2 & 3 & | & 1 & 0 & 0 \\ -3 & -2 & -1 & | & 0 & 1 & 0 \\ -1 & 0 & 1 & | & 0 & 0 & 1 \end{bmatrix}$

$3R_1 + R_2 \rightarrow R_2$ $\begin{bmatrix} 1 & 2 & 3 & | & 1 & 0 & 0 \\ 0 & 4 & 8 & | & 3 & 1 & 0 \\ 0 & 2 & 4 & | & 1 & 0 & 1 \end{bmatrix}$
$R_1 + R_3 \rightarrow R_3$

$-2R_1 + R_2 \rightarrow R_1$ $\begin{bmatrix} -2 & 0 & 2 & | & 1 & 1 & 0 \\ 0 & 4 & 8 & | & 3 & 1 & 0 \\ 0 & 0 & 0 & | & 1 & 1 & -2 \end{bmatrix}$
$-2R_3 + R_2 \rightarrow R_3$

23. Let $A = \begin{bmatrix} 2 & 4 & 6 \\ -1 & -4 & -3 \\ 0 & 1 & -1 \end{bmatrix}$.

$[A \mid I] = \begin{bmatrix} 2 & 4 & 6 & | & 1 & 0 & 0 \\ -1 & -4 & -3 & | & 0 & 1 & 0 \\ 0 & 1 & -1 & | & 0 & 0 & 1 \end{bmatrix}$

$R_1 + 2R_2 \rightarrow R_2$ $\begin{bmatrix} 2 & 4 & 6 & | & 1 & 0 & 0 \\ 0 & -4 & 0 & | & 1 & 2 & 0 \\ 0 & 1 & -1 & | & 0 & 0 & 1 \end{bmatrix}$

$R_1 + R_2 \rightarrow R_1$ $\begin{bmatrix} 2 & 0 & 6 & | & 2 & 2 & 0 \\ 0 & -4 & 0 & | & 1 & 2 & 0 \\ 0 & 0 & -4 & | & 1 & 2 & 4 \end{bmatrix}$
$R_2 + 4R_3 \rightarrow R_3$

$4R_1 + 6R_3 \rightarrow R_1$ $\begin{bmatrix} 8 & 0 & 0 & | & 14 & 20 & 24 \\ 0 & -4 & 0 & | & 1 & 2 & 0 \\ 0 & 0 & -4 & | & 1 & 2 & 4 \end{bmatrix}$

$\frac{1}{8}R_1 \rightarrow R_1$

$-\frac{1}{4}R_2 \rightarrow R_2$ $\begin{bmatrix} 1 & 0 & 0 & | & \frac{7}{4} & \frac{5}{2} & 3 \\ 0 & 1 & 0 & | & -\frac{1}{4} & -\frac{1}{2} & 0 \\ 0 & 0 & 1 & | & -\frac{1}{4} & -\frac{1}{2} & -1 \end{bmatrix}$

$-\frac{1}{4}R_3 \rightarrow R_3$

$A^{-1} = \begin{bmatrix} \frac{7}{4} & \frac{5}{2} & 3 \\ -\frac{1}{4} & -\frac{1}{2} & 0 \\ -\frac{1}{4} & -\frac{1}{2} & -1 \end{bmatrix}$

25. Let $A = \begin{bmatrix} 1 & -2 & 3 & 0 \\ 0 & 1 & -1 & 1 \\ -2 & 2 & -2 & 4 \\ 0 & 2 & -3 & 1 \end{bmatrix}$.

$[A \mid I] = \begin{bmatrix} 1 & -2 & 3 & 0 & | & 1 & 0 & 0 & 0 \\ 0 & 1 & -1 & 1 & | & 0 & 1 & 0 & 0 \\ -2 & 2 & -2 & 4 & | & 0 & 0 & 1 & 0 \\ 0 & 2 & -3 & 1 & | & 0 & 0 & 0 & 1 \end{bmatrix}$

$2R_1 + R_3 \rightarrow R_3$ $\begin{bmatrix} 1 & -2 & 3 & 0 & | & 1 & 0 & 0 & 0 \\ 0 & 1 & -1 & 1 & | & 0 & 1 & 0 & 0 \\ 0 & -2 & 4 & 4 & | & 2 & 0 & 1 & 0 \\ 0 & 2 & -3 & 1 & | & 0 & 0 & 0 & 1 \end{bmatrix}$

$$2R_2 + R_1 \to R_1 \quad \begin{bmatrix} 1 & 0 & 1 & 2 & | & 1 & 2 & 0 & 0 \\ 0 & 1 & -1 & 1 & | & 0 & 1 & 0 & 0 \\ 0 & 0 & 2 & 6 & | & 2 & 2 & 1 & 0 \\ 0 & 0 & -1 & -1 & | & 0 & -2 & 0 & 1 \end{bmatrix}$$

$$\begin{matrix} 2R_2 + R_3 \to R_3 \\ -2R_2 + R_4 \to R_4 \end{matrix}$$

$$\begin{matrix} R_3 + (-2)R_1 \to R_1 \\ R_3 + 2R_2 \to R_2 \\ \\ R_3 + 2R_4 \to R_4 \end{matrix} \quad \begin{bmatrix} -2 & 0 & 0 & 2 & | & 0 & -2 & 1 & 0 \\ 0 & 2 & 0 & 8 & | & 2 & 4 & 1 & 0 \\ 0 & 0 & 2 & 6 & | & 2 & 2 & 1 & 0 \\ 0 & 0 & 0 & 4 & | & 2 & -2 & 1 & 2 \end{bmatrix}$$

$$\begin{matrix} -2R_1 + R_4 \to R_1 \\ R_2 + (-2)R_4 \to R_2 \\ 2R_3 + (-3)R_4 \to R_3 \\ \\ \end{matrix} \quad \begin{bmatrix} 4 & 0 & 0 & 0 & | & 2 & 2 & -1 & 2 \\ 0 & 2 & 0 & 0 & | & -2 & 8 & -1 & -4 \\ 0 & 0 & 4 & 0 & | & -2 & 10 & -1 & -6 \\ 0 & 0 & 0 & 4 & | & 2 & -2 & 1 & 2 \end{bmatrix}$$

$$\begin{matrix} \frac{1}{4}R_1 \to R_1 \\ \\ \frac{1}{2}R_2 \to R_2 \\ \\ \frac{1}{4}R_3 \to R_3 \\ \\ \frac{1}{4}R_4 \to R_4 \end{matrix} \quad \begin{bmatrix} 1 & 0 & 0 & 0 & | & \frac{1}{2} & \frac{1}{2} & -\frac{1}{4} & \frac{1}{2} \\ 0 & 1 & 0 & 0 & | & -1 & 4 & -\frac{1}{2} & -2 \\ 0 & 0 & 1 & 0 & | & -\frac{1}{2} & \frac{5}{2} & -\frac{1}{4} & -\frac{3}{2} \\ 0 & 0 & 0 & 1 & | & \frac{1}{2} & -\frac{1}{2} & \frac{1}{4} & \frac{1}{2} \end{bmatrix}$$

$$A^{-1} = \begin{bmatrix} \frac{1}{2} & \frac{1}{2} & -\frac{1}{4} & \frac{1}{2} \\ -1 & 4 & -\frac{1}{2} & -2 \\ -\frac{1}{2} & \frac{5}{2} & -\frac{1}{4} & -\frac{3}{2} \\ \frac{1}{2} & -\frac{1}{2} & \frac{1}{4} & \frac{1}{2} \end{bmatrix}$$

27. $2x + 3y = 10$

$\quad\;\; x - y = -5$

First write the system in matrix form.

$$\begin{bmatrix} 2 & 3 \\ 1 & -1 \end{bmatrix} \begin{bmatrix} x \\ y \end{bmatrix} = \begin{bmatrix} 10 \\ -5 \end{bmatrix}$$

Let $A = \begin{bmatrix} 2 & 3 \\ 1 & -1 \end{bmatrix}$, $X = \begin{bmatrix} x \\ y \end{bmatrix}$, $B = \begin{bmatrix} 10 \\ -5 \end{bmatrix}$.

The system is (in matrix form) $AX = B$.
Now use row operations to find A^{-1}.

$$[A \mid I] = \begin{bmatrix} 2 & 3 & | & 1 & 0 \\ 1 & -1 & | & 0 & 1 \end{bmatrix}$$

$$R_1 + (-2)R_2 \to R_2 \quad \begin{bmatrix} 2 & 3 & | & 1 & 0 \\ 0 & 5 & | & 1 & -2 \end{bmatrix}$$

$$5R_1 + (-3)R_2 \to R_1 \quad \begin{bmatrix} 10 & 0 & | & 2 & 6 \\ 0 & 5 & | & 1 & -2 \end{bmatrix}$$

$$\begin{matrix} \frac{1}{10}R_1 \to R_1 \\ \\ \frac{1}{5}R_2 \to R_2 \end{matrix} \quad \begin{bmatrix} 1 & 0 & | & \frac{1}{5} & \frac{3}{5} \\ 0 & 1 & | & \frac{1}{5} & -\frac{2}{5} \end{bmatrix}$$

$$A^{-1} = \begin{bmatrix} \frac{1}{5} & \frac{3}{5} \\ \frac{1}{5} & -\frac{2}{5} \end{bmatrix}.$$

Next, find the product $A^{-1}B$.

$$A^{-1}B = \begin{bmatrix} \frac{1}{5} & \frac{3}{5} \\ \frac{1}{5} & -\frac{2}{5} \end{bmatrix} \begin{bmatrix} 10 \\ -5 \end{bmatrix} = \begin{bmatrix} -1 \\ 4 \end{bmatrix}$$

Since $X = A^{-1}B$,

$$X = \begin{bmatrix} x \\ y \end{bmatrix} = \begin{bmatrix} -1 \\ 4 \end{bmatrix}.$$

Thus, the solution of the system is $(-1, 4)$.

29. $2x + y = 5$

$\quad\;\; 5x + 3y = 13$

Let $A = \begin{bmatrix} 2 & 1 \\ 5 & 3 \end{bmatrix}$, $X = \begin{bmatrix} x \\ y \end{bmatrix}$, $B = \begin{bmatrix} 5 \\ 13 \end{bmatrix}$.

Use row operations to obtain

$$A^{-1} = \begin{bmatrix} 3 & -1 \\ -5 & 2 \end{bmatrix}.$$

$$X = A^{-1}B = \begin{bmatrix} 3 & -1 \\ -5 & 2 \end{bmatrix} \begin{bmatrix} 5 \\ 13 \end{bmatrix} = \begin{bmatrix} 2 \\ 1 \end{bmatrix}$$

The solution of the system is $(2, 1)$.

31. $-x + y = 1$

$\quad\;\; 2x - y = 1$

Let $A = \begin{bmatrix} -1 & 1 \\ 2 & -1 \end{bmatrix}$, $X = \begin{bmatrix} x \\ y \end{bmatrix}$, $B = \begin{bmatrix} 1 \\ 1 \end{bmatrix}$.

Use row operations to obtain

$$A^{-1} = \begin{bmatrix} 1 & 1 \\ 2 & 1 \end{bmatrix}.$$

$$X = A^{-1}B = \begin{bmatrix} 1 & 1 \\ 2 & 1 \end{bmatrix} \begin{bmatrix} 1 \\ 1 \end{bmatrix} = \begin{bmatrix} 2 \\ 3 \end{bmatrix}$$

The solution of the system is $(2, 3)$.

33. $-x - 8y = 12$

$3x + 24y = -36$

Let $A = \begin{bmatrix} -1 & -8 \\ 3 & 24 \end{bmatrix}$, $X = \begin{bmatrix} x \\ y \end{bmatrix}$, $B = \begin{bmatrix} 12 \\ -36 \end{bmatrix}$.

Using row operations on $\begin{bmatrix} A \mid I \end{bmatrix}$ leads to the matrix

$$\begin{bmatrix} 1 & -8 & -1 & 0 \\ 0 & 0 & 3 & 1 \end{bmatrix},$$

but the zeros in the second row indicate that matrix A does not have an inverse. We cannot complete the solution by this method.

Since the second equation is a multiple of the first, the equations are dependent.

Solve the first equation of the system for x.

$$-x - 8y = 12$$
$$-x = 8y + 12$$
$$x = -8y - 12$$

The solution of the system is $(-8y - 12, y)$.

35. $-x - y - z = 1$

$4x + 5y = -2$

$y - 3z = 3$

has coefficient matrix

$$A = \begin{bmatrix} -1 & -1 & -1 \\ 4 & 5 & 0 \\ 0 & 1 & 3 \end{bmatrix}.$$

In Exercise 19, it was found that

$$A^{-1} = \begin{bmatrix} -1 & -1 & -1 \\ 4 & 5 & 0 \\ 0 & 1 & 3 \end{bmatrix}^{-1}$$

$$= \begin{bmatrix} 15 & 4 & -5 \\ -12 & -3 & 4 \\ -4 & -1 & 1 \end{bmatrix}.$$

Since $X = A^{-1}B$,

$$\begin{bmatrix} x \\ y \\ z \end{bmatrix} = \begin{bmatrix} 15 & 4 & -5 \\ -12 & -3 & 4 \\ -4 & -1 & 1 \end{bmatrix} \begin{bmatrix} 1 \\ -2 \\ 3 \end{bmatrix} = \begin{bmatrix} -8 \\ 6 \\ 1 \end{bmatrix}.$$

The solution of the system is $(-8, 6, 1)$.

37. $2x + 4y + 6z = 4$

$-x - 4y - 3z = 8$

$y - z = -4$

has coefficient matrix

$$A = \begin{bmatrix} 2 & 4 & 6 \\ -1 & -4 & -3 \\ 0 & 1 & -1 \end{bmatrix}.$$

In Exercise 23, it was found that

$$A^{-1} = \begin{bmatrix} 2 & 4 & 6 \\ -1 & -4 & -3 \\ 0 & 1 & -1 \end{bmatrix}^{-1}$$

$$= \begin{bmatrix} \frac{7}{4} & \frac{5}{2} & 3 \\ -\frac{1}{4} & -\frac{1}{2} & 0 \\ -\frac{1}{4} & -\frac{1}{2} & -1 \end{bmatrix}.$$

Since $X = A^{-1}B$,

$$\begin{bmatrix} x \\ y \\ z \end{bmatrix} = \begin{bmatrix} \frac{7}{4} & \frac{5}{2} & 3 \\ -\frac{1}{4} & -\frac{1}{2} & 0 \\ -\frac{1}{4} & -\frac{1}{2} & -1 \end{bmatrix} \begin{bmatrix} 4 \\ 8 \\ -4 \end{bmatrix} = \begin{bmatrix} 15 \\ -5 \\ -1 \end{bmatrix}.$$

The solution of the system is $(15, -5, -1)$.

39. $2x - 2y = 5$

$4y + 8z = 7$

$x + 2z = 1$

has coefficient matrix

$$A = \begin{bmatrix} 2 & -2 & 0 \\ 0 & 4 & 8 \\ 1 & 0 & 2 \end{bmatrix}.$$

However, using row operations on $[A \mid I]$ shows that A does not have an inverse, so another method must be used.

Try the Gauss–Jordon method. The augmented matrix is

$$\left[\begin{array}{ccc|c} 2 & -2 & 0 & 5 \\ 0 & 4 & 8 & 7 \\ 1 & 0 & 2 & 1 \end{array}\right].$$

After several row operations, we obtain the matrix

$$\left[\begin{array}{ccc|c} 1 & 0 & 2 & \frac{17}{4} \\ 0 & 1 & 2 & \frac{7}{4} \\ 0 & 0 & 0 & 13 \end{array}\right].$$

The bottom row of this matrix shows that the system has no solution, since $0 = 13$ is a false statement.

41.
$$\begin{array}{rcr} x - 2y + 3z + & = & 4 \\ y - z + w & = & -8 \\ -2x + 2y - 2z + 4w & = & 12 \\ 2y - 3z + w & = & -4 \end{array}$$

has coefficient matrix

$$A = \left[\begin{array}{cccc} 1 & -2 & 3 & 0 \\ 0 & -1 & -1 & 1 \\ -2 & 2 & -2 & 4 \\ 0 & 2 & -3 & 1 \end{array}\right].$$

In Exercise 25, it was found that

$$A^{-1} = \left[\begin{array}{cccc} \frac{1}{2} & \frac{1}{2} & -\frac{1}{4} & \frac{1}{2} \\ -1 & 4 & -\frac{1}{2} & -2 \\ -\frac{1}{2} & \frac{5}{2} & -\frac{1}{4} & -\frac{3}{2} \\ \frac{1}{2} & -\frac{1}{2} & \frac{1}{4} & \frac{1}{2} \end{array}\right].$$

Since $X = A^{-1}B$,

$$\left[\begin{array}{c} x \\ y \\ z \\ w \end{array}\right] = \left[\begin{array}{cccc} \frac{1}{2} & \frac{1}{2} & -\frac{1}{4} & \frac{1}{2} \\ -1 & 4 & -\frac{1}{2} & -2 \\ -\frac{1}{2} & \frac{5}{2} & -\frac{1}{4} & -\frac{3}{2} \\ \frac{1}{2} & -\frac{1}{2} & \frac{1}{4} & \frac{1}{2} \end{array}\right]\left[\begin{array}{c} 4 \\ -8 \\ 12 \\ -4 \end{array}\right] = \left[\begin{array}{c} -7 \\ -34 \\ -19 \\ 7 \end{array}\right].$$

The solution of the system is $(-7, -34, -19, 7)$.

43.
$$A = \left[\begin{array}{cc} a & b \\ c & d \end{array}\right]$$

$$IA = \left[\begin{array}{cc} 1 & 0 \\ 0 & 1 \end{array}\right]\left[\begin{array}{cc} a & b \\ c & d \end{array}\right] = \left[\begin{array}{cc} a & b \\ c & d \end{array}\right] = A$$

Thus, $IA = A$.

45.
$$A = \left[\begin{array}{cc} a & b \\ c & d \end{array}\right], \quad 0 = \left[\begin{array}{cc} 0 & 0 \\ 0 & 0 \end{array}\right]$$

$$A \cdot 0 = \left[\begin{array}{cc} a & b \\ c & d \end{array}\right]\left[\begin{array}{cc} 0 & 0 \\ 0 & 0 \end{array}\right] = \left[\begin{array}{cc} 0 & 0 \\ 0 & 0 \end{array}\right] = 0$$

Thus, $A \cdot 0 = 0$

47. $A = \left[\begin{array}{cc} a & b \\ c & d \end{array}\right]$

In Exercise 46, it was found that

$$A^{-1} = \frac{1}{ad - bc}\left[\begin{array}{cc} d & -b \\ -c & a \end{array}\right].$$

$$A^{-1}A = \left(\frac{1}{ad - bc}\left[\begin{array}{cc} d & -b \\ -c & a \end{array}\right]\right)\left[\begin{array}{cc} a & b \\ c & d \end{array}\right]$$

$$= \frac{1}{ad - bc}\left(\left[\begin{array}{cc} d & -b \\ -c & a \end{array}\right]\left[\begin{array}{cc} a & b \\ c & d \end{array}\right]\right)$$

$$= \frac{1}{ad - bc}\left[\begin{array}{cc} ad - bc & 0 \\ 0 & ad - bc \end{array}\right]$$

$$= \left[\begin{array}{cc} 1 & 0 \\ 0 & 1 \end{array}\right] = I$$

Thus, $A^{-1}A = I$.

49.
$$AB = 0$$
$$A^{-1}(AB) = A^{-1} \cdot 0$$
$$(A^{-1}A)B = 0$$
$$I \cdot B = 0$$
$$B = 0$$

Thus, if $AB = 0$ and A^{-1} exists, then $B = 0$.

51. (a) The matrix is $B = \begin{bmatrix} 72 \\ 48 \\ 60 \end{bmatrix}$.

(b) The matrix equation is

$$\begin{bmatrix} 2 & 4 & 2 \\ 2 & 1 & 2 \\ 2 & 1 & 3 \end{bmatrix} \begin{bmatrix} x_1 \\ x_2 \\ x_3 \end{bmatrix} = \begin{bmatrix} 72 \\ 48 \\ 60 \end{bmatrix}.$$

(c) To solve the system, begin by using row operations to find A^{-1}.

$$[A \mid I] = \begin{bmatrix} 2 & 4 & 2 & | & 1 & 0 & 0 \\ 2 & 1 & 2 & | & 0 & 1 & 0 \\ 2 & 1 & 3 & | & 0 & 0 & 1 \end{bmatrix}$$

$$\begin{matrix} \\ R_1 - 1R_2 \to R_2 \\ R_1 - 1R_3 \to R_3 \end{matrix} \begin{bmatrix} 2 & 4 & 2 & | & 1 & 0 & 0 \\ 0 & 3 & 0 & | & 1 & -1 & 0 \\ 0 & 3 & -1 & | & 1 & 0 & -1 \end{bmatrix}$$

$$\begin{matrix} 3R_1 - 4R_2 \to R_1 \\ \\ R_2 - 1R_3 \to R_3 \end{matrix} \begin{bmatrix} 6 & 0 & 6 & | & -1 & 4 & 0 \\ 0 & 3 & 0 & | & 1 & -1 & 0 \\ 0 & 0 & 1 & | & 0 & -1 & 1 \end{bmatrix}$$

$$R_1 - 6R_3 \to R_1 \begin{bmatrix} 6 & 0 & 0 & | & -1 & 10 & -6 \\ 0 & 3 & 0 & | & 1 & -1 & 0 \\ 0 & 0 & 1 & | & 0 & -1 & 1 \end{bmatrix}$$

$$\begin{matrix} \frac{1}{6}R_1 \to R_1 \\ \frac{1}{3}R_2 \to R_2 \\ \\ \end{matrix} \begin{bmatrix} 1 & 0 & 0 & | & -\frac{1}{6} & \frac{5}{3} & -1 \\ 0 & 1 & 0 & | & \frac{1}{3} & -\frac{1}{3} & 0 \\ 0 & 0 & 1 & | & 0 & -1 & 1 \end{bmatrix}$$

The inverse matrix is

$$A^{-1} = \begin{bmatrix} -\frac{1}{6} & \frac{5}{3} & -1 \\ \frac{1}{3} & -\frac{1}{3} & 0 \\ 0 & -1 & 1 \end{bmatrix}.$$

Since $X = A^{-1}B$,

$$\begin{bmatrix} x_1 \\ x_2 \\ x_3 \end{bmatrix} = \begin{bmatrix} -\frac{1}{6} & \frac{5}{3} & -1 \\ \frac{1}{3} & -\frac{1}{3} & 0 \\ 0 & -1 & 1 \end{bmatrix} \begin{bmatrix} 72 \\ 48 \\ 60 \end{bmatrix} = \begin{bmatrix} 8 \\ 8 \\ 12 \end{bmatrix}.$$

There are 8 daily orders for type I, 8 for type II, and 12 for type III.

53. Let x_1 = amount invested in AAA bonds;
x_2 = amount invested in A bonds; and
x_3 = amount invested in B bonds.

The total investment is

$$x_1 + x_2 + x_3.$$

The annual return is

$$.06x_1 + .07x_2 + .10x_3.$$

Since twice as much is invested in AAA bonds as in B bonds,

$$x_1 = 2x_3.$$

(a) The system to be solved is

$$\begin{aligned} x_1 + x_2 + x_3 &= 25{,}000 \\ .06x_1 + .07x_2 + .10x_3 &= 1810 \\ x_1 \qquad\quad - 2x_3 &= 0. \end{aligned}$$

Let $A = \begin{bmatrix} 1 & 1 & 1 \\ .06 & .07 & .10 \\ 1 & 0 & -2 \end{bmatrix}$, $X = \begin{bmatrix} x_1 \\ x_2 \\ x_3 \end{bmatrix}$,

$$B = \begin{bmatrix} 25{,}000 \\ 1800 \\ 0 \end{bmatrix}.$$

Use row operations to obtain

$$A^{-1} = \begin{bmatrix} -14 & 200 & 3 \\ 22 & -300 & -4 \\ -7 & 100 & 1 \end{bmatrix}.$$

Since $X = A^{-1}B$,

$$\begin{bmatrix} x_1 \\ x_2 \\ x_3 \end{bmatrix} = \begin{bmatrix} -14 & 200 & 3 \\ 22 & -300 & -4 \\ -7 & 100 & 1 \end{bmatrix} \begin{bmatrix} 25{,}000 \\ 1810 \\ 0 \end{bmatrix} = \begin{bmatrix} 12{,}000 \\ 7000 \\ 6000 \end{bmatrix}.$$

$12,000 should be invested in AAA bonds at 6%, $7000 in A bonds at 7%, and $6000 in B bonds at 10%.

(b) The matrix of constants is changed to

$$B = \begin{bmatrix} 30,000 \\ 2150 \\ 0 \end{bmatrix}.$$

$$X = A^{-1}B = \begin{bmatrix} -14 & 200 & 3 \\ 22 & -300 & -4 \\ -7 & 100 & 1 \end{bmatrix}\begin{bmatrix} 30,000 \\ 2150 \\ 0 \end{bmatrix}$$

$$= \begin{bmatrix} 10,000 \\ 15,000 \\ 5000 \end{bmatrix}$$

$10,000 should be invested in AAA bonds at 6%, $15,000 in A bonds at 7%, and $5000 in B bonds at 10%.

(c) The matrix of constants is changed to

$$B = \begin{bmatrix} 40,000 \\ 2900 \\ 0 \end{bmatrix}.$$

$$X = A^{-1}B = \begin{bmatrix} -14 & 200 & 3 \\ 22 & -300 & -4 \\ -7 & 100 & 1 \end{bmatrix}\begin{bmatrix} 40,000 \\ 2900 \\ 0 \end{bmatrix}$$

$$= \begin{bmatrix} 20,000 \\ 10,000 \\ 10,000 \end{bmatrix}$$

$20,000 should be invested in AAA bonds at 6%, $10,000 in A bonds at 7%, and $10,000 in B bonds at 10%.

55. Use a computer to perform this calculation. With entries rounded to 6 places, the answer is

$$(CD)^{-1} = \begin{bmatrix} .010146 & -.011883 & .002772 & .020724 & -.012273 \\ .006353 & .014233 & -.001861 & -.029146 & .019225 \\ -.000638 & .006782 & -.004823 & -.022658 & .019344 \\ -.005261 & .003781 & .006192 & .004837 & -.006910 \\ -.012252 & -.001177 & -.006126 & .006744 & .002792 \end{bmatrix}$$

57. Use a computer to determine that no, $C^{-1}D^{-1}$ and $(CD)^{-1}$ are not equal.

59. Use a computer to obtain $X = \begin{bmatrix} .62963 \\ .148148 \\ .259259 \end{bmatrix}$.

61. Use a computer to obtain $X = \begin{bmatrix} .489558 \\ 1.00104 \\ 2.11853 \\ -1.20793 \\ -.961346 \end{bmatrix}$.

Section 2.6

1. $A = \begin{bmatrix} .5 & .4 \\ .25 & .2 \end{bmatrix}$, $D = \begin{bmatrix} 2 \\ 4 \end{bmatrix}$

    To find the production matrix, we first calculate $I - A$.

    $$I - A = \begin{bmatrix} 1 & 0 \\ 0 & 1 \end{bmatrix} - \begin{bmatrix} .5 & .4 \\ .25 & .2 \end{bmatrix}$$

    $$= \begin{bmatrix} .5 & -.4 \\ -.25 & .8 \end{bmatrix}$$

    Using row operations, find the inverse of $I - A$.

    $$[I - A \mid I] = \begin{bmatrix} .5 & -.4 & 1 & 0 \\ -.25 & .8 & 0 & 1 \end{bmatrix}$$

    $$\begin{matrix} 10R_1 \rightarrow R_1 \\ 100R_2 \rightarrow R_2 \end{matrix} \begin{bmatrix} 5 & -4 & 10 & 0 \\ -25 & 80 & 0 & 100 \end{bmatrix}$$

    $$5R_1 + R_2 \rightarrow R_2 \begin{bmatrix} 5 & -4 & 10 & 0 \\ 0 & 60 & 50 & 100 \end{bmatrix}$$

    $$15R_1 + R_2 \rightarrow R_1 \begin{bmatrix} 75 & 0 & 200 & 100 \\ 0 & 60 & 50 & 100 \end{bmatrix}$$

    $$\begin{matrix} \frac{1}{75}R_1 \rightarrow R_1 \\ \frac{1}{60}R_2 \rightarrow R_2 \end{matrix} \begin{bmatrix} 1 & 0 & \frac{8}{3} & \frac{4}{3} \\ 0 & 1 & \frac{5}{6} & \frac{5}{3} \end{bmatrix}$$

    $$(I - A)^{-1} = \begin{bmatrix} \frac{8}{3} & \frac{4}{3} \\ \frac{5}{6} & \frac{5}{3} \end{bmatrix} \approx \begin{bmatrix} 2.67 & 1.33 \\ .83 & 1.67 \end{bmatrix}$$

    Since $X = (I - A)^{-1}D$, the production matrix is

    $$X = \begin{bmatrix} 2.67 & 1.33 \\ .83 & 1.67 \end{bmatrix} \begin{bmatrix} 2 \\ 4 \end{bmatrix} = \begin{bmatrix} 10.67 \\ 8.33 \end{bmatrix}.$$

3. $A = \begin{bmatrix} .1 & .03 \\ .07 & .6 \end{bmatrix}$, $D = \begin{bmatrix} 5 \\ 10 \end{bmatrix}$

    First, calculate $I - A$.

    $$I - A = \begin{bmatrix} .9 & -.03 \\ -.07 & .4 \end{bmatrix}$$

Use row operations to find the inverse of $I - A$, which is

$$(I - A)^{-1} \approx \begin{bmatrix} 1.118 & .084 \\ .195 & 2.515 \end{bmatrix}.$$

Since $X = (I - A)^{-1}D$, the production matrix is

$$X = \begin{bmatrix} 1.118 & .084 \\ .195 & 2.515 \end{bmatrix} \begin{bmatrix} 5 \\ 10 \end{bmatrix} = \begin{bmatrix} 6.43 \\ 26.12 \end{bmatrix}.$$

5. $A = \begin{bmatrix} .4 & 0 & .3 \\ 0 & .8 & .1 \\ 0 & .2 & .4 \end{bmatrix}$, $D = \begin{bmatrix} 1 \\ 3 \\ 2 \end{bmatrix}$

    First, calculate $I - A$.

    $$I - A = \begin{bmatrix} 1 & 0 & 0 \\ 0 & 1 & 0 \\ 0 & 0 & 1 \end{bmatrix} - \begin{bmatrix} .4 & 0 & .3 \\ 0 & .8 & .1 \\ 0 & .2 & .4 \end{bmatrix}$$

    $$= \begin{bmatrix} .6 & 0 & -.3 \\ 0 & .2 & -.1 \\ 0 & -.2 & .6 \end{bmatrix}$$

    Now use row operations to find $(I - A)^{-1}$.

    $$[I - A \mid I] = \begin{bmatrix} .6 & 0 & -.3 & 1 & 0 & 0 \\ 0 & .2 & -.1 & 0 & 1 & 0 \\ 0 & -.2 & .6 & 0 & 0 & 1 \end{bmatrix}$$

    $$R_2 + R_3 \rightarrow R_3 \begin{bmatrix} .6 & 0 & -.3 & 1 & 0 & 0 \\ 0 & .2 & -.1 & 0 & 1 & 0 \\ 0 & 0 & .5 & 0 & 1 & 1 \end{bmatrix}$$

    $$\begin{matrix} 5R_1 + 3R_3 \rightarrow R_1 \\ 5R_2 + R_3 \rightarrow R_2 \end{matrix} \begin{bmatrix} 3 & 0 & 0 & 5 & 3 & 3 \\ 0 & 1 & 0 & 0 & 6 & 1 \\ 0 & 0 & .5 & 0 & 1 & 1 \end{bmatrix}$$

    $$\begin{matrix} \frac{1}{3}R_1 \rightarrow R_1 \\ \\ 2R_3 \rightarrow R_3 \end{matrix} \begin{bmatrix} 1 & 0 & 0 & \frac{5}{3} & 1 & 1 \\ 0 & 1 & 0 & 0 & 6 & 1 \\ 0 & 0 & 1 & 0 & 2 & 2 \end{bmatrix}$$

    $$(I - A)^{-1} \approx \begin{bmatrix} 1.67 & 1 & 1 \\ 0 & 6 & 1 \\ 0 & 2 & 2 \end{bmatrix}.$$

    Since $X = (I - A)^{-1}D$, the production matrix is

    $$X = \begin{bmatrix} 1.67 & 1 & 1 \\ 0 & 6 & 1 \\ 0 & 2 & 2 \end{bmatrix} \begin{bmatrix} 1 \\ 3 \\ 2 \end{bmatrix} = \begin{bmatrix} 6.67 \\ 20 \\ 10 \end{bmatrix}.$$

$$\begin{array}{cc} & \begin{array}{ccc} A & B & C \end{array} \\ 7. & \begin{array}{c} A \\ B \\ C \end{array} \left[\begin{array}{ccc} .3 & .1 & .8 \\ .5 & .6 & .1 \\ .2 & .3 & .1 \end{array} \right] = A \end{array}$$

$$I - A = \left[\begin{array}{ccc} .7 & -.1 & -.8 \\ -.5 & .4 & -.1 \\ -.2 & -.3 & .9 \end{array} \right]$$

Set $(I - A)X = 0$ to obtain the following.

$$\left[\begin{array}{ccc} .7 & -.1 & -.8 \\ -.5 & .4 & -.1 \\ -.2 & -.3 & .9 \end{array} \right] \left[\begin{array}{c} x_1 \\ x_2 \\ x_3 \end{array} \right] = \left[\begin{array}{c} 0 \\ 0 \\ 0 \end{array} \right]$$

$$\left[\begin{array}{c} .7x_1 - .1x_2 - .8x_3 \\ -.5x_1 + .4x_2 - .1x_3 \\ -.2x_1 - .3x_2 + .9x_3 \end{array} \right] \left[\begin{array}{c} x_1 \\ x_2 \\ x_3 \end{array} \right] = \left[\begin{array}{c} 0 \\ 0 \\ 0 \end{array} \right]$$

Rewrite this matrix equation as a system of equations.

$$.7x_1 - .1x_2 - .8x_3 = 0$$
$$-.5x_1 + .4x_2 - .1x_3 = 0$$
$$-.2x_1 - .3x_2 + .9x_3 = 0$$

Rewrite the equations without decimals.

$$7x_1 - x_2 - 8x_3 = 0 \quad (1)$$
$$-5x_1 + 4x_2 - x_3 = 0 \quad (2)$$
$$-2x_1 - 3x_2 + 9x_3 = 0 \quad (3)$$

Use row operations to solve this system of equations. Begin by eliminating x_1 in equations (2) and (3).

$$7x_1 - x_2 - 8x_3 = 0 \quad (1)$$
$$5R_1 + 7R_2 \rightarrow R_2 \qquad 23x_2 - 47x_3 = 0 \quad (4)$$
$$2R_1 + 7R_3 \rightarrow R_3 \qquad -23x_2 + 47x_3 = 0 \quad (5)$$

Eliminate x_2 in equations (1) and (5).

$$23R_1 + R_2 \rightarrow R_1 \quad 161x_1 \qquad - 231x_3 = 0 \quad (6)$$
$$23x_2 - 47x_3 = 0 \quad (4)$$
$$0 = 0 \quad (7)$$

The true statement in equation (7) indicates that the equations are dependent. Solve equation (6) for x_1 and equation (4) for x_2, each in terms of x_3.

$$x_1 = \frac{231}{161}x_3 = \frac{33}{23}x_3$$

$$x_2 = \frac{47}{23}x_3$$

The solution of the system is $\left(\frac{33}{23}x_3, \frac{47}{23}x_3, x_3 \right)$.

If $x_3 = 23$, then $x_1 = 33$ and $x_2 = 47$, so the production of the three commodities should be in the ratio $33:47:23$.

9. In Example 4, it was found that

$$(I - A)^{-1} \approx \left[\begin{array}{cc} 1.39 & .13 \\ .51 & 1.17 \end{array} \right].$$

Since $X = (I - A)^{-1}D$, the production matrix is

$$X = \left[\begin{array}{cc} 1.39 & .13 \\ .51 & 1.17 \end{array} \right] \left[\begin{array}{c} 690 \\ 920 \end{array} \right] = \left[\begin{array}{c} 1078.7 \\ 1428.3 \end{array} \right].$$

Thus, about 1079 metric tons of wheat and 1428 metric tons of oil should be produced.

11. In Example 3, it was found that

$$(I - A)^{-1} \approx \left[\begin{array}{ccc} 1.40 & .50 & .59 \\ .84 & 1.36 & .62 \\ .56 & .47 & 1.30 \end{array} \right].$$

Since $X = (I - A)^{-1}D$, the production matrix is

$$X = \left[\begin{array}{ccc} 1.40 & .50 & .59 \\ .84 & 1.36 & .62 \\ .56 & .47 & 1.30 \end{array} \right] \left[\begin{array}{c} 516 \\ 516 \\ 516 \end{array} \right]$$

$$= \left[\begin{array}{c} 1284.8 \\ 1455.1 \\ 1202.3 \end{array} \right].$$

Thus, about 1285 units of agriculture, 1455 units of manufacturing, and 1202 units of transportation should be produced.

13. From the given data, we get the input-output matrix

$$A = \begin{bmatrix} 0 & \frac{1}{2} & \frac{1}{4} \\ \frac{1}{4} & 0 & \frac{1}{4} \\ \frac{1}{2} & \frac{1}{4} & 0 \end{bmatrix}.$$

$$I - A = \begin{bmatrix} 1 & -\frac{1}{2} & -\frac{1}{4} \\ -\frac{1}{4} & 1 & -\frac{1}{4} \\ -\frac{1}{2} & -\frac{1}{4} & 1 \end{bmatrix}$$

Use row operations to find the inverse of $I - A$, which is

$$(I - A)^{-1} \approx \begin{bmatrix} 1.538 & .923 & .615 \\ .615 & 1.436 & .513 \\ .923 & .821 & 1.436 \end{bmatrix}.$$

Since $X = (I - A)^{-1}D$, the production matrix is

$$X = \begin{bmatrix} 1.538 & .923 & .615 \\ .615 & 1.436 & .513 \\ .923 & .821 & 1.436 \end{bmatrix} \begin{bmatrix} 1000 \\ 1000 \\ 1000 \end{bmatrix}$$

$$\approx \begin{bmatrix} 3077 \\ 2564 \\ 3179 \end{bmatrix}.$$

Thus, the production should be about 3077 units of agriculture, 2564 units of manufacturing, and 3179 units of transportation.

15. From the given data, we get the input-output matrix

$$A = \begin{bmatrix} \frac{1}{4} & \frac{1}{6} \\ \frac{1}{2} & 0 \end{bmatrix}.$$

$$I - A = \begin{bmatrix} \frac{3}{4} & -\frac{1}{6} \\ -\frac{1}{2} & 1 \end{bmatrix}$$

Use row operations to find the inverse of $I - A$, which is

$$(I - A)^{-1} = \begin{bmatrix} \frac{3}{2} & \frac{1}{4} \\ \frac{3}{4} & \frac{9}{8} \end{bmatrix}.$$

(a) The production matrix is

$$X = (I - A)^{-1}D = \begin{bmatrix} \frac{3}{2} & \frac{1}{4} \\ \frac{3}{4} & \frac{9}{8} \end{bmatrix} \begin{bmatrix} 1 \\ 1 \end{bmatrix} = \begin{bmatrix} \frac{7}{4} \\ \frac{15}{8} \end{bmatrix}.$$

Thus, $\frac{7}{4}$ bushels of yams and $\frac{15}{8} \approx 2$ pigs should be produced.

(b) The production matrix is

$$X = (I - A)^{-1}D = \begin{bmatrix} \frac{3}{2} & \frac{1}{4} \\ \frac{3}{4} & \frac{9}{8} \end{bmatrix} \begin{bmatrix} 100 \\ 70 \end{bmatrix} = \begin{bmatrix} 167.5 \\ 153.75 \end{bmatrix}.$$

Thus, 167.5 bushels of yams and $153.75 \approx 154$ pigs should be produced.

17. Use a computer to find the production matrix $X = (I - A)^{-1}D$. The answer is

$$X = \begin{bmatrix} 2930 \\ 3570 \\ 2300 \\ 580 \end{bmatrix}.$$

19. $A = \begin{bmatrix} .1 & .2 & .1 \\ .2 & .1 & .05 \\ 0 & .05 & .1 \end{bmatrix}$, $D = \begin{bmatrix} 1000 \\ 1000 \\ 1000 \end{bmatrix}$

Use a computer to find the production matrix $X = (I - A)^{-1}D$. The answer is

$$X = \begin{bmatrix} 1583.91 \\ 1529.54 \\ 1196.09 \end{bmatrix}.$$

Chapter 2 Review Exercises

3. $2x + 3y = 10$ *(1)*

$-3x + y = 18$ *(2)*

Eliminate x in equation (2).

$$2x + 3y = 10 \quad (1)$$

$3R_2 + 2R_1 \rightarrow R_2$ $11y = 66$ *(3)*

Make each leading coefficient equal 1.

$\frac{1}{2}R_1 \rightarrow R_1$ $x + \frac{3}{2}y = 5$ *(4)*

$\frac{1}{11}R_2 \rightarrow R_2$ $y = 6$ *(5)*

Substitute 6 for y in equation (4) to get x = -4.

The solution of the system is (-4, 6).

5. $2x - 3y + z = -5$ *(1)*

$x + 4y + 2z = 13$ *(2)*

$5x + 5y + 3z = 14$ *(3)*

Eliminate x in equations (2) and (3).

$$2x - 3y + z = -5 \quad (1)$$

$-2R_2 + R_1 \rightarrow R_2$ $-11y - 3z = -31$ *(4)*

$5R_1 + (-2)R_3 \rightarrow R_3$ $-25y - z = -53$ *(5)*

Eliminate y in equation (5).

$$2x - 3y + z = -5 \quad (1)$$

$-11y - 3z = -31$ *(6)*

$-25R_2 + R_3 \rightarrow R_3$ $64z = 192$ *(7)*

Make each leading coefficient equal 1.

$\frac{1}{2}R_1 \rightarrow R_1$ $x - \frac{3}{2}y + \frac{1}{2}z = -\frac{5}{2}$ *(8)*

$-\frac{1}{11}R_2 \rightarrow R_2$ $y + \frac{3}{11}z = \frac{31}{11}$ *(9)*

$\frac{1}{64}R_3 \rightarrow R_3$ $z = 3$ *(10)*

Substitute 3 for z in equation (9) to get y = 2. Substitute 3 for z and 2 for y in equation (8) to get x = -1. The solution of the system is (-1, 2, 3).

7. $2x + 4y = -6$

$-3x - 5y = 12$

Write the augmented matrix and use row operations.

$$\begin{bmatrix} 2 & 4 & -6 \\ -3 & -5 & 12 \end{bmatrix}$$

$3R_1 + 2R_2 \rightarrow R_2$ $\begin{bmatrix} 2 & 4 & -6 \\ 0 & 2 & 6 \end{bmatrix}$

$-2R_2 + R_1 \rightarrow R_1$ $\begin{bmatrix} 2 & 0 & -18 \\ 0 & 2 & 6 \end{bmatrix}$

$\frac{1}{2}R_1 \rightarrow R_1$ $\begin{bmatrix} 1 & 0 & -9 \\ 0 & 1 & 3 \end{bmatrix}$

$\frac{1}{2}R_2 \rightarrow R_2$

The solution of the system is (-9, 3).

9. $x - y + 3z = 13$

$4x + y + 2z = 17$

$3x + 2y + 2z = 1$

Write the augmented matrix and use row operations.

$$\begin{bmatrix} 1 & -1 & 3 & 13 \\ 4 & 1 & 2 & 17 \\ 3 & 2 & 2 & 1 \end{bmatrix}$$

$-4R_1 + R_2 \rightarrow R_2$ $\begin{bmatrix} 1 & -1 & 3 & 13 \\ 0 & 5 & -10 & -35 \\ 0 & 5 & -7 & -38 \end{bmatrix}$
$-3R_1 + R_3 \rightarrow R_3$

$5R_1 + R_2 \rightarrow R_1$ $\begin{bmatrix} 5 & 0 & 5 & 30 \\ 0 & 5 & -10 & -35 \\ 0 & 0 & 3 & -3 \end{bmatrix}$
$-1R_2 + R_3 \rightarrow R_3$

$-3R_1 + 5R_3 \rightarrow R_1$ $\begin{bmatrix} -15 & 0 & 0 & -105 \\ 0 & 15 & 0 & -135 \\ 0 & 0 & 3 & -3 \end{bmatrix}$
$3R_2 + 10R_3 \rightarrow R_2$

$-\frac{1}{15}R_1 \rightarrow R_1$ $\begin{bmatrix} 1 & 0 & 0 & 7 \\ 0 & 1 & 0 & -9 \\ 0 & 0 & 1 & -1 \end{bmatrix}$
$\frac{1}{15}R_2 \rightarrow R_2$

$\frac{1}{3}R_3 \rightarrow R_3$

The solution of the system is (7, -9, -1).

11. $3x - 6y + 9z = 12$

$-x + 2y - 3z = -4$

$x + y + 2z = 7$

Write the augmented matrix and use row operations.

$$\begin{bmatrix} 3 & -6 & 9 & | & 12 \\ -1 & 2 & -3 & | & -4 \\ 1 & 1 & 2 & | & 7 \end{bmatrix}$$

$$\begin{matrix} \\ R_1 + 3R_2 \to R_2 \\ -1R_1 + 3R_3 \to R_3 \end{matrix} \begin{bmatrix} 3 & -6 & 9 & | & 12 \\ 0 & 0 & 0 & | & 0 \\ 0 & 9 & -3 & | & 9 \end{bmatrix}$$

The zero in row 2, column 2 is an obstacle. To proceed, interchange the second and third rows.

$$\begin{bmatrix} 3 & -6 & 9 & | & 12 \\ 0 & 9 & -3 & | & 9 \\ 0 & 0 & 0 & | & 0 \end{bmatrix}$$

$$3R_1 + 2R_2 \to R_1 \begin{bmatrix} 9 & 0 & 21 & | & 54 \\ 0 & 9 & -3 & | & 9 \\ 0 & 0 & 0 & | & 0 \end{bmatrix}$$

$$\begin{matrix} \frac{1}{9}R_1 \to R_1 \\ \\ \frac{1}{9}R_2 \to R_2 \\ \\ \end{matrix} \begin{bmatrix} 1 & 0 & \frac{7}{3} & | & 6 \\ 0 & 1 & -\frac{1}{3} & | & 1 \\ 0 & 0 & 0 & | & 0 \end{bmatrix}$$

The row of zeros indicates dependent equations. Solve the first two equations respectively for x and y in terms of z to obtain

$$x = 6 - \frac{7}{3}z \quad \text{and} \quad y = 1 + \frac{1}{3}z.$$

The solution of the system is

$$\left(6 - \frac{7}{3}z, \ 1 + \frac{1}{3}z, \ z\right).$$

13. $$\begin{bmatrix} 2 & x \\ y & 6 \\ 5 & z \end{bmatrix} = \begin{bmatrix} a & -1 \\ 4 & 6 \\ p & 7 \end{bmatrix}$$

The size of these matrices is 3 × 2. For matrices to be equal, corresponding elements must be equal, so $a = 2$, $x = -1$, $y = 4$, $p = 5$, and $z = 7$.

15. $$\begin{bmatrix} a+5 & 3b & 6 \\ 4c & 2+d & -3 \\ -1 & 4p & q-1 \end{bmatrix} = \begin{bmatrix} -7 & b+2 & 2k-3 \\ 3 & 2d-1 & 4\ell \\ m & 12 & 8 \end{bmatrix}$$

These are 3 × 3 square matrices. Since corresponding elements must be equal,

$a + 5 = -7$, so $\quad a = -12$;

$3b = b + 2$, so $\quad b = 1$;

$6 = 2k - 3$, so $\quad k = \frac{9}{2}$;

$4c = 3$, so $\quad c = \frac{3}{4}$;

$2 + d = 2d - 1$, so $d = 3$;

$-3 = 4\ell$, so $\quad \ell = -\frac{3}{4}$;

$\quad m = -1$;

$4p = 12$, so $\quad p = 3$; and

$q - 1 = 8$, so $\quad q = 9$.

17. $2G - 4F = 2\begin{bmatrix} 2 & 5 \\ 1 & 6 \end{bmatrix} - 4\begin{bmatrix} -1 & 4 \\ 3 & 7 \end{bmatrix}$

$= \begin{bmatrix} 4 & 10 \\ 2 & 12 \end{bmatrix} + \begin{bmatrix} 4 & -16 \\ -12 & -28 \end{bmatrix}$

$= \begin{bmatrix} 8 & -6 \\ -10 & -16 \end{bmatrix}$

19. Since B is a 3 × 3 matrix, and A is a 3 × 2 matrix, the calculation of B − A is not possible.

21. A has size 3 × 2

and

F has size 2 × 2,

so AF will have size 3 × 2.

$$AF = \begin{bmatrix} 4 & 10 \\ -2 & -3 \\ 6 & 9 \end{bmatrix}\begin{bmatrix} -1 & 4 \\ 3 & 7 \end{bmatrix}$$

$$= \begin{bmatrix} 26 & 86 \\ -7 & -29 \\ 21 & 87 \end{bmatrix}$$

23. D has size 3 × 1
and
E has size 1 × 3,
so DE will have size 3 × 3.

$$DE = \begin{bmatrix} 6 \\ 1 \\ 0 \end{bmatrix} \begin{bmatrix} 1 & 3 & -4 \end{bmatrix}$$

$$= \begin{bmatrix} 6 & 18 & -24 \\ 1 & 3 & -4 \\ 0 & 0 & 0 \end{bmatrix}$$

25. B has size 3 × 3
and
D has size 3 × 1,
so BD will have size 3 × 1.

$$BD = \begin{bmatrix} 2 & 3 & -2 \\ 2 & 4 & 0 \\ 0 & 1 & 2 \end{bmatrix} \begin{bmatrix} 6 \\ 1 \\ 0 \end{bmatrix} = \begin{bmatrix} 15 \\ 16 \\ 1 \end{bmatrix}$$

27. $F = \begin{bmatrix} -1 & 4 \\ 3 & 7 \end{bmatrix}$

$$[F \mid I] = \begin{bmatrix} -1 & 4 & 1 & 0 \\ 3 & 7 & 0 & 1 \end{bmatrix}$$

$$3R_1 + R_2 \to R_2 \begin{bmatrix} -1 & 4 & 1 & 0 \\ 0 & 19 & 3 & 1 \end{bmatrix}$$

$$-19R_1 + 4R_2 \to R_1 \begin{bmatrix} 19 & 0 & -7 & 4 \\ 0 & 19 & 3 & 1 \end{bmatrix}$$

$$\frac{1}{19}R_1 \to R_1 \begin{bmatrix} 1 & 0 & -\frac{7}{19} & \frac{4}{19} \\ 0 & 1 & \frac{3}{19} & \frac{1}{19} \end{bmatrix}$$
$$\frac{1}{19}R_2 \to R_2$$

$$F^{-1} = \begin{bmatrix} -\frac{7}{19} & \frac{4}{19} \\ \frac{3}{19} & \frac{1}{19} \end{bmatrix}$$

29. A and C are 3 × 2 matrices, so their sum A + C is a 3 × 2 matrix. Only square matrices have inverses. Therefore, $(A + C)^{-1}$ does not exist.

31. Let $A = \begin{bmatrix} -4 & 2 \\ 0 & 3 \end{bmatrix}$.

$$[A \mid I] = \begin{bmatrix} -4 & 2 & 1 & 0 \\ 0 & 3 & 0 & 1 \end{bmatrix}$$

$$-3R_1 + 2R_2 \to R_1 \begin{bmatrix} 12 & 0 & -3 & 2 \\ 0 & 3 & 0 & 1 \end{bmatrix}$$

$$\frac{1}{12}R_1 \to R_1 \begin{bmatrix} 1 & 0 & -\frac{1}{4} & \frac{1}{6} \\ 0 & 1 & 0 & \frac{1}{3} \end{bmatrix}$$
$$\frac{1}{3}R_2 \to R_2$$

The inverse is $\begin{bmatrix} -\frac{1}{4} & \frac{1}{6} \\ 0 & \frac{1}{3} \end{bmatrix}$.

33. Let $A = \begin{bmatrix} 6 & 4 \\ 3 & 2 \end{bmatrix}$.

$$[A \mid I] = \begin{bmatrix} 6 & 4 & 1 & 0 \\ 3 & 2 & 0 & 1 \end{bmatrix}$$

$$R_1 + (-2)R_2 \to R_2 \begin{bmatrix} 6 & 4 & 1 & 0 \\ 0 & 0 & 1 & -2 \end{bmatrix}$$

The zeros in the second row indicate that the original matrix has no inverse.

35. Let $A = \begin{bmatrix} 2 & 0 & 4 \\ 1 & -1 & 0 \\ 0 & 1 & -2 \end{bmatrix}$.

$$[A \mid I] = \begin{bmatrix} 2 & 0 & 4 & 1 & 0 & 0 \\ 1 & -1 & 0 & 0 & 1 & 0 \\ 0 & 1 & -2 & 0 & 0 & 1 \end{bmatrix}$$

$$-2R_2 + R_1 \to R_2 \begin{bmatrix} 2 & 0 & 4 & 1 & 0 & 0 \\ 0 & 2 & 4 & 1 & -2 & 0 \\ 0 & 1 & -2 & 0 & 0 & 1 \end{bmatrix}$$

$$-2R_3 + R_2 \to R_3 \begin{bmatrix} 2 & 0 & 4 & 1 & 0 & 0 \\ 0 & 2 & 4 & 1 & -2 & 0 \\ 0 & 0 & 8 & 1 & -2 & -2 \end{bmatrix}$$

$$-1R_3 + 2R_1 \to R_1 \begin{bmatrix} 4 & 0 & 0 & 1 & 2 & 2 \\ 0 & 4 & 0 & 1 & -2 & 2 \\ 0 & 0 & 8 & 1 & -2 & -2 \end{bmatrix}$$
$$-1R_3 + 2R_2 \to R_2$$

$\frac{1}{4}R_1 \rightarrow R_1$
$\frac{1}{4}R_2 \rightarrow R_2$
$\frac{1}{8}R_3 \rightarrow R_3$
$\begin{bmatrix} 1 & 0 & 0 & \frac{1}{4} & \frac{1}{2} & \frac{1}{2} \\ 0 & 1 & 0 & \frac{1}{4} & -\frac{1}{2} & \frac{1}{2} \\ 0 & 0 & 1 & \frac{1}{8} & -\frac{1}{4} & -\frac{1}{4} \end{bmatrix}$

The inverse is $\begin{bmatrix} \frac{1}{4} & \frac{1}{2} & \frac{1}{2} \\ \frac{1}{4} & -\frac{1}{2} & \frac{1}{2} \\ \frac{1}{8} & -\frac{1}{4} & -\frac{1}{4} \end{bmatrix}$.

37. Let $A = \begin{bmatrix} 2 & 3 & 5 \\ -2 & -3 & 5 \\ 1 & 4 & 2 \end{bmatrix}$.

$[A \mid I] = \begin{bmatrix} 2 & 3 & 5 & 1 & 0 & 0 \\ -2 & -3 & -5 & 0 & 1 & 0 \\ 1 & 4 & 2 & 0 & 0 & 1 \end{bmatrix}$

$\begin{matrix} R_1 + R_2 \rightarrow R_2 \\ -2R_3 + R_1 \rightarrow R_3 \end{matrix} \begin{bmatrix} 2 & 3 & 5 & 1 & 0 & 0 \\ 0 & 0 & 0 & 1 & 1 & 0 \\ 0 & -5 & 1 & 1 & 0 & -2 \end{bmatrix}$

The zeros in the second row to the left of the vertical bar indicate that the original matrix has no inverse.

39. $A = \begin{bmatrix} 1 & 2 \\ 2 & 4 \end{bmatrix}$, $B = \begin{bmatrix} 5 \\ 10 \end{bmatrix}$

Row operations may be used to see that matrix A has no inverse. The matrix equation AX = B may be written as the system of equations

$$x + 2y = 5 \quad (1)$$
$$2x + 4y = 10. \quad (2)$$

Use the elimination method to solve this system. Begin by eliminating x in equation (2).

$$x + 2y = 5 \quad (1)$$
$-2R_1 + R_2 \rightarrow R_2 \qquad 0 = 0 \quad (3)$

The true statement in equation (3) indicates that the equations are dependent. Solve equation (1) for x in terms of y.

$$x = -2y + 5$$

The solution of the system is $(-2y + 5, y)$.

41. $A = \begin{bmatrix} 2 & 4 & 0 \\ 1 & -2 & 0 \\ 0 & 0 & 3 \end{bmatrix}$, $B = \begin{bmatrix} 72 \\ -24 \\ 48 \end{bmatrix}$

Use row operations to find the inverse of A, which is

$$A^{-1} = \begin{bmatrix} \frac{1}{4} & \frac{1}{2} & 0 \\ \frac{1}{8} & -\frac{1}{4} & 0 \\ 0 & 0 & \frac{1}{3} \end{bmatrix}.$$

Since $X = A^{-1}B$,

$$X = \begin{bmatrix} \frac{1}{4} & \frac{1}{2} & 0 \\ \frac{1}{8} & -\frac{1}{4} & 0 \\ 0 & 0 & \frac{1}{3} \end{bmatrix} \begin{bmatrix} 72 \\ -24 \\ 48 \end{bmatrix} = \begin{bmatrix} 6 \\ 15 \\ 16 \end{bmatrix}.$$

43. $5x + 10y = 80$
$3x - 2y = 120$

Let $A = \begin{bmatrix} 5 & 10 \\ 3 & -2 \end{bmatrix}$, $X = \begin{bmatrix} x \\ y \end{bmatrix}$, $B = \begin{bmatrix} 80 \\ 120 \end{bmatrix}$.

Use row operations to find the inverse of A, which is

$$A^{-1} = \begin{bmatrix} \frac{1}{20} & \frac{1}{4} \\ \frac{3}{40} & -\frac{1}{8} \end{bmatrix}.$$

Since $X = A^{-1}B$,

$$\begin{bmatrix} x \\ y \end{bmatrix} = \begin{bmatrix} \frac{1}{20} & \frac{1}{4} \\ \frac{3}{40} & -\frac{1}{8} \end{bmatrix} \begin{bmatrix} 80 \\ 120 \end{bmatrix} = \begin{bmatrix} 34 \\ -9 \end{bmatrix}.$$

The solution of the system is $(34, -9)$.

45. $A = \begin{bmatrix} .01 & .05 \\ .04 & .03 \end{bmatrix}, \quad D = \begin{bmatrix} 200 \\ 300 \end{bmatrix}$

$I - A = \begin{bmatrix} 1 & 0 \\ 0 & 1 \end{bmatrix} - \begin{bmatrix} .01 & .05 \\ .04 & .03 \end{bmatrix}$

$= \begin{bmatrix} .99 & -.05 \\ -.04 & .97 \end{bmatrix}$

Use row operations to find the inverse of I - A, which is

$(I - A)^{-1} \approx \begin{bmatrix} 1.0122 & .0522 \\ .0417 & 1.0331 \end{bmatrix}.$

Since $X = (I - A)^{-1}D$, the production matrix is

$X = \begin{bmatrix} 1.0122 & .0522 \\ .0417 & 1.0331 \end{bmatrix} \begin{bmatrix} 200 \\ 300 \end{bmatrix}$

$= \begin{bmatrix} 218.1 \\ 318.2 \end{bmatrix}$

47. Use a chart to organize the information.

	Standard	Extra Large	Time Available
Hours Cutting	$\frac{1}{4}$	$\frac{1}{3}$	4
Hours Shaping	$\frac{1}{2}$	$\frac{1}{3}$	6

Let x = the number of standard paper clips (in thousands); and

y = the number of extra large paper clips (in thousands).

The given information leads to the system

$\frac{1}{4}x + \frac{1}{3}y = 4$

$\frac{1}{2}x + \frac{1}{3}y = 6.$

Solve this system by any method to get x = 8, y = 6.

The manufacturer can make 8 thousand standard and 6 thousand extra large paper clips.

49. Let x_1 = the number of blankets;

x_2 = the number of rugs; and

x_3 = the number of skirts.

The given information leads to the system

$24x_1 + 30x_2 + 12x_3 = 306 \quad (1)$

$4x_1 + 5x_2 + 3x_3 = 59 \quad (2)$

$15x_1 + 18x_2 + 9x_3 = 201 \quad (3)$

Simplify equations (1) and (3).

$\frac{1}{6}R_1 \to R_1 \quad 4x_1 + 5x_2 + 2x_3 = 51 \quad (4)$

$4x_1 + 5x_2 + 3x_3 = 59 \quad (5)$

$\frac{1}{3}R_3 \to R_3 \quad 5x_1 + 6x_2 + 3x_3 = 67 \quad (6)$

Solve this system by the Gauss–Jordan method. Write the augmented matrix and use row operations.

$\begin{bmatrix} 4 & 5 & 2 & | & 51 \\ 4 & 5 & 3 & | & 59 \\ 5 & 6 & 3 & | & 67 \end{bmatrix}$

$\begin{matrix} \\ -1R_1 + R_2 \to R_2 \\ -4R_3 + 5R_1 \to R_3 \end{matrix} \begin{bmatrix} 4 & 5 & 2 & | & 51 \\ 0 & 0 & 1 & | & 8 \\ 0 & 1 & -2 & | & -13 \end{bmatrix}$

Interchange the second and third rows.

$\begin{bmatrix} 4 & 5 & 2 & | & 51 \\ 0 & 1 & -2 & | & -13 \\ 0 & 0 & 1 & | & 8 \end{bmatrix}$

$-5R_2 + R_1 \to R_1 \begin{bmatrix} 4 & 0 & 12 & | & 116 \\ 0 & 1 & -2 & | & -13 \\ 0 & 0 & 1 & | & 8 \end{bmatrix}$

$\begin{matrix} -12R_3 + R_1 \to R_1 \\ R_2 + 2R_3 \to R_2 \end{matrix} \begin{bmatrix} 4 & 0 & 0 & | & 20 \\ 0 & 1 & 0 & | & 3 \\ 0 & 0 & 1 & | & 8 \end{bmatrix}$

$\frac{1}{4}R_1 \to R_1$ $\begin{bmatrix} 1 & 0 & 0 & | & 5 \\ 0 & 1 & 0 & | & 3 \\ 0 & 0 & 1 & | & 8 \end{bmatrix}$

The solution of the system is $x = 5$, $y = 3$, $z = 8$.

5 blankets, 3 rugs, and 8 skirts can be made.

51. The 4×5 matrix of stock reports is

$$\begin{bmatrix} 5 & 7 & 2532 & 52\frac{3}{8} & -\frac{1}{4} \\ 3 & 9 & 1464 & 56 & \frac{1}{8} \\ 2.50 & 5 & 4974 & 41 & -1\frac{1}{2} \\ 1.36 & 10 & 1754 & 18\frac{7}{8} & \frac{1}{2} \end{bmatrix}.$$

53.   (a)   The input-output matrix is

$$A = \begin{bmatrix} 0 & \frac{1}{2} \\ \frac{2}{3} & 0 \end{bmatrix}.$$

(b) $I - A = \begin{bmatrix} 1 & -\frac{1}{2} \\ -\frac{2}{3} & 1 \end{bmatrix}$, $D = \begin{bmatrix} 400 \\ 800 \end{bmatrix}$

Use row operations to find the inverse of $I - A$, which is

$$(I - A)^{-1} = \begin{bmatrix} \frac{3}{2} & \frac{3}{4} \\ 1 & \frac{3}{2} \end{bmatrix}.$$

Since $X = (I - A)^{-1}D$,

$$X = \begin{bmatrix} \frac{3}{2} & \frac{3}{4} \\ 1 & \frac{3}{2} \end{bmatrix} \begin{bmatrix} 400 \\ 800 \end{bmatrix} = \begin{bmatrix} 1200 \\ 1600 \end{bmatrix}.$$

The production required is 1200 units of cheese and 1600 units of goats.

55.

$$\begin{aligned} x + 2y + z &= 7 \quad (1) \\ 2x - y - z &= 2 \quad (2) \\ 3x - 3y + 2z &= -5 \quad (3) \end{aligned}$$

(a)   To solve the system by elimination, begin by eliminating x in equations (2) and (3).

$$\begin{array}{l} \ \ x + 2y + z = \ \ \ 7 \quad (1) \\ -2R_1 + R_2 \to R_2 \quad -5y - 3z = -12 \quad (4) \\ -3R_1 + R_3 \to R_3 \quad -9y - z = -26 \quad (5) \end{array}$$

Eliminate y in equation (5).

$$\begin{array}{l} \ \ x + 2y + z = \ \ \ 7 \quad (1) \\ \ \ \ \ \ \ \ -5y - 3z = -12 \quad (4) \\ -9R_2 + 5R_3 \to R_3 \quad \quad \ \ 22z = -22 \quad (6) \end{array}$$

Make each leading coefficient equal 1.

$$\begin{array}{l} \phantom{-\frac{1}{5}R_2 \to R_2} \ \ x + 2y + z = 7 \quad (1) \\ -\frac{1}{5}R_2 \to R_2 \quad \quad y + \frac{3}{5}z = \frac{12}{5} \quad (7) \\ \frac{1}{22}R_3 \to R_3 \quad \quad \quad \ \ z = -1 \quad (8) \end{array}$$

Substitute -1 for z in equation (7) to get $y = 3$. Substitute -1 for z and 3 for y in equation (1) to get $x = 2$. The solution of the system is $(2, 3, -1)$.

(b)   The same system is to be solved using the Gaussian method. Write the augmented matrix and use row operations.

$$\begin{bmatrix} 1 & 2 & 1 & | & 7 \\ 2 & -1 & -1 & | & 2 \\ 3 & -3 & 2 & | & -5 \end{bmatrix}$$

$$\begin{matrix} -2R_1 + R_2 \to R_2 \\ -3R_1 + R_3 \to R_3 \end{matrix} \begin{bmatrix} 1 & 2 & 1 & | & 7 \\ 0 & -5 & -3 & | & -12 \\ 0 & -9 & -1 & | & -26 \end{bmatrix}$$

$$-9R_2 + 5R_3 \to R_3 \begin{bmatrix} 1 & 2 & 1 & | & 7 \\ 0 & -5 & -3 & | & -12 \\ 0 & 0 & 22 & | & -22 \end{bmatrix}$$

$$-\frac{1}{5}R_2 \to R_2 \quad \begin{bmatrix} 1 & 2 & 1 & 7 \\ 0 & 1 & \frac{3}{5} & \frac{12}{5} \\ 0 & 0 & 1 & -1 \end{bmatrix}$$
$$\frac{1}{22}R_3 \to R_3$$

The corresponding system is

$$x + 2y + z = 7$$
$$y + \frac{3}{5}z = \frac{12}{5}$$
$$z = -1.$$

Use back-substitution to see again that the solution of the system is $(2, 3, -1)$.

(c) The same system is to be solved using the Gauss–Jordan method. Write the augmented matrix and use row operations.

$$\begin{bmatrix} 1 & 2 & 1 & 7 \\ 2 & -1 & -1 & 2 \\ 3 & -3 & 2 & -5 \end{bmatrix}$$

$$\begin{matrix} -2R_1 + R_2 \to R_2 \\ -3R_1 + R_3 \to R_3 \end{matrix} \begin{bmatrix} 1 & 2 & 1 & 7 \\ 0 & -5 & -3 & -12 \\ 0 & -9 & -1 & -26 \end{bmatrix}$$

$$\begin{matrix} 5R_1 + 2R_2 \to R_1 \\ \\ -9R_2 + 5R_3 \to R_3 \end{matrix} \begin{bmatrix} 5 & 0 & -1 & 11 \\ 0 & -5 & -3 & -12 \\ 0 & 0 & 22 & -22 \end{bmatrix}$$

$$\begin{matrix} 22R_1 + R_3 \to R_1 \\ 22R_2 + 3R_3 \to R_2 \end{matrix} \begin{bmatrix} 110 & 0 & 0 & 220 \\ 0 & -110 & 0 & -330 \\ 0 & 0 & 22 & -22 \end{bmatrix}$$

$$\begin{matrix} \frac{1}{110}R_1 \to R_1 \\ -\frac{1}{110}R_2 \to R_2 \\ \frac{1}{22}R_3 \to R_3 \end{matrix} \begin{bmatrix} 1 & 0 & 0 & 2 \\ 0 & 1 & 0 & 3 \\ 0 & 0 & 1 & -1 \end{bmatrix}$$

Once again, the solution of the system is $(2, 3, -1)$.

(d) The system can be written as a matrix equation $AX = B$ by writing

$$\begin{bmatrix} 1 & 2 & 1 \\ 2 & -1 & -1 \\ 3 & -3 & 2 \end{bmatrix} \begin{bmatrix} x \\ y \\ z \end{bmatrix} = \begin{bmatrix} 7 \\ 2 \\ -5 \end{bmatrix}.$$

(e) The inverse of the coefficient matrix A can be found by using row operations.

$$[A \mid I] = \begin{bmatrix} 1 & 2 & 1 & 1 & 0 & 0 \\ 2 & -1 & -1 & 0 & 1 & 0 \\ 3 & -3 & 2 & 0 & 0 & 1 \end{bmatrix}$$

$$\begin{matrix} -2R_1 + R_2 \to R_2 \\ -3R_1 + R_3 \to R_3 \end{matrix} \begin{bmatrix} 1 & 2 & 1 & 1 & 0 & 0 \\ 0 & -5 & -3 & -2 & 1 & 0 \\ 0 & -9 & -1 & -3 & 0 & 1 \end{bmatrix}$$

$$\begin{matrix} 5R_1 + 2R_2 \to R_1 \\ \\ -9R_2 + 5R_3 \to R_3 \end{matrix} \begin{bmatrix} 5 & 0 & -1 & 1 & 2 & 0 \\ 0 & -5 & -3 & -2 & 1 & 0 \\ 0 & 0 & 22 & 3 & -9 & 5 \end{bmatrix}$$

$$\begin{matrix} 22R_1 + R_3 \to R_1 \\ 22R_2 + 3R_3 \to R_2 \end{matrix} \begin{bmatrix} 110 & 0 & 0 & 25 & 35 & 5 \\ 0 & -110 & 0 & -35 & -5 & 15 \\ 0 & 0 & 22 & 3 & -9 & 5 \end{bmatrix}$$

$$\begin{matrix} \frac{1}{110}R_1 \to R_1 \\ -\frac{1}{110}R_2 \to R_2 \\ \frac{1}{22}R_3 \to R_3 \end{matrix} \begin{bmatrix} 1 & 0 & 0 & \frac{5}{22} & \frac{7}{22} & \frac{1}{22} \\ 0 & 1 & 0 & \frac{7}{22} & \frac{1}{22} & -\frac{3}{22} \\ 0 & 0 & 1 & \frac{3}{22} & -\frac{9}{22} & \frac{5}{22} \end{bmatrix}$$

The inverse of matrix A is

$$A^{-1} = \begin{bmatrix} \frac{5}{22} & \frac{7}{22} & \frac{1}{22} \\ \frac{7}{22} & \frac{1}{22} & -\frac{3}{22} \\ \frac{3}{22} & -\frac{9}{22} & \frac{5}{22} \end{bmatrix}$$

$$\approx \begin{bmatrix} .23 & .32 & .05 \\ .32 & .05 & -.14 \\ .14 & -.41 & .23 \end{bmatrix}.$$

(f) Since $X = A^{-1}B$,

$$\begin{bmatrix} x \\ y \\ z \end{bmatrix} = \begin{bmatrix} \frac{5}{22} & \frac{7}{22} & \frac{1}{22} \\ \frac{7}{22} & \frac{1}{22} & -\frac{3}{22} \\ \frac{3}{22} & -\frac{9}{22} & \frac{5}{22} \end{bmatrix} \begin{bmatrix} 7 \\ 2 \\ -5 \end{bmatrix} = \begin{bmatrix} 2 \\ 3 \\ -1 \end{bmatrix}.$$

Once again, the solution of the system is $(2, 3, -1)$.

Extended Application

1. **(a)** $A = \begin{bmatrix} .245 & .102 & .051 \\ .099 & .291 & .279 \\ .433 & .372 & .011 \end{bmatrix}$, $D = \begin{bmatrix} 2.88 \\ 31.45 \\ 30.91 \end{bmatrix}$, $X = \begin{bmatrix} x_1 \\ x_2 \\ x_3 \end{bmatrix}$

(b) $I - A = \begin{bmatrix} 1 & 0 & 0 \\ 0 & 1 & 0 \\ 0 & 0 & 1 \end{bmatrix} - \begin{bmatrix} .245 & .102 & .051 \\ .099 & .291 & .279 \\ .433 & .372 & .011 \end{bmatrix} = \begin{bmatrix} .755 & -.102 & -.051 \\ -.099 & .709 & -.279 \\ -.433 & -.372 & .989 \end{bmatrix}$

(c) $(I - A)^{-1}(I - A) = \begin{bmatrix} 1.454 & .291 & .157 \\ .533 & 1.763 & .525 \\ .837 & .791 & 1.278 \end{bmatrix}\begin{bmatrix} .755 & -.102 & -.051 \\ -.099 & .709 & -.279 \\ -.433 & -.372 & .989 \end{bmatrix}$

$= \begin{bmatrix} 1.00098 & .0004 & .00007 \\ .00055 & 1.0003 & .00017 \\ .00025 & .00003 & 1.00057 \end{bmatrix} \approx \begin{bmatrix} 1 & 0 & 0 \\ 0 & 1 & 0 \\ 0 & 0 & 1 \end{bmatrix}$

(d) $X = (I - A)^{-1}D = \begin{bmatrix} 1.454 & .291 & .157 \\ .533 & 1.763 & .525 \\ .837 & .791 & 1.278 \end{bmatrix}\begin{bmatrix} 2.88 \\ 31.45 \\ 30.91 \end{bmatrix} \approx \begin{bmatrix} 18.2 \\ 73.2 \\ 66.8 \end{bmatrix}$

(Each entry of X has been rounded to three significant digits.)

(e) $18.2 billion of agriculture, $73.2 billion of manufacturing, and $66.8 billion of household would be required to support a demand of $2.88 billion, $31.45 billion, and $30.91 billion, respectively.

2. **(a)** $A = \begin{bmatrix} .293 & 0 & 0 \\ .014 & .207 & .017 \\ .044 & .010 & .216 \end{bmatrix}$, $D = \begin{bmatrix} 138,213 \\ 17,597 \\ 1786 \end{bmatrix}$

(b) $I - A = \begin{bmatrix} 1 & 0 & 0 \\ 0 & 1 & 0 \\ 0 & 0 & 1 \end{bmatrix} - \begin{bmatrix} .293 & 0 & 0 \\ .014 & .207 & .017 \\ .044 & .010 & .216 \end{bmatrix} = \begin{bmatrix} .707 & 0 & 0 \\ -.014 & .793 & -.017 \\ -.044 & -.010 & .784 \end{bmatrix}$

(c) $(I - A)^{-1}(I - A) = \begin{bmatrix} 1.414 & 0 & 0 \\ .027 & 1.261 & .027 \\ .080 & .016 & 1.276 \end{bmatrix}\begin{bmatrix} .707 & 0 & 0 \\ -.014 & .793 & -.017 \\ -.044 & -.010 & .784 \end{bmatrix}$

$= \begin{bmatrix} .9997 & 0 & 0 \\ .0002 & .9997 & -.0003 \\ .0002 & -.00007 & 1.0001 \end{bmatrix} \approx \begin{bmatrix} 1 & 0 & 0 \\ 0 & 1 & 0 \\ 0 & 0 & 1 \end{bmatrix}$

(d) $X = (I - A)^{-1}D = \begin{bmatrix} 1.414 & 0 & 0 \\ .027 & 1.261 & .027 \\ .080 & .016 & 1.276 \end{bmatrix}\begin{bmatrix} 138,213 \\ 17,597 \\ 1786 \end{bmatrix} \approx \begin{bmatrix} 195,000 \\ 26,000 \\ 13,600 \end{bmatrix}$

Since the entries in D were in thousands of pounds, this means that 195 million pounds of agricultural products, 26 million pounds of manufactured goods, and 13.6 million pounds of energy are required.

CHAPTER 2   TEST

1. Solve the system below using the echelon method.

$$3x + y = 11$$
$$x - 2y = -8$$

2. Solve using the Gauss–Jordan method.

$$x + 2y + 3z = 5$$
$$2x - y + z = 5$$
$$x + y + z = 2$$

3. Find the values of the variables.

$$\begin{bmatrix} 6-x & 2y+1 \\ 3m & 5p+2 \end{bmatrix} = \begin{bmatrix} 8 & 10 \\ -5 & 3p-1 \end{bmatrix}$$

4. Given the matrices below, perform the indicated operations, if possible.

$$A = \begin{bmatrix} 1 & 2 & -1 \\ 0 & 1 & 1 \\ 1 & 0 & 1 \end{bmatrix} \qquad B = \begin{bmatrix} 1 & -2 \\ 1 & 1 \\ 0 & 1 \end{bmatrix} \qquad C = \begin{bmatrix} 2 & 1 & 3 \\ 0 & 4 & 1 \\ 1 & 1 & 1 \end{bmatrix}$$

    (a) AB   **(b)** 2A − C   **(c)** A + 2B

5. Find the inverse of each matrix which has an inverse.

$$A = \begin{bmatrix} 1 & 0 & -1 \\ 2 & 1 & 1 \\ 1 & 1 & 5 \end{bmatrix} \qquad B = \begin{bmatrix} 2 & -1 & 1 \\ 0 & 2 & 4 \\ 2 & 1 & 5 \end{bmatrix}$$

6. For $A = \begin{bmatrix} 1 & 0 & 1 \\ 1 & 1 & 1 \\ 2 & 1 & 3 \end{bmatrix}$, $A^{-1} = \begin{bmatrix} 2 & 1 & -1 \\ -1 & 1 & 0 \\ -1 & -1 & 1 \end{bmatrix}$.

    Use this inverse to solve the equation AX = B, where $B = \begin{bmatrix} 1 \\ 2 \\ -1 \end{bmatrix}$.

7. Solve the system below using the inverse of the coefficient matrix.

$$x - 2y = 3$$
$$x + 3y = 5$$

8. Find the production matrix, given the following input–output and demand matrices.

$$A = \begin{bmatrix} .2 & .1 & .3 \\ .1 & 0 & .2 \\ 0 & 0 & .4 \end{bmatrix} \qquad D = \begin{bmatrix} 1000 \\ 2000 \\ 5000 \end{bmatrix}$$

CHAPTER 2   TEST ANSWERS

1.   (2, 5)

2.   (1, −1, 2)

3. $x = -2$, $y = \dfrac{9}{2}$, $m = -\dfrac{5}{3}$, $p = -\dfrac{3}{2}$

4.   (a) $\begin{bmatrix} 3 & -1 \\ 1 & 2 \\ 1 & -1 \end{bmatrix}$ (b) $\begin{bmatrix} 0 & 3 & -5 \\ 0 & -2 & 1 \\ 1 & -1 & 1 \end{bmatrix}$ (c) Not possible

5. $A^{-1} = \begin{bmatrix} \dfrac{4}{3} & -\dfrac{1}{3} & \dfrac{1}{3} \\ -3 & 2 & -1 \\ \dfrac{1}{3} & -\dfrac{1}{3} & \dfrac{1}{3} \end{bmatrix}$; B^{-1} does not exist.

6. $X = \begin{bmatrix} 5 \\ 1 \\ -4 \end{bmatrix}$

7. $\left(\dfrac{19}{5}, \dfrac{2}{5} \right)$

8. $\begin{bmatrix} 4895 \\ 4156 \\ 8333 \end{bmatrix}$

CHAPTER 3   LINEAR PROGRAMMING: THE GRAPHICAL METHOD

Section 3.1

For Exercises 1–39, see the answer graphs in the back of the textbook.

1. $x + y \leq 2$

First graph the boundary line $x + y = 2$ using the points $(2, 0)$ and $(0, 2)$. Since the points on this line satisfy $x + y \leq 2$, draw a solid line. To find the correct region to shade, choose any point not on the line. If $(0, 0)$ is used as the test point, we have

$$x + y \leq 2$$
$$0 + 0 \leq 2$$
$$0 \leq 2,$$

which is a true sttement.
Shade the side of the line which includes $(0, 0)$, which is the region below the line.

3. $x \geq 3 + y$

First graph the boundary line $x = 3 + y$ using the points $(0, -3)$ and $(3, 0)$. This will be a solid line. Choose $(0, 0)$ as a test point, and we have

$$x \geq 3 + y$$
$$0 \geq 3 + 0$$
$$0 \geq 3,$$

which is a false statement.

Shade the side which does not include $(0, 0)$, which is the region below the line.

5. $4x - y < 6$

Graph $4x - y = 6$ as a dashed line, since the points on the line are not part of the solution; the line passes through the points $(0, -6)$ and $\left(\frac{3}{2}, 0\right)$. Using the test point $(0, 0)$, we get $0 - 0 < 6$ or $0 < 6$, a true statement. Shade the side of the boundary line that includes the origin $(0, 0)$, which is the region above the line.

7. $3x + y < 6$

Graph $3x + y = 6$ as a dashed line through $(2, 0)$ and $(0, 6)$. Using the test point $(0, 0)$, we get $3(0) + 0 < 6$ or $0 < 6$, a true statement. Shade the side that contains the origin, which is the region below the line.

9. $x + 3y \geq -2$

The graph includes the line $x + 3y = -2$, whose intercepts are the points $(0, -2/3)$ and $(-2, 0)$. Graph $x + 3y = -2$ as a solid line and use the origin as a test point. Since $0 + 3(0) \geq -2$ is true, shade the side which includes the origin, which is the region above the line.

11. $4x + 3y > -3$

Graph $4x + 3y = -3$ as a dashed line through the points $(0, -1)$ and $(-3/4, 0)$. Use the origin as a test point. Since $4(0) + 3(0) > -3$ is true, the origin will be included in the region, so we shade the half-plane above the line.

13. $2x - 4y < 3$

Graph $2x - 4y = 3$ as a dashed line through the points $(3/2, 0)$ and $(0, -3/4)$. Use the origin as a test point. $2(0) - 4(0) < 3$ is true, so the region above the line, which includes the origin, is the correct half-plane to shade.

15. $x \leq 5y$

Graph $x = 5y$ as a solid line through the points $(0, 0)$ and $(5, 1)$. Since this line contains the origin, some point other than $(0, 0)$ must be used as a test point. If we use the point $(1, 2)$, we obtain $1 \leq 5(2)$ or $1 \leq 10$, a true statement. Shade the side of the line containing $(1, 2)$, that is, the region above the line.

17. $-3x < y$

Graph $y = -3x$ as a dashed line through the points $(0, 0)$ and $(1, -3)$. Since this line contains the origin, use some point other than $(0, 0)$ as a test point. If $(1, 1)$ is used as a test point, we obtain $-3 < 3$, a true statement.

Shade the region containing $(1, 1)$, which is the half-plane above the line.

19. $x + y \leq 0$

Graph $x + y = 0$ as a solid line through the points $(0, 0)$ and $(1, -1)$. This line contains $(0, 0)$. If we use $(-1, 0)$ as a test point, we obtain $-1 + 0 \leq 0$ or $-1 \leq 0$, a true statement. Shade the region containing $(-1, 0)$, which is the half-plane below the line.

21. $y < x$

Graph $y = x$ as a dashed line through the points $(0, 0)$ and $(1, 1)$. Since this line contains the origin, choose a point other than $(0, 0)$ as a test point. If we use $(2, 3)$, we obtain $3 < 2$, which is false. Shade the region that does not contain the test point, which is the half-plane below the line.

23. $x < 4$

Graph $x = 4$ as a dashed line. This is the vertical line crossing the x-axis at the point $(4, 0)$. Using $(0, 0)$ as a test point, we obtain $0 < 4$, which is true. Shade the region containing the origin, which is the half-plane to the left of the line.

25. $y \leq -2$

Graph $y = -2$ as a solid horizontal line through the point $(0, -2)$. Using the origin as a test point, we obtain $0 \leq -2$, which is false. Shade the region which does not contain the origin, which is the half-plane below the line.

27. $x + y \leq 1$
$x - y \geq 2$

Graph each inequality separately, using solid boundary lines. In both cases, the graph is the region below the line. Shade the overlapping part of these two half-planes, which is the region below both lines. The shaded region is the feasible region for this system.

29. $2x - y < 1$
$3x + y < 6$

Graph $2x - y < 1$ as the half-plane above the dashed line $2x - y = 1$. Graph $3x + y < 6$ as the half-plane below the dashed line $3x + y = 6$. Shade the overlapping part of these two half-planes to show the feasible region for this system.

31. $-x - y < 5$
$2x - y < 4$

Graph $-x - y < 5$ as the half-plane above the dashed line $-x - y = 5$.

Graph $2x - y < 4$ as the half-plane above the dashed line $2x - y = 4$. Shade the overlapping part of these two half-planes to show the feasible region.

33. $x + y \leq 4$
$x - y \leq 5$
$4x + y \leq -4$

The graph of $x + y \leq 4$ consists of the solid line $x + y = 4$ and all the points below it. The graph of $x - y \leq 5$ consists of the solid line $x - y = 5$ and all the points above it. The graph of $4x + y \leq -4$ consists of the solid line $4x + y = -4$ and all the points below it. The feasible region is the overlapping part of these three half-planes.

35. $-2 < x < 3$
$-1 \leq y \leq 5$
$2x + y < 6$

The graph of $-2 < x < 3$ is the region between the vertical lines $x = -2$ and $x = 3$, but not including the lines themselves (so the two vertical boundaries are drawn as dashed lines). The graph of $-1 \leq y \leq 5$ is the region between the horizontal lines $y = -1$ and $y = 5$, including the lines (so the two horizontal boundaries are drawn as solid lines). The graph of

2x + y < 6 is the region below the line 2x + y = 6 (so the boundary is drawn as a dashed line). Shade the region common to all three graphs to show the feasible region.

37.   2y + x ≥ -5

   y ≤ 3 + x

   x ≥ 0

   y ≥ 0

The graph of 2y + x ≥ -5 contains of the boundary line 2y + x = 5 and the region above it. The graph of y ≤ 3 + x consists of the boundary line y = 3 + x and the region below it. The inequalities x ≥ 0 and y ≤ 0 restrict the feasible region to the first quadrant. Shade the region in the first quadrant where the first two graphs overlap to show the feasible region.

39.   3x + 4y > 12

   2x - 3y < 6

    0 ≤ y ≤ 2

      x ≥ 0

3x + 4y > 12 is the set of points above the dashed line 3x + 4y = 12; 2x - 3y < 6 is the set of points above the dashed line 2x - 3y = 6; 0 ≤ y ≤ 2 is the rectangular strip of points lying on or between the horizontal lines y = 0 and y = 2; and x ≥ 0 consists of all the points on and to the right of the y-axis. Shade the feasible region, which is the triangular region satisfying all of the inequalities.

41.   (a)

	Number Made	Time on Wheel	Time in Kiln
Glazed	x	1/2	1
Unglazed	y	1	6
Maximum Time Available		8	20

(b) On the wheel, x glazed pots require $\frac{1}{2} \cdot x = \frac{1}{2}x$ hr and y unglazed pots require $1 \cdot y = y$ hr. Since the wheel is available for at most 8 hr per day,

$$\frac{1}{2}x + y \le 8.$$

In the kiln, x glazed pots require $1 \cdot x = x$ hr and y unglazed pots require $6 \cdot y = 6y$ hr. Since the kiln is available for at most 20 hr per day,

$$x + 6y \le 20.$$

Since it is not possible to produce a negative number of pots,

$$x \ge 0 \text{ and } y \ge 0.$$

Thus, we have the system

$$\frac{1}{2}x + y \le 8$$

$$x + 6y \le 20$$

$$x \ge 0$$

$$y \ge 0.$$

See the graph of the feasible region in the back of the textbook.

(c) Yes, 5 glazed and 2 unglazed planters can be made, since the point (5, 2) lies within the feasible region.

From the graph, it looks like the point (10, 2) might lie right on a boundary of the feasible region. However, (10, 2) does not satisfy the inequality $x + 6y \le 20$, so the point is definitely outside the feasible region. Therefore, no, 10 glazed and 2 unglazed planters cannot be made.

43. **(a)**
$$x \ge 3000$$
$$y \ge 5000$$
$$x + y \le 10,000$$

(b) The first inequality gives the half-plane to the right of the vertical line x = 3000, including the points on the line. The second inequality gives the half-plane above the horizontal line y = 5000, including the points on the line. The third inequality gives the half-plane below the line $x + y \le 10,000$, including the points on the line. Shade the region where the three half-planes overlap to show the feasible region. (See the graph of the feasible region in the back of the textbook.)

45. **(a)**
$$x \ge 1000$$
$$y \ge 800$$
$$x + y \le 2400$$

(b) The first inequality gives the set of points on and to the right of the vertical line x = 1000. The second inequality gives the set of points on and above the horizontal line y = 800. The third inequality gives the set of points on and below the line x + y = 2400. Shade the region where the three graphs overlap to show the feasible region. (See the graph of the feasible region in the back of the textbook.)

Section 3.2

1. Make a table indicating the value of the objective function z = 3x + 5y at each corner point.

Corner Point	Value of z = 3x + 5y
(1, 1)	3(1) + 5(1) = 8 Minimum
(2, 7)	3(2) + 5(7) = 41
(5, 10)	3(5) + 5(10) = 65 Maximum
(6, 3)	3(6) + 5(3) = 33

The maximum value of 65 occurs at (5, 10). The minimum value of 8 occurs at (1, 1).

3.

Corner Point	Value of z = .40x + .75y
(0, 0)	.40(0) + .75(0) = 0
(0, 12)	.40(0) + .75(12) = 9
(4, 8)	.40(4) + .75(8) = 7.6
(7, 3)	.40(7) + .75(3) = 5.05
(8, 0)	.40(8) + .75(0) = 3.2

The maximum value is 9 at (0, 12);
the minimum value is 0 at (0, 0).

5.   (a)

Corner Point	Value of z = 4x + 2y
(0, 8)	4(0) + 2(8) = 16
(3, 4)	4(3) + 2(4) = 20
(13/2, 2)	4(13/2) + 2(2) = 30
(12, 0)	4(12) + 2(0) = 48

The minimum value is 16 at (0, 8).
Since the feasible region is unbounded, there is no maximum value.

(b)

Corner Point	Value of z = 2x + 3y
(0, 8)	2(0) + 3(8) = 24
(3, 4)	2(3) + 3(4) = 18
(13/2, 2)	2(13/2) + 3(2) = 19
(12, 0)	2(12) + 3(0) = 24

The minimum value is 18 at (3, 4);
there is no maximum value.

(c)

Corner Point	Value of z = 2x + 4y
(0, 8)	2(0) + 4(8) = 32
(3, 4)	2(3) + 4(4) = 22
(13/2, 2)	2(13/2) + 4(2) = 21
(12, 0)	2(12) + 4(0) = 24

The minimum value is 21 at
(13/2, 2); there is no maximum
value.

(d)

Corner Point	Value of z = x + 4y
(0, 8)	0 + 4(8) = 32
(3, 4)	3 + 4(4) = 19
(13/2, 2)	13/2 + 4(2) = 29/2
(12, 0)	12 + 4(0) = 12

The minimum value is 12 at (12, 0);
there is no maximum value.

7.   Maximize z = 5x + 2y

    subject to: 2x + 3y ≤ 6
                  4x + y ≥ 6
                     x ≥ 0
                     y ≥ 0.

Sketch the feasible region.

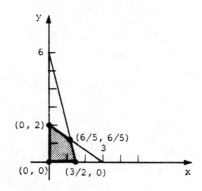

The graph shows that the feasible region is bounded.

The corner points are (0, 0), (0, 2), (3/2, 0), and (6/5, 6/5), which is the intersection of $2x + 3y = 6$ and $4x + y = 6$. Use the corner points to find the maximum value of the objective function.

Corner Point	Value of $z = 5x + 2y$
(0, 0)	$5(0) + 2(0) = 0$
(0, 2)	$5(0) + 2(2) = 4$
$\left(\frac{6}{5}, \frac{6}{5}\right)$	$5\left(\frac{6}{5}\right) + 2\left(\frac{6}{5}\right) = \frac{42}{5}$
$\left(\frac{3}{2}, 0\right)$	$5\left(\frac{3}{2}\right) + 2(0) = \frac{15}{2}$

The maximum value of $z = 5x + 2y$ is 42/5 at the corner point (6/5, 6/5).

9. Maximize $z = 2x + 2y$

   subject to: $3x - y \le 12$
   $x + y \le 15$
   $x \ge 2$
   $y \ge 5.$

   Sketch the feasible region.

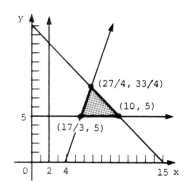

The graph shows that the feasible region is bounded. The corner points are: (17/3, 5), which is the intersection of $y = 5$ and $3x - y = 12$; (27/4, 33/4), which is the intersection of $3x - y = 12$ and $x + y = 15$; and (10, 5), which is the intersection of $y = 5$ and $x + y = 15$.

Use the corner points to find the maximum value of the objective function.

Corner Point	Value of $z = 2x + 2y$
$\left(\frac{17}{3}, 5\right)$	$2\left(\frac{17}{3}\right) + 2(5) = \frac{64}{3}$
$\left(\frac{27}{4}, \frac{33}{4}\right)$	$2\left(\frac{27}{4}\right) + 2\left(\frac{33}{4}\right) = 30$
(10, 5)	$2(10) + 2(5) = 30$

The maximum value of $z = 2x + 2y$ is 30 at (27/4, 33/4) and it is also 30 at (10, 5), as well as at all points on the line segment joining (27/4, 33/4) and (10, 5).

11. Maximize $z = 4x + 2y$

   subject to: $x - y \le 10$
   $5x + 3y \le 75$
   $x + y \le 20$
   $x \ge 0$
   $y \ge 0.$

Sketch the feasible region.

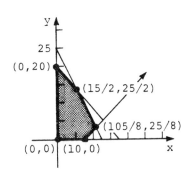

The region is bounded, with corner points (0, 0), (0, 20), (15/2, 25/2), (105/8, 25/8), and (10, 0).

Corner Point	Value of z = 4x + 2y
(0, 0)	$4(0) + 2(0) = 0$
(0, 20)	$4(0) + 2(20) = 40$
$\left(\frac{15}{2}, \frac{25}{2}\right)$	$4\left(\frac{15}{2}\right) + 2\left(\frac{25}{2}\right) = 55$
$\left(\frac{105}{8}, \frac{25}{8}\right)$	$4\left(\frac{105}{8}\right) + 2\left(\frac{25}{8}\right) = \frac{235}{4}$
(10, 0)	$4(10) + 2(0) = 40$

The maximum value of z = 4x + 2y is 235/4 at (105/8, 25/8).

13. **(a)** $x + y \leq 20$

$x + 3y \leq 24$

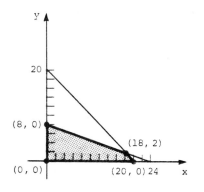

Corner Point	Value of z = 10x + 12y
(0, 0)	$10(0) + 12(0) = 0$
(0, 8)	$10(0) + 12(8) = 96$
(18, 2)	$10(18) + 12(2) = 204$
(20, 0)	$10(20) + 12(0) = 200$

The maximum value of 204 occurs 0 when x = 18 and y = 2.

(b) $3x + y \leq 15$

$x + 2y \leq 18$

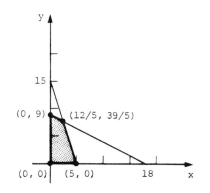

Corner Point	Value of z = 10x + 12y
(0, 0)	$10(0) + 12(0) = 0$
(0, 9)	$10(0) + 12(9) = 108$
$\left(\frac{12}{5}, \frac{39}{5}\right)$	$10\left(\frac{12}{5}\right) + 12\left(\frac{39}{5}\right) = \frac{588}{5}$
(5, 0)	$10(5) + 12(0) = 50$

The maximum value of 588/5 occurs when x = 12/5 and y = 39/5.

(c) $2x + 5y \geq 22$

$4x + 3y \leq 28$

$2x + 2y \leq 17$

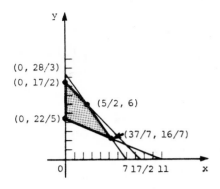

Corner Point	Value of $z = 10x + 12y$
$\left(0, \frac{22}{5}\right)$	$10(0) + 12\left(\frac{22}{5}\right) = \frac{264}{5}$
$\left(0, \frac{17}{2}\right)$	$10(0) + 12\left(\frac{17}{2}\right) = 102$
$\left(\frac{5}{2}, 6\right)$	$10\left(\frac{5}{2}\right) + 12(6) = 97$
$\left(\frac{37}{7}, \frac{16}{7}\right)$	$10\left(\frac{37}{7}\right) + 12\left(\frac{16}{7}\right) = \frac{562}{7}$

The minimum value of 102 occurs when $x = 0$ and $y = 17/2$.

15. Maximize $z = c_1 x_1 + c_2 x_2$
Subject to: $2x_1 + x_2 \leq 11$
$-x_1 + 2x_2 \leq 2$
$x_1 \geq 0, x_2 \geq 0.$

Sketch the feasible region.

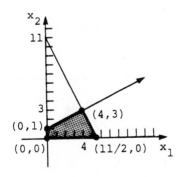

The region is bounded, with corner points $(0, 0)$ $(0, 1)$, $(4, 3)$, and $(11/2, 0)$.

Corner Point	Value of $z = c_1 x_1 + c_2 x_2$
$(0, 0)$	$c_1(0) + c_2(0) = 0$
$(0, 1)$	$c_1(0) + c_2(1) = c_2$
$(4, 3)$	$c_1(4) + c_2(3) = 4c_1 + 3c_2$
$\left(\frac{11}{2}, 0\right)$	$c_1\left(\frac{11}{2}\right) + c_2(0) = \frac{11}{2}c_1$

If we are to have $(x_1, x_2) = (4, 3)$ as an optimal solution, then it must be true that both $4c_1 + 3c_2 \geq c_2$ and $4c_1 + 3c_2 \geq \frac{11}{2}c_1$, because the value of z at $(4, 3)$ cannot be smaller than the other values of z in the table. Manipulate the symbols in these two inequalities in order to isolate $\frac{c_1}{c_2}$ in each; keep in mind the given information that $c_2 > 0$ when performing division by c_2. First,

$$4c_1 + 3c_2 \geq c_2$$
$$4c_1 \geq -2c_2$$
$$\frac{4c_1}{4c_2} \geq \frac{-2c_2}{4c_2}$$
$$\frac{c_1}{c_2} \geq -\frac{1}{2}.$$

Then,

$$4c_1 + 3c_2 \geq \frac{11}{2}c_1$$
$$-\frac{3}{2}c_1 + 3c_2 \geq 0$$
$$3c_1 - 6c_2 \leq 0$$
$$3c_1 \leq 6c_2$$
$$\frac{3c_1}{3c_2} \leq \frac{6c_2}{3c_2}$$
$$\frac{c_1}{c_2} \leq 2.$$

Since $\frac{c_1}{c_2} \geq -\frac{1}{2}$ and $\frac{c_1}{c_2} \leq 2$, the desired range for c_1/c_2 is $[-1/2, 2]$, which corresponds to choice (b).

Section 3.3

1. Let x represent the number of product A made, and y represent the number of product B. Each item of A uses 2 hr on the machine, so 2x represents the total hours required for x items of product A. Similarly, 3y represents the total hours used for product B. There are only 45 hr available, so

$$2x + 3y \leq 45.$$

3. Let x represent the number of green pills and y represent the number of red pills. Then 4x represents the number of vitamin units provided by the green pills, and y represents the vitamin units provided by the red ones. Since at least 25 units are needed per day,

$$4x + y \geq 25.$$

5. Let x represent the number of pounds of \$6 coffee and y represent the number of pounds of \$5 coffee. Since the mixture must weigh at least 50 lb,

$$x + y \geq 50.$$

(Notice that the price per pound is not used in setting up this inequality.)

For Exercises 7–21, the graphs of the feasible regions appear in the answer section of the textbook.

7. Let x represent the number of engines sent to plant I and y represent the number of engines sent to plant II.

Minimize z = 20x + 35y

subject to: $x \geq 50$

$\qquad\qquad y \geq 27$

$\qquad x + y \leq 85$

$\qquad\qquad x \geq 0$

$\qquad\qquad y \geq 0.$

The corner points are found by solving the following systems of equations.

The solution of the system

$$x = 50$$
$$y = 27$$

is (50, 27).

The solution of the system

$$x = 50$$
$$x + y = 85$$

is (50, 35).

The solution of the system

$$y = 27$$
$$x + y = 85$$

is (58, 27).

The feasible region is the triangular region shown in the graph in the back of the textbook. The corner points are (50, 27), (50, 35), and (58, 27).

Use the corner points to find the minimum value of the objective function.

Corner Point	Value of z = 20z + 35y
(50, 27)	20(50) + 35(27) = 1945
(50, 35)	20(50) + 35(35) = 2225
(58, 27)	20(58) + 35(27) = 2105

The minimum value of z occurs at the corner point (50, 27).
50 engines should be shipped to plant I and 27 engines should be shipped to plant II. The minimum cost is $1945.

9. **(a)** Let x represent the number of units of policy A to purchase and y represent the number of units of policy B to purchase.
Minimize z = 50x + 40y
subject to: 10x + 15y ≥ 100
80x + 120y ≥ 1000
x ≥ 0
y ≥ 0.

In the first quadrant, the graph of 80x + 120y = 100 lies entirely above the graph of 10x + 15y = 100, so the corner points are the x- and y-intercepts of 80x + 120y = 1000. These are (0, 25/3) and (25/2, 0). Use the corner points to find the minimum value of the objective function.

Corner Point	Value of z = 50x + 40y
(0, 25/3)	1000/3 = 333 1/3 Minimum
(25/2, 0)	625

In order to minimize the premium costs, 0 units of policy A and 25/3 (or 8 1/3) units of policy B should be purchased for a minimum premium cost of $333.33.

(b) The objective function to be minimized is altered to
z = 25x + 40y, but the corner points of the feasible region remain the same since none of the constraints are altered.

Corner Point	Value of z = 25x + 40y
(0, 25/3)	1000/3 = 333 1/3
(25/2, 0)	625/2 = 312 1/2 Minimum

In order to minimize the altered premium costs, 25/2 (or 12 1/2) units of policy A and 0 units of policy B should be purchased for a minimum premium cost of $312.50.

11. Let x represent the number of type 1 bolts and y represent the number of type 2 bolts.
Maximize z = .10x + .12y
subject to: .1x + .1y ≤ 240
.1x + .4y ≤ 720
.1x + .5y ≤ 160
x ≥ 0
y ≥ 0.

The first two inequalities do not affect the solution. The feasible region is the triangular region bounded by the x- and y-axes and the line .1x + .5y = 160. The corner points are (0, 0), (0, 320), and (1600, 0).

Use the corner points to find the maximum value of the objective function.

Corner Point	Value of z = .10x + .12y
(0, 0)	0
(0, 320)	38.40
(1600, 0)	160.00 Maximum

1600 type 1 bolts and 0 type 2 bolts should be manufactured for a maximum revenue of $160 per day.

13.   (a) Let x represent the number of kilograms of the half-and-half mix and y represent the number of kilograms of the other mix. The problem requests maximum revenue, so we need a revenue function: z = 6x + 4.8y. The constraints are on the available nuts and raisins that make up the mixes.

Maximize z = 6x + 4.8y

subject to: $\frac{1}{2}x + \frac{1}{3}y \leq 100$

$\frac{1}{2}x + \frac{2}{3}y \leq 125$

$x \geq 0$

$y \geq 0.$

Three of the corner points are (0, 0), (0, 187.5), (200, 0). The fourth corner point is the point of intersection of

$\frac{1}{2}x + \frac{1}{3}y = 100$ and $\frac{1}{2}x + \frac{2}{3}y = 125,$

which is (150, 75).

Corner Point	Value of z = 6x + 4.8y
(0, 0)	0
(0, 187.5)	900
(150, 75)	1260 Maximum
(200, 0)	1200

The company should prepare 150 kg of the half-and-half mix and 75 kg of the other mix for a maximum revenue of $1260.

(b) The objective function to be maximized is altered to

z = 8x + 4.8y,

but the corner points remain the same.

Corner Point	Value of z = 8x + 4.8y
(0, 0)	0
(0, 187.5)	900
(150, 75)	1560
(200, 0)	1600 Maximum

In order to maximize the revenue under the altered conditions, the company should prepare 200 kg of the half-and-half mix and 0 kg of the other mix for a maximum revenue of $1600.

15. Let x represent the number of gallons of milk from Dairy II and y
represent the number of gallons of
milk from Dairy I.

Maximize z = .032x + .037y

subject to: x ≤ 80

            y ≤ 50

      x + y ≤ 100

           x ≥ 0

           y ≥ 0.

The corner points are (0, 0),
(80, 0), (80, 20), (50, 50),
and (0, 50).

Corner Point	Value of z = .032x + .037y
(0, 0)	0
(80, 0)	2.56
(80, 20)	3.30
(50, 50)	3.45 Maximum
(0, 50)	1.85

50 gal from Dairy I and 50 gal from
Dairy II should be used to get milk
with a maximum butterfat of 3.45
gal.

17. Let x represent the amount invested
in bonds and y represent the amount
invested in mutual funds (in millions of dollars).

The amount of annual interest is

.12x + .08y.

Maximize z = .12x + .08y

subject to: x ≥ 20

            y ≥ 15

      x + y ≤ 40.

The corner points are (20, 15),
(20, 20), and (25, 15).

Corner Point	Value of z = .12x + .08y
(20, 15)	3.6
(20, 20)	4.0
(25, 15)	4.2 Maximum

He should invest $25 million in
bonds and $15 million in mutual
funds for maximum annual interest
of $4.20 million.

19. Let x represent the number of
species I prey and y represent
the number of species II prey.

	Protein	Fat
Species I	5	2
II	3	4

Minimize z = 2x + 3y

subject to: 5x + 3y ≥ 10

           2x + 4y ≥ 8

                x ≥ 0

                y ≥ 0.

The corner points are (0, 10/3),
(4, 0), and the intersection of
5x + 3y = 10 and 2x + 4y = 8,
which is (8/7, 10/7).

Corner Point	Value of z = 2x + 3y
(0, 10/3)	10
(4, 0)	8
(8/7, 10/7)	46/7 Minimum

The minimum value of z is
$46/7 \approx 6.57$. 8/7 units of species
I and 10/7 units of species II will
meec the daily food requirements
with the least expenditure of
energy. However, a predator prob-
ably can catch and digest only whole
numbers of prey. This problem shows
that it is important to consider
whether a model produces a real-
istic answer to a problem.

21. Let x represent the number of servings of product A and y represent
the number of servings of product
B. The cost function (in dollars)
is z = .25x + .40y.
Minimize z = .25x + .40y
subject to: $3x + 2y \geq 15$
$2x + 4y \geq 15$
$x \geq 0$
$y \geq 0$.

The corner points are (0, 15/2),
(15/4, 15/8), and (15/2, 0).

Corner Point	Value of z = .25x + .40y
(0, 15/2)	3
(15/4, 15/8)	1.6875 Minimum
(15/2, 0)	1.875

15/4 (or 3 3/4) servings of A and
15/8 (or 1 7/8) servings of B will
satisfy the requirements at a minimum cost of $1.69 (rounded to the
nearest cent).

25. Beta is limited to 400 units per
day, so Beta $\leq$ 400. The correct
answer is choice (a).

Chapter 3 Review Exercises

For Exercises 3-13, see the answer graphs
in the back of the textbook.

3. $y \geq 2x + 3$

Graph y = 2x + 3 as a solid line,
using the intercepts (0, 3) and
(-3/2, 0). Using the origin as a
test point, we get $0 \geq 2(0) + 3$ or
$0 \geq 3$, which is false. Shade the
region that does not contain the
origin, that is, the half-plane
above the line.

5. $3x + 4y \leq 12$

Graph 3x + 4y = 12 as a solid line,
using the intercepts (0, 3) and
(4, 0). Using the origin as a test
point, we get $0 \leq 12$, which is true.
Shade the region that contains the
origin, that is, the half-plane
below the line.

7. $y \geq x$

Graph y = x as a solid line. Since
this line contains the origin,
choose a point other than (0, 0) as
a test point. If we use (1, 4), we
get $4 \geq 1$, which is true. Shade the
region that contains the test point,
that is, the half-plane above the
line.

9. $x + y \le 6$

$2x - y \ge 3$

$x + y \le 6$ is the half-plane on or below the line $x + y = 6$; $2x - y \ge 3$ is the half-plane on or below the line $2x - y = 3$. Shade the overlapping part of these two half-planes, which is the region below both lines. The only corner point is the intersection of the two boundary lines, the point $(3, 3)$.

11. $-4 \le x \le 2$

$-1 \le y \le 3$

$x + y \le 4$

$-4 \le x \le 2$ is the rectangular region lying on or between the two vertical lines, $x = -4$ and $x = 2$; $-1 \le y \le 3$ is the rectangular region lying on or between the two horizontal lines, $y = -1$ and $y = 3$; $x + y \le 4$ is the half-plane lying on or below the line $x + y = 4$. Shade the overlapping part of these three regions. The corner points are $(-4, -1)$, $(-4, 3)$, $(1, 3)$, $(2, 2)$, and $(2, -1)$.

13. $x + 3y \ge 6$

$4x - 3y \le 12$

$x \ge 0$

$y \ge 0$

$x + 3y \ge 6$ is the half-plane on or above the line $x + 3y = 6$; $4x - 3y \le 12$ is the half-plane on or above the line $4x - 3y = 12$; $x \ge 0$ and $y \ge 0$ together restrict the graph to the first quadrant. Shade

the portion of the first quadrant where the half-planes overlap. The corner points are $(0, 2)$ and $(18/5, 4/5)$.

15.

Corner Point	Value of $z = 2x + 4y$
$(1, 6)$	$2(1) + 4(6) = 26$
$(6, 7)$	$2(6) + 4(7) = 40$ Maximum
$(7, 3)$	$2(7) + 4(3) = 26$
$\left(1, 2\frac{1}{2}\right)$	$2(1) + 4\left(2\frac{1}{2}\right) = 12$
$(2, 1)$	$2(2) + 4(1) = 8$ Minimum

The maximum value of 40 occurs at $(6, 7)$, and the minimum value of 8 occurs at $(2, 1)$.

17. Maximize $z = 2x + 4y$

subject to: $3x + 2y \le 12$

$5x + y \ge 5$

$x \ge 0$

$y \ge 0.$

Sketch the feasible region.

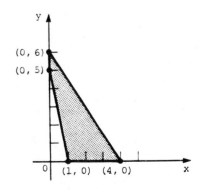

The corner points are $(0, 5)$, $(0, 6)$, $(4, 0)$, and $(1, 0)$.

Corner Point	Value of z = 2x + 4y
(0, 5)	20
(0, 6)	24 Maximum
(4, 0)	8
(1, 0)	2

The maximum value of z = 2x + 4y is 24 at (0, 6).

19. Minimize z = 4x + 2y

subject to: x + y ≤ 50

2x + y ≥ 20

x + 2y ≥ 30

x ≥ 0

y ≥ 0.

Sketch the feasible region.

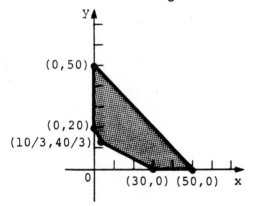

The corner points are (0, 20), (10/3, 40/3), (30, 0), (50, 0), and (0, 50).

Corner Point	Value of z = 4x + 2y
(0, 20)	40 Minimum
$\left(\frac{10}{3}, \frac{40}{3}\right)$	40 Minimum
(30, 0)	120
(50, 0)	200
(0, 50)	100

Thus, the minimum value of 4x + 2y is 40 and occurs at every point on the line segment joining (0, 20) and (10/3, 40/3).

23. Let x represent the number of batches of cakes and y represent the number of batches of cookies. Then we have the following inequalities.

$$2x + \frac{3}{2}y \le 15 \quad \text{(oven time)}$$

$$3x + \frac{2}{3}y \le 13 \quad \text{(decorating)}$$

$$x \ge 0$$

$$y \ge 0$$

See the answer graph in the back of the textbook.

25. From the graph for Exercise 23, the corner points are (0, 10), (3, 6), (13/3, 0) and (0, 0). Since x was the number of batches of cakes and y the number of batches of cookies, the revenue function is

z = 30x + 20y.

Evaluate this objective function at each corner point.

Corner Point	Value of z = 30x + 20y
(0, 0)	200
(3, 6)	210 Maximum
$\left(\frac{13}{3}, 0\right)$	130
(0, 0)	0

Therefore, 3 batches of cakes and 6 batches of cookies should be made to produce a maximum profit of $210.

27. Let x represent the number of hours Charles should spend with his math tutor and y represent the number of hours he should spend with his accounting tutor. The number of points he expects to get on the two tests combined is $z = 3x + 5y$. The given information translates into the following problem.

Maximize $z = 3x + 5y$

subject to:

$$20x + 40y \leq 220 \quad \text{(finances)}$$
$$x + 3y \leq 16 \quad \text{(aspirin)}$$
$$x + 3y \leq 15 \quad \text{(sleep)}$$
$$x \geq 0$$
$$y \geq 0$$

Graph the feasible region.

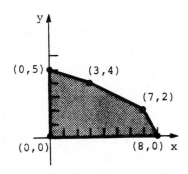

The corner points of the feasible region are (0, 0), (0, 5), (3, 4), (7, 2), and (8, 0). Evaluate the objective function at each corner point.

Corner Point	Value of $z = 3x + 5y$
(0, 0)	0
(0, 5)	25
(3, 4)	29
(7, 2)	31 Maximum
(8, 0)	24

Therefore, Charles should spend 7 hr with the math tutor and 2 hr with the accounting tutor in order to earn a maximum of 31 points.

29. Maximize $z = 4x + 2y$

subject to: $x - y \leq 10$
$$5x + 3y \leq 75$$
$$x + y \leq 20$$
$$x \geq 0$$
$$y \geq 0.$$

The feasible region is the same as it was when we solved this problem previously. Add the line $z = 40$, which is the same as $4x + 2y = 40$. (40 is chosen because the numbers work out simply and the line will show up clearly in our graph, but the chosen value of z is arbitrary.)

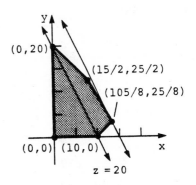

Also drawn in this graph is a line
parallel to the line z = 40 that is
as far from the origin as possible
but still touches the feasible region; this second line passes
through the corner point
(105/8, 25/8) and so, as before,
the objective function has a maximum value of 235/4 when x = 105/8,
y = 25/8.

CHAPTER 3   TEST

1. Graph the inequality $2x + y \leq 4$.

2. Graph the solution to the system of inequalities below. Find all corner
    points.

$$x + y \leq 4$$
$$2x + y \geq 6$$
$$x \geq 0$$
$$y \geq 0$$

3. Maximize the objective function $z = 3x + 2y$ for the region sketched below.

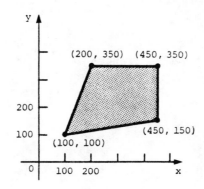

4. Use the graphical method to solve the following linear programming problem.

$$\text{Maximize } z = 4x + 3y$$
$$\text{subject to:}\quad 3x + y \leq 12$$
$$x + y \geq 3$$
$$x \geq 0$$
$$y \geq 0.$$

5. The Gigantic Zipper Company manufactures two kinds of zippers. Type I
    zippers require 2 min on machine I and 3 min on machine II. Type II
    zippers require 1 min on machine I and 4 min on machine II. Machine I
    is available for 20 min, while machine II is available for 12 min. The
    profit on each type I zipper is \$.30 and on each type II zipper is \$.20.
    How many of each type of zipper should be manufactured to ensure the
    maximum profit?

CHAPTER 3   TEST ANSWERS

1.

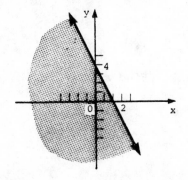

2.

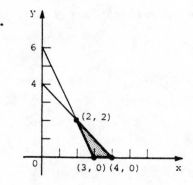

Corner points: (3, 0), (4, 0), (2, 2)

3. Maximum of 2050 at (450, 350)

4. Maximum of 36 at (0, 12)

5. 4 type I zippers and 0 type II zippers

CHAPTER 4 LINEAR PROGRAMMING: THE SIMPLEX METHOD

Section 4.1

1. $x_1 + 2x_2 \le 6$ becomes

   $x_1 + 2x_2 + x_3 = 6.$

3. $2x_1 + 4x_2 + 3x_3 \le 100$ becomes

   $2x_1 + 4x_2 + 3x_3 + x_4 = 100.$

5. **(a)** Since there are three constraints to be converted into equations, we need 3 slack variables.

   (b) The original problem uses x_1 and x_2, so we use x_3, x_4, x_5 for the slack variables.

   (c) The equations are

   $$4x_1 + 2x_2 + x_3 \qquad\qquad = 20$$
   $$5x_1 + \;\; x_2 \qquad + x_4 \qquad = 50$$
   $$2x_1 + 3x_2 \qquad\qquad + x_5 = 25.$$

7. **(a)** There are two constraints to be converted into equations, so we must introduce 2 slack variables.

   (b) x_1, x_2, and x_3 are already used in the problem, so call the slack variables x_4 and x_5.

   (c) The equations are

   $$7x_1 + 6x_2 + 8x_3 + x_4 \qquad = 118$$
   $$4x_1 + 5x_2 + 10x_3 \qquad + x_5 = 220.$$

9.

x_1	x_2	x_3	x_4	x_5	z	
2	2	0	3	1	0	15
3	4	1	6	0	0	20
-2	-1	0	1	0	1	10

The variables x_3 and x_5 are basic variables, because the columns for these variables have all zeros except for one nonzero entry. If the remaining variables x_1, x_2, and x_4 are zero, then $x_3 = 20$ and $x_5 = 15$. From the bottom row, $z = 10$. The basic feasible solution is

$x_1 = 0$, $x_2 = 0$, $x_3 = 20$, $x_4 = 0$, $x_5 = 15$, and $z = 10$.

11.

x_1	x_2	x_3	x_4	x_5	x_6	z	
6	2	2	3	0	0	0	16
2	2	0	1	0	5	0	35
2	1	0	3	1	0	0	6
-3	-2	0	2	0	0	3	36

The basic variables are x_3, x_5 and x_6. If x_1, x_2, and x_4 are zero, then $2x_3 = 16$, so $x_3 = 8$. Similarly, $x_5 = 6$ and $5x_6 = 35$, so $x_6 = 7$. From the bottom row, $3z = 36$, so $z = 12$. The basic feasible solution is

$x_1 = 0$, $x_2 = 0$, $x_3 = 8$, $x_4 = 0$, $x_5 = 6$, $x_6 = 7$, and $z = 12$.

13.

x_1	x_2	x_3	x_4	x_5	z	
1	2	4	1	0	0	56
2	②	1	0	1	0	40
-1	-3	-2	0	0	1	0

$$-R_2 + 2R_1 \rightarrow R_1 \quad \begin{array}{ccccccc} x_1 & x_2 & x_3 & x_4 & x_5 & z \\ \left[\begin{array}{cccccc|c} -1 & 0 & 3 & 1 & -1 & 0 & 16 \\ 2 & ② & 1 & 0 & 1 & 0 & 40 \\ \hline -1 & -3 & -2 & 0 & 0 & 1 & 0 \end{array}\right] \end{array}$$

$$\quad \begin{array}{ccccccc} & x_1 & x_2 & x_3 & x_4 & x_5 & z \\ & \left[\begin{array}{cccccc|c} -1 & 0 & 3 & 1 & -1 & 0 & 16 \\ & 2 & 2 & 1 & 0 & 1 & 0 & 40 \\ \hline 3R_2 + 2R_3 \rightarrow R_3 & 4 & 0 & -1 & 0 & 3 & 2 & 120 \end{array}\right] \end{array}$$

The solution is $x_1 = 0$, $x_2 = 20$, $x_3 = 0$, $x_4 = 16$, $x_5 = 0$, and $z = 60$.

15.

$$\begin{array}{ccccccc} x_1 & x_2 & x_3 & x_4 & x_5 & x_6 & z \\ \left[\begin{array}{ccccccc|c} 2 & 2 & ① & 1 & 0 & 0 & 0 & 12 \\ 1 & 2 & 3 & 0 & 1 & 0 & 0 & 45 \\ 3 & 1 & 1 & 0 & 0 & 1 & 0 & 20 \\ \hline -2 & -1 & -3 & 0 & 0 & 0 & 1 & 0 \end{array}\right] \end{array}$$

$$\begin{array}{l} \\ -3R_1 + R_2 \rightarrow R_2 \\ -R_1 + R_3 \rightarrow R_3 \\ \\ 3R_1 + R_4 \rightarrow R_4 \end{array} \begin{array}{ccccccc} x_1 & x_2 & x_3 & x_4 & x_5 & x_6 & z \\ \left[\begin{array}{ccccccc|c} 2 & 2 & 1 & 1 & 0 & 0 & 0 & 12 \\ -5 & -4 & 0 & -3 & 1 & 0 & 0 & 9 \\ 1 & -1 & 0 & -1 & 0 & 1 & 0 & 8 \\ \hline 4 & 5 & 0 & 3 & 0 & 0 & 1 & 36 \end{array}\right] \end{array}$$

The solution is $x_1 = 0$, $x_2 = 0$, $x_3 = 12$, $x_4 = 0$, $x_5 = 9$, $x_6 = 8$, and $z = 36$.

17.

$$\begin{array}{ccccccc} x_1 & x_2 & x_3 & x_4 & x_5 & x_6 & z \\ \left[\begin{array}{ccccccc|c} 1 & 1 & 1 & 1 & 0 & 0 & 0 & 60 \\ 3 & 1 & ② & 0 & 1 & 0 & 0 & 100 \\ 1 & 2 & 3 & 0 & 0 & 1 & 0 & 200 \\ \hline -1 & -1 & -2 & 0 & 0 & 0 & 1 & 0 \end{array}\right] \end{array}$$

$$\begin{array}{l} 2R_1 - R_2 \rightarrow R_1 \\ \\ -3R_2 + 2R_3 \rightarrow R_3 \\ \\ R_2 + R_4 \rightarrow R_4 \end{array} \begin{array}{ccccccc} x_1 & x_2 & x_3 & x_4 & x_5 & x_6 & z \\ \left[\begin{array}{ccccccc|c} -1 & 1 & 0 & 2 & -1 & 0 & 0 & 20 \\ 3 & 1 & 2 & 0 & 1 & 0 & 0 & 100 \\ -7 & 1 & 0 & 0 & -3 & 2 & 0 & 100 \\ \hline 2 & 0 & 0 & 0 & 1 & 1 & 1 & 100 \end{array}\right] \end{array}$$

The solution is $x_1 = 0$, $x_2 = 0$, $x_3 = 50$, $x_4 = 10$, $x_5 = 0$, $x_6 = 50$, and $z = 100$.

19. Find $x_1 \geq 0$ and $x_2 \geq 0$ such that

$$2x_1 + 3x_2 \leq 6$$
$$4x_1 + x_2 \leq 6$$

and $z = 5x_1 + x_2$ is maximized. We need two slack variables, x_3 and x_4. Then the problem can be restated as: Find $x_1 \geq 0$, $x_2 \geq 0$, $x_3 \geq 0$, and $x_4 \geq 0$ such that

$$2x_1 + 3x_2 + x_3 \qquad = 6$$
$$4x_1 + x_2 \qquad + x_4 = 6$$

and $z = 5x_1 + x_2$ is maximized. Rewrite the objective function as $-5x_1 - x_2 + z = 0$. The initial simple tableau is

$$\begin{array}{ccccc} x_1 & x_2 & x_3 & x_4 & z \\ \left[\begin{array}{cccc|c} 2 & 3 & 1 & 0 & 0 & 6 \\ 4 & 1 & 0 & 1 & 0 & 6 \\ \hline -5 & -1 & 0 & 0 & 1 & 0 \end{array}\right] \end{array}.$$

21. Find $x_1 \geq 0$ and $x_2 \geq 0$ such that

$$x_1 + x_2 \leq 10$$
$$5x_1 + 2x_2 \leq 20$$
$$x_1 + 2x_2 \leq 36$$

and $z = x_1 + 3x_2$ is maximized.

Using slack variables x_3, x_4, and x_5, the problem can be restricted as:

Find $x_1 \geq 0$, $x_2 \geq 0$, $x_3 \geq 0$, $x_4 \geq 0$, and $x_5 \geq 0$ such that

$$
\begin{aligned}
x_1 + x_2 + x_3 &= 10 \\
5x_1 + 2x_2 + x_4 &= 20 \\
x_1 + 2x_2 + x_5 &= 36
\end{aligned}
$$

and $z = x_1 + 3x_2$ is maximized.

Rewrite the objective function as $-x_1 - 3x_2 + z = 0$.

The initial simplex tableau is

$$
\begin{array}{cccccc}
x_1 & x_2 & x_3 & x_4 & x_5 & z
\end{array}
$$

$$
\left[
\begin{array}{cccccc|c}
1 & 1 & 1 & 0 & 0 & 0 & 10 \\
5 & 2 & 0 & 1 & 0 & 0 & 20 \\
1 & 2 & 0 & 0 & 1 & 0 & 36 \\
\hline
-1 & -3 & 0 & 0 & 0 & 1 & 0
\end{array}
\right].
$$

23. Find $x_1 \geq 0$ and $x_2 \geq 0$ such that

$$
\begin{aligned}
3x_1 + x_2 &\leq 12 \\
x_1 + x_2 &\leq 15
\end{aligned}
$$

and $z = 2x_1 + x_2$ is maximized.
Using slack variables x_3 and x_4, the problem can be restated as:
Find $x_1 \geq 0$, $x_2 \geq 0$, $x_3 \geq 0$, and $x_4 \geq 0$ such that

$$
\begin{aligned}
3x_1 + x_2 + x_3 &= 12 \\
x_1 + x_2 + x_4 &= 15
\end{aligned}
$$

and $z = 2x_1 + x_2$ is maximized.
Rewrite the objective function as $-2x_1 - x_2 + z = 0$.
The initial simplex tableau is

$$
\begin{array}{ccccc}
x_1 & x_2 & x_3 & x_4 & z
\end{array}
$$

$$
\left[
\begin{array}{ccccc|c}
3 & 1 & 1 & 0 & 0 & 12 \\
1 & 1 & 0 & 1 & 0 & 15 \\
\hline
-2 & -1 & 0 & 0 & 1 & 0
\end{array}
\right].
$$

25. Let x_1 represent the number of prams, x_2 the number of runabouts, and x_3 the number of trimarans. Organize some of the information in a table.

	Pram	Run-about	Tri-maran	Available Time
Sec A	1	2	3	6240
Sec B	2	5	4	10,800
Profit	$75	$90	$100	

Using this information, together with the fact that the total number of boats cannot exceed 3000, the problem may be stated as:
Find $x_1 \geq 0$, $x_2 \geq 0$, and $x_3 \geq 0$ such that

$$
\begin{aligned}
x_1 + 2x_2 + 3x_3 &\leq 6240 \\
2x_1 + 5x_2 + 4x_3 &\leq 10,800 \\
x_1 + x_2 + x_3 &\leq 3000
\end{aligned}
$$

and $z = 75x_1 + 90x_2 + 100x_3$ is maximized.

Introduce slack variables x_4, x_5, and x_6, and the problem can be restated as:
Find $x_1 \geq 0$, $x_2 \geq 0$, $x_3 \geq 0$, $x_4 \geq 0$, $x_5 \geq 0$, and $x_6 \geq 0$ such that

$$
\begin{aligned}
x_1 + 2x_2 + 3x_3 + x_4 &= 6240 \\
2x_1 + 5x_2 + 4x_3 + x_5 &= 10,800 \\
x_1 + x_2 + x_3 + x_6 &= 3000
\end{aligned}
$$

and $z = 75x_1 + 90x_2 + 100x_3$ is maximized.

Rewrite the objective as

$$-75x_1 - 90x_2 - 100x_3 + z = 0.$$

The initial simplex tableau is

$$
\begin{array}{ccccccc}
x_1 & x_2 & x_3 & x_4 & x_5 & x_6 & z
\end{array}
$$

$$
\left[
\begin{array}{ccccccc|c}
1 & 2 & 3 & 1 & 0 & 0 & 0 & 6240 \\
2 & 5 & 4 & 0 & 1 & 0 & 0 & 10,800 \\
1 & 1 & 1 & 0 & 0 & 1 & 0 & 3000 \\
\hline
-75 & -90 & -100 & 0 & 0 & 0 & 1 & 0
\end{array}
\right]
$$

27. Let x_1 represent the number of simple figures, x_2 the number of figures with additions, and x_3 the number of computer–drawn sketches. Organize some of the information in a table.

	Simple Figures	Figures with Additions	Computer– Drawn Sketches	Maximum Allowed
Cost	20	35	60	2200
Royalties	95	200	325	

The cost constraint is

$$20x_1 + 35x_2 + 60x_3 \le 2200.$$

The limit of 400 figures leads to the constraint $x_1 + x_2 + x_3 \le 400$. The other stated constraints are $x_3 \le x_1 + x_2$ and $x_1 \ge 2x_2$, and these can be rewritten in standard form as $-x_1 - x_2 + x_3 \le 0$ and $-x_1 + 2x_2 \le 0$ respectively.

The problem may be stated as:

Find $x_1 \ge 0$, $x_2 \ge 0$, and $x_3 \ge 0$ such that

$$
\begin{aligned}
20x_1 + 35x_2 + 60x_3 &\le 2200 \\
x_1 + x_2 + x_3 &\le 400 \\
-x_1 - x_2 + x_3 &\le 0 \\
-x_1 + 2x_2 &\le 0
\end{aligned}
$$

and $z = 95x_1 + 200x_2 + 325x_3$ is maximized.

Introduce slack variables x_4, x_5, x_6, and x_7, and the problem can be restated as:

Find $x_1 \ge 0$, $x_2 \ge 0$, $x_3 \ge 0$, $x_4 \ge 0$, $x_5 \ge 0$, $x_6 \ge 0$, and $x_7 \ge 0$ such that

$$
\begin{aligned}
20x_1 + 35x_2 + 60x_3 + x_4 \phantom{{}+ x_5 + x_6 + x_7} &= 2200 \\
x_1 + x_2 + x_3 \phantom{{}+ x_4} + x_5 \phantom{{}+ x_6 + x_7} &= 400 \\
-x_1 - x_2 + x_3 \phantom{{}+ x_4 + x_5} + x_6 \phantom{{}+ x_7} &= 0 \\
-x_1 + 2x_2 \phantom{{}+ x_3 + x_4 + x_5 + x_6} + x_7 &= 0
\end{aligned}
$$

and $z = 95x_1 + 200x_2 + 325x_3$ is maximized.

Rewrite the objective function as

$-95x_1 - 200x_2 - 325x_3 + z = 0.$

The initial simplex tableau is

$$
\begin{array}{cccccccc|c}
x_1 & x_2 & x_3 & x_4 & x_5 & x_6 & x_7 & z & \\
\hline
20 & 35 & 60 & 1 & 0 & 0 & 0 & 0 & 2200 \\
1 & 1 & 1 & 0 & 1 & 0 & 0 & 0 & 400 \\
-1 & -1 & 1 & 0 & 0 & 1 & 0 & 0 & 0 \\
-1 & 2 & 0 & 0 & 0 & 0 & 1 & 0 & 0 \\
\hline
-95 & -200 & -325 & 0 & 0 & 0 & 0 & 1 & 0
\end{array}
$$

29. Let x_1 represent the number of one-speed bicycles, x_2 the number of three-speed bicycles, and x_3 the number of ten-speed bicycles.

Organize the information in a table.

	One-Speed	Three-Speed	Ten-Speed	Amount Available
Steel	17	27	34	91,800
Aluminum	12	21	15	42,000
Profit	$8	$12	$22	

Using this information, the problem may be stated as:

Find $x_1 \ge 0$, $x_2 \ge 0$, and $x_3 \ge 0$ such that

$$17x_1 + 27x_2 + 34x_3 \le 91,800$$
$$12x_1 + 21x_2 + 15x_3 \le 42,000$$

and $z = 8x_1 + 12x_2 + 22x_3$ is maximized.

Introduce slack variables x_4 and x_5 and the problem can be restated as:

Find $x_1 \ge 0$, $x_2 \ge 0$, $x_3 \ge 0$, $x_4 \ge 0$, and $x_5 \ge 0$ such that

$$17x_1 + 27x_2 + 34x_3 + x_4 \qquad = 91,800$$
$$12x_1 + 21x_2 + 15x_3 \qquad + x_5 = 42,000$$

and $z = 8x_1 + 12x_2 + 22x_3$ is maximized.

Rewrite the objective function as

$-8x_1 - 12x_2 - 22x_3 + z = 0.$

The initial simplex tableau is

$$
\begin{array}{cccccc|c}
x_1 & x_2 & x_3 & x_4 & x_5 & z & \\
\hline
17 & 27 & 34 & 1 & 0 & 0 & 91,800 \\
12 & 21 & 15 & 0 & 1 & 0 & 42,000 \\
\hline
-8 & -12 & -22 & 0 & 0 & 1 & 0
\end{array}
$$

Section 4.2

1. The initial simplex tableau is as follows.

$$
\begin{array}{cccccc|c}
x_1 & x_2 & x_3 & x_4 & x_5 & z & \\
\hline
1 & 2 & 4 & 1 & 0 & 0 & 8 \\
2 & 2 & 1 & 0 & 1 & 0 & 10 \\
\hline
-2 & -5 & -1 & 0 & 0 & 1 & 0
\end{array}
$$

The most negative indicator is -5, in the second column. Find the quotients $8/2 = 4$ and $10/2 = 5$; since 4 is the smallest quotient, 2 in the first row, second column, is the pivot.

$$
\begin{array}{c}
8/2 = 4 \\
10/2 = 5 \\
\\
\end{array}
\begin{array}{cccccc|c}
x_1 & x_2 & x_3 & x_4 & x_5 & z & \\
\hline
1 & ② & 4 & 1 & 0 & 0 & 8 \\
2 & 2 & 1 & 0 & 1 & 0 & 10 \\
\hline
-2 & -5 & -1 & 0 & 0 & 1 & 0
\end{array}
$$

Performing row transformations, we get the following tableau.

$$
\begin{array}{c}
\\
-R_1 + R_2 \to R_2 \\
5R_1 + 2R_3 \to R_3
\end{array}
\begin{array}{cccccc|c}
x_1 & x_2 & x_3 & x_4 & x_5 & z & \\
\hline
1 & 2 & 4 & 1 & 0 & 0 & 8 \\
1 & 0 & -3 & -1 & 1 & 0 & 2 \\
1 & 0 & 18 & 5 & 0 & 2 & 40
\end{array}
$$

All of the numbers in the last row are nonnegative, so we are finished pivoting.

Create a 1 in the columns corresponding to x_2 and z.

$$
\begin{array}{c}
\\
\frac{1}{2}R_1 \to R_1 \\
\\
\\
\frac{1}{2}R_3 \to R_3
\end{array}
\begin{array}{cccccc}
x_1 & x_2 & x_3 & x_4 & x_5 & z \\
\end{array}
\left[
\begin{array}{cccccc|c}
\frac{1}{2} & 1 & 2 & \frac{1}{2} & 0 & 0 & 4 \\
1 & 0 & -3 & -1 & 1 & 0 & 2 \\
\hline
\frac{1}{2} & 0 & 9 & \frac{5}{2} & 0 & 1 & 20
\end{array}
\right]
$$

The maximum value is 20 and occurs when $x_1 = 0$, $x_2 = 4$, $x_3 = 0$, $x_4 = 0$, and $x_5 = 2$.

3.

$$
\begin{array}{cccccc}
x_1 & x_2 & x_3 & x_4 & x_5 & z \\
\end{array}
\left[
\begin{array}{cccccc|c}
1 & 3 & 1 & 0 & 0 & 0 & 12 \\
2 & 1 & 0 & 1 & 0 & 0 & 10 \\
1 & 1 & 0 & 0 & 1 & 0 & 4 \\
\hline
-2 & -1 & 0 & 0 & 0 & 1 & 0
\end{array}
\right]
$$

The most negative indicator is -2, in the first column. Find the quotients $12/1 = 12$, $10/2 = 5$, and $4/1 = 4$; since 4 is the smallest quotient, 1 in the third row, first column, is the pivot.

$$
\begin{array}{cccccc}
x_1 & x_2 & x_3 & x_4 & x_5 & z \\
\end{array}
\left[
\begin{array}{cccccc|c}
1 & 3 & 1 & 0 & 0 & 0 & 12 \\
2 & 1 & 0 & 1 & 0 & 0 & 10 \\
① & 1 & 0 & 0 & 1 & 0 & 4 \\
\hline
-2 & -1 & 0 & 0 & 0 & 1 & 0
\end{array}
\right]
$$

$$
\begin{array}{c}
\\
-R_3+R_1 \to R_1 \\
-2R_3+R_2 \to R_2 \\
\\
2R_3+R_4 \to R_4
\end{array}
\begin{array}{ccccc}
x_1 & x_2 & x_3 & x_4 & x_5 \quad z \\
\end{array}
\left[
\begin{array}{ccccc|cc}
0 & 2 & 1 & 0 & -1 & 0 & 8 \\
0 & -1 & 0 & 1 & -2 & 0 & 2 \\
1 & 1 & 0 & 0 & 1 & 0 & 4 \\
\hline
0 & 1 & 0 & 0 & 2 & 1 & 8
\end{array}
\right]
$$

This is a final tableau, since all of the numbers in the last row are

nonnegative. The maximum value is 8 when $x_1 = 4$, $x_2 = 0$, $x_3 = 8$, $x_4 = 2$, and $x_5 = 0$.

5.

$$
\begin{array}{ccccccc}
x_1 & x_2 & x_3 & x_4 & x_5 & x_6 & z \\
\end{array}
\left[
\begin{array}{ccccccc|c}
2 & 2 & 8 & 1 & 0 & 0 & 0 & 40 \\
4 & -5 & 6 & 0 & 1 & 0 & 0 & 60 \\
2 & -2 & 6 & 0 & 0 & 1 & 0 & 24 \\
\hline
-14 & -10 & -12 & 0 & 0 & 0 & 1 & 0
\end{array}
\right]
$$

The most negative indicator is -14, in the first column. Find the quotients $40/2 = 20$, $60/4 = 15$, and $24/2 = 12$; since 12 is the smallest quotient, 2 in the third row, first column, is the pivot.

$$
\begin{array}{ccccccc}
x_1 & x_2 & x_3 & x_4 & x_5 & x_6 & z \\
\end{array}
\left[
\begin{array}{ccccccc|c}
2 & 2 & 8 & 1 & 0 & 0 & 0 & 40 \\
4 & -5 & 6 & 0 & 1 & 0 & 0 & 60 \\
② & -2 & 6 & 0 & 0 & 1 & 0 & 24 \\
\hline
-14 & -10 & -12 & 0 & 0 & 0 & 1 & 0
\end{array}
\right]
$$

Performing row transformations, we get the following tableau.

$$
\begin{array}{c}
-R_3+R_1 \to R_1 \\
-2R_3+R_2 \to R_2 \\
\\
7R_3+R_4 \to R_4
\end{array}
\begin{array}{ccccccc}
x_1 & x_2 & x_3 & x_4 & x_5 & x_6 & z \\
\end{array}
\left[
\begin{array}{ccccccc|c}
0 & ④ & 2 & 1 & 0 & -1 & 0 & 16 \\
0 & -1 & -6 & 0 & 1 & -2 & 0 & 12 \\
2 & -2 & 6 & 0 & 0 & 1 & 0 & 24 \\
\hline
0 & -24 & 30 & 0 & 0 & 7 & 1 & 168
\end{array}
\right]
$$

Since there is still a negative indicator, we must repeat the process. The second pivot is the 4 in the second column, since $16/4$ is the only nonnegative quotient in the only column with a negative indicator.

Performing row transformations again, we get the following tableau.

$$
\begin{array}{c}
\\
R_1+4R_2\to R_2\\
R_1+2R_3\to R_3\\
\\
6R_1+R_4\to R_4
\end{array}
\begin{array}{c}
\begin{array}{ccccccc}
x_1 & x_2 & x_3 & x_4 & x_5 & x_6 & z
\end{array}\\
\left[\begin{array}{ccccccc|c}
0 & 4 & 2 & 1 & 0 & -1 & 0 & 16\\
0 & 0 & -22 & 1 & 4 & -9 & 0 & 64\\
4 & 0 & 14 & 1 & 0 & 1 & 0 & 64\\
\hline
0 & 0 & 42 & 6 & 0 & 1 & 1 & 264
\end{array}\right]
\end{array}
$$

All of the numbers in the last row are nonnegative, so we are finished pivoting. Create a 1 in the columns corresponding to x_1, x_2, and x_5.

$$
\begin{array}{c}
\frac{1}{4}R_1\to R_1\\
\frac{1}{4}R_2\to R_2\\
\frac{1}{4}R_3\to R_3\\
\\
\end{array}
\begin{array}{c}
\begin{array}{ccccccc}
x_1 & x_2 & x_3 & x_4 & x_5 & x_6 & z
\end{array}\\
\left[\begin{array}{ccccccc|c}
0 & 1 & \frac{1}{2} & \frac{1}{4} & 0 & -\frac{1}{4} & 0 & 4\\
0 & 0 & -\frac{11}{2} & \frac{1}{4} & 1 & -\frac{9}{4} & 0 & 16\\
1 & 0 & \frac{7}{2} & \frac{1}{4} & 0 & \frac{1}{4} & 0 & 16\\
\hline
0 & 0 & 42 & 6 & 0 & 1 & 1 & 264
\end{array}\right]
\end{array}
$$

The maximum value is 264 and occurs when $x_1 = 16$, $x_2 = 4$, $x_3 = 0$, $x_4 = 0$, $x_5 = 16$, and $x_6 = 0$.

7. Maximize $z = 4x_1 + 3x_2$
   subject to: $2x_1 + 3x_2 \leq 11$
   $x_1 + 2x_2 \leq 6$
   with $x_1 \geq 0$, $x_2 \geq 0$.

Two slack variables need to be introduced, x_3 and x_4.
The problem can be restated as:

Maximize $z = 4x_1 + 3x_2$
subject to: $2x_1 + 3x_2 + x_3 = 11$
$x_1 + 2x_2 + x_4 = 6$
with $x_1 \geq 0$, $x_2 \geq 0$,
$x_3 \geq 0$, $x_4 \geq 0$.

Rewrite the objective function as $-4x_1 - 3x_2 + z = 0$.

The initial simplex tableau follows.

$$
\begin{array}{c}
\begin{array}{ccccc}
x_1 & x_2 & x_3 & x_4 & z
\end{array}\\
\left[\begin{array}{ccccc|c}
2 & 3 & 1 & 0 & 0 & 11\\
1 & 2 & 0 & 1 & 0 & 6\\
\hline
-4 & -3 & 0 & 0 & 1 & 0
\end{array}\right]
\end{array}
$$

The most negative indicator is -4, in column 1; to select the pivot from column 1, find the quotients $11/2$ and $6/1$. The smallest is $11/2$, so 2 is the pivot.

$$
\begin{array}{c}
\begin{array}{ccccc}
x_1 & x_2 & x_3 & x_4 & z
\end{array}\\
\left[\begin{array}{ccccc|c}
\boxed{2} & 3 & 1 & 0 & 0 & 11\\
1 & 2 & 0 & 1 & 0 & 6\\
\hline
-4 & -3 & 0 & 0 & 1 & 0
\end{array}\right]
\end{array}
$$

$$
\begin{array}{c}
\\
-R_1 + 2R_2 \to R_2\\
\\
2R_1 + R_3 \to R_3
\end{array}
\begin{array}{c}
\begin{array}{ccccc}
x_1 & x_2 & x_3 & x_4 & z
\end{array}\\
\left[\begin{array}{ccccc|c}
2 & 3 & 1 & 0 & 0 & 11\\
0 & 1 & -1 & 2 & 0 & 1\\
\hline
0 & 3 & 2 & 0 & 1 & 22
\end{array}\right]
\end{array}
$$

All of the indicators are nonnegative. Create a 1 in the columns corresponding to x_1 and x_4

$$
\begin{array}{c}
\frac{1}{2}R_1 \to R_1\\
\\
\frac{1}{2}R_2 \to R_2\\
\\
\end{array}
\begin{array}{c}
\begin{array}{ccccc}
x_1 & x_2 & x_3 & x_4 & z
\end{array}\\
\left[\begin{array}{ccccc|c}
1 & \frac{3}{2} & \frac{1}{2} & 0 & 0 & \frac{11}{2}\\
0 & \frac{1}{2} & -\frac{1}{2} & 1 & 0 & \frac{1}{2}\\
\hline
0 & 3 & 2 & 0 & 1 & 22
\end{array}\right]
\end{array}
$$

The maximum value is 22 when $x_1 = 5.5$, $x_2 = 0$, $x_3 = 0$, and $x_4 = .5$.

9. Maximize $z = 10x_1 + 12x_2$

    subject to: $4x_1 + 2x_2 \leq 20$

    $\qquad\qquad 5x_1 + x_2 \leq 50$

    $\qquad\qquad 2x_1 + 2x_2 \leq 24$

    with $x_1 \geq 0,\ x_2 \geq 0.$

    Three slack variables need to be introduced, x_3, x_4, and x_5. Then the initial tableau is as follows.

    $$\begin{array}{c} x_1 \quad x_2 \quad x_3 \quad x_4 \quad x_5 \quad z \\ \left[\begin{array}{cccccc|c} 4 & 2 & 1 & 0 & 0 & 0 & 20 \\ 5 & 1 & 0 & 1 & 0 & 0 & 50 \\ 2 & 2 & 0 & 0 & 1 & 0 & 24 \\ \hline -10 & -12 & 0 & 0 & 0 & 1 & 0 \end{array}\right] \end{array}$$

    The most negative indicator is -12, in column 2. The quotients are $20/2 = 10$, $50/1 = 50$, and $24/2 = 12$; the smallest is 10, so 2 in the first row, second column, is the pivot.

    $$\begin{array}{c} x_1 \quad x_2 \quad x_3 \quad x_4 \quad x_5 \quad z \\ \left[\begin{array}{cccccc|c} 4 & ② & 1 & 0 & 0 & 0 & 20 \\ 5 & 1 & 0 & 1 & 0 & 0 & 50 \\ 2 & 2 & 0 & 0 & 1 & 0 & 24 \\ \hline -10 & -12 & 0 & 0 & 0 & 1 & 0 \end{array}\right] \end{array}$$

    $$\begin{array}{l} \\ \\ -R_1 + 2R_2 \rightarrow R_2 \\ -R_1 + R_3 \rightarrow R_3 \\ 6R_1 + R_4 \rightarrow R_4 \end{array} \begin{array}{c} x_1 \ x_2 \ x_3 \ \ x_4 \ x_5 \ z \\ \left[\begin{array}{cccccc|c} 4 & 2 & 1 & 0 & 0 & 0 & 20 \\ 6 & 0 & -1 & 2 & 0 & 0 & 80 \\ -2 & 0 & -1 & 0 & 1 & 0 & 4 \\ \hline 14 & 0 & 6 & 0 & 0 & 1 & 120 \end{array}\right] \end{array}$$

    All of the indicators are nonnegative, so we are finished pivoting. Create a 1 in the columns corresponding to x_2 and x_4.

    $$\begin{array}{l} \frac{1}{2}R_1 \rightarrow R_1 \\ \frac{1}{2}R_2 \rightarrow R_2 \\ \\ \\ \end{array} \begin{array}{c} x_1 \quad x_2 \ x_3 \quad\ x_4 \ x_5 \ z \\ \left[\begin{array}{cccccc|c} 2 & 1 & \frac{1}{2} & 0 & 0 & 0 & 10 \\ 3 & 0 & -\frac{1}{2} & 1 & 0 & 0 & 40 \\ -2 & 0 & -1 & 0 & 1 & 0 & 4 \\ \hline 14 & 0 & 6 & 0 & 0 & 1 & 120 \end{array}\right] \end{array}$$

    The maximum value is 120 when $x_1 = 0$, $x_2 = 10$, $x_3 = 0$, $x_4 = 40$, and $x_5 = 4$.

11. Maximize $z = 8x_1 + 3x_2 + x_3$

    subject to: $x_1 + 6x_2 + 8x_3 \leq 118$

    $\qquad\qquad x_1 + 5x_2 + 10x_3 \leq 220$

    with $x_1 \geq 0,\ x_2 \geq 0,\ x_3 \geq 0.$

    Two slack variables need to be introduced, x_4 and x_5. The initial simplex tableau is as follows.

    $$\begin{array}{c} x_1 \quad x_2 \quad x_3 \quad x_4 \ x_5 \ z \\ \left[\begin{array}{cccccc|c} ① & 6 & 8 & 1 & 0 & 0 & 118 \\ 1 & 5 & 10 & 0 & 1 & 0 & 220 \\ \hline -8 & -3 & -1 & 0 & 0 & 1 & 0 \end{array}\right] \end{array}$$

    The most negative indicator is -8, in the first column. The quotients are $118/1 = 118$ and $220/1 = 220$; since 118 is the smallest, 1 in the first row, first column is the pivot. Performing row transformations, we get the following tableau.

    $$\begin{array}{l} \\ -R_1 + R_2 \rightarrow R_2 \\ \\ 8R_1 + R_3 \rightarrow R_3 \end{array} \begin{array}{c} x_1 \ x_2 \ \ x_3 \quad x_4 \ x_5 \ z \\ \left[\begin{array}{cccccc|c} 1 & 6 & 8 & 1 & 0 & 0 & 118 \\ 0 & -1 & 2 & -1 & 1 & 0 & 102 \\ \hline 0 & 45 & 63 & 8 & 0 & 1 & 944 \end{array}\right] \end{array}$$

    All of the indicators are nonnegative, so we are finished pivoting.

The maximum value is 944 when $x_1 = 118$, $x_2 = 0$, $x_3 = 0$, $x_4 = 0$, and $x_5 = 102$.

13. Maximize $z = x_1 + 2x_2 + x_3 + 5x_4$ subject to:

$$x_1 + 2x_2 + x_3 + x_4 \leq 50$$
$$3x_1 + x_2 + 2x_3 + x_4 \leq 100$$

with $x_1 \geq 0$, $x_2 \geq 0$,
$x_3 \geq 0$, $x_4 \geq 0$.

Two slack variables need to be introduced, x_5 and x_6. Then the initial simplex tableau is as follows.

$$
\begin{array}{ccccccc}
x_1 & x_2 & x_3 & x_4 & x_5 & x_6 & z \\
\end{array}
$$
$$
\left[
\begin{array}{ccccccc|c}
1 & 2 & 1 & ① & 1 & 0 & 0 & 50 \\
3 & 1 & 2 & 1 & 0 & 1 & 0 & 100 \\
\hline
-1 & -2 & -1 & -5 & 0 & 0 & 1 & 0 \\
\end{array}
\right]
$$

In the column with the most negative indicator, -5, the quotients are $50/1 = 50$ and $100/1 = 100$; the smallest is 50, so 1 in the first row, fourth column, is the pivot.

$$
\begin{array}{ccccccc}
x_1 & x_2 & x_3 & x_4 & x_5 & x_6 & z \\
\end{array}
$$
$$
\begin{array}{l}
 \\
-R_1+R_2\to R_2 \\
5R_1+R_3\to R_3
\end{array}
\left[
\begin{array}{ccccccc|c}
1 & 2 & 1 & 1 & 1 & 0 & 0 & 50 \\
2 & -1 & 1 & 0 & -1 & 1 & 0 & 50 \\
4 & 8 & 4 & 0 & 5 & 0 & 1 & 250 \\
\end{array}
\right]
$$

This is a final tableau, since all of the indicators are nonnegative. The maximum value is 250 when $x_1 = 0$, $x_2 = 0$, $x_3 = 0$, $x_4 = 50$, $x_5 = 0$, and $x_6 = 50$.

17. Organize the information in a table.

	Church group	Labor Union	Maximum Time Available
Letter Writing	2	2	16
Follow-up	1	3	12
Money Raised	$100	$200	

Let x_1 and x_2 be the number of church groups and labor unions contacted respectively. We need 2 slack variables, x_3 and x_4. We want to maximize $z = 100x_1 + 200x_2$ subject to:

$$2x_1 + 2x_2 + x_3 = 16$$
$$x_1 + 3x_2 + x_4 = 12$$

with $x_1 \geq 0$, $x_2 \geq 0$, $x_3 \geq 0$, $x_4 \geq 0$.

The initial simplex tableau is as follows.

$$
\begin{array}{ccccc}
x_1 & x_2 & x_3 & x_4 & z \\
\end{array}
$$
$$
\left[
\begin{array}{ccccc|c}
2 & 2 & 1 & 0 & 0 & 16 \\
1 & ③ & 0 & 1 & 0 & 12 \\
\hline
-100 & -200 & 0 & 0 & 1 & 0 \\
\end{array}
\right]
$$

Pivot on the 3 in the second row, second column.

$$
\begin{array}{ccccc}
x_1 & x_2 & x_3 & x_4 & z \\
\end{array}
$$
$$
\begin{array}{l}
3R_1-2R_2\to R_1 \\
 \\
200R_2+3R_3\to R_3
\end{array}
\left[
\begin{array}{ccccc|c}
④ & 0 & 3 & -2 & 0 & 24 \\
1 & 3 & 0 & 1 & 0 & 12 \\
\hline
-100 & 0 & 0 & 200 & 3 & 2400 \\
\end{array}
\right]
$$

Pivot on the 4 in the first row, first column.

$$\begin{array}{c} \\ 4R_2 - R_1 \rightarrow R_2 \\ \\ 25R_1 + R_4 \rightarrow R_4 \end{array} \begin{array}{cccccc} x_1 & x_2 & x_3 & x_4 & z & \\ \left[\begin{array}{ccccc|c} 4 & 0 & 3 & -2 & 0 & 24 \\ 0 & 12 & -3 & 6 & 0 & 24 \\ \hline 0 & 0 & 75 & 150 & 3 & 3000 \end{array}\right] \end{array}$$

This is a final tableau, since all of the indicators are nonnegative.

The maximum value is $\dfrac{3000}{3} = 1000$ when $x_1 = \dfrac{24}{4} = 6$, $x_2 = \dfrac{24}{12} = 2$, $x_3 = 0$, and $x_4 = 0$.

She should contact 6 churches and 2 labor unions to raise a maximum of $1000 per month.

19. Organize the information in a table.

	Recording	Mixing	Editing	Profit
Jazz	4	2	6	$.80
Blues	4	8	2	$.60
Reggae	10	4	6	$1.20
Maximum Available	80	52	54	

Let x_1, x_2, x_3 represent the number of jazz, blues and reggae albums respectively. We need 3 slack variables, x_4, x_5, and x_6.

We want to maximize $z = .8x_1 + .6x_2 + 1.2x_3$
subject to:
$$4x_1 + 4x_2 + 10x_3 + x_4 \qquad\qquad = 80$$
$$2x_1 + 8x_2 + 4x_3 \qquad + x_5 \qquad = 52$$
$$6x_1 + 2x_2 + 6x_3 \qquad\qquad + x_6 = 54$$
with $x_1 \geq 0$, $x_2 \geq 0$, $x_3 \geq 0$, $x_4 \geq 0$, $x_5 \geq 0$, and $x_6 \geq 0$.

The initial simplex tableau is as follows.

$$\begin{array}{ccccccc} x_1 & x_2 & x_3 & x_4 & x_5 & x_6 & z \\ \left[\begin{array}{ccccccc|c} 4 & 4 & ⑩ & 1 & 0 & 0 & 0 & 80 \\ 2 & 8 & 4 & 0 & 1 & 0 & 0 & 52 \\ 6 & 2 & 6 & 0 & 0 & 1 & 0 & 54 \\ \hline -.8 & -.6 & -1.2 & 0 & 0 & 0 & 1 & 0 \end{array}\right] \end{array}$$

Pivot on the 10 in the first row, third column.

$$
\begin{array}{c}
\\
-2R_1 + 5R_2 \to R_2 \\
-3R_1 + 5R_3 \to R_3 \\
\\
3R_1 + 25R_4 \to R_4
\end{array}
\begin{array}{ccccccc|c}
x_1 & x_2 & x_3 & x_4 & x_5 & x_6 & z & \\
4 & 4 & 10 & 1 & 0 & 0 & 0 & 80 \\
2 & 32 & 0 & -2 & 5 & 0 & 0 & 100 \\
\boxed{18} & -2 & 0 & -3 & 0 & 5 & 0 & 30 \\
-8 & -3 & 0 & 3 & 0 & 0 & 25 & 240
\end{array}
$$

Pivot on the 18 in the third row, first column.

$$
\begin{array}{c}
9R_1 - 2R_3 \to R_1 \\
9R_2 - R_3 \to R_2 \\
\\
4R_3 + 9R_4 \to R_4
\end{array}
\begin{array}{ccccccc|c}
x_1 & x_2 & x_3 & x_4 & x_5 & x_6 & z & \\
0 & 40 & 90 & 15 & 0 & -10 & 0 & 660 \\
0 & \boxed{290} & 0 & -15 & 45 & -5 & 0 & 870 \\
18 & -2 & 0 & -3 & 0 & 5 & 0 & 30 \\
0 & -35 & 0 & 15 & 0 & 20 & 225 & 2280
\end{array}
$$

Pivot on the 290 in the second row, second column.

$$
\begin{array}{c}
29R_1 - 4R_2 \to R_1 \\
\\
R_2 + 145R_3 \to R_3 \\
7R_2 + 58R_4 \to R_4
\end{array}
\begin{array}{ccccccc|c}
x_1 & x_2 & x_3 & x_4 & x_5 & x_6 & z & \\
0 & 0 & 2610 & 495 & -180 & -270 & 0 & 15,660 \\
0 & 290 & 0 & -15 & 45 & -5 & 0 & 870 \\
2610 & 0 & 0 & -450 & 45 & 720 & 0 & 5220 \\
0 & 0 & 0 & 765 & 315 & 1125 & 13,050 & 138,330
\end{array}
$$

This is a final tableau, since all of the indicators are nonnegative. The maximum value is

$$
\frac{138,330}{13,050} = 10.60
$$

when $x_1 = \dfrac{5220}{2610} = 2$, $x_2 = \dfrac{870}{290} = 3$, $x_3 = \dfrac{15,660}{2610} = 6$, $x_4 = 0$, $x_5 = 0$, and $x_6 = 0$.

The company should produce 2 jazz albums, 3 blues albums, and 6 reggae albums for a maximum weekly profit of $10.60.

21. Let x_1 represent the number of one- speed bicycles, x_2 the number of three-speed bicycles, and x_3 the number of ten-speed bicycles.

From Exercise 29 in Section 4.1, the initial simplex tableau is as follows.

$$
\begin{array}{cccccc|c}
x_1 & x_2 & x_3 & x_4 & x_5 & z & \\
17 & 27 & \boxed{34} & 0 & 1 & 0 & 91,800 \\
12 & 21 & 15 & 1 & 0 & 0 & 42,000 \\
\hline
-8 & -12 & -22 & 0 & 0 & 1 & 0
\end{array}
$$

Pivot on the 34 in the first row, third column.

$$
\begin{array}{r}
\\
-15R_1+34R_2\rightarrow R_2 \\
\\
11R_1+17R_3\rightarrow R_3
\end{array}
\begin{array}{c}
\begin{array}{cccccc}
x_1 & x_2 & x_3 & x_4 & x_5 & z
\end{array} \\
\left[\begin{array}{cccccc|c}
17 & 27 & 34 & 0 & 1 & 0 & 91{,}800 \\
153 & 309 & 0 & 34 & -15 & 0 & 51{,}000 \\
\hline
51 & 93 & 0 & 0 & 11 & 17 & 1{,}009{,}800
\end{array}\right]
\end{array}
$$

This is a final tableau, since all of the indicators are nonnegative. The maximum value is

$$\frac{1{,}009{,}800}{17} = 59{,}400 \text{ when}$$

$$x_1 = 0, \ x_2 = 0, \ x_3 = \frac{91{,}800}{34} = 2700,$$

$$x_4 = \frac{51{,}000}{34} = 1500, \text{ and } x_5 = 0.$$

The company should make no one-speed or three-speed bicycles and 2700 ten-speed bicycles for a maximum profit of $59,400.

23. Let x_1 represent the amount of half nuts and half raisins mix, and x_2 be the amount of 1/3 nuts and 2/3 raisins mix. From Exercise 13 in Section 3.3, the problem is to Maximize $z = 6x_1 + 4.8x_2$ subject to:

$$\frac{1}{2}x_1 + \frac{1}{3}x_2 \leq 100$$

$$\frac{1}{2}x_1 + \frac{2}{3}x_2 \leq 125$$

with $x_1 \geq 0$ and $x_2 \geq 0$.

We need 2 slack variables, x_3 and x_4. Then the initial simplex tableau is as follows.

$$
\begin{array}{c}
\begin{array}{ccccc}
x_1 & x_2 & x_3 & x_4 & z
\end{array} \\
\left[\begin{array}{ccccc|c}
\boxed{\tfrac{1}{2}} & \tfrac{1}{3} & 1 & 0 & 0 & 100 \\
\tfrac{1}{2} & \tfrac{2}{3} & 0 & 1 & 0 & 125 \\
\hline
-6 & -4.8 & 0 & 0 & 1 & 0
\end{array}\right]
\end{array}
$$

Pivot on the $\frac{1}{2}$ in the first row, first column.

$$
\begin{array}{r}
\\
-R_1+R_2\rightarrow R_2 \\
\\
12R_1+R_3\rightarrow R_3
\end{array}
\begin{array}{c}
\begin{array}{ccccc}
x_1 & x_2 & x_3 & x_4 & z
\end{array} \\
\left[\begin{array}{ccccc|c}
\tfrac{1}{2} & \tfrac{1}{3} & 1 & 0 & 0 & 100 \\
0 & \boxed{\tfrac{1}{3}} & -1 & 1 & 0 & 25 \\
\hline
0 & -.8 & 12 & 0 & 1 & 1200
\end{array}\right]
\end{array}
$$

Pivot on the $\frac{1}{3}$ in the second row, second column.

$$
\begin{array}{r}
-R_2+R_1\rightarrow R_1 \\
\\
\\
12R_2+5R_3\rightarrow R_3
\end{array}
\begin{array}{c}
\begin{array}{ccccc}
x_1 & x_2 & x_3 & x_4 & z
\end{array} \\
\left[\begin{array}{ccccc|c}
\tfrac{1}{2} & 0 & 2 & -1 & 0 & 75 \\
0 & \tfrac{1}{3} & -1 & 1 & 0 & 25 \\
\hline
0 & 0 & 48 & 12 & 5 & 6300
\end{array}\right]
\end{array}
$$

All of the indicators are non-negative, so we are finished pivoting. Create a 1 in the columns corresponding to x_1, x_2, and z.

$$
\begin{array}{r}
2R_1\rightarrow R_1 \\
3R_2\rightarrow R_2 \\
\\
\tfrac{1}{5}R_3\rightarrow R_3
\end{array}
\begin{array}{c}
\begin{array}{ccccc}
x_1 & x_2 & x_3 & x_4 & z
\end{array} \\
\left[\begin{array}{ccccc|c}
1 & 0 & 4 & -2 & 0 & 150 \\
0 & 1 & -3 & 3 & 0 & 75 \\
\hline
0 & 0 & \tfrac{48}{5} & \tfrac{12}{5} & 1 & 1260
\end{array}\right]
\end{array}
$$

The maximum value is 1260 when $x_1 = 150$, $x_2 = 75$, $x_3 = 0$, and $x_4 = 0$.

The maximum revenue is $1260 when 150 kg of the half-and-half mix and 75 kg of the other mix are prepared.

25. (a) Look at the first chart, which has to do with the profits.

The profit-maximization formula is

$$\$2A + \$5B + \$4C = X,$$

so the answer is choice 1.

(b) Look at the Painting row of the second chart.

The Painting constraint is

$$1A + 2B + 2C \leq 38,000,$$

so the answer is choice 3.

27. Let x_1 represent the number of minutes for the senator, x_2 the number of minutes for the congresswoman, and x_3 the number of minutes for the governor. Of the half–hour show's time, only $30 - 3 = 27$ min are available to be allotted to the politicans.

The given information leads to the equation $x_1 + x_2 + x_3 = 27$ and the inequalities

$$x_1 \geq 2x_3 \quad \text{and} \quad x_1 + x_3 \geq 2x_2,$$

and we are to maximize the objective function $z = 40x_1 + 60x_2 + 50x_3$.

Rewrite the equation as $x_3 = 27 - x_1 - x_2$ and the inequalities as

$$x_1 - 2x_3 \geq 0 \quad \text{and} \quad x_1 - 2x_2 + x_3 \geq 0.$$

Substitute $27 - x_1 - x_2$ for x_3 in the objective function and the inequalites, and the problem is as follows.

Maximize $z = 40x_1 + 60x_2 + 50(27 - x_1 - x_2)$ subject to:

$$x_1 \qquad - 2(27 - x_1 - x_2) \geq 0$$
$$x_1 - 2x_2 + (27 - x_1 - x_2) \geq 0$$

with $x_1 \geq 0$, $x_2 \geq 0$, and

$x_3 = 27 - x_1 - x_2 \geq 0$.

Simplify to obtain the following.

Maximize $z = 1350 - 10x_1 + 10x_2$

subject to: $3x_1 + 2x_2 \geq 54$

$$x_2 \leq 9$$

with $x_1 \geq 0$, $x_2 \geq 0$, and

$x_3 = 27 - x_1 - x_2 \geq 0$.

This linear programming problem is not in standard maximum form, so we will attempt to use reason rather than the simplex algorithm. (The graphical method of solution examined in Chapter 3 would also be acceptable.)

Decreasing x_2 would increase x_1 (since $x_1 + x_2 + x_3$ must remain equal to 27) and would decrease z (since $z = 1350 - 10x_1 + 10x_2$). Therefore, choose x_2 to be as large as possible, which means $x_2 = 9$. The constraint $3x_1 + 2x_2 \geq 54$ can be rewritten as $x_1 \geq \dfrac{54 - 2x_2}{3}$.

Since we have chosen $x_2 = 9$, this means $x_1 \geq \dfrac{54 - 2(9)}{3}$ or $x_1 \geq 12$.

Increasing x_1 would decrease z, so choose x_1 to be as small as possible. This means $x_1 = 12$.

With $x_1 = 12$ and $x_2 = 9$, we have $x_3 = 27 - 12 - 9$ or $x_3 = 6$, and $z = 1350 - 10(12) + 10(9)$ or $z = 1320$. For a maximum of $1,320,000$ viewers, the time allotments should be 12 min for the senator, 9 min for the congresswoman, and 6 min for the governor.

29. **(a)** Let x_1 represent the number of species A,

x_2 represent the number of species B, and

x_3 represent the number of species C.

We are asked to

maximize $z = 1.62x_1 + 2.14x_2 + 3.01x_3$

subject to:

$$1.32x_1 + 2.1x_2 + .86x_3 \leq 490$$
$$2.9x_1 + .95x_2 + 1.52x_3 \leq 897$$
$$1.75x_1 + .6x_2 + 2.01x_3 \leq 653$$

with $x_1 \geq 0$, $x_2 \geq 0$, and $x_3 \geq 0$.

Use a computer to solve this problem and find that the answer is to stock none of species A, 114 of species B, and 291 of species C for a maximum combined weight of 1119.72 kg.

(b) Many answers are possible. The idea is to choose average weights for species B and C that are considerably smaller than the average weight chosen for species A, so that species A dominates the objective function.

(c) Many answers are possible. The idea is to choose average weights for species A and B that are considerably smaller than the average weight chosen for species C.

Section 4.3

1. $2x_1 + 3x_2 \leq 8$
    $x_1 + 4x_2 \geq 7$

    Introduce the slack variable x_3 and the surplus variable x_4 to obtain the following equations.

    $$2x_1 + 3x_2 + x_3 \quad\quad = 8$$
    $$x_1 + 4x_2 \quad\quad - x_4 = 7$$

3. $x_1 + x_2 + x_3 \leq 100$
    $x_1 + x_2 + x_3 \geq 75$
    $x_1 + x_2 \quad\quad \geq 27$

    Introduce the slack variable x_4 and the surplus variables x_5 and x_6 to obtain the following equations.

    $$x_1 + x_2 + x_3 + x_4 \quad\quad\quad = 100$$
    $$x_1 + x_2 + x_3 \quad\quad - x_5 \quad\quad = 75$$
    $$x_1 + x_2 \quad\quad\quad\quad - x_6 = 27$$

5. Minimize $w = 4y_1 + 3y_2 + 2y_3$
    subject to: $y_1 + y_2 + y_3 \geq 5$
    $$y_1 + y_2 \quad\quad \geq 4$$
    $$2y_1 + y_2 + 3y_3 \geq 15$$
    with $y_1 \geq 0, \ y_2 \geq 0, \ y_3 \geq 0$.

    Change this to a maximization problem by letting $z = -w$. The problem can now be stated as equivalently as follows.
    Maximize $z = -4y_1 - 3y_2 - 2y_3$
    subject to: $y_1 + y_2 + y_3 \geq 5$
    $$y_1 + y_2 \quad\quad \geq 4$$
    $$2y_1 + y_2 + 3y_3 \geq 15$$
    with $y_1 \geq 0, \ y_2 \geq 0, \ y_3 \geq 0$.

7. Minimize $w = y_1 + 2y_2 + y_3 + 5y_4$
    subject to: $y_1 + y_2 + y_3 + y_4 \geq 50$
    $$3y_1 + y_2 + 2y_3 + y_4 \geq 100$$
    with $y_1 \geq 0, \ y_2 \geq 0, \ y_3 \geq 0,$
    $$y_4 \geq 0.$$

    Change this to a maximization problem by letting $z = -w$. The problem can now be stated equivalently as follows.
    Maximize $z = -y_1 - 2y_2 - y_3 - 5y_4$
    subject to: $y_1 + y_2 + y_3 + y_4 \geq 50$
    $$3y_1 + y_2 + 2y_3 + y_4 \geq 100$$
    with $y_1 \geq 0, \ y_2 \geq 0, \ y_3 \geq 0,$
    $$y_4 \geq 0.$$

9. Find $x_1 \geq 0$ and $x_2 \geq 0$ such that
    $$x_1 + 2x_2 \geq 24$$
    $$x_1 + x_2 \leq 40$$
    and $z = 12x_1 + 10x_2$ is maximized.
    Subtracting the surplus variable x_3 and adding the slack variable x_4 leads to the equations

$$x_1 + 2x_2 - x_3 \quad\quad = 24$$
$$x_1 + \quad x_2 \quad\quad + x_4 = 40.$$

The initial simplex tableau is as follows.

$$\begin{array}{ccccc} x_1 & x_2 & x_3 & x_4 & z \\ \end{array}$$
$$\left[\begin{array}{ccccc|c} ① & 2 & -1 & 0 & 0 & 24 \\ 1 & 1 & 0 & 1 & 0 & 40 \\ \hline -12 & -10 & 0 & 0 & 1 & 0 \end{array}\right]$$

The initial basic solution is not feasible, since $x_3 = -24$ is negative, so row transformations must be used. We pivot on the 1 in the first row, first column, since it is the positive entry that is farthest to the left in the first row (the row containing the -1) and since, in the first column, $24/1 = 24$ is a smaller quotient than $40/1 = 40$. After row transformations, we obtain the following tableau.

$$\begin{array}{ccccc} x_1 & x_2 & x_3 & x_4 & z \\ \end{array}$$
$$\begin{array}{c} \\ -R_1 + R_2 \rightarrow R_2 \\ \\ 12R_1 + R_3 \rightarrow R_3 \end{array} \left[\begin{array}{ccccc|c} 1 & 2 & -1 & 0 & 0 & 24 \\ 0 & -1 & ① & 1 & 0 & 16 \\ \hline 0 & 14 & -12 & 0 & 1 & 288 \end{array}\right]$$

The basic solution is now feasible, but the problem is not yet finished since there is a negative indicator. Continue now in the usual way. The 1 in the third column is the next pivot. After row transformations, we get the following tableau.

$$\begin{array}{ccccc} x_1 & x_2 & x_3 & x_4 & z \\ \end{array}$$
$$\begin{array}{c} R_1 + R_2 \rightarrow R_1 \\ \\ \\ 12R_2 + R_3 \rightarrow R_3 \end{array} \left[\begin{array}{ccccc|c} 1 & 1 & 0 & 1 & 0 & 40 \\ 0 & -1 & 1 & 1 & 0 & 16 \\ \hline 0 & 2 & 0 & 12 & 1 & 480 \end{array}\right]$$

This is a final tableau, since the entries in the last row are all non-negative. The maximum value is 480 and occurs when $x_1 = 40$, $x_2 = 0$, $x_3 = 16$, and $x_4 = 0$.

11. Find $x_1 \geq 0$, $x_2 \geq 0$, and $x_3 \geq 0$ such that

$$x_1 + x_2 + x_3 \leq 150$$
$$x_1 + x_2 + x_3 \geq 100$$

and $z = 2x_1 + 5x_2 + 3x_3$ is maximized.

The initial tableau is as follows.

$$\begin{array}{cccccc} x_1 & x_2 & x_3 & x_4 & x_5 & z \\ \end{array}$$
$$\left[\begin{array}{cccccc|c} 1 & 1 & 1 & 1 & 0 & 0 & 150 \\ 1 & ① & 1 & 0 & -1 & 0 & 100 \\ \hline -2 & -5 & -3 & 0 & 0 & 1 & 0 \end{array}\right]$$

Note that x_4 is a slack variable, while x_5 is a surplus variable. The initial basic solution is not feasible, since $x_5 = -100$ is negative. Pivot on the 1 in the second row, second column. (Note that any column with a non-negative entry in the second row could be used as the pivot column.)

$$\begin{array}{cccccc} x_1 & x_2 & x_3 & x_4 & x_5 & z \\ \end{array}$$
$$\begin{array}{c} -R_2 + R_1 \rightarrow R_1 \\ \\ 5R_2 + R_3 \rightarrow R_3 \end{array} \left[\begin{array}{cccccc|c} 0 & 0 & 0 & 1 & ① & 0 & 50 \\ 1 & 1 & 1 & 0 & -1 & 0 & 100 \\ \hline 3 & 0 & 2 & 0 & -5 & 1 & 500 \end{array}\right]$$

The basic solution is now feasible, but there is a negative indicator. Pivot on the 1 in the first row, fifth column.

$$
\begin{array}{c}
\\
R_1 + R_2 \rightarrow R_2 \\
\\
5R_1 + R_3 \rightarrow R_3
\end{array}
\begin{array}{cccccc}
x_1 & x_2 & x_3 & x_4 & x_5 & z \\
\end{array}
\left[
\begin{array}{cccccc|c}
0 & 0 & 0 & 1 & 1 & 0 & 50 \\
1 & 1 & 1 & 1 & 0 & 0 & 150 \\
\hline
3 & 0 & 2 & 5 & 0 & 1 & 750
\end{array}
\right]
$$

This is a final tableau. The maximum value is 750 when $x_1 = 0$, $x_2 = 150$, $x_3 = 0$, $x_4 = 0$, and $x_5 = 50$.

13. Find $x_1 \geq 0$ and $x_2 \geq 0$ such that

$$
\begin{aligned}
x_1 + x_2 &\leq 100 \\
x_1 + x_2 &\geq 50 \\
2x_1 + x_2 &\leq 110
\end{aligned}
$$

and $z = -2x_1 + 3x_2$ is maximized. The initial tableau is as follows.

$$
\begin{array}{cccccc}
x_1 & x_2 & x_3 & x_4 & x_5 & z \\
\end{array}
\left[
\begin{array}{cccccc|c}
1 & 1 & 1 & 0 & 0 & 0 & 100 \\
1 & ① & 0 & -1 & 0 & 0 & 50 \\
2 & 1 & 0 & 0 & 1 & 0 & 110 \\
\hline
2 & -3 & 0 & 0 & 0 & 1 & 0
\end{array}
\right]
$$

The initial basic solution is not feasible since $x_4 = -50$.
Pivot on the 1 in the second row, second column.

$$
\begin{array}{c}
-R_2 + R_1 \rightarrow R_1 \\
\\
-R_2 + R_3 \rightarrow R_3 \\
\\
3R_2 + R_4 \rightarrow R_4
\end{array}
\begin{array}{cccccc}
x_1 & x_2 & x_3 & x_4 & x_5 & z \\
\end{array}
\left[
\begin{array}{cccccc|c}
0 & 0 & 1 & ① & 0 & 0 & 50 \\
1 & 1 & 0 & -1 & 0 & 0 & 50 \\
1 & 0 & 0 & 1 & 1 & 0 & 60 \\
\hline
5 & 0 & 0 & -3 & 0 & 1 & 150
\end{array}
\right]
$$

The basic solution is now feasible, but there is a negative indicator. Pivot on the 1 in the first row, fourth column.
(It would also be acceptable to pivot on the 1 in the second row, first column.)

$$
\begin{array}{c}
-R_1 + R_2 \rightarrow R_2 \\
\\
-R_1 + R_3 \rightarrow R_3 \\
\\
3R_1 + R_4 \rightarrow R_4
\end{array}
\begin{array}{cccccc}
x_1 & x_2 & x_3 & x_4 & x_5 & z \\
\end{array}
\left[
\begin{array}{cccccc|c}
0 & 0 & 1 & 1 & 0 & 0 & 50 \\
1 & 1 & 1 & 0 & 0 & 0 & 100 \\
1 & 0 & -1 & 0 & 1 & 0 & 10 \\
\hline
5 & 0 & 3 & 0 & 0 & 1 & 300
\end{array}
\right]
$$

This is a final tableau. The maximum value is 300 when $x_1 = 0$, $x_2 = 100$, $x_3 = 0$, $x_4 = 50$, and $x_5 = 10$.

15. Find $y_1 \geq 0$, $y_2 \geq 0$ such that

$$
\begin{aligned}
10y_1 + 5y_2 &\geq 100 \\
20y_1 + 10y_2 &\geq 150
\end{aligned}
$$

and $w = 4y_1 + 5y_2$ is minimized. Let $z = -w$, and then $z = -4y_1 - 5y_2$ is to be maximized. The initial tableau is as follows.

$$
\begin{array}{ccccc}
y_1 & y_2 & y_3 & y_4 & z \\
\end{array}
\left[
\begin{array}{ccccc|c}
10 & 5 & -1 & 0 & 0 & 100 \\
㉀ & 10 & 0 & -1 & 0 & 150 \\
\hline
4 & 5 & 0 & 0 & 1 & 0
\end{array}
\right]
$$

The initial basic solution is not feasible since $y_3 = -100$ and $y_4 = -150$. Pivot on the 20 in the second row, first column.

$$
\begin{array}{c}
2R_1 - R_2 \rightarrow R_1 \\
\\
5R_3 - R_2 \rightarrow R_3
\end{array}
\begin{array}{ccccc}
y_1 & y_2 & y_3 & y_4 & z \\
\end{array}
\left[
\begin{array}{ccccc|c}
0 & 0 & -2 & ① & 0 & 50 \\
20 & 10 & 0 & -1 & 0 & 150 \\
\hline
0 & 15 & 0 & 1 & 5 & -150
\end{array}
\right]
$$

The basic solution is still not feasible. Pivot on the 1 in the first row, fourth column.

$$
\begin{array}{c}
\\
R_1 + R_2 \rightarrow R_2 \\
\\
-R_1 + R_3 \rightarrow R_3
\end{array}
\begin{array}{ccccc}
y_1 & y_2 & y_3 & y_4 & z \\
\end{array}
\left[
\begin{array}{ccccc|c}
0 & 0 & -2 & 1 & 0 & 50 \\
20 & 10 & -2 & 0 & 0 & 200 \\
\hline
0 & 15 & 2 & 0 & 5 & -200
\end{array}
\right]
$$

The basic solution is now feasible. Additionally, all of the indicators are nonnegative, so we are finished pivoting. For ease in reading the basic solution, create a 1 in the columns corresponding to y_1 and z.

$$\begin{array}{c} \\ \frac{1}{20}R_2 \to R_2 \\ \\ \frac{1}{5}R_3 \to R_3 \end{array} \begin{array}{cccccc} y_1 & y_2 & y_3 & y_4 & z & \\ \left[\begin{array}{ccccc|c} 0 & 0 & -2 & 1 & 0 & 50 \\ 1 & .5 & -.1 & 0 & 0 & 10 \\ \hline 0 & 3 & .4 & 0 & 1 & -40 \end{array}\right] \end{array}$$

This is a final tableau. The maximum value of $z = -w$ is -40. Therefore, the minimum value of w is 40, and it occurs when $y_1 = 10$, $y_2 = 0$, $y_3 = 0$, and $y_4 = 50$.

17. Minimize $w = 2y_1 + y_2 + 3y_3$

subject to: $y_1 + y_2 + y_3 \geq 100$
$$2y_1 + y_2 \qquad \geq 50$$

with $\qquad y_1 \geq 0$, $y_2 \geq 0$, $y_3 \geq 0$.

Let $z = -w$, and then $w = -2y_1 - y_2 - 3y_3$ is to be maximized.

The initial tableau is as follows.

$$\begin{array}{cccccc} y_1 & y_2 & y_3 & y_4 & y_5 & z \\ \left[\begin{array}{cccccc|c} 1 & 1 & 1 & -1 & 0 & 0 & 100 \\ ② & 1 & 0 & 0 & -1 & 0 & 50 \\ \hline 2 & 1 & 3 & 0 & 0 & 1 & 0 \end{array}\right] \end{array}$$

The initial basic solution is not feasible. Pivot on the 2 in the second row, first column.

$$\begin{array}{c} 2R_1 - R_2 \to R_1 \\ \\ \\ -R_2 + R_3 \to R_3 \end{array} \begin{array}{cccccc} y_1 & y_2 & y_3 & y_4 & y_5 & z \\ \left[\begin{array}{cccccc|c} 0 & 1 & 2 & -2 & 1 & 0 & 150 \\ 2 & ① & 0 & 0 & -1 & 0 & 50 \\ \hline 0 & 0 & 3 & 0 & 1 & 1 & -50 \end{array}\right] \end{array}$$

The basic solution is still not feasible. Pivot on the 1 in the second row, second column.

$$-R_2 + R_1 \to R_1 \begin{array}{cccccc} y_1 & y_2 & y_3 & y_4 & y_5 & z \\ \left[\begin{array}{cccccc|c} -2 & 0 & ② & -2 & 2 & 0 & 100 \\ 2 & 1 & 0 & 0 & -1 & 0 & 50 \\ \hline 0 & 0 & 3 & 0 & 1 & 1 & -50 \end{array}\right] \end{array}$$

The basic solution is still not feasible. Pivot on the 2 in the first row, third column.

$$-3R_1 + 2R_3 \to R_3 \begin{array}{cccccc} y_1 & y_2 & y_3 & y_4 & y_5 & z \\ \left[\begin{array}{cccccc|c} -2 & 0 & 2 & -2 & ② & 0 & 100 \\ 2 & 1 & 0 & 0 & -1 & 0 & 50 \\ \hline 6 & 0 & 0 & 6 & -4 & 2 & -400 \end{array}\right] \end{array}$$

The basic solution is now feasible, but there is a negative indicator. Pivot on the 2 in the first row, fifth column.

$$\begin{array}{c} \\ R_1 + 2R_2 \to R_2 \\ \\ 2R_1 + R_3 \to R_3 \end{array} \begin{array}{cccccc} y_1 & y_2 & y_3 & y_4 & y_5 & z \\ \left[\begin{array}{cccccc|c} -2 & 0 & 2 & -2 & 2 & 0 & 100 \\ 2 & 2 & 2 & -2 & 0 & 0 & 200 \\ \hline 2 & 0 & 4 & 2 & 0 & 2 & -200 \end{array}\right] \end{array}$$

All of the indicators are nonnegative, so we are finished pivoting. For ease in reading the basic solution, create a 1 in the columns corresponding to y_2, y_5, and z.

$$\begin{array}{c} \frac{1}{2}R_1 \to R_1 \\ \\ \frac{1}{2}R_2 \to R_2 \\ \\ \frac{1}{2}R_3 \to R_3 \end{array} \begin{array}{cccccc} y_1 & y_2 & y_3 & y_4 & y_5 & z \\ \left[\begin{array}{cccccc|c} -1 & 0 & 1 & -1 & 1 & 0 & 50 \\ 1 & 1 & 1 & -1 & 0 & 0 & 100 \\ \hline 1 & 0 & 2 & 1 & 0 & 1 & -100 \end{array}\right] \end{array}$$

This is a final tableau. The maximum value of $z = -w$ is -100. Therefore, the minimum value of w is 100, and it occurs when $y_1 = 0$, $y_2 = 100$, $y_3 = 0$, $y_4 = 0$, and $y_5 = 50$.

19. Maximize $z = 3x_1 + 2x_2$

subject to: $x_1 + x_2 = 50$

$\qquad\quad 4x_1 + 2x_2 \geq 120$

$\qquad\quad 5x_1 + 2x_2 \leq 200$

with $x_1 \geq 0,\ x_2 \geq 0$.

The artificial variable a_1 is used to rewrite $x_1 + x_2 = 50$ as $x_1 + x_2 + a_1 = 50$; note that a_1 must equal 0 for this equation to be a true statement. Also, the surplus variable x_3 and the slack variable x_4 are needed. The initial tableau is a follows.

$$\begin{array}{cccccc} x_1 & x_2 & a_1 & x_3 & x_4 & z \\ \left[\begin{array}{cccccc|c} 1 & 1 & 1 & 0 & 0 & 0 & 50 \\ ④ & 2 & 0 & -1 & 0 & 0 & 120 \\ 5 & 2 & 0 & 0 & 1 & 0 & 200 \\ \hline -3 & -2 & 0 & 0 & 0 & 1 & 0 \end{array}\right] \end{array}$$

The initial basic solution is not feasible. Pivot on the 4 in the second row, first column.

$$\begin{array}{c} \\ 4R_1 - R_2 \to R_1 \\ \\ -5R_2 + 4R_3 \to R_3 \\ \\ 3R_2 + 4R_4 \to R_4 \end{array} \begin{array}{cccccc} x_1 & x_2 & a_1 & x_3 & x_4 & z \\ \left[\begin{array}{cccccc|c} 0 & 2 & 4 & 1 & 0 & 0 & 80 \\ 4 & 2 & 0 & -1 & 0 & 0 & 120 \\ 0 & -2 & 0 & ⑤ & 2 & 0 & 200 \\ \hline 2 & -2 & 0 & -3 & 0 & 4 & 360 \end{array}\right] \end{array}$$

The basic solution is now feasible, but there are negative indicators. Pivot on the 5 in the third row, fourth column (which is the column with the most negative indicator and the row with the smallest nonnegative quotient).

$$\begin{array}{c} \\ 5R_1-R_3 \to R_1 \\ 5R_2+R_3 \to R_2 \\ \\ 3R_3+5R_4 \to R_4 \end{array} \begin{array}{cccccc} x_1 & x_2 & a_1 & x_3 & x_4 & z \\ \left[\begin{array}{cccccc|c} 0 & ⑫ & 20 & 0 & -2 & 0 & 200 \\ 20 & 8 & 0 & 0 & 2 & 0 & 800 \\ 0 & -2 & 0 & 5 & 2 & 0 & 200 \\ \hline 0 & -16 & 0 & 0 & 6 & 20 & 2400 \end{array}\right] \end{array}$$

Pivot on the 12 in the first row, second column.

$$\begin{array}{c} \\ \\ -2R_1+3R_2 \to R_2 \\ R_1+6R_3 \to R_3 \\ \\ 4R_1+3R_4 \to R_4 \end{array} \begin{array}{cccccc} x_1 & x_2 & a_1 & x_3 & x_4 & z \\ \left[\begin{array}{cccccc|c} 0 & 12 & 20 & 0 & -2 & 0 & 200 \\ 60 & 0 & -40 & 0 & 10 & 0 & 2000 \\ 0 & 0 & 20 & 30 & 10 & 0 & 1400 \\ \hline 0 & 0 & 80 & 0 & 10 & 60 & 8000 \end{array}\right] \end{array}$$

We now have $a_1 = 0$, so drop the a_1 column.

$$\begin{array}{ccccc} x_1 & x_2 & x_3 & x_4 & z \\ \left[\begin{array}{ccccc|c} 0 & 12 & 0 & -2 & 0 & 200 \\ 60 & 0 & 0 & 10 & 0 & 2000 \\ 0 & 0 & 30 & 10 & 0 & 1400 \\ \hline 0 & 0 & 0 & 10 & 60 & 8000 \end{array}\right] \end{array}$$

We are finished pivoting. Create a 1 in the columns corresponding to x_1, x_2, x_3, and z.

$$\begin{array}{c} \frac{1}{12}R_1 \to R_1 \\ \frac{1}{60}R_2 \to R_2 \\ \frac{1}{30}R_3 \to R_3 \\ \\ \frac{1}{60}R_4 \to R_4 \end{array} \begin{array}{ccccc} x_1 & x_2 & x_3 & x_4 & z \\ \left[\begin{array}{ccccc|c} 0 & 1 & 0 & -\frac{1}{6} & 0 & 16\frac{2}{3} \\ 1 & 0 & 0 & \frac{1}{6} & 0 & 33\frac{1}{3} \\ 0 & 0 & 1 & \frac{1}{3} & 0 & 46\frac{2}{3} \\ \hline 0 & 0 & 0 & \frac{1}{6} & 1 & 133\frac{1}{3} \end{array}\right] \end{array}$$

The maximum value is $133\frac{1}{3}$ when $x_1 = 33\frac{1}{3}$, $x_2 = 16\frac{2}{3}$, $x_3 = 46\frac{2}{3}$, and $x_4 = 0$.

21. Minimize $w = 15y_1 + 12y_2$

subject to: $y_1 + 2y_2 \leq 12$

$\qquad\quad 3y_1 + y_2 \geq 18$

$\qquad\quad y_1 + y_2 = 10$

with $y_1 \geq 0,\ y_2 \geq 0$.

Let $z = -w$, and then $z = -15y_1 - 12y_2$ is to be maximized. Introduce the slack variable y_3, the surplus variable y_4, and the artificial variable a_1. The initial tableau is as follows.

$$\begin{array}{cccccc} y_1 & y_2 & y_3 & y_4 & a_1 & z \\ \end{array}$$
$$\left[\begin{array}{cccccc|c} 1 & 2 & 1 & 0 & 0 & 0 & 12 \\ ③ & 1 & 0 & -1 & 0 & 0 & 18 \\ 1 & 1 & 0 & 0 & 1 & 0 & 10 \\ \hline 15 & 12 & 0 & 0 & 0 & 1 & 0 \end{array}\right]$$

The initial basic solution is not feasible. Pivot on the 3 in the second row, first column.

$$\begin{array}{cccccc} & y_1 & y_2 & y_3 & y_4 & a_1 & z \\ \end{array}$$
$$\begin{array}{l} 3R_1 - R_2 \to R_1 \\ \\ \\ 3R_3 - R_2 \to R_3 \\ \\ -5R_2 + R_4 \to R_4 \end{array} \left[\begin{array}{cccccc|c} 0 & 5 & 3 & 1 & 0 & 0 & 18 \\ 3 & 1 & 0 & -1 & 0 & 0 & 18 \\ 0 & 2 & 0 & ① & 3 & 0 & 12 \\ \hline 0 & 7 & 0 & 5 & 0 & 1 & -90 \end{array}\right]$$

The basic solution is now feasible, but the artificial variable still remains, so we are not finished. Try pivoting on the 1 in the third row, fourth column.

$$\begin{array}{cccccc} & y_1 & y_2 & y_3 & y_4 & a_1 & z \\ \end{array}$$
$$\begin{array}{l} -R_3 + R_1 \to R_1 \\ R_3 + R_2 \to R_2 \\ \\ -5R_3 + R_4 \to R_4 \end{array} \left[\begin{array}{cccccc|c} 0 & 3 & 3 & 0 & -3 & 0 & 6 \\ 3 & 3 & 0 & 0 & 3 & 0 & 30 \\ 0 & 2 & 0 & 1 & 3 & 0 & 12 \\ \hline 0 & -3 & 0 & 0 & -15 & 1 & -150 \end{array}\right]$$

We now have $a_1 = 0$, so drop the a_1 column.

$$\begin{array}{ccccc} y_1 & y_2 & y_3 & y_4 & z \\ \end{array}$$
$$\left[\begin{array}{ccccc|c} 0 & ③ & 3 & 0 & 0 & 6 \\ 3 & 3 & 0 & 0 & 0 & 30 \\ 0 & 2 & 0 & 1 & 0 & 12 \\ \hline 0 & -3 & 0 & 0 & 1 & -150 \end{array}\right]$$

Pivot on the 3 in the first row, second column.

$$\begin{array}{ccccc} y_1 & y_2 & y_3 & y_4 & z \\ \end{array}$$
$$\begin{array}{l} \\ -R_1 + R_2 \to R_2 \\ -2R_1 + 3R_3 \to R_3 \\ \\ R_1 + R_4 \to R_4 \end{array} \left[\begin{array}{ccccc|c} 0 & 3 & 3 & 0 & 0 & 6 \\ 3 & 0 & -3 & 0 & 0 & 24 \\ 0 & 0 & -6 & 3 & 0 & 24 \\ \hline 0 & 0 & 3 & 0 & 1 & -144 \end{array}\right]$$

We are finished pivoting. Create a 1 in the columns corresponding to y_1, y_2, and y_4.

$$\begin{array}{ccccc} y_1 & y_2 & y_3 & y_4 & z \\ \end{array}$$
$$\begin{array}{l} \frac{1}{3}R_1 \to R_1 \\ \\ \frac{1}{3}R_2 \to R_2 \\ \\ \frac{1}{3}R_3 \to R_3 \end{array} \left[\begin{array}{ccccc|c} 0 & 1 & 1 & 0 & 0 & 2 \\ 1 & 0 & -1 & 0 & 0 & 8 \\ 0 & 0 & -2 & 1 & 0 & 8 \\ \hline 0 & 0 & 3 & 0 & 1 & -144 \end{array}\right]$$

This is a final tableau. The maximum value of $z = -w$ is -144. Therefore, the minimum value of w is 144, and it occurs when $y_1 = 8$, $y_2 = 2$, $y_3 = 0$, and $y_4 = 8$.

25. Let y_1 = the number of barrels from S_1 to D_1;

y_2 = the number of barrels from S_2 to D_1;

y_3 = the number of barrels from S_1 to D_2; and

y_4 = the number of barrels from S_2 to D_2.

The problem is to minimize

$$w = 30y_1 + 25y_2 + 20y_3 + 22y_4$$

subject to:

$$y_1 + y_2 \geq 3000$$
$$y_3 + y_4 \geq 5000$$
$$y_1 + y_3 \leq 5000$$
$$y_2 + y_4 \leq 5000$$

with $y_1 \geq 0$, $y_2 \geq 0$, $y_3 \geq 0$, $y_4 \geq 0$.

Let z = -w, and maximize z.   The initial tableau is as follows.

y_1	y_2	y_3	y_4	y_5	y_6	y_7	y_8	z	
①	1	0	0	-1	0	0	0	0	3000
0	0	1	1	0	-1	0	0	0	5000
1	0	1	0	0	0	1	0	0	5000
0	1	0	1	0	0	0	1	0	5000
30	25	20	22	0	0	0	0	1	0

The initial basic solution is not feasible.   Pivot on the 1 in the first row, first column.

	y_1	y_2	y_3	y_4	y_5	y_6	y_7	y_8	z	
	1	1	0	0	-1	0	0	0	0	3000
	0	0	1	1	0	-1	0	0	0	5000
$-R_1 + R_3 \to R_3$	0	-1	①	0	1	0	1	0	0	2000
	0	1	0	1	0	0	0	1	0	5000
$-20R_1 + R_5 \to R_5$	0	-5	20	22	30	0	0	0	1	-90,000

The basic solution is still not feasible.   Pivot on the 1 in the third row, third column.

	y_1	y_2	y_3	y_4	y_5	y_6	y_7	y_8	z	
	1	1	0	0	-1	0	0	0	0	3000
$-R_3 + R_2 \to R_2$	0	①	0	1	-1	-1	-1	0	0	3000
	0	-1	1	0	1	0	1	0	0	2000
	0	1	0	1	0	0	0	1	0	5000
$-20R_3 + R_5 \to R_5$	0	15	0	22	10	0	-20	0	1	-130,000

The basic solution is still not feasible.   Pivot on the 1 in the second row, second column.

	y_1	y_2	y_3	y_4	y_5	y_6	y_7	y_8	z	
$-R_2 + R_1 \to R_1$	1	0	0	-1	0	①	1	0	0	0
	0	1	0	1	-1	-1	-1	0	0	3000
$R_2 + R_3 \to R_3$	0	0	1	1	0	-1	0	0	0	5000
$-R_2 + R_4 \to R_4$	0	0	0	0	1	1	1	1	0	2000
$-15R_2 + R_5 \to R_5$	0	0	0	7	25	15	-5	0	1	-175,000

The basic solution is now feasible, but there is a negative indicator.   Pivot on the 1 in the first row, seventh column.

$$
\begin{array}{c}
\\
\\
R_1 + R_2 \rightarrow R_2 \\
\\
-R_1 + R_4 \rightarrow R_4 \\
5R_1 + R_5 \rightarrow R_5
\end{array}
\begin{array}{cccccccccc}
y_1 & y_2 & y_3 & y_4 & y_5 & y_6 & y_7 & y_8 & z & \\
\left[\begin{array}{ccccccccc|c}
1 & 0 & 0 & -1 & 0 & 1 & 1 & 0 & 0 & 0 \\
1 & 1 & 0 & 0 & -1 & 0 & 0 & 0 & 0 & 3000 \\
0 & 0 & 1 & 1 & 0 & -1 & 0 & 0 & 0 & 5000 \\
-1 & 0 & 0 & 1 & 1 & 0 & 0 & 1 & 0 & 2000 \\
\hline
5 & 0 & 0 & 2 & 25 & 20 & 0 & 0 & 1 & -175{,}000
\end{array}\right]
\end{array}
$$

This is a final tableau. The minimum value of w is 175,000 when $y_1 = 0$, $y_2 = 3000$, $y_3 = 5000$, $y_4 = 0$, $y_5 = 0$, $y_6 = 0$, $y_7 = 0$, and $y_8 = 2000$. To minimize the shipping cost, ship 5000 barrels of oil from supplier S_1 to distributor D_2, and ship 3000 barrels of oil from supplier S_2 to distributor D_1. The minimum cost is \$175,000.

27. Let x_1 = the number of kilograms of sauce and
    and x_2 = the number of kilograms of whole tomatoes.

    The problem is to

$$
\begin{aligned}
&\text{minimize } w = 3.25y_1 + 4y_2 \\
&\text{subject to: } \quad y_1 + y_2 \le 3{,}000{,}000 \\
&\qquad\qquad\qquad y_1 \quad\;\ge \quad\;\; 80{,}000 \\
&\qquad\qquad\qquad\quad\; y_2 \ge \quad\; 800{,}000 \\
&\text{with} \qquad\qquad y_1 \ge 0,\; y_2 \ge 0.
\end{aligned}
$$

Let $z = -w$, and maximize z.

The initial tableau is as follows.

$$
\begin{array}{ccccccc}
y_1 & y_2 & y_3 & y_4 & y_5 & z & \\
\left[\begin{array}{ccccc|c}
1 & 1 & 1 & 0 & 0 & 0 & 3{,}000{,}000 \\
\textcircled{1} & 0 & 0 & -1 & 0 & 0 & 80{,}000 \\
0 & 1 & 0 & 0 & -1 & 0 & 800{,}000 \\
\hline
3.25 & 4 & 0 & 0 & 0 & 1 & 0
\end{array}\right]
\end{array}
$$

The initial basic solution is not feasible. Pivot on the 1 in the second row, first column.

$$
\begin{array}{c}
\\
-R_2 + R_1 \rightarrow R_1 \\
\\
\\
-3.25R_2 + R_4 \rightarrow R_4
\end{array}
\begin{array}{ccccccc}
y_1 & y_2 & y_3 & y_4 & y_5 & z & \\
\left[\begin{array}{ccccc|c}
0 & 1 & 1 & 1 & 0 & 0 & 2{,}920{,}000 \\
1 & 0 & 0 & -1 & 0 & 0 & 80{,}000 \\
0 & \textcircled{1} & 0 & 0 & -1 & 0 & 800{,}000 \\
\hline
0 & 4 & 0 & 3.25 & 0 & 1 & -260{,}000
\end{array}\right]
\end{array}
$$

The basic solution is still not feasible. Pivot on the 1 in the third row, second column.

$$
\begin{array}{c}
\\
-R_3 + R_1 \rightarrow R_1 \\
\\
\\
4R_3 + R_4 \rightarrow R_4
\end{array}
\begin{array}{ccccccc}
y_1 & y_2 & y_3 & y_4 & y_5 & z & \\
\end{array}
\left[
\begin{array}{cccccc|r}
0 & 0 & 1 & 1 & 1 & 0 & 2,120,000 \\
1 & 0 & 0 & -1 & 0 & 0 & 80,000 \\
0 & 1 & 0 & 0 & -1 & 0 & 800,000 \\
\hline
0 & 0 & 0 & 3.25 & 4 & 1 & -3,460,000
\end{array}
\right]
$$

The basic solution is now feasible. Additionally, there are no negative indicators, so this is a final tableau. The minimum value of w is 3,460,000 when y_1 = 80,000, y_2 = 800,000, y_3 = 2,120,000, y_4 = 0, and y_5 = 0.

Brand X should use 800,000 kg of tomatoes for whole tomatoes and 80,000 kg of tomatoes for sauce for a minimum cost of $3,460,000.

29. Let y_1 = the number of small tubes

and y_2 = the number of large tubes.

The problem is to minimize $.15y_1 + .12y_2$

$$
\begin{aligned}
\text{subject to: }\quad y_1 &\geq 800 \\
y_2 &\geq 500 \\
y_1 + y_2 &\geq 1500 \\
y_1 &\geq 2y_2 \\
\text{with }\quad y_1 \geq 0,\ y_2 &\geq 0.
\end{aligned}
$$

The last constraint can be rewitten as $y_1 - 2y_2 \geq 0$.

Let z = -w, and maximize z. The initial tableau is as follows.

$$
\begin{array}{ccccccc}
y_1 & y_2 & y_3 & y_4 & y_5 & y_6 & z \\
\end{array}
\left[
\begin{array}{ccccccc|r}
①& 0 & -1 & 0 & 0 & 0 & 0 & 800 \\
0 & 1 & 0 & -1 & 0 & 0 & 0 & 500 \\
1 & 1 & 0 & 0 & -1 & 0 & 0 & 1500 \\
1 & -2 & 0 & 0 & 0 & -1 & 0 & 0 \\
\hline
.15 & .12 & 0 & 0 & 0 & 0 & 1 & 0
\end{array}
\right]
$$

Continue by pivoting on the circled entry in the above tableau.

$$
\begin{array}{c}
\\
\\
-R_1 + R_3 \rightarrow R_3 \\
-R_1 + R_4 \rightarrow R_4 \\
\\
-.15R_1 + R_5 \rightarrow R_5
\end{array}
\begin{array}{ccccccc}
y_1 & y_2 & y_3 & y_4 & y_5 & y_6 & z \\
\end{array}
\left[
\begin{array}{ccccccc|r}
1 & 0 & -1 & 0 & 0 & 0 & 0 & 800 \\
0 & ① & 0 & -1 & 0 & 0 & 0 & 500 \\
0 & 1 & 1 & 0 & -1 & 0 & 0 & 700 \\
0 & -2 & 1 & 0 & 0 & -1 & 0 & -800 \\
\hline
0 & .12 & .15 & 0 & 0 & 0 & 1 & -120
\end{array}
\right]
$$

$$\begin{array}{c} \\ \\ -R_2 + R_3 \to R_3 \\ 2R_2 + R_4 \to R_4 \\ \\ -.12R_2 + R_5 \to R_5 \end{array} \begin{array}{cccccccc} y_1 & y_2 & y_3 & y_4 & y_5 & y_6 & z & \\ \begin{bmatrix} 1 & 0 & -1 & 0 & 0 & 0 & 0 & 800 \\ 0 & 1 & 0 & -1 & 0 & 0 & 0 & 500 \\ 0 & 0 & 1 & 1 & -1 & 0 & 0 & 200 \\ 0 & 0 & ① & -2 & 0 & -1 & 0 & 200 \\ \hline 0 & 0 & .15 & .12 & 0 & 0 & 1 & -180 \end{bmatrix} \end{array}$$

$$\begin{array}{c} R_1 + R_4 \to R_1 \\ \\ -R_4 + R_3 \to R_3 \\ \\ -.15R_4 + R_5 \to R_5 \end{array} \begin{array}{cccccccc} y_1 & y_2 & y_3 & y_4 & y_5 & y_6 & z & \\ \begin{bmatrix} 1 & 0 & 0 & -2 & 0 & -1 & 0 & 1000 \\ 0 & 1 & 0 & -1 & 0 & 0 & 0 & 500 \\ 0 & 0 & 0 & 3 & -1 & 1 & 0 & 0 \\ 0 & 0 & 1 & -2 & 0 & -1 & 0 & 200 \\ \hline 0 & 0 & 0 & .42 & 0 & .15 & 1 & -210 \end{bmatrix} \end{array}$$

This is a final tableau. The minimum value of w is 210 when $y_1 = 1000$, $y_2 = 500$, $y_3 = 200$, $y_4 = 0$, $y_5 = 0$, and $y_6 = 0$.

1000 small and 500 large test tubes should be ordered for a minimum cost of \$210.

31. Let x_1 = the amount to invest in securities;

   x_2 = the amount to invest in bonds; and

   x_3 = the amount to invest in mutual funds.

The problem is to maximize $\quad .07x_1 + .06x_2 + .10x_3$

$$\begin{array}{rl} \text{subject to: } & x_1 + x_2 + x_3 \le 100,000 \\ & x_1 \qquad\qquad \ge 40,000 \\ & \qquad x_2 + x_3 \ge 50,000 \end{array}$$

with $x_1 \ge 0$, $x_2 \ge 0$, $x_3 \ge 0$.

The initial tableau is as follows.

$$\begin{array}{ccccccc} x_1 & x_2 & x_3 & x_4 & x_5 & x_6 & z \\ \begin{bmatrix} 1 & 1 & 1 & 1 & 0 & 0 & 0 & 100,000 \\ 1 & 0 & 0 & 0 & -1 & 0 & 0 & 40,000 \\ 0 & 1 & 1 & 0 & 0 & -1 & 0 & 50,000 \\ \hline -.07 & -.06 & -.1 & 0 & 0 & 0 & 1 & 0 \end{bmatrix} \end{array}$$

Continue by pivoting on the circled entry in each tableau.

$$
\begin{array}{c}
\\
-R_2 + R_1 \rightarrow R_1 \\
\\
\\
.07R_2 + R_4 \rightarrow R_4
\end{array}
\begin{array}{ccccccc}
x_1 & x_2 & x_3 & x_4 & x_5 & x_6 & z \\
\end{array}
\left[
\begin{array}{ccccccc|c}
0 & 1 & 1 & 1 & 1 & 0 & 0 & 60{,}000 \\
1 & 0 & 0 & 0 & -1 & 0 & 0 & 40{,}000 \\
0 & ① & 1 & 0 & 0 & -1 & 0 & 50{,}000 \\
\hline
0 & -.06 & -.1 & 0 & -.07 & 0 & 1 & 2800
\end{array}
\right]
$$

$$
\begin{array}{c}
\\
-R_3 + R_1 \rightarrow R_1 \\
\\
\\
.06R_3 + R_4 \rightarrow R_4
\end{array}
\begin{array}{ccccccc}
x_1 & x_2 & x_3 & x_4 & x_5 & x_6 & z \\
\end{array}
\left[
\begin{array}{ccccccc|c}
0 & 0 & 0 & 1 & ① & 0 & 0 & 10{,}000 \\
1 & 0 & 0 & 0 & -1 & 0 & 0 & 40{,}000 \\
0 & 1 & 1 & 0 & 0 & -1 & 0 & 50{,}000 \\
\hline
0 & 0 & -.04 & 0 & -.07 & -.06 & 1 & 5800
\end{array}
\right]
$$

$$
\begin{array}{c}
\\
\\
R_1 + R_2 \rightarrow R_2 \\
\\
.07R_1 + R_4 \rightarrow R_4
\end{array}
\begin{array}{ccccccc}
x_1 & x_2 & x_3 & x_4 & x_5 & x_6 & z \\
\end{array}
\left[
\begin{array}{ccccccc|c}
0 & 0 & 0 & 1 & 1 & 1 & 0 & 10{,}000 \\
1 & 0 & 0 & 1 & 0 & 1 & 0 & 50{,}000 \\
0 & 1 & ① & 0 & 0 & -1 & 0 & 50{,}000 \\
\hline
0 & 0 & -.04 & .07 & 0 & .01 & 1 & 6500
\end{array}
\right]
$$

$$
\begin{array}{c}
\\
\\
\\
\\
.04R_3 + R_4 \rightarrow R_4
\end{array}
\begin{array}{ccccccc}
x_1 & x_2 & x_3 & x_4 & x_5 & x_6 & z \\
\end{array}
\left[
\begin{array}{ccccccc|c}
0 & 0 & 0 & 1 & 1 & ① & 0 & 10{,}000 \\
1 & 0 & 0 & 1 & 0 & 1 & 0 & 50{,}000 \\
0 & 1 & 1 & 0 & 0 & -1 & 0 & 50{,}000 \\
\hline
0 & .04 & 0 & .07 & 0 & -.03 & 1 & 8500
\end{array}
\right]
$$

$$
\begin{array}{c}
\\
\\
-R_1 + R_2 \rightarrow R_2 \\
R_1 + R_3 \rightarrow R_3 \\
.03R_1 + R_4 \rightarrow R_4
\end{array}
\begin{array}{ccccccc}
x_1 & x_2 & x_3 & x_4 & x_5 & x_6 & z \\
\end{array}
\left[
\begin{array}{ccccccc|c}
0 & 0 & 0 & 1 & 1 & 1 & 0 & 10{,}000 \\
1 & 0 & 0 & 0 & -1 & 0 & 0 & 40{,}000 \\
0 & 1 & 1 & 1 & 1 & 0 & 0 & 60{,}000 \\
\hline
0 & .04 & 0 & .1 & .03 & 0 & 1 & 8800
\end{array}
\right]
$$

This is a final tableau. The maximum value is 8800 when $x_1 = 40{,}000$, $x_2 = 0$, $x_3 = 60{,}000$, $x_4 = 0$, $x_5 = 0$, and $x_6 = 10{,}000$.

$40,000 should be invested in government securities and $60,000 in mutual funds for a maximum annual interest of $8800.

33. Let y_1 = the number of #1 pills

and y_2 = the number of #2 pills to be purchased.

Organize the given information in a table.

	Vitamin A	Vitamin B_1	Vitamin C	Cost
#1	8	1	2	$.10
#2	2	1	7	$.20
Total Needed	16	5	20	

The problem is to minimize $w = .1y_1 + .2y_2$

subject to:
$$8y_1 + 2y_2 \geq 16$$
$$y_1 + y_2 \geq 5$$
$$2y_1 + 7y_2 \geq 20$$

with $y_1 \geq 0$, $y_2 \geq 0$.

Let $z = -w$, and maximize z. The initial tableau is as follows.

$$
\begin{array}{cccccc|c}
y_1 & y_2 & y_3 & y_4 & y_5 & z & \\
\hline
⑧ & 2 & -1 & 0 & 0 & 0 & 16 \\
1 & 1 & 0 & -1 & 0 & 0 & 5 \\
2 & 7 & 0 & 0 & -1 & 0 & 20 \\
\hline
.1 & .2 & 0 & 0 & 0 & 1 & 0
\end{array}
$$

Continue by pivoting on the circled entry in each tableau.

$$
\begin{array}{l}
 \\
8R_2 - R_1 \rightarrow R_2 \\
4R_3 - R_1 \rightarrow R_3 \\
 \\
80R_4 - R_1 \rightarrow R_4
\end{array}
\begin{array}{cccccc|c}
y_1 & y_2 & y_3 & y_4 & y_5 & z & \\
\hline
8 & 2 & -1 & 0 & 0 & 0 & 16 \\
0 & 6 & 1 & -8 & 0 & 0 & 24 \\
0 & ㉖ & 1 & 0 & -4 & 0 & 64 \\
\hline
0 & 14 & 1 & 0 & 0 & 80 & -16
\end{array}
$$

$$
\begin{array}{l}
13R_1 - R_3 \rightarrow R_1 \\
13R_2 - 3R_3 \rightarrow R_2 \\
 \\
-7R_3 + 13R_1 \rightarrow R_4
\end{array}
\begin{array}{cccccc|c}
y_1 & y_2 & y_3 & y_4 & y_5 & z & \\
\hline
104 & 0 & -14 & 0 & 4 & 0 & 144 \\
0 & 0 & ⑩ & -104 & 12 & 0 & 120 \\
0 & 26 & 1 & 0 & -4 & 0 & 64 \\
\hline
0 & 0 & 6 & 0 & 28 & 1040 & -656
\end{array}
$$

$$
\begin{array}{l}
5R_1 + 7R_2 \rightarrow R_1 \\
 \\
10R_3 - R_2 \rightarrow R_3 \\
-3R_3 + 5R_4 \rightarrow R_4
\end{array}
\begin{array}{cccccc|c}
y_1 & y_2 & y_3 & y_4 & y_5 & z & \\
\hline
520 & 0 & 0 & -728 & 144 & 0 & 1560 \\
0 & 0 & 10 & -104 & 12 & 0 & 120 \\
0 & 260 & 0 & 104 & 52 & 0 & 520 \\
\hline
0 & 0 & 0 & 312 & 104 & 5200 & 3640
\end{array}
$$

Create a 1 in the columns corresponding to y_1, y_2, y_3, and z.

$$
\begin{array}{c}
\frac{1}{520}R_1 \to R_1 \\
\frac{1}{10}R_2 \to R_2 \\
\frac{1}{260}R_3 \to R_3 \\
\frac{1}{5200}R_4 \to R_4
\end{array}
\begin{array}{ccccccc}
y_1 & y_2 & y_3 & y_4 & y_5 & z & \\
\end{array}
\left[
\begin{array}{cccccc|c}
1 & 0 & 0 & -1.4 & .28 & 0 & 3 \\
0 & 0 & 1 & -10.4 & 1.2 & 0 & 12 \\
0 & 1 & 0 & .4 & .2 & 0 & 2 \\
\hline
0 & 0 & 0 & .06 & .02 & 1 & -.7
\end{array}
\right]
$$

This is a final tableau. The minimum value is .7 when y_1 = 3, y_2 = 2, y_3 = 12, y_4 = 0, and y_5 = 0.

Mark should buy 3 of pill #1 and 2 of pill #2 for a minimum cost of 70¢.

35. Let y_1 = the number of units of ingredient I;
     y_2 = the number of units of ingredient II; and
     y_3 = the number of units of ingredient III.

The problem is to minimize w = $4y_1$ + $7y_2$ + $5y_3$

$$
\begin{aligned}
\text{subject to: }\quad 4y_1 + \ \ y_2 + 10y_3 &\ge 10 \\
3y_1 + 2y_2 + \ \ y_3 &\ge 12 \\
4y_2 + \ \ 5y_3 &\ge 20
\end{aligned}
$$

with $y_1 \ge 0$, $y_2 \ge 0$, $y_3 \ge 0$.

Let z = -w, and maximize z.   The initial tableau is as follows.

$$
\begin{array}{ccccccc}
y_1 & y_2 & y_3 & y_4 & y_5 & y_6 & z \\
\end{array}
\left[
\begin{array}{ccccccc|c}
④ & 1 & 10 & -1 & 0 & 0 & 0 & 10 \\
3 & 2 & 1 & 0 & -1 & 0 & 0 & 12 \\
0 & 4 & 5 & 0 & 0 & -1 & 0 & 20 \\
\hline
4 & 7 & 5 & 0 & 0 & 0 & 1 & 0
\end{array}
\right]
$$

Continue by pivoting on the circled entry in each tableau.

$$
\begin{array}{c}
\\
-3R_1 + 4R_2 \to R_2 \\
\\
-R_1 + R_4 \to R_4
\end{array}
\begin{array}{ccccccc}
y_1 & y_2 & y_3 & y_4 & y_5 & y_6 & z \\
\end{array}
\left[
\begin{array}{ccccccc|c}
4 & 1 & 10 & -1 & 0 & 0 & 0 & 10 \\
0 & ⑤ & -26 & 3 & -4 & 0 & 0 & 18 \\
0 & 4 & 5 & 0 & 0 & -1 & 0 & 20 \\
\hline
0 & 6 & -5 & 1 & 0 & 0 & 1 & -10
\end{array}
\right]
$$

	y_1	y_2	y_3	y_4	y_5	y_6	z	
$5R_1 - R_2 \to R_1$	20	0	76	−8	4	0	0	32
	0	5	−26	3	−4	0	0	18
$-4R_2 + 5R_3 \to R_3$	0	0	129	−12	(16)	−5	0	28
$-6R_2 + 5R_4 \to R_4$	0	0	131	−13	24	0	5	−158

	y_1	y_2	y_3	y_4	y_5	y_6	z	
$4R_1 - R_3 \to R_1$	80	0	175	−20	0	5	0	100
$4R_2 + R_3 \to R_2$	0	20	25	0	0	−5	0	100
	0	0	(129)	−12	16	−5	0	28
$-3R_3 + 2R_4 \to R_4$	0	0	−125	10	0	15	10	−400

	y_1	y_2	y_3	y_4	y_5	y_6	z	
$129R_1 - 175R_3 \to R_1$	10,320	0	0	−480	−2800	1520	0	8000
$129R_2 - 25R_3 \to R_2$	0	2580	0	(300)	−400	−520	0	12,200
	0	0	129	−12	16	−5	0	28
$125R_3 + 129R_4 \to R_4$	0	0	0	−210	2000	259,310	1290	−48,100

	y_1	y_2	y_3	y_4	y_5	y_6	z	
$5R_1 + 8R_2 \to R_1$	51,600	20,640	0	0	−17,200	3440	0	137,600
	0	2580	0	300	−400	−520	0	12,200
$R_2 + 25R_3 \to R_3$	0	2580	3225	0	0	−645	0	12,900
$7R_2 + 10R_4 \to R_4$	0	18,060	0	0	17,200	2,589,460	12,290	−395,600

Create a 1 in the columns corresponding to y_1, y_3, y_4, and z.

	y_1	y_2	y_3	y_4	y_5	y_6	z	
$\frac{1}{51,600}R_1 \to R_1$	1	.4	0	0	−.33	.067	0	$2\frac{2}{3}$
$\frac{1}{300}R_2 \to R_2$	0	8.6	0	1	−1.33	−1.733	0	$40\frac{2}{3}$
$\frac{1}{3225}R_3 \to R_3$	0	.8	1	0	0	−.2	0	4
$\frac{1}{12,900}R_4 \to R_4$	0	1.4	0	0	1.33	201	1	$-30\frac{2}{3}$

This is a final tableau. The minimum value is $30\frac{2}{3}$ when $y_1 = 2\frac{2}{3}$, $y_2 = 0$, $y_3 = 4$, $y_4 = 40\frac{2}{3}$, $y_5 = 0$, and $y_6 = 0$.

The biologist can meet his needs at a minimum cost of \$30.67 by using $2\frac{2}{3}$ units of ingredient I and 4 units of ingredient III. (Ingredient II should not be used at all.)

37. Let y_1 = the number of ounces of ingredient I;

y_2 = the number of ounces of ingredient II; and

y_3 = the number of ounces of ingredient III.

The problem is to

minimize $w = .30y_1 + .09y_2 + .27y_3$

subject to: $y_1 + y_2 + y_3 \geq 10$

$\qquad\qquad y_1 + y_2 + y_3 \leq 15$

$\qquad\qquad y_1 \geq \frac{1}{4}y_2$

$\qquad\qquad y_3 \geq y_1$

with $\qquad y_1 \geq 0,\ y_2 \geq 0,\ y_3 \geq 0.$

Use a computer to find that the

minimum is $w = 1.55$ when $y_1 = 1\frac{2}{3}$,

$y_2 = 6\frac{2}{3}$, and $y_3 = 1\frac{2}{3}$.

The additive should consist of $1\frac{2}{3}$ oz

of ingredient I, $6\frac{2}{3}$ oz of ingredient

II, and $1\frac{2}{3}$ oz of ingredient III, for

a minimum cost of \$1.55. The amount

of additive that should be used per

gallon of gasoline is $1\frac{2}{3} + 6\frac{2}{3} + 1\frac{2}{3} =$

10 oz.

Section 4.4

1. To find the transpose of a matrix, the rows of the original matrix are written as the columns of the transpose.

The transpose of $\begin{bmatrix} 1 & 2 & 3 \\ 3 & 2 & 1 \\ 1 & 10 & 0 \end{bmatrix}$

is $\begin{bmatrix} 1 & 3 & 1 \\ 2 & 2 & 10 \\ 3 & 1 & 0 \end{bmatrix}$.

3. The transpose of $\begin{bmatrix} -1 & 4 & 6 & 12 \\ 13 & 25 & 0 & 4 \\ -2 & -1 & 11 & 3 \end{bmatrix}$

is $\begin{bmatrix} -1 & 13 & -2 \\ 4 & 25 & -1 \\ 6 & 0 & 11 \\ 12 & 4 & 3 \end{bmatrix}$.

5. Maximize $\qquad z = 4x_1 + 3x_2 + 2x_3$

subject to: $x_1 + x_2 + x_3 \leq 5$

$\qquad\qquad x_1 + x_2 \qquad \leq 4$

$\qquad\qquad 2x_1 + x_2 + 3x_3 \leq 15$

with $\qquad\qquad x_1 \geq 0,\ x_2 \geq 0,\ x_3 \geq 0.$

Begin by writing the augmented matrix for the given problem.

$$\left[\begin{array}{ccc|c} 1 & 1 & 1 & 5 \\ 1 & 1 & 0 & 4 \\ 2 & 1 & 3 & 15 \\ \hline 4 & 3 & 2 & 0 \end{array}\right]$$

Find the transpose of this matrix.

$$\left[\begin{array}{ccc|c} 1 & 1 & 2 & 4 \\ 1 & 1 & 1 & 3 \\ 1 & 0 & 3 & 2 \\ \hline 5 & 4 & 15 & 0 \end{array}\right]$$

The dual problem is stated from this second matrix as follows (using y instead of x).

Minimize $w = 5y_1 + 4y_2 + 15y_3$

subject to: $y_1 + y_2 + 2y_3 \geq 4$

$\qquad\qquad y_1 + y_2 + y_3 \geq 3$

$\qquad\qquad y_1 \qquad + 3y_3 \geq 2$

with $\qquad y_1 \geq 0,\ y_2 \geq 0,\ y_3 \geq 0.$

7. Minimize $\qquad w = y_1 + 2y_2 + y_3 + 5y_4$

subject to: $y_1 + y_2 + y_3 + y_4 \geq 50$

$\qquad\qquad 3y_1 + y_2 + 2y_3 + y_4 \geq 100$

with $y_1 \geq 0,\ y_2 \geq 0,\ y_3 \geq 0,\ y_4 \geq 0.$

Write the augmented matrix for the given problem.

$$\left[\begin{array}{cccc|c} 1 & 1 & 1 & 1 & 50 \\ 3 & 1 & 2 & 1 & 100 \\ \hline 1 & 2 & 1 & 5 & 0 \end{array}\right]$$

Find the transpose of this matrix.

$$\left[\begin{array}{cc|c} 1 & 3 & 1 \\ 1 & 1 & 2 \\ 1 & 2 & 1 \\ 1 & 1 & 5 \\ \hline 50 & 100 & 0 \end{array}\right]$$

The dual problem is stated from this second matrix as follows (using x instead of y).

Maximize $z = 50x_1 + 100x_2$

subject to: $x_1 + 3x_2 \leq 1$

$x_1 + x_2 \leq 2$

$x_1 + 2x_2 \leq 1$

$x_1 + x_2 \leq 5$

with $x_1 \geq 0,\ x_2 \geq 0.$

9. Find $y_1 \geq 0$ and $y_2 \geq 0$ such that

$$2y_1 + 3y_2 \geq 6$$
$$2y_1 + y_2 \geq 7$$

and $w = 5y_1 + 2y_2$ is minimized.

Write the augmented matrix for this problem.

$$\left[\begin{array}{cc|c} 2 & 3 & 6 \\ 2 & 1 & 7 \\ \hline 5 & 2 & 0 \end{array}\right]$$

Find the transpose of this matrix.

$$\left[\begin{array}{cc|c} 2 & 2 & 5 \\ 3 & 1 & 2 \\ \hline 6 & 7 & 0 \end{array}\right]$$

Use this matrix to write the dual problem.

Find $x_1 \geq 0$ and $x_2 \geq 0$ such that

$$2x_1 + 2x_2 \leq 5$$
$$3x_1 + x_2 \leq 2$$

and $z = 6x_1 + 7x_2$ is maximized. Introduce slack variables x_3 and x_4. The initial tableau is as follows.

x_1	x_2	x_3	x_4	z	
2	2	1	0	0	5
3	①	0	1	0	2
-6	-7	0	0	1	0

Pivot on the 1 in the second row, second column, since that column has the most negative indicator and that row has the smallest nonnegative quotient.

	x_1	x_2	x_3	x_4	z	
$-2R_2 + R_1 \rightarrow R_1$	-4	0	1	-2	0	1
	3	1	0	1	0	2
$7R_2 + R_3 \rightarrow R_3$	15	0	0	7	1	14

The minimum value of w is the same as the maximum value of z. The minimum value of w is 14 when $y_1 = 0$ and $y_2 = 7$.

(Note that the values of y_1 and y_2 are given by the entries in the bottom row of the columns corresponding to the slack variables in the final tableau.)

11. Find $y_1 \geq 0$ and $y_2 \geq 0$ such that

$$10y_1 + 5y_2 \geq 100$$
$$20y_1 + 10y_2 \geq 150$$

and $w = 4y_1 + 5y_2$ is minimized.

The dual problem is as follows.
Find $x_1 \geq 0$ and $x_2 \geq 0$ such that

$$10x_1 + 20x_2 \leq 4$$
$$5x_1 + 10x_2 \leq 5$$

and $z = 100x_1 + 150x_2$ is maximized.
The initial simplex tableau is as
follows.

$$
\begin{array}{ccccc}
x_1 & x_2 & x_3 & x_4 & z \\
\end{array}
$$

$$
\left[
\begin{array}{ccccc|c}
10 & ⟨20⟩ & 1 & 0 & 0 & 4 \\
5 & 10 & 0 & 1 & 0 & 5 \\
\hline
-100 & -150 & 0 & 0 & 1 & 0 \\
\end{array}
\right]
$$

Pivot on the 20 in the first row,
second column.

$$
\begin{array}{ccccc}
x_1 & x_2 & x_3 & x_4 & z \\
\end{array}
$$

$$
\begin{array}{r}
\\
2R_2 - R_1 \rightarrow R_2 \\
\\
15R_1 + 2R_3 \rightarrow R_3
\end{array}
\left[
\begin{array}{ccccc|c}
⟨10⟩ & 20 & 1 & 0 & 0 & 4 \\
0 & 0 & -1 & 2 & 0 & 6 \\
\hline
-50 & 0 & 15 & 0 & 2 & 60 \\
\end{array}
\right]
$$

Pivot on the 10 in the first row,
first column.

$$
\begin{array}{ccccc}
x_1 & x_2 & x_3 & x_4 & z \\
\end{array}
$$

$$
\begin{array}{r}
\\
\\
5R_1 + R_3 \rightarrow R_3
\end{array}
\left[
\begin{array}{ccccc|c}
10 & 20 & 1 & 0 & 0 & 4 \\
0 & 0 & -1 & 2 & 0 & 6 \\
\hline
0 & 100 & 20 & 0 & 2 & 80 \\
\end{array}
\right]
$$

Create a 1 in the columns corresponding to x_1, x_4, and z.

$$
\begin{array}{ccccc}
x_1 & x_2 & x_3 & x_4 & z \\
\end{array}
$$

$$
\begin{array}{r}
\frac{1}{10}R_1 \rightarrow R_1 \\[4pt]
\frac{1}{2}R_2 \rightarrow R_2 \\[4pt]
\frac{1}{2}R_3 \rightarrow R_3
\end{array}
\left[
\begin{array}{ccccc|c}
1 & 2 & \frac{1}{10} & 0 & 0 & \frac{2}{5} \\
0 & 0 & -\frac{1}{2} & 1 & 0 & 3 \\
\hline
0 & 50 & 10 & 0 & 1 & 40 \\
\end{array}
\right]
$$

The minimum value of w is 40 when
$y_1 = 10$ and $y_2 = 0$. (These values
of y_1 and y_2 are read from the last

row of the columns corresponding to
x_3 and x_4 in the final tableau.)

13. Minimize $w = 2y_1 + y_2 + 3y_3$
     subject to: $y_1 + y_2 + y_3 \geq 100$
     $\qquad\qquad\quad 2y_1 + y_2 \qquad \geq 50$
     with $\quad y_1 \geq 0, \; y_2 \geq 0, \; y_3 \geq 0.$

The dual problem is as follows.
Maximize $z = 100x_1 + 50x_2$
subject to: $x_1 + 2x_2 \leq 2$
$\qquad\qquad\quad x_1 + \;\; x_2 \leq 1$
$\qquad\qquad\quad x_1 \qquad\quad \leq 3$
with $\quad x_1 \geq 0, \; x_2 \geq 0, \; x_3 \geq 0.$

The initial simplex tableau is as
follows.

$$
\begin{array}{cccccc}
x_1 & x_2 & x_3 & x_4 & x_5 & z \\
\end{array}
$$

$$
\left[
\begin{array}{cccccc|c}
1 & 2 & 1 & 0 & 0 & 0 & 2 \\
⟨1⟩ & 1 & 0 & 1 & 0 & 0 & 1 \\
1 & 0 & 0 & 0 & 1 & 0 & 3 \\
\hline
-100 & -50 & 0 & 0 & 0 & 1 & 0 \\
\end{array}
\right]
$$

Pivot on the 1 in the second row,
first column.

$$
\begin{array}{cccccc}
x_1 & x_2 & x_3 & x_4 & x_5 & z \\
\end{array}
$$

$$
\begin{array}{r}
-R_2 + R_1 \rightarrow R_1 \\
\\
-R_2 + R_3 \rightarrow R_3 \\
100R_2 + R_4 \rightarrow R_4
\end{array}
\left[
\begin{array}{cccccc|c}
0 & 1 & 1 & -1 & 0 & 0 & 1 \\
1 & 1 & 0 & 1 & 0 & 0 & 1 \\
0 & -1 & 0 & -1 & 1 & 0 & 2 \\
\hline
0 & 50 & 0 & 100 & 0 & 1 & 100 \\
\end{array}
\right]
$$

The minimum value of w is 100 when
$y_1 = 0$, $y_2 = 100$, and $y_3 = 0$.

15. Minimize $z = x_1 + 2x_2$
     subject to: $-2x_1 + \;\; x_2 \geq 1$
     $\qquad\qquad\quad x_1 - 2x_2 \geq 1$
     with $\quad x_1 \geq 0, \; x_2 \geq 0.$

A quick sketch of the constraints $-2x_1 + x_2 \geq 1$ and $x_1 - 2x_2 \geq 1$ will verify that the two corresponding half planes do not overlap in the first quadrant of the x_1x_2-plane. Therefore, this problem (P) has no feasible solutions. The dual of the given problem is as follows.

Maximize $w = y_1 + y_2$
subject to: $-2y_1 + y_2 \leq 1$
$\qquad\qquad\quad y_1 - 2y_2 \leq 2$
with $\qquad\quad y_1 \geq 0, \ y_2 \geq 0.$

A quick sketch here will verify that there is a feasible region in the y_1y_2-plane, and it is unbounded. Therefore, there is no maximum value of w in this problem (D). (P) has no feasible solutions and the objective function of (D) is unbounded is choice (a).

17. Maximize $x_1 + 1.5x_2 = z$
subject to: $x_1 + 2x_2 \leq 200$
$\qquad\qquad\quad 4x_1 + 3x_2 \leq 600$
$\qquad\qquad\qquad\qquad x_2 \leq 90$
with $\qquad\quad x_1 \geq 0, \ x_2 \geq 0.$

The final simplex tableau of this problem is as follows.

$$
\begin{array}{cccccc}
x_1 & x_2 & x_3 & x_4 & x_5 & z \\
\left[\begin{array}{cccccc|c}
0 & 1 & .8 & -.2 & 0 & 0 & 40 \\
1 & 0 & -.6 & .4 & 0 & 0 & 120 \\
0 & 0 & -.8 & .2 & 1 & 0 & 50 \\
\hline
0 & 0 & .6 & .1 & 0 & 1 & 180
\end{array}\right]
\end{array}
$$

(a) The corresponding dual problem is as follows.

Minimize $w = 200y_1 + 600y_2 + 90y_3$
subject to:

$$y_1 + 4y_2 \qquad \geq \ 1$$
$$2y_1 + 3y_2 + y_3 \geq 1.5$$

with $y_1 \geq 0, \ y_2 \geq 0, \ y_3 \geq 0.$

(b) From the given final tableau, the optimal solution to the dual problem is $y_1 = .6$, $y_2 = .1$, $y_3 = 0$, and $w = 180$.

(c) The shadow value for felt is .6; an increase in supply of 10 units of felt will increase profit to $180 + .6(10) = \$186$.

(d) The shadow values are .1 for stuffing and 0 for trim. If stuffing and trim are each decreased by 10 units, the profit will be $180 - .1(10) - 0(10) = \$179$.

19. Let $y_1 =$ the number of packages of Sun Hill and

$\qquad y_2 =$ the number of packages of Bear Valley.

The problem is to
minimize $w = 3y_1 + 2y_2$
subject to: $10y_1 + 2y_2 \geq 20$
$\qquad\qquad\quad 4y_1 + 4y_2 \geq 24$
$\qquad\qquad\quad 2y_1 + 8y_2 \geq 24$
with $\qquad\quad y_1 \geq 0, \ y_2 \geq 0.$

The constraints can be simplified and the problem restated as follows.

Minimize $w = 3y_1 + 2y_2$
subject to: $5y_1 + y_2 \geq 10$
$\qquad\qquad\quad y_1 + y_2 \geq 6$
$\qquad\qquad\quad y_1 + 4y_2 \geq 12$
with $\qquad\quad y_1 \geq 0, \ y_2 \geq 0.$

The dual problem is as follows.

Maximize $z = 10x_1 + 6x_2 + 12x_3$

subject to: $5x_1 + x_2 + x_3 \leq 3$

$\qquad\qquad x_1 + x_2 + 4x_3 \leq 2$

with $\qquad\quad x_1 \geq 0,\ x_2 \geq 0.$

The initial simplex tableau is as follows.

$$
\begin{array}{cccccc}
x_1 & x_2 & x_3 & x_4 & x_5 & z \\
\end{array}
$$

$$
\left[\begin{array}{ccccccc|c}
5 & 1 & 1 & 1 & 0 & 0 & 3 \\
1 & 1 & ④ & 0 & 1 & 0 & 2 \\
\hline
-10 & -6 & -12 & 0 & 0 & 1 & 0
\end{array}\right]
$$

Continue by pivoting on the circled entry in each tableau.

$$
\begin{array}{cccccc}
x_1 & x_2 & x_3 & x_4 & x_5 & z
\end{array}
$$

$4R_1 - R_2 \to R_1$
$$
\left[\begin{array}{cccccc|c}
⑲ & 3 & 0 & 4 & -1 & 0 & 10 \\
1 & 1 & 4 & 0 & 1 & 0 & 2 \\
\hline
\end{array}\right.
$$
$3R_2 + R_3 \to R_3 \quad \left.\begin{array}{cccccc|c} -7 & -3 & 0 & 0 & 3 & 1 & 6 \end{array}\right]$

$$
\begin{array}{cccccc}
x_1 & x_2 & x_3 & x_4 & x_5 & z
\end{array}
$$

$$
\left[\begin{array}{cccccc|c}
19 & 3 & 0 & 4 & -1 & 0 & 10 \\
\end{array}\right.
$$
$19R_2 - R_1 \to R_2 \quad \left[\begin{array}{cccccc|c} 0 & ⑯ & 76 & -4 & 20 & 0 & 28 \end{array}\right.$

$7R_1 + 19R_3 \to R_3 \quad \left[\begin{array}{cccccc|c} 0 & -36 & 0 & 28 & 50 & 19 & 184 \end{array}\right]$

$$
\begin{array}{cccccc}
x_1 & x_2 & x_3 & x_4 & x_5 & z
\end{array}
$$

$16R_1 - 3R_2 \to R_1 \quad \left[\begin{array}{cccccc|c} 304 & 0 & -228 & 76 & -76 & 0 & 76 \end{array}\right.$

$\left[\begin{array}{cccccc|c} 0 & 16 & 76 & -4 & 20 & 0 & 28 \end{array}\right.$

$9R_2 - 4R_3 \to R_3 \quad \left[\begin{array}{cccccc|c} 0 & 0 & 684 & 76 & 380 & 76 & 988 \end{array}\right]$

We are finished pivoting. Create a 1 in the columns corresponding to x_1, x_3, and z.

$$
\begin{array}{cccccc}
x_1 & x_2 & x_3 & x_4 & x_5 & z
\end{array}
$$

$\frac{1}{304}R_1 \to R_1 \quad \left[\begin{array}{cccccc|c} 1 & 0 & -\frac{3}{4} & \frac{1}{4} & -\frac{1}{4} & 0 & \frac{1}{4} \end{array}\right.$

$\frac{1}{16}R_2 \to R_2 \quad \left[\begin{array}{cccccc|c} 0 & 1 & \frac{19}{4} & -\frac{1}{4} & \frac{5}{4} & 0 & \frac{7}{4} \end{array}\right.$

$\frac{1}{76}R_3 \to R_3 \quad \left[\begin{array}{cccccc|c} 0 & 0 & 9 & 1 & 5 & 1 & 13 \end{array}\right]$

The minimum value of w is 13 when $y_1 = 1$ and $y_2 = 5$.

You should buy 1 package of Sun Hill and 5 packages of Bear Valley, for a minimum cost of $13.

21. Organize the information in a table.

	Units of Nutrient A (per bag)	Units of Nutrient B (per bag)	Cost (per bag)
Feed 1	1	2	$3
Feed 2	3	1	$2
Minimum	7	4	

Let y_1 = the number of bags of feed 1 and y_2 = the number of bags of feed 2.

(a) We want the cost to equal $7 for 7 units of A and 4 units of B exactly. Therefore, use a system of equations rather than a system of inequalities.

$$3y_1 + 2y_2 = 7$$
$$y_1 + 3y_2 = 7$$
$$2y_1 + y_2 = 4$$

Use Gauss–Jordan elimination to solve this system of equations.

$$
\left[\begin{array}{cc|c}
3 & 2 & 7 \\
1 & 3 & 7 \\
2 & 1 & 4
\end{array}\right]
$$

$3R_2 - R_1 \to R_2$
$-2R_1 + 3R_3 \to R_3$
$$
\left[\begin{array}{cc|c}
3 & 2 & 7 \\
0 & 7 & 14 \\
0 & -1 & -2
\end{array}\right]
$$

$7R_1 - 2R_2 \to R_1$
$R_2 + 7R_3 \to R_3$
$$
\left[\begin{array}{cc|c}
21 & 0 & 21 \\
0 & 7 & 14 \\
0 & 0 & 0
\end{array}\right]
$$

$\frac{1}{21}R_1 \to R_1$
$\frac{1}{7}R_2 \to R_2$
$$
\left[\begin{array}{cc|c}
1 & 0 & 1 \\
0 & 1 & 2 \\
0 & 0 & 0
\end{array}\right]
$$

Thus, $y_1 = 1$ and $y_2 = 2$, so use 1 bag of feed 1 and 2 bags of feed 2. The cost will be $3(1) + 2(2) = \$7$ as desired. The number of units of A is $1(1) + 3(2) = 7$ and the number of units of B is $2(1) + 1(2) = 4$.

(b)

	Units of Nutrient A (per bag)	Units of Nutrient B (per bag)	Cost (per bag)
Feed 1	1	2	\$3
Feed 2	3	1	\$2
Minimum	5	4	

The problem is to

minimize $w = 3y_1 + 2y_2$

subject to: $y_1 + 3y_2 \geq 5$

$\qquad\qquad 2y_1 + y_2 \geq 4$

with $\qquad y_1 \geq 0, y_2 \geq 0.$

The dual problem is as follows.

Maximize $z = 5x_1 + 4x_2$

subject to: $x_1 + 2x_2 \leq 3$

$\qquad\qquad 3x_1 + x_2 \leq 2$

with $\qquad x_1 \geq 0, x_2 \geq 0.$

The initial simplex tableau is as follows.

$$\begin{array}{ccccc} x_1 & x_2 & x_3 & x_4 & z \end{array}$$
$$\begin{bmatrix} 1 & 2 & 1 & 0 & 0 & | & 3 \\ ③ & 1 & 0 & 1 & 0 & | & 2 \\ \hline -5 & -4 & 0 & 0 & 1 & | & 0 \end{bmatrix}$$

Pivot as indicated.

$$\begin{array}{cccccc} & x_1 & x_2 & x_3 & x_4 & z \end{array}$$
$$\begin{array}{c} 3R_1-R_2 \to R_1 \\ \\ 5R_2+3R_3 \to R_3 \end{array} \begin{bmatrix} 0 & ⑤ & 3 & -1 & 0 & | & 7 \\ 3 & 1 & 0 & 1 & 0 & | & 2 \\ 0 & -7 & 0 & 5 & 3 & | & 10 \end{bmatrix}$$

$$\begin{array}{ccccc} & x_1 & x_2 & x_3 & x_4 & z \end{array}$$
$$\begin{array}{c} 5R_2-R_1 \to R_2 \\ \\ 7R_1+5R_3 \to R_3 \end{array} \begin{bmatrix} 0 & 5 & 3 & -1 & 0 & | & 7 \\ 15 & 0 & -3 & 6 & 0 & | & 3 \\ \hline 0 & 0 & 21 & 18 & 15 & | & 99 \end{bmatrix}$$

$$\begin{array}{ccccc} & x_1 & x_2 & x_3 & x_4 & z \end{array}$$
$$\begin{array}{c} \frac{1}{5}R_1 \to R_1 \\ \frac{1}{15}R_2 \to R_2 \\ \\ \frac{1}{15}R_3 \to R_3 \end{array} \begin{bmatrix} 0 & 1 & \frac{3}{5} & -\frac{1}{5} & 0 & | & \frac{7}{5} \\ 1 & 0 & -\frac{1}{5} & \frac{2}{5} & 0 & | & \frac{1}{5} \\ \hline 0 & 0 & \frac{7}{5} & \frac{6}{5} & 1 & | & \frac{33}{5} \end{bmatrix}$$

Reading from the final column of the final tableau, $x_2 = \$1.40$ is the cost of nutrient B and $x_1 = \$.20$ is the cost of nutrient A. With 5 units of A and 4 units of B, this gives a minimum cost of $5(\$.20) + 4(\$1.40) = \$6.60$ as given in the lower right corner. 1.4 (or 7/5) bags of feed 1 and 1.2 (or 6/5) bags of feed 2 should be used.

23. Let $y_1 =$ the number of ounces of ingredient I;

$\qquad y_2 =$ the number of ounces of ingredient II; and

$\qquad y_3 =$ the number of ounces of ingredient III.

As seen in Exercise 37 in Section 4.3, the problem is to

minimize $w = .30y_1 + .09y_2 + .27y_3$

subject to:

$\qquad\qquad y_1 + y_2 + y_3 \geq 10$

$\qquad\qquad y_1 + y_2 + y_3 \leq 15$

$\qquad\qquad\qquad y_1 \geq \frac{1}{4}y_2$

$\qquad\qquad\qquad y_3 \geq y_1$

with $\qquad y_1 \geq 0, y_2 \geq 0, y_3 \geq 0.$

Two of the constraints need to be rearranged.   The problem can be restated as follows.

Minimize $w = .30y_1 + .09y_2 + .27y_3$

subject to:

$$y_1 + y_2 + y_3 \geq 10$$
$$-y_1 - y_2 - y_3 \geq -15$$
$$4y_1 - y_2 \qquad \geq 0$$
$$-y_1 \qquad + y_3 \geq 0$$

with $y_1 \geq 0, \ y_2 \geq 0, \ y_3 \geq 0.$

The dual problem is as follows.

Maximize $z = 10x_1 - 15x_2$

subject to:

$$x_1 - x_2 + 4x_3 - x_4 \leq .30$$
$$x_1 - x_2 - x_3 \qquad \leq .09$$
$$x_1 - x_2 \qquad - x_4 \leq .27$$

with $x_1 \geq 0, \ x_2 \geq 0, \ x_3 \geq 0, \ x_4 \geq 0.$
The initial simplex tableau is as follows.

x_1	x_2	x_3	x_4	x_5	x_6	x_7	z	
1	-1	4	-1	1	0	0	0	.30
1	-1	-1	0	0	1	0	0	.09
1	-1	0	-1	0	0	1	0	.27
-10	15	0	0	0	0	0	1	0

Use a computer to perform the simplex algorithm and find that the minimum is $w = 1.55$ when $y_1 = \frac{5}{3}$, $y_2 = \frac{20}{3}$, and $y_3 = \frac{5}{3}$. The additive should consist of $1\frac{2}{3}$ oz of ingredient I, $6\frac{2}{3}$ oz ingredient II, and $1\frac{2}{3}$ oz of ingredient III for a minimum cost of $1.55.   The amount of

additive that should be used per gallon of gasoline is

$$\frac{5}{3} + \frac{20}{3} + \frac{5}{3} = 10 \text{ oz.}$$

Chapter 4   Review Exercises

3.   Maximize $z = 5x_1 + 3x_2$

subject to: $2x_1 + 5x_2 \leq 50$
$$x_1 + 3x_2 \leq 25$$
$$4x_1 + x_2 \leq 18$$
$$x_1 + x_2 \leq 12$$
with $x_1 \geq 0, \ x_2 \geq 0.$

(a) Adding the slack variables x_3, x_4, x_5, and x_6, we obtain the following equations.

$$2x_1 + 5x_2 + x_3 \qquad\qquad = 50$$
$$x_1 + 3x_2 \qquad + x_4 \qquad\qquad = 25$$
$$4x_1 + x_2 \qquad\qquad + x_5 \qquad = 18$$
$$x_1 + x_2 \qquad\qquad\qquad + x_6 = 12$$

(b) The initial tableau is

x_1	x_2	x_3	x_4	x_5	x_6	z	
2	5	1	0	0	0	0	50
1	3	0	1	0	0	0	25
4	1	0	0	1	0	0	18
1	1	0	0	0	1	0	12
-5	-3	0	0	0	0	1	0

5.   Maximize $z = 5x_1 + 8x_2 + 6x_3$

subject to: $x_1 + x_2 + x_3 \leq 90$
$$2x_1 + 5x_2 + x_3 \leq 120$$
$$x_1 + 3x_2 \geq 80$$
with $x_1 \geq 0, \ x_2 \geq 0, \ x_3 \geq 0.$

(a) Adding the slack variables x_4 and x_5 and subtracting the surplus variable x_6, we obtain the following equations.

$$x_1 + x_2 + x_3 + x_4 \qquad\qquad = 90$$
$$2x_1 + 5x_2 + x_3 \qquad + x_5 \qquad = 120$$
$$x_1 + 3x_2 \qquad\qquad\qquad - x_6 = 80$$

(b) The initial tableau is

	x_1	x_2	x_3	x_4	x_5	x_6	z	
	1	1	1	1	0	0	0	90
	2	5	1	0	1	0	0	120
	1	3	0	0	0	-1	0	80
	-5	-8	-6	0	0	0	1	0

7.

	x_1	x_2	x_3	x_4	x_5	z	
	1	2	3	1	0	0	28
	②	4	1	0	1	0	32
	-5	-2	-3	0	0	1	0

The most negative entry in the last row is -5 and the smaller of the two quotients is $32/2 = 16$. Hence, the 2 in the second row, first column, is the first pivot. Performing row transformations leads to the following tableau.

	x_1	x_2	x_3	x_4	x_5	z	
$2R_1-R_2\to R_1$	0	0	⑤	2	-1	0	24
	2	4	1	0	1	0	32
$5R_2+2R_3\to R_3$	0	16	-1	0	5	2	160

Pivot on the 5 in the first row, third column.

	x_1	x_2	x_3	x_4	x_5	z	
$5R_2-R_1\to R_2$	0	0	5	2	-1	0	24
	10	20	0	-2	6	0	136
$5R_3+R_1\to R_3$	0	80	0	2	24	10	824

Create a 1 in the columns corresponding to x_1, x_3, and z.

	x_1	x_2	x_3	x_4	x_5	z	
$\frac{1}{5}R_1\to R_1$	0	0	1	.4	-.2	0	4.8
$\frac{1}{10}R_2\to R_2$	1	2	0	-.2	.6	0	13.6
$\frac{1}{10}R_3\to R_3$	0	8	0	.2	2.4	1	82.4

The maximum value is 82.4 when $x_1 = 13.6$, $x_2 = 0$, $x_3 = 4.8$, $x_4 = 0$, and $x_5 = 0$.

9.

	x_1	x_2	x_3	x_4	x_5	x_6	z	
	1	2	2	1	0	0	0	50
	③	1	0	0	1	0	0	20
	1	0	2	0	0	-1	0	15
	-5	-3	-2	0	0	0	1	0

The initial basic solution is not feasible since $x_6 = -15$. In the third row, where the negative coefficient appears, the nonnegative entry that appears farthest to the left is the 1 in the first column; in the first column, the smallest nonnegative quotient is $\frac{20}{3}$. Pivot on the 3 in the second row, first column.

$$
\begin{array}{c}
\\
\\
3R_1-R_2\rightarrow R_1 \\
\\
3R_3-R_2\rightarrow R_3 \\
\\
5R_2+3R_4\rightarrow R_4
\end{array}
\begin{array}{cccccccc}
x_1 & x_2 & x_3 & x_4 & x_5 & x_6 & z & \\
\end{array}
\left[\begin{array}{ccccccc|c}
0 & 5 & 6 & 3 & 0 & 0 & 0 & 130 \\
3 & 1 & 0 & 0 & 1 & 0 & 0 & 20 \\
0 & -1 & ⑥ & 0 & 0 & -3 & 0 & 25 \\
\hline
0 & -4 & -6 & 0 & 5 & 0 & 3 & 100
\end{array}\right]
$$

Continue by pivoting on each circled entry.

$$
\begin{array}{c}
\\
R_1-R_3\rightarrow R_1 \\
\\
\\
\\
R_3+R_4\rightarrow R_4
\end{array}
\begin{array}{cccccccc}
x_1 & x_2 & x_3 & x_4 & x_5 & x_6 & z & \\
\end{array}
\left[\begin{array}{ccccccc|c}
0 & ⑥ & 0 & 3 & 0 & 3 & 0 & 105 \\
3 & 1 & 0 & 0 & 1 & 0 & 0 & 20 \\
0 & -1 & 6 & 0 & 0 & -3 & 0 & 25 \\
\hline
0 & -5 & 0 & 0 & 5 & -3 & 3 & 125
\end{array}\right]
$$

The basic solution is now feasible, but there are negative indicators. Continue pivoting.

$$
\begin{array}{c}
\\
6R_2-R_1\rightarrow R_2 \\
6R_3+R_1\rightarrow R_3 \\
\\
5R_1+6R_4\rightarrow R_4
\end{array}
\begin{array}{cccccccc}
x_1 & x_2 & x_3 & x_4 & x_5 & x_6 & z & \\
\end{array}
\left[\begin{array}{ccccccc|c}
0 & 6 & 0 & 3 & 0 & ③ & 0 & 105 \\
18 & 0 & 0 & -3 & 6 & -3 & 0 & 15 \\
0 & 0 & 36 & 3 & 0 & -15 & 0 & 255 \\
\hline
0 & 0 & 0 & 15 & 30 & -3 & 18 & 1275
\end{array}\right]
$$

$$
\begin{array}{c}
\\
R_1+R_2\rightarrow R_2 \\
5R_1+R_3\rightarrow R_3 \\
\\
R_1+R_4\rightarrow R_4
\end{array}
\begin{array}{cccccccc}
x_1 & x_2 & x_3 & x_4 & x_5 & x_6 & z & \\
\end{array}
\left[\begin{array}{ccccccc|c}
0 & 6 & 0 & 3 & 0 & 3 & 0 & 105 \\
18 & 6 & 0 & 0 & 6 & 0 & 0 & 120 \\
0 & 30 & 36 & 18 & 0 & 0 & 0 & 780 \\
\hline
0 & 6 & 0 & 18 & 30 & 0 & 18 & 1380
\end{array}\right]
$$

Create a 1 in the columns corresponding to x_1, x_3, x_6 and z.

$$
\begin{array}{c}
\\
\frac{1}{3}R_1\rightarrow R_1 \\
\\
\frac{1}{18}R_2\rightarrow R_2 \\
\\
\frac{1}{36}R_3\rightarrow R_3 \\
\\
\frac{1}{18}R_4\rightarrow R_4
\end{array}
\begin{array}{cccccccc}
x_1 & x_2 & x_3 & x_4 & x_5 & x_6 & z & \\
\end{array}
\left[\begin{array}{ccccccc|c}
0 & 2 & 0 & 1 & 0 & 1 & 0 & 35 \\
1 & .33 & 0 & 0 & .33 & 0 & 0 & 6.67 \\
0 & .83 & 1 & .5 & 0 & 0 & 0 & 21.67 \\
\hline
0 & .33 & 0 & 1 & 1.67 & 0 & 1 & 76.67
\end{array}\right]
$$

The maximum value is about 76.67 when $x_1 \approx 6.67$, $x_2 = 0$, $x_3 \approx 21.67$, $x_4 = 0$, $x_5 = 0$, and $x_6 = 35$.

11. Minimize $w = 10y_1 + 15y_2$

subject to: $y_1 + y_2 \geq 17$

$\quad\quad\quad\quad 5y_1 + 8y_2 \geq 42$

with $y_1 \geq 0$, $y_2 \geq 0$.

Let $z = -w$, and maximize z. The problem can be restated as follows.
Maximize $z = -10y_1 - 15y_2$
subject to: $y_1 + y_2 \geq 17$

$\quad\quad\quad\quad 5y_1 + 8y_2 \geq 42$

with $y_1 \geq 0$, $y_2 \geq 0$.

13. Minimize $w = 7y_1 + 2y_2 + 3y_3$

subject to: $y_1 + y_2 + 2y_3 \geq 48$

$\quad\quad\quad\quad y_1 + y_2 \quad\quad \geq 12$

$\quad\quad\quad\quad\quad\quad\quad y_3 \geq 10$

$\quad\quad 3y_1 \quad\quad + y_3 \geq 30$

with $y_1 \geq 0$, $y_2 \geq 0$, $y_3 \geq 0$.

Let $z = -w$, and maximize z.
The problem can be restated as follows.
Maximize $z = -7y_1 - 2y_2 - 3y_3$
subject to: $y_1 + y_2 + 2y_3 \geq 48$

$\quad\quad\quad\quad y_1 + y_2 \quad\quad \geq 12$

$\quad\quad\quad\quad\quad\quad\quad y_3 \geq 10$

$\quad\quad 3y_1 \quad\quad + y_3 \geq 30$

with $y_1 \geq 0$, $y_2 \geq 0$, $y_3 \geq 0$.

15.
$$
\begin{array}{ccccccc}
y_1 & y_2 & y_3 & y_4 & y_5 & y_6 & z \\
\end{array}
\left[\begin{array}{ccccccc|c}
0 & 0 & 3 & 0 & 1 & 1 & 0 & 2 \\
1 & 0 & -2 & 0 & 2 & 0 & 0 & 8 \\
0 & 1 & 7 & 0 & 0 & 0 & 0 & 12 \\
0 & 0 & 1 & 1 & -4 & 0 & 0 & 1 \\
\hline
0 & 0 & 5 & 0 & 8 & 0 & 1 & -62
\end{array}\right]
$$

From this final tableau, read that the maximum value of $z = -w$ is -62 when $y_1 = 8$, $y_2 = 12$, $y_3 = 0$, $y_4 = 1$, $y_5 = 0$, and $y_6 = 2$. Therefore the minimum value of w is 62 when $y_1 = 8$, $y_2 = 12$, $y_3 = 0$, $y_4 = 1$, $y_5 = 0$, and $y_6 = 2$.

19.
$$\left[\begin{array}{cccccc|c} 4 & 2 & 3 & 1 & 0 & 0 & 9 \\ 5 & 4 & 1 & 0 & 1 & 0 & 10 \\ \hline -6 & -7 & -5 & 0 & 0 & 1 & 0 \end{array}\right]$$

(a) The 1 in column 4 and the 1 in column 5 indicate that the constraints involve $\leq$. The problem being solved with this tableau is to

maximize $z = 6x_1 + 7x_2 + 5x_3$

subject to: $4x_1 + 2x_2 + 3x_3 \leq 9$
$5x_1 + 4x_2 + x_3 \leq 10$

with $x_1 \geq 0, x_2 \geq 0, x_3 \geq 0.$

(b) If the 1 in row 1, column 4 were -1 rather than 1, then the first constraint would have a surplus variable rather than a slack variable, which means the first constraint would be $4x_1 + 2x_2 + 3x_3 \geq 9$ instead of $4x_1 + 2x_2 + 3x_3 \leq 9$.

(c)
x_1	x_2	x_3	x_4	x_5	z	
3	0	5	2	-1	0	8
11	10	0	-1	3	0	21
47	0	0	13	11	10	227

From this tableau, the solution is

$x_1 = 0$, $x_2 = \dfrac{21}{10} = 2.1$, $x_3 = \dfrac{8}{5} = 1.6$,

and $z = \dfrac{227}{10} = 22.7.$

(d) The dual of the original problem is as folows.

Minimize $w = 9y_1 + 10y_2$
subject to: $4y_1 + 5y_2 \geq 6$
$2y_1 + 4y_2 \geq 7$
$3y_1 + y_2 \geq 5$
with $y_1 \geq 0, y_2 \geq 0.$

(e) From the tableau in part (c), the solution of the dual in part (d) is $y_1 = \dfrac{13}{10} = 1.3$, $y_2 = \dfrac{11}{10} = 1.1$, and

$z = \dfrac{227}{10} = 22.7.$

21. (a) Let x_1 = the amount invested in the oil leases;

x_2 = the amount invested in bonds;

and x_3 = the amount invested in stock.

(b) We want to maximize

$z = .15x_1 + .09x_2 + .05x_3.$

(c) The constraints are as follows.

$x_1 + x_2 + x_3 \leq 50,000$
$x_1 + x_2 \leq 15,000$
$x_1 + x_3 \leq 25,000$

23. (a) Let x_1 = the number of 5-gallon bags;

x_2 = the number of 10-gallon bags; and

x_3 = the number of 20-gallon bags.

(b) We want to maximize

$z = x_1 + .9x_2 + .95x_3.$

(c) The constraints are as follows.

$x_1 + 1.1x_2 + 1.5x_3 \leq 8$
$x_1 + 1.2x_2 + 1.3x_3 \leq 8$
$2x_1 + 3x_2 + 4x_3 \leq 8$

25. Based on the information given in Exercise 21, the initial tableau is as follows.

$$
\begin{array}{ccccccc}
x_1 & x_2 & x_3 & x_4 & x_5 & x_6 & z \\
\end{array}
$$

$$
\left[
\begin{array}{ccccccc|c}
1 & 1 & 1 & 1 & 0 & 0 & 0 & 50{,}000 \\
① & 1 & 0 & 0 & 1 & 0 & 0 & 15{,}000 \\
1 & 0 & 1 & 0 & 0 & 1 & 0 & 25{,}000 \\
\hline
-.15 & -.09 & -.05 & 0 & 0 & 0 & 1 & 0
\end{array}
\right]
$$

Continue by pivoting on each circled entry.

$$
\begin{array}{ccccccc}
x_1 & x_2 & x_3 & x_4 & x_5 & x_6 & z \\
\end{array}
$$

$R_1 - R_2 \rightarrow R_1$

$R_3 - R_2 \rightarrow R_2$

$.15R_2 + R_4 \rightarrow R_4$

$$
\left[
\begin{array}{ccccccc|c}
0 & 0 & 1 & 1 & -1 & 0 & 0 & 35{,}000 \\
1 & 1 & 0 & 0 & 1 & 0 & 0 & 15{,}000 \\
0 & -1 & ① & 0 & -1 & 1 & 0 & 10{,}000 \\
\hline
0 & .06 & -.05 & 0 & .15 & 0 & 1 & 2250
\end{array}
\right]
$$

$$
\begin{array}{ccccccc}
x_1 & x_2 & x_3 & x_4 & x_5 & x_6 & z \\
\end{array}
$$

$R_1 - R_3 \rightarrow R_1$

$.05R_3 + R_4 \rightarrow R_4$

$$
\left[
\begin{array}{ccccccc|c}
0 & 1 & 0 & 1 & 0 & -1 & 0 & 25{,}000 \\
1 & 1 & 0 & 0 & 1 & 0 & 0 & 15{,}000 \\
0 & -1 & 1 & 0 & -1 & 1 & 0 & 10{,}000 \\
\hline
0 & .01 & 0 & 0 & .1 & .05 & 1 & 2750
\end{array}
\right]
$$

The maximum value is z = 2750 when

$$x_1 = 15{,}000,$$
$$x_2 = 0,$$
$$x_3 = 10{,}000,$$
$$x_4 = 25{,}000,$$
$$x_5 = 0, \text{ and}$$
$$x_6 = 0.$$

He should invest \$15,000 in oil leases and \$10,000 in stock for a maximum return of \$2750.

27. Based on the information in Exercise 23, the initial tableau is as follows.

$$
\begin{array}{ccccccc}
x_1 & x_2 & x_3 & x_4 & x_5 & x_6 & z \\
\end{array}
$$

$$
\left[
\begin{array}{ccccccc|c}
1 & 1.1 & 1.5 & 1 & 0 & 0 & 0 & 8 \\
1 & 1.2 & 1.3 & 0 & 1 & 0 & 0 & 8 \\
② & 3 & 4 & 0 & 0 & 1 & 0 & 8 \\
\hline
-1 & -.9 & -.95 & 0 & 0 & 0 & 1 & 0
\end{array}
\right]
$$

Pivot on the circled entry.

$$
\begin{array}{c}
 \\
2R_1 - R_3 \rightarrow R_1 \\
2R_2 - R_3 \rightarrow R_2 \\
 \\
 \\
2R_4 + R_3 \rightarrow R_4
\end{array}
\begin{array}{ccccccc}
x_1 & x_2 & x_3 & x_4 & x_5 & x_6 & z \\
\left[\begin{array}{ccccccc|c}
0 & -.8 & -1 & 2 & 0 & -1 & 0 & 8 \\
0 & -.6 & -1.4 & 0 & 2 & -1 & 0 & 8 \\
2 & 3 & 4 & 0 & 0 & 1 & 0 & 8 \\
\hline
0 & 1.2 & 2.1 & 0 & 0 & 1 & 2 & 8
\end{array}\right]
\end{array}
$$

The maximum value is $z = \dfrac{8}{2} = 4$

when $x_1 = \dfrac{8}{2} = 4$, $x_2 = 0$, $x_3 = 0$,

$x_4 = \dfrac{8}{2} = 4$, $x_5 = \dfrac{8}{2} = 4$, and $x_6 = 0$.

For a maximum profit of $4 per unit, 4 units of 5-gallon bags (and none of the others) should be made.

Extended Application 1

1. **(a)** We obtain

(i) $.4x_1 + .23x_2 + .805x_3 + .998x_5$
$+ .04x_6 + .5x_{10} + .625x_{11} \geq 14$

(ii) $.054x_1 + .069x_2 + .025x_4 + .078x_6$
$+ .28x_7 + .97x_8 \geq 6$

(iii) $.707x_9 + .1x_{10} \geq 16$

(iv) $.35x_{10} + .315x_{11} \geq .35$

(v) $x_1 + x_2 + x_3 + x_4 + x_5 + x_6 + x_7$
$+ x_8 + x_9 + x_{10} + x_{11} \leq 99.6$

Solving on a computer gives a minimum cost of $12.55 when $x_2 = 58.6957$, $x_8 = 2.01031$, $x_9 = 22.4895$, $x_{10} = 1$, $x_{12} = 1$, and $x_{13} = 1$.

(b) Replace 14 with 17 in inequality (i) and 16 with 16.5 in inequality (iii).

The minimum cost is $15.05 when $x_2 = 71.7391$, $x_8 = 3.14433$, $x_9 = 23.1966$, $x_{10} = 1$, $x_{12} = 1$, and $x_{13} = 1$.

2. Use a computer to solve the problem in the text.

The minimum cost is $14.31 when $x_2 = 67.394$, $x_8 = 3.459$, $x_9 = 22.483$, $x_{10} = 1$, $x_{12} = .25$, and $x_{13} = .15$.

Extended Application 2

1. $w_1 = \dfrac{570}{600} = .95$

$w_2 = \dfrac{500}{600} = .83$

$w_3 = \dfrac{450}{600} = .75$

$w_4 = \dfrac{600}{600} = 1.00$

$w_5 = \dfrac{520}{600} = .87$

$w_6 = \dfrac{565}{600} = .94$

2. Use the simplex method with a computer to find the merit increase for each employee. The correct answers are $x_1 = 100$, $x_2 = 0$, $x_3 = 0$, $x_4 = 90$, $x_5 = 0$, and $x_6 = 210$.

CHAPTER 4   TEST

1. For the maximization problem below:

   (a) Determine the number of slack variables needed.

   (b) Determine the number of surplus variables needed.

   (c) Convert each constraint into a linear equation.

   Maximize $z = 50x_1 + 80x_2$

   subject to: $x_1 + 2x_2 \leq 32$

   $\qquad\qquad 3x_1 + 4x_2 \leq 84$

   $\qquad\qquad\qquad x_2 \leq 12$

   with $\qquad x_1 \geq 0, \; x_2 \geq 0$.

2. For the maximization problem below:

   (a) Set up the initial simplex tableau.

   (b) Determine the initial basic solution.

   (c) Find the first pivot element and justify your choice.

   Maximize $z = 10x_1 + 5x_2$

   subject to: $6x_1 + 2x_2 \leq 36$

   $\qquad\qquad 2x_1 + 4x_2 \leq 32$

   with $\qquad x_1 \geq 0, \; x_2 \geq 0$.

3. Solve the problem with given initial tableau.

$$
\begin{array}{ccccccc|c}
x_1 & x_2 & x_3 & x_4 & x_5 & x_6 & z & \\
\hline
1 & 1 & 1 & 1 & 0 & 0 & 0 & 1000 \\
40 & 20 & 30 & 0 & 1 & 0 & 0 & 3200 \\
1 & 2 & 1 & 0 & 0 & 1 & 0 & 160 \\
\hline
-100 & -300 & -200 & 0 & 0 & 0 & 1 & 0
\end{array}
$$

4. Use the simplex method to solve the following problem.

   Mammoth Micros markets computers with single-sided and double-sided disk drives. They obtain these drives from Large Disks, Inc. and Double Drives Are Us. Large Disk charges $250 for a single-sided and $350 for a double-sided disk. Double Drives charges $290 and $320. Each month Large Disks can supply at most 1000 drives in all. Double Drives can supply at most 2000. Mammoth needs at least 1200 single and 1600 double drives. How many of each type should they buy from each company to minimize their total costs? What is the minimum cost?

5. Maximize $z = 6x_1 - 2x_2$

   subject to: $x_1 + x_2 \leq 10$

   $\qquad\qquad 3x_1 + 2x_2 \geq 24$

   with $\qquad x_1 \geq 0, \ x_2 \geq 0.$

6. For the following minimization problem,

   (a) state the dual problem, and

   (b) solve the problem using the simplex method.

   Minimize $w = 5y_1 + 7y_2$

   subject to: $y_1 \qquad\quad \geq 4$

   $\qquad\qquad y_1 + y_2 \geq 8$

   $\qquad\qquad y_1 + 2y_2 \geq 10$

   with $\qquad y_1 \geq 0, \ y_2 \geq 0.$

CHAPTER 4   TEST ANSWERS

1. **(a)**   3 slack variables are needed.

    (b)   0 surplus variables are needed.

    (c) $x_1 + 2x_2 + x_3 \qquad\qquad = 32$

        $3x_1 + 4x_2 \qquad + x_4 \qquad = 84$

             $x_2 \qquad\qquad + x_5 = 12$

2. **(a)**

x_1	x_2	x_3	x_4	z	
6	2	1	0	0	36
2	4	0	1	0	32
-10	-5	0	1	1	0

    (b) $x_3 = 36$, $x_4 = 32$, $z = 0$

    (c)   Column one has the most negative indicator. The smallest nonnegative quotient occurs in row 1, since $\dfrac{36}{6}$ is smaller than $\dfrac{32}{2}$. Pivot on the 6.

3. The final tableau is

x_1	x_2	x_3	x_4	x_5	x_6	z	
1	0	1	2	0	-1	0	40
10	0	0	-40	1	10	0	800
0	1	0	-1	0	1	0	60
100	0	0	100	0	100	1	26,000

   The maximum value is 26,000 when $x_1 = 0$, $x_2 = 60$, and $x_3 = 40$.

4. They should buy 1000 single-sided from Large Disks, and they should buy 200 single-sided and 1600 double-sided from Doubles Are Us for a minimum cost of $820,000.

5. The maximum value is 60 when $x_1 = 10$ and $x_2 = 0$.

6. **(a)**   Maximize $z = 4x_1 + 8x_2 + 10x_3$

       subject to: $x_1 + x_2 + \quad x_3 \le 5$

                  $x_2 + 2x_3 \le 7$

       with $x_1 \ge 0$, $x_2 \ge 0$, $x_3 \ge 0$.

    (b) The minimum value is 44 when $y_1 = 6$ and $y_2 = 2$.

CHAPTER 5   SETS AND PROBABILITY

Section 5.1

1. $3 \in \{2, 5, 7, 9, 10\}$

   False. The number 3 is not an element of the set.

3. $9 \notin \{2, 1, 5, 8\}$

   True. 9 is not an element of the set.

5. $\{2, 5, 8, 9\} = \{2, 5, 9, 8\}$

   True. The sets contain exactly the same elements, so they are equal.

7. $\{\text{all whole numbers greater than 7 and less than 10}\} = \{8, 9\}$

   True. 8 and 9 are the only such numbers.

9. $\{x \mid x \text{ is an odd integer}, 6 \leq x \leq 18\}$
   $= \{7, 9, 11, 15, 17\}$

   False. The number 13 should be included.

11. $\emptyset$ and $\{\ \}$ are notations that represent the empty set, and they correspond to choices (b) and (c).
    (The other two choices do not represent the empty set since $\{\emptyset\}$ contains an element and 0 is not a set.)

13. $A \subseteq U$

    Every element of A is also an element of U.

15. $A \nsubseteq E$

    A contains elements that do not belong to E, namely 2, 4, 8, 10, and 12.

17. $\emptyset \subseteq A$

    The empty set is a subset of every set.

19. $D \subseteq B$

    Every element of D is also an element of B.

21. There are exactly $2^6 = 64$ subsets of A. A set with n distinct elements has 2^n subsets, and A has $n = 6$ elements.

23. There are exactly $2^3 = 8$ subsets of C. A set with n distinct elements has 2^n subsets, and C has $n = 3$ elements.

25. $\{4, 5, 6\}$

    Since the set has 3 elements, there are $2^3 = 8$ subsets.

27. $\{x \mid x \text{ is a counting number between 6 and 12}\}$

    The set contains five elements: 7, 8, 9, 10, and 11. Therefore, there are $2^5 = 32$ subsets.

31. $\{8, 11, 15\} \cap \{8, 11, 19, 20\} = \{8, 11\}$

$\{8, 11\}$ is the set of all elements belonging to both of the first two sets, so it is the intersection of those sets.

33. $\{6, 12, 14, 16\} \cap \{6, 14, 19\} = \{6, 14\}$

$\{6, 14\}$ is the set of all elements belonging to both of the first two sets, so it is the intersection of those sets.

35. $\{3, 5, 9, 10\} \cup \emptyset = \{3, 5, 9, 10\}$

The empty set contains no elements, so the union of any set with the empty set will result in an answer set that is identical to the original set. (On the other hand, $\{3, 5, 9, 10\} \cap \emptyset = \emptyset$.)

37. $\{1, 2, 4\} \cup \{1, 2\} = \{1, 2, 4\}$

The answer set $\{1, 2, 4\}$ consists of all elements belonging to the first set, to the second set, or to both sets, and therefore it is the union of the first two sets. (On the other hand,
$$\{1, 2, 4\} \cap \{1, 2\} = \{1, 2\}.)$$

39. $X \cap Y = \{2, 3, 4, 5\} \cap \{3, 5, 7, 9\}$
$$= \{3, 5\}$$

41. X′ consists of those elements of U which are not in X.
$$X' = \{7, 9\}$$

43. X′ ∩ Y′

$X' = \{7, 9\}$ and $Y' = \{2, 4\}$ so $X' \cap Y' = \emptyset$.

45. X ∪ (Y ∩ Z)

$X = \{2, 3, 4, 5\}$ and $Y \cap Z = \{5, 7, 9\}$.

Hence, the union of these two sets is $X \cup (Y \cap Z) = \{2, 3, 4, 5, 7, 9\}$
$$= U.$$

47. M′ consists of all students in U who art not in M, so M′ consists of all students in this school not taking this course.

49. N ∩ P is the set of all students in this school taking both accounting and zoology.

51. $A = \{2, 4, 6, 8, 10, 12\}$, $B = \{2, 4, 8, 10\}$, $C = \{4, 8, 12\}$, $D = \{2, 10\}$, $E = \{6\}$, $U = \{2, 4, 6, 8, 10, 12, 14\}$

A pair of sets is disjoint if the two sets have no elements in common. The pairs of these sets that are disjoint are B and E, C and E, D and E, and C and D.

53. Since B is the set of all stocks in the list with a price-to-earnings ratio of a least 13, B′ is the set of all stocks on the list whose price-to-earnings ratio is less than 13.

However, none of the stocks in the list has a price-to-earnings ratio that is less than 13, so B′ = Ø.

55. A ∩ B is the set of all stocks with a dividend greater than $3 and a price-to-earnings ratio of at least 13. A ∩ B = {IBM, Mobil}, and (A ∩ B)′ is the set of all stocks in the list that are not elements of A ∩ B. Therefore,
(A ∩ B)′ = {ATT, GE, Hershey, Nike}.

57. F = {networks with more than 5000 subscribers}
= {HBO, Showtime, Cinemax}

59. H = {networks that show movies}
= {HBO, Showtime, Cinemax, Disney Channel, Movie Channel}

61. H ∩ G = {networks that show movies and non-cartoon comedy}
= {Showtime, Cinemax}

63. U = {s, d, c, g, i, m, h}
and O = {i, m, h, g}, so
O′ = {s, d, c}.

65. N ∩ O = {s, d, c, g} ∩ {i, m, h, g}
= {g}

67. N ∩ O′ = {s, d, c, g} ∩ {s, d, c}
= {s, d, c}

Section 5.2

1. B ∩ A′ is the set of all elements in B *and* not in A.
See the Venn diagram in the back of the textbook.

3. A′ ∪ B is the set of all elements which do not belong to *or* which do belong to B, or both.
See the Venn diagram in the back of the textbook.

5. B′ ∪ (A′ ∩ B′)
First find A′ ∩ B′, the set of elements not in A *and* not in B.

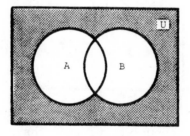

For the union, we want those elements in B′ *or* (A′ ∩ B′), or both.

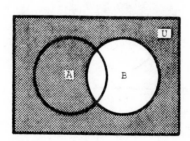

7. U′ is the empty set ∅.

See the Venn diagram in the back of the textbook.

9. Three sets divide the universal set into 8 regions. (Examples of this situation will be seen in Exercises 11–17.)

11. (A ∩ C′) ∪ B

First find A ∩ C′, the region in A *and* not in C.

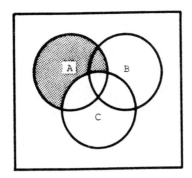

For the union, we want the region in (A ∩ C′) *or* in B, or both.

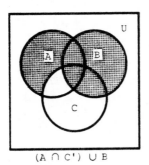

(A ∩ C') ∪ B

13. A′ ∩ (B ∩ C)

First find A′, the region not in A.

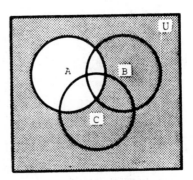

Then find B ∩ C, the region where B and C overlap.

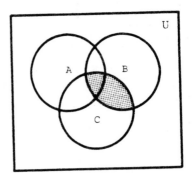

Now intersect these regions.

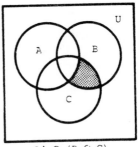

A' ∩ (B ∩ C)

15.   (A ∩ B′) ∪ C

First find A ∩ B′, the region in A *and* not in B.

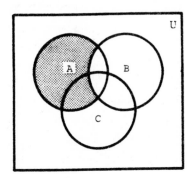

For the union, we want the region in (A ∩ B′) *or* in C, or both.

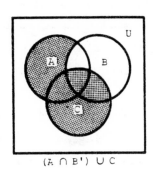

(A ∩ B′) ∪ C

17.   A′ ∩ (B′ ∪ C)

First find A′.

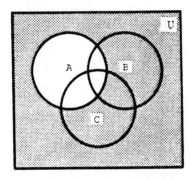

Then find B′ ∪ C, the region not in B *or* in C, or both.

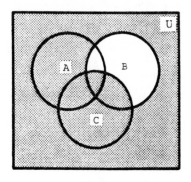

Now intersect these regions.

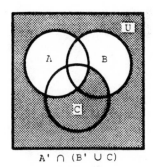

A′ ∩ (B′ ∪ C)

19.   n(A ∪ B) = n(A) + n(B) − (A ∩ B)
　　　　　 = 5 + 8 − 4
　　　　　 = 9

21.   n(A ∪ B) = n(A) + n(B) − n(A ∩ B)
　　　　　 20 = n(A) + 7 − 3
　　　　　 20 = n(A) + 4
　　　　　 16 = n(A)

23.   n(U) = 38
　　　 n(A) = 16
　　　 n(A ∩ B) = 12
　　　 n(B′) = 20

First put 12 in A ∩ B. Since
n(A) = 16, and 12 are in A ∩ B,
there must be 4 elements in A that
are not in A ∩ B. n(B′) = 20, so
there are 20 not in B. We already
have 4 not in B (but in A), so there
must be another 16 outside B *and*
outside A. So far we have accounted
for 32, and n(U) = 38, so 6 must be
in B but not in any region yet
identified. Thus n(A′ ∩ B) = 6.
See the Venn diagram in the back of
the textbook.

25.   n(A ∪ B) = 17
      n(A ∩ B) = 3
      n(A) = 8
      n(A′ ∪ B′) = 21

Start with n(A ∩ B) = 3. Since
n(A) = 8, there must be 5 more in A
not in B. n(A ∪ B) = 17; we already
have 8, so 9 more must be in B not
yet counted. A′ ∪ B′ consists of
all the region not in A ∩ B, where
we have 3. So far 5 + 9 = 14 are in
this region, so another 21 − 14 = 7
must be outside both A and B.
See the Venn diagram in the back of
the textbook.

27.   n(A) = 28
      n(B) = 34
      n(C) = 25
      n(A ∩ B) = 14
      n(B ∩ C) = 15
      n(A ∩ C) = 11
      n(A ∩ B ∩ C) = 9
      n(U) = 59

We start with n(A ∩ B ∩ C) = 9.
If n(A ∩ B) = 14, an additional 5
are in A ∩ B but not in A ∩ B ∩ C.
Similarly, N(B ∩ C) = 15, so
15 − 9 = 6 are in B ∩ C but not in
A ∩ B ∩ C. Also, n(A ∩ C) = 11, so
11 − 9 = 2 are in A ∩ C but not in
A ∩ B ∩ C. Now we turn our attention to n(A) = 28. So far we have
2 + 9 + 5 = 16 in A; there must be
another 28 − 16 = 12 in A not yet
counted. Similarly, n(B) = 34; we
have 5 + 9 + 6 = 20 so far, and
34 − 20 = 14 more must be put in B.
For C, n(C) = 25; we have
2 + 9 + 6 = 17 counted so far. Then
there must be 8 more in C not yet
counted. The count now stands at
56, and n(U) = 59, so 3 must be
outside the three sets.
See the Venn diagram in the back of
the textbook.

29.   n(A ∩ B) = 6
      n(A ∩ B ∩ C) = 4
      n(A ∩ C) = 7
      n(B ∩ C) = 4
      n(A ∩ C′) = 11
      n(B ∩ C′) = 8
      n(C) = 15
      n(A′ ∩ B′ ∩ C′) = 5

Start with n(A ∩ B) = 6 and
n(A ∩ B ∩ C) = 4 to get 6 − 4 = 2
in that portion of A ∩ B outside of
C. From n(B ∩ C) = 4, there are
4 − 4 = 0 elements in that portion

of B ∩ C outside of A. Use
n(A ∩ C) = 7 to get 7 − 4 = 3 elements in that portion of A ∩ C outside of B. Since n(A ∩ C') = 11,
there are 11 − 2 = 9 elements in
that part of A outside of B and C.
Use n(B ∩ C') = 8 to get 8 − 2 = 6
elements in that part of B outside
of A and C. Since n(C) = 15, there
are 15 − 3 − 4 − 0 = 8 elements in C
outside of A and B. Finally, 5 must
be outside all three sets, since
n(A' ∩ B' ∩ C') = 5.
See the Venn diagram in the back of
the textbook.

For A' ∩ B', first find A' and B'
individually.

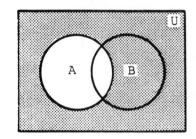 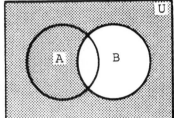

Then A' ∩ B' is the region where A'
and B' overlap, which is the entire
region outside A ∪ B (the same
result as in the second diagram).
Therefore, (A ∪ B)' = A' ∩ B'.

31. (A ∪ B)' = A' ∩ B'

For (A ∪ B)', first find A ∪ B.

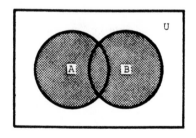

Now find (A ∪ B)', the region outside A ∪ B.

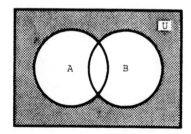

33. A ∩ (B ∪ C) = (A ∩ B) ∪ (A ∩ C)

First find A and B ∪ C individually.

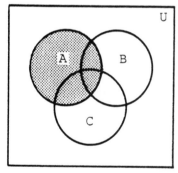 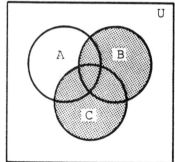

Then A ∩ (B ∪ C) is the region where
the above two diagrams overlap.

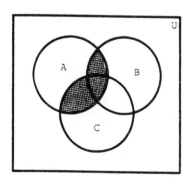

Next find A ∩ B and A ∩ C
individually.

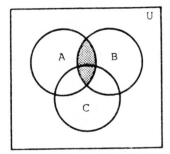 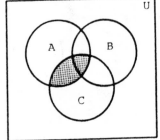

Then (A ∩ B) ∪ (A ∩ C) is the union
of the above two diagrams.

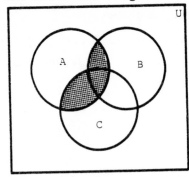

The Venn diagram for A ∩ (B ∪ C) is
identical to the Venn diagram for
(A ∩ B) ∪ (A ∩ C), so conclude that
A ∩ (B ∪ C) = (A ∩ B) ∪ (A ∩ C).

35. Let M be the set of those who use a
microwave oven. E be the set of
those who use an electric range, and
G be the set of those who use a gas
range. We are given the following
information.

n(U) = 140

n(M) = 58

N(E) = 63

N(G) = 58

n(M ∩ E) = 19

n(M ∩ G) = 17

n(G ∩ E) = 4

n(M ∩ G ∩ E) = 1

n(M′ ∩ G′ ∩ E′) = 2

Since n(M ∩ G ∩ E) = 1, there is
1 element in the region where the
three sets overlap.
Since n(M ∩ E) = 19, there are
19 − 1 = 18 elements in M ∩ E but
not in M ∩ E ∩ G.
Since n(M ∩ G) = 17, there are
17 − 1 = 16 elements in M ∩ G but
not in M ∩ E ∩ G.
Since n(G ∩ E) = 4, there are
4 − 1 = 3 elements in G ∩ E but
not in M ∩ E ∩ G.
Now consider n(M) = 58. So far we
have 16 + 1 + 18 = 35 in M; there
must be another 58 − 35 = 23 in M
not yet counted.
Similarly, n(E) = 63; we have
18 + 1 + 13 = 32 counted so far.
There must be 63 − 32 = 31 more in
E not yet counted.
Also, n(G) = 58; we have
16 + 1 + 13 = 30 counted so far.
There must be 58 − 30 = 28 more
in G not yet counted.
Lastly, n(M′ ∩ G′ ∩ E′) = 2 indicates that there are 2 elements outside of all three sets.

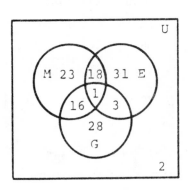

Note that the numbers in the Venn diagram add up to 142 even though $n(U) = 140$. Jeff has made some error and he should definitely be reassigned.

37. **(a)** $n(Y \cap R) = 40$

since 40 is the number in the table where the Y row and R column meet.

(b) $n(M \cap D) = 30$

since 30 is the number in the table where the M row and the D column meet.

(c) $n(D \cap Y) = 15$

$n(M) = 80$ since that is the total in the M row.

$m(M \cap (D \cap Y)) = 0$ since no person can simultaneously have an age in the range 21-25 *and* have an age in the range 26-35. By the union rule for sets,

$n(M \cup (D \cap Y))$
$= n(M) + n(D \cap Y) - n(M \cap (D \cap Y))$
$= 80 + 15 - 0$
$= 95.$

(d) $Y' \cap (D \cup N)$ consists of all people in the D column or in the N column who are at the same time not in the Y row. Therefore,

$n(Y' \cap (D \cup N)) = 30 + 50 + 20 + 10$
$\qquad\qquad\qquad = 110.$

(e) $n(N) = 45$
  $n(0) = 70$
  $n(0') = 220 - 70 = 150$
  $n(0' \cap N) = 15 + 20 = 35$

By the union rule,

$n(0' \cup N) = n(0') + n(N) - n(0' \cap N)$
$\qquad\qquad = 150 + 45 - 35$
$\qquad\qquad = 160.$

(f) $M' \cap (R' \cap N')$ consists of all people who are not in the R column and not in the N column and who are at the same time not in the M row. Therefore,

$n(M' \cap (R' \cap N')) = 15 + 50 = 65.$

(g) $M \cup (D \cap Y)$ consists of all people age 21-25 who drink diet cola *or* anyone age 26-35.

39. Let T be the set of tall pea plants, G be the set of plants with green peas, and S be the set of plants with smooth peas. We are given the following information.

$n(T) = 22$
$n(G) = 25$
$n(S) = 39$
$n(T \cap G) = 9$
$n(T \cap S) = 17$
$n(G \cap S) = 20$
$n(T \cap G \cap S) = 6$
$n(T' \cap G' \cap S') = 4$

Start with the last two restricted regions, $T \cap G \cap S$ and $T' \cap G' \cap S'$. With $n(G \cap S) = 20$, there are $20 - 6 = 14$ yet to be labeled; $n(T \cap S) = 17$ puts $17 - 6 = 11$ in $T \cap S \cap G'$; also, $n(T \cap G) = 9$ puts $9 - 6 = 3$ in $T \cap G \cap S'$. Now fill in $T \cap (G' \cap S') = 22 - 20 = 2$,

$G \cap (T' \cap S') = 25 - 23 = 2$, and

$S \cap (T' \cap G') = 39 - 31 = 8$.

$N(U) = 2 + 3 + 2 + 11 + 6 + 14 + 8 + 4$

$\qquad = 50$

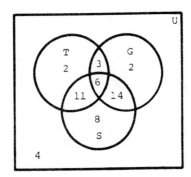

(a) $n(U) = 50$

(b) $n(T \cap S' \cap G') = 2$

(c) $n(T' \cap S \cap G) = 14$

41.   First fill in the Venn diagram, starting with the region common to all three sets.

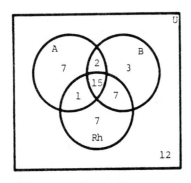

(a) The total of these numbers in the diagram is 54.

(b) $7 + 3 + 7 = 17$ had only one antigen.

(c) $1 + 2 + 7 = 10$ had exactly two antigens.

(d) A person with O-positive blood has only the Rh antigen, so this number is 7.

(e) A person with AB-positive blood has all three antigens, so this number is 15.

(f) A person with B-negative blood has only the B antigen, so this number is 3.

(g) A person with O-negative blood has none of the antigens. There are 12 such people.

(h) A person with A-positive blood has the A and Rh antigens, but not the B-antigen. The number is 1.

43.   Let W be the set of women, C the set of those who speak Cantonese, and F the set of those who set off firecrackers. We are given the following information.

$$n(W) = 120$$
$$n(C) = 150$$
$$n(F) = 170$$
$$n(W' \cap C) = 108$$
$$n(W' \cap F') = 100$$
$$n(W \cap C' \cap F) = 18$$
$$n(W' \cap C' \cap F') = 78$$
$$n(W \cap C \cap F) = 30$$

Note that

$n(W' \cap C \cap F')$

$\quad = n(W' \cap F') - n(W' \cap C' \cap F')$

$\quad = 100 - 78 = 22.$

Furthermore,

$n(W' \cap C \cap F)$
  $= n(W' \cap C) - n(W' \cap C \cap F')$
  $= 108 - 22 = 86.$

We now have

$n(W \cap C \cap F')$
  $= n(C) - n(W' \cap C \cap F)$
    $- n(W \cap C \cap F) - n(W' \cap C \cap F')$
  $= 150 - 86 - 30 - 22 = 12.$

With all of the overlaps of W, C, and F determined, we can now compute $n(W \cap C' \cap F') = 60$ and $n(W' \cap C' \cap F) = 36.$

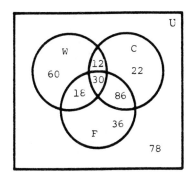

(a) Adding up the disjoint components, we find the total attendance to be

$60 + 12 + 18 + 30 + 22 + 86 + 36 + 78 = 342.$

(b) $n(C') = 342 - n(C)$
          $= 342 - 150 = 192$

(c) $n(W \cap F') = 60 + 12 = 72$

(d) $n(W' \cap C \cap F) = 86$

Section 5.3

3. The sample space is the set of the twelve months, {January, February, March, ..., December}.

5. The possible number of points earned could be any whole number from 0 to 80. The sample space is the set
    {0, 1, 2, 3, ..., 80}.

7. The possible decisions are to go ahead with a new oil shale plant or to cancel it. The sample space is the set {go ahead, cancel}.

9. Let h = heads and t = tails for the coin; the die can display 6 different numbers. There are 12 possible outcomes in the sample space, which is the set
    {(h, 1), (t, 1), (h, 2), (t, 2),
     (h, 3), (t, 3), (h, 4), (t, 4),
     (h, 5), (t, 5), (h, 6), (t, 6)}.

13. Use the first letter of each name. The sample space is the set

    {AB, AC, AD, AE, BC,
     BD, BE, CD, CE, DE}.

    (a) One of the committee members must be Chinn. This event is {AC, BC, CD, CE}.

    (b) Alam, Bartolini, and Chinn may be on any committee; Dickson and Ellsberg may not be on the same committee. This event is {AB, AC, AD, AE, BC, BD, BE, CD, CE}.

(c) Both Alam and Chinn are on the committee. This event is $\{AC\}$.

15. Each outcome consists of two of the numbers 1, 2, 3, 4, and 5, without regard for order. For example, let (2, 5) represent the outcome that the slips of paper marked with 2 and 5 are drawn. There are 10 pairs in this sample space, which is

$$\{(1, 2), (1, 3), (1, 4), (1, 5),$$
$$(2, 3), (2, 4), (2, 5), (3, 4),$$
$$(3, 5), (4, 5)\}.$$

(a) Both numbers in the outcome pair are even. This event is $\{(2, 4)\}$, which is called a simple event since it consists of only one outcome.

(b) One number in the pair is even and the other number is odd. This event is

$$\{(1, 2), (1, 4), (2, 3),$$
$$(2, 5), (3, 4), (4, 5)\}.$$

(c) Each slip of paper has a different number written on it, so it is not possible to draw two slips marked with the same number. This event is $\emptyset$, which is called an impossible event since it contains no outcomes.

17. There are 6 possibilities for the first number and 6 for the second number. For example, let (2, 5) represent the outcome of observing 2 on the first toss and 5 on the

second toss. There are 36 ordered pairs in this sample space, which is

$$\{(1, 1), (1, 2), (1, 3), (1, 4),$$
$$(1, 5), (1, 6), (2, 1), (2, 2),$$
$$(2, 3), (2, 4), (2, 5), (2, 6),$$
$$(3, 1), (3, 2), (3, 3), (3, 4),$$
$$(3, 5), (3, 6), (4, 1), (4, 2),$$
$$(4, 3), (4, 4), (4, 5), (4, 6),$$
$$(5, 1), (5, 2), (5, 3), (5, 4),$$
$$(5, 5), (5, 6), (6, 1), (6, 2),$$
$$(6, 3), (6, 4), (6, 5), (6, 6)\}.$$

(a) There are six outcomes that have 3 as the first number of the ordered pair. This event is

$$\{(3, 1), (3, 2), (3, 3),$$
$$(3, 4), (3, 5), (3, 6)\}.$$

(b) There are five outcomes in which the sum of the two numbers is 8. This event is

$$\{(2, 6), (3, 5), (4, 4),$$
$$(5, 3), (6, 2)\}.$$

(c) The sum never exceeds 12, so this event is impossible. The event is the empty set, which we write as $\emptyset$.

19. $S = \{1, 2, 3, 4, 5, 6\}$, so $n(S) = 6$. $E = \{2\}$, so $n(E) = 1$.

If all the outcomes in a sample space S are equally likely, then the probability of an event E is

$$P(E) = \frac{n(E)}{n(S)}.$$

In this problem,

$$P(E) = \frac{n(E)}{n(S)} = \frac{1}{6}.$$

21. $S = \{1, 2, 3, 4, 5, 6\}$, so $n(S) = 6$.
$E = \{1, 2, 3, 4\}$, so $n(E) = 4$.

$$P(E) = \frac{4}{6} = \frac{2}{3}.$$

23. $S = \{1, 2, 3, 4, 5, 6\}$, so $n(S) = 6$.
$E = \{3, 4\}$, so $n(E) = 2$.

$$P(E) = \frac{2}{6} = \frac{1}{3}.$$

25. $n(S) = 52$

Let E be the event "a 9 is drawn."
There are four 9's in the deck, so
$n(E) = 4$.

$$P(9) = P(E) = \frac{n(E)}{n(S)} = \frac{4}{52} = \frac{1}{13}$$

27. Let F be the event "a black 9 is
drawn." There are two black 9's in
the deck, so $n(F) = 2$. As before,
$n(S) = 52$.

$$P(\text{black } 9) = P(F) = \frac{n(F)}{n(S)} = \frac{2}{52} = \frac{1}{26}$$

29. Let G be the event "a 9 of hearts
is drawn." There is only one 9 of
hearts in a deck of 52 cards, so
$n(G) = 1$. Again, $n(S) = 52$.

$$P(9 \text{ of hearts}) = P(G) = \frac{n(G)}{n(S)} = \frac{1}{52}$$

31. Let H be the event "a 2 or a queen is
drawn." There are four 2's and four
queens in the deck, so $n(H) = 8$.
Also, $n(S) = 52$.

$$P(2 \text{ or queen}) = P(H) = \frac{n(H)}{n(S)} = \frac{8}{52} = \frac{2}{13}$$

33. Let J be the event "a red face card
is drawn." There are three face
cards (jack, queen, and king) in each
of the two red suits (hearts and
diamonds), so $n(J) = 6$. Also,
$n(S) = 52$.

$$P(\text{red face card}) = P(J) = \frac{n(J)}{n(S)}$$

$$= \frac{6}{52} = \frac{3}{26}$$

35. Since there are $2 + 3 + 5 + 8 = 18$
marbles in the jar and the experiment
consists of drawing one of them at
random, $n(S) = 18$. 2 of the marbles
are white, so

$$P(\text{white}) = \frac{2}{18} = \frac{1}{9}.$$

37. 5 of the marbles are yellow and
$n(S) = 18$, so

$$P(\text{yellow}) = \frac{5}{18}.$$

39. $2 + 3 + 5 = 10$ of the marbles are not
black and $n(S) = 18$, so

$$P(\text{not black}) = \frac{10}{18} = \frac{5}{9}.$$

41. E: worker is female
F: worker has worked less than 5 yr
G: worker contributes to a voluntary
retirement plan

(a) E′ occurs when E does not, so E
is the event "worker is male."

(b) E ∩ F occurs when both E and F
occur, so E ∩ F is the event "worker
is female and has worked less than
5 yr."

(c) $E \cup G'$ is the event "worker is female or does not contribute to a voluntary retirement plan, or both."

43. First, calculate that

$$9.0 + 1.2 + 2.7 + 1.0 + 1.1 = 15.0$$

is the total amount of funding (in billions of dollars).

(a) P(federal government)

$$= \frac{9.0}{15.0} = \frac{9}{15} = \frac{3}{5}$$

(b) $P(\text{industry}) = \frac{1.0}{15.0} = \frac{1}{15}$

(c) P(institutional)

$$= \frac{2.7}{15.0} = \frac{27}{150} = \frac{9}{50}$$

45. E: person smokes

F: person has a family of heart disease

G: person is overweight

(a) $E \cup F$ occurs when E or F or both occur, so $E \cup F$ is the event "person smokes or has a family history of heart disease, or both."

(b) $E' \cap F$ occurs when E does not occur and F does occur, so $E' \cap F$ is the event "person does not smoke and has a family history of heart disease."

(c) $F' \cup G'$ is the event "person does not have a family history of heart disease or is not overweight, or both."

Section 5.4

3. No, it is possible to wear both glasses and sandals. The events are not mutually exclusive.

5. Yes, teenagers cannot be over 30. The events are mutually exclusive.

7. No, a postal worker can be male. The events are not mutually exclusive.

9. When the two dice are rolled, there are 36 equally likely outcomes. Let 5–3 represent the outcome "the first die shows a 5 and the second die shows a 3," and so on.

(a) Rolling a sum of 8 occurs when the outcome is 2–6, 3–5, 4–4, 5–3, or 6–2. Therefore, since there are five such outcomes, the probability of thi event is

$$P(\text{sum is 8}) = \frac{5}{36}.$$

(b) A sum of 9 occurs when the outcome is 3–6, 4–5, 5–4, or 6–3, so

$$P(\text{sum is 9}) = \frac{4}{36} = \frac{1}{9}.$$

(c) A sum of 10 occurs when the outcome is 4–6, 5–5, or 6–4, so

$$P(\text{sum is 10}) = \frac{3}{36} = \frac{1}{12}.$$

(d) A sum of 13 does not occur in any of the 36 outcomes, so

$$P(\text{sum is } 16) = \frac{0}{36} = 0.$$

11. **(a)** P(sum is not more than 5)

$$= P(2) + P(3) + P(4) + P(5)$$

$$= \frac{1}{36} + \frac{2}{36} + \frac{3}{36} + \frac{4}{36}$$

$$= \frac{10}{36} = \frac{5}{18}$$

(b) P(sum is not less than 8)

$$= P(8) + P(9) + P(10)$$
$$+ P(11) + (12)$$

$$= \frac{5}{36} + \frac{4}{36} + \frac{3}{36} + \frac{2}{36} + \frac{1}{36}$$

$$= \frac{15}{36} = \frac{5}{12}$$

(c) P(sum is between 3 and 7)

$$= P(4) + P(5) + P(6)$$

$$= \frac{3}{36} + \frac{4}{36} + \frac{5}{36}$$

$$= \frac{12}{36} = \frac{1}{3}$$

13. **(a)** There are a total of 12 aces, 2's and 3's in a deck of 52, so

$$P(A \text{ or } 2 \text{ or } 3) = \frac{12}{52} = \frac{3}{13}.$$

(b) There are 13 diamonds plus three 7's in other suits, so

$$P(D \text{ or } 7) = \frac{16}{52} = \frac{4}{13}.$$

Alternatively, using the union rule for probability,

$$P(D) + P(7) - P(7 \text{ of diamonds})$$

$$= \frac{13}{52} + \frac{4}{52} - \frac{1}{52} = \frac{16}{52} = \frac{4}{13}.$$

(c) There are 26 black cards plus 2 red aces, so

$$P = \frac{28}{52} = \frac{7}{13}.$$

(d) There are 13 hearts plus 3 additional jacks in other suits, so

$$P = \frac{16}{52} = \frac{4}{13}.$$

15. **(a)** There are 3 uncles plus 2 cousins out of 10, so

$$P = \frac{5}{10} = \frac{1}{2}.$$

(b) There are 3 uncles, 2 brothers, and 2 cousins, for a total of 7 out of 10, so

$$P = \frac{7}{10}.$$

(c) There are 2 aunts, 2 cousins, and 1 mother, for a total of 5 out of 10, so

$$P = \frac{5}{10} = \frac{1}{2}.$$

17. **(a)** The sample space for this experiment is listed in part (b) below. The only outcomes in which both numbers are even are (2, 4) and (4, 2), so

$$P = \frac{2}{20} = \frac{1}{10}.$$

(b) The sample space is

$$\{(\underline{1, 2}), (1, 3), (\underline{1, 4}), (\underline{1, 5}),$$
$$(\underline{2, 1}), (\underline{2, 3}), (\underline{2, 4}), (\underline{2, 5}),$$
$$(3, 1), (\underline{3, 2}), (\underline{3, 4}), (\underline{3, 5}),$$
$$(\underline{4, 1}), (\underline{4, 2}), (\underline{4, 3}), (\underline{4, 5}),$$
$$(\underline{5, 1}), (\underline{5, 2}), (\underline{5, 3}), (\underline{5, 4})\}.$$

The 18 underlined pairs are the outcomes in which one number is even or greater than 3, so

$$P = \frac{18}{20} = \frac{9}{10}.$$

(c) The sum is 5 in the outcomes (1, 4), (2, 3), (3, 2), and (4, 1). The second draw is 2 in the outcomes (1, 2), (3, 2), (4, 2), and (5, 2). There are 7 distinct outcomes out of 20, so

$$P = \frac{7}{20}.$$

19. $P(Z) = .42$, $P(Y) = .38$, $P(Z \cup Y) = .61$

Begin by using the union rule for probability.

$$P(Z \cup Y) = P(Z) + P(Y) - P(Z \cap Y)$$
$$.61 = .42 + .38 - P(Z \cap Y)$$
$$.61 = .80 - P(Z \cap Y)$$
$$-.19 = -P(Z \cap Y)$$
$$.19 = P(Z \cap Y)$$

This gives the first value to be labelled in the Venn diagram. Then the part of Z outside Y must contain $.42 - .19 = .23$, and the part of Y outside Z must contain $.38 - .19 = .19$. Observe that $.23 + .19 + .19 = .61$, which agrees with the given information that $P(Z \cup Y) = .61$. The part of U outside both Y and Z must contain $1 - .61 = .39$.

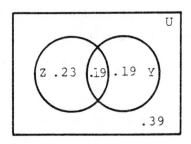

This Venn diagram may now be used to find the following probabilities.

(a) $Z' \cap Y'$ is the event represented by the part of the Venn diagram that is outside Z and outside Y.

$$P(Z' \cap Y') = .39$$

(b) $Z' \cup Y'$ is everything outside Z or outside Y or both, which is all of U except $Z \cap Y$.

$$P(Z' \cup Y') = .39 + .19 + .23$$
$$= .81$$

(c) $Z' \cup Y$ is everything outside Z or inside Y or both.

$$P(Z' \cup Y) = .19 + .19 + .39$$
$$= .77$$

(d) $Z \cap Y'$ is everything inside Z and outside Y.

$$P(Z \cap Y') = .23$$

21. Let E be the event "a 5 is rolled."

$$P(E) = \frac{1}{6} \text{ and } P(E') = \frac{5}{6}.$$

The odds in favor of rolling a 5 are

$$\frac{P(E)}{P(E')} = \frac{1/6}{5/6} = \frac{1}{5},$$

which is written "1 to 5."

23. Let E be the event "1, 2, 3, or 4 is rolled." Here P(E) = 4/6 = 2/3 and P(E′) = 1/3. The odds in favor of E are

$$\frac{P(E)}{P(E')} = \frac{2/3}{1/3} = \frac{2}{1},$$

which is written "2 to 1."

25. **(a)** Yellow: there are 3 ways to win and 12 ways to lose. The odds in favor of drawing yellow are 3 to 12, or 1 to 4.

    (b) Blue: there are 8 ways to win and 7 ways to lose; the odds in favor of drawing blue are 8 to 7.

    (c) White: there are 4 ways to win and 11 ways to lose; the odds in favor of drawing white are 4 to 11.

27. We are given that P(profit) = .74, so P(no profit) = .26. Hence, the odds against making a profit are .26 to .74 or 13 to 37.

29. The odds of winning are 3 to 2; that means there are 3 ways to win and 2 ways to lose, out of a total of 2 + 3 = 5 ways altogether. Hence, the probability of losing is $\frac{2}{5}$.

31. It is possible to establish an exact probability for this event, so this is not an empirical probability.

33. It is not possible to establish an exact probability for this event, so this is an empirical probability.

35. It is not possible to establish an exact probability for this event, so this is an empirical probability.

37. The gambler's claim is a mathematical fact, so this is not an empirical proability.

41. **(a)** The event "in the 70s" corresponds to two rows in the given probability distribution table.

    P(in the 70s)
    = P(70 – 74) + P(75 – 79)
    = .28 + .22
    = .50

    (b) P(not in the 60s)
    = 1 – P(in the 60s)
    = 1 – (.08 + .15)
    = 1 – .23 = .77

    (c) P(not in the 60s or 70s)
    = P(below 60) + P(80 – 84)
    + P(85 – 89) + P(90 – 94)
    + P(95 – 99) + P(100 or more)
    = .01 + .08 + .06 + .04
    + .02 + .06
    = .27

43. **(a)** P(less than $350)
    = 1 – P($350 or more)
    = 1 – (.08 + .03)
    = 1 – .11 = .89

(b) P($75 or more)

= P($75 – $99.99)

+ P($100 – $199.99)

+ P($200 – $349.99)

+ P($350 – $499.99)

+ P($500 or more)

= .11 + .09 + .07 + .08 + .03

= .38

(c) P($200 or more)

= P($200 – $340.99)

+ P($350 – $499.99)

+ P($500 or more)

= .07 + .08 + .03

= .18

45. The probability assignment is possible because the probability of each outcome is a number between 0 and 1, and the sum of the probabilities of all the outcomes is 1.

47. The probability assignment is not possible. All of the probabilities are between 0 and 1, but the sum of the probabilities is $\frac{13}{12}$ which is larger than 1.

49. The probability assignment is not possible. One of the probabilities is negative instead of being between 0 and 1, and the sum of the probabilities is not 1.

51. (a) P($500 or more)

= 1 – P(less than $500)

= 1 – (.31 + .18)

= 1 – .49 = .51

(b) P(less than $1000)

= .31 + .18 + .18

= .67

(c) P($500 to $2999)

= .18 + .13 + .08

= .39

(d) P($3000 or more)

= .05 + .06 + .01

= .12

53. P(C) = .049, P(M ∩ C) = .042, P(M ∪ C) = .534

Place the given information in a Venn diagram by starting with .042 in the intersection of the regions for M and C.

Since P(C) = .049,

.049 – .042 = .007

goes inside region C, but outside the intersection of C and M. Thus,

P(C ∩ M′) = .007.

Since P(M ∪ C) = .534,

.534 – .042 – .007 = .485

goes inside region M, but outside the intersection of C and M. Thus, P(M ∩ C′) = .485. The labeled regions have probability

.485 + .042 + .007 = .534.

Since the entire region of the Venn diagram must have probability 1, the region outside M and C, or M′ ∩ C′, has probability

$$1 - .534 = .466.$$

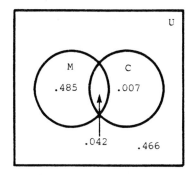

(a) P(C′) = 1 − P(C)

  = 1 − .049

  = .951

(b)   P(M) = .485 + .042

  = .527

(c) P(M′) = 1 − P(M)

  = 1 − .527

  = .473

(d) P(M′ ∩ C′) = .466

(e) P(C ∩ M′) = .007

(f) P(C ∪ M′)

  = P(C) + P(M′) − P(C ∩ M′)

  = .049 + .473 − .007

  = .515

55. (a) Now red is no longer dominant, and RW or WR results in pink, so P(red) = P(RR) = 1/4.

(b) P(pink) = P(RW) + P(WR)

  $= \dfrac{1}{4} + \dfrac{1}{4} = \dfrac{1}{2}$

(c) P(white) = P(WW) = 1/4

57. Since 55 of the workers were women, 130 − 55 = 75 were men. Since 3 of the women earned more than $40,000, 55 − 3 = 52 of them earned less than $40,000. Since 62 of the men earned less than $40,000, 75 − 62 = 13 earned more than $40,000.

These data for the 130 workers can be summarized in the following table.

	Men	Women
Under $40,000	62	52
Over $40,000	13	3

(a) P(a woman earning less than $40,000)

  $= \dfrac{52}{130} = .4$

(b) P(a man earning more than $40,000)

  $= \dfrac{13}{130} = .1$

(c) P(a man or is earning more than $40,000)

  $= \dfrac{62 + 13 + 3}{130}$

  $= \dfrac{78}{130} = .6$

(d) P(a woman or is earning less $40,000)

  than

  $= \dfrac{52 + 3 + 62}{130}$

  $= \dfrac{117}{130} = .9$

59. Let A be the set of refugees who came to escape abject poverty and B be the set of refugees who came to escape political oppression. Then $P(A) = .80$, $P(B) = .90$, and $P(A \cap B) = .70$.

$$P(A \cup B) = P(A) + P(B) - P(A \cap B)$$
$$= .80 + .90 - .70 = 1$$
$$P(A' \cap B') = 1 - P(A \cap B)$$
$$= 1 - 1 = 0$$

The probability that a refugee in the camp was not poor nor seeking political asylum is 0.

61. This exercise should be solved by computer methods. The solution will vary according to the computer program that is used. The theoretical probabilities may be found in the following way.

When a coin is tossed 5 times, the sample space is

S = {hhhhh, hhhht, hhhth, hhthh,
      hthhh, thhhh, hhhtt, hhtht,
      hthht, thhht, hhtth, hthth,
      thhth, ththh, tthhh, htthh,
      hhttt, hthtt, httht, httth,
      ttthh, tthth, tthht, ththt,
      thhtt, thtth, tttth, tttht,
      tthtt, thttt, htttt, ttttt}.

(a) P(4 heads) $= \dfrac{5}{32} = .15625$

(b) P(hhthh) $= \dfrac{1}{32} = .03125$

63. This exercise should be solved by computer methods. The solution will vary according to the computer program that is used. The answer is .632.

Section 5.5

1. Let A be the event

"the number is 2" and B be the event "the number is odd."

The problem seeks the conditional probability $P(A|B)$. Use the definition

$$P(A|B) = \frac{P(A \cap B)}{P(B)}.$$

Here, $P(A \cap B) = 0$ and $P(B) = 1/2$. Thus,

$$P(A|B) = \frac{0}{1/2} = 0.$$

3. Let A be the event "the number is even" and B be the event "the number is 6." Then

$$P(A|B) = \frac{P(A \cap B)}{P(B)} = \frac{1/6}{1/6} = 1.$$

5. Let A be the event "sum of 6" and B be the event "double." 6 of the 36 ordered pairs have a sum of 6, so $P(B) = 6/36 = 1/6$. There is only one outcome, 3-3, in $A \cap B$, so $P(A \cap B) = 1/36$. Thus

$$P(A|B) = \frac{1/36}{1/6} = \frac{6}{36} = \frac{1}{6}.$$

7. Use a reduced sample space. After the first card drawn is a heart, there remain 51 cards, of which 12 are hearts. Thus,

P(heart on 2nd|heart on 1st)

$= \dfrac{12}{51} = \dfrac{4}{17}.$

9. Use a reduced sample space. After the first card drawn is a jack, there remain 51 cards, of which 11 are face cards. Thus,

P(face card on 2nd|jack on 1st)

$= \dfrac{11}{51}.$

11. P(a jack and a 10)

   = P(jack followed by 10)

   + P(10 followed by jack)

   $= \dfrac{4}{52} \cdot \dfrac{4}{51} + \dfrac{4}{52} \cdot \dfrac{4}{51}$

   $= \dfrac{16}{2652} + \dfrac{16}{2652}$

   $= \dfrac{32}{2652} \approx .012$

13. P(two black cards)

   = P(black on 1st)

   • P(black on 2nd|black on 1st)

   $= \dfrac{26}{52} \cdot \dfrac{25}{51}$

   $= \dfrac{650}{2652} \approx .245$

19. First draw the tree diagram.

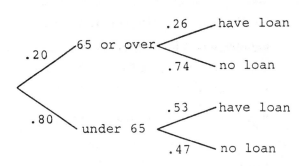

(a) P(person is 65 or over and has a loan)

   = P(65 or over)

   • P(has loan|65 or over)

   = (.20)(.26) = .052

(b) P(person has a loan)

   = P(65 or over and has loan)

   + P(under 65 and has loan)

   = (.20)(.26) + (.80)(.53)

   = .052 + .424 = .476

21. Draw the tree diagram.

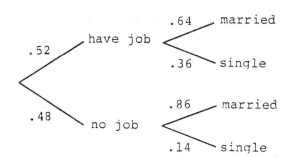

(a) P(married)

   = P(job and married)

   + P(no job and married)

   = (.52)(.64) + (.48)(.86)

   = .3328 + .4128 = .7456

(b) P(job and single)
$$= (.52)(.36) = .1872$$

23. Draw the tree diagram.

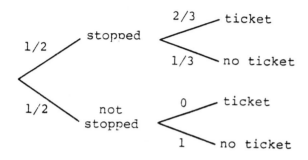

(a) P(no ticket)

$$= \text{P(stopped and no ticket)}$$
$$+ \text{P(not stopped and no ticket)}$$

$$= \left(\frac{1}{2}\right)\left(\frac{1}{3}\right) + \left(\frac{1}{2}\right)(1)$$

$$= \frac{1}{6} + \frac{1}{2} = \frac{2}{3}$$

(b) If his activities on the 3 consecutive weekends are independent, then

P(no ticket in 3 weekends)

$$= \text{P(no ticket)} \cdot \text{P(no ticket)}$$
$$\cdot \text{P(no ticket)}$$

$$= \frac{2}{3} \cdot \frac{2}{3} \cdot \frac{2}{3}$$

$$= \frac{8}{27} \approx .296$$

25. P(component fails) = .03

(a) Let n represent the number of these components to be connected in parallel; that is, the component has n − 1 backup components.

P(at least one component works)

$$= 1 - \text{P(no component works)}$$
$$= 1 - \text{P(all n components fail)}$$
$$= 1 - (.03)^n$$

If this probability is to be at least .999999, then

$$1 - (.03)^n \geq .999999$$
$$-(.03)^n \geq -.000001$$
$$(.03)^n \leq .000001,$$

and the smallest whole number value of n for which this inequality holds true is n = 4.

4 − 1 = 3 backup components must be used.

27. Let A be the event "student studies" and B be the event "student gets a good grade." We are told that P(A) = .6, P(B) = .7, and P(A ∩ B) = .52.

$$\text{P(A)} \cdot \text{P(B)} = (.6)(.7) = .42$$

Since P(A) · P(B) is not equal to P(A ∩ B), A and B are not independent. Rather, they are dependent events.

29. The probability that a customer cashing a check will fail to make a deposit is

$$P(D' | C) = \frac{n(D' \cap C)}{n(C)} = \frac{30}{80} = \frac{3}{8}.$$

31. The probability that a customer making a deposit will not cash a check is

$$P(C' | D) = \frac{n(C' \cap D)}{n(D)} = \frac{20}{70} = \frac{2}{7}.$$

33. **(a)** Since the separate flights are independent, the probability of 3 flights in a row is

$$(.98)(.98)(.98) = .941192 \approx .94.$$

35. Since 60% of production comes from assembly line B, $P(A) = .40$ (the remaining 40%). Also $P(\text{pass inspection}|A) = .95$, so $P(\text{not pass}|A) = .05$. Therefore,

$$P(A \cap \text{not pass})$$
$$= P(A) \cdot P(\text{not pass}|A)$$
$$= (.40)(.05) = .02.$$

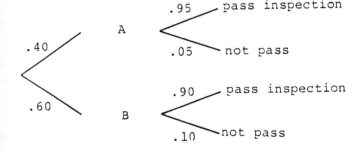

37. The sample space is $\{RW, WR, RR, WW\}$. The event "red" is $\{RW, WR, RR\}$ and the event "mixed" is $\{RW, WR\}$.

$$P(\text{mixed}|\text{red}) = \frac{n(\text{mixed and red})}{n(\text{red})}$$
$$= \frac{2}{3}.$$

39. $P(3 \text{ girls}|3\text{rd is a girl})$
$$= \frac{P(3 \text{ girls and 3rd is girl})}{P(3\text{rd is girl})}$$
$$= \frac{1/8}{1/2} = \frac{1}{4}$$

41. $P(3 \text{ girls}|\text{at least 2 girls})$
$$= \frac{P(3 \text{ girls and at least 2 girls})}{P(\text{at least 2 girls})}$$
$$= \frac{P(3 \text{ girls})}{P(\text{at least 2 girls})}$$
$$= \frac{P(3 \text{ girls})}{P(2 \text{ girls}) + P(3 \text{ girls})}$$
$$= \frac{1/8}{(3/8) + (1/8)}$$
$$= \frac{1/8}{4/8} = \frac{1}{4}$$

(Note that $P(3 \text{ girls}) = P(GGG)$
$$= \frac{1}{2} \cdot \frac{1}{2} \cdot \frac{1}{2} = \frac{1}{8}$$

and

$P(2 \text{ girls}) = P(GGB)$
$$+ P(BGG)$$
$$+ P(GBG)$$
$$= \frac{1}{8} + \frac{1}{8} + \frac{1}{8} = \frac{3}{8}.)$$

43. $P(M) = .527$, the total of the M column.

45. P(M ∩ C) = .042, the entry in the M column and C row.

47. $P(M|C) = \dfrac{P(M \cap C)}{P(C)} = \dfrac{.042}{.049} \approx .857$

(The exact value is $\dfrac{.042}{.049} = \dfrac{42}{49} = \dfrac{6}{7}$.)

49. $P(M'|C) = \dfrac{P(M' \cap C)}{P(C)}$

$= \dfrac{.007}{.049} \approx .143$

(The exact value is $\dfrac{.007}{.049} = \dfrac{7}{49} = \dfrac{1}{7}$.)

51. From the table,

P(C ∩ D) = .0004 and

P(C) • P(D) = (.0800)(.0050)

= .0004.

Since P(C ∩ D) = P(C) • P(D), C and D are independent events; color blindness and deafness are independent events.

53. First draw the tree diagram.

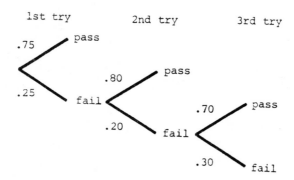

(a) P(fails both 1st and 2nd tests) = P(fails 1st) • P(fails 2nd|fails 1st)

= .25(.20)

= .05

(b) P(fails three times in a row) = (.25)(.2)(.3) = .015

(c) P(requires at least 2 tries) = P(does not pass on 1st try) = .25

55. There are 6 males of which 1 is a beagle and 1 a cocker spaniel, so there are 4 male poodles and hence 2 female poodles. The table is as follows.

	Beagle	Cocker Spaniel	Poodle	Totals
Male	1	1	4	6
Female	2	0	2	4
Totals	3	1	6	10

$$P(\text{beagle}|\text{male}) = \frac{n(\text{beagle and male})}{n(\text{male})} = \frac{1}{6}$$

57. $$P(\text{cocker spaniel}|\text{female}) = \frac{n(\text{cocker spaniel and female})}{n(\text{female})} = \frac{0}{4} = 0$$

59. $$P(\text{female}|\text{beagle}) = \frac{n(\text{female and beagle})}{n(\text{beagle})} = \frac{2}{3}$$

Section 5.6

1. Use Bayes' theorem with two possibilities M and M′.

$$P(M|N) = \frac{P(M) \cdot P(N|M)}{P(M) \cdot P(N|M) + P(M') \cdot P(N|M')} = \frac{(.4)(.3)}{(.4)(.3) + (.6)(.4)}$$

$$= \frac{.12}{.12 + .24} = \frac{.12}{.36} = \frac{12}{36} = \frac{1}{3}$$

3. Using Bayes' theorem,

$$P(R_1|Q) = \frac{P(R_1) \cdot P(Q|R_1)}{P(R_1) \cdot P(Q|R_1) + P(R_2) \cdot P(Q|R_2) + P(R_3) \cdot P(Q|R_3)}$$

$$= \frac{.05(.40)}{.05(.40) + .6(.30) + .35(.60)} = \frac{.02}{.41} = \frac{2}{41}.$$

5. Using Bayes' theorem,

$$P(R_3|Q) = \frac{P(R_3) \cdot P(Q|R_3)}{P(R_1) \cdot P(Q|R_1) + P(R_2) \cdot P(Q|R_2) + P(R_3) \cdot P(Q|R_3)}$$

$$= \frac{.35(.60)}{.05(.40) + .6(.30) + .35(.60)} = \frac{.21}{.41} = \frac{21}{41}.$$

7. We first draw the tree diagram and determine the probabilities as indicated below.

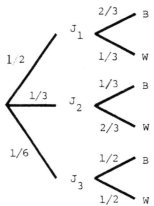

We want to determine the probability that if a white ball is drawn, it came from the second jar. This $P(J_2|W)$. Use Bayes' theorem.

$$P(J_2|W) = \frac{P(J_2) \cdot P(W|J_2)}{P(J_2) \cdot P(W|J_2) + P(J_1) \cdot P(W|J_1) + P(J_3) \cdot P(W|J_3)}$$

$$= \frac{\dfrac{1}{3} \cdot \dfrac{2}{3}}{\dfrac{1}{3} \cdot \dfrac{2}{3} + \dfrac{1}{2} \cdot \dfrac{1}{3} + \dfrac{1}{6} \cdot \dfrac{1}{2}} = \frac{\dfrac{2}{9}}{\dfrac{2}{9} + \dfrac{1}{6} + \dfrac{1}{12}} = \frac{\dfrac{2}{9}}{\dfrac{17}{36}} = \frac{8}{17}$$

9. Let G represent "good worker," B represent "bad worker," S represent "passes the test," and F represent "fails the test."

The given information is

$P(G) = .70$, $P(B) = P(G') = .30$, $P(S|G) = .80$ (and therefore $P(F|G) = .20$), and $P(S|B) = .40$ (and therefore $P(F|B) = .60$).

If passing the test is made a requirement for employment, then the percent of the new hires that will turn out to be good workers is

$$P(G|S) = \frac{P(G) \cdot P(S|G)}{P(G) \cdot P(S|G) + P(B) \cdot P(S|B)} = \frac{.70(.80)}{.70(.80) + .30(.40)} = \frac{.56}{.56 + .12}$$

$$= \frac{.56}{.68} \approx .824 \quad \text{or} \quad 82.4\%.$$

11. Let Q represent "qualified" and M represent "approved by the manager." Set up the tree diagram.

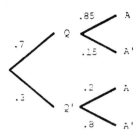

$$P(Q' \mid M) = \frac{P(Q') \cdot P(M \mid Q')}{P(Q) \cdot P(M \mid Q) + P(Q') \cdot P(M \mid Q')} = \frac{.30(.20)}{.70(.85) + .30(.20)} = \frac{.06}{.655} \approx .092$$

13. Let D represent "damaged," A represent "qualified," and B represent "unqualifed."
Set up the tree diagram.

$$P(B \mid D) = \frac{P(B) \cdot P(D \mid B)}{P(B) \cdot P(D \mid B) + P(A) \cdot P(D \mid A)} = \frac{.30(.05)}{.30(.05) + .70(.10)}$$

$$= \frac{.015}{.015 + .07} = \frac{.015}{.085} \approx .176$$

15. Let A represent "slow pay" and L represent "large down payment." Set up the tree diagram.

$$P(S' \mid L) = \frac{P(S') \cdot P(L \mid S')}{P(S) \cdot P(L \mid S) + P(S') \cdot P(L \mid S')} = \frac{.98(.5)}{.02(.14) + .98(.5)} = \frac{.49}{.4928} \approx .994$$

17. Start with the tree diagram, where the first stage refers to the companies and the second to a defective appliance.

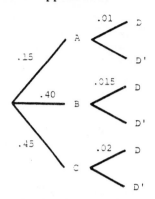

$$P(B|D) = \frac{P(B) \cdot P(D|B)}{P(A) \cdot P(D|A) + P(B(\cdot P(D|B) + P(C) \cdot P(D|C)}$$

$$= \frac{.40(.015)}{.15(.01) + .40(.015) + .45(.02)} = \frac{.0060}{.0165} \approx .364$$

19. Let that PF mean professional football, CF mean college football, B mean baseball, and HR mean high ratings. Draw the tree diagram.

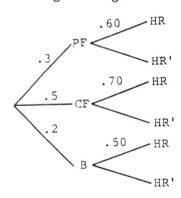

$$P(PF|HR) = \frac{P(PF) \cdot P(HR|PF)}{P(CF) \cdot P(HR|CF) + P(B) \cdot P(HR|B) + P(PF) \cdot P(HR|PF)}$$

$$= \frac{.3(.60)}{.5(.70) + .2(.50) + .3(.60)} = \frac{.18}{.35 + .10 + .18} = \frac{.18}{.63} = \frac{18}{63} = \frac{2}{7}$$

21. Let R mean recession, M mean mediocre, B mean booming, and H mean huge profits. Draw the tree diagram.

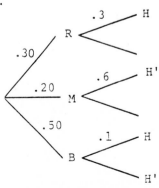

$$P(R|H) = \frac{P(R) \cdot P(H|R)}{P(R) \cdot P(H|R) + P(M) \cdot P(H|M) + P(B) \cdot P(H|B)}$$

$$= \frac{.30(.3)}{.30(.3) + .20(.6) + .50(.1)} = \frac{.09}{.26} = \frac{9}{26}$$

23. There are a total of 1260 + 700 + 560 + 280 = 2800 mortgages being studied at the bank.

(a) $P(5\% \text{ down}|\text{default}) =$

$$\frac{.05\left(\frac{1260}{2800}\right)}{.05\left(\frac{1260}{2800}\right) + .03\left(\frac{700}{2800}\right) + .02\left(\frac{560}{2800}\right) + .01\left(\frac{280}{2800}\right)}$$

$$= \frac{.05(.45)}{.05(.45) + .03(.25) + .02(.2) + .01(.1)}$$

$$= \frac{.0225}{.0225 + .0075 + .004 + .001} = \frac{.0225}{.035} \approx .643$$

(b) A mortgage being paid to maturity is the complement of a mortgage being defaulted.

$P(10\% \text{ down}|\text{paid to maturity}) = P(10\% \text{ down}|\text{not default})$

$$= \frac{.97\left(\frac{700}{2800}\right)}{.95\left(\frac{1260}{2800}\right) + .97\left(\frac{700}{2800}\right) + .98\left(\frac{560}{2800}\right) + .99\left(\frac{280}{2800}\right)}$$

$$= \frac{.97(.25)}{.95(.45) + .97(.25) + .98(.2) + .99(.1)}$$

$$= \frac{.2425}{.4275 + .2425 + .196 + .099} = \frac{.2425}{.965} \approx .251$$

25. (a) Draw the tree diagram.

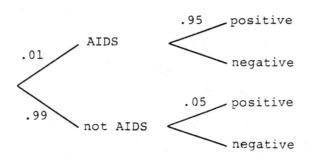

$$P(\text{AIDS}|\text{positive}) = \frac{.01(.95)}{.01(.95) + .99(.05)} = \frac{.0095}{.0095 + .0495} = \frac{.0095}{.059} \approx .161$$

27. P(30 - 34|man who never married)

$$= \frac{.126(.250)}{.151(.875) + .126(.433) + .126(.250) + .110(.140) + .487(.054)}$$

$$= \frac{.0315}{.259881} \approx .121$$

29. P(35 - 39|woman who never married)

$$= \frac{.103(.090)}{.142(.752) + .117(.295) + .116(.161) + .103(.090) + .522(.033)}$$

$$= \frac{.00927}{.186471} \approx .050$$

31. Draw the tree diagram.

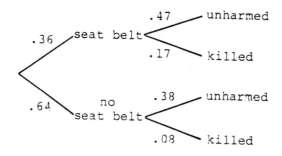

P(no seat belt|unharmed) $= \dfrac{.64(.38)}{.36(.47) + .64(.38)} = \dfrac{.2432}{.1692 + .2432} = \dfrac{.2432}{.4124} \approx .590$

33. Let V represent voting age and 65 represent 65 or over. Draw the tree diagram.

```
                        .48   V
              18-21 <
         .11          \  V'
              22-24 < .53   V
         .076         \  V'
         .376   .68  V
              25-44 <
         .283         \  V'
              45-64 < .64   V
         .155         \  V'
              65  < .74   V
                      \  V'
```

$$P(65|V) = \frac{.155(.74)}{.11(.48) + .076(.53) + .376(.68) + .283(.64) + .155(.74)} = \frac{.1147}{.64458} \approx .178$$

Chapter 5 Review Exercises

1. $9 \in \{8, 4, -3, -9, 6\}$

   False. 9 is not an element of the set.

3. $2 \notin \{0, 1, 2, 3, 4\}$

   False. 2 is an element of the set.

5. $\{3, 4, 5\} \subset \{2, 3, 4, 5, 6\}$

   True. Every element of $\{3, 4, 5\}$ is an element of $\{2, 3, 4, 5, 6\}$.

7. $\{3, 6, 9, 10\} \subset \{3, 9, 11, 13\}$

   False. 10 is an element of $\{3, 6, 9, 10\}$, but 10 is not an element of $\{3, 9, 11, 13\}$. Therefore, $\{3, 6, 9, 10\}$ is not a subset of $\{3, 9, 11, 13\}$.

9. $\{2, 8\} \not\subset \{2, 4, 6, 8\}$

   False. Since both 2 and 8 are elements of $\{2, 4, 6, 8\}$, $\{2, 8\}$ is a subset of $\{2, 4, 6, 8\}$.

11. $K = \{c, d, f, g\}$

    K has 4 elements, so it has $2^4 = 16$ subsets.

13. K' (the complement of K) is the set of all elements of U that do *not* belong to K.

    $K' = \{a, b, e\}$

15. $K \cap R$ (the intersection of K and R) is the set of all elements belonging to both set K and set R.

    $K \cap R = \{c, d, g\}$

17. $(K \cap R)' = \{a, b, e, f\}$

19. $\emptyset' = U$

21. $A \cap C$ is the set of all employees in the KO Brown Company who are in the accounting department *and* have at least 10 years in the company.

23. $A \cup D$ is the set of all employees in the KO Brown Company who are in the accounting department *or* have MBA degrees.

25. $B' \cap C'$ is the set of employees who are not in the sales department *and* have worked less than 10 years with the company.

27. $A \cup B'$ is the set of all elements which belong to A or do not belong to B, or both.

    See the Venn diagram in the back of the textbook.

29. $(A \cap B) \cup C$

    First find $A \cap B$.

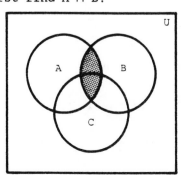

Now find the union of this region with C.

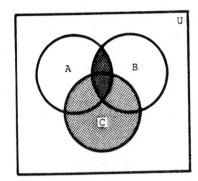

31. The sample space for rolling a die is
$$S = \{1, 2, 3, 4, 5, 6\}.$$

33. The sample space of the possible weights is
$$S = \{0, .5, 1, 1.5, 2, \ldots, 299.5, 300\}.$$

35. The sample space consists of all ordered pairs (a, b) where a can be 3, 5, 7, 9, or 11, and b is either R(red) or G(green). Thus,
$$S = \{(3, R), (3, G), (5, R), (5, G),$$
$$(7, R), (7, G), (9, R), (9, G),$$
$$(11, R), (11, G)\}.$$

37. The event that the second ball is green is
$$F = \{(3, G), (5, G), (7, G),$$
$$(9, G), (11, G)\}.$$

39. There are 13 hearts out of 52 cards in a deck. Thus,
$$P(\text{heart}) = \frac{13}{52} = \frac{1}{4}.$$

41. There are three face cards in each suit (jack, queen, and king) and there are four suits, so there are 3 · 4 = 12 face cards out of the 52 cards. Thus,
$$P(\text{face card}) = \frac{12}{52} = \frac{3}{13}.$$

43. There are 4 queens of which 2 are red, so
$$P(\text{red}|\text{queen}) = \frac{n(\text{red and queen})}{n(\text{queen})}$$
$$= \frac{2}{4} = \frac{1}{2}.$$

45. There are 4 kings of which all 4 are face cards. Thus,
$$P(\text{face card}|\text{king})$$
$$= \frac{n(\text{face card and king})}{n(\text{king})}$$
$$= \frac{4}{4} = 1.$$

51. Marilyn vos Savant's answer is that the contestant should switch doors. To understand why, recall that the puzzle begins with the contestant choosing door 1 and then the host opening door 3 to reveal a goat. When the host opens door 3 and shows the goat, that does not affect the probability of the car being behind door 1; the contestant had a $\frac{1}{3}$ probability of being correct to begin with, and he still has a $\frac{1}{3}$ probability after the host opens door 3.

The contestant knew that the host would open another door regardless of what was behind door 1, so opening either other door gives no new information about door 1. The probability of the car being behind door 1 is still $\frac{1}{3}$; with the goat behind door 3, the only other place the car could be is behind door 2, so the probability that the car is behind door 2 is now $\frac{2}{3}$. By switching to door 2, the contestant can double his chances of winning the car.

53. Let E represent the event "draw a black jack." $P(E) = 2/52 = 1/26$ and then $P(E') = 25/26$. The odds in favor of drawing a black jack are

$$\frac{P(E)}{P(E')} = \frac{1/26}{25/26} = \frac{1}{25},$$

or "1 to 25."

55. The sum is 8 for each of the 5 outcomes 2-6, 3-5, 4-4, 5-3, and 6-2. There are 36 outcomes in all in the sample space.

$$P(\text{sum is } 8) = \frac{5}{36} \approx .139$$

57. $P(\text{sum is at least } 10)$

$= P(\text{sum is } 10) + P(\text{sum is } 11)$

$\quad + P(\text{sum is } 12)$

$= \frac{3}{36} + \frac{2}{36} + \frac{1}{36}$

$= \frac{6}{36} = \frac{1}{6} \approx .167$

59. The sum can be 9 or 11.

$P(\text{sum is } 9) = 4/36$ and

$P(\text{sum is } 11) = 2/36$.

$P(\text{sum is odd number greater than } 8)$

$= \frac{4}{36} + \frac{2}{36} = \frac{6}{36} = \frac{1}{6} \approx .167$

61. Consider the reduced sample space of the 11 outcomes in which at least one die is a four. Of these, 2 have a sum of 7, 3-4 and 4-3. Therefore,

$P(\text{sum is } 7 \,|\, \text{at least one die is a four})$

$= \frac{2}{11} \approx .182$.

63. $P(E) = .51.$ $P(F) = .37,$

$P(E \cap F) = .22$

(a) $P(E \cup F) = P(E) + P(F) - P(E \cap F)$

$= .51 + .37 - .22$

$= .66$

(b) Draw a Venn diagram.

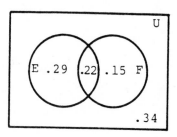

$E \cap F'$ is the portion of the diagram that is inside E and outside F.

$P(E \cap F') = .29$

(c) $E' \cup F$ is outside E or inside F, or both.

$P(E' \cup F) = .22 + .15 + .34 = .71$

(d) $E' \cap F'$ is outside E and outside F.

$P(E' \cap F') = .34$

65. The probability that the ball came from box B, given that it is red, is

$P(B|\text{red})$

$$= \frac{P(B)P(\text{red}|B)}{P(B)P(\text{red}|B) + P(A)P(\text{red}|A)}$$

$$= \frac{\frac{5}{8}\left(\frac{2}{5}\right)}{\frac{5}{8}\left(\frac{2}{5}\right) + \frac{3}{8}\left(\frac{5}{6}\right)}$$

$$= \frac{4}{9} \approx .444$$

67. Let C represent "competent shop" and R represent "able to repair appliance."

Draw a tree diagram and label the given information.

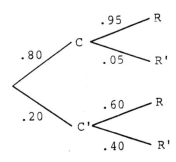

The probability that an appliance that was repaired correctly was repaired by an incompetent shop is

$P(C'|R) = \dfrac{P(C' \cap R)}{P(R)}$

$$= \frac{P(C') \cdot P(R|C')}{P(C) \cdot P(R|C) + P(C') \cdot P(R|C')}$$

$$= \frac{.20(.60)}{.80(.95) + .20(.60)}$$

$$= \frac{.12}{.76 + .12} = \frac{.12}{.88} = \frac{12}{88}$$

$$= \frac{3}{22} \approx .136.$$

69. The probability that an appliance that was repaired incorrectly was repaired by an incompetent shop is

$P(C'|R') = \dfrac{P(C' \cap R')}{P(R')}$

$$= \frac{.20(.40)}{.20(.40) + .80(.05)}$$

$$= \frac{.08}{.12} = \frac{8}{12} = \frac{2}{3}.$$

71. (a) P(no more than 3 defects)

$$= P(0) + P(1) + P(2) + P(3)$$

$$= .31 + .25 + .18 + .12$$

$$= .86$$

(b) P(at least 3 defects)

$$= P(3) + P(4) + P(5)$$

$$= .12 + .08 + .06$$

$$= .26$$

73. (a)

	N_2	T_2
N_1	$N_1 N_2$	$N_1 T_2$
T_1	$T_1 N_2$	$T_1 T_2$

Since the four combinations are equally likely, each has probability 1/4.

(b) P(two trait cells)

$$= P(T_1 T_2) = \frac{1}{4}$$

(c) P(one normal cell and one trait cell)

$$= P(N_1 T_2) + P(T_1 N_2)$$

$$= \frac{1}{4} + \frac{1}{4} = \frac{1}{2}$$

(d) P(not a carrier and does not have the disease)

$$= P(N_1 N_2) = \frac{1}{4}$$

75. **(a)** P(answer yes) = P(answer B) • P(answer yes|answer B)

         + P(answer A) • P(answer yes|answer A)

Divide by P(answer B).

$$\frac{P(\text{answer yes})}{P(\text{answer B})} = P(\text{answer yes|answer B}) + \frac{P(\text{answer A}) \cdot P(\text{yes|A})}{P(\text{answer B})}$$

Solve for P(answer yes|answer B).

$$P(\text{answer yes|answer B}) = \frac{P(\text{answer yes}) - P(\text{answer A}) \cdot P(\text{yes|A})}{P(\text{answer B})}$$

(b) Using the formula from part (a),

$$\frac{(.6) - \left(\frac{1}{2}\right)\left(\frac{1}{2}\right)}{\left(\frac{1}{2}\right)} = \frac{7}{10}.$$

77. **(a)**

Car Type	Satisfied	Not Satisfied	Totals
New	300	100	400
Used	450	150	600
Totals	750	250	1000

(b) 1000 buyers were surveyed.

(c) 300 bought a new car and were satisfied.

(d) 250 were not satisfied.

(e) 600 bought used cars.

(f) 150 who were not satisfied had bought a used car.

(g) The event is "the buyer purchased a used car given that the buyer is not satisfied."

(h) P(used car|not satisfied) = $\dfrac{n(\text{used car and not satisfied})}{n(\text{not satisfied})}$

          = $\dfrac{150}{250} = \dfrac{3}{5}$

(i) P(used car and not satisfied) = $\dfrac{n(\text{used car and not satisfied})}{n(\text{buyers})}$

          = $\dfrac{150}{1000} = \dfrac{3}{20}$

Extended Application

1. Using Bayes' theorem,

$$P(H_2|C_1) = \frac{P(C_1|H_2) \cdot P(H_2)}{P(C_1|H_1) \cdot P(H_1) + P(C_1|H_2) \cdot P(H_2) + P(C_1|H_3) \cdot P(H_3)}$$

$$= \frac{.4(.15)}{.9(.8) + .4(.15) + .1(.05)} = \frac{.06}{.785} \approx .076.$$

2. Using Bayes' theorem,

$$P(H_1|C_2) = \frac{P(C_2|H_1) \cdot P(H_1)}{P(C_2|H_1) \cdot P(H_1) + P(C_2|H_2) \cdot P(H_2) + P(C_2|H_3) \cdot P(H_3)}$$

$$= \frac{(.2)(.8)}{(.2)(.8) + (.8)(.15) + (.3)(.05)} \approx .542.$$

3. Using Bayes' theorem,

$$P(H_3|C_2) = \frac{P(C_2|H_3) \cdot P(H_3)}{P(C_2|H_1) \cdot P(H_1) + P(C_2|H_2) \cdot P(H_2) + P(C_2|H_3) \cdot P(H_3)}$$

$$= \frac{.3(.05)}{.2(.8) + .8(.15) + .3(.05)} = \frac{.015}{.295} \approx .051.$$

CHAPTER 5 TEST

1. Write true or false for each statement.

    (a) $3 \in \{1, 5, 7, 9\}$ **(b)** $\{1, 3\} \not\subset \{0, 1, 2, 3, 4\}$ **(c)** $\emptyset \subset \{2\}$

    (d) A set of 6 distinct elements has exactly 64 subsets.

2. Let $U = \{1, 2, 3, 4, 5, 6, 7, 8, 9\}$, $A = \{1, 3, 4, 5\}$, $B = \{2, 4, 5\}$, and
    $C = \{1, 3, 5, 7\}$.

    Find each of the following sets.

    (a) $A \cap B'$ **(b)** $A \cap (B \cup C')$

3. Draw a Venn diagram and shade the region that represents $A \cap (B \cup C')$.

4. Draw a Venn diagram and fill in regions given that $n(U) = 25$, $n(A) = 11$,
    $n(B \cap A') = 9$, and $n(A \cap B) = 6$.

5. In the Mellonville Social Club, bridge, poker, and rummy are the 3 most
    popular card games. A recent survey of the members found that

                    60 play poker;
                    63 play bridge;
                    37 play rummy;
                    22 play poker and rummy;
                    19 play bridge and rummy;
                    23 play bridge only;
                    10 play all three games;
                     7 play none of the games.

    (a) Draw Venn diagram and fill in the numbers in each region.

    (b) How many members do not play poker?

    (c) How may members play bridge or poker?

    (d) How many members play poker but not rummy?

6. Suppose that for events A and B, $P(A) = .4$, $P(B) = .3$ and $P(A \cup B) = .68$. Find each of the following probabilities.

   (a) $P(A \cap B)$ (b) $P(A')$ (c) $P(A \cap B')$ (d) $P(A' \cap B')$

7. An urn contains 4 red, 3 blue, and 2 yellow marbles. A single marble is drawn.

   (a) Find the odds in favor of drawing a red marble.

   (b) Find the probability that a red or a blue marble is drawn.

8. Three cards are drawn without replacement from a standard deck of 52.

   (a) What is the probability that all three are spades?

   (b) What is the probability that all three spades, given that the first card drawn is a spade?

9. The probability of passing the University of Waterloo's physical fitness test is .3. If you fail the first time, your chances of passing on the second try drop to .1. Draw a tree diagram and compute the probability that a person will pass on the first or second try.

10. The Magnum Opus Publishing Company uses three printers to put out its lengthy tomes. Printer A produces 40% of their books with a 20% failure rate. Printer B produces 25% with a 10% failure rate. Printer C produces the remainder with a 40% failure rate. Given that a book is badly printed, what is the probability that it was printed by Printer C?

CHAPTER 5 TEST ANSWERS

1. **(a)** False   **(b)** False   **(c)** True   **(d)** True

2. **(a)** $\{1, 3\}$ **(b)** $\{4, 5\}$

3.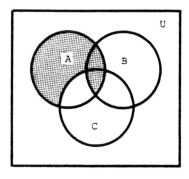

4.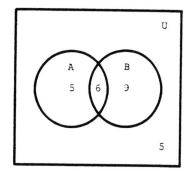

5. **(a)** 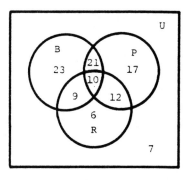   **(b)** 45   **(c)** 92   **(d)** 38

6. **(a)** .02   **(b)** .6   **(c)** .38   **(d)** .32

7. **(a)** 4 to 5   **(b)** 7/9

8. **(a)** $\left(\frac{32}{52}\right)\left(\frac{12}{51}\right)\left(\frac{11}{50}\right) \approx .013$ **(b)** $\left(\frac{12}{51}\right)\left(\frac{11}{50}\right) \approx .052$

9.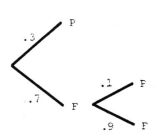

10. $\dfrac{(.35)(.4)}{(.35)(.4) + (.4)(.2) + (.25)(.1)} \approx .57$

P(pass) = .3 + .07 = .37

CHAPTER 6 COUNTING PRINCIPLES; FURTHER PROBABILITY TOPICS

Section 6.1

1. $P(4, 2) = \dfrac{4!}{(4 - 2)!} = \dfrac{4!}{2!}$

    $= \dfrac{4 \cdot 3 \cdot 2!}{2!} = 12$

3. $6! = 6 \cdot 5 \cdot 4 \cdot 3 \cdot 2 \cdot 1 = 720$

5. $0! = 1$

    This is the definition of $0!$.

7. $4! = 4 \cdot 3 \cdot 2 \cdot 1 = 24$

9. $P(13, 2) = \dfrac{13!}{(13 - 2)!} = \dfrac{13!}{11!}$

    $= \dfrac{13 \cdot 12 \cdot 11!}{11!}$

    $= 13 \cdot 12 = 156$

11. $P(25, 12) = \dfrac{25!}{(25 - 12)!}$

    $= \dfrac{25!}{13!}$

    $= 2.490952 \times 10^{15}$

13. $P(14, 5) = \dfrac{14!}{(14 - 5)!}$

    $= \dfrac{14!}{9!}$

    $= \dfrac{14 \cdot 13 \cdot 12 \cdot 11 \cdot 10 \cdot 9!}{9!}$

    $= 14 \cdot 13 \cdot 12 \cdot 11 \cdot 10$

    $= 240,240$

15. By the multiplication principle, there will be $5 \cdot 3 \cdot 2 = 30$ different home types available.

17. By the multiplication principle, there will be $3 \cdot 8 \cdot 5 = 120$ different meals possible.

19. **(a)** The number of ways 5 works can be arranged is

    $$P(5, 5) = 5! = 120.$$

    (b) If one of the 2 overtures must be chosen first, followed by arrangements of the 4 remaining pieces, then

    $$P(2, 1) \cdot P(4, 4) = 2 \cdot 24 = 48$$

    is the number of ways the program can be arranged.

21. The number of possible arrangements is

    $$P(10, 7) = \dfrac{10!}{3!} = 604,800.$$

23. Pick any 4 of the 380 nonmathematical courses.

    The number of possible schedules is

    $P(380, 4) = \dfrac{380!}{376!}$

    $= 380 \cdot 379 \cdot 378 \cdot 377$

    $= 20,523,714,120.$

    In scientific notation, this answer would be written as 2.05237×10^{10}.

25. The number of possible batting orders is

    $$P(20, 9) = \dfrac{20!}{11!}$$

    $$= 60,949,324,800.$$

    In scientific notation, this answer would be written as 6.09493×10^{10}.

27. **(a)** Pick one of the 5 traditional numbers followed by an arrangement of the remaining total of 7. The program can be arranged in

$P(5, 1) \cdot P(7, 7) = 5 \cdot 7! = 25,200$ different ways.

(b) Pick one of the 3 original Cajun compositions to play last, preceded by an arrangement of the remaining total of 7. This program can be arranged in

$P(7, 7) \cdot P(3, 1) = 7! \cdot 3 = 15,120$ different ways.

29. By the multiplication principle, a person could schedule the evening of television viewing in $8 \cdot 5 \cdot 6 = 240$ different ways.

31. There is exactly one 3-letter subset of the letters A, B, and C, namely A, B, and C.

33. **(a)** initial

This word contains 3 i's, 1 n, 1 t, 1 a, and 1 ℓ. Use the formula for distinguishable permutations with $n = 7$, $n_1 = 3$, $n_2 = 1$, $n_3 = 1$, $n_4 = 1$ and $n_5 = 1$.

$$\frac{7!}{3!1!1!1!1!} = \frac{7 \cdot 6 \cdot 5 \cdot 4 \cdot 3!}{3!} = 840$$

There are 840 distinguishable permutations of the letters.

(b) little

Use the formula for distinguishable permutations with $n = 6$, $n_1 = 2$, $n_2 = 1$, $n_3 = 2$, and $n_4 = 1$.

$$\frac{6!}{2!1!2!1!} = \frac{6!}{2!2!}$$
$$= \frac{6 \cdot 5 \cdot 4 \cdot 3 \cdot 2 \cdot 1}{2 \cdot 1 \cdot 2 \cdot 1}$$

There are 180 distinguishable permutations.

(c) decreed

Use the formula for distinguishable permutations with $n = 7$, $n_1 = 2$, $n_2 = 3$, $n_3 = 1$, and $n_4 = 1$.

$$\frac{7!}{2!3!1!1!} = \frac{7!}{2!3!}$$
$$= \frac{7 \cdot 6 \cdot 5 \cdot 4 \cdot 3!}{2 \cdot 1 \cdot 3!}$$
$$= 420$$

There are 420 distinguishable permutations.

35. **(a)** The 9 books can be arranged in $P(9, 9) = 9! = 362,880$ ways.

(b) The blue books can be arranged in 4! ways, the green books can be arranged in 3! ways, and the red books can be arranged in 2! ways. There are 3! ways to choose the order of the 3 groups of books. Therefore, using the multiplication principle, the number of possible arrangements is

$$4!3!2!3! = 24 \cdot 6 \cdot 2 \cdot 6$$
$$= 1728.$$

(c) Use the formula for distinguishable permutations with $n = 9$, $n_1 = 4$, $n_2 = 3$, and $n_3 = 2$.

The number of distinguishable arrangements is

$$\frac{9!}{4!3!2!} = \frac{9 \cdot 8 \cdot 7 \cdot 6 \cdot 5 \cdot 4!}{4! \cdot 6 \cdot 2}$$
$$= 1260.$$

37. 5 of the 11 drugs can be administered in

$$P(11, 5) = \frac{11!}{(11 - 5)!} = \frac{11!}{6!}$$
$$= \frac{11 \cdot 10 \cdot 9 \cdot 8 \cdot 7 \cdot 6!}{6!}$$
$$= 55,440$$

different sequences.

39. (a) There are 4 tasks to be performed in selecting 4 letters for the call letters. The first task may be done in 2 ways, the second in 25, the third in 24, and the fourth in 23. By the multiplication principle, there will be $2 \cdot 25 \cdot 24 \cdot 23 = 27,600$ different call letter names possible.

(b) With repeats possible, there will be $2 \cdot 26 \cdot 26 \cdot 26 = 2 \cdot 26^3$ or 35,152 call letter names possible.

(c) To start with W or K, make no repeats, and end in R, there will be $2 \cdot 24 \cdot 23 \cdot 1 = 1104$ possible call letter names.

41. (a) Our number system has ten digits, which are 1 through 9 and 0.

There are 3 tasks to be performed in selecting 3 digits for the area code. The first task may be done in 8 ways, the second in 2, and the third in 10. By multiplication principle, there will be

$$8 \cdot 2 \cdot 10 = 160$$

different area codes possible.

There are 7 tasks to be performed in selecting 7 digits for the telephone number. The first task may be done in 8 ways, and the other 6 tasks may each be done in 10 ways. By the multiplication principle, there will be

$$8 \cdot 10^6 = 8,000,000$$

different telephone numbers possible within each area code.

(b) Some numbers, such as 911, 800, and 900 are reserved for special purposes and therefore unavailable for use as an area code.

43. There would be 4 tasks to be performed in selecting 4 digits for this new type of area code. The first task could be done in 8 ways, the second in 2, the third in 10, and the fourth in 10.
By the multiplication principle, there would be

$$8 \cdot 2 \cdot 10^2 = 1600$$

different area codes possible with this plan.

45. **(a)** There were $26^3 \cdot 10^3 = 17{,}576{,}000$ license plates possible that had 3 letters followed by 3 digits.

(b) There were $10^3 \cdot 26^3 = 17{,}576{,}000$ new license plates possible when plates were also issued having 3 digits followed by 3 numbers.

(c) There were
$26 \cdot 10^3 \cdot 26^3 = 456{,}976{,}000$ new license plates possible when plates were also issued having 1 letter followed by 3 digits and then 3 letters.

47. If there are no restrictions on the digits used, there would be $10^5 = 100{,}000$ different 5-digit zip codes possible.
If the first digit is not allowed to be 0, there would be $9 \cdot 10^4 = 90{,}000$ zip codes possible.

Section 6.2

1. To evaluate $\binom{8}{3}$, use the formula

$$\binom{n}{r} = \frac{n!}{(n-r)!\,r!}$$ with $n = 8$ and $r = 3$.

$$\binom{8}{3} = \frac{8!}{(8-3)!\,3!}$$
$$= \frac{8!}{5!\,3!}$$
$$= \frac{8 \cdot 7 \cdot 6 \cdot 5!}{5! \cdot 3 \cdot 2 \cdot 1} = 56$$

3. $\binom{12}{5} = \frac{12!}{(12-5)!\,5!} = \frac{12!}{7!\,5!}$
$$= \frac{12 \cdot 11 \cdot 10 \cdot 9 \cdot 8 \cdot 7!}{7! \cdot 5 \cdot 4 \cdot 3 \cdot 2 \cdot 1} = 792$$

5. $\binom{6}{0} = \frac{6!}{(6-0)!\,0!}$
$$= \frac{6!}{6!\,0!} = 1$$

7. $\binom{21}{10} = \frac{21!}{(21-10)!\,10!}$
$$= \frac{21!}{11!\,10!}$$
$$= 352{,}716$$

9. $\binom{25}{16} = \frac{25!}{(25-16)!\,16!}$
$$= \frac{25!}{9!\,16!}$$
$$= 2{,}042{,}975$$

11. $\binom{27}{10} = \frac{27!}{(27-10)!\,10!}$
$$= \frac{27!}{17!\,10!}$$
$$= 8{,}436{,}285$$

13. Pick 13 from 52. There are
$$\binom{52}{13} = \frac{52!}{13!\,39!}$$
$$= 635{,}013{,}559{,}600$$
different 13-card bridge hands possible.
In scientific notation, this answer would be written as 6.3501×10^{11}.

15. **(a)** There are
$$\binom{5}{2} = \frac{5!}{3!\,2!} = \frac{5 \cdot 4 \cdot 3!}{3! \cdot 2 \cdot 1} = 10$$
different 2-card combinations possible.

(b) The 10 possible hands are

$\{1, 2\}, \{2, 3\}, \{3, 4\}, \{4, 5\},$
$\{1, 3\}, \{2, 4\}, \{3, 5\}, \{1, 4\},$
$\{2, 5\}, \{1, 5\}.$

Of these, 7 contain a card numbered less than 3.

17. To have *at least* 2 good hitters among the 3 chosen, there will either be *exactly* 2 good hitters or 3 good hitters. The number of ways the coach can choose exactly 2 good hitters (and hence 1 poor hitter) is

$$\binom{5}{2} \cdot \binom{4}{1} = 10 \cdot 4 = 40.$$

The number of ways the coach can choose 3 good hitters is

$$\binom{5}{3} = 10.$$

The total number of ways to select at least 2 good hitters is

$$40 + 10 = 50.$$

19. Choose 2 letters from $\{L, M, N\}$; order is important.

(a)

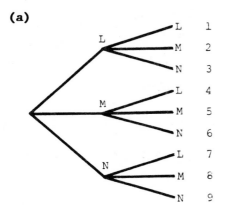

There are 9 ways to choose 2 letters if repetition is allowed.

(b)

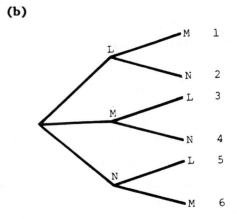

There are 6 ways to choose 2 letters if no repeats are allowed.

(c) $\binom{3}{2} = \dfrac{3!}{1!2!} = 3$

This answer differs from both (a) and (b).

23. Order does not matter in choosing members of a committee, so use combinations rather than permutations.

(a) The number of committees whose members are all men is

$$\binom{8}{5} = \frac{8!}{3!5!} = \frac{8 \cdot 7 \cdot 6 \cdot 5!}{3 \cdot 2 \cdot 1 \cdot 5!}$$
$$= 56.$$

(b) The number of committees whose members are all women is

$$\binom{11}{5} = \frac{11!}{6!5!} = \frac{11 \cdot 10 \cdot 9 \cdot 8 \cdot 7 \cdot 6!}{6! \cdot 5 \cdot 4 \cdot 3 \cdot 2 \cdot 1}$$
$$= 462.$$

(c) The 3 men can be chosen in

$$\binom{8}{3} = \frac{8!}{5!3!} = \frac{8 \cdot 7 \cdot 6 \cdot 5!}{5! \cdot 3 \cdot 2 \cdot 1}$$
$$= 56 \text{ ways}.$$

The 2 women can be chosen in

$$\binom{11}{2} = \frac{11!}{9!2!} = \frac{11 \cdot 10 \cdot 9!}{9! \cdot 2 \cdot 1}$$
$$= 55 \text{ ways}.$$

Using the multiplication principle, a committee of 3 men and 2 women can be chosen in

$$56 \cdot 55 = 3080 \text{ ways}.$$

25. Order does not matter, so use combinations.

(a) The 3 students who will take part in the course can be chosen in

$$\binom{12}{3} = \frac{12!}{9!3!}$$
$$= \frac{12 \cdot 11 \cdot 10 \cdot 9!}{9! \cdot 3 \cdot 2 \cdot 1}$$
$$= 220 \text{ ways}.$$

(b) The 9 students who will not take part in the course can be chosen in

$$\binom{12}{9} = \frac{12!}{3!9!} = 220 \text{ ways}.$$

27. Order is important, so use permutations.
The number of ways in which the children can find seats is

$$P(12, 11) = \frac{12!}{(12 - 11)!} = \frac{12!}{1!}$$
$$= 12!$$
$$= 479,001,600.$$

In scientific notation, this answer would be written as 4.79×10^8.

29. Since order is important, use a permutation. (Each secretary is being assigned to a manager, which is essentially the same as putting them in numbered slots.)
The secretaries can be selected in $P(7, 3) = 7 \cdot 6 \cdot 5 = 210$ different ways.

31. Since order does not matter, use combinations.

(a) There are $\binom{25}{3} = 2300$ possible samples of 3 apples.

(b) There are $\binom{5}{3} = 10$ possible samples of 3 rotten apples.

(c) There are $\binom{5}{1}\binom{20}{2} = 950$ possible samples with exactly 1 rotten apple.

33. Since order is important, use a permutation. The plants can be arranged in

$$P(9, 5) = 9 \cdot 8 \cdot 7 \cdot 6 \cdot 5 = 15,120$$
different ways.

35. Since order does not matter, use combinations.

(a) There are $\binom{20}{5} = 15,504$ different ways to select five of the orchids.

(b) If 2 special orchids must be included in the show, that leaves 18 other orchids from which the other 3 orchids for the show must be chosen. This can be done in $\binom{18}{3}$ = 816 different ways.

37. $\binom{n}{n-1} = \dfrac{n!}{[n-(n-1)]!(n-1)!}$

$\qquad = \dfrac{n!}{1!(n-1)!}$

$\qquad = \dfrac{n \cdot (n-1)!}{1 \cdot (n-1)!}$

$\qquad = n$

39. There are 2 types of meat and 6 types of extras. Order does not matter here, so use combinations.

(a) There are $\binom{2}{1}$ ways to choose one type of meat and $\binom{6}{3}$ ways to choose exactly three extras. By the multiplication principle, there are

$$\binom{2}{1} \cdot \binom{6}{3} = 2 \cdot 20 = 40$$

different ways to order a hamburger with exactly three extras.

(b) There are $\binom{6}{3}$ = 20 different ways to choose exactly three extras.

(c) "At least five extras" means "5 extras or 6 extras." There are $\binom{6}{5}$ different ways to choose exactly 5 extras and $\binom{6}{6}$ ways to choose

exactly 6 extras, so there are $\binom{6}{5} + \binom{6}{6}$ = 6 + 1 = 7 different ways to choose at least five extras.

41. Select 8 of the 16 smokers and 8 of the 20 nonsmokers; order does not matter in the group, so use combinations. There are

$$\binom{16}{8} \cdot \binom{20}{8} = 1,621,233,900$$

different ways to select the study group. In scientific notation, this answer would be written as 1.6212×10^9.

43. Order does not matter in choosing a delegation, so use combinations. This committee has 5 + 4 = 9 members.

(a) There are

$$\binom{9}{3} = \frac{9!}{6!3!}$$

$$= \frac{9 \cdot 8 \cdot 7 \cdot 6!}{6!3 \cdot 2 \cdot 1}$$

$$= 84 \text{ possible delegations.}$$

(b) To have all Democrats, the number of possible delegations is

$$\binom{5}{3} = 10.$$

(c) To have 2 Democrats and 1 Republican, the number of possible delegations is

$$\binom{5}{2} \cdot \binom{4}{1} = 10 \cdot 4 = 40.$$

(d) We have previously calculated that there are 84 possible delegations, of which 10 consist of all Democrats. Those 10 delegations are the only ones with no Republicans, so the remaining 84 − 10 = 74 delegations include at least one Republican.

45. In the lottery, 6 different numbers are to be chosen from the 99 numbers.

(a) There are

$$\binom{99}{6} = \frac{99!}{93!6!} = 1,120,529,256$$

different ways to choose 6 numbers if order is not important. In scientific notation, this answer would be written as 1.1205×10^9.

(b) There are

$$P(99, \ 6) = \frac{99!}{93!} = 806,781,064,320$$

different ways to choose 6 numbers if order matters. In scientific notation, this answer would be written as 8.0678×10^{11}.

Section 6.3

1. There are $\binom{10}{3}$ samples of 3 apples.

$$\binom{10}{3} = \frac{10 \cdot 9 \cdot 8}{3 \cdot 2 \cdot 1} = 120$$

There are $\binom{6}{3}$ samples of 3 red apples.

$$\binom{6}{3} = \frac{6 \cdot 5 \cdot 4}{3 \cdot 2 \cdot 1} = 20$$

Thus,

$$P(\text{all red apples}) = \frac{20}{120} = \frac{1}{6}.$$

3. There are $\binom{4}{2}$ samples of 2 yellow apples.

$$\binom{4}{2} = \frac{4 \cdot 3}{2 \cdot 1} = 6$$

There are $\binom{6}{1} = 6$ samples of 1 red apple. Thus, there are 6 · 6 = 36 samples of 3 in which 2 are yellow and 1 red. Thus,

$$P(\text{2 yellow and 1 red apple}) = \frac{36}{120}$$
$$= \frac{3}{10}.$$

5. The number of 2-card hands is

$$\binom{52}{2} = \frac{52 \cdot 51}{2 \cdot 1} = 1326.$$

7. There are $\binom{52}{2} = 1326$ different 2-card hands. The number of 2-card hands with exactly one ace is

$$\binom{4}{1}\binom{48}{1} = 4 \cdot 48 = 192.$$

The number of 2-card hands with two aces is

$$\binom{4}{2} = 6.$$

Thus there are 198 hands with at least one ace. Therefore,

P(the 2-card hand contains an ace)

$$= \frac{198}{1326} = \frac{33}{221} \approx .149.$$

9. There are $\binom{52}{2} = 1326$ different 2-card hands. There are $\binom{13}{2} = 78$ ways to get a 2-card hand where both cards are of a single named suit, but there are 4 suits to choose from. Thus,

P(two cards of same suit)

$$= \frac{4 \cdot \binom{13}{2}}{\binom{52}{2}} = \frac{312}{1326} = \frac{52}{221} \approx .235.$$

11. There are $\binom{52}{2} = 1326$ different 2-card hands. There are 12 face cards in a deck, so there are 40 cards that are not face cards.

Thus,

P(no face cards)

$$= \frac{\binom{40}{2}}{\binom{52}{2}} = \frac{780}{1326} = \frac{130}{221} \approx .588.$$

13. There are 26 choices for each slip pulled out, and there are 5 slips pulled out, so there are

$$26^5 = 11,881,376$$

different "words" that can be formed from the letters. If the "word" must be "chuck," there is only one choice for each of the 5 letters

(the first slip must contain a "c," the second an "h," and so on). Thus,

P(word is "chuck")

$$= \frac{1^5}{26^5} = \left(\frac{1}{26}\right)^5 \approx 8.42 \times 10^{-8}.$$

15. There are $26^5 = 11,881,376$ different "words" that can be formed. If the "word" is to have no repetition of letters, then there are 26 choices for the first letter, but only 25 choices for the second (since the letters must all be different), 24 choices for the third, and so on. Thus,

P(all different letters)

$$= \frac{26 \cdot 25 \cdot 24 \cdot 23 \cdot 22}{26^5}$$

$$= \frac{1 \cdot 25 \cdot 24 \cdot 23 \cdot 22}{26^4}$$

$$= \frac{303,600}{456,976}$$

$$= \frac{18,975}{28,561} \approx .664.$$

17. There are $\binom{52}{5}$ different 5-card poker hands. There are 4 royal flushes, one for each suit. Thus,

P(royal flush)

$$= \frac{4}{\binom{52}{5}} = \frac{4}{2,598,960}$$

$$= \frac{1}{649,740}$$

$$\approx 1.54 \times 10^{-6}.$$

19. The four of a kind can be chosen in 13 ways and then is matched with 1 of the remaining 48 cards to make a 5-card hand containing 4 of a kind, Thus there are $13 \cdot 48 = 624$ poker hands with 4 of a kind. It follows that

P(four of a kind)

$$= \frac{624}{\binom{52}{5}} = \frac{624}{2,598,960}$$

$$= \frac{1}{4165} \approx 2.40 \times 10^{-4}.$$

21. There are $\binom{52}{13}$ different 13-card bridge hands. Since there are only 13 hearts, there is exactly one way to get a bridge hand containing only hearts. Thus,

$$P(\text{only hearts}) = \frac{1}{\binom{52}{13}}.$$

23. There are $\binom{4}{3}$ ways to obtain 3 aces, $\binom{4}{3}$ ways to obtain 3 kings, and $\binom{44}{7}$ ways to obtain the 7 remaining cards. Thus,

P(exactly 3 aces and exactly 3 kings)

$$= \frac{\binom{4}{3}\binom{4}{3}\binom{44}{7}}{\binom{52}{13}}.$$

25. There are 21 books, so the number of selections of any 6 books is

$$\binom{21}{6} = 54,264.$$

(a) The probability that the selection consisted of 3 Hughes and 3 Morrison books is

$$\frac{\binom{9}{3}\binom{7}{3}}{\binom{21}{6}} = \frac{84 \cdot 35}{54,264} = \frac{2940}{54,264} \approx .054.$$

(b) A selection containing exactly 4 Baldwin books will contain 2 of the 16 books by the other authors, so the probability is

$$\frac{\binom{5}{4}\binom{16}{2}}{\binom{21}{6}} = \frac{5 \cdot 120}{54,264} = \frac{600}{54,264} \approx .011.$$

(c) The probability of a selection consisting of 2 Hughes, 3 Baldwin, and 1 Morrison book is

$$\frac{\binom{9}{2}\binom{5}{3}\binom{7}{1}}{\binom{21}{6}} = \frac{36 \cdot 10 \cdot 7}{54,264}$$

$$= \frac{2520}{54,274} \approx .046.$$

(d) A selection consisting of at least 4 Hughes books may contain 4, 5, or 6 Hughes books, with any remaining books by the other authors. Therefore, the probability is

$$\frac{\binom{9}{4}\binom{12}{2} + \binom{9}{5}\binom{12}{1} + \binom{9}{6}}{\binom{21}{6}}$$

$$= \frac{126 \cdot 66 + 126 \cdot 12 + 84}{54,264}$$

$$= \frac{8316 + 1512 + 84}{54,264}$$

$$= \frac{9912}{54,264} \approx .183.$$

(e) Since there are 9 Hughes books and 5 Baldwin books, there are 14 books written by males. The probability of a selection with exactly 4 books written by males is

$$\frac{\binom{14}{4}\binom{7}{2}}{\binom{21}{6}} = \frac{1001 \cdot 21}{54,264}$$

$$= \frac{21,021}{54,264} \approx .387.$$

(f) A selection with no more than 2 books written by Baldwin may contain 0, 1, or 2 books by Baldwin, with the remaining books by the other authors. Therefore, the probability is

$$\frac{\binom{16}{6} + \binom{5}{1}\binom{16}{5} + \binom{5}{2}\binom{16}{4}}{\binom{21}{6}}$$

$$= \frac{8008 + 5 \cdot 4368 + 10 \cdot 1820}{54,264}$$

$$= \frac{8008 + 21,840 + 18,200}{54,264}$$

$$= \frac{48,048}{54,264} \approx .885.$$

29. P(at least 2 presidents have
the same birthday)

= 1 - P(no 2 presidents have
the same birthday)

The number of ways that 41 people can have the same or different birthdays is $(365)^{41}$. The number of ways that 41 people can have all different birthdays is the number of permutations of 365 things taken 41 at a time or P(365, 41). Thus,

P(at least 2 presidents have
the same birthday)

$$= 1 - \frac{P(365, 41)}{365^{41}}.$$

(Be careful to realize that the symbol P is sometimes used to indicate permutations and sometimes used to indicate probability; in this solution, the symbol is used both ways.)

31. Since there are 435 members of the House of Representatives, and there are only 365 days in a year, it is a certain event that at least 2 people will have the same birthday. Thus,

P(at least 2 members have
the same birthday) = 1.

33. Each of the four people can choose to get off at any one of the seven floors, so there are 7^4 ways the four people can leave the elevator. The number of ways the people can leave at different floors is the number of permutations of 7 things (floors) taken 4 at a time or

P(7, 4) = 7 · 6 · 5 · 4 = 840.

Thus, the probability that no two passengers leave at the same floor is

$$\frac{P(7, 4)}{7^4} = \frac{840}{2401} = \frac{120}{343} \approx .3499.$$

(Note the similarity of this problem and the "birthday problem.")

35. There are 9 ways to choose 1 typewriter from the shipment of 9. Since 2 of the 9 are defective, there are 7 ways to choose 1 nondefective typewriter. Thus,

P(1 drawn from the 9 is not defective)

= 7/9.

37. There are $\binom{9}{3}$ ways to choose 3 typewriters.

$$\binom{9}{3} = \frac{9!}{3!6!} = \frac{9 \cdot 8 \cdot 7}{3 \cdot 2 \cdot 1} = 84$$

There are $\binom{7}{3}$ ways to choose 3 nondefective typewriters.

$$\binom{7}{3} = \frac{7!}{3!4!} = \frac{7 \cdot 6 \cdot 5}{3 \cdot 2 \cdot 1} = 35$$

Thus,

P(3 drawn from the 9 are non-defective)

$$= \frac{35}{84} = \frac{5}{12}.$$

39. There are $\binom{12}{4} = 495$ different ways to choose 4 engines for testing from the crate of 12. A crate will not be shipped if any of the 4 in the sample is defective. If there are 2 defectives in the crate, then there are $\binom{10}{4} = 210$ ways of choosing a sample with no defectives. Thus,

P(shipping a crate with 2 defectives)

$$= \frac{210}{495} \approx .424.$$

41. There are $\binom{99}{6} = 1,120,529,256$ different ways to pick 6 numbers from 1 to 99, but there is only 1 way to win; the 6 numbers you pick must exactly match the 6 winning numbers, without regard to order. Thus,

P(win the big prize)

$$= \frac{1}{1,120,529,256}$$

$$\approx 8.9 \times 10^{-10}.$$

43. **(a)** There were 28 games played in the season, since the numbers in the "Won" column have a sum of 28 (and the numbers in the "Lost" column have a sum of 28). Each of those 28 games had 2 possible outcomes; either Team A won and Team B lost, or else Team A lost and Team B won. By the multiplication principle, this means that there were 2^{28} different win/lose progressions possible. Any one of the 8 teams could have been the one that won all of its games, any one of the remaining 7 teams could have been the one that won all but one of its games, and so on, until there is only one team left and it is the one that lost all of its games. By the multiplication principle, this means that there were

$$8 \cdot 7 \cdot 6 \cdot 5 \cdot 4 \cdot 3 \cdot 2 \cdot 1 = 8!$$

different "perfect progressions" possible.

Thus,

P("perfect progression" in an
  8-team league)

$$= \frac{8!}{2^{28}} \approx .000150.$$

(b) If there are n teams in the
league, then the "Won" column will
begin with n − 1, followed by n − 2,
then n − 3, and so on down to 0. It
can be shown that the sum of these n

numbers is $\frac{n(n-1)}{2}$, so there are

$2^{n(n-1)/2}$ different win/lose
progressions possible. The n teams
can be ordered in n! different ways,
so there are n! different "perfect
progressions" possible. Thus,

P("perfect progression" in an
  n-team league)

$$= \frac{n!}{2^{n(n-1)/2}}.$$

45. This exercise should be solved by
using the Monte Carlo method. The
answers may vary from run to run
since the random numbers will vary
from run to run. By non-computer
methods, the answers are as follows.

(a) P(no aces) $= \dfrac{\binom{48}{13}}{\binom{52}{13}} \approx .3038$

(b) P(2 kings and 2 aces)

$$= \frac{\binom{4}{2}\binom{4}{2}\binom{44}{9}}{\binom{52}{13}} \approx .0402$$

(c) P(only 3 suits are represented)

$$= \frac{4 \cdot \binom{39}{13}}{\binom{52}{13}} \approx .6651$$

Section 6.4

1. This is a Bernoulli trial problem
with P(success) = P(girl) = 1/2.
The probability of exactly x
successes in n trials is

$$\binom{n}{x} p^x (1 - p)^{n-x},$$

where p is the probability of
success in a single trial. We
have n = 5, x = 2, and p = 1/2.

Note that $1 - p = 1 - \frac{1}{2} = \frac{1}{2}$.

P(exactly 2 girls)

$$= \binom{5}{2}\left(\frac{1}{2}\right)^2\left(\frac{1}{2}\right)^3$$

$$= \frac{10}{32} = \frac{5}{16} \approx .313$$

3. We have n = 5, x = 0, p = 1/2, and
1 − p = 1/2.

$$\text{P(no girls)} = \binom{5}{0}\left(\frac{1}{2}\right)^0\left(\frac{1}{2}\right)^5$$

$$= \frac{1}{32} \approx .031$$

5. "At least 4 girls" means either 4 or
5 girls.

P(at least 4 girls)

$$= \binom{5}{4}\left(\frac{1}{2}\right)^4\left(\frac{1}{2}\right)^1 + \binom{5}{5}\left(\frac{1}{2}\right)^5\left(\frac{1}{2}\right)^0$$

$$= \frac{5}{32} + \frac{1}{32} = \frac{6}{32} = \frac{3}{16} \approx .188$$

7. P(no more than 3 boys)

$$= 1 - P(\text{at least 4 boys})$$

$$= 1 - P(\text{4 boys or 5 boys})$$

$$= 1 - [P(\text{4 boys}) + P(\text{5 boys})]$$

$$= 1 - \left(\frac{5}{32} + \frac{1}{32}\right)$$

$$= 1 - \frac{6}{32}$$

$$= 1 - \frac{3}{16} = \frac{13}{16} \approx .813$$

9. On one roll, $P(1) = \frac{1}{6}$. We have

$n = 12$, $x = 12$, and $p = 1/6$.

Note that $1 - p = 5/6$. Thus,

P(exactly twelve 1's)

$$= \binom{12}{12}\left(\frac{1}{6}\right)^{12}\left(\frac{5}{6}\right)^{0}$$

$$\approx 4.6 \times 10^{-10}.$$

11. P(exactly one 1)

$$= \binom{12}{1}\left(\frac{1}{6}\right)^{1}\left(\frac{5}{6}\right)^{11} \approx .269$$

13. "No more than three 1's" means 0, 1, 2, or 3 ones. Thus,

P(no more than three 1's)

$$= P(\text{zero 1's}) + P(\text{one 1})$$

$$+ P(\text{two 1's}) + P(\text{three 1's})$$

$$= \binom{12}{0}\left(\frac{1}{6}\right)^{0}\left(\frac{5}{6}\right)^{12} + \binom{12}{1}\left(\frac{1}{6}\right)^{1}\left(\frac{5}{6}\right)^{11}$$

$$+ \binom{12}{2}\left(\frac{1}{6}\right)^{2}\left(\frac{5}{6}\right)^{10} + \binom{12}{3}\left(\frac{1}{6}\right)^{3}\left(\frac{5}{6}\right)^{9}$$

$$\approx .875.$$

15. Each time the coin is tossed

$P(\text{head}) = \frac{1}{2}$. We have $n = 5$,

$x = 5$, $p = 1/2$, and $1 - p = 1/2$.

Thus,

P(all heads)

$$= \binom{5}{5}\left(\frac{1}{2}\right)^{5}\left(\frac{1}{2}\right)^{0}$$

$$= \frac{1}{32} \approx .031.$$

17. P(no more than 3 heads)

$$= P(\text{0 heads}) + P(\text{1 head})$$

$$+ P(\text{2 heads}) + P(\text{3 heads})$$

$$= \binom{5}{0}\left(\frac{1}{2}\right)^{0}\left(\frac{1}{2}\right)^{5} + \binom{5}{1}\left(\frac{1}{2}\right)^{1}\left(\frac{1}{2}\right)^{4}$$

$$+ \binom{5}{2}\left(\frac{1}{2}\right)^{2}\left(\frac{1}{2}\right)^{3} + \binom{5}{3}\left(\frac{1}{2}\right)^{3}\left(\frac{1}{2}\right)^{2}$$

$$= \frac{26}{32} = \frac{13}{16} \approx .813$$

21. We have $n = 10$, $x = 2$, $p = .2$, and $1 - p = .8$, so

$$P(x = 2) = \binom{10}{2}(.2)^{2}(.8)^{8} \approx .302.$$

23. We have

P(x ≤ 3)

$$= P(x = 0) + P(x = 1)$$

$$+ P(x = 2) + P(x = 3)$$

$$= \binom{10}{0}(.2)^{0}(.8)^{10} + \binom{10}{1}(.2)^{1}(.8)^{9}$$

$$+ \binom{10}{2}(.2)^{2}(.8)^{8} + \binom{10}{3}(.2)^{3}(.8)^{7}$$

$$\approx .879.$$

25. We have $n = 10$, $x = 10$, $p = .9$, and $1 - p = .1$, so

$$P(x = 10) = \binom{10}{10}(.9)^{10}(.1)^{0}$$

$$\approx .349.$$

27. We have

$$P(x \geq 9)$$
$$= P(x = 9) + P(x = 10)$$
$$= \binom{10}{9}(.9)^9(.1)^1 + \binom{10}{10}(.9)^{10}(.1)^0$$
$$\approx .387 + .349$$
$$= .736.$$

29. We have $n = 6$, $x = 2$, $p = 1/5$, and $1 - p = 4/5$. Thus,

P(exactly 2 correct)
$$= \binom{6}{2}\left(\tfrac{1}{5}\right)^2\left(\tfrac{4}{5}\right)^4$$
$$\approx .246.$$

31. We have

P(at least 4 correct)
$$= P(4 \text{ correct}) + P(5 \text{ correct})$$
$$\quad + P(6 \text{ correct})$$
$$= \binom{6}{4}\left(\tfrac{1}{5}\right)^4\left(\tfrac{4}{5}\right)^2 + \binom{6}{5}\left(\tfrac{1}{5}\right)^5\left(\tfrac{4}{5}\right)^1$$
$$\quad + \binom{6}{6}\left(\tfrac{1}{5}\right)^6\left(\tfrac{4}{5}\right)^0$$
$$\approx .017.$$

33. We have $n = 3$, $x = 1$, $p = .1$, and $1 - p = .9$. Thus,

P(exactly 1 loses everything)
$$= \binom{3}{1}(.1)^1(.9)^2 \approx .243.$$

35. We have $n = 20$, $p = .05$, and $1 - p = .95$. Thus,

P(at most 2 defective transistors)
$$= P(x \leq 2)$$
$$= P(x = 0) + P(x = 1) + P(x = 2)$$

$$= \binom{20}{0}(.05)^0(.95)^{20} + \binom{20}{1}(.05)^1(.95)^{19}$$
$$\quad + \binom{20}{2}(.05)^2(.95)^{18}$$
$$\approx .925.$$

37. We have $n = 20$, $x = 17$, $p = .7$, and $1 - p = .3$. Thus,

P(exactly 17 cured)
$$= \binom{20}{17}(.7)^{17}(.3)^3$$
$$= (1140)(.7)^{17}(.3)^3$$
$$\approx .072.$$

39. P(at least 18 cured)
$$= P(\text{exactly 18 cured})$$
$$\quad + P(\text{exactly 19 cured})$$
$$\quad + P(20 \text{ cured})$$
$$= \binom{20}{18}(.7)^{18}(.3)^2 + \binom{20}{19}(.7)^{19}(.3)^1$$
$$\quad + \binom{20}{20}(.7)^{20}(.3)^0$$
$$= 190(.7)^{18}(.3)^2 + 20(.7)^{19}(.3)$$
$$\quad + (.7)^{20}$$
$$\approx .035$$

41. We have $n = 100$, $p = .012$, and $1 - p = .988$. Thus,

$$P(x \leq 2)$$
$$= P(x = 0) + P(x = 1) + P(x = 2)$$
$$= \binom{100}{0}(.012)^0(.988)^{100}$$
$$\quad + \binom{100}{1}(.012)^1(.988)^{99}$$
$$\quad + \binom{100}{2}(.012)^2(.988)^{98}$$
$$= (.012)^0(.988)^{100}$$
$$\quad + 100(.012)^1(.988)^{99}$$
$$\quad + 4950(.012)^2(.988)^{98}$$
$$\approx .881.$$

43. We have n = 6, x = 3, p = .70, and 1 − p = .30. Thus,

P(exactly 3 recover)

$$= \binom{6}{3}(.7)^3(.3)^3$$

$$\approx .185.$$

45. P(no more than 3 recover)

= P(0 recover) + P(1 recovers)

  + P(2 recover) + P(3 recover)

$$= \binom{6}{0}(.7)^0(.3)^6 + \binom{6}{1}(.7)^1(.3)^5$$

$$+ \binom{6}{2}(.7)^2(.3)^4 + \binom{6}{3}(.7)^3(.3)^3$$

$$\approx .256$$

47. We have n = 10,000,

p = 2.5 × 10⁻⁷ = .00000025, and

1 − p = .99999975. Thus,

P(at least 1 mutation occurs)

  = 1 − P(none occur)

$$= 1 - \binom{10,000}{0}(p)^0(1 - p)^{10,000}$$

$$= 1 - (.99999975)^{10,000} \approx .0025.$$

49. We have n = 5, p = 1/3, and 1 − p = 2/3. Thus,

P(x ≤ 3)

  = P(x = 0) + P(x = 1) + P(x = 2)

   + P(x = 3)

$$= \binom{5}{0}\left(\tfrac{1}{3}\right)^0\left(\tfrac{2}{3}\right)^5 + \binom{5}{1}\left(\tfrac{1}{3}\right)^1\left(\tfrac{2}{3}\right)^4$$

$$+ \binom{5}{2}\left(\tfrac{1}{3}\right)^2\left(\tfrac{2}{3}\right)^3 + \binom{5}{3}\left(\tfrac{1}{3}\right)^3\left(\tfrac{2}{3}\right)^2$$

$$= \frac{232}{243} \approx .955.$$

51. We have n = 10, x = 7, p = .2, and 1 − p = .8. Thus,

$$P(x = 7) = \binom{10}{7}(.2)^7(.8)^3 \approx .00079.$$

53. We have

P(x < 8)

  = 1 − P(x ≥ 8)

  = 1 − [P(x = 8) + P(x = 9) + P(x = 10)]

  = 1 − P(x = 8) − P(x = 9) − P(x = 10)

$$= 1 - \binom{10}{8}(.2)^8(.8)^2 - \binom{10}{9}(.2)^9(.8)^1$$

$$- \binom{10}{10}(.2)^{10}(.8)^0$$

  = 1 − .000074 − .000004 − .0000001

  = .999922.

55. Let success mean not getting the flu. Then we have n = 134, p = .80, and 1 − p = .20.

(a) If exactly 10 of the people get the flu, then x = 134 − 10 = 124 since we have let success mean *not* getting the flu. Thus,

$$P(x = 124) = \binom{134}{124}(.80)^{124}(.20)^{10}$$

$$\approx .000036.$$

(b) If no more than 10 people get the flu, then more than 10 people did *not* get the flu, so we are interested in x ≥ 10. Thus,

P(x ≥ 10)

  = 1 − P(x < 10)

  = 1 − P(x = 9) − P(x = 8) − P(x = 7)

   − P(x = 6) − P(x = 5) − P(x = 4)

   − P(x = 3) − P(x = 2) − P(x = 1)

   − P(x = 0)

  ≈ .000054.

(c) If none of the people get the flu, then x = 134.   Thus,

P(x = 134)

$$= \binom{134}{134}(.80)^{134}(.20)^0$$

$$\approx .000000000000103$$

$$\approx 1.0 \times 10^{-13}.$$

57. Let success mean producing a defective item.   Then we have n = 75, p = .05, and 1 − p = .95.

(a) If there are exactly 5 defective items, then x = 5.   Thus,

$$P(x = 5) = \binom{75}{5}(.05)^5(.95)^{70}$$

$$\approx .148774.$$

(b) If there are no defective items, then x = 0.   Thus,

$$P(x = 0) = \binom{75}{0}(.05)^0(.95)^{75}$$

$$\approx .021344.$$

(c) If there is at least 1 defective item, then we are interested in x ≥ 1.   We have

$$P(x \geq 1) = 1 - P(x = 0)$$

$$\approx 1 - .021344$$

$$= .978656.$$

Section 6.5

1. **(a)** P(1 germinated) $= \frac{0}{10} = 0$,

P(2 germinated) $= \frac{1}{10} = .1$,

and so on.   The probability distribution is as follows.

Number Germinated	0	1	2	3	4	5
Probability	0	0	.1	.3	.4	.2

(b) The histogram corresponding to this probability distribution consists of 6 rectangles, the first two of which have no height.   See the histogram in the back of the textbook.

3. **(a)** P(0 bullseyes) $= \frac{0}{25} = 0$,

P(1 bullseye) $= \frac{1}{25} = .04$,

and so on.

Number of Bullseyes	0	1	2	3	4	5	6
Probability	0	.04	0	.16	.40	.32	.08

(b) See the histogram in the back of the textbook.

5. **(a)** P(0 with disease) $= \frac{3}{20} = .15$,

P(1 with disease) $= \frac{5}{20} = .25$,

and so on.

Number with Disease	0	1	2	3	4	5
Probability	.15	.25	.3	.15	.1	.05

(b) See the histogram in the back of the textbook.

7. Let x denote the number of heads observed. Then x can take on 0, 1, 2, 3, 4 as values. The probabilities are as follows.

$$P(x = 0) = \binom{4}{0}\left(\frac{1}{2}\right)^0\left(\frac{1}{2}\right)^4 = \frac{1}{16}$$

$$P(x = 1) = \binom{4}{1}\left(\frac{1}{2}\right)^1\left(\frac{1}{2}\right)^3 = \frac{4}{16} = \frac{1}{4}$$

$$P(x = 2) = \binom{4}{2}\left(\frac{1}{2}\right)^2\left(\frac{1}{2}\right)^2 = \frac{6}{16} = \frac{3}{8}$$

$$P(x = 3) = \binom{4}{3}\left(\frac{1}{2}\right)^3\left(\frac{1}{2}\right)^1 = \frac{4}{16} = \frac{1}{4}$$

$$P(x = 4) = \binom{4}{4}\left(\frac{1}{2}\right)^4\left(\frac{1}{2}\right)^0 = \frac{1}{16}$$

Therefore, the probability distribution is as follows.

Number of Heads	0	1	2	3	4
Probability	$\frac{1}{16}$	$\frac{1}{4}$	$\frac{3}{8}$	$\frac{1}{4}$	$\frac{1}{16}$

9. Let x denote the number of aces drawn. Then x can take on values 0, 1, 2, 3. The probabilities are as follows.

$$P(x = 0) = \binom{3}{0}\left(\frac{48}{52}\right)\left(\frac{47}{51}\right)\left(\frac{46}{50}\right) \approx .783$$

$$P(x = 1) = \binom{3}{1}\left(\frac{4}{52}\right)\left(\frac{48}{51}\right)\left(\frac{47}{50}\right) \approx .204$$

$$P(x = 2) = \binom{3}{2}\left(\frac{4}{52}\right)\left(\frac{3}{51}\right)\left(\frac{48}{50}\right) \approx .013$$

$$P(x = 3) = \binom{3}{3}\left(\frac{4}{52}\right)\left(\frac{3}{51}\right)\left(\frac{2}{50}\right) \approx .0002$$

Therefore, the probability distribution is as follows.

Number of Aces	0	1	2	3
Probability	.783	.204	.013	.0002

For Exercises 11–15, see the histograms in the back of the textbook.

11. Draw a histogram with 5 rectangles, corresponding to x = 0, x = 1, x = 2, x = 3, and x = 4. $P(x \leq 2)$ corresponds to

$$P(x = 0) + P(x = 1) + P(x = 2),$$

so shade the first 3 rectangles in the histogram.

13. Draw a histogram with 4 rectangles, corresponding to x = 0, x = 1, x = 2, and x = 3.

P(at least one ace) = $P(x \geq 1)$

corresponds to

$$P(x = 1) + P(x = 2) + P(x = 3),$$

so shade the last 3 rectangles.

15. Draw a histogram with 5 rectangles, the same histogram as in Exercise 11. P(x = 2 or x = 3) corresponds to P(x = 2) + P(x = 3), so shade those 2 rectangles.

17. E(x) = 2(.1) + 3(.4) + 4(.3) + 5(.2)
     = 3.6

19. $E(z) = 9(.14) + 12(.22) + 15(.36)$
    $\qquad + 18(.18) + 21(.10)$
    $\quad = 14.64$

21. It is possible (but not necessary) to begin by writing the histogram's data as a probability distribution, which would look as follows.

x	1	2	3	4
P(x)	.2	.3	.1	.4

The expected value of x is

$E(x) = 1(.2) + 2(.3) + 3(.1) + 4(.4)$
$\qquad = 2.7.$

23. The expected value of x is

$E(x) = 6(.1) + 12(.2) + 18(.4)$
$\qquad + 24(.2) + 30(.1)$
$\qquad = 18.$

25. Using the data from Example 6, the expected winnings for Mary are

$E(x) = (-.4)\left(\frac{1}{4}\right) + .4\left(\frac{1}{4}\right)$
$\qquad + .4\left(\frac{1}{4}\right) + (-.4)\left(\frac{1}{4}\right)$
$\qquad = 0.$

Yes, it is still a fair game if Mary tosses and Donna calls.

27. Below is the probability distribution of x, which stands for the person's net winnings.

x	$99	$39	−$1
P(x)	$\frac{1}{500} = .002$	$\frac{2}{500} = .004$	$\frac{497}{500} = .994$

The expected value of the person's winnings is

$E(x) = 99(.002) + 39(.004)$
$\qquad + (-1)(.994)$
$\qquad \approx -\$.64 \quad \text{or} \quad -64\text{¢}.$

Since the expected value of the winnings is not 0, this is not a fair game.

29. The number of possible samples is

$$\binom{7}{3} = \frac{7!}{3!4!} = \frac{7 \cdot 6 \cdot 5}{3 \cdot 2 \cdot 1} = 35.$$

The number of samples containing no yellows and therefore 3 whites is $\binom{4}{3} = 4$, so the probability of drawing a sample containing no yellow is 4/35.

The number of samples containing 1 yellow and therefore 2 whites is $\binom{3}{1}\binom{4}{2} = 3 \cdot 6 = 18$, so the probability of drawing a sample containing 1 yellow is 18/35.

Similarly, the probability of drawing a sample containing 2 yellows is

$$\frac{\binom{3}{2}\binom{4}{1}}{\binom{7}{3}} = \frac{12}{35},$$

and the probability of drawing a sample containing 3 yellows is

$$\frac{\binom{3}{3}}{\binom{7}{3}} = \frac{1}{35}.$$

Let x denote the number of yellow marbles drawn. The probability distribution of x is as follows.

x	0	1	2	3
P(x)	$\frac{4}{35}$	$\frac{18}{35}$	$\frac{12}{35}$	$\frac{1}{35}$

The expected value is

$$E(x) = 0\left(\frac{4}{35}\right) + 1\left(\frac{18}{35}\right) + 2\left(\frac{12}{35}\right) + 3\left(\frac{1}{35}\right)$$

$$= \frac{45}{35} = \frac{9}{7} \approx 1.3 \text{ yellow marbles.}$$

31. The probability that the delegation contains no liberals and 3 conservatives is

$$\frac{\binom{5}{0}\binom{4}{3}}{\binom{9}{3}} = \frac{1 \cdot 4}{84} = \frac{4}{84}.$$

Similarly, use combinations to calculate the remaining probabilities for the probability distribution.

(a) Let x denote the number of liberals on the delegation. The probability distribution of x is as follows.

x	0	1	2	3
P(x)	$\frac{4}{84}$	$\frac{30}{84}$	$\frac{40}{84}$	$\frac{10}{84}$

The expected value is

$$E(x) = 0\left(\frac{4}{84}\right) + 1\left(\frac{30}{84}\right) + 2\left(\frac{40}{84}\right) + 3\left(\frac{10}{84}\right)$$

$$= \frac{140}{84} = \frac{5}{3} \approx 1.67 \text{ liberals.}$$

(b) Let y denote the number of conservatives on the committee. The probability distribution of y is as follows.

y	0	1	2	3
P(y)	$\frac{10}{84}$	$\frac{40}{84}$	$\frac{30}{84}$	$\frac{4}{84}$

The expected value is

$$E(y) = 0\left(\frac{10}{84}\right) + 1\left(\frac{40}{84}\right) + 2\left(\frac{30}{84}\right) + 3\left(\frac{4}{84}\right)$$

$$= \frac{112}{84} = \frac{4}{3} \approx 1.33 \text{ conservatives.}$$

33. Let x represent the number of junior members on the committee. Use combinations to find the probabilities of 0, 1, 2, and 3 junior members. The probability distribution of x is as follows.

x	0	1	2	3
P(x)	$\frac{57}{203}$	$\frac{95}{203}$	$\frac{45}{203}$	$\frac{6}{203}$

The expected value is

$$E(x) = 0\left(\frac{57}{203}\right) + 1\left(\frac{95}{203}\right) + 2\left(\frac{45}{203}\right) + 3\left(\frac{6}{203}\right)$$

$$= 1 \text{ junior member.}$$

35. The probability of drawing 2 diamonds is

$$\frac{\binom{13}{2}}{\binom{52}{2}} = \frac{78}{1326},$$

and the probability of not drawing 2 diamonds is

$$1 - \frac{78}{1326} = \frac{1248}{1326}.$$

Let x denote your net winnings. Then the expected value of the game is

$$E(x) = 4.5\left(\frac{78}{1326}\right) + (-.5)\left(\frac{1248}{1326}\right) = -\frac{273}{1326} \approx \$-.21 \quad \text{or} \quad -21\cancel{c}.$$

The game is not fair since your expected winnings are not zero.

37. The probability of getting exactly 3 of the 4 selections correct and winning this game is $4\left(\frac{1}{13}\right)^3\left(\frac{12}{13}\right) \approx .001681$. The probability of losing is .998319. If you win, your winnings are \$199. Otherwise, you lose \$1 (win -\$1). If x denotes your winnings, then the expected value is

$$E(x) = 199(.001681) + (-1)(.998319)$$

$$= .334519 - .998319 = -.6638$$
$$\approx -\$.66 \quad \text{or} \quad -66\cancel{c}.$$

39. In this form of roulette, $P(\text{even}) = \frac{18}{37}$ and $P(\text{noneven}) = \frac{19}{37}$.

If an even number comes up, you win \$1. Otherwise, you lose \$1 (win -\$1.) If x denotes your winnings, then the expected value is

$$E(x) = 1\left(\frac{18}{37}\right) + (-1)\left(\frac{19}{37}\right) = -\frac{1}{37} \approx -2.7\cancel{c}.$$

41. In this form of the game Keno,

$$P(\text{your number comes up}) = \frac{20}{80} = \frac{1}{4} \text{ and } P(\text{your number doesn't come up}) = \frac{60}{80} = \frac{3}{4}.$$

If your number comes up, you win \$2.20. Otherwise, you lose \$1 (win -\$1). If x denotes your winings, then the expected value is

$$E(x) = 2.20\left(\frac{1}{4}\right) - 1\left(\frac{3}{4}\right) = .55 - .75 = -\$.20 \quad \text{or} \quad -20\cancel{c}.$$

45. The expected value is

$$E(x) = 0(.01) + 1(.05) + 2(.15) + 3(.26) + 4(.33) + 5(.14) + 6(.06)$$
$$= 3.51 \text{ complaints.}$$

Account number	Existing volume	Potential add volume	Probability of getting it	Expected value	Exist. vol. + exp. value	Class
1	$15,000	$10,000	.25	$2500	$17,500	C
2	$40,000	$0	---	-----	$40,000	C
3	$20,000	$10,000	.20	$2000	$22,000	C
4	$50,000	$10,000	.10	$1000	$51,000	B
5	$5000	$50,000	.50	$25,000	$30,000	C
6	$0	$100,000	.60	$60,000	$60,000	A
7	$30,000	$20,000	.80	$16,000	$46,000	B

47. (table above)

49. **(a)** Let x denote the amount of damage in millions of dollars. For seeding, the expected value is

$$E(x) = (.038)(335.8) + (.143)(191.1) + (.392)(100) + (.255)(46.7) + (.172)(16.3)$$
$$\approx \$94.0 \text{ million}$$

For not seeding, the expected value is

$$E(x) = (.054)(335.8) + (.206)(191.1) + (.480)(100) + (.206)(46.7) + (.054)(16.3)$$
$$\approx \$116.0 \text{ million}.$$

(b) Seed, since the total expected damage is less with that option.

51. Let x denote the winnings. The probability distribution is as follows.

x	P(x)
100,000	1/2,000,000
40,000	2/2,000,000
10,000	2/2,000,000
0	1,999,995/2,000,000

The expected value is

$$E(x) = 100,000\left(\frac{1}{2,000,000}\right) + 40,000\left(\frac{2}{2,000,000}\right) + 10,000\left(\frac{2}{2,000,000}\right) + 0\left(\frac{1,999,995}{2,000,000}\right)$$
$$= .05 + .04 + .01 + 0 = \$.10 = 10¢$$

Since the expecting winnings are 10¢, if entering the contest costs 50¢ then it would not be worth it to enter.

Chapter 6   Review Exercises

1. 6 shuttle vans can line up at the airport in

$$P(6, 6) = 6! = 720$$

different ways.

3. 3 oranges can be taken from a bag of 12 in

$$\binom{12}{3} = \frac{12!}{9!3!} = \frac{12 \cdot 11 \cdot 10}{3 \cdot 2 \cdot 1} = 220$$

different ways.

5. 2 pictures from a group of 5 different pictures can be arranged in

$$P(5, 2) = 5 \cdot 4 = 20$$

different ways.

7. (a) There are 2! ways to arrange the landscapes, 3! ways to arrange the puppies, and 2 choices whether landscapes or puppies come first. Thus, the pictures can be arranged in

$$(2!)(3!) \cdot 2 = 24$$

different ways.

(b) The pictures must be arranged puppy, landscape, puppy, landscape, puppy. Arrange the puppies in 3! or 6 ways. Arrange the landscapes in 2! or 2 ways. In this scheme, the pictures can be arranged in $6 \cdot 2 = 12$ different ways.

9. (a) There are $7 \cdot 5 \cdot 4 = 140$ different groups of 3 representatives possible.

(b) $7 \cdot 5 \cdot 4 = 140$ is the number of groups with 3 representatives. For 2 representatives, the number of groups is

$$7 \cdot 5 + 7 \cdot 4 + 5 \cdot 4 = 83.$$

For 1 representative, the number of groups is

$$7 + 5 + 4 = 16.$$

The total number of these groups is

$$140 + 83 + 16 = 239$$

groups.

13. It is impossible to draw 3 blue balls, since there are only 2 blue balls in the basket; hence,

$$P(\text{all blue balls}) = 0.$$

15. $P(\text{exactly 2 black balls})$

$$= \frac{\binom{4}{2}\binom{7}{1}}{\binom{11}{3}} = \frac{42}{165} = \frac{14}{55} \approx .255$$

17. $P(\text{2 green balls and 1 blue ball})$

$$= \frac{\binom{5}{2}\binom{2}{1}}{\binom{11}{3}} = \frac{20}{165} = \frac{4}{33} \approx .121$$

19. Let x denote the number of girls. We have $n = 6$, $x = 6$, $p = 1/2$, and $1 - p = 1/2$, so

$$P(\text{all girls}) = \binom{6}{6}\left(\frac{1}{2}\right)^6\left(\frac{1}{2}\right)^0$$

$$= \frac{1}{64} \approx .016.$$

21. Let x denote the number of boys, and then p = 1/2 and 1 − p = 1/2.

We have

P(no more than 2 boys)

$= P(x \leq 2)$

$= P(x = 0) + P(x = 1) + P(x = 2)$

$= \binom{6}{0}\left(\frac{1}{2}\right)^{0}\left(\frac{1}{2}\right)^{6} + \binom{6}{1}\left(\frac{1}{2}\right)^{1}\left(\frac{1}{2}\right)^{5}$

$\quad + \binom{6}{2}\left(\frac{1}{2}\right)^{2}\left(\frac{1}{2}\right)^{4}$

$= \frac{11}{32} \approx .344.$

23. $P(2 \text{ spades}) = \dfrac{\binom{13}{2}}{\binom{52}{2}} = \dfrac{78}{1326}$

$\qquad\qquad\qquad = \dfrac{1}{17} \approx .059$

25. P(exactly 1 face card)

$= \dfrac{\binom{12}{1}\binom{40}{1}}{\binom{52}{2}} = \dfrac{480}{1326} = \dfrac{80}{221} \approx .362$

27. P(at most 1 queen)

$= P(0 \text{ queens}) + P(1 \text{ queen})$

$= \dfrac{\binom{48}{2}}{\binom{52}{2}} + \dfrac{\binom{4}{1}\binom{48}{1}}{\binom{52}{2}}$

$= \dfrac{1128}{1326} + \dfrac{192}{1326}$

$= \dfrac{1320}{1326} = \dfrac{220}{221} \approx .9955$

29. Add up the frequencies to obtain n = 23. Divide the frequencies by n to obtain the probabilities.

(a) The probability distribution is as follows.

x	8	9	10	11	12	13	14
P(x)	$\frac{1}{23}$	0	$\frac{2}{23}$	$\frac{5}{23}$	$\frac{8}{23}$	$\frac{4}{23}$	$\frac{3}{23}$

(b) The histogram consists of 7 rectangles, the second of which has no height.

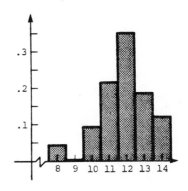

(c) The expected value is

$E(x) = 8\left(\frac{1}{23}\right) + 9(0) + 10\left(\frac{2}{23}\right) + 11\left(\frac{5}{23}\right)$

$\quad + 12\left(\frac{8}{23}\right) + 13\left(\frac{4}{23}\right) + 14\left(\frac{3}{23}\right)$

$= \dfrac{273}{23} \approx 11.87.$

31. (a) There are n = 36 possible outcomes. Let x represent the sum of the dice, and note that the possible values of x are the whole numbers from 2 to 12. The probability distribution is as follows.

x	2	3	4	5	6
P(x)	$\frac{1}{36}$	$\frac{2}{36}$	$\frac{3}{36}$	$\frac{4}{36}$	$\frac{5}{36}$

x	7	8	9	10	11	12
P(x)	$\frac{6}{36}$	$\frac{5}{36}$	$\frac{4}{36}$	$\frac{3}{36}$	$\frac{2}{36}$	$\frac{1}{36}$

(b) The histogram consists of 11 rectangles.

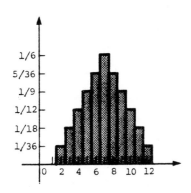

(c) The expected value is

$$E(x) = 2\left(\tfrac{1}{36}\right) + 3\left(\tfrac{2}{36}\right) + 4\left(\tfrac{3}{36}\right) + 5\left(\tfrac{4}{36}\right)$$

$$+ 6\left(\tfrac{5}{36}\right) + 7\left(\tfrac{6}{36}\right) + 8\left(\tfrac{5}{36}\right) + 9\left(\tfrac{4}{36}\right)$$

$$+ 10\left(\tfrac{3}{36}\right) + 11\left(\tfrac{2}{36}\right) + 12\left(\tfrac{1}{36}\right)$$

$$= \frac{252}{36} = 7.$$

33. The probability that corresponds to the shaded region of the histogram is the total of the shaded areas, which is

$$1(.3) + 1(.2) + 1(.1) = .6.$$

35. The probability of rolling a 6 is 1/6, and your net winnings would be $2. The probability of rolling a 5 is 1/6, and your net winnings would be $1. The probability of rolling something else is 4/6, and your net winnings would be −$2. Let x represent your winnings. The expected value is

$$E(x) = 2\left(\tfrac{1}{6}\right) + 1\left(\tfrac{1}{6}\right) + (-2)\left(\tfrac{4}{6}\right)$$

$$= -\frac{5}{6} \approx -\$.833 \quad \text{or} \quad -83.3¢.$$

This is not a fair game since the expected value is not 0.

37. Let x denote the number of girls. The probability distribution is as follows.

x	0	1	2	3	4	5
P(x)	$\tfrac{1}{32}$	$\tfrac{5}{32}$	$\tfrac{10}{32}$	$\tfrac{10}{32}$	$\tfrac{5}{32}$	$\tfrac{1}{32}$

The expected value is

$$E(x) = 0\left(\tfrac{1}{32}\right) + 1\left(\tfrac{5}{32}\right) + 2\left(\tfrac{10}{32}\right) + 3\left(\tfrac{10}{32}\right)$$

$$+ 4\left(\tfrac{5}{32}\right) + 5\left(\tfrac{1}{32}\right)$$

$$= \frac{80}{32} = 2.5 \text{ girls.}$$

39. $P(3 \text{ clubs}) = \dfrac{\binom{13}{3}}{\binom{52}{3}} = \dfrac{286}{22,100} \approx .0129$

Thus,

P(win) = .0129 and

P(lose) = 1 − .0129 = .9871.

Let x represent the amount you should pay.

Your net winnings are 100 − x if you win and −x if you lose.

If it is a fair game, your expected winnings will be 0. Thus, E(x) = 0 becomes

$$.0129(100 - x) + .9871(-x) = 0$$
$$1.29 - .0129x - .9871x = 0$$
$$1.29 - x = 0$$
$$x = 1.29.$$

You should pay $1.29.

41. We have n = 20, x = 3, p = .01, and 1 − p = .99, so

$$P(x = 3) = \binom{20}{3}(.01)^3(.99)^{17}$$
$$\approx .00096.$$

43. P(x ≥ 12)

$$= P(x = 12) + P(x = 13) + \cdots$$
$$+ P(x = 20)$$
$$= \binom{20}{12}(.01)^{12}(.99)^8$$
$$+ \binom{20}{13}(.01)^{13}(.99)^7$$
$$+ \cdots + \binom{20}{20}(.01)^{20}(.99)^0$$

45. Let x denote your winnings. Reduce the $1000 prize by 62¢ for the cost of entering. (This cost covers the stamped envelope for mailing the entry and the self-addressed stamped envelope enclosed with the entry.) Then, the expected value is

$$E(x) = 999.38\left(\frac{1}{4000}\right) + (-.62)\left(\frac{3999}{4000}\right)$$
$$\approx -.37,$$

which is a loss of 37¢.

47. If a box is good (probability .9) and the merchant samples an excellent piece of fruit from that box (probability .80), then he will accept the box and earn a $200 profit on it. If a box is bad (probability .1) and he samples an excellent piece of fruit from the box (probability .30), then he will accept the box and earn a −$1000 profit on it. If the merchant ever samples a non-excellent piece of fruit, he will not accept the box. In this case he pays nothing and earns nothing, so the profit will be $0.

Let x denote the merchant's earnings. Note that   .9(.80) = .72,

$$.1(.30) = .03,$$

and 1 − (.72 + .03) = .25.

The probability distribution is as follows.

x	200	−1000	0
P(x)	.72	.03	.25

The expected value when the merchant samples the fruit is

$$E(x) = 200(.72) + (-1000)(.03) + 0(.25)$$
$$= 144 - 30 + 0$$
$$= \$114.$$

We must also consider the case in which the merchant does not sample the fruit. Let x again denote the merchant's earnings. The probability distribution is as follows.

x	200	−1000
P(x)	.9	.1

The expected value when the merchant does not sample the fruit is

$$E(x) = 200(.9) + (-1000)(.1)$$
$$= 180 - 100$$
$$= \$80.$$

Combining these two results, the expected value of the right to sample is $114 - $80 = $34, which corresponds to choice (c).

Extended Application

1. **(a)** $C(M_0)$

$$= NL[1 - (1-p_1)(1-p_2)(1-p_3)]$$
$$= 3(54)[1 - (.91)(.76)(.83)]$$
$$= \$69.01$$

(b) $C(M_2)$

$$= H_2 + NL[1 - (1 - p_1)(1 - p_3)]$$
$$= 40 + 3(54)[1 - (.91)(.83)]$$
$$= \$79.64$$

(c) $C(M_3)$

$$= H_3 + NL[1 - (1 - p_1)(1 - p_2)]$$
$$= 9 + 3(54)[1 - (.91)(.76)]$$
$$= \$58.96$$

(d) $C(M_{12})$

$$= H_1 + H_2 + NL[1 - (1 - p_3)]$$
$$= 15 + 40 + 3(54)[1 - .83]$$
$$= \$82.54$$

(e) $C(M_{13})$

$$= H_1 + H_3 + NL[1 - (1 - p_2)]$$
$$= 15 + 9 + 3(54)[1 - .76]$$
$$= \$62.88$$

(f) $C(M_{123})$

$$= H_1 + H_2 + H_3 + NL[1 - 1]$$
$$= 15 + 40 + 9$$
$$= \$64.00$$

2. Policy M_3, stocking only part 3 on the truck, leads to the lowest expected cost.

3. It is not necessary for the probabilities to add up to 1 because it is possible that no parts will be needed. That is, the events of needing parts 1, 2, and 3 are not the only events in the sample space.

4. For 3 different parts we have 8 different policies: 1 with no parts, 3 with 1 part, 3 with 2 parts, and 1 with 3 parts. The number of different policies, $8 = 2^3$, is the number of subsets of a set containing 3 distinct elements. If there are n different parts, the number of policies is the number of subsets of a set containing n distinct elements which we showed in Chapter 5 to be 2^n.

CHAPTER 6 TEST

1. Evaluate the following factorials, permutations, and combinations.

   (a) 7!   **(b)** $\binom{8}{3}$ **(c)** P(6, 4)

2. The 24 members of the 3rd grade Mitey Mites hockey team must select a head basher, a second basher, and a designated tripper for their team. How many ways can this be done?

3. A basketball team consists of 7 good shooters and 5 poor shooters.

   (a) In how many ways can a 5-person team be selected?

   (b) In how many ways can a team consisting of only good shooters be selected?

   (c) In how many ways can a team consisting of 3 good and 2 poor shooters be selected?

4. In Ohio, most license plates consist of three letters followed by three digits. How many different plates are possible in the following situations?

   (a) Repeats of letters and digits are allowed.

   (b) Repeats of letters, but not digits are allowed.

   (c) The first two letters must be AR, and repeats are allowed.

5. A coin and a die are tossed. List the sample space for this experiment.

6. A three-card hand is drawn from a standard deck of 52 cards. Set up the probability of each of the following events.

   (a) The hand contains exactly 2 hearts.

   (b) The hand contains fewer than 2 hearts.

7. A shipment of bolts has 20% of the bolts defective.
   (a) What is the probability of finding 2 or fewer defective bolts in a sample of 15?

   (b) What is the probability that 9 bolts will have to be sampled before 2 defectives are found?

   (c) How many bolts must be sampled to ensure a probability of at least .6 that a defective bolt is found?

8. A nickel, a dime, and a quarter are tossed simultaneously. Let the random variable x denote the number of heads observed in the experiment, and prepare a probability distribution.

9. Two dice are rolled, and the total number of points is recoreded. Find the expected value.

10. There is a game called Double or Nothing, and it costs $1 to play. If you draw the ace of spades, you are paid $2, but you are paid nothing if you draw any other card. Is this a fair game?

CHAPTER 6   TEST ANSWERS

1. **(a)** 5040   **(b)** 56   **(c)** 360   2. 12,144 ways   3. **(a)** 792 ways

   (b) 21 ways   (c) 350 ways   4. **(a)** $26 \cdot 26 \cdot 26 \cdot 10 \cdot 10 \cdot 10 \cdot = 17,576,000$

   (b) $26 \cdot 26 \cdot 26 \cdot 10 \cdot 9 \cdot 8 = 12,654,720$ **(c)** $1 \cdot 1 \cdot 26 \cdot 10 \cdot 10 \cdot 10 = 26,000$

5. $\{(h, 1), (h, 2), (h, 3), (h, 4), (h, 5), (h, 6),$
   $(t, 1), (t, 2), (t, 3), (t, 4), (t, 5), (t, 6)\}$

6. **(a)** $\dfrac{\binom{13}{2}\binom{39}{1}}{\binom{52}{3}}$ **(b)** $\dfrac{\binom{13}{0}\binom{39}{3} + \binom{13}{1}\binom{39}{2}}{\binom{52}{3}}$

7. **(a)** $\binom{15}{0}(.2)^0(.8)^{15} + \binom{15}{1}(.2)^1(.8)^{14} + \binom{15}{2}(.2)^2(.8)^{13} \approx .3980$

   (b) $\binom{8}{1}(.2)^2(.8)^7 \approx .0671$ **(c)** 5

8.

x	0	1	2	3
P(x)	$\frac{1}{8}$	$\frac{3}{8}$	$\frac{3}{8}$	$\frac{1}{8}$

9. 7

10. No, it is not a fair game.

CHAPTER 7   STATISTICS

Section 7.1

1. **(a) – (b)** Since 0 – 24 is to be the
    first interval and there are 25 num-
    bers between 0 and 24 inclusive, we
    will let all six intervals be of size
    25.   The other five intervals are
    25 – 49, 50 – 74, 75 – 99, 100 – 124,
    and 125 – 149.   Keeping a tally of
    how many data values lie in each
    interval leads to the following fre-
    quency distribution.

Interval	Frequency
0 – 24	4
25 – 49	3
50 – 74	6
75 – 99	3
100 – 124	5
125 – 149	9

    (c) Draw the histogram.   It consists
    of 6 bars of equal width and having
    heights as determined by the fre-
    quency of each interval.   See the
    histogram that appears in the back of
    the textbook.

    (d) To construct the frequency poly-
    gon, join consecutive midpoints of
    the tops of the histogram bars with
    straight line segments.   See the back
    of the textbook.

3. **(a) – (b)** Since 70 – 74 is to be the
    first interval, we let all the inter-
    vals be of size 5.   The largest data
    value is 111, so the last interval
    that will be needed is 110 – 114.
    The frequency distribution is as fol-
    lows.

Interval	Frequency
70 – 74	2
75 – 79	1
80 – 84	3
85 – 89	2
90 – 94	6
95 – 99	5
100 – 104	6
105 – 109	4
110 – 114	2

    (c) Draw the histogram.   It consists
    of 7 bars of equal width and having
    heights as determined by the fre-
    quency of each interval.   See the
    histogram that appears in the back of
    the textbook.

    (d) Construct the frequency polygon
    by joining consecutive midpoints of
    the tops of the histogram bars with
    straight line segments.   See the back
    of the textbook.

7. The mean of the 5 numbers is

    $$\bar{x} = \frac{\Sigma x}{5}$$

    $$= \frac{8 + 10 + 16 + 21 + 25}{5}$$

    $$= \frac{80}{5} = 16.$$

9. $\Sigma x = 21,900 + 22,850 + 24,930$
$ + 29,710 + 28,340 + 40,000$
$ = 167,730$

The mean of the 6 numbers is

$\bar{x} = \dfrac{\Sigma x}{6} = \dfrac{167,730}{6} = 27,955.$

11. $\Sigma x = 9.4 + 11.3 + 10.5 + 7.4 + 9.1$
$ + 8.4 + 9.7 + 5.2 + 1.1 + 4.7$
$ = 76.8$

The mean of the 10 numbers is

$\bar{x} = \dfrac{\Sigma x}{10} = \dfrac{76.8}{10} = 7.68 \approx 7.7.$

13. Add to the frequency distribution a new column, "Value × Frequency."

Value	Frequency	Value × Frequency
3	4	3 · 4 = 12
5	2	5 · 2 = 10
9	1	9 · 1 = 9
12	3	12 · 3 = 36
	Total: 10	Total: 67

The means is $\bar{x} = \dfrac{67}{10} = 6.7.$

15.

Value	Frequency	Value × Frequency
12	4	12 · 4 = 48
13	2	13 · 2 = 26
15	5	15 · 5 = 75
19	3	19 · 3 = 57
22	1	22 · 1 = 22
23	5	23 · 5 = 115
	Total: 20	Total: 343

The mean is $\bar{x} = \dfrac{343}{20} = 17.15 \approx 17.2.$

17. 12, 18, 32, 51, 58, 92, 106

The median is the middle number, 51.

19. 100, 114, 125, 135, 150, 172

The median is the mean of the 2 middle numbers, which is

$$\frac{125 + 135}{2} = \frac{260}{2} = 130.$$

21. First arrange the numbers in numerical order, from smallest to largest.

3.4, 9.1, 22.6, 28.4, 29.8, 32.1, 47.6, 59.8

There are 8 numbers here; the median is the mean of the 2 middle numbers, which is

$$\frac{28.4 + 29.8}{2} = \frac{58.2}{2} = 29.1.$$

23. 4, 9, 8, 6, 9, 2, 1, 3

The mode is the number that occurs most often. Here, the mode is 9.

25. 74, 68, 68, 68, 75, 75, 74, 74, 70

The mode is the number that occurs most often. Here, there are two modes, 68 and 74, since they both appear three times.

27. 6.8, 6.3, 6.3, 6.9, 6.7, 6.4, 6.1, 6.0

The mode is 6.3.

31.

Interval	Midpoint, x	Frequency, f	Product, xf
70 – 74	72	2	144
75 – 79	77	1	77
80 – 84	82	3	246
85 – 89	87	2	174
90 – 94	92	6	552
95 – 99	97	5	485
100 – 104	102	6	612
105 – 109	107	4	428
110 – 114	112	2	224
		Total: 31	Total: 2942

The mean of this collection of grouped data is

$$\bar{x} = \frac{2942}{31} \approx 94.9.$$

The intervals 90 – 94 and 100 – 104 each contain the most data values, 6, so they are the modal classes.

33. Draw the histogram. It consists of 5 bars of equal width and having heights as determined by the frequency of each interval (the fourth bar has no height). Construct the frequency polygon by joining the consecutive midpoints of the tops of the histogram bars with straight line segments. See the answer graph in the back of the textbook.

35. $\Sigma x = 2200 + 2000 + 1750 + 2200 + 2400 + 2800 + 2800 + 2450 + 2600 + 2750$
    $= 23,950$

The mean of the wheat production is

$$\bar{x} = \frac{\Sigma x}{10} = \frac{23,950}{10} = 2395 \text{ million bushels.}$$

37. (a) The height of the 10 – 19 histogram bar looks like about 17.5, so about 17.5% of the population was between 10 and 19 years old.

(b) The 60 – 69 bar has a height of about 8, so about 8% of the population was between 60 and 69 years old.

(c) The 20 – 29 bar appears to have the greatest height, so that age group had the largest percentage of the population.

39. **(a)** $\bar{x} = \dfrac{\Sigma x}{n} = \dfrac{666}{12} = 55.5$

The mean of the maximum temperatures is 55.5°F.

(b) $\bar{x} = \dfrac{\Sigma x}{n} = \dfrac{347}{12} \approx 28.9$

The mean of the minimum temperatures is about 28.9°F.

Section 7.2

1. The standard deviation of a sample of numbers is the square root of the variance of the sample.

3. The range is $74 - 29 = 53$, the difference of the highest and lowest numbers in the set.

The mean is $\bar{x} = \dfrac{1}{7}(42 + 38 + 29 + 74 + 82 + 71 + 35) = \dfrac{371}{7} = 53$.

To prepare for calculating the standard deviation, construct a table of the set of numbers, their deviations from the mean $\bar{x}$, and the squares of these deviations.

x	$x - \bar{x}$	$(x - \bar{x})^2$
42	−11	121
38	−15	225
29	−24	576
74	21	441
82	29	841
71	18	324
35	−18	324
		Total: 2852

The standard deviation is $s = \sqrt{\dfrac{2852}{7 - 1}} \approx \sqrt{475.3} \approx 21.8$.

5. The range is 287 − 241 = 46. The mean is

$$\overline{x} = \frac{1}{6}(241 + 248 + 251 + 257 + 252 + 287) = 256.$$

x	x − $\overline{x}$	(x − $\overline{x}$)²
241	−15	225
248	−8	64
251	−5	25
257	1	1
252	−4	16
287	31	961
		Total: 1292

The standard deviation is $s = \sqrt{\dfrac{1292}{6 - 1}} = \sqrt{258.4} \approx 16.1.$

7. The range is 27 − 3 = 24. The mean is

$$\overline{x} = \frac{1}{10}(3 + 7 + 4 + 12 + 15 + 18 + 19 + 27 + 24 + 11) = 14.$$

x	x − $\overline{x}$	(x − $\overline{x}$)²
3	−11	121
7	−7	49
4	−10	100
12	−2	4
15	1	1
18	4	16
19	5	25
27	13	169
24	10	100
11	−3	9
		Total: 594

The standard deviation is $s = \sqrt{\dfrac{594}{10 - 1}} = \sqrt{66} \approx 8.1.$

9. Expand the table to include columns for the midpoint x of each interval, and for fx, x^2, and fx^2.

Interval	f	x	fx	x^2	fx^2
0 - 24	4	12	48	144	576
25 - 49	3	37	111	1369	4107
50 - 74	6	62	372	3844	23,064
75 - 99	3	87	261	7569	22,707
100 - 124	5	112	560	12,544	62,720
125 - 149	9	137	1233	18,769	168,921
Totals:	30		2585		282,095

The mean of the grouped data is

$$\overline{x} = \frac{\Sigma fx}{n} = \frac{2585}{30} \approx 86.2.$$

The standard deviation for the grouped data is

$$s = \sqrt{\frac{\Sigma fx^2 - n\overline{x}^2}{n - 1}} = \sqrt{\frac{282,095 - 30(86.2)^2}{30 - 1}} \approx \sqrt{2046.7} \approx 45.2.$$

11. Start with the frequency distribution that was the answer to Exercise 3 of Section 7.1, and expand the table to include columns for the midpoint x of each interval, and for fx, x^2, and fx^2.

Interval	f	x	fx	x^2	fx^2
70 - 74	2	72	144	5184	10,368
75 - 79	1	77	77	5929	5929
80 - 84	3	82	246	6724	20,172
85 - 89	2	87	174	7569	15,138
90 - 94	6	92	552	8464	50,784
95 - 99	5	97	485	9409	47,045
100 - 104	6	102	612	10,404	62,424
105 - 109	4	107	428	11,449	45,796
110 - 114	2	112	224	12,544	25,088
Totals:	31		2942		282,744

The mean of the grouped data is

$$\bar{x} = \frac{\Sigma fx}{n} = \frac{2942}{31} \approx 94.9.$$

The standard deviation for the grouped data is

$$s = \sqrt{\frac{\Sigma fx^2 - n\bar{x}^2}{n - 1}}$$

$$= \sqrt{\frac{282,744 - 31(94.9)^2}{31 - 1}}$$

$$\approx \sqrt{118.0} \approx 10.9.$$

13. Use $k = 2$ in Chebyshev's theorem.

$$P(\mu - 2\sigma \le x \le \mu + 2\sigma) \ge 1 - \frac{1}{2^2} = \frac{3}{4},$$

so at least 3/4 of the distribution is within 2 standard deviations of the mean.

15. Use $k = 5$ in Chebyshev's theorem.

$$P(\mu - 5\sigma \le x \le \mu + 5\sigma) \ge 1 - \frac{1}{5^2} = \frac{24}{25},$$

so at least 24/25 of the distribution is within 5 standard deviations of the mean.

17. Here $32 = 50 - 3 \cdot 6 = \mu - 3\sigma$ and $68 = 50 + 3 \cdot 6 = \mu + 3\sigma$, so Chebyshev's theorem applies with $k = 3$; hence at least

$$1 - \frac{1}{k^2} = \frac{8}{9} \approx 88.9\%$$

of the numbers lie between 32 and 68.

19. The answer here is the complement of the answer to Exercise 16. It was found that at least 75% of the distribution of numbers are between 38 and 62, so at most 100% – 75% = 25% of the numbers are less than 38 or more than 62.

21. 300, 320, 380, 420, 500, 2000

(a) $\mu = \frac{1}{6}(300 + 320 + 380 + 420$

$$+ 500 + 2000)$$

$$= \frac{1}{6}(3920) \approx 653.33.$$

The mean of the salaries is \$653.33.

x	x^2
300	90,000
320	102,400
380	144,400
420	176,400
500	250,000
2000	4,000,000
	Total: 4,763,200

$$\sigma = \sqrt{\frac{\Sigma x^2 - n\bar{\mu}^2}{n - 1}}$$

$$= \sqrt{\frac{4,763,200 - 6(653.33)^2}{6 - 1}}$$

$$= \sqrt{440,426.67} \approx 663.65$$

The standard deviation of the salaries is \$663.65.

(b) $653.33 - 663.65 = -10.32$ is 1 standard deviation below the mean, and $653.33 + 663.65 = 1316.98$ is 1 standard deviation above the mean.

5 of the data values lie between these two numbers, so 5 of the workers earn wages within 1 standard deviation of the mean.

(c) 653.33 − 2(663.65) = −673.97 is 2 standard deviations below the mean, and 653.33 + 2(663.65) = 1980.63 is 2 standard deviations above the mean.

5 of the data values lie between these two numbers, so 5 of the workers earn wages within 2 standard deviations of the mean.

(d) Use k = 2 in Chebyshev's theorem.

$$P(\mu - 2\sigma \leq x \leq \mu + 2\sigma) \geq 1 - \frac{1}{2^2} = \frac{3}{4}$$

Chebyshev's theorem guarantees that at least 3/4 of a distribution will lie within 2 standard deviations of the mean, which agrees with our findings in part (c) above.

23.   15, 18, 19, 23, 25, 25, 28, 30, 34, 38

(a) $\bar{x} = \frac{1}{10}(15 + 18 + 19 + 23 + 25 + 25 + 28 + 30 + 34 + 38) = \frac{1}{10}(255) = 25.5$

The mean life of the sample of Brand X batteries is 25.5 hr.

x	$x - \bar{x}$	$(x - \bar{x})^2$
15	−10.5	110.25
18	−7.5	56.25
19	−6.5	42.25
23	−2.5	6.25
25	−.5	.25
25	−.5	.25
28	2.5	6.25
30	4.5	20.25
34	8.5	72.25
38	12.5	156.25
		Total: 470.50

$$s = \sqrt{\frac{470.50}{10 - 1}} \approx \sqrt{52.28} \approx 7.2$$

The standard deviation of the Brand X battery lives is 7.2 hr.

(b) Forever Power has a smaller standard deviation (4.1 hr, as opposed to 7.2 hr for Brand X), which indicates a more uniform life.

(c) Forever Power has a higher mean (26.2 hr, as opposed to 25.5 hr for Brand X), which indicates a longer average life.

25.

	Sample Number									
	1	2	3	4	5	6	7	8	9	10
	2	3	-2	-3	-1	3	0	-1	2	0
	-2	-1	0	1	2	2	1	2	3	0
	1	4	1	2	4	2	2	3	2	2
(a) $\bar{x}$	$\frac{1}{3}$	2	$-\frac{1}{3}$	0	$\frac{5}{3}$	$\frac{7}{3}$	1	$\frac{4}{3}$	$\frac{7}{3}$	$\frac{2}{3}$
(b) s	2.1	2.6	1.5	2.6	2.5	.6	1	2.1	.6	1.2

(c) $\bar{X} = \dfrac{\Sigma \bar{x}}{n} \approx \dfrac{11.3}{10} = 1.13$

(d) $\bar{s} = \dfrac{\Sigma s}{n} = \dfrac{16.8}{10} = 1.68$

(e) The upper control limit for the sample means is

$$\mu + k_1 \bar{s} = 1.13 + (1.954)(1.68) \approx 4.41.$$

The lower control limit for the sample means is

$$\mu - k_1 s = 1.13 - (1.954)(1.68) \approx -2.15.$$

(f) The upper control limit for the sample standard deviations is

$$k_2 \bar{s} = 2.568(1.68) \approx 4.31.$$

The lower control limit for the sample standard deviations is

$$k_3 \bar{s} = 0(1.68) = 0.$$

27. 8.54, 8.31, 8.24, 7.43, 6.70, 6.53, 6.87

x	x^2
8.54	72.9316
8.31	69.0561
8.24	67.8976
7.43	55.2049
6.70	44.89
6.53	42.6409
6.87	47.1969
Totals: 52.62	399.818

(a) The mean is

$$\mu = \frac{\Sigma x}{n} = \frac{52.62}{7} \approx 7.5171 \approx 7.52.$$

1987 with 7.43 thousand unemployed is the year that has unemployment closest to the mean.

(b) The standard deviation is

$$\sigma = \sqrt{\frac{\Sigma x^2 - n\mu^2}{n - 1}}$$
$$= \sqrt{\frac{399.818 - 7(7.5171)^2}{7 - 1}}$$
$$\approx \sqrt{.7117} \approx .84.$$

(c) 7.52 - .84 = 6.68 is 1 standard deviation below the mean, and 7.52 + .84 = 8.36 is 1 standard deviation above the mean. 8.31, 8.24, 7.43, 6.70, and 6.87 are all between 6.68 and 8.36, so unemployment is within 1 standard deviation of the mean in 5 of the years.

29. This exercise should be solved using a computer or a calculator with a standard deviation key. The answers are 1.8158 mm for the mean and .4451 mm for the standard deviation.

Section 7.3

1. The peak in a normal curve occurs directly above the mean.

3. For normal distributions where $\mu \neq 0$ or $\sigma \neq 1$, z-scores are found by using the formula $z = \frac{x - \mu}{\sigma}$.

5. Use the table, "Area Under a Normal Curve to the Left of z", in the Appendix. To find the percent of the area under a normal curve between the mean and 2.50 standard deviations from the mean, subtract the table entry for z = 0 (representing the mean) from the table entry for z = 2.5.

$$.9938 - .5000 = .4938$$

Therefore, 49.38% of the area lies between μ and $\mu + 2.5\sigma$.

7. Subtract the table entry for z = -1.71 from the table entry for z = 0.

$$.5000 - .0436 = .4564$$

45.64% of the area lies between μ and $\mu - 1.71\sigma$.

9. $P(1.41 \le z \le 2.83)$

    $= P(z \le 2.83) - P(z \le 1.41)$

    $= .9977 - .9207$

    $= .077$

    7.7% of the total area under the standard normal curve lies between z = 1.41 and z = 2.83.

11. $P(-2.48 \le z \le -.05)$

    $= P(z \le -.05) - P(z \le -2.48)$

    $= .4801 - .0066$

    $= .4735$

    47.35% of the area lies between z = -2.48 and z = -.05.

13. $P(-3.11 \le z \le 1.44)$

    $= P(z \le 1.44) - P(z \le -3.11)$

    $= .9251 - .0009$

    $= .9242$

    92.42% of the area lies between z = -3.11 and z = 1.44.

15. $P(-.42 \le z \le .42)$

    $= P(z \le .42) - P(z \le -.42)$

    $= .6628 - .3372$

    $= .3256$

    32.56% of the area lies between z = -.42 and z = .42.

17. 5% of the total area is to the left of z.

    Use the table backwards.
    Look in the body of the table for an area of .05, and find the corresponding z using the left column and top column of the table.

The closest values to .05 in the body of the table are .0505, which corresponds to z = -1.64, and .0495, which corresponds to z = -1.65.

19. 15% of the area is to the right of z. If 15% of the area is to the right of z, then 85% of the area is to the left of z. The closest value to .85 in the body of the table is .8504, which corresponds to z = 1.04.

21. Let x represent the bolt diameter.

    $\mu = .25, \ \sigma = .02$

    First, find the probability that a bolt has a diameter less than or equal to .3 in, that is, $P(x \le .3)$. The z-score corresponding to x = .3 is

    $$z = \frac{x - \mu}{\sigma}$$

    $$= \frac{.3 - .25}{.02} = 2.5.$$

    Using the table, find the area to the left of z = 2.5. This gives us

    $P(x \le .3) = P(z \le 2.5) = .9938.$

    Then

    $$P(x > .3) = 1 - P(x \le .3)$$

    $$= 1 - .9938$$

    $$= .0062.$$

23. Let x represent a grocery bill.

    $\mu = 52.25, \ \sigma = 19.50$

    The middle 50% of the grocery bills have cutoffs at 25% below the mean and 25% above the mean.

At 25% below the mean, the area to the left is .2500, which corresponds to about $z = -.67$. At 25% above the mean, the area to the left is .7500, which corresponds to about $z = .67$. Find the x-value that corresponds to each z-score.

For $z = -.67$,

$$-.67 = \frac{x - 52.25}{19.50} \quad \text{or} \quad x \approx 39.19.$$

For $z = .67$,

$$.67 = \frac{x - 52.25}{19.50} \quad \text{or} \quad x \approx 65.32.$$

The middle 50% of the customers spend between \$39.19 and \$65.32.

25. Let x represent the length of a fish.

$$\mu = 12.3, \ \sigma = 4.1$$

Find the z-score for $x = 18$.

$$z = \frac{18 - 12.3}{4.1} \approx 1.39$$

$$\begin{aligned} P(x > 18) &= 1 - P(x \le 18) \\ &= 1 - P(z \le 1.39) \\ &= 1 - .9177 = .0823 \end{aligned}$$

27. Let x represent the life of a light bulb.

$$\mu = 500, \ \sigma = 100$$

At least 500 hr means $x \ge 500$.

$$P(x \ge 500) = 1 - P(x < 500)$$

Find the z-score that corresponds to $x = 500$.

$$z = \frac{500 - 500}{100} = 0$$

$$P(x \ge 500) = P(z \ge 0)$$
$$= 1 - .5000 = .5000$$

Thus, $10,000(.5) = 5000$ bulbs can be expected to last at least 500 hr.

29. Let x represent the life of a light bulb.

$$\mu = 500, \ \sigma = 100$$

Between 650 and 780 hr means the following.

For $x = 650$, $z = \dfrac{650 - 500}{100} = 1.5$.

For $x = 780$, $z = \dfrac{780 - 500}{100} = 2.8$.

$$\begin{aligned} P(650 \le x \le 780) \\ = P(1.5 \le z \le 2.8) \\ = P(z \le 2.8) - P(z \le 1.5) \\ = .9974 - .9332 \\ = .0642 \end{aligned}$$

Thus, $10,000(.0642) = 642$ bulbs can be expected to last between 650 and 780 hr.

31. Let x represent the life of a light bulb.

$$\mu = 500, \ \sigma = 100$$

Less than 740 hr means $x < 740$.

For $x = 740$, $z = \dfrac{740 - 500}{100} = 2.4$.

$$P(x < 740) = P(z < 2.4) = .9918$$

Thus, $10,000(.9918) = 9918$ bulbs can be expected to last at least 740 hr.

33. Let x represent the weight of a package.

$$\mu = 16.5, \ \sigma = .5$$

For x = 16, $z = \dfrac{16 - 16.5}{.5} = -1.$

$P(x < 16) = P(z < -1) = .1587$

is the fraction of the boxes that are
underweight.

35. Let x represent the weight of a package.

$$\mu = 16.5, \ \sigma = .2$$

For x = 16, $z = \dfrac{16 - 16.5}{.2} = -2.5$

$P(x < 16) = P(z < -2.5) = .0062$

is the fraction of the boxes that are
underweight.

37. Let x represent the weight of a
chicken.

$$\mu = 1850, \ \sigma = 150$$

More than 1700 g means x > 1700.

For x = 1700, $z = \dfrac{1700 - 1850}{150} = -1.0.$

$$\begin{aligned} P(x > 1700) &= 1 - P(x \le 1700) \\ &= 1 - P(z \le -1.0) \\ &= 1 - .1587 \\ &= .8413 \end{aligned}$$

Thus, 84.13% of the chickens will
weigh more than 1700 g.

39. Let x represent the weight of a
chicken.

$$\mu = 1850, \ \sigma = 150$$

Between 1750 and 1900 g means
$1750 \le x \le 1900.$

For x = 1750, $z = \dfrac{1750 - 1850}{150} = -.67.$

For x = 1900, $z = \dfrac{1900 - 1850}{150} = -.33.$

$$\begin{aligned} P(1750 &\le x \le 1900) \\ &= P(-.67 \le z \le -.33) \\ &= P(z \le -.33) - P(z \le -.67) \\ &= .6293 - .2514 \\ &= .3779 \end{aligned}$$

Thus, 37.79% of the chicken will
weigh between 1750 and 1900 g.

41. Let x represent the weight of a
chicken.

$$\mu = 1850, \ \sigma = 150$$

Less than 1550 g means x < 1550.

For x = 1550, $z = \dfrac{1550 - 1850}{150} = -2.0.$

$$\begin{aligned} P(x < 1550) &= P(z < -2.0) \\ &= .0228 \end{aligned}$$

Thus, 2.28% of the chickens will
weigh less than 1550 g.

43. Let x represent the amount of
vitamins a person needs. Then

$$\begin{aligned} P(x \le \mu + 2.5\sigma) &= P(z \le 2.5) \\ &= .9938. \end{aligned}$$

99.38% of the people will receive
adequate amounts of vitamins.

45. The Recommended Daily Allowance is

$$\mu + 2.5\sigma = 159 + (2.5)(12)$$
$$= 189 \text{ units.}$$

47. Let x represent a driving speed.

$$\mu = 50, \; \sigma = 10$$

At the 85th percentile, the area to the left is .8500, which corresponds to about z = 1.04.
Find the x-value that corresponds to this z-score.

$$z = \frac{x - \mu}{\sigma}$$
$$1.04 = \frac{x - 50}{10}$$
$$10.4 = x - 50$$
$$60.4 = x$$

The 85th percentile speed for this road is 60.4 mph.

49. Let x stand for a student's total points.

$$P\left(x \geq \mu + \frac{3}{2}\sigma\right) = P(z \geq 1.5)$$
$$= 1 - P(z \leq 1.5)$$
$$= 1 - .9332$$
$$= .0668$$

Thus, 6.68% of the students receive A's.

51. Let x stand for a student's total points.

$$P\left(\mu - \frac{1}{2}\sigma \leq x \leq \mu + \frac{1}{2}\sigma\right)$$
$$= P(-.5 \leq z \leq .5)$$
$$= P(z \leq .5) - P(z \leq -.5)$$
$$= .6915 - .3085$$
$$= .383$$

Thus, 38.3% of the students receive C's.

53. Let x represent a student's test score.

$$\mu = 74, \; \sigma = 6$$

Since the top 8% get A's, we want to find the number a for which

$$P(x \geq a) = .08,$$
or $P(x \leq a) = .92.$

Read the table backwards to find the z-score for an area of .92, which is 1.41. Find the value of x that corresponds to z = 1.41.

$$z = \frac{x - \mu}{\sigma}$$
$$1.41 = \frac{x - 74}{6}$$
$$8.46 = x - 74$$
$$82.46 = x$$

The bottom cutoff score for an A is 82.

55. Let x represent a student's test score.

$$\mu = 74, \; \sigma = 6$$

23% of the students will receive D's and F's, so to find the bottom cut-off score for a C we need to find the number c for which

$$P(x \leq c) = .23.$$

Read the table backwards to find the z-score for an area of .23, which is -.74. Find the value of x that corresponds to z = -.74.

$$-.74 = \frac{x - 74}{6}$$

$$-4.44 = x - 74$$

$$69.56 = x$$

The bottom cutoff score for a C is 70.

Exercises 57–63 should be completed using a computer. The solution may vary according to the computer program that is used. The answers to these exercises are given below.

57. $P(1.372 \le z \le 2.548) = .0796$

59. $P(z > -2.476) = .9933$

61. $P(12.275 < x < 28.432) = .1527$

63. $P(x < 17.462) = .0051$

Section 7.4

3. **(a)** Use the formula

$$P(x) = \binom{n}{x} p^x (1 - p)^{n-x}$$

to calculate the probabilities.

$$P(x) = \binom{6}{x} \left(\frac{1}{6}\right)^x \left(\frac{5}{6}\right)^{6-x};$$

$$x = 0, 1, 2, \ldots, 6$$

$$P(0) = \binom{6}{0} \left(\frac{1}{6}\right)^0 \left(\frac{5}{6}\right)^6 \approx .335$$

$$P(6) = \binom{6}{6} \left(\frac{1}{6}\right)^6 \left(\frac{5}{6}\right)^0 \approx .00002 \approx .000$$

Find $P(1)$ through $P(5)$ similarly.

The distribution is as follows.

x	0	1	2	3
P(x)	.335	.402	.201	.054

x	4	5	6
P(x)	.008	.001	.000

(b) The mean is $\mu = np = (6)\left(\frac{1}{6}\right)$

$$= 1.00.$$

(c) The standard deviation is

$$\sigma = \sqrt{np(1 - p)}$$

$$= \sqrt{6\left(\frac{1}{6}\right)\left(\frac{5}{6}\right)}$$

$$= \sqrt{\frac{5}{6}} \approx .91.$$

5. **(a)**

$$P(x) = \binom{3}{x}(.02)^x(.98)^{3-x};$$

$$x = 0, 1, 2, 3$$

$$P(0) = \binom{3}{0}(.02)^0(.98)^3 \approx .941$$

$$P(3) = \binom{3}{3}(.02)^3(.98)^0$$

$$= .000008 \approx .000$$

x	0	1	2	3
P(x)	.941	.058	.001	.000

(b) $\mu = np = 3(.02) = .06$

(c) $\sigma = \sqrt{np(1 - p)}$

$$= \sqrt{3(.02)(.98)} \approx .24$$

7. **(a)**

$$P(x) = \binom{4}{x}(.7)^x(.3)^{4-x};$$

$$x = 0, 1, 2, 3, 4$$

$$P(0) = \binom{4}{0}(.7)^0(.3)^4 = .0081$$

$$P(4) = \binom{4}{4}(.7)^4(.3)^0 = .2401$$

x	0	1	2	3	4
P(x)	.0081	.0756	.2646	.4116	.2401

(b) $\mu = np = (4)(.7) = 2.8$

(c) $\sigma = \sqrt{np(1 - p)}$

$$= \sqrt{4(.7)(.3)}$$

$$\approx .92$$

9. $n = 500, \ p = .025$

$\mu = np = (500)(.025) = 12.5$

$\sigma = \sqrt{np(1 - p)} = \sqrt{(500)(.025)(.975)}$

$\quad = \sqrt{12.1875} \approx 3.49$

11. $n = 64, \ p = .80$

$\mu = np = 64(.80) = 51.2$

$\sigma = \sqrt{np(1 - p)} = \sqrt{64(.80)(.20)}$

$\quad = \sqrt{10.24} \approx 3.2$

13. The normal distribution can be used to approximate a binomial distribution as long as $np \geq 5$ and $n(1 - p) \geq 5$.

15. 16 coins are tossed, which means $n = 16$ and $p = 1/2$. To approximate the binomial distribution, use a normal distribution with

$\mu = np = 16\left(\frac{1}{2}\right) = 8$ and

$\sigma = \sqrt{np(1 - p)} = \sqrt{16\left(\frac{1}{2}\right)\left(\frac{1}{2}\right)} = 2.$

Exactly 7 heads corresponds to the area under the normal curve between $x = 6.5$ and $x = 7.5$. The corresponding z-scores are

$$z = \frac{6.5 - 8}{2} = -.75$$

and $\quad z = \frac{7.5 - 8}{2} = -.25.$

P(exactly 7 heads)

$\quad = P(6.5 \leq x \leq 7.5)$

$\quad = P(-.75 \leq z \leq -.25)$

$\quad = P(z \leq -.25) - P(z \leq -.75)$

$\quad = .4013 - .2266$

$\quad = .1747$

17. $n = 16, \ p = \frac{1}{2}$

As in Exercise 15, $\mu = 8$ and $\sigma = 2$. Less than 12 tails corresponds to the area under the normal curve to the left of $x = 11.5$. The corresponding z-score is

$$z = \frac{11.5 - 8}{2} = 1.75.$$

P(less than 12 tails)

$\quad = P(x \leq 11.5)$

$\quad = P(z \leq 1.75)$

$\quad = .9599$

19. $n = 1000, \ p = \frac{1}{2}$

We have

$\mu = np = 1000\left(\frac{1}{2}\right) = 500$ and

$\sigma = \sqrt{np(1 - p)} = \sqrt{1000\left(\frac{1}{2}\right)\left(\frac{1}{2}\right)}$

$$= \sqrt{250} \approx 15.81.$$

Exactly 510 heads corresponds to the area under the normal curve between $x = 509.5$ and $x = 510.5$. The corresponding z-scores are

$$z = \frac{509.5 - 500}{15.81} \approx .60$$

and

$$z = \frac{510.5 - 500}{15.81} \approx .66.$$

P(exactly 510 heads)

$= P(509.5 \leq x \leq 510.5)$

$= P(.60 \leq z \leq .66)$

$= P(z \leq .66) - P(z \leq .60)$

$= .7454 - .7257$

$= .0197$

21. $n = 1000$, $p = \frac{1}{2}$

As in Exercise 19, $\mu = 500$ and $\sigma = 15.81$.

Less than 470 tails corresponds to the area under the normal curve to the left of $x = 469.5$. The corresponding z-score is

$$z = \frac{469.5 - 500}{15.8} = -1.93.$$

P(less than 470 tails)

$= P(x \leq 469.5)$

$= P(z \leq -1.93)$

$= .0268$

23. $n = 120$, $p = \frac{1}{6}$

We have $\mu = np = 120\left(\frac{1}{6}\right) = 20$ and

$$\sigma = \sqrt{np(1 - p)}$$

$$= \sqrt{120\left(\frac{1}{6}\right)\left(\frac{5}{6}\right)}$$

$$\approx 4.08.$$

We want the area between $x = 23.5$ and $x = 24.5$. The corresponding z-scores are .86 and 1.103.

P(exactly twenty-four 6's)

$= P(23.5 \leq x \leq 24.5)$

$= P(.86 \leq z \leq 1.103)$

$= P(z \leq 1.103) - P(z \leq .86)$

$= .8643 - .8051$

$= .0592$

25. $n = 120$, $p = \frac{1}{6}$

As in Exercise 23, $\mu = 20$ and $\sigma = 4.08$.

We want the area to the left of $x = 21.5$. The corresponding z-score is .37.

P(fewer than twenty-two 6's)

$= P(x \leq 21.5)$

$= P(z \leq .37)$

$= .6443$

27. We have $n = 10,000$ and $p = .02$, so $\mu = np = 200$ and $\sigma = \sqrt{np(1 - p)} = 14$. We want the area to the right of $x = 222.5$. The corresponding z-score value is 1.61.

P(more than 222 defects)

$= P(x \geq 222.5)$

$= P(z \geq 1.61)$

$= 1 - P(z \leq 1.61)$

$= 1 - .9463$

$= .0537$

29. $n = 120$, $p = .6$

We have $\mu = np = 120(.6) = 72$ and
$\sigma = \sqrt{np(1 - p)} = \sqrt{120(.6)(.4)} \approx 5.4$.

(a) P(exactly 80 units)

$\quad = P(79.5 \leq x \leq 80.5)$

$\quad = P(1.39 \leq z \leq 1.57)$

$\quad = P(z \leq 1.57) - P(z \leq 1.39)$

$\quad = .9418 - .9177$

$\quad = .0241$

(b) P(at least 70 units)

$\quad = P(x \geq 69.5)$

$\quad = P(z \geq -.46)$

$\quad = 1 - P(z \leq -.46)$

$\quad = 1 - .3228$

$\quad = .6772$

31. $n = 25$, $p = .80$

We have $\mu = np = 25(.80) = 20$ and
$\sigma = \sqrt{np(1 - p)} = \sqrt{25(.80)(.20)} = 2$.

P(exactly 20 cured)

$\quad = P(19.5 \leq x \leq 20.5)$

$\quad = P(-.25 \leq z \leq .25)$

$\quad = P(z \leq .25) - P(z \leq -.25)$

$\quad = .5987 - .4013$

$\quad = .1974$

33. $P(x = 0) = \binom{25}{0}(.80)^0(.20)^{25}$

$\quad\quad\quad = (.20)^{25}$

$\quad\quad\quad = 3.36 \times 10^{-18}$

$\quad\quad\quad \approx 0$

35. $n = 1400$, $p = .55$

We have $\mu = np = 1400(.55) = 770$ and
$\sigma = \sqrt{np(1 - p)} = \sqrt{1400(.55)(.45)}$

$\quad \approx 18.6$.

P(at least 750 people)

$\quad = P(x \geq 749.5)$

$\quad = P(z \geq -1.10)$

$\quad = 1 - P(z \leq -1.10)$

$\quad = 1 - .1357$

$\quad = .8643$

Exercises 37 and 39 should be completed using a computer. The solution may vary according to the computer program that is used. The answers to these exercises are given below.

37. **(a)** P(exactly 5 are color-blind)

$\quad = .0472$

(b) P(no more than 5 are color-blind)

$\quad = .9875$

(c) P(at least 1 is color-blind)

$\quad = .8814$

39. **(a)** P(all 58 like it)

$\quad = 1.04 \times 10^{-9} \approx 0$

(b) P(exactly 28, 29, or 30 like it)

$\quad = .0018$

Chapter 7 Review Exercises

3. **(a)** Since 450 – 474 is to be the first interval, we will let all the intervals be
of size 25. The largest data value is 566, so the last interval that will be
needed is 550 – 574. The frequency distribution is as follows.

Interval	Frequency
450 – 474	5
475 – 499	6
500 – 524	5
525 – 549	2
550 – 574	2

(b) Draw the histogram. It consists of 5 bars of equal width and having heights
as determined by the frequency of each interval. See the histogram that appears
in the back of the textbook.

(c) Construct the frequency polygon by joining consecutive midpoints of the tops
of the histogram bars with straight line segments. See the back of the textbook.

5. $\Sigma x = 41 + 60 + 67 + 68 + 72 + 74 + 78 + 83 + 90 + 97 = 730$

The mean of the 10 numbers is

$$\overline{x} = \frac{\Sigma x}{10} = \frac{730}{10} = 73.$$

7.

Interval	Midpoint, x	Frequency, f	Product, xf
10 – 19	14.5	6	87
20 – 29	24.5	12	294
30 – 39	34.5	14	483
40 – 49	44.5	10	445
50 – 59	54.5	8	436
		Total: 50	Total: 1745

The mean of this collection of grouped data is

$$\overline{x} = \frac{1745}{50} = 34.9.$$

11. Arrange the numbers in numerical order, from smallest to largest.

   35, 36, 36, 38, 38, 42, 44, 48

There are 8 numbers here; the median is the mean of the 2 middle numbers, which is

$$\frac{38 + 38}{2} = \frac{76}{2} = 38.$$

The mode is the number that occurs most often. Here, there are two modes, 36 and 38, since they both appear twice.

13. The modal class for the distribution of Exercise 8 is the interval 55 - 59, since it contains more data values than any of the other intervals.

17. The range is 93 - 26 = 67, the difference of the highest and lowest numbers in the distribution.

The mean is $\bar{x} = \frac{1}{10}(26 + 43 + 51 + 29 + 37 + 56 + 29 + 82 + 74 + 93) = \frac{520}{10} = 52.$

Construct a table with the values of x, $x - \bar{x}$, and $(x - \bar{x})^2$.

x	$x - \bar{x}$	$(x - \bar{x})^2$
26	−26	676
43	−9	81
51	−1	1
29	−23	529
37	−15	225
56	4	16
29	−23	529
82	30	900
74	22	484
93	41	1681
	Total:	5122

The standard deviation is $s = \sqrt{\dfrac{5122}{10 - 1}} \approx \sqrt{569.1} \approx 23.9.$

19. Start with the frequency distribution that was the answer to Exercise 8, and expand the table to include columns for the midpoint x of each interval, and for fx, x^2, and fx^2.

Interval	f	x	fx	x^2	fx^2
40 – 44	2	42	84	1764	3528
45 – 49	5	47	235	2209	11,045
50 – 54	7	52	364	2704	18,928
55 – 59	10	57	570	3249	32,490
60 – 64	4	62	248	3844	15,376
65 – 69	1	67	67	4489	4489
Totals:	29		1568		85,856

The mean of the grouped data is

$$\bar{x} = \frac{\Sigma fx}{n} = \frac{1568}{29} \approx 54.07.$$

The standard deviation for the grouped data is

$$s = \sqrt{\frac{\Sigma fx^2 - n\bar{x}^2}{n - 1}} = \sqrt{\frac{85,856 - 29(54.07)^2}{29 - 1}} \approx \sqrt{38.3} \approx 6.2.$$

21. A skewed distribution has the largest frequency at one end rather than in the middle.

23. **(a)** Using the standard normal curve table,

$$P(-2.5 \le z \le 2.5) = .9938 - .0062 = .9876,$$

so 98.76% of the distribution is within 2.5 standard deviations of the mean.

(b) According to Chebyshev's theorem,

$$P(-2.5 \le z \le 2.5) \ge 1 - \frac{1}{(2.5)^2} \approx .84,$$

so at least 84% of the distribution is within 2.5 standard deviations of the mean.

25. Using the standard normal curve table,

$$P(z < .41) = .6591.$$

27. $P(1.53 \leq z \leq 2.82)$

$= P(z \leq 2.82) - P(z \leq 1.53)$

$= .9976 - .9370$

$= .0606$

29. The normal distribution is not a good approximation of a binomial distribution that has a value of p close to 0 or 1 because the histogram of such a binomial distribution is skewed and therefore not close to the shape of a normal distribution.

31. $P(x < 32)$

$= P\left(z < \dfrac{32 - 32.1}{.1}\right)$

$= P(z < -1)$

$= .1587$ or 15.87%

33. $n = 500$, $p = .06$

To approximate the binomial distribution, use a normal distribution with

$\mu = np = 500(.06) = 30$ and
$\sigma = \sqrt{np(1 - p)}$

$= \sqrt{28.2}$

$\approx 5.31.$

(a) 25 or fewer overstuffed corresponds to the area under the normal curve to the left of $x = 25.5$. The corresponding z–score is

$$z = \frac{25.5 - 30}{5.3} = -.85.$$

$P(25$ or fewer overstuffed)

$= P(x \leq 25.5)$

$= P(z < -.85)$

$= .1977$

(b) P(exactly 30 overstuffed)

$= P(29.5 \leq x \leq 30.5)$

$= P(-.09 \leq z \leq .09)$

$= .5359 - .4641$

$= .0718$

(c) P(more than 40 overstuffed)

$= P(x \geq 40.5)$

$= P(z \geq 1.98)$

$= 1 - .9761$

$= .0239$

35. The table below records the mean and standard deviation for diet A and for diet B.

	μ	σ
diet A	2.7	2.14
diet B	1.3	.9

(a) Diet A had the greater mean gain, since the mean for diet A is larger.

(b) Diet B had a more consistent gain, since diet B has a smaller standard deviation.

37. $n = 100$, $p = .98$

We have $\mu = np = 100(.98) = 98$ and

$\sigma = \sqrt{np(1 - p)} = \sqrt{100(.98)(.02)}$

$= 1.4.$

At least 95% or .95(100) = 95 of the flies corresponds to the area under the normal curve to the right of x = 94.5. The corresponding z-score is

$$z = \frac{94.5 - 98}{1.4} = -2.5.$$

$$
\begin{aligned}
P(\text{at least 95 flies are killed}) &= P(x \geq 94.5) \\
&= P(z \geq -2.5) \\
&= 1 - P(z \leq -2.5) \\
&= 1 - .0062 \\
&= .9938
\end{aligned}
$$

39. n = 100, p = .98

As in Exercise 37, we have μ = 98 and σ = 1.4. All the flies, or 100 flies, corresponds to the area under the normal curve between x = 99.5 and x = 100.5. The corresponding z-scores are

$$z = \frac{99.5 - 98}{1.4} \approx 1.07 \quad \text{and}$$

$$z = \frac{100.5 - 98}{1.4} \approx 1.79.$$

$$
\begin{aligned}
P(\text{all the flies are killed}) &= P(99.5 \leq x \leq 100.5) \\
&= P(1.07 \leq z \leq 1.79) \\
&= P(z \leq 1.79) - P(z \leq 1.07) \\
&= .9633 - .8577 \\
&= .1056
\end{aligned}
$$

41. μ = 42, σ = 12

Find the z-score for z = 35.

$$z = \frac{35 - 42}{12} \approx -.58$$

$$P(x \leq 35) = P(z \leq -.58) = .2810$$

28.10% of the residents commute no more than 35 min per day.

43. $\mu = 42$, $\sigma = 12$

For x = 38, z ≈ -.33.

For x = 60, z = 1.5.

$$
\begin{aligned}
P(38 \leq x \leq 60) &= P(-.33 \leq z \leq 1.5) \\
&= P(z \leq 1.5) - P(z \leq -.33) \\
&= .9332 - .3707 \\
&= .5625
\end{aligned}
$$

56.25% of the residents commute between 38 and 60 min per day.

45.

Interval	x	Tally	f	x • f
1 - 3	2	|||||		6	12
4 - 6	5	|||||	5	25
7 - 9	8	||||| |||||		11	88
10 - 12	11	||||| ||||| ||||| |||||	20	220
13 - 15	14	|||||		6	84
16 - 18	17	||	2	34
Totals			50	463

x	P(x)	x • p(x)
0	.001	0
1	.010	.010
2	.044	.088
3	.117	.351
4	.205	.820
5	.246	1.230
6	.205	1.230
7	.117	.819
8	.044	.352
9	.010	.090
10	.001	.010
Totals: 1.000		5.000

(a) The mean of the frequency distribution is

$$\bar{x} = \frac{\Sigma(x \cdot f)}{n} = \frac{463}{50} = 9.26 \approx 9.3.$$

The expected value of the probability distribution is

$$E(x) = \Sigma(x \cdot P(x)) = 5.$$

(b) The standard deviation of the frequency distribution is

$$s = \sqrt{\frac{\Sigma f x^2 - n\bar{x}^2}{n - 1}} = \sqrt{\frac{5027 - 50(9.26)^2}{50 - 1}} \approx 3.9.$$

The standard deviation of the probability distribution is

$$s = \sqrt{np(1 - p)} = \sqrt{10(.5)(.5)} \approx 1.58.$$

(c) Use $k = 2$ in Chebyshev's theorem since $1 - \frac{1}{2^2} = \frac{3}{4} = 75\%$. For the frequency distribution, the interval $\bar{x} - 2s \leq x \leq \bar{x} + 2s$ becomes

$$9.3 - 2(3.9) \leq x \leq 9.3 + 2(3.9)$$
$$\text{or} \qquad 1.5 \leq x \leq 17.1.$$

For the probability distribution, the interval is

$$5 - 2(1.58) \leq x \leq 5 + 2(1.58) \quad \text{or} \quad 1.84 \leq x \leq 8.16.$$

(d) 95.44% of the area under the normal approximation of the binomial probability distribution will lie between $z = -2$ and $z = 2$.

$$z = -2 \text{ means } -2 = \frac{x - 5}{1.58} \quad \text{or} \quad 1.84 = x.$$

$$z = 2 \text{ means } 2 = \frac{x - 5}{1.58} \quad \text{or} \quad 8.16 = x.$$

The interval is $1.84 \leq x \leq 8.16$.

(e) The normal distribution can't be used to answer probability questions about the frequency distribution because the histogram of the frequency distribution is not close enough to the shape of a normal curve.

CHAPTER 7 TEST

1. A certain species of fruit fly is 40% red-winged and 60% black-winged. Let x represent the number of red-winged flies. Compute the probability distribution for x if 4 flies are chosen at random.

2. The following table gives the probability distribution for a random variable x. Find the expected value for x.

x	3	4	5	7
P(x)	.1	.3	.5	.1

3. Find the mean and standard deviation for the set of numbers given below. Also find the median and the mode.

$$3, \ 3, \ 4, \ 4, \ 4, \ 5, \ 6, \ 6, \ 6, \ 6$$

4. Find the mean and standard deviation of a binomial distribution with n = 56 and p = .4.

5. A distribution of 100 incomes has μ = $10,500 and σ = $1000. Use Chebyshev's theorem to find:

   (a) A range of incomes which would include at least 75 of the 100 incomes.

   (b) The minimum number of incomes we would expect to find in the interval from $7500 to $13,500.

6. Using the normal curve table, find the percent of the total area under the normal curve in each case:

   (a) z between -1.13 and 2.14.   (b) z greater than 2.1.

7. Al's Quick Photo store develops an average roll of film in 2.3 min. The standard deviation in time is .6 min. Assume a normal distribution.

(a) Out of 1000 rolls of film, how many will be developed in less than 3.5 min?

(b) What percentage of film is developed in between 2 and 3 min?

8. A loaded coin with P(head) = .6 is tossed 100 times. Using the normal curve approximation to the binomial distribution, find the probability of getting each of the following:

(a) at least 55 heads.   (b) exactly 61 heads.

(c) between 55 and 65 heads (inclusive).

CHAPTER 7 TEST ANSWERS

1.

x	0	1	2	3	4
P(x)	.1296	.3456	.3456	.1536	.0256

2. E(x) = 4.7   3. μ = 4.7, $\sigma \approx$ 1.25   4. μ = 22.4, $\sigma \approx$ 3.67

Median = 4.5, mode = 6

5. **(a)** $8500 to $12,500   **(b)** 88   6. **(a)** .8546   **(b)** .0179

7. **(a)** 977   **(b)** 57%   8. **(a)** .8686   **(b)** .0819   **(c)** .7372

CHAPTER 8   MARKOV CHAINS AND GAME THEORY

Section 8.1

1. $\begin{bmatrix} \frac{2}{3} & \frac{1}{2} \end{bmatrix}$ could not be a probability vector because the sum of the entries in the row is not equal to 1.

3. [0   1] could be a probability vector since it is a matrix of only one row, having nonnegative entries whose sum is 1.

5. [.4   .2   0] could not be a probability vector because the sum of the entries in the row is not equal to 1.

7. [.07   .04   .37   .52] could be a probability vector. It is a matrix of only one row, having nonnegative entries whose sum is 1.

9. [0,   −.2   .6   .6] could not be a probability vector because it has a negative entry.

11.
$$\begin{array}{cc} & \begin{array}{cc} A & B \end{array} \\ \begin{array}{c} A \\ B \end{array} & \begin{bmatrix} \frac{2}{3} & \frac{1}{3} \\ 1 & 0 \end{bmatrix} \end{array}$$

This could be a transition matrix since it is a square matrix, all entries are between 0 and 1, inclusive, and the sum of the entries in each row is 1.

To draw the transition diagram, give names to the two states (such as A and B) and label the probabilities of going from one state to another.   In this case,

$$P_{AA} = \frac{2}{3}, \qquad P_{AB} = \frac{1}{3},$$
$$P_{BA} = 1, \text{ and } P_{BB} = 0.$$

See the transition diagram in the back of the textbook.

13. $\begin{bmatrix} \frac{1}{4} & \frac{3}{4} & 0 \\ 2 & 0 & 1 \\ 1 & \frac{2}{3} & 3 \end{bmatrix}$

This could not be a transition matrix because it has entries that are greater than 1.

15. $\begin{bmatrix} \frac{1}{3} & \frac{1}{2} & 1 \\ 0 & 1 & 0 \\ \frac{1}{2} & \frac{1}{2} & 1 \end{bmatrix}$

This could not be a transition matrix because the sum of the entries in the first row and in the third row is more than 1.

17. The transition diagram provides the information

$$P_{AA} = .9, \ P_{AB} = .1, \ P_{AC} = 0,$$
$$P_{BA} = .1, \ P_{BB} = .6, \ P_{BC} = .3,$$
$$P_{CA} = 0, \ P_{CB} = .3, \text{ and } P_{CC} = .7.$$

The transition matrix associated with this diagram is

$$
\begin{array}{c} \\ A \\ B \\ C \end{array}
\begin{array}{ccc} A & B & C \end{array}
\begin{bmatrix} .9 & .1 & 0 \\ .1 & .6 & .3 \\ 0 & .3 & .7 \end{bmatrix}.
$$

19. $A = \begin{bmatrix} 1 & 0 \\ .8 & .2 \end{bmatrix}$

$$
A^2 = \begin{bmatrix} 1 & 0 \\ .8 & .2 \end{bmatrix} \begin{bmatrix} 1 & 0 \\ .8 & .2 \end{bmatrix} = \begin{bmatrix} 1 & 0 \\ .96 & .04 \end{bmatrix}
$$

$$
A^3 = \begin{bmatrix} 1 & 0 \\ .8 & .2 \end{bmatrix} \begin{bmatrix} 1 & 0 \\ .96 & .04 \end{bmatrix}
$$

$$
= \begin{bmatrix} 1 & 0 \\ .992 & .008 \end{bmatrix}
$$

The entry in row 1, column 2 of A^3 gives the probability that state 1 changes to state 2 after 3 repetitions of the experiment. This probability is 0.

21. $C = \begin{bmatrix} .5 & .5 \\ .72 & .28 \end{bmatrix}$

$$
C^2 = \begin{bmatrix} .5 & .5 \\ .72 & .28 \end{bmatrix} \begin{bmatrix} .5 & .5 \\ .72 & .28 \end{bmatrix}
$$

$$
= \begin{bmatrix} .61 & .39 \\ .5616 & .4384 \end{bmatrix}
$$

$$
C^3 = \begin{bmatrix} .5 & .5 \\ .72 & .28 \end{bmatrix} \begin{bmatrix} .61 & .39 \\ .5616 & .4384 \end{bmatrix}
$$

$$
= \begin{bmatrix} .5858 & .4142 \\ .596448 & .403552 \end{bmatrix}
$$

The probability that state 1 changes to state 2 after 3 repetitions is .4142, since that is the entry in row 1, column 2 of C^3.

23. $E = \begin{bmatrix} .8 & .1 & .1 \\ .3 & .6 & .1 \\ 0 & 1 & 0 \end{bmatrix}$

$$
E^2 = \begin{bmatrix} .8 & .1 & .1 \\ .3 & .6 & .1 \\ 0 & 1 & 0 \end{bmatrix} \begin{bmatrix} .8 & .1 & .1 \\ .3 & .6 & .1 \\ 0 & 1 & 0 \end{bmatrix}
$$

$$
= \begin{bmatrix} .67 & .24 & .09 \\ .42 & .49 & .09 \\ .3 & .6 & .1 \end{bmatrix}
$$

$$
E^3 = \begin{bmatrix} .8 & .1 & .1 \\ .3 & .6 & .1 \\ 0 & 1 & 0 \end{bmatrix} \begin{bmatrix} .67 & .24 & .09 \\ .42 & .49 & .09 \\ .3 & .6 & .1 \end{bmatrix}
$$

$$
= \begin{bmatrix} .608 & .301 & .091 \\ .483 & .426 & .091 \\ .42 & .49 & .09 \end{bmatrix}
$$

The probability that state 1 changes to state 2 after 3 repetitions is .301, since that is the entry in row 1, column 2 of E^3.

25. (a) We are asked to show that $I(P^n) = (\cdots(((IP)P)P)\cdots P)$ is true for any natural number n, where the expression on the right side of the equation has a total of n factors of P. This may be proven by mathematical induction on n.

When n = 1, the statement becomes $I(P) = (IP)$, which is obviously true. When n = 2, the statement becomes $(P^2) = (IP)P$, or $I(PP) = (IP)P$, which is true since matrix multiplication is associative. Next, assume the nth statement is true in order to show that (n + 1)st statement is true. That is, assume that

$$I(P^n) = (\cdots(((IP)P)P)\cdots P)$$

is true. Associativity plays a role here also.

$$I(P^{n+1}) = I(P^n \cdot P)$$
$$= (IP^n)P$$
$$= (\cdots(((IP)P)P)\cdots P)P$$

Conclude that

$$I(P^n) = (\cdots(((IP)P)P)\cdots P)$$

is true for any natural number n.

27. The probability vector is [.4   .6].

(a) $[.4 \quad .6]\begin{bmatrix} .8 & .2 \\ .35 & .65 \end{bmatrix}$

$= [.53 \quad .47]$

Thus, after 1 week Johnson has 53%
market share, and Northclean has a
47% share.

(b) $C^2 = \begin{bmatrix} .8 & .2 \\ .35 & .65 \end{bmatrix}\begin{bmatrix} .8 & .2 \\ .35 & .65 \end{bmatrix}$

$= \begin{bmatrix} .71 & .29 \\ .5075 & .4925 \end{bmatrix}$

$[.4 \quad .6]\begin{bmatrix} .71 & .29 \\ .5075 & .4925 \end{bmatrix}$

$= [.5885 \quad .4115]$

After 2 weeks, Johnson has a 58.85%
market share, and Northclean has a
41.15% share.

(c) $C^3 = \begin{bmatrix} .8 & .2 \\ .35 & .65 \end{bmatrix}\begin{bmatrix} .71 & .29 \\ .5075 & .4925 \end{bmatrix}$

$= \begin{bmatrix} .6695 & .3305 \\ .5784 & .4216 \end{bmatrix}$

$[.4 \quad .6]\begin{bmatrix} .6695 & .3305 \\ .5784 & .4216 \end{bmatrix}$

$= [.6148 \quad .3852]$

After 3 weeks, the shares are 61.48%
and 38.52%, respectively.

(d) $C^4 = \begin{bmatrix} .8 & .2 \\ .35 & .65 \end{bmatrix}\begin{bmatrix} .6695 & .3305 \\ .5784 & .4216 \end{bmatrix}$

$= \begin{bmatrix} .6513 & .3487 \\ .6103 & .3897 \end{bmatrix}$

$[.4 \quad .6]\begin{bmatrix} .6513 & .3487 \\ .6103 & .3897 \end{bmatrix}$

$= [.62666 \quad .37334]$

After 4 weeks the shares are 62.666%
and 37.334%, respectively.

29. The transition matrix P is

$$\begin{array}{ccc} & G_0 & G_1 & G_2 \end{array}$$
$$\begin{array}{c} G_0 \\ G_1 \\ G_2 \end{array}\begin{bmatrix} .85 & .1 & .05 \\ 0 & .8 & .2 \\ 0 & 0 & 1 \end{bmatrix}.$$

We have 50,000 new policy holders,
all in G_0. The probability vector
for these people is

$$\begin{array}{ccc} G_0 & G_1 & G_2 \end{array}$$
$$[1 \quad 0 \quad 0].$$

(a) After 1 yr, the distribution of
people in each group is

$[1 \quad 0 \quad 0]\begin{bmatrix} .85 & .1 & .05 \\ 0 & .8 & .2 \\ 0 & 0 & 1 \end{bmatrix}$

$= [.85 \quad .1 \quad .05].$

There are

(.85)(50,000) = 42,500 people in G_0,

(.1)(50,000) = 5000 people in G_1, and

(.05)(50,000) = 2500 people in G_2.

(b)

$$P^2 = \begin{bmatrix} .85 & .1 & .05 \\ 0 & .8 & .2 \\ 0 & 0 & 1 \end{bmatrix} \begin{bmatrix} .85 & .1 & .05 \\ 0 & .8 & .2 \\ 0 & 0 & 1 \end{bmatrix}$$

$$= \begin{bmatrix} .7225 & .165 & .1125 \\ 0 & .64 & .36 \\ 0 & 0 & 1 \end{bmatrix}$$

After 2 yr, the distribution of people in each group is

$$[1 \quad 0 \quad 0] \begin{bmatrix} .7225 & .165 & .1125 \\ 0 & .64 & .36 \\ 0 & 0 & 1 \end{bmatrix}$$

$$= [.7225 \quad .165 \quad .1125].$$

There are

$(.7225)(50,000) = 36,125$ in G_0,

$(.165)(50,000) = 8250$ in G_1, and

$(.1125)(50,000) = 5625$ in G_2.

(c)

$$P^3 = \begin{bmatrix} .85 & .1 & .05 \\ 0 & .8 & .2 \\ 0 & 0 & 1 \end{bmatrix} \begin{bmatrix} .7225 & .165 & .1125 \\ 0 & .64 & .36 \\ 0 & 0 & 1 \end{bmatrix}$$

$$= \begin{bmatrix} .61413 & .20425 & .18163 \\ 0 & .512 & .488 \\ 0 & 0 & 1 \end{bmatrix}$$

After 3 yr, the distribution is

$$[1 \quad 0 \quad 0] \begin{bmatrix} .61413 & .20425 & .18163 \\ 0 & .512 & .488 \\ 0 & 0 & 1 \end{bmatrix}$$

$$= [.61413 \quad .20425 \quad .18163].$$

There are

$(50,000)(.61413) = 30,706$ in G_0,

$(50,000)(.20425) = 10,213$ in G_1, and

$(50,000)(.18163) = 9081$ in G_2.

(d)

$$P^4 = \begin{bmatrix} .85 & .1 & .05 \\ 0 & .8 & .2 \\ 0 & 0 & 1 \end{bmatrix} \begin{bmatrix} .6141 & .2043 & .1816 \\ 0 & .512 & .488 \\ 0 & 0 & 1 \end{bmatrix}$$

$$\begin{bmatrix} .52201 & .22481 & .25318 \\ 0 & .4096 & .5904 \\ 0 & 0 & 1 \end{bmatrix}$$

The probabilities are

$$[1 \quad 0 \quad 0] \begin{bmatrix} .52201 & .22481 & .25318 \\ 0 & .4096 & .5904 \\ 0 & 0 & 1 \end{bmatrix}$$

$$= [.52201 \quad .22481 \quad .25318].$$

There are

$(50,000)(.52201) = 26,100$ in G_0,

$(50,000)(.22481) = 11,241$ in G_1, and

$(50,000)(.25318) = 12,659$ in G_2.

31. $$P = \begin{bmatrix} .825 & .175 & 0 \\ .060 & .919 & .021 \\ .049 & 0 & .951 \end{bmatrix}$$

The initial probability vector is

$$[.26 \quad .6 \quad .14].$$

(a) The share held by each type is given by

$$[.26 \quad .6 \quad .14] \begin{bmatrix} .825 & .175 & 0 \\ .060 & .919 & .021 \\ .049 & 0 & .951 \end{bmatrix}$$

$$= [.257 \quad .597 \quad .146].$$

(b)

$$P^2 = \begin{bmatrix} .825 & .175 & 0 \\ .060 & .919 & .021 \\ .049 & 0 & .951 \end{bmatrix} \begin{bmatrix} .825 & .175 & 0 \\ .060 & .919 & .021 \\ .049 & 0 & .951 \end{bmatrix}$$

$$= \begin{bmatrix} .691 & .305 & .004 \\ .106 & .855 & .039 \\ .087 & .009 & .904 \end{bmatrix}$$

The share held by each is

$$[.26 \quad .6 \quad .14] \begin{bmatrix} .691 & .305 & .004 \\ .106 & .855 & .039 \\ .087 & .009 & .904 \end{bmatrix}$$

$$= [.255 \quad .594 \quad .151].$$

(c)

$$P^3 = \begin{bmatrix} .825 & .175 & 0 \\ .060 & .919 & .021 \\ .049 & 0 & .951 \end{bmatrix} \begin{bmatrix} .691 & .305 & .004 \\ .106 & .855 & .039 \\ .087 & .009 & .904 \end{bmatrix}$$

$$= \begin{bmatrix} .589 & .401 & .01 \\ .141 & .804 & .055 \\ .117 & .023 & .860 \end{bmatrix}$$

The share held by each after 3 yr is

$$[.26 \quad .6 \quad .14] \begin{bmatrix} .589 & .401 & .01 \\ .141 & .804 & .055 \\ .117 & .023 & .860 \end{bmatrix}$$

$$= [.254 \quad .590 \quad .156].$$

33. (a) If there is no one in line, then after 1 min there will be either 0, 1, or 2 people in line with probabilities $p_{00} = 1/2$, $p_{01} = 1/3$, and $p_{02} = 1/6$. If there is one person in line, then that person will be served and either 0, 1, or 2 new people will join the line, with probabilities $p_{10} = 1/2$, $p_{11} = 1/3$, and $p_{12} = 1/6$. If there are two people in line, then one of them will be served and either 1 or 2 new people will join the line, with probabilities $p_{21} = 1/2$ and $p_{22} = 1/2$; it is impossible for both people in line to be served, so $p_{20} = 0$. Therefore,

the transition matrix is

$$A = \begin{bmatrix} \frac{1}{2} & \frac{1}{3} & \frac{1}{6} \\ \frac{1}{2} & \frac{1}{3} & \frac{1}{6} \\ 0 & \frac{1}{2} & \frac{1}{2} \end{bmatrix}.$$

(b) The transition matrix for a two-minute period is

$$A^2 = \begin{bmatrix} \frac{1}{2} & \frac{1}{3} & \frac{1}{6} \\ \frac{1}{2} & \frac{1}{3} & \frac{1}{6} \\ 0 & \frac{1}{2} & \frac{1}{2} \end{bmatrix} \begin{bmatrix} \frac{1}{2} & \frac{1}{3} & \frac{1}{6} \\ \frac{1}{2} & \frac{1}{3} & \frac{1}{6} \\ 0 & \frac{1}{2} & \frac{1}{2} \end{bmatrix}$$

$$= \begin{bmatrix} \frac{5}{12} & \frac{13}{36} & \frac{2}{9} \\ \frac{5}{12} & \frac{13}{36} & \frac{2}{9} \\ \frac{1}{4} & \frac{5}{12} & \frac{1}{3} \end{bmatrix}.$$

(c) The probability that a queue with no one in line has two people in line 2 min later is 2/9, since that is the entry in row 1, column 3 of A^2.

35. (a) We use L for liberal, C for conservative, and I for independent. The transition matrix P is give by

$$\begin{array}{c} \\ L \\ C \\ I \end{array} \begin{array}{ccc} L & C & I \\ \begin{bmatrix} .80 & .15 & .05 \\ .20 & .70 & .10 \\ .20 & .20 & .60 \end{bmatrix} \end{array}.$$

(b) The initial distribution is given by

$$I = [.40 \quad .45 \quad .15].$$

(c) On month later in July,

$$IP = [.44 \quad .405 \quad .155],$$

so there will be 44% liberals, 40.5% conservatives, and 15.5% independents.

(d) Two months later in August,

$$IP^2 = (IP)P = [.464 \quad .3805 \quad .1555],$$

so there will be 46.4% liberals, 38.05% conservatives, and 15.55% independents.

(e) Three months later in September,

$$IP^3 = (IP^2)P = [.4784 \quad .36705 \quad .15455],$$

so there will be 47.84% liberals, 36.705% conservatives, and 15.455% independents.

(f) Four months later in October,

$$IP^4 = (IP^3)P = [.48704 \quad .359605 \quad .153355],$$

so there will be 48.704% liberals, 35.9605% conservatives, and 15.3355% independents.

37. This exercise should be solved by computer methods. The solution will vary according to the computer program that is used. The answers are as follows. The first five powers of the transition matrix are

$$A = \begin{bmatrix} .1 & .2 & .2 & .3 & .2 \\ .2 & .1 & .1 & .2 & .4 \\ .2 & .1 & .4 & .2 & .1 \\ .3 & .1 & .1 & .2 & .3 \\ .1 & .3 & .1 & .1 & .4 \end{bmatrix}, \quad A^2 = \begin{bmatrix} .2 & .15 & .17 & .19 & .29 \\ .16 & .2 & .15 & .18 & .31 \\ .19 & .14 & .24 & .21 & .22 \\ .16 & .19 & .16 & .2 & .29 \\ .16 & .19 & .14 & .17 & .34 \end{bmatrix},$$

$$A^3 = \begin{bmatrix} .17 & .178 & .171 & .191 & .29 \\ .171 & .178 & .161 & .185 & .305 \\ .18 & .163 & .191 & .197 & .269 \\ .175 & .174 & .164 & .187 & .3 \\ .167 & .184 & .158 & .182 & .309 \end{bmatrix}, \quad A^4 = \begin{bmatrix} .1731 & .175 & .1683 & .188 & .2956 \\ .1709 & .1781 & .1654 & .1866 & .299 \\ .1748 & .1718 & .1753 & .1911 & .287 \\ .1712 & .1775 & .1667 & .1875 & .2971 \\ .1706 & .1785 & .1641 & .1858 & .301 \end{bmatrix},$$

and

$$A^5 = \begin{bmatrix} .17193 & .17643 & .1678 & .18775 & .29609 \\ .17167 & .17689 & .16671 & .18719 & .29754 \\ .17293 & .17488 & .17007 & .18878 & .29334 \\ .17192 & .17654 & .16713 & .18741 & .297 \\ .17142 & .17726 & .16629 & .18696 & .29807 \end{bmatrix}.$$

The probability that state 2 changes to state 4 after 5 repetitions of the experiment is .18719, since that is the entry in row 2, column 4 of A^5.

39. This exercise should be solved by computer methods. The solution will vary according to the computer program that is used. The answers are as follows.

   (a) The proportion is
   .847423 or about 85%.

   (b) The proportions are as listed in the probability vector

   $$[.0128 \quad .0513 \quad .0962 \quad .8397].$$

Section 8.2

1. Let $A = \begin{bmatrix} .2 & .8 \\ .9 & .1 \end{bmatrix}$.

   A is a regular transition matrix since $A^1 = A$ contains all positive entries.

3. Let $B = \begin{bmatrix} 1 & 0 \\ .6 & .4 \end{bmatrix}$.

   $$B^2 = \begin{bmatrix} 1 & 0 \\ .6 & .4 \end{bmatrix} \begin{bmatrix} 1 & 0 \\ .6 & .4 \end{bmatrix}$$

   $$= \begin{bmatrix} 1 & 0 \\ .84 & .16 \end{bmatrix}$$

   B is not regular since any power of B will have [1   0] as its first row and thus cannot have all positive entries.

5. Let $P = \begin{bmatrix} 0 & 1 & 0 \\ .4 & .2 & .4 \\ 1 & 0 & 0 \end{bmatrix}$.

   $$P^2 = \begin{bmatrix} 0 & 1 & 0 \\ .4 & .2 & .4 \\ 1 & 0 & 0 \end{bmatrix} \begin{bmatrix} 0 & 1 & 0 \\ .4 & .2 & .4 \\ 1 & 0 & 0 \end{bmatrix}$$

   $$= \begin{bmatrix} .4 & .2 & .4 \\ .48 & .44 & .08 \\ 0 & 1 & 0 \end{bmatrix}$$

   $$P^3 = \begin{bmatrix} 0 & 1 & 0 \\ .4 & .2 & .4 \\ 1 & 0 & 0 \end{bmatrix} \begin{bmatrix} .4 & .2 & .4 \\ .48 & .44 & .08 \\ 0 & 1 & 0 \end{bmatrix}$$

   $$= \begin{bmatrix} .48 & .44 & .08 \\ .256 & .568 & .176 \\ .4 & .2 & .4 \end{bmatrix}$$

   P is a regular transition matrix since P^3 contains all positive entries.

7. Let $P = \begin{bmatrix} \frac{1}{4} & \frac{3}{4} \\ \frac{1}{2} & \frac{1}{2} \end{bmatrix}$ and let V be the probability vector $[v_1 \quad v_2]$.
   We want to find V such that
   $$VP = V,$$

   or $[v_1 \quad v_2] \begin{bmatrix} \frac{1}{4} & \frac{3}{4} \\ \frac{1}{2} & \frac{1}{2} \end{bmatrix} = [v_1 \quad v_2]$.

   Use matrix multiplication on the left and obtain

   $$\left[\frac{1}{4}v_1 + \frac{1}{2}v_2 \quad \frac{3}{4}v_1 + \frac{1}{2}v_2\right] = [v_1 \quad v_2].$$

Set corresponding entries from the two matrices equal to get

$$\frac{1}{4}v_1 + \frac{1}{2}v_2 = v_1$$

$$\frac{3}{4}v_1 + \frac{1}{2}v_2 = v_2.$$

Multiply both equations by 4 to eliminate fractions.

$$v_1 + 2v_2 = 4v_1$$
$$3v_1 + 2v_2 = 4v_2$$

Simplify both equations.

$$-3v_1 + 2v_2 = 0$$
$$3v_1 - 2v_2 = 0$$

This is a dependent system. To find the values of v_1 and v_2, an additional equation is needed. Since $V = [v_1 \quad v_2]$ is a probability vector,

$$v_1 + v_2 = 1.$$

To find v_1 and v_2, solve the system

$$-3v_1 + 2v_2 = 0 \quad (1)$$
$$v_1 + v_2 = 1. \quad (2)$$

From equation (2), $v_1 = 1 - v_2$. Substitute $1 - v_2$ for v_1 in equation (1) to obtain

$$-3(1 - v_2) + 2v_2 = 0$$
$$-3 + 3v_2 + 2v_2 = 0$$
$$-3 + 5v_2 = 0$$
$$v_2 = \frac{3}{5}.$$

Since $v_1 = 1 - v_2$, $v_1 = 2/5$, and the equilibrium vector is

$$V = \begin{bmatrix} \frac{2}{5} & \frac{3}{5} \end{bmatrix}.$$

9. Let $P = \begin{bmatrix} .3 & .7 \\ .4 & .6 \end{bmatrix}$ and let V be the probability vector $[v_1 \quad v_2]$.
We want to find V such that

$$VP = V,$$

or $[v_1 \quad v_2] \begin{bmatrix} .3 & .7 \\ .4 & .6 \end{bmatrix} = [v_1 \quad v_2].$

By matrix multiplication and equality of matrices,

$$.3v_1 + .4v_2 = v_1$$
$$.7v_1 + .6v_2 = v_2.$$

Simplify these equations to get the dependent system

$$-.7v_1 + .4v_2 = 0$$
$$.7v_1 - .4v_2 = 0.$$

Since V is a probability vector,

$$v_1 + v_2 = 1.$$

To find v_1 and v_2, solve the system

$$.7v_1 - .4v_2 = 0$$
$$v_1 + v_2 = 0$$

by the substitution method.
Observe that $v_2 = 1 - v_1$.

$$.7v_1 - .4(1 - v_1) = 0$$
$$1.1v_1 - .4 = 0$$
$$v_1 = \frac{.4}{1.1} = \frac{4}{11}$$
$$v_2 = 1 - \frac{4}{11} = \frac{7}{11}$$

The equilibrium vector is

$$\begin{bmatrix} \frac{4}{11} & \frac{7}{11} \end{bmatrix}.$$

11. Let V be the probability vector
$[v_1 \quad v_2 \quad v_3]$.

$$[v_1 \quad v_2 \quad v_3] \begin{bmatrix} .1 & .1 & .8 \\ .4 & .4 & .2 \\ .1 & .2 & .7 \end{bmatrix}$$

$$= [v_1 \quad v_2 \quad v_3]$$

$.1v_1 + .4v_2 + .1v_3 = v_1$

$.1v_1 + .4v_2 + .2v_3 = v_2$

$.8v_1 + .2v_2 + .7v_3 = v_3$

Simplify these equations to get the dependent system

$-.9v_1 + .4v_2 + .1v_3 = 0$

$.1v_1 - .6v_2 + .2v_3 = 0$

$.8v_1 + .2v_2 - .3v_3 = 0.$

Since V is a probability vector,

$$v_1 + v_2 + v_3 = 1.$$

Solving the above system of 4 equations using the Gauss–Jordan method, we obtain

$$v_1 = \frac{14}{83}, \quad v_2 = \frac{19}{83}, \quad v_3 = \frac{50}{83}.$$

Thus, the equilibrium vector is

$$\left[\frac{14}{83} \quad \frac{19}{83} \quad \frac{50}{83}\right].$$

13. Let V be the probability vector
$[v_1 \quad v_2 \quad v_3]$.

$$[v_1 \quad v_2 \quad v_3] \begin{bmatrix} .25 & .35 & .4 \\ .1 & .3 & .6 \\ .55 & .4 & .05 \end{bmatrix}$$

$$= [v_1 \quad v_2 \quad v_3]$$

$.25v_1 + .1v_2 + .55v_3 = v_1$

$.35v_1 + .3v_2 + .4v_3 = v_2$

$.4v_1 + .6v_2 + .05v_3 = v_3$

Simplify these equations to get the dependent system

$-.75v_1 + .1v_2 + .55v_3 = 0$

$.35v_1 - .7v_2 + .4v_3 = 0$

$.4v_1 + .6v_2 - .95v_3 = 0.$

Since V is a probability vector,

$$v_1 + v_2 + v_3 = 1.$$

Solving this system we obtain

$$v_1 = \frac{170}{563}, \quad v_2 = \frac{197}{563}, \quad v_3 = \frac{196}{563}.$$

Thus, the equilibrium vector is

$$\left[\frac{170}{563} \quad \frac{197}{563} \quad \frac{196}{563}\right].$$

15. $[v_1 \quad v_2 \quad v_3] \begin{bmatrix} .85 & .1 & .05 \\ 0 & .8 & .2 \\ 0 & 0 & 1 \end{bmatrix}$

$$= [v_1 \quad v_2 \quad v_3]$$

$.85v_1 = v_1$

$.1v_1 + .8v_2 = v_2$

$.05v_1 + .2v_2 + v_3 = v_3$

We also have $v_1 + v_2 + v_3 = 1$.
Solving this system, we obtain

$$v_1 = 0, \ v_2 = 0, \ v_3 = 1.$$

The equilibrium vector is

$$[0 \quad 0 \quad 1].$$

17. $[v_1 \quad v_2 \quad v_3] \begin{bmatrix} .825 & .175 & 0 \\ .060 & .919 & .021 \\ .049 & 0 & .951 \end{bmatrix}$

$$= [v_1 \quad v_2 \quad v_3]$$

$.825v_1 + .060v_2 + .049v_3 = v_1$

$.175v_1 + .919v_2 \qquad = v_2$

$.021v_2 + .951v_3 = v_3$

Also, $v_1 + v_2 + v_3 = 1$.

Solving this system, we obtain

$$v_1 = .244, \quad v_2 = .529, \quad v_3 = .227.$$

The equilibrium vector is

$$[.244 \quad .529 \quad .227].$$

19. $[v_1 \quad v_2 \quad v_3] \begin{bmatrix} .80 & .15 & .05 \\ 0 & .90 & .10 \\ .10 & .20 & .70 \end{bmatrix}$

$= [v_1 \quad v_2 \quad v_3]$

$.80v_2 \qquad\qquad + .10v_3 = v_1$

$.15v_1 + .90v_2 + .20v_3 = v_2$

$.05v_1 + .10v_2 + .70v_3 = v_3$

Also, $v_1 + v_2 + v_3 = 1$.

Solving this system, we obtain

$$v_1 = \frac{1}{2}, \quad v_2 = \frac{7}{20}, \quad v_3 = \frac{3}{20}.$$

The equilibrium vector is

$$\left[\frac{1}{2} \quad \frac{7}{20} \quad \frac{3}{20}\right].$$

21. Let V be the probability vector $[x_1 \quad x_2]$. We want to find V such that

$$V\begin{bmatrix} p & 1 - p \\ 1 - q & q \end{bmatrix} = V.$$

The system of equations is

$$px_1 + (1 - q)x_2 = x_1$$
$$(1 - p)x_1 + qx_2 = x_2.$$

Collecting like terms and simplifying leads to

$$(p - 1)x_1 + (1 - q)x_2 = 0,$$

so $\qquad\qquad x_1 = \frac{1 - q}{1 - p}x_2.$

Substituting this into $x_1 + x_2 = 1$, we obtain

$$\frac{1 - q}{1 - p}x_2 + x_2 = 1$$

or $\qquad \frac{2 - p - q}{1 - p}x_2 = 1;$

therefore,

$$x_2 = \frac{1 - p}{2 - p - q}$$

and $\qquad x_1 = \frac{1 - q}{2 - p - q},$

so

$$V = \left[\frac{1 - q}{2 - p - q} \quad \frac{1 - p}{2 - p - q}\right].$$

Since $0 < p < 1$ and $0 < q < 1$, the matrix is always regular.

23. Let V be the probability vector $[x_1 \quad x_2]$.

We have $P = \begin{bmatrix} a_{11} & a_{12} \\ a_{21} & a_{22} \end{bmatrix},$

where $a_{11} + a_{21} = 1$

and $\quad a_{12} + a_{22} = 1.$

The resulting equations are

$$a_{11}x_1 + a_{21}x_2 = x_1$$
$$a_{12}x_1 + a_{22}x_2 = x_2,$$

which we simplify to

$$(a_{11} - 1)x_1 + a_{21}x_2 = 0$$
$$a_{12}x_1 + (a_{22} - 1)x_2 = 0.$$

Hence, $x_1 = \frac{a_{21}}{1 - a_{11}}x_2$

$$= \frac{a_{21}}{a_{21}}x_2 = x_2,$$

which we substitute into $x_1 + x_2 = 1$, obtaining

$$x_2 + x_2 = 1$$
$$2x_2 = 1$$
$$x_2 = \frac{1}{2}$$

and therefore $x_1 = 1 - \frac{1}{2} = \frac{1}{2}$.

The equilibrium vector is $[\frac{1}{2} \quad \frac{1}{2}]$.

25. The transition matrix for the given information is

$$\begin{array}{c} \\ \text{Works} \\ \text{Dosen't Work} \end{array} \begin{array}{cc} \text{Works} & \begin{array}{c}\text{Doesn't}\\\text{work}\end{array} \\ \begin{bmatrix} .9 & .1 \\ .7 & .3 \end{bmatrix} \end{array}.$$

Let V be the probability vector $[v_1 \quad v_2]$.

$$[v_1 \quad v_2]\begin{bmatrix} .9 & .1 \\ .7 & .3 \end{bmatrix} = [v_1 \quad v_2]$$

$$.9v_1 + .7v_2 = v_1$$
$$.1v_1 + .3v_2 = v_2$$

Simplify these equations to get the dependent system

$$-.1v_1 + .7v_2 = 0$$
$$.1v_1 - .7v_2 = 0.$$

Also, $v_1 + v_2 = 1$, so $v_1 = 1 - v_2$.

$$.1(1 - v_2) - .7v_2 = 0$$
$$.1 - .8v_2 = 0$$

$$v_2 = 1/8, \quad v_1 = 7/8$$

The equilibrium vector is

$$[\frac{7}{8} \quad \frac{1}{8}].$$

The long-range probability that the line will work correctly is 7/8.

27. The transition matrix is

	Low	Medium	High
Low	.5	.4	.1
Medium	.25	.45	.3
High	.05	.4	.55

Let V be the probability vector $[v_1 \quad v_2 \quad v_3]$.

$$[v_1 \quad v_2 \quad v_3]\begin{bmatrix} .5 & .4 & .1 \\ .25 & .45 & .3 \\ .05 & .4 & .55 \end{bmatrix}$$

$$= [v_1 \quad v_2 \quad v_3]$$

$$.5v_1 + .25v_2 + .05v_3 = v_1$$
$$.4v_1 + .45v_2 + .4v_3 = v_2$$
$$.1v_1 + .3v_2 + .55v_3 = v_3$$

Also, $v_1 + v_2 + v_3 = 1$.

Solving this system, we obtain

$$v_1 = \frac{51}{209}, \quad v_2 = \frac{88}{209}, \quad v_3 = \frac{70}{209}.$$

The equilibrium vector is

$$[\frac{51}{209} \quad \frac{88}{209} \quad \frac{70}{209}].$$

Thus, the long-range proportions for low, medium, and high producers are 51/209, 88/209, and 70/209, respectively.

29. Let F, C, and R represent fair, cloudy, and rainy, respectively, and let $[x_1 \quad x_2 \quad x_3]$ be a probability vector.

The transition matrix is

	F	C	R
F	.60	.25	.15
C	.40	.35	.25
R	.35	.40	.25

and the resulting system of equations is

$$.60x_1 + .40x_2 + .35x_3 = x_1$$
$$.25x_1 + .35x_2 + .40x_3 = x_2$$
$$.15x_1 + .25x_2 + .25x_3 = x_3.$$

Also, $x_1 + x_2 + x_3 = 1$.
Solving this system, we obtain

$$x_1 = \frac{155}{318}, \quad x_2 = \frac{99}{318}, \quad x_3 = \frac{32}{159}.$$

The equilibrium vector is

$$\left[\frac{155}{318} \quad \frac{99}{318} \quad \frac{32}{159}\right].$$

Over the long term, the proportion of days that are expected to be fair, cloudy, and rainy are $\frac{155}{318} \approx 48.7\%$, $\frac{99}{318} \approx 31.1\%$, and $\frac{32}{159} \approx 20.1\%$, respectively.

31. The transition matrix is

$$\begin{bmatrix} p & 1-p \\ 1-p & p \end{bmatrix}.$$

The columns sum to $p + (1-p) = 1$, so by Exercise 23 equilibrium vector is $[\frac{1}{2} \quad \frac{1}{2}]$.

The long-range prediction for the fraction of the people who will hear the decision correctly is 1/2.

33. The transition matrix is

$$P = \begin{bmatrix} .12 & .88 \\ .54 & .46 \end{bmatrix}.$$ Let V be the probability vector $[v_1 \quad v_2]$.

$$[v_1 \quad v_2]\begin{bmatrix} .12 & .88 \\ .54 & .46 \end{bmatrix} = [v_1 \quad v_2]$$

$$.12v_1 + .54v_2 = v_1$$
$$.88v_1 + .46v_2 = v_2$$

Simplify these equations to get the dependent system

$$-.88v_1 + .54v_2 = 0$$
$$.88v_1 - .54v_2 = 0.$$

Also, $v_1 + v_2 = 1$.
Solving this system, we obtain

$$v_1 = \frac{27}{71} \quad \text{and} \quad v_2 = \frac{44}{71}, \text{ and note}$$

that $\frac{27}{71} \approx .38 = 38\%$.

About 38% of letters in English text are expected to be vowels.

35. This exercise should be solved by computer methods. The solution will vary according to the computer program that is used. The answer is

V = [.171898 .176519 .167414
     .187526 .296644]

37. This exercise should be solved by computer methods. The solution will vary according to the computer program that is used. The answers are as follows.

(a)

[.4 .6]; [.53 .47]; [.5885 .4115];
[.614825 .385175]; [.626671 .373329];
[.632002 .367998]; [.634401 .365599];
[.635480 .364520]; [.635966 .364034];
[.636185 .363815]

(b) .24; .11; .048; .022; .0097;
.0044; .0020; .00088; .00040; .00018

(c) The ratio is roughly .45 for each week.

(d) Each week, the difference between the probability vector and the equilibrium vector is slightly less than half of what it was the previous week.

(e) [.75 .25]; [.6875 .3125]; [.659375 .340625]; [.641023 .358977]; [.638461 .361539]; [.637307 .362693]; [.636788 .363212]; [.636555 .363445]; [.636450 .363550];

.11; .051; .023; .010; .0047; .0021; .00094; .00042; .00019; .000086

The ratio is roughly .45 for each week, which is the same conclusion as before.

Section 8.3

1. **(a)** Buy speculative; she thinks the market will go up and $30,000 is the highest possible net amount.

   (b) Buy blue-chip; she thinks the market will go down, and $18,000 is better than $11,000.

   (c) If there is a .7 probability the market will go up, then there is a .3 probability it will go down. Find her expected profit for each strategy.

Blue-chip:

$(.7)(25,000) + (.3)(18,000) = \$22,900$

Speculative:

$(.7)(30,000) + (.3)(11,000) = \$24,300$

Therefore she should buy speculative and her expected profit is $24,300.

(d) Find her expected profit for each strategy.

Blue-chip:

$(.2)(25,000) + (.8)(18,000) = \$19,400$

Speculative:

$(.2)(30,000) + (.8)(11,000) = \$14,800$

Therefore she should buy blue-chip and her expected profit is $19,400.

3. **(a)** Set up in the stadium; she doesn't think it will rain and $1500 is the highest possible net profit.

   (b) Set up in the gym; the worst that can happen is a profit of $1000.

   (c) If there is a .6 probability of rain, then there is a .4 probability of no rain. Find her expected profit for each strategy.

Stadium: $.6(-1550) + .4(1500) = -\$330$
  Gym: $.6(1000) + .4(1000) = \$1000$
  Both: $.6(750) + .4(1400) = \$1010$

She should set up both for a maximum expected profit of $1010.

5. (a) The payoff matrix is as follows.

	Better	Not Better
Market	$50,000	-$25,000
Don't Market	-$40,000	-$10,000

(b) Find the expected profit under the 2 strategies.

Market product:

(.4)(50,000) + (.6)(-25,000)

= $5000

Don't market:

(.4)(-40,000) + (.6)(-10,000)

= -$22,000

They should market the product and make a profit of $5000 since that is better than losing $22,000.

7. (a) The payoff matrix is as follows.

	Strike	No Strike
Bid $30,000	-$5500	$4500
Bid $40,000	$4500	$0

(b) Find his expected earnings under each stategy.

Bid $30,000:

(.6)(-5500) + (.4)(4500) = -$1500

Bid $40,000:

(.6)(4500) + (.4)(0) = $2700

He should bid $40,000 and make a profit of $2700, since that is better than losing $1500.

9. Find the expected utility under each strategy.

Jobs:

(.35)(25) + (.65)(-10) = 2.25

Environment:

(.35)(-15) + (.65)(30) = 14.25

She should emphasize the environment. The expected utility of this strategy is 14.25.

Section 8.4

1. $$A \begin{array}{c} 1 \\ 2 \\ 3 \end{array} \begin{bmatrix} 6 & -4 & 0 \\ 3 & -2 & 6 \\ -1 & 5 & 11 \end{bmatrix}$$

with B columns 1, 2, 3.

Consider the strategy (1, 1).
The first row, first column entry is 6, indicating a payoff of $6 from B to A.

3. Consider the strategy (2, 2).
The second row, second column entry is -2, indicating a payoff of $2 from A to B.

5. Consider the strategy (3, 1).
The third row, first column entry is -1, indicating a payoff of $1 from A to B.

7. Yes, each entry in column 2 is smaller than the corresponding entry in column 3, so column 2 dominates column 3.

9. $\begin{bmatrix} 0 & -2 & 8 \\ 3 & -1 & -9 \end{bmatrix}$

Column 2 dominates column 1, so remove column 1 to obtain

$$\begin{bmatrix} -2 & 8 \\ -1 & -9 \end{bmatrix}.$$

11. $\begin{bmatrix} 1 & 4 \\ 4 & -1 \\ 3 & 5 \\ -4 & 0 \end{bmatrix}$

Row 3 dominates rows 1 and 4, so remove rows 1 and 4 to obtain

$$\begin{bmatrix} 4 & -1 \\ 3 & 5 \end{bmatrix}.$$

13. $\begin{bmatrix} 8 & 12 & -7 \\ -2 & 1 & 4 \end{bmatrix}$

Column 1 dominates column 2, so remove column 2 to obtain

$$\begin{bmatrix} 8 & -7 \\ -2 & 4 \end{bmatrix}.$$

15. $\begin{bmatrix} ③ & ⑤ \\ 2 & \underline{-5} \end{bmatrix}$

Underline the smallest number in each row and circle the largest value in each column.
The 3 at (1, 1) is the smallest number in its row and the largest number in its column, so the saddle point is 3 at (1, 1). This game is strictly determined and its value is 3.

17. $\begin{bmatrix} 3 & \underline{-4} & ① \\ 5 & ③ & \underline{-2} \end{bmatrix}$

Underline the smallest number in each row and circle the largest value in each column; in this matrix, the two categorizations do not overlap.
There is no saddle point. This game is not strictly determined.

19. $\begin{bmatrix} \underline{-6} & 2 \\ -1 & \underline{-10} \\ ③ & ⑤ \\ \underline{} & \end{bmatrix}$

The 3 at (3, 1) is the smallest number in its row and the largest number in its column, so the saddle point is 3 at (3, 1). This game is strictly determined and its value is 3.

21. $\begin{bmatrix} 2 & 3 & ① \\ -1 & ④ & \underline{-7} \\ 5 & 2 & \underline{0} \\ ⑧ & \underline{-4} & -1 \end{bmatrix}$

The 1 at (1, 3) is the smallest number in its row and the largest number in its column, so the saddle point is 1 at (1, 3). This game is strictly determined and its value is 1.

23. $$\begin{bmatrix} \underline{-6} & 1 & ④ & ② \\ ⑨ & ③ & \underline{-8} & -7 \end{bmatrix}$$

There is no saddle point. This game is not strictly determined.

25. Focus on any single column of the payoff matrix. In that column, suppose the row 1 entry is x_1, the row 2 entry is x_2, and the row 3 entry is x_3.

Since row 1 dominates row 2, $x_1 > x_2$, and since row 2 dominates row 3, $x_2 > x_3$. By transitivity, it follows that $x_1 > x_3$. The same phenomenon involving these three rows occurs in every column, so row 1 dominates row 3.

We have shown that, whenever row 1 dominates row 2 and row 2 dominates row 3, it must also be true that row 1 dominates row 3.

27.

		B		
		City 1	City 2	City 3
	City 1	5	−2	6
A	City 2	7	⑤	9
	City 3	3	−3	5

To get the entries in the above matrix, look, for example, at the entry in row 2, column 1. If merchant A locates in city 2 and merchant B in city 1, then merchant A will get 80% of the business in city 2, 20% in city 3, and 60% in city 1.

Taking into account the fraction of the population living in each city, we get

.80(.45) + .20(.30) + .60(.25) = .57.

Thus, merchant A gets 57% of the total business. Now 57% is 7 percentage points above 50%, so the entry in row 2, column 1 is +7. Likewise for row 3, column 1, we get

.80(.25) + .20(.30) + .60(.45) = .53 = 53%,

which is 3 percentage points above 50%. The other entries are found in a similar manner. (Note that all diagonal entries are 5 since 55% is 5 percentage points above 50%.) The 5 at (2, 2) is the smallest entry in its row and the largest in its column, so the saddle point is the 5 at (2, 2) and the value of the game is 5.

29. $$\begin{bmatrix} ③ & -8 & \boxed{-9} \\ 0 & ⑥ & \underline{-12} \\ -8 & 4 & \underline{-10} \end{bmatrix}$$

Underline the smallest number in each row and circle the largest value in each column.

−9 is the smallest entry in its row and the largest in its column. The saddle point is −9 at (1, 3), and the value of the game is −9.

31. The payoff matrix is as follows.

	Stone	Scissors	Paper
Stone	0	①	-1
Scissors	-1	0	①
Paper	①	-1	0

Underline the smallest number in each row and circle the largest value in each column; in this matrix, the two categorizations do not overlap.
The game is not strictly determined since it does not have a saddle point.

Section 8.5

1. **(a)** $[.5 \quad .5]\begin{bmatrix} 3 & -4 \\ -5 & 2 \end{bmatrix}\begin{bmatrix} .3 \\ .7 \end{bmatrix}$

$= [.5 \quad .5]\begin{bmatrix} -1.9 \\ -.1 \end{bmatrix}$

$= [-.95 - .05]$

$= [-1]$

The expected value is -1.

(b) $[.1 \quad .9]\begin{bmatrix} 3 & -4 \\ -5 & 2 \end{bmatrix}\begin{bmatrix} .3 \\ .7 \end{bmatrix}$

$= [.1 \quad .9]\begin{bmatrix} -1.9 \\ -.1 \end{bmatrix} = [-.28]$

The expected value is -.28.

(c) $[.8 \quad .2]\begin{bmatrix} 3 & -4 \\ -5 & 2 \end{bmatrix}\begin{bmatrix} .3 \\ .7 \end{bmatrix}$

$= [.8 \quad .2]\begin{bmatrix} -1.9 \\ -.1 \end{bmatrix} = [-1.54]$

The expected value is -1.54.

(d) $[.2 \quad .8]\begin{bmatrix} 3 & -4 \\ -5 & 2 \end{bmatrix}\begin{bmatrix} .3 \\ .7 \end{bmatrix}$

$= [.2 \quad .8]\begin{bmatrix} -1.9 \\ -.1 \end{bmatrix} = [-.46]$

The expected value is -.46.

3. $\begin{bmatrix} 5 & 1 \\ 3 & 4 \end{bmatrix}$

There are no saddle points. For A, the optimum strategy is

$$p_1 = \frac{4 - 3}{5 - 3 - 1 + 4} = \frac{1}{5},$$

$$p_2 = 1 - \frac{1}{5} = \frac{4}{5}.$$

For player B, the optimum strategy is

$$q_1 = \frac{4 - 1}{5 - 3 - 1 + 4} = \frac{3}{5},$$

$$q_2 = 1 - \frac{3}{5} = \frac{2}{5}.$$

The value of the game is

$$\frac{(5)(4) - (3)(1)}{5 - 3 - 1 + 4}$$

$$= \frac{20 - 3}{5} = \frac{17}{5}.$$

5. $\begin{bmatrix} -2 & 0 \\ 3 & -4 \end{bmatrix}$

There are no saddle points. For player A, the optimum strategy is

$$p_1 = \frac{-4 - 3}{-2 - 3 - 0 - 4} = \frac{-7}{-9} = \frac{7}{9},$$

$$p_2 = 1 - \frac{7}{9} = \frac{2}{9}.$$

For player B, the optimum strategy is

$$q_1 = \frac{-4 - 0}{-2 - 3 - 0 - 4} = \frac{-4}{-9} = \frac{4}{9},$$

$$q_2 = 1 - \frac{4}{9} = \frac{5}{9}.$$

The value of the game is

$$\frac{-2(-4) - (0)(3)}{-2 - 3 - 0 - 4} = -\frac{8}{9}.$$

7. $\begin{bmatrix} 4 & -3 \\ -1 & 7 \end{bmatrix}$

There are no saddle points. For player A, the optimum strategy is

$$p_1 = \frac{7 - (-1)}{4 - (-1) - (-3) + 7} = \frac{8}{15},$$

$$p_2 = 1 - \frac{8}{15} = \frac{7}{15}.$$

For player B, the optimum strategy is

$$q_1 = \frac{7 - (-3)}{15} = \frac{10}{15} = \frac{2}{3},$$

$$q_2 = 1 - \frac{2}{3} = \frac{1}{3}.$$

The value of the game is

$$\frac{(4)(7) - (-3)(-1)}{15} = \frac{25}{15} = \frac{5}{3}.$$

9. $\begin{bmatrix} -2 & \frac{1}{2} \\ 0 & -3 \end{bmatrix}$

There are no saddle points. For player A, the optimum strategy is

$$p_1 = \frac{-3 - 0}{-2 - 0 - \frac{1}{2} - 3} = \frac{-3}{\frac{-11}{2}}$$

$$= \frac{6}{11},$$

$$p_2 = 1 - \frac{6}{11} = \frac{5}{11}.$$

For player B, the optimum strategy is

$$q_1 = \frac{-3 - \frac{1}{2}}{-2 - 0 - \frac{1}{2} - 3} = \frac{-\frac{7}{2}}{\frac{-11}{2}} = \frac{7}{11},$$

$$q_2 = 1 - \frac{7}{11} = \frac{4}{11}.$$

The value of the game is

$$\frac{-2(-3) - \left(\frac{1}{2}\right)(0)}{-2 - 0 - \frac{1}{2} - 3} = \frac{6}{\frac{-11}{2}} = -\frac{12}{11}.$$

11. $\begin{bmatrix} \frac{8}{3} & -\frac{1}{2} \\ \frac{3}{4} & -\frac{5}{12} \end{bmatrix}$

The game is strictly determined since it has a saddle point at $(2, 2)$. The value of the game is $-\frac{5}{12}$.

13. $\begin{bmatrix} -1 & 2 \\ 3 & 1 \end{bmatrix}$

There are no saddle points. For player A, the optimum strategy is

$$p_1 = \frac{1 - 3}{-1 - 3 - 2 + 1} = \frac{-2}{-5} = \frac{2}{5},$$

$$p_2 = 1 - \frac{2}{5} = \frac{3}{5}.$$

For player B, the optimum strategy is

$$q_1 = \frac{1 - 2}{1 - 3 - 2 + 1} = \frac{-1}{-5} = \frac{1}{5},$$

$$q_2 = 1 - \frac{1}{5} = \frac{4}{5}.$$

The value of the game is

$$\frac{-1(1) - (2)(3)}{-1 - 3 - 2 + 1} = \frac{-7}{-5} = \frac{7}{5}.$$

15. $\begin{bmatrix} -4 & 9 \\ 3 & -5 \\ 8 & 7 \end{bmatrix}$

Row 3 dominates row 2, so remove row 2. This gives the matrix

$$\begin{bmatrix} -4 & 9 \\ 8 & 7 \end{bmatrix}.$$

For player A, the optimum strategy is

$$p_1 = \frac{7 - 8}{-4 - 8 - 9 + 7} = \frac{-1}{-14} = \frac{1}{14},$$

$p_2 = 0$ (row 2 was removed),

$$p_3 = 1 - \frac{1}{14} = \frac{13}{14}.$$

For player B, the optimum strategy is

$$q_1 = \frac{7 - 9}{-4 - 8 - 9 + 7} = \frac{-2}{-14} = \frac{1}{7},$$

$$q_2 = 1 - \frac{1}{7} = \frac{6}{7}.$$

The value of the game is

$$\frac{-4(7) - (9)(8)}{-4 - 8 - 9 + 7} = \frac{-100}{-14} = \frac{50}{7}.$$

17. $\begin{bmatrix} 8 & 6 & 3 \\ -1 & -2 & 4 \end{bmatrix}$

Column 2 dominates column 1, so remove column 1. This gives the matrix

$$\begin{bmatrix} 6 & 3 \\ -2 & 4 \end{bmatrix}.$$

For player A, the optimum strategy is

$$p_1 = \frac{4 - (-2)}{6 + 2 - 3 + 4} = \frac{6}{9} = \frac{2}{3},$$

$$p_2 = 1 - \frac{2}{3} = \frac{1}{3}.$$

For player B, the optimum strategy is

$$q_1 = 0 \text{ (column 1 was removed)},$$

$$q_2 = \frac{4 - 3}{6 + 2 - 3 + 4} = \frac{1}{9},$$

$$q_3 = 1 - \frac{1}{9} = \frac{8}{9}.$$

The value of the game is

$$\frac{6(4) - (3)(-2)}{6 + 2 - 3 + 4} = \frac{30}{9} = \frac{10}{3}.$$

19. $\begin{bmatrix} 9 & -1 & 6 \\ 13 & 11 & 8 \\ 6 & 0 & 9 \end{bmatrix}$

Row 2 dominates row 1, so remove row 1. This gives the matrix

$$\begin{bmatrix} 13 & 11 & 8 \\ 6 & 0 & 9 \end{bmatrix}.$$

Now, column 2 dominates column 1, so remove column 1. This gives the matrix

$$\begin{bmatrix} 11 & 8 \\ 0 & 9 \end{bmatrix}.$$

For player A, the optimum strategy is

$$p_1 = 0 \text{ (row 1 was removed)},$$

$$p_2 = \frac{9 - 0}{11 - 0 - 8 + 9} = \frac{9}{12} = \frac{3}{4},$$

$$p_3 = 1 - \frac{3}{4} = \frac{1}{4}.$$

For player B, the optimum strategy is

$$q_1 = 0, \text{ (column 1 was removed)},$$

$$q_2 = \frac{9 - 8}{11 - 0 - 8 + 9} = \frac{1}{12},$$

$$q_3 = 1 - \frac{1}{12} = \frac{11}{12}.$$

The value of the game is

$$\frac{11(9) - (8)(0)}{11 - 0 - 8 + 9} = \frac{99}{12} = \frac{33}{4}.$$

21. In a non-strictly-determined game, there is no saddle point. Let

$M = \begin{bmatrix} a_{11} & a_{12} \\ a_{21} & a_{22} \end{bmatrix}$ be the payoff matrix

of the game. Assume that player B chooses column 1 with probability p_1. The expected value for B, assuming A plays row 1, is E_1, where

$$E_1 = a_{11} \cdot p_1 + a_{12} \cdot (1 - p_1).$$

The expected value for B if A plays row 2 is E_2, where

$$E_2 = a_{21} \cdot p_1 + a_{22} \cdot (1 - p_1).$$

The optimum strategy for player B is found by letting $E_1 = E_2$.

$a_{11} \cdot p_1 + a_{12} \cdot (1 - p_1)$
$= a_{21} \cdot p_1 + a_{22}(1 - p_1)$
$a_{11} \cdot p_1 + a_{12} - a_{12} \cdot p_1$
$= a_{21} \cdot p_1 + a_{22} - a_{22} \cdot p_1$
$a_{11} \cdot p_1 - a_{21} \cdot p_1 - a_{12} \cdot p_1 + a_{22} \cdot p_1$
$= a_{22} - a_{12}$
$p_1(a_{11} - a_{21} - a_{12} + a_{22})$
$= a_{22} - a_{12}$

$$p_1 = \frac{a_{22} - a_{12}}{a_{11} - a_{21} - a_{12} + a_{22}}$$

Since $p_2 = 1 - p_1$,

$$p_2 = 1 - \frac{a_{22} - a_{12}}{a_{11} - a_{21} - a_{12} + a_{22}}$$

$$= \frac{a_{11} - a_{21} - a_{12} + a_{22} - (a_{22} - a_{12})}{a_{11} - a_{21} - a_{12} + a_{22}}$$

$$= \frac{a_{11} - a_{21}}{a_{11} - a_{21} - a_{12} + a_{22}}.$$

These are the formulas given in the text for p_1 and p_2.

25.

$$\begin{array}{cc} & Bates \\ & \begin{array}{cc} \text{T.V.} & \text{Radio} \end{array} \\ Allied \begin{array}{c} \text{T.V.} \\ \text{Radio} \end{array} & \begin{bmatrix} 1.0 & -.7 \\ -.5 & .5 \end{bmatrix} \end{array}$$

The optimum strategy for Allied is

$$p_1 = \frac{.5 - (-.5)}{1 - (-.5) - (-.7) + (.5)}$$

$$= \frac{1}{2.7} = \frac{10}{27},$$

$$p_2 = 1 - \frac{10}{27} = \frac{17}{27}.$$

Allied should use T.V. with probability 10/27 and use radio with probability 17/27.
The value of the game is

$$\frac{(1)(.5) - (-.7)(-.5)}{2.7} = \frac{.15}{2.7} = \frac{1}{18},$$

which represents increased sales of

$$1,000,000\left(\frac{1}{18}\right) \approx 1,000,000(.055556)$$

$$= \$55,556.$$

27.

$$\begin{array}{cc} & Selling\ during: \\ & \begin{array}{cc} \text{Rain} & \text{Shine} \end{array} \\ Buying\ for: \begin{array}{c} \text{Rain} \\ \text{Shine} \end{array} & \begin{bmatrix} 250 & -150 \\ -150 & 350 \end{bmatrix} \end{array}$$

The best mixed strategy for Merrill is

$$p_1 = \frac{350 - (-150)}{250 - (-150) - (-150) + 350}$$

$$= \frac{500}{900} = \frac{5}{9},$$

$$p_2 = 1 - \frac{5}{9} = \frac{4}{9}.$$

He should invest in rainy day goods 5/9 of the time and in sunny day goods 4/9 of the time.

The value of the game is

$$\frac{(250)(350) - (-150)^2}{900} = \frac{65,000}{900}$$

$$= \frac{650}{9}$$

$$\approx \$72.22$$

Thus, his profit is $72.22.

29. The payoff matrix is as follows.

Jamie

		Pounce	Freeze
Euclid	Pounce	3	1
	Freeze	-2	2

The optimum strategy for Euclid is

$$p_1 = \frac{2 - (-2)}{3 - (-2) - 1 + 2} = \frac{4}{6} = \frac{2}{3},$$

$$p_2 = 1 - \frac{2}{3} = \frac{1}{3}.$$

Euclid should pounce 2/3 of the time and freeze 1/3 of the time.

The optimum strategy for Jamie is

$$q_1 = \frac{2 - 1}{3 - (-2) - 1 + 2} = \frac{1}{6},$$

$$q_2 = 1 - \frac{1}{6} = \frac{5}{6}.$$

Jamie should pounce 1/6 of the time and freeze 5/6 of the time.

The value of the game is

$$\frac{3(2) - (-2)(1)}{3 - (-2) - 1 + 2} = \frac{8}{6} = \frac{4}{3}.$$

31. (a) The payoff matrix is as follows.

Number of Fingers
B Shows

		1	2
Number of Fingers A Shows	1	2	-3
	2	-3	4

(b) For player A, the optimum strategy is

$$p_1 = \frac{4 - (-3)}{2 - (-3) - (-3) + 4} = \frac{7}{12},$$

$$p_2 = 1 - \frac{7}{12} = \frac{5}{12}.$$

For player B, the optimum strategy is

$$q_1 = \frac{4 - (-3)}{2 - (-3) - (-3) + 4} = \frac{7}{12},$$

$$q_2 = 1 - \frac{7}{12} = \frac{5}{12}.$$

This means that each player should show 1 finger 7/12 of the time and 2 fingers 5/12 of the time. The value of the game is

$$\frac{2(4) - (-3)(-3)}{2 - (-3) - (-3) + 4} = -\frac{1}{12}.$$

Section 8.6

1. $\begin{bmatrix} 1 & 2 \\ 3 & 1 \end{bmatrix}$

To find the optimum strategy for player A,

minimize $w = x + y$

subject to: $x + 3y \geq 1$

$2x + y \geq 1$

with $x \geq 0, y \geq 0.$

Solve this linear programming problem by the graphical method. Sketch the feasible region.

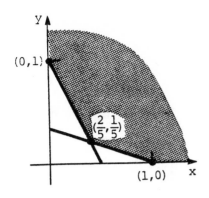

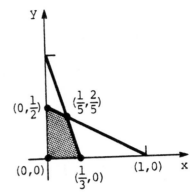

The region is unbounded, with corner points $(0, 1)$, $\left(\frac{2}{5}, \frac{1}{5}\right)$, and $(1, 0)$.

Corner Point	Value of $w = x + y$
$(0, 1)$	$0 + 1 = 1$
$\left(\frac{2}{5}, \frac{1}{5}\right)$	$\frac{2}{5} + \frac{1}{5} = \frac{3}{5}$
$(1, 0)$	$1 + 0 = 1$

The minimum value is $w = \frac{3}{5}$ at the point where $x = \frac{2}{5}$, $y = \frac{1}{5}$. Thus, the value of the game is $g = \frac{1}{w} = \frac{5}{3}$, and the optimum strategy for A is

$$p_1 = gx = \left(\frac{5}{3}\right)\left(\frac{2}{5}\right) = \frac{2}{3},$$

$$p_2 = gy = \left(\frac{5}{3}\right)\left(\frac{1}{5}\right) = \frac{1}{3}.$$

To find the optimum strategy for player B,

maximize $z = x + y$

subject to: $x + 2y \leq 1$

$\qquad\qquad 3x + y \leq 1$

with $x \geq 0,\ y \geq 0.$

Corner Point	Value of $z = x + y$
$\left(0, \frac{1}{2}\right)$	$0 + \frac{1}{2} = \frac{1}{2}$
$\left(\frac{1}{5}, \frac{2}{5}\right)$	$\frac{1}{5} + \frac{2}{5} = \frac{3}{5}$
$\left(\frac{1}{3}, 0\right)$	$\frac{1}{3} + 0 = \frac{1}{3}$
$(0, 0)$	$0 + 0 = 0$

The maximum value is $z = \frac{3}{5}$ at the point where $x = \frac{1}{5}$, $y = \frac{2}{5}$. The value of the game is $g = \frac{1}{z} = \frac{5}{3}$ (agreeing with our earlier findings), and the optimum strategy for B is

$$q_1 = gx = \left(\frac{5}{3}\right)\left(\frac{1}{5}\right) = \frac{1}{3} \text{ and}$$

$$q_2 = gy = \left(\frac{5}{3}\right)\left(\frac{2}{5}\right) = \frac{2}{3}.$$

3. $\begin{bmatrix} 4 & -2 \\ -1 & 6 \end{bmatrix}$

Get rid of negative numbers by adding 2 to each entry, to obtain

$$\begin{bmatrix} 6 & 0 \\ 1 & 8 \end{bmatrix}.$$

(The 2 will have to be subtracted away later, after the calculations have been performed.)

To find the optimum strategy for player A,

minimize   w = x + y

subject to: 6x + y ≥ 1

                8y ≥ 1

with   x ≥ 0, y ≥ 0.

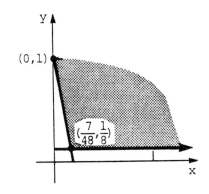

Corner Point	Value of w = x + y
(0, 1)	$0 + 1 = 1$
$\left(\frac{7}{48}, \frac{1}{8}\right)$	$\frac{7}{48} + \frac{1}{8} = \frac{13}{48}$

The minimum value is $w = \frac{13}{48}$ at

$\left(\frac{7}{48}, \frac{1}{8}\right)$. Thus, $g = \frac{1}{w} = \frac{48}{13}$, and the

optimum strategy for A is

$$p_1 = gx = \left(\frac{48}{13}\right)\left(\frac{7}{48}\right) = \frac{7}{13},$$

$$p_2 = gy = \left(\frac{48}{13}\right)\left(\frac{1}{8}\right) = \frac{6}{13}.$$

The value of the game is $\frac{48}{13} - 2 = \frac{22}{13}$.

To find the optimum strategy for player B,

maximize   z = x + y

subject to: 6x   ≤ 1

             x + 8y ≤ 1

with   x ≥ 0, y ≥ 0.

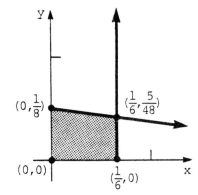

Corner Point	Value of z = x + y
$\left(0, \frac{1}{8}\right)$	$0 + \frac{1}{8} = \frac{1}{8}$
$\left(\frac{1}{6}, \frac{25}{48}\right)$	$\frac{1}{6} + \frac{5}{48} = \frac{13}{48}$
$\left(\frac{1}{6}, 0\right)$	$\frac{1}{6} + 0 = \frac{1}{6}$
(0, 0)	$0 + 0 = 0$

The maximum value is $z = \frac{13}{48}$ at

$\left(\frac{1}{6}, \frac{5}{48}\right)$. The optimum strategy for B

is

$$q_1 = gx = \left(\frac{48}{13}\right)\left(\frac{1}{6}\right) = \frac{8}{13},$$

$$q_2 = gy = \left(\frac{48}{13}\right)\left(\frac{5}{48}\right) = \frac{5}{13}.$$

5. $\begin{bmatrix} 8 & -7 \\ -2 & 4 \end{bmatrix}$

Add 7 to each entry, to obtain

$\begin{bmatrix} 15 & 0 \\ 5 & 11 \end{bmatrix}$.

To find the optimum strategy for player A,

minimize $w = x + y$

subject to: $15x + 5y \geq 1$

$\qquad\qquad\quad 11y \geq 1$

with $x \geq 0, \; y \geq 0.$

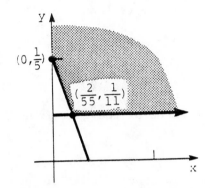

Corner Point	Value of $w = x + y$
$\left(0, \dfrac{1}{5}\right)$	$0 + \dfrac{1}{5} = \dfrac{1}{5}$
$\left(\dfrac{2}{55}, \dfrac{1}{11}\right)$	$\dfrac{2}{55} + \dfrac{1}{11} = \dfrac{7}{55}$

The minimum value is $w = \dfrac{7}{55}$ at

$\left(\dfrac{2}{55}, \dfrac{1}{11}\right)$. Thus, $g = \dfrac{1}{w} = \dfrac{55}{7}$, and

the optimum strategy for A is

$$p_1 = gx = \left(\frac{55}{7}\right)\left(\frac{2}{55}\right) = \frac{2}{7},$$

$$p_2 = gy = \left(\frac{55}{7}\right)\left(\frac{1}{11}\right) = \frac{5}{7}.$$

The value of the game is $\dfrac{55}{7} - 7 = \dfrac{6}{7}$.

To find the optimum strategy for player B,

maximize $z = x + y$

subject to: $15x \qquad\quad \leq 1$

$\qquad\qquad\quad 5x + 11y \leq 1$

with $x \geq 0, \; y \geq 0.$

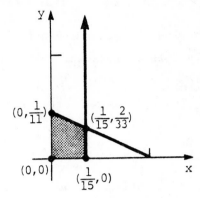

Corner Point	Value of $z = x + y$
$\left(0, \dfrac{1}{11}\right)$	$0 + \dfrac{1}{11} = \dfrac{1}{11}$
$\left(\dfrac{1}{15}, \dfrac{2}{33}\right)$	$\dfrac{1}{15} + \dfrac{2}{33} = \dfrac{21}{165} = \dfrac{7}{55}$
$\left(\dfrac{1}{15}, 0\right)$	$\dfrac{1}{15} + 0 = \dfrac{1}{15}$
$(0, 0)$	$0 + 0 = 0$

The maximum value is $z = \dfrac{7}{55}$ at

$\left(\dfrac{1}{15}, \dfrac{2}{33}\right)$.

The optimum strategy for B is

$$q_1 = gx = \left(\frac{55}{7}\right)\left(\frac{1}{15}\right) = \frac{11}{21},$$

$$q_2 = gy = \left(\frac{55}{7}\right)\left(\frac{2}{33}\right) = \frac{10}{21}.$$

7. $\begin{bmatrix} 3 & -4 & 1 \\ 5 & 3 & -2 \end{bmatrix}$

Column 1 is dominated by the other two columns, so delete it.

$\begin{bmatrix} -4 & 1 \\ 3 & -2 \end{bmatrix}$

Add 4 to each entry, to obtain

$\begin{bmatrix} 0 & 5 \\ 7 & 2 \end{bmatrix}.$

The linear programming problem to be solved is as follows.

Maximize $z = x_2 + x_3$
subject to: $5x_3 \leq 1$
            $7x_2 + 2x_3 \leq 1$
with $x_2 \geq 0, \ x_3 \geq 0$.

Use the simplex method to solve the problem.

The initial tableau is

$$
\begin{array}{ccccc}
x_2 & x_3 & x_4 & x_5 & z \\
\end{array}
$$
$$
\left[
\begin{array}{ccccc|c}
0 & 5 & 1 & 0 & 0 & 1 \\
⑦ & 2 & 0 & 1 & 0 & 1 \\
\hline
-1 & -1 & 0 & 0 & 1 & 0
\end{array}
\right].
$$

Pivot on the circled entry.

$$
\begin{array}{ccccc}
x_2 & x_3 & x_4 & x_5 & z \\
\end{array}
$$
$$
\begin{array}{c}
\\
\\
7R_3 + R_2 \rightarrow R_3
\end{array}
\left[
\begin{array}{ccccc|c}
0 & ⑤ & 1 & 0 & 0 & 1 \\
7 & 2 & 0 & 1 & 0 & 1 \\
\hline
0 & -5 & 0 & 1 & 7 & 1
\end{array}
\right]
$$

Pivot again.

$$
\begin{array}{ccccc}
x_2 & x_3 & x_4 & x_5 & z \\
\end{array}
$$
$$
\begin{array}{c}
\\
-2R_1 + 5R_2 \rightarrow R_2 \\
R_1 + R_3 \rightarrow R_3
\end{array}
\left[
\begin{array}{ccccc|c}
0 & 5 & 1 & 0 & 0 & 1 \\
35 & 0 & -2 & 5 & 0 & 3 \\
\hline
0 & 0 & 1 & 1 & 7 & 2
\end{array}
\right]
$$

Create a 1 in the columns corresponding to x_2, x_3, and z.

$$
\begin{array}{ccccc}
x_2 & x_3 & x_4 & x_5 & z \\
\end{array}
$$
$$
\begin{array}{c}
\frac{1}{5}R_1 \rightarrow R_1 \\[4pt]
\frac{1}{35}R_2 \rightarrow R_2 \\[4pt]
\frac{1}{7}R_3 \rightarrow R_3
\end{array}
\left[
\begin{array}{ccccc|c}
0 & 1 & \frac{1}{5} & 0 & 0 & \frac{1}{5} \\[4pt]
1 & 0 & -\frac{2}{35} & \frac{1}{7} & 0 & \frac{3}{35} \\[4pt]
0 & 0 & \frac{1}{7} & \frac{1}{7} & 1 & \frac{2}{7}
\end{array}
\right]
$$

From this final tableau, we have

$x_2 = \dfrac{3}{35}$, $x_3 = \dfrac{1}{5}$, $y_1 = \dfrac{1}{7}$, $y_2 = \dfrac{1}{7}$,

$z = \dfrac{2}{7}$. Note that $g = \dfrac{1}{z} = \dfrac{7}{2}$.

The optimum strategy for player A is

$$p_1 = gy_1 = \left(\frac{7}{2}\right)\left(\frac{1}{7}\right) = \frac{1}{2},$$

$$p_2 = gy_2 = \left(\frac{7}{2}\right)\left(\frac{1}{7}\right) = \frac{1}{2}.$$

The optimum strategy for player B is

$q_1 = 0$ (column 1 was removed),

$$q_2 = gx_2 = \left(\frac{7}{2}\right)\left(\frac{3}{35}\right) = \frac{3}{10},$$

$$q_3 = gx_3 = \left(\frac{7}{2}\right)\left(\frac{1}{5}\right) = \frac{7}{10}.$$

The value of the game is $\dfrac{7}{2} - 4 = -\dfrac{1}{2}$.

9. $\begin{bmatrix} -1 & 2 & 4 \\ 3 & -2 & 0 \end{bmatrix}$

Delete column 3, since it is dominated by column 2.

$\begin{bmatrix} -1 & 2 \\ 3 & -2 \end{bmatrix}$

Add 2 to each entry, to obtain

$\begin{bmatrix} 1 & 4 \\ 5 & 0 \end{bmatrix}$.

The linear programming problem to be solved is as follows.

Maximize $z = x_1 + x_2$
subject to: $x_1 + 4x_2 \leq 1$
            $5x_1 \qquad \leq 1$
with $x_1 \geq 0, \ x_2 \geq 0$.

The initial tableau is

$$
\begin{array}{ccccc}
x_1 & x_2 & x_4 & x_5 & z \\
\end{array}
$$
$$
\left[
\begin{array}{ccccc|c}
1 & 4 & 1 & 0 & 0 & 1 \\
⑤ & 0 & 0 & 1 & 0 & 1 \\
\hline
-1 & -1 & 0 & 0 & 1 & 0
\end{array}
\right].
$$

Pivot on each circled entry.

$5R_1 - R_2 \to R_1$

$5R_3 + R_2 \to R_3$

x_1	x_2	x_4	x_5	z	
0	⑳	5	-1	0	4
5	0	0	1	0	1
0	-5	0	1	5	1

x_1	x_2	x_4	x_5	z	
0	20	5	-1	0	4
5	0	0	1	0	1
0	0	5	3	20	8

$4R_3 + R_1 \to R_3$

Create a 1 in the columns corresponding to x_1, x_2, and z.

$\frac{1}{20}R_1 \to R_1$

$\frac{1}{5}R_2 \to R_2$

$\frac{1}{20}R_3 \to R_3$

x_1	x_2	x_4	x_5	z	
0	1	$\frac{1}{4}$	$-\frac{1}{20}$	0	$\frac{1}{5}$
1	0	0	$\frac{1}{5}$	0	$\frac{1}{5}$
0	0	$\frac{1}{4}$	$\frac{3}{20}$	1	$\frac{2}{5}$

From this final tableau, we have

$x_1 = \frac{1}{5}$, $x_2 = \frac{1}{5}$, $y_1 = \frac{1}{4}$. $y_2 = \frac{3}{20}$, and

$z = \frac{2}{5}$.

Note that $g = \frac{1}{z} = \frac{5}{2}$.

The optimum strategy for player A is

$$p_1 = gy_1 = \left(\frac{5}{2}\right)\left(\frac{1}{4}\right) = \frac{5}{8},$$

$$p_2 = gy_2 = \left(\frac{5}{2}\right)\left(\frac{3}{20}\right) = \frac{3}{8}.$$

The optimum strategy for player B is

$$q_1 = gx_1 = \left(\frac{5}{2}\right)\left(\frac{1}{5}\right) = \frac{1}{2},$$

$$q_2 = gx_2 = \left(\frac{5}{2}\right)\left(\frac{1}{5}\right) = \frac{1}{2},$$

$$q_3 = 0 \quad \text{(column 3 was removed)}.$$

The value of the game is $\frac{5}{2} - 2 = \frac{1}{2}$.

11. $\begin{bmatrix} 1 & 0 & -1 \\ -1 & 0 & 1 \\ 2 & -1 & 2 \end{bmatrix}$

Add 1 to each entry, to obtain

$\begin{bmatrix} 2 & 1 & 0 \\ 0 & 1 & 2 \\ 3 & 0 & 3 \end{bmatrix}$.

The linear programming problem to be solved is as follows.

Maximize $z = x_1 + x_2 + x_3$

subject to: $2x_1 + x_2 \quad\quad \leq 1$

$\quad\quad\quad\quad\quad x_2 + 2x_3 \leq 1$

$\quad\quad\quad 3x_1 + \quad\quad 3x_3 \leq 1$

with $\quad\quad x_1 \geq 0,\; x_2 \geq 0,\; x_3 \geq 0.$

The initial tableau is

x_1	x_2	x_3	x_4	x_5	x_6	z	
2	1	0	1	0	0	0	1
0	1	2	0	1	0	0	1
③	0	3	0	0	1	0	1
-1	-1	-1	0	0	0	1	0

Pivot on each circled entry.

$3R_1 - 2R_3 \to R_1$

$3R_4 + R_3 \to R_4$

x_1	x_2	x_3	x_4	x_5	x_6	z	
0	③	-6	3	0	-2	0	1
0	1	2	0	1	0	0	1
3	0	3	0	0	1	0	1
0	-3	0	0	0	1	3	1

$3R_2 - R_1 \to R_2$

$R_1 + R_4 \to R_4$

x_1	x_2	x_3	x_4	x_5	x_6	z	
0	3	-6	3	0	-2	0	1
0	0	⑫	-3	3	2	0	2
3	0	3	0	0	1	0	1
0	0	-6	3	0	-1	3	2

$2R_1 + R_2 \to R_1$

$4R_3 - R_2 \to R_3$

$2R_4 + R_2 \to R_4$

x_1	x_2	x_3	x_4	x_5	x_6	z	
0	6	0	3	3	-2	0	4
0	0	12	-3	3	2	0	2
12	0	0	3	-3	2	0	2
0	0	0	3	3	0	6	6

Create a 1 in the columns corresponding to x_1, x_2, x_3, and z.

$$
\begin{array}{c}
\frac{1}{6}R_1 \to R_1 \\[4pt]
\frac{1}{12}R_2 \to R_2 \\[4pt]
\frac{1}{12}R_3 \to R_3 \\[4pt]
\frac{1}{6}R_4 \to R_4
\end{array}
\begin{array}{c}
\begin{array}{ccccccc}
x_1 & x_2 & x_3 & x_4 & x_5 & x_6 & z
\end{array} \\
\left[
\begin{array}{ccccccc|c}
0 & 1 & 0 & \frac{1}{2} & \frac{1}{2} & -\frac{1}{3} & 0 & \frac{2}{3} \\[4pt]
0 & 0 & 1 & -\frac{1}{4} & \frac{1}{4} & \frac{1}{6} & 0 & \frac{1}{6} \\[4pt]
1 & 0 & 0 & \frac{1}{4} & -\frac{1}{4} & \frac{1}{6} & 0 & \frac{1}{6} \\[4pt]
0 & 0 & 0 & \frac{1}{2} & \frac{1}{2} & 0 & 1 & 1
\end{array}
\right]
\end{array}
$$

From this final tableau, we have

$$x_1 = \frac{1}{6}, \ x_2 = \frac{2}{3}, \ x_3 = \frac{1}{6}, \ y_1 = \frac{1}{2},$$

$$y_2 = \frac{1}{2}, \ y_3 = 0, \text{ and } z = 1.$$

Note that $g = \frac{1}{z} = 1$.

The optimum strategy for player A is

$$p_1 = gy_1 = (1)\left(\frac{1}{2}\right) = \frac{1}{2},$$

$$p_2 = gy_2 = (1)\left(\frac{1}{2}\right) = \frac{1}{2},$$

$$p_3 = gy_3 = (1)(0) = 0.$$

The optimum strategy for player B is

$$q_1 = gx_1 = (1)\left(\frac{1}{6}\right) = \frac{1}{6},$$

$$q_2 = gx_2 = (1)\left(\frac{2}{3}\right) = \frac{2}{3},$$

$$q_3 = gx_3 = (1)\left(\frac{1}{6}\right) = \frac{1}{6}.$$

The value of the game is $1 - 1 = 0$.

13.   The payoff matrix is as follows.

$$
\begin{array}{c}
\text{Bid \$30,000} \\
\text{Bid \$40,000}
\end{array}
\begin{array}{c}
\begin{array}{cc}
\text{Strike} & \text{No Strike}
\end{array} \\
\left[
\begin{array}{cc}
-\$5500 & \$4500 \\
\$4500 & \$0
\end{array}
\right]
\end{array}
$$

Add $5500 to each entry, to obtain

$$\begin{bmatrix} 0 & 10,000 \\ 10,000 & 5500 \end{bmatrix}.$$

To find the optimum strategy for the contractor,

minimize $w = x + y$

subject to: $10,000y \geq 1$

$10,000x + 5500y \geq 1$

with $x \geq 0, \ y \geq 0.$

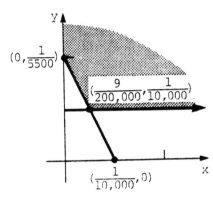

Corner Point	Value of $w = x + y$
$\left(0, \ \dfrac{1}{5500}\right)$	$0 + \dfrac{1}{5500} = \dfrac{1}{5500}$
$\left(\dfrac{9}{200,000}, \ \dfrac{1}{10,000}\right)$	$\dfrac{9}{200,000} + \dfrac{1}{10,000} = \dfrac{29}{200,000}$

The minimum value is $w = \dfrac{29}{200,000}$ at $\left(\dfrac{9}{200,000}, \ \dfrac{1}{10,000}\right)$. Thus, $g = \dfrac{1}{w} = \dfrac{200,000}{29}$,

and the optimum strategy for the contractor is

$$p_1 = gx = \left(\frac{200,000}{29}\right)\left(\frac{9}{200,000}\right) = \frac{9}{29},$$

$$p_2 = gy = \left(\frac{200,000}{29}\right)\left(\frac{1}{10,000}\right) = \frac{20}{29}.$$

That is, the contractor should bid $30,000 with probability $\dfrac{9}{29}$ and bid $40,000

with probability $\dfrac{20}{29}$. The value of the game is

$$\frac{200,000}{29} - 5500 \approx \$1396.55.$$

15. (a) The payoff matrix is as follows.

$$\text{Original Imitators}$$

General Items
$$\begin{array}{cc} & \begin{array}{ccc} \text{Atlanta} & \text{Boston} & \text{Cleveland} \end{array} \\ \begin{array}{c} \text{Atlanta} \\ \text{Boston} \\ \text{Cleveland} \end{array} & \begin{bmatrix} \$5000 & \$10{,}000 & \$10{,}000 \\ \$8000 & \$4000 & \$8000 \\ \$6000 & \$6000 & \$3000 \end{bmatrix} \end{array}$$

Note that $5000 = \frac{1}{2}(10{,}000)$, $4000 = \frac{1}{2}(8000)$, and $3000 = \frac{1}{2}(6000)$ are the reduced profits for General Items when the two companies run ads in the same city.

(b) The linear programming problem to be solved is as follows.

Maximize $z = x_1 + x_2 + x_3$

subject to: $5000x_1 + 10{,}000x_2 + 10{,}000x_3 \le 1$

$8000x_1 + \quad 4000x_2 + \quad 8000x_3 \le 1$

$6000x_1 + \quad 6000x_2 + \quad 3000x_3 \le 1$

with $x_1 \ge 0,\ x_2 \ge 0,\ x_3 \ge 0.$

The initial tableau is

x_1	x_2	x_3	x_4	x_5	x_6	z	
5000	10,000	10,000	1	0	0	0	1
(8000)	4000	8000	0	1	0	0	1
6000	6000	3000	0	0	1	0	1
−1	−1	−1	0	0	0	1	0

Pivot on each circled entry.

	x_1	x_2	x_3	x_4	x_5	x_6	z	
$-5R_2 + 8R_1 \to R_1$	0	(60,000)	40,000	8	−5	0	0	3
	8000	4000	8000	0	1	0	0	1
$-3R_2 + 4R_3 \to R_3$	0	12,000	−12,000	0	−3	4	0	1
$R_2 + 8000R_4 \to R_4$	0	−4000	0	0	1	1	8000	1

	x_1	x_2	x_3	x_4	x_5	x_6	z	
	0	60,000	40,000	8	−5	0	0	3
$-R_1 + 15R_2 \to R_2$	120,000	0	80,000	−8	20	0	0	12
$-R_1 + 5R_3 \to R_3$	0	0	−100,000	−8	−10	20	0	2
$R_1 + 15R_4 \to R_4$	0	0	40,000	8	10	0	120,000	18

Create a 1 in the columns corresponding to x_1, x_2, x_6, and z.

$$
\begin{array}{c}
\\
\frac{1}{60,000}R_1 \to R_1 \\
\frac{1}{120,000}R_2 \to R_2 \\
\frac{1}{20}R_3 \to R_3 \\
\frac{1}{120,000}R_4 \to R_4
\end{array}
\begin{array}{cccccccc}
x_1 & x_2 & x_3 & x_4 & x_5 & x_6 & z & \\
0 & 1 & \frac{2}{3} & \frac{1}{7500} & -\frac{1}{12,000} & 0 & 0 & \frac{1}{20,000} \\
0 & 0 & \frac{2}{3} & -\frac{1}{15,000} & \frac{1}{6000} & 0 & 0 & \frac{1}{10,000} \\
0 & 0 & -5000 & -\frac{2}{5} & -\frac{1}{2} & 1 & 0 & \frac{1}{10} \\
\hline
0 & 0 & \frac{1}{3} & \frac{1}{15,000} & \frac{1}{12,000} & 0 & 1 & \frac{3}{20,000}
\end{array}
$$

From this final tableau, we have

$$x_1 = \frac{1}{10,000}, \quad x_2 = \frac{1}{20,000}, \quad x_3 = 0, \quad y_1 = \frac{1}{15,000}, \quad y_2 = \frac{1}{12,000}, \quad y_3 = 0, \text{ and}$$

$$z = \frac{3}{20,000}.$$

Note that $g = \frac{1}{z} = \frac{20,000}{3} \approx 6666.67$, so the value of the game is $6666.67.

The optimum strategy for General Items is

$$p_1 = gy_1 = \left(\frac{20,000}{3}\right)\left(\frac{1}{15,000}\right) = \frac{4}{9},$$

$$p_2 = gy_2 = \left(\frac{20,000}{3}\right)\left(\frac{1}{12,000}\right) = \frac{5}{9},$$

$$p_3 = gy_3 = \left(\frac{20,000}{3}\right)(0) = 0.$$

That is, General Items should advertise in Atlanta with probability $\frac{4}{9}$, in Boston with probability $\frac{5}{9}$, and never in Cleveland.

The optimum strategy for Original Imitators is

$$q_1 = gx_1 = \left(\frac{20,000}{3}\right)\left(\frac{1}{10,000}\right) = \frac{2}{3},$$

$$q_2 = gx_2 = \left(\frac{20,000}{3}\right)\left(\frac{1}{20,000}\right) = \frac{1}{3},$$

$$q_3 = gx_3 = \left(\frac{20,000}{3}\right)(0) = 0.$$

That is, Original Imitators should advertise in Atlanta with probability $\frac{2}{3}$, in Boston with probability $\frac{1}{3}$, and never in Cleveland.

17. $\begin{bmatrix} 50 & 0 \\ 10 & 40 \end{bmatrix}$

To find the student's optimum strategy,

minimize $w = x + y$

subject to: $50x + 10y \geq 1$

$\qquad\qquad\quad 40y \geq 1$

with $x \geq 0,\ y \geq 0$.

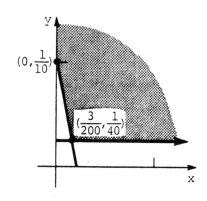

Corner Point	Value of $w = x + y$
$(0, \frac{1}{10})$	$0 + \frac{1}{10} = \frac{1}{10}$
$(\frac{3}{200}, \frac{1}{40})$	$\frac{3}{200} + \frac{1}{40} = \frac{1}{25}$

The minimum value is $w = \frac{1}{25}$ at

$\left(\frac{3}{200}, \frac{1}{40}\right)$.

Thus, the value of the game is

$g = \frac{1}{w} = 25$, and the optimum strategy

for the student is

$$p_1 = gx = (25)\left(\frac{3}{200}\right) = \frac{3}{8},$$

$$p_2 = gy = (25)\left(\frac{1}{40}\right) = \frac{5}{8}.$$

That is, the student should choose

the calculator with probability $\frac{3}{8}$

and the book with probability $\frac{5}{8}$.

19. (a) The payoff matrix is as follows.

$$\begin{array}{c} \\ \text{Rock} \\ \text{Scissors} \\ \text{Paper} \end{array} \begin{array}{ccc} \text{Rock} & \text{Scissors} & \text{Paper} \end{array}$$

$$\begin{bmatrix} 0 & 1 & -1 \\ -1 & 0 & 1 \\ 1 & -1 & 0 \end{bmatrix}$$

Add 1 to each entry, to obtain

$$\begin{bmatrix} 1 & 2 & 0 \\ 0 & 1 & 2 \\ 2 & 0 & 1 \end{bmatrix}.$$

The linear programming problem to be solved is

maximize $w = x_1 + x_2 + x_3$

subject to: $x_1 + 2x_2 \qquad\quad \leq 1$

$\qquad\qquad\quad x_2 + 2x_3 \leq 1$

$\qquad 2x_1 \qquad\quad + x_3 \leq 1$

with $x_1 \geq 0,\ x_2 \geq 0,\ x_3 \geq 0$.

The initial tableau is

$$\begin{array}{ccccccc} x_1 & x_2 & x_3 & x_4 & x_5 & x_6 & z \end{array}$$

$$\begin{bmatrix} 1 & ② & 0 & 1 & 0 & 0 & 0 & 1 \\ 0 & 1 & 2 & 0 & 1 & 0 & 0 & 1 \\ 2 & 0 & 1 & 0 & 0 & 1 & 0 & 1 \\ -1 & -1 & -1 & 0 & 0 & 0 & 1 & 0 \end{bmatrix}.$$

Pivot on each circled entry.

$$\begin{array}{ccccccc} x_1 & x_2 & x_3 & x_4 & x_5 & x_6 & z \end{array}$$

$$\begin{array}{c} \\ 2R_2 - R_1 \to R_2 \\ \\ 2R_4 + R_1 \to R_4 \end{array} \begin{bmatrix} 1 & 2 & 0 & 1 & 0 & 0 & 0 & 1 \\ -1 & 0 & ④ & -1 & 2 & 0 & 0 & 1 \\ 2 & 0 & 1 & 0 & 0 & 1 & 0 & 1 \\ -1 & 0 & -2 & 1 & 0 & 0 & 2 & 1 \end{bmatrix}$$

$$\begin{array}{ccccccc} x_1 & x_2 & x_3 & x_4 & x_5 & x_6 & z \end{array}$$

$$\begin{array}{c} \\ \\ 4R_3 - R_2 \to R_3 \\ 2R_4 + R_2 \to R_4 \end{array} \begin{bmatrix} 1 & 2 & 0 & 1 & 0 & 0 & 0 & 1 \\ -1 & 0 & 4 & -1 & 2 & 0 & 0 & 1 \\ ⑨ & 0 & 0 & 1 & -2 & 4 & 0 & 3 \\ -3 & 0 & 0 & 1 & 2 & 0 & 4 & 3 \end{bmatrix}$$

$$
\begin{array}{r}
9R_1 - R_3 \to R_1 \\
9R_2 + R_3 \to R_2 \\
\\
3R_4 + R_3 \to R_4
\end{array}
\begin{array}{ccccccc|c}
x_1 & x_2 & x_3 & x_4 & x_5 & x_6 & z & \\
0 & 18 & 0 & 8 & 2 & -4 & 0 & 6 \\
0 & 0 & 36 & -8 & 16 & 4 & 0 & 12 \\
9 & 0 & 0 & 1 & -2 & 4 & 0 & 3 \\
0 & 0 & 0 & 4 & 4 & 4 & 12 & 12
\end{array}
$$

Create a 1 in the columns corresponding to x_1, x_2, x_3, and z.

$$
\begin{array}{r}
\frac{1}{18}R_1 \to R_1 \\[4pt]
\frac{1}{36}R_2 \to R_2 \\[4pt]
\frac{1}{9}R_3 \to R_3 \\[4pt]
\frac{1}{12}R_4 \to R_4
\end{array}
\begin{array}{ccccccc|c}
x_1 & x_2 & x_3 & x_4 & x_5 & x_6 & z & \\
0 & 1 & 0 & \frac{4}{9} & \frac{1}{9} & -\frac{2}{9} & 0 & \frac{1}{3} \\[4pt]
0 & 0 & 1 & -\frac{2}{9} & \frac{4}{9} & \frac{1}{9} & 0 & \frac{1}{3} \\[4pt]
1 & 0 & 0 & \frac{1}{9} & -\frac{2}{9} & \frac{4}{9} & 0 & \frac{1}{3} \\[4pt]
0 & 0 & 0 & \frac{1}{3} & \frac{1}{3} & \frac{1}{3} & 1 & 1
\end{array}
$$

From this final tableau, we have

$$x_1 = \frac{1}{3}, \ x_2 = \frac{1}{3}, \ x_3 = \frac{1}{3},$$

$$y_1 = \frac{1}{3}, \ y_2 = \frac{1}{3}, \ y_3 = \frac{1}{3}, \text{ and } z = 1.$$

Note that $g = \frac{1}{z} = 1$.

The optimum strategy for player A is

$$p_1 = gy_1 = (1)\left(\frac{1}{3}\right) = \frac{1}{3},$$

$$p_2 = gy_2 = (1)\left(\frac{1}{3}\right) = \frac{1}{3},$$

$$p_3 = gy_3 = (1)\left(\frac{1}{3}\right) = \frac{1}{3}.$$

The optimum strategy for player B is

$$q_1 = gx_1 = (1)\left(\frac{1}{3}\right) = \frac{1}{3},$$

$$q_2 = gx_2 = (1)\left(\frac{1}{3}\right) = \frac{1}{3},$$

$$q_3 = gx_3 = (1)\left(\frac{1}{3}\right) = \frac{1}{3}.$$

The value of the game is $1 - 1 = 0$.

(b) The game is symmetric in that neither player has an advantage, and each choice is as strong as every other choice.

21. (a) The payoff matrix is as follows.

Kije

		(3, 0)	(0, 3)	(2, 1)	(1, 2)
	(4, 0)	4	0	2	1
	(0, 4)	0	4	1	2
Blotto	(3, 1)	1	−1	3	0
	(1, 3)	−1	1	0	3
	(2, 2)	−2	−2	2	2

Note that the strategy (3, 1) for Colonel Blotto means that he sends 3 regiments to the first post and 1 regiment to the second post; the strategy (0, 3) for Captain Kije means that he sends 0 regiments to the first post and 3 regiments to the second post.

Row 1, column 1 of the payoff matrix means (4, 0) for Blotto and (3, 0) for Kije. Blotto sends more regiments to the first post, so he wins 1 point, plus 3 points for capturing Kije's 3 regiments. Neither leader sends any regiments to the second post, so no points are won there. The payoff here is 1 + 3 = 4 points to Blotto.

Row 4, column 2 of the matrix means (1, 3) for Blotto and (0, 3) for Kije. Blotto earns a point for capturing the first post, but Kije sends no regiments there so there are no points for capturing regiments. Both leaders send 3 regiments to the second post, so there is a standoff and no additional points. The payoff here is 1 point to Blotto.

Row 5, column 1 of the matrix means (2, 2) for Blotto and (3, 0) for Kije. Kije earns 1 post–capturing point and 2 regiment–capturing points for the first post, while Blotto earns 1 post–capturing point for the second post. The net result is a payoff of 2 points to Kije, represented in the matrix as −2.

Continue in this manner to obtain all of the entries of the payoff matrix.

(b) Add 2 to each entry of the payoff matrix, to obtain

$$\begin{bmatrix} 6 & 2 & 4 & 3 \\ 2 & 6 & 3 & 4 \\ 3 & 1 & 5 & 2 \\ 1 & 3 & 2 & 5 \\ 0 & 0 & 4 & 4 \end{bmatrix}.$$

The linear programming problem to be solved is

maximize $z = x_1 + x_2 + x_3 + x_4$

subject to:
$$6x_1 + 2x_2 + 4x_3 + 3x_4 \le 1$$
$$2x_1 + 6x_2 + 3x_3 + 4x_4 \le 1$$
$$3x_1 + x_2 + 5x_3 + 2x_4 \le 1$$
$$x_1 + 3x_2 + 2x_3 + 5x_4 \le 1$$
$$4x_3 + 4x_4 \le 1$$

with $x_1 \ge 0$, $x_2 \ge 0$, $x_3 \ge 0$, $x_4 \ge 0$.

The initial tableau is

x_1	x_2	x_3	x_4	x_5	x_6	x_7	x_8	x_9	z	
⑥	2	4	3	1	0	0	0	0	0	1
2	6	3	4	0	1	0	0	0	0	1
3	1	5	2	0	0	1	0	0	0	1
1	3	2	5	0	0	0	1	0	0	1
0	0	4	4	0	0	0	0	1	0	1
−1	−1	−1	−1	0	0	0	0	0	1	0

Pivot on each circled entry.

	x_1	x_2	x_3	x_4	x_5	x_6	x_7	x_8	x_9	z	
	6	2	4	3	1	0	0	0	0	0	1
$3R_2 - R_1 \to R_2$	0	⑯	5	9	−1	3	0	0	0	0	2
$2R_3 - R_1 \to R_3$	0	0	6	1	−1	0	2	0	0	0	1
$6R_4 - R_1 \to R_4$	0	16	8	27	−1	0	0	6	0	0	5
	0	0	4	4	0	0	0	0	1	0	1
$6R_6 + R_1 \to R_6$	0	−4	−2	−3	1	0	0	0	0	6	1

	x_1	x_2	x_3	x_4	x_5	x_6	x_7	x_8	x_9	z	
$8R_1 - R_2 \to R_1$	48	0	27	15	9	−3	0	0	0	0	6
	0	16	5	9	−1	3	0	0	0	0	2
	0	0	⑥	1	−1	0	2	0	0	0	1
$R_4 - R_2 \to R_4$	0	0	3	18	0	−3	0	6	0	0	3
	0	0	4	4	0	0	0	0	1	0	1
$4R_6 + R_2 \to R_6$	0	0	−3	−3	3	3	0	0	0	24	6

$$
\begin{array}{c}
\begin{array}{cccccccccc}
x_1 & x_2 & x_3 & x_4 & x_5 & x_6 & x_7 & x_8 & x_9 & z
\end{array}
\end{array}
$$

	x_1	x_2	x_3	x_4	x_5	x_6	x_7	x_8	x_9	z	
$2R_1 - 9R_3 \rightarrow R_1$	96	0	0	21	27	-6	-18	0	0	0	3
$6R_2 - 5R_3 \rightarrow R_2$	0	96	0	49	-1	18	-10	0	0	0	7
	0	0	6	1	-1	0	2	0	0	0	1
$2R_4 - R_3 \rightarrow R_4$	0	0	0	35	1	-6	-2	12	0	0	5
$3R_5 - 2R_3 \rightarrow R_5$	0	0	0	⑩	2	0	-4	0	3	0	1
$2R_6 + R_3 \rightarrow R_6$	0	0	0	-5	5	6	2	0	0	48	13

	x_1	x_2	x_3	x_4	x_5	x_6	x_7	x_8	x_9	z	
$10R_1 - 21R_5 \rightarrow R_1$	960	0	0	0	228	-60	-96	0	-63	0	9
$10R_2 - 49R_5 \rightarrow R_2$	0	960	0	0	-108	180	96	0	-147	0	21
$10R_3 - R_5 \rightarrow R_3$	0	0	60	0	-12	0	24	0	-3	0	9
$2R_4 - 7R_5 \rightarrow R_4$	0	0	0	0	-12	-12	24	24	-21	0	3
	0	0	0	10	2	0	-4	0	3	0	1
$2R_6 + R_5 \rightarrow R_6$	0	0	0	0	12	12	0	0	3	96	27

Create a 1 in the columns corresponding to x_1, x_2, x_3, x_4, x_8, and z.

	x_1	x_2	x_3	x_4	x_5	x_6	x_7	x_8	x_9	z	
$\frac{1}{960}R_1 \rightarrow R_1$	1	0	0	0	$\frac{19}{80}$	$-\frac{1}{16}$	$-\frac{1}{10}$	0	$-\frac{21}{320}$	0	$\frac{3}{320}$
$\frac{1}{960}R_2 \rightarrow R_2$	0	1	0	0	$-\frac{9}{80}$	$\frac{3}{16}$	$\frac{1}{10}$	0	$-\frac{49}{320}$	0	$\frac{7}{320}$
$\frac{1}{60}R_3 \rightarrow R_3$	0	0	1	0	$-\frac{1}{5}$	0	$\frac{2}{5}$	0	$-\frac{1}{20}$	0	$\frac{3}{20}$
$\frac{1}{24}R_4 \rightarrow R_4$	0	0	0	0	$-\frac{1}{2}$	$-\frac{1}{2}$	1	1	$-\frac{7}{8}$	0	$\frac{1}{8}$
$\frac{1}{10}R_5 \rightarrow R_5$	0	0	0	1	$\frac{1}{5}$	0	$-\frac{2}{5}$	0	$\frac{3}{10}$	0	$\frac{1}{10}$
$\frac{1}{96}R_6 \rightarrow R_6$	0	0	0	0	$\frac{1}{8}$	$\frac{1}{8}$	0	0	$\frac{1}{32}$	1	$\frac{9}{32}$

From this final tableau, we have

$$x_1 = \frac{3}{320}, \ x_2 = \frac{7}{320}, \ x_3 = \frac{3}{20}, \ x_4 = \frac{1}{10},$$

$$y_1 = \frac{1}{8}, \ y_2 = \frac{1}{8}, \ y_3 = 0, \ y_4 = 0, \ y_5 = \frac{1}{32}, \text{ and } z = \frac{9}{32}.$$

Note that $g = \frac{1}{z} = \frac{32}{9}$.

The optimum strategy for Colonel Blotto is

$$p_1 = gy_1 = \left(\frac{32}{9}\right)\left(\frac{1}{8}\right) = \frac{4}{9},$$

$$p_2 = gy_2 = \left(\frac{32}{9}\right)\left(\frac{1}{8}\right) = \frac{4}{9},$$

$$p_3 = gy_3 = \left(\frac{32}{9}\right)(0) = 0,$$

$$p_4 = gy_4 = \left(\frac{32}{9}\right)(0) = 0,$$

$$p_5 = gy_5 = \left(\frac{32}{9}\right)\left(\frac{1}{32}\right) = \frac{1}{9}.$$

That is, Blotto uses strategies (4, 0) and (0, 4) with probability 4/9 each, strategy (2, 2) with probability 1/9, and never sends 3 regiments to one post and 1 to the other.

The optimum strategy for Captain Kije is

$$q_1 = gx_1 = \left(\frac{32}{9}\right)\left(\frac{3}{320}\right) = \frac{1}{30},$$

$$q_2 = gx_2 = \left(\frac{32}{9}\right)\left(\frac{7}{320}\right) = \frac{7}{90},$$

$$q_3 = gx_3 = \left(\frac{32}{9}\right)\left(\frac{3}{20}\right) = \frac{8}{15},$$

$$q_4 = gx_4 = \left(\frac{32}{9}\right)\left(\frac{1}{10}\right) = \frac{16}{45}.$$

That is, Kije uses strategy (3, 0) with probability 1/30, strategy (0, 3) with probability 7/90, strategy (2, 1) with probability 8/15, and strategy (1, 2) with probability 16/45.

The value of the game is

$$\frac{32}{9} - 2 = \frac{14}{9}.$$

(c) Let $B = \begin{bmatrix} q_1 \\ q_2 \\ q_3 \\ q_4 \end{bmatrix}$ be any stragegy that Captain Kije could use, which means that it is a probability vector and $q_1 + q_2 + q_3 + q_4 = 1$. If Colonel Blotto uses the strategy $A = \begin{bmatrix} \frac{4}{9} & \frac{4}{9} & 0 & 0 & \frac{1}{9} \end{bmatrix}$ that was found in part (b), then

$$AMB = \begin{bmatrix} \frac{4}{9} & \frac{4}{9} & 0 & 0 & \frac{1}{9} \end{bmatrix} \begin{bmatrix} 4 & 0 & 2 & 1 \\ 0 & 4 & 1 & 2 \\ 1 & -1 & 3 & 0 \\ -1 & 1 & 0 & 3 \\ -2 & -2 & 2 & 2 \end{bmatrix} \begin{bmatrix} q_1 \\ q_2 \\ q_3 \\ q_4 \end{bmatrix}$$

$$= \begin{bmatrix} \frac{14}{9} & \frac{14}{9} & \frac{14}{9} & \frac{14}{9} \end{bmatrix} \begin{bmatrix} q_1 \\ q_2 \\ q_3 \\ q_4 \end{bmatrix}$$

$$= \begin{bmatrix} \frac{14}{9}q_1 + \frac{14}{9}q_2 + \frac{14}{9}q_3 + \frac{14}{9}q_4 \end{bmatrix}$$

$$= \begin{bmatrix} \frac{14}{9}(q_1 + q_2 + q_3 + q_4) \end{bmatrix}$$

$$= \begin{bmatrix} \frac{14}{9}(1) \end{bmatrix}$$

$$= \begin{bmatrix} \frac{14}{9} \end{bmatrix}.$$

Therefore, the value of the game is $\frac{14}{9}$ regardless of what strategy Captain Kije uses.

Chapter 8   Review Exercises

3. $\begin{bmatrix} .4 & .6 \\ 1 & 0 \end{bmatrix}$

This could be a transition matrix since it is a square matrix, all entries are between 0 and 1, inclusive, and the sum of the entries in each row is 1.

5. $\begin{bmatrix} .8 & .2 & 0 \\ 0 & 1 & 0 \\ .1 & .4 & .5 \end{bmatrix}$

This could be a transition matrix.

7. (a) $C = \begin{bmatrix} .6 & .4 \\ 1 & 0 \end{bmatrix}$

$C^2 = \begin{bmatrix} .6 & .4 \\ 1 & 0 \end{bmatrix}\begin{bmatrix} .6 & .4 \\ 1 & 0 \end{bmatrix}$

$= \begin{bmatrix} .76 & .24 \\ .6 & .4 \end{bmatrix}$

$C^3 = \begin{bmatrix} .6 & .4 \\ 1 & 0 \end{bmatrix}\begin{bmatrix} .76 & .24 \\ .6 & .4 \end{bmatrix}$

$= \begin{bmatrix} .696 & .304 \\ .76 & .24 \end{bmatrix}$

(b) The probability that state 2 changes to state 1 after 3 repetitions is .76, since that is the entry in row 2, column 1 of C^3.

9. (a) $E = \begin{bmatrix} .2 & .5 & .3 \\ .1 & .8 & .1 \\ 0 & 1 & 0 \end{bmatrix}$

$E^2 = \begin{bmatrix} .2 & .5 & .3 \\ .1 & .8 & .1 \\ 0 & 1 & 0 \end{bmatrix}\begin{bmatrix} .2 & .5 & .3 \\ .1 & .8 & .1 \\ 0 & 1 & 0 \end{bmatrix}$

$= \begin{bmatrix} .09 & .8 & .11 \\ .1 & .79 & .11 \\ .1 & .8 & .1 \end{bmatrix}$

$E^3 = \begin{bmatrix} .2 & .5 & .3 \\ .1 & .8 & .1 \\ 0 & 1 & 0 \end{bmatrix}\begin{bmatrix} .09 & .8 & .11 \\ .1 & .79 & .11 \\ .1 & .8 & .1 \end{bmatrix}$

$= \begin{bmatrix} .098 & .795 & .107 \\ .099 & .792 & .109 \\ .1 & .79 & .11 \end{bmatrix}$

(b) The probability that state 2 changes to state 1 after 3 repetitions is .099, since that is the entry in row 2, column 1 of E^3.

11. $T^2 = \begin{bmatrix} .4 & .6 \\ .5 & .5 \end{bmatrix}\begin{bmatrix} .4 & .6 \\ .5 & .5 \end{bmatrix}$

$= \begin{bmatrix} .46 & .54 \\ .45 & .55 \end{bmatrix}$

The distribution after 2 repetitions is

$[.3 \quad .7]\begin{bmatrix} .46 & .54 \\ .45 & .55 \end{bmatrix} = [.453 \quad .547].$

To predict the long-range distribution, let V be the probability vector $[v_1 \quad v_2]$.

$[v_1 \quad v_2]\begin{bmatrix} .46 & .54 \\ .45 & .55 \end{bmatrix} = [v_1 \quad v_2]$

$.46v_1 + .45v_2 = v_1$
$.54v_1 + .55v_2 = v_2$
$v_1 + \quad v_2 = 1$

$$-.54v_1 + .45v_2 = 0$$
$$.54v_1 - .45v_2 = 0$$
$$v_2 = 1 - v_1$$

$$.54v_1 - .45(1 - v_1) = 0$$
$$.99v_1 = .45$$

$$v_1 = \frac{45}{99} = \frac{5}{11}$$
$$\approx .455$$
$$v_2 = \frac{54}{99} = \frac{6}{11}$$
$$\approx .545$$

The long-range distribution is

$$\begin{bmatrix} \frac{5}{11} & \frac{6}{11} \end{bmatrix} \quad \text{or} \quad [.455 \quad .545].$$

13. $T^2 = \begin{bmatrix} .6 & .2 & .2 \\ .3 & .3 & .4 \\ .5 & .4 & .1 \end{bmatrix} \begin{bmatrix} .6 & .2 & .2 \\ .3 & .3 & .4 \\ .5 & .4 & .1 \end{bmatrix}$

$$= \begin{bmatrix} .52 & .26 & .22 \\ .47 & .31 & .22 \\ .47 & .26 & .27 \end{bmatrix}$$

The distribution after 2 repetitions is

$$[.2 \quad .4 \quad .4] \begin{bmatrix} .52 & .26 & .22 \\ .47 & .31 & .22 \\ .47 & .26 & .27 \end{bmatrix}$$

$$= [.48 \quad .28 \quad .24].$$

To predict the long-range distribution, let V be the probability vector $[v_1 \quad v_2 \quad v_3]$.

$$[v_1 \quad v_2 \quad v_3] \begin{bmatrix} .6 & .2 & .2 \\ .3 & .3 & .4 \\ .5 & .4 & .1 \end{bmatrix}$$

$$= [v_1 \quad v_2 \quad v_3]$$

$$.6v_1 + .3v_2 + .5v_3 = v_1$$
$$.2v_1 + .3v_2 + .4v_3 = v_2$$
$$.2v_1 + .4v_2 + .1v_3 = v_3$$

Also, $v_1 + v_2 + v_3 = 1$.
Solving this system by the Gauss-Jordan method gives

$$v_1 = \frac{47}{95} \approx .495$$
$$v_2 = \frac{26}{95} \approx .274$$
$$v_3 = \frac{22}{95} \approx .232.$$

The long-range distribution is

$$\begin{bmatrix} \frac{47}{95} & \frac{26}{95} & \frac{22}{95} \end{bmatrix} \quad \text{or} \quad [.495 \quad .274$$
$$.232].$$

17. $A = \begin{bmatrix} .4 & .2 & .4 \\ 0 & 1 & 0 \\ .6 & .3 & .1 \end{bmatrix}$

$$A^2 = \begin{bmatrix} .4 & .4 & .2 \\ 0 & 1 & 0 \\ .3 & .45 & .25 \end{bmatrix}$$

$$A^3 = \begin{bmatrix} .28 & .54 & .18 \\ 0 & 1 & 0 \\ .27 & .585 & .145 \end{bmatrix}$$

Note that the second row will always have zeros; hence, the matrix is not regular.

21. $\begin{bmatrix} -2 & 5 & -6 & 3 \\ 0 & -1 & 7 & 5 \\ 2 & 6 & -4 & 4 \end{bmatrix}$

The entry at (1, 1) is −2, indicating that the payoff is $2 from A to B.

23. The entry at (2, 3) is 7, indicating that the payoff is $7 from B to A.

25. Row 3 dominates row 1 and column 1 dominates column 4.

27. $\begin{bmatrix} -11 & 6 & 8 & 9 \\ -10 & -12 & 3 & 2 \end{bmatrix}$

Column 1 dominates both column 3 and column 4. Remove the dominated columns to obtain

$$\begin{bmatrix} -11 & 6 \\ -10 & -12 \end{bmatrix}.$$

29. $\begin{bmatrix} -2 & 4 & 1 \\ 3 & 2 & 7 \\ -8 & 1 & 6 \\ 0 & 3 & 9 \end{bmatrix}$

Row 2 dominates row 3. Remove row 3 to obtain

$$\begin{bmatrix} -2 & 4 & 1 \\ 3 & 2 & 7 \\ 0 & 3 & 9 \end{bmatrix}.$$

Column 1 dominates column 3. Remove column 3 to obtain

$$\begin{bmatrix} -2 & 4 \\ 3 & 2 \\ 0 & 3 \end{bmatrix}.$$

31. $\begin{bmatrix} \boxed{-2} & 3 \\ \underline{-4} & ⑤ \end{bmatrix}$

Underline the smallest number in each row and circle the largest value in each column. The -2 at

(1, 1) is the smallest number in its row and the largest number in its column, so the saddle point is -2 at (1, 1). The value of the game is -2.

33. $\begin{bmatrix} \underline{-4} & -1 \\ 6 & ⓪ \\ ⑧ & \underline{3} \end{bmatrix}$

The 0 at (2, 2) is the smallest number in its row and the largest number in its column, so the saddle point is 0 at (2, 2). The value of the game is 0, and so it is a fair game.

35. $\begin{bmatrix} ⑧ & 1 & \underline{-7} & 2 \\ -1 & ④ & \underline{-3} & ③ \end{bmatrix}$

The -3 at (2, 3) is the smallest number in its row and the largest number in its column, so the saddle point is -3 at (2, 3). The value of the game is -3.

37. $\begin{bmatrix} 1 & 0 \\ -2 & 3 \end{bmatrix}$

The optimum strategy for player A is

$$p_1 = \frac{3 - (-2)}{1 - (-2) - 0 + 3} = \frac{5}{6},$$

$$p_2 = 1 - \frac{5}{6} = \frac{1}{6}.$$

The optimum strategy for player B is

$$q_1 = \frac{3 - (0)}{1 - (-2) - 0 + 3} = \frac{3}{6} = \frac{1}{2},$$

$$q_2 = 1 - \frac{1}{2} = \frac{1}{2}.$$

The value of the game is

$$\frac{(1)(3) - (0)(-2)}{6} = \frac{1}{2}.$$

39. $\begin{bmatrix} -3 & 5 \\ 1 & 0 \end{bmatrix}$

The optimum strategy for player A is

$$p_1 = \frac{0 - 1}{-3 - 1 - 5 + 0} = \frac{-1}{-9} = \frac{1}{9},$$

$$p_2 = 1 - \frac{1}{9} = \frac{8}{9}.$$

The optimum strategy for player B is

$$q_1 = \frac{0 - 5}{-3 - 1 - 5 + 0} = \frac{-5}{-9} = \frac{5}{9},$$

$$q_2 = 1 - \frac{5}{9} = \frac{4}{9}.$$

The value of the game is

$$\frac{-3(0) - (5)(1)}{-3 - 1 - 5 + 0} = \frac{-5}{-9} = \frac{5}{9}.$$

41. $\begin{bmatrix} -4 & 8 & 0 \\ -2 & 9 & -3 \end{bmatrix}$

Column 1 dominates column 2. Remove column 2, to obtain

$$\begin{bmatrix} -4 & 0 \\ -2 & -3 \end{bmatrix}.$$

The optimum strategy for player A is

$$p_1 = \frac{-3 + 2}{-4 + 2 - 0 - 3} = \frac{-1}{-5} = \frac{1}{5},$$

$$p_2 = 1 - \frac{1}{5} = \frac{4}{5}.$$

The optimum strategy for player B is

$$q_1 = \frac{-3 - 0}{-4 + 2 - 0 - 3} = \frac{-3}{-5} = \frac{3}{5},$$

$q_2 = 0$ (column 2 was removed),

$$q_3 = 1 - \frac{3}{5} = \frac{2}{5}.$$

The value of the game is

$$\frac{-4(-3) - (0)(-2)}{-4 + 2 - 0 - 3} = \frac{12}{-5} = -\frac{12}{5}.$$

43. $\begin{bmatrix} 2 & -1 \\ -4 & 5 \\ -1 & -2 \end{bmatrix}$

Row 1 dominates row 3. Remove row 3 to obtain

$$\begin{bmatrix} 2 & -1 \\ -4 & 5 \end{bmatrix}.$$

The optimum strategy for player A is

$$p_1 = \frac{5 + 4}{2 + 4 + 1 + 5} = \frac{9}{12} = \frac{3}{4},$$

$$p_2 = 1 - \frac{3}{4} = \frac{1}{4},$$

$p_3 = 0$ (row 3 was removed).

The optimum strategy for player B is

$$q_1 = \frac{5 + 1}{2 + 4 + 1 + 5} = \frac{6}{12} = \frac{1}{2},$$

$$q_2 = 1 - \frac{1}{2} = \frac{1}{2}.$$

The value of the game is

$$\frac{2(5) - (-1)(-4)}{2 + 4 + 1 + 5} = \frac{6}{12} = \frac{1}{2}.$$

45.   (a) The distribution after the campaign is

$$[.35 \quad .65]\begin{bmatrix} .8 & .2 \\ .4 & .6 \end{bmatrix} = [.54 \quad .46].$$

(b) $P^3 = \begin{bmatrix} .688 & .312 \\ .624 & .376 \end{bmatrix}$

The distribution after 3 campaigns is

$$[.35 \quad .65]\begin{bmatrix} .688 & .312 \\ .624 & .376 \end{bmatrix}$$

$$= [.6464 \quad .3536].$$

47. The distribution after one month is

$$[.4 \quad .4 \quad .2]\begin{bmatrix} .8 & .15 & .05 \\ .25 & .55 & .2 \\ .04 & .21 & .75 \end{bmatrix}$$

$$= [.428 \quad .332 \quad .25].$$

49. $P^2 = \begin{bmatrix} .8 & .15 & .05 \\ .25 & .55 & .2 \\ .04 & .21 & .75 \end{bmatrix}\begin{bmatrix} .8 & .15 & .05 \\ .25 & .55 & .2 \\ .04 & .21 & .75 \end{bmatrix}$

$$= \begin{bmatrix} .6795 & .213 & .1075 \\ .3455 & .382 & .2725 \\ .1145 & .279 & .6065 \end{bmatrix}$$

$P^3 = \begin{bmatrix} .8 & .15 & .05 \\ .25 & .55 & .2 \\ .04 & .21 & .75 \end{bmatrix}\begin{bmatrix} .6795 & .213 & .1075 \\ .3455 & .382 & .2725 \\ .1145 & .279 & .6065 \end{bmatrix}$

$$= \begin{bmatrix} .60115 & .24165 & .1572 \\ .3828 & .31915 & .29805 \\ .18561 & .29799 & .5164 \end{bmatrix}$$

The distribution after 3 months is

$$[.4 \quad .4 \quad .2]\begin{bmatrix} .60115 & .24165 & .1572 \\ .3828 & .31915 & .29805 \\ .18561 & .29799 & .5164 \end{bmatrix}$$

$$= [.431 \quad .284 \quad .285].$$

51.

$$\begin{array}{cc} & \textit{Management} \\ & \begin{array}{cc} \text{Friendly} & \text{Hostile} \end{array} \\ \textit{Labor}\ \begin{array}{c} \text{Friendly} \\ \text{Hostile} \end{array} & \begin{bmatrix} \$600 & \$800 \\ \$400 & \$950 \end{bmatrix} \end{array}$$

Be hostile; then he has a chance at the $950 wage increase, which is the largest possible increase.

53. If there is a .7 chance that the company will be hostile, then there is a .3 chance that it will be friendly. Find the expected payoff for each strategy.

Friendly:

$$(.3)(600) + (.7)(800) = \$740$$

Hostile:

$$(.3)(400) + (.7)(950) = \$785$$

Therefore, he should be hostile. The expected payoff is $785.

55.

$$\begin{bmatrix} \boxed{600} & 800 \\ 400 & \boxed{950} \end{bmatrix}$$

The 600 at (1, 1) is the smallest number in its row and the largest number in its column, so it is a saddle point and the game is strictly determined. Labor and management should both always be friendly, and the value of the game is 600.

57. The original transition matrix is

$$\begin{array}{c} \\ A = \end{array} \begin{array}{c} \\ \begin{array}{c} T \\ N \\ O \end{array} \end{array} \begin{array}{ccc} T & N & O \\ \begin{bmatrix} .3 & .5 & .2 \\ .2 & .6 & .2 \\ .1 & .5 & .4 \end{bmatrix} \end{array}.$$

A grandson is 2 generations down, so we need to look at A^2.

$$A^2 = \begin{array}{c} \\ T \\ N \\ O \end{array} \begin{array}{ccc} T & N & O \\ \begin{bmatrix} .21 & .55 & .24 \\ .2 & .56 & .24 \\ .17 & .55 & .28 \end{bmatrix} \end{array}$$

The probability that the grandson of a normal man is thin is given by the entry in row 2, column 1 of A^2. This probability is .2.

59. We can read this probability directly from matrix A: the probability that an overweight man has an overweight son is the entry in row 3, column 3 of A. This probability is .4.

61. $A^3 = \begin{array}{c} \\ T \\ N \\ O \end{array} \begin{array}{ccc} T & N & O \\ \begin{bmatrix} .197 & .555 & .248 \\ .196 & .556 & .248 \\ .189 & .555 & .256 \end{bmatrix} \end{array}$

is the transition matrix over 3 generations. The probability of an overweight man having an overweight great-grandson is the entry in row 3, column 3 of A^3. This probability is .256.

63. Let D = [.2 .55 .25], the initial distribution. From Exercises 36, we have

$$A^2 = \begin{bmatrix} .21 & .55 & .24 \\ .2 & .56 & .24 \\ .17 & .55 & .28 \end{bmatrix}.$$

The distribution after 2 generations is

$$DA^2 = [.2 \quad .55 \quad .25] \begin{bmatrix} .21 & .55 & .24 \\ .2 & .56 & .24 \\ .17 & .55 & .28 \end{bmatrix}$$

$$= [.1945 \quad .5555 \quad .25].$$

65. $A = \begin{bmatrix} .3 & .5 & .2 \\ .2 & .6 & .2 \\ .1 & .5 & .4 \end{bmatrix}$

To find the long-term distribution, we use the system

$$v_1 + v_2 + v_3 = 1$$
$$.3v_1 + .2v_2 + .1v_3 = v_1$$
$$.5v_1 + .6v_2 + .5v_3 = v_2$$
$$.2v_1 + .2v_2 + .4v_3 = v_3.$$

Simplify these equations to obtain the system

$$v_1 + v_2 + v_3 = 1$$
$$-.7v_1 + .2v_2 + .1v_3 = 0$$
$$.5v_1 - .4v_2 + .5v_3 = 0$$
$$.2v_1 + .2v_2 - .6v_3 = 0.$$

Solve this system by the Gauss-Jordan method to obtain $v_1 = 7/36$, $v_2 = 5/9$, and $v_3 = 1/4$. The long-range distribution is

$$\left[\frac{7}{36} \quad \frac{5}{9} \quad \frac{1}{4}\right].$$

67. The distribution of men by weight after 3 generations is

$$[.2 \quad .55 \quad .25] \begin{bmatrix} .197 & .555 & .248 \\ .196 & .556 & .248 \\ .189 & .555 & .256 \end{bmatrix}$$

$$= [.194 \quad .556 \quad .25].$$

69. If the offspring both carry genes AA, then so must their offspring; hence, state 1 ends up in state 1 with prob- ability 1. If the offspring both carry genes aa, then so must their offspring; hence, state 6 ends up in state 6 with probability 1. If AA mates with aa, then the offspring will carry genes Aa; hence, state 3 ends up in state 4 with probability 1. If AA mates with Aa, there are four possible outcomes for a pair of offspring; AA and AA is one of the outcomes, so state 2 ends up in state 1 with probability 1/4, AA and Aa can happen two ways, so state 2 ends up in state 2 with probability 2/4 or 1/2, and Aa and Aa is the last pos- sible outcome, so state 2 ends up in state 4 with probability 1/4. If Aa mates with Aa, there are sixteen possible outcomes for a pair of off-spring; state 4 ends up in states 1, 2, 3, 4, 5, 6 with respective probabilities 1/16, 1/4, 1/8, 1/4, 1/4, and 1/16. If Aa mates with aa, there are four possible outcomes for a pair of offspring, corresponding to three of the possible states; state 5 ends up in states 4, 5, 6 with respective probabilities 1/4, 1/2, and 1/4. This verifies that the transition matrix for this mating experiment is

$$
\begin{array}{c}
\\ 1\\ 2\\ 3\\ 4\\ 5\\ 6
\end{array}
\begin{array}{cccccc}
\;1 & \;2 & \;3 & \;4 & \;5 & \;6\\
\left[\begin{array}{cccccc}
1 & 0 & 0 & 0 & 0 & 0\\
\frac{1}{4} & \frac{1}{2} & 0 & \frac{1}{4} & 0 & 0\\
0 & 0 & 0 & 1 & 0 & 0\\
\frac{1}{16} & \frac{1}{4} & \frac{1}{8} & \frac{1}{4} & \frac{1}{4} & \frac{1}{16}\\
0 & 0 & 0 & \frac{1}{4} & \frac{1}{2} & \frac{1}{4}\\
0 & 0 & 0 & 0 & 0 & 1
\end{array}\right].
\end{array}
$$

Extended Application

1. This verification should be accom- plished by computer methods.

2. The entry in row 1, column 1 of the fundamental matrix indicates that there will be about 6.6 visits until there is some tooth decay. Each visit represents 6 months, so 6.6 visits = 39.6 months = 3.3 yr.

3. Use as a computer to show that

$$
FR = \begin{bmatrix}
.768332 & .231668\\
.814382 & .185618\\
.775304 & .224696\\
.807692 & .192308\\
.5 & .5\\
.5 & .5\\
.5 & .5\\
.5 & .5
\end{bmatrix}.
$$

4. The probability that a healthy tooth is eventually lost is .231668 $\approx$ .23, since that is the entry in row 1, column 2 of FR.

5. The four .5 entries in column 2 of
   FR each correspond to a five-digit
   number whose first digit is 1, which
   corresponds to decay on the occlusal
   surface.

CHAPTER 8   TEST

1. Decide which of the following matrices could be transition matrices. Justify your answer.

   (a) $\begin{bmatrix} .1 & .9 \\ 0 & 1 \end{bmatrix}$ **(b)** $\begin{bmatrix} .1 & .2 & .5 & .2 \\ 0 & .8 & .3 & -.1 \\ .1 & .8 & .1 & 0 \\ 0 & 0 & 0 & 1 \end{bmatrix}$ **(c)** $\begin{bmatrix} .21 & 0 & .88 \\ 0 & .45 & .55 \\ .79 & .03 & .18 \end{bmatrix}$

2. Let A be the transition matrix

$$\begin{bmatrix} \frac{1}{2} & \frac{1}{2} \\ \frac{1}{3} & \frac{2}{3} \end{bmatrix}.$$

   (a) Find the first 3 powers of A.

   (b) Find the probability that state 1 changes to state 2 after 2 repetitions of the experiment.

   (c) Find the probability that state 2 changes to state 1 after 3 repetitions of the experiment.

3. A survey conducted for General Motors revealed that 40% of GM buyers would buy GM again while 75% of Toyota buyers would buy Toyotas again. Suppose the initial distribution vector is [100   50] with GM coming first.

   (a) Write the transition matrix for this problem.

   (b) Find the distribution vector after 1 time period.

   (c) Find the distribution vector after 2 time periods.

4. Find the equilibrium vector for the regular Markov matrix $\begin{bmatrix} .2 & .8 \\ .4 & .6 \end{bmatrix}$.

5. In the small Illinois town of Red Bud, there are three political parties: the Corn Party, the Wheat Party, and the Tea Party. Each year 50% of the Corn Party members switch to the Wheat Party while the rest remain loyal. Among Wheat Party members, 25% switch to Corn, and another 25% switch to Tea, and the rest remain loyal. 50% of the Tea Party members switch to Wheat and the rest remain loyal.

   (a) Draw the transition diagram for this problem.

   (b) Write the transition matrix.

   (c) Show that the matrix is regular.

   (d) Find the long-range prediction for the proportion of voters in each party.

6. A politician is faced with three possible levels of opposition to his policies: Light, Medium, and Heavy. He must decide on three levels of advertising to combat this opposition: Low, Medium, or High. His payoff matrix is given below.

$$
\begin{array}{c}
\text{Low} \\
\text{Medium} \\
\text{High}
\end{array}
\begin{array}{ccc}
\text{Light} & \text{Medium} & \text{Heavy} \\
\begin{bmatrix}
100 & -30 & -60 \\
-10 & 75 & -70 \\
-50 & -20 & 50
\end{bmatrix}
\end{array}
$$

   (a) What would an optimist do?

   (b) What would a pessimist do?

   (c) The probability of heavy opposition is .3, and of medium opposition is .4. What is the best course of action?

7. Remove any dominated strategies from the payoff matrix.

$$
\begin{bmatrix}
2 & 3 & 0 & -1 \\
1 & 2 & -2 & -5 \\
0 & 3 & 1 & 0 \\
-1 & 0 & 0 & 1
\end{bmatrix}.
$$

8. Find the saddle point, if it exists, in the matrix

$$
\begin{bmatrix}
1 & 20 & 3 \\
-1 & 1 & 0 \\
-3 & 0 & 1
\end{bmatrix}.
$$

9. For each game below, find the optimum strategy for each player and the value
    of the game.

    (a) $\begin{bmatrix} 1 & -2 \\ 2 & 3 \end{bmatrix}$ **(b)** $\begin{bmatrix} 6 & -3 \\ 5 & 9 \end{bmatrix}$

10. Consider the game

    $$\begin{bmatrix} -3 & 2 \\ 4 & -5 \end{bmatrix}.$$

    (a) Find the optimum strategy for player A using the graphical method.

    (b) Find the optimum strategy for player B using the simplex method.

    (c) Find the value of the game.

CHAPTER 8   TEST ANSWERS

1. **(a)** This could be a transition matrix since it is a square matrix with all entries between 0 and 1, inclusive, and the sum of the entries in each row is 1.

   (b) This could not be a transition matrix because it has a negative entry.

   (c) This could not be a transition matrix because the sum of the entries in the first row is not 1.

2. **(a)** $A = \begin{bmatrix} \frac{1}{2} & \frac{1}{2} \\ \frac{1}{3} & \frac{2}{3} \end{bmatrix}$, $A^2 = \begin{bmatrix} \frac{5}{12} & \frac{7}{12} \\ \frac{7}{18} & \frac{11}{18} \end{bmatrix}$ **(b)** 7/12   **(c)** 43/108

   $A^3 = \begin{bmatrix} \frac{29}{72} & \frac{43}{72} \\ \frac{43}{108} & \frac{33}{54} \end{bmatrix}$

3. **(a)** $A = \begin{bmatrix} .4 & .6 \\ .25 & .75 \end{bmatrix}$ **(b)** [52.5   97.5]   **(c)** [45.375   104.625]

4. [1/3   2/3]

5. **(a)**

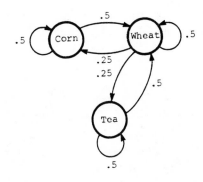

   (b)
	Corn	Wheat	Tea
Corn	.5	.5	0
Wheat	.25	.5	.25
Tea	0	.5	.5

   (c) $A^2 = \begin{bmatrix} .375 & .5 & .125 \\ .25 & .5 & .25 \\ .125 & .25 & .375 \end{bmatrix}$, which has all positive entries.

   (d) [.25   .5   .25]

6. **(a)** Low advertising   **(b)** High advertising   **(c)** Medium advertising

7. $\begin{bmatrix} 2 & 0 & -1 \\ 0 & 1 & 0 \\ -1 & 0 & 1 \end{bmatrix}$

8. Saddle point 1 at (1, 1)

9. **(a)** Player A: $p_1 = 0$, $p_2 = 1$

    Player B: $q_1 = 1$, $q_2 = 0$

    Value of game = 2

    (b) Player A: $p_1 = \dfrac{4}{13}$, $p_2 = \dfrac{9}{13}$

    Player B: $q_1 = \dfrac{12}{13}$, $q_2 = \dfrac{1}{13}$

    Value of game = $\dfrac{69}{13}$

10. **(a)** Player A: $p_1 = \dfrac{9}{14}$, $p_2 = \dfrac{5}{14}$

    (b) Player B: $q_1 = \dfrac{1}{2}$, $q_2 = \dfrac{1}{2}$

    (c) Value of game = $-\dfrac{1}{2}$

CHAPTER 9 MATHEMATICS OF FINANCE

Section 9.1

1. The interest rate and time period determine the amount of interest on a fixed principal. In the formula $I = Prt$, if P remains constant then the value of I will be affected by the values of r and t.

3. $I = Prt = 3850(.09)\left(\frac{8}{12}\right) = \$231.00$

5. $I = Prt = 3724(.084)\left(\frac{11}{12}\right) = \$286.75$

7. $I = Prt = 2930.42(.119)\left(\frac{123}{360}\right)$
$= \$119.15$

9. There are $14 + 30 + 30 + 30 + 30 = 134$ days, since a 30-day month is assumed.

$I = Prt = 5408(.12)\left(\frac{134}{360}\right) = \$241.56$

11. There are $9 + 31 + 30 + 31 = 101$ days.

$I = Prt = 11,000(.10)\left(\frac{101}{365}\right) = \$304.38$

13. There are $18 + 31 + 30 + 31 + 31 + 28 + 31 + 30 + 31 + 30 + 30 = 321$ days.

$I = Prt = 37,098(.112)\left(\frac{321}{365}\right)$
$= \$3654.10$

17. $P = \dfrac{A}{1 + rt}$

$= \dfrac{48,000}{1 + .05\left(\frac{9}{12}\right)}$

$= \$46,265.06$

19. $P = \dfrac{A}{1 + rt}$

$= \dfrac{29,764}{1 + .072\left(\frac{310}{360}\right)}$

$= \$28,026.37$

21. $P = A(1 - rt)$

$= 9450\left[1 - .10\left(\frac{7}{12}\right)\right]$

$\approx 9450(.9417)$

$\approx \$8898.75$

23. $P = A(1 - rt)$

$= 50,900\left[1 - .082\left(\frac{238}{360}\right)\right]$

$\approx 50,900(.9458)$

$\approx \$48,140.65$

25. The discount is

$6200(.10)\left(\frac{8}{12}\right) \approx 413.33.$

The proceeds are found by subtracting the discount from the original amount.

$6200 - 413.33 = 5786.67$

Use $I = Prt$ with $I = 413.33$, $P = 5786.67$, and $t = \frac{8}{12}$, and solve for r.

$413.33 = 5786.67(r)\left(\frac{8}{12}\right)$

$.107 \approx r$

The interest rate paid by the borrower is about 10.7%.

27. The discount is

$$58,000(.108)\left(\frac{9}{12}\right) = 4698$$

and the proceeds are

$$58,000 - 4698 = 53,302.$$

Use $I = Prt$ with $I = 4698$, $P = 53,302$, and $t = \frac{9}{12}$, and solve for r.

$$4698 = 53,302(r)\left(\frac{9}{12}\right)$$

$$.118 \approx r$$

The interest rate paid by the borrower is about 11.8%.

29. Use the formula for future value.

$$A = P(1 + rt)$$

$$= 25,900\left[1 + (.084)\left(\frac{11}{12}\right)\right]$$

$$= 25,900(1.077)$$

$$= \$27,894.30$$

She repaid $27,894.30.

31. The interest is

$$101,133.33 - 100,000 = 1133.33.$$

Use the formula for simple interest.

$$I = Prt$$

$$1133.33 = 100,000(r)\left(\frac{60}{360}\right)$$

$$.068 \approx r$$

The interest rate was about 6.8%.

33. The future value is $7(5104) = 35,728$. Use the formula for present value.

$$P = \frac{A}{1 + rt}$$

$$= \frac{35,728}{1 + (.0642)\left(\frac{7}{12}\right)}$$

$$\approx 34,438.29.$$

They should deposit about $34,438.29.

35. The proceeds are $6100. Use the formula for proceeds and solve for A.

$$P = A(1 - rt)$$

$$A = \frac{P}{1 - rt}$$

$$= \frac{6100}{1 - .188\left(\frac{7}{12}\right)}$$

$$\approx 6550.92$$

The amount of the loan is about $6550.92.

37. The interest per share is

$$(24 - 22) + .50 = 2.50.$$

Use the simple interest formula with $I = 2.50$, $P = 22$, and $t = 1$, and solve for r.

$$I = Prt$$

$$2.50 = 22(r)(1)$$

$$.114 \approx r$$

The interest rate is about 11.4%.

39. Find the maturity value of the loan, the amount the contractor must pay to the plumber.

$$A = P(1 + rt)$$
$$= 13,500\left[1 + .09\left(\frac{9}{12}\right)\right]$$
$$= 13,500(1.0675)$$
$$= 14,411.25$$

Three months after the note is signed, there are six more months before the loan is payable. The bank will apply its discount rate to the 14,411.25, to obtain

Amount of discount
$$= 14,411.25(.101)\left(\frac{6}{12}\right)$$
$$\approx 727.77.$$

The plumber will receive from the bank

$14,411.25 - \$727.77 = \$13,683.48$, which will be enough to pay the $13,582 bill.

Section 9.2

3. The formula for the compound amount is

$$A = P(1 + i)^n.$$

To find A, substitute P = 1000, i = .06, and n = 8.

$$A = 1000(1 + .06)^8$$
$$= 1000(1.06)^8$$
$$\approx 1000(1.59385)$$
$$= 1593.85$$

The compound amount is $1593.85.

5. To find the compound amount A, substitute P = 470, $i = \frac{.10}{2} = .05$, and n = 12 · 2 = 24 in the formula for the compound amount.

$$A = P(1 + i)^n$$
$$= 470(1.05)^{24}$$
$$\approx 470(3.22510)$$
$$= \$1515.80$$

7. P = 6500, $i = \frac{.12}{4}$, n = 6 · 4

$$A = P(1 + i)^n$$
$$= 6500\left(1 + \frac{.12}{4}\right)^{6(4)}$$
$$= 6500(1.03)^{24}$$
$$\approx 6500(2.03279)$$
$$= \$13,213.14$$

9. First find the compound amount.

$$A = P(1 + i)^n$$
$$= 6000(1 + .08)^8$$
$$= 6000(1.08)^8$$
$$\approx 6000(1.85093)$$
$$= \$11,105.58$$

To find the amount of interest earned, subtract the initial deposit from the compound amount.

Amount of interest
$$= A - P$$
$$= \$11,105.58 - \$6000$$
$$= \$5105.58$$

11. $A = P(1 + i)^n$
$$= 43,000\left(1 + \frac{.10}{2}\right)^{2(9)}$$
$$= 43,000(1.05)^{18}$$
$$\approx 43,000(2.40662)$$
$$= \$103,484.66$$

Amount of interest

$$= \$103{,}484.66 - \$43{,}000$$

$$= \$60{,}484.66$$

13. $A = P(1 + i)^n$

$$A = 2196.58\left(1 + \frac{.108}{4}\right)^{4(4)}$$

$$= 2196.58(1.027)^{16}$$

$$\approx 2196.58(1.53153)$$

$$\approx 3364.14$$

Amount of interest

$$= \$3364.14 - \$2196.58$$

$$= \$1167.56$$

15. $A = 4500$, $i = .08$, $n = 9$

Substitute these values in the formula for present value of an amount at compound interest.

$$P = \frac{A}{(1 + i)^n}$$

$$= \frac{4500}{(1.08)^9}$$

$$\approx \frac{4500}{1.99900}$$

$$\approx \$2251.12$$

17. $P = \frac{A}{(1 + i)^n}$

$$= \frac{15{,}902.74}{(1 + .098)^7}$$

$$\approx \frac{15{,}902.74}{1.92405}$$

$$\approx \$8265.24$$

19. $P = \frac{A}{(1 + i)^n}$

$$= \frac{2000}{\left(1 + \frac{.09}{2}\right)^{8(2)}}$$

$$= \frac{2000}{(1.045)^{16}}$$

$$\approx \frac{2000}{2.02237}$$

$$\approx \$988.94$$

21. $P = \frac{A}{(1 + i)^n}$

$$= \frac{8000}{\left(1 + \frac{.10}{4}\right)^{5(4)}}$$

$$= \frac{8800}{(1.025)^{20}}$$

$$\approx \$5370.38$$

25. Substitute $r = .04$ and $m = 2$ in the formula for effective rate.

$$r_e = \left(1 + \frac{r}{m}\right)^m - 1$$

$$= \left(1 + \frac{.04}{2}\right)^2 - 1$$

$$= (1.02)^2 - 1$$

$$= 1.0404 - 1$$

$$= .0404$$

The effective rate is 4.04%.

27. $r = .08$, $m = 2$

$$r_e = \left(1 + \frac{r}{m}\right)^m - 1$$

$$= \left(1 + \frac{.08}{2}\right)^2 - 1$$

$$= (1.04)^2 - 1$$

$$= .0816$$

The effective rate is 8.16%.

29. $r = .12$, $m = 2$

$$r_e = \left(1 + \frac{r}{m}\right)^m - 1$$

$$= \left(1 + \frac{.12}{2}\right)^2 - 1$$

$$= (1.06)^2 - 1$$

$$= .1236$$

The effective rate is 12.36%.

31. Substitute $P = 3000$, $i = \frac{.06}{4}$, and $n = 18(4)$ in the formula for the compound amount.

$$A = P(1 + i)^n$$

$$= 3000\left(1 + \frac{.06}{4}\right)^{18(4)}$$

$$= 3000(1.015)^{72}$$

$$\approx \$8763.47$$

33. Substitute $P = 78,000$, $i = .03$, and $n = 12$ in the formula for the compound amount.

$$A = P(1 + i)^n$$

$$= 78,000(1.03)^{12}$$

$$\approx 111,000$$

The average price will be about $111,000.

35. $P = 50,000$, $i = \frac{.12}{12}$, $n = 12(4)$

First find the compound amount.

$$A = P(1 + i)^n$$

$$= 50,000\left(1 + \frac{.12}{12}\right)^{12(4)}$$

$$= 50,000(1.01)^{48}$$

$$\approx \$80,611.30$$

Amount of interest

$$= A - P$$

$$= \$80,611.30 - \$50,000$$

$$= \$30,611.30$$

37. Substitute $A = 2.9$ million, $i = \frac{.08}{12}$, and $n = 12(5)$ in the formula for present value with compound interest.

$$P = \frac{A}{(1 + i)^n}$$

$$= \frac{2.9}{\left(1 + \frac{.08}{12}\right)^{12(5)}}$$

$$\approx \frac{2.9}{(1.00667)^{60}}$$

$$\approx 1.946$$

They should invest about $1.946 million now.

39. Substitute $P = 10,000$, $i = \frac{.06}{2}$, and $n = 2(3)$ in the formula for the compound amount.

$$A = P(1 + i)^n$$

$$= 10,000\left(1 + \frac{.06}{2}\right)^{2(3)}$$

$$= 10,000(1.03)^6$$

$$\approx 11,940.52$$

She should contribute about $11,940.52 in 3 yr.

41. To find the number of years it will take prices to double at 4% annual inflation, find n in the equation

$$2 = (1 + .04)^n,$$

which simplifies to

$$2 = (1.04)^n.$$

By trying various values of n, find that n = 18 is approximately correct, because

$$1.04^{18} \approx 2.0258 \approx 2.$$

Prices will double in about 18 yr.

43. To find the number of years it will be until the generating capacity will need to be doubled, find n in the equation

$$2 = (1 + .06)^n,$$

which simplifies to

$$2 = (1.06)^n.$$

By trying various values of n, find that n = 12 is approximately correct, because $1.06^{12} \approx 2.0122 \approx 2$. The generating capacity will need to be doubled in about 12 yr.

45. P = 150,000, i = −.024, n = 2

$$\begin{aligned}
A &= P(1 + i)^n \\
&= 150,000(1 - .024)^2 \\
&= 150,000(.976)^2 \\
&= \$142,886.40
\end{aligned}$$

47. P = 150,000, i = −.024, n = 8

$$\begin{aligned}
A &= P(1 + i)^n \\
&= 150,000(1 - .024)^8 \\
&= 150,000(.976)^8 \\
&= \$123,506.50
\end{aligned}$$

49. First consider the case of earning interest at a rate of k per annum compounded quarterly for all eight years and earning $2203.76 interest on the $1000 investment.

$$2203.76 = 1000\left(1 + \frac{k}{4}\right)^{8(4)}$$

$$2.20376 = \left(1 + \frac{k}{4}\right)^{32}$$

Use a calculator to raise both sides to the power $\frac{1}{32}$.

$$1.025 = 1 + \frac{k}{4}$$

$$.025 = \frac{k}{4}$$

$$.1 = k$$

Next consider the actual investments. The $1000 was invested for the first five years at a rate of j per annum compounded semiannually.

$$A = 1000\left(1 + \frac{j}{2}\right)^{5(2)}$$

$$A = 1000\left(1 + \frac{j}{2}\right)^{10}$$

This amount was then invested for the remaining three years at k = .1 per annum compounded quarterly for a final compound amount of $1990.76.

$$1990.76 = A\left(1 + \frac{.1}{4}\right)^{3(4)}$$

$$1990.76 = A(1.025)^{12}$$

$$1480.24 \approx A$$

Recall that $A = 1000\left(1 + \frac{j}{2}\right)^{10}$ and substitute this value into the above equation.

$$1480.24 = 1000\left(1 + \frac{j}{2}\right)^{10}$$

$$1.48024 = \left(1 + \frac{j}{2}\right)^{10}$$

Use a calculator to raise both sides to the power $\frac{1}{10}$.

$$1.04 \approx 1 + \frac{j}{2}$$

$$.04 = \frac{j}{2}$$

$$.08 = j$$

The ratio of k to j is

$$\frac{k}{j} = \frac{.1}{.08} = \frac{10}{8} = \frac{5}{4}.$$

Section 9.3

1. $a = 3$, $r = 2$

The first five terms are

$$3,\ 3(2),\ 3(2)^2,\ 3(2)^3,\ 3(2)^4$$

or $3,\ 6,\ 12,\ 24,\ 48$.

The fifth term is 48.

3. $a = -8$, $r = 3$

The first five terms are

$$-8,\ -8(3),\ -8(3)^2,\ -8(3)^3,\ -8(3)^4$$

or $-8,\ -24,\ -72,\ -216,\ -648$.

The fifth term is -648.

5. $a = 1$, $r = -3$

The first five terms are

$$1,\ 1(-3),\ 1(-3)^2,\ 1(-3)^3,\ 1(-3)^4$$

or $1,\ -3,\ 9,\ -27,\ 81$.

The fifth term is 81.

7. $a = 1024$, $r = \frac{1}{2}$

The first five terms are

$$1024,\ 1024\left(\tfrac{1}{2}\right),\ 1024\left(\tfrac{1}{2}\right)^2,\ 1024\left(\tfrac{1}{2}\right)^3,$$

$$1024\left(\tfrac{1}{2}\right)^4$$

or $1024,\ 512,\ 256,\ 128,\ 64$.

The fifth term is 64.

9. $a = 1$, $r = 2$, $n = 4$

To find the sum of the first 4 terms, S_4, use the formula for the sum of the first n terms of geometric sequence.

$$S_n = \frac{a(r^n - 1)}{r - 1}$$

$$S_4 = \frac{1(2^4 - 1)}{2 - 1}$$

$$= \frac{16 - 1}{1} = 15$$

11. $a = 5$, $r = \frac{1}{5}$, $n = 4$

$$S_n = \frac{a(r^n - 1)}{r - 1}$$

$$S_4 = \frac{5[\left(\tfrac{1}{5}\right)^4 - 1]}{\tfrac{1}{5} - 1}$$

$$= \frac{5\left(-\tfrac{624}{625}\right)}{-\tfrac{4}{5}}$$

$$= \frac{-\tfrac{624}{125}}{-\tfrac{4}{5}}$$

$$= \left(-\tfrac{624}{125}\right)\left(-\tfrac{5}{4}\right)$$

$$= \frac{156}{25}$$

13. $a = 128$, $r = -\frac{3}{2}$, $n = 4$

$$S_n = \frac{a(r^n - 1)}{r - 1}$$

$$S_4 = \frac{128\left[\left(-\frac{3}{2}\right)^4 - 1\right]}{-\frac{3}{2} - 1}$$

$$= \frac{128\left(\frac{65}{16}\right)}{-\frac{5}{2}}$$

$$= -208$$

15. $s_{\overline{n}|i} = \frac{(1 + i)^n - 1}{i}$

$$s_{\overline{12}|.05} = \frac{(1 + .05)^{12} - 1}{.05}$$

$$\approx 15.91713$$

17. $s_{\overline{n}|i} = \frac{(1 + i)^n - 1}{i}$

$$s_{\overline{16}|.04} = \frac{(1.04)^{16} - 1}{.04}$$

$$\approx 21.82453$$

19. $s_{\overline{n}|i} = \frac{(1 + i)^n - 1}{i}$

$$s_{\overline{20}|.01} = \frac{(1.01)^{20} - 1}{.01}$$

$$\approx 22.01900$$

23. $R = 1000$, $i = .06$, $n = 5$

Use the formula for the future value of an ordinary annuity.

$$S = R \cdot s_{\overline{n}|i}$$

$$= 1000 \cdot s_{\overline{5}|.06}$$

$$= 1000 \cdot \frac{(1.06)^5 - 1}{.06}$$

$$\approx 1000(5.63709)$$

$$= \$5637.09$$

25. $R = 29,500$, $i = .05$, $n = 15$

$$S = R \cdot s_{\overline{n}|i}$$

$$= 29,500 \cdot s_{\overline{15}|.05}$$

$$\approx \$636,567.63$$

27. $R = 3700$, $i = \frac{.08}{2} = .04$,

$n = 2 \cdot 11 = 22$

$$S = R \cdot s_{\overline{n}|i}$$

$$= 3700 \cdot s_{\overline{22}|.04}$$

$$\approx \$126,717.49$$

29. $R = 4600$, $i = \frac{.08}{4} = .02$,

$n = 9 \cdot 4 = 36$

$$S = R \cdot s_{\overline{n}|i}$$

$$= 4600 \cdot s_{\overline{36}|.02}$$

$$\approx \$239,174.10$$

31. $R = 42,000$, $i = \frac{.10}{2} = .05$,

$n = 12 \cdot 2 = 24$

$$S = R \cdot s_{\overline{n}|i}$$

$$= 42,000 \cdot s_{\overline{24}|.05}$$

$$\approx \$1,869,084.00$$

33. $R = 1400$, $i = .08$, $n = 10$

To find the value of an annuity due, use the formula for the future value of an ordinary annuity, but include one additional time period and subtract the amount of one payment.

$$S = R \cdot s_{\overline{n+1}|i} - R$$

$$= 1400 \cdot s_{\overline{11}|.08} - 1400$$

$$\approx \$21,903.68$$

35. $R = 4000$, $i = .06$, $n = 11$

$$S = R \cdot s_{\overline{n+1}|i} - R$$

$$= 4000 \cdot s_{\overline{12}|.06} - 4000$$

$$\approx \$63,479.76$$

37. $R = 750$, $i = \dfrac{.06}{12} = .005$,

$n = 15 \cdot 12 = 180$

$$S = R \cdot s_{\overline{n+1}|i} - R$$

$$= 750 \cdot s_{\overline{181}|.005} - 750$$

$$\approx \$219,204.60$$

39. $R = 1500$, $i = \dfrac{.06}{2} = .03$,

$n = 11 \cdot 2 = 22$

$$S = R \cdot s_{\overline{n+1}|i} - R$$

$$= 1500 \cdot s_{\overline{23}|.03} - 1500$$

$$\approx \$47,179.32$$

41. $S = 100,000$, $i = \dfrac{.08}{2} = .04$,

$n = 9 \cdot 2 = 18$

Let R represent the amount to be
deposited into the sinking fund each
year. Solve the formula $S = R \cdot s_{\overline{n}|i}$
for R and proceed.

$$R = \frac{S}{s_{\overline{n}|i}}$$

$$= \frac{100,000}{s_{\overline{18}|.04}}$$

$$\approx \$3899.32$$

43. $S = 8500$, $i = .08$, $n = 7$

$$R = \frac{S}{s_{\overline{n}|i}}$$

$$= \frac{8500}{s_{\overline{7}|.08}}$$

$$\approx \$952.62$$

45. $S = 75,000$, $i = \dfrac{.06}{2} = .03$,

$n\left(4\tfrac{1}{2}\right)(2) = 9$

$$R = \frac{S}{s_{\overline{n}|i}}$$

$$= \frac{75,000}{s_{\overline{9}|.03}}$$

$$\approx \$7382.54$$

47. $S = 25,000$, $i = \dfrac{.05}{4} = .0125$,

$n = \left(3\tfrac{1}{2}\right)(4) = 14$

$$R = \frac{S}{s_{\overline{n}|i}}$$

$$= \frac{25,000}{s_{\overline{14}|.0125}}$$

$$\approx \$1645.13$$

49. $S = 9000$, $i = \dfrac{.12}{12} = .01$,

$n = \left(2\dfrac{1}{2}\right)(12) = 30$

$R = \dfrac{S}{s_{\overline{n}|i}}$

$= \dfrac{9000}{s_{\overline{30}|}.01}$

$\approx \$258.73$

51. $R = 12{,}000$, $i = .08$, $n = 9$

(a) $S = R \cdot s_{\overline{n}|i}$

$= 12{,}000 \cdot s_{\overline{9}|}.08$

$\approx \$149{,}850.69$

(b) $R = 12{,}000$, $i = .06$, $n = 9$

$S = R \cdot s_{\overline{n}|i}$

$= 12{,}000 \cdot s_{\overline{9}|}.06$

$\approx \$137{,}895.79$

(c) The amount that would be lost is the difference of the above two amounts, which is

$\$149{,}850.69 - \$137{,}895.79$

$= \$11{,}954.90.$

53. $R = 80$, $i = \dfrac{.075}{12} = .00625$,

$n = 3 \cdot 12 + 9 = 45$

Because the deposits are made at the beginning of each month, this is an annuity due.

$S = R \cdot s_{\overline{n+1}|i} - R$

$= 80 \cdot s_{\overline{46}|}.00625 - R$

≈ 4168.30

There will be about $\$4168.30$ in the account.

55. For the first 15 yr, we have an ordinary annuity with $R = 1000$,

$i = \dfrac{.08}{4} = .02$, and $n = 15 \cdot 4 = 60$.

The amount on deposit after 15 yr is

$S = R \cdot s_{\overline{n}|i}$

$= 1000 \cdot s_{\overline{60}|}.02$

$\approx 114{,}051.54.$

For the remaining 5 yr, this amount earns compound interest at 8% compounded quarterly.

To find the final amount on deposit, use the formula for the compound amount, with $P = 114{,}051.54$,

$i = \dfrac{.08}{4} = .02$, and $n = 5 \cdot 4 = 20$.

$A = P(1 + i)^n$

$= 114{,}051.54(1.02)^{20}$

$\approx 169{,}474.59$

The man will have about $\$169{,}474.59$ in the account when he retires.

57. For the first 8 yr, we have an annuity due, with $R = 2435$, $i = \dfrac{.06}{2} = .03$,

and $n = 8 \cdot 2 = 16$. The amount on deposit after 8 yr is

$S = R \cdot s_{\overline{n+1}|i} - R$

$= 2435 \cdot s_{\overline{17}|}.03 - 2435$

$\approx 2435(21.76159) - 2435$

$= 52{,}989.47 - 2435$

$= \$50{,}554.47.$

For the remaining 5 yr, this amount earns compound interest at 6% compounded semiannually. To find the final amount on deposit, use the formula for the compound amount, with

$P = 50,554.47$, $i = \dfrac{.06}{2} = .03$, and

$n = 5 \cdot 2 = 10$.

$$A = P(1 + i)^n$$
$$= 50,554.47(1.03)^{10}$$
$$\approx 67,940.98$$

The final amount on deposit will be about $67,940.98.

59.   **(a)** $S = 10,000$, $i = \dfrac{.08}{4} = .02$,

$n = 8 \cdot 4 = 32$

Let R represent the amount of each payment.

$$S = R \cdot s_{\overline{n}|i}$$
$$10,000 = R \cdot s_{\overline{32}|.02}$$
$$R \approx \frac{10,000}{44.22703}$$
$$\approx 226.11$$

If the money is deposited at 8% compounded quarterly, Berkowitz's quarterly deposit will need to be about $226.11.

(b) $S = 10,000$, $i = \dfrac{.06}{4} = .015$,

$n = 8 \cdot 4 = 32$

Let R represent the amount of each payment.

$$S = R \cdot s_{\overline{n}|i}$$
$$10,000 = R \cdot s_{\overline{32}|.015}$$
$$R \approx \frac{10,000}{40.68829}$$
$$\approx 245.77$$

If the money is deposited at 6% compounded quarterly, Berkowitz's quarterly deposit will need to be about $245.77.

61. $S = 18,000$, $i = \dfrac{.05}{4} = .0125$,

$n = 6 \cdot 4 = 24$

Let R represent the amount of each payment.

$$S = R \cdot s_{\overline{n}|i}$$
$$18,000 = R \cdot s_{\overline{24}|.0125}$$
$$R = \frac{18,000}{s_{\overline{24}|.0125}}$$
$$\approx \frac{18,000}{27.78808}$$
$$\approx 647.76$$

She must deposit about $647.76 at the end of each quarter.

63. $R = \dfrac{2000}{2} = 1000$, $i = \dfrac{.08}{2} = .04$,

$n = 25 \cdot 2 = 50$

$$S = R \cdot s_{\overline{n}|i}$$
$$= 1000 \cdot s_{\overline{50}|.04}$$
$$\approx 152,667.08$$

There will be about $152,667.08 in the IRA.

65. $R = \dfrac{2000}{2} = 1000$, $i = \dfrac{.10}{2} = .05$,

$n = 25 \cdot 2 = 50$

$$S = R \cdot s_{\overline{n}|i}$$

$$= 1000 \cdot s_{\overline{50}|.05}$$

$$\approx 209{,}348.00$$

There will be about \$209,348 in the IRA.

67. This exercise should be solved by computer methods. The solution will vary according to the computer program that is used. The answers are as follows.

(a) The buyer's quarterly interest payment will be \$1200.

(b) The buyer's semiannual payments into the sinking fund will be \$3511.58 for each of the first 13 payments and \$3511.59 for the last payment.

(c) A table showing the amount in the sinking fund after each deposit is as follows.

Payment Number	Amount of Deposit	Interest Earned	Total
1	\$3511.58	\$0	\$3511.58
2	\$3511.58	\$105.35	\$7128.51
3	\$3511.58	\$213.86	\$10,853.95
4	\$3511.58	\$325.62	\$14,691.15
5	\$3511.58	\$440.73	\$18,643.46
6	\$3511.58	\$559.30	\$22,714.34
7	\$3511.58	\$681.43	\$26,907.35
8	\$3511.58	\$807.22	\$31,226.15
9	\$3511.58	\$936.78	\$35,674.51
10	\$3511.58	\$1070.24	\$40,256.33
11	\$3511.58	\$1207.69	\$44,975.60
12	\$3511.58	\$1349.27	\$49,836.45
13	\$3511.58	\$1495.09	\$54,843.12
14	\$3411.59	\$1645.29	\$60,000.00

Section 9.4

1. $$\dfrac{1 - (1 + i)^{-n}}{i}$$

is represented by $a_{\overline{n}|i}$, and it is choice (c).

3. $$a_{\overline{n}|i} = \dfrac{1 - (1 + i)^{-n}}{i}$$

$$a_{\overline{15}|.06} = \dfrac{1 - (1 + .06)^{-15}}{.06}$$

$$= \dfrac{1 - (1.06)^{-15}}{.06}$$

$$\approx 9.71255$$

5. $$a_{\overline{n}|i} = \dfrac{1 - (1 + i)^{-n}}{i}$$

$$a_{\overline{18}|.04} = \dfrac{1 - (1.04)^{-18}}{.04}$$

$$\approx 12.65930$$

7. $$a_{\overline{n}|i} = \dfrac{1 - (1 + i)^{-n}}{i}$$

$$a_{\overline{16}|.01} = \dfrac{1 - (1.01)^{-16}}{.01}$$

$$\approx 14.71787$$

11. $R = 1400$, $i = .08$, $n = 8$

Use the formula for the present value of an annuity.

$$P = R \cdot a_{\overline{n}|i}$$

$$P = 1400 \cdot a_{\overline{8}|.08}$$

$$\approx 1400(5.746639)$$

$$\approx \$8045.30$$

13. $R = 50,000$, $i = \dfrac{.08}{4} = .02$,

$n = 10 \cdot 4 = 40$

$$P = R \cdot a_{\overline{n}|i}$$
$$P = 50,000 \cdot a_{\overline{40}|}.02$$
$$\approx 50,000(27.35548)$$
$$\approx \$1,367,774.00$$

15. $R = 18,579$, $i = \dfrac{.094}{2} = .047$,

$n = 8 \cdot 2 = 16$

$$P = R \cdot a_{\overline{n}|i}$$
$$P = 18,579 \cdot a_{\overline{16}|}.047$$
$$\approx 18,579(11.072953)$$
$$\approx \$205,724.40$$

17. The lump sum is the same as the present value of the annuity.

$R = 10,000$, $i = .05$, $n = 15$

$$P = R \cdot a_{\overline{n}|i}$$
$$P = 10,000 \cdot a_{\overline{15}|}.05$$
$$\approx 10,000(10.37966)$$
$$\approx \$103,796.60$$

21. $P = 41,000$, $i = \dfrac{.10}{2} = .05$, $n = 10$

Let R be the amount of each payment.

$$P = R \cdot a_{\overline{n}|i}$$
$$R = \frac{P}{a_{\overline{n}|i}}$$
$$R = \frac{41,000}{a_{\overline{10}|}.05}$$
$$\approx \frac{41,000}{7.72173}$$
$$\approx \$5309.69$$

23. $P = 140,000$, $i = \dfrac{.12}{4} = .03$, $n = 15$

Let R be the amount of each payment.

$$R = \frac{P}{a_{\overline{n}|i}}$$
$$R = \frac{140,000}{a_{\overline{15}|}.03}$$
$$\approx \frac{140,000}{11.93794}$$
$$\approx \$11,727.32$$

25. $P = 5500$, $i = \dfrac{.12}{12} = .01$, $n = 24$

Let R be the amount of each payment.

$$R = \frac{P}{a_{\overline{n}|i}}$$
$$R = \frac{5500}{a_{\overline{24}|}.01}$$
$$\approx \frac{5500}{21.24339}$$
$$\approx \$258.90$$

27. The Portion to Principal column of the table indicates that $87.10 of the 11th payment of $88.85 is used to reduce the debt.

29. The amount of interest paid in the last 4 months of the loan is

$$3.47 + 2.61 + 1.75 + .88 = \$8.71.$$

31. $4000 deposited every 6 months for 10 yr at 6% compounded semiannually will be worth $4000 \cdot s_{\overline{20}|}.03 \approx \$107,481.48$.

(Note that $i = \frac{.06}{2}$ and

$n = 10 \cdot 2 = 20$.) For the lump sum investment of x dollars, use

$i = \frac{.08}{4} = .02$ and $n = 10 \cdot 4 = 40$ in

the formula $A = P(1 + i)^n$.
Our unknown amount x will be worth $x(1.02)^{40}$, so

$$x(1.02)^{40} = 107,481.48$$
$$x \approx 48,677.34.$$

About \$48,677.34 should be invested today.

33. $P = 170,892$, $i = \frac{.0811}{12} = .006758$,

$n = 30 \cdot 12 = 360$

$$R = \frac{P}{a_{\overline{n}|i}}$$

$$R = \frac{170,892}{a_{\overline{360}|}.006758}$$

$$\approx \frac{170,892}{134.87644}$$

$$\approx \$1267.07$$

35. $P = 96,511$, $i = \frac{.0957}{12} = .007975$,

$n = 25 \cdot 12 = 300$

$$R = \frac{P}{a_{\overline{n}|i}}$$

$$R = \frac{96,511}{a_{\overline{300}|}.007975}$$

$$\approx \frac{96,511}{113.82165}$$

$$\approx \$847.91$$

37.   **(a)** $P = 6000$, $i = \frac{.12}{12} = .01$,

$n = 4 \cdot 12 = 48$

Let R be the amount of each payment.

$$R = \frac{P}{a_{\overline{n}|i}}$$

$$R = \frac{6000}{a_{\overline{48}|}.01}$$

$$\approx \frac{6000}{37.97396}$$

$$\approx \$158.00$$

(b) 48 payments of \$158.00 are made, and 48(\$158.00) = \$7584. The total amount of interest Le will pay is \$7584 - \$6000 = \$1584.00.

39.   This is just the present value of the annuity with $R = 50,000$, $i = .06$, and $n = 20$.

$$P = R \cdot a_{\overline{n}|i}$$
$$P = 50,000 \cdot a_{\overline{20}|}.06$$

$$\approx 50,000(11.46992)$$

$$\approx 573,496.00$$

The lump sum that the management must invest is about \$573,496.00.

41. $P = 72,000$, $i = \frac{.10}{2} = .05$, $n = 9$

$R = \frac{72,000}{a_{\overline{9}|}.05} \approx \$10,129.69$ is the amount

of each payment.

Payment Number	Amount of Payment	Interest for Period	Portion to Principal	Principal at End of Period
0	———	———	———	$72,000.00
1	$10,129.69	$3600.00	$6529.69	$65,470.31
2	$10,129.69	$3273.52	$6856.17	$58,614.14
3	$10,129.69	$2930.71	$7198.98	$51,415.16
4	$10,129.69	$2570.76	$7558.93	$43,856.23

43. The loan is for 14,000 + 7200 − 1200 = $20,000.

$P = 20,000$, $i = \dfrac{.12}{2} = .06$, $n = 5 \cdot 2 = 10$

$R = \dfrac{20,000}{a_{\overline{10}|.06}} \approx \$2717.36$ is the amount of each payment.

Payment Number	Amount of Payment	Interest for Period	Portion to Principal	Principal at End of Period
0	———	———	———	$20,000.00
1	$2717.36	$1200.00	$1517.36	$18,482.64
2	$2717.36	$1108.96	$1608.40	$16,874.24
3	$2717.36	$1012.45	$1704.91	$15,169.33
4	$2717.36	$910.16	$1807.20	$13,362.13

45. This is an amortization problem with $P = 25,000$. R represents the amount of each annual withdrawal.

(a) $i = .06$, $n = 8$

$$R = \frac{P}{a_{\overline{n}|i}}$$

$$R = \frac{25,000}{a_{\overline{8}|.06}} \approx \frac{25,000}{6.20979} \approx 4025.90$$

He will be able to withdraw about $4025.90 per month for the 8 yr.

(b) $i = .06$, $n = 12$

$$R = \frac{P}{a_{\overline{n}|i}}$$

$$R = \frac{25,000}{a_{\overline{12}|.06}}$$

$$\approx \frac{25,000}{8.38384}$$

$$\approx 2981.93$$

He will be able to withdraw about \$2981.93 per month for the 12 yr.

47. This exercise should be solved by computer methods. The solution will vary according to the computer program that is used. The amortization schedule is as follows.

Payment Number	Amount of Payment	Interest for Period	Portion to Principal	Principal at End of Period
1	\$5783.49	\$3225.54	\$2557.95	\$35,389.55
2	\$5783.49	\$3008.11	\$2775.38	\$32,614.17
3	\$5783.49	\$2772.20	\$3011.29	\$29,602.88
4	\$5783.49	\$2516.24	\$3267.25	\$26,335.63
5	\$5783.49	\$2238.53	\$3544.96	\$22,790.67
6	\$5783.49	\$1937.21	\$3846.28	\$18,944.39
7	\$5783.49	\$1610.27	\$4173.22	\$14,771.17
8	\$5783.49	\$1255.55	\$4527.94	\$10,243.22
9	\$5783.49	\$870.67	\$4912.82	\$5,330.41
10	\$5783.49	\$453.08	\$5530.41	\$0

Chapter 9   Review Exercises

1. $I = Prt$

$= 15,903(.08)\left(\frac{8}{12}\right)$

$= \$848.16$

3. $I = Prt$

$= 42,368(.0522)\left(\frac{5}{12}\right)$

$\approx \$921.50$

7. $P = \dfrac{A}{1 + rt}$

$= \dfrac{459.57}{1 + .045\left(\frac{7}{12}\right)}$

$\approx \$447.81$

9. $P = A(1 - rt)$

$= 802.34\left[1 - (.086)\left(\frac{11}{12}\right)\right]$

$\approx \$739.09$

11. For a given amount of money at a given interest rate for a given time period greater than 1, compound interest produces more interest than simple interest.

13. $A = P(1 + i)^n$

$= 19,456.11\left(1 + \frac{.12}{2}\right)^{2(7)}$

$= 19,456.11(1.06)^{14}$

$\approx 19,456.11(2.26090)$

$\approx \$43,988.40$

15. $A = P(1 + i)^n$

$= 57,809.34\left(1 + \frac{.12}{4}\right)^{4(5)}$

$= 57,809.34(1.03)^{20}$

$\approx 57,809.34(1.80611)$

$\approx \$104,410.10$

17. $A = P(1 + i)^n$

$= 12,699.36\left(1 + \frac{.10}{2}\right)^{2(7)}$

$= 12,699.36(1.05)^{14}$

$\approx 12,699.36(1.97993)$

$\approx \$25,143.86$

The interest earned is

$\$25,143.86 - \$12,699.36$

$= \$12,444.50.$

19. $A = P(1 + i)^n$

$= 34,677.23\left(1 + \frac{.0972}{12}\right)^{32}$

$\approx 34,677.23(1.29454)$

$\approx \$44,891.08$

The interest earned is

$\$44,891.08 - \$34,677.23$

$= \$10,213.85.$

21. $P = \dfrac{A}{(1 + i)^n}$

$= \dfrac{17,650}{\left(1 + \frac{.08}{4}\right)^{4(4)}}$

$= \dfrac{17,650}{(1.02)^{16}}$

$\approx \dfrac{17,650}{1.37279}$

$\approx \$12,857.07$

23. $P = \dfrac{A}{(1 + i)^n}$

$= \dfrac{2388.90}{\left(1 + \dfrac{.0593}{12}\right)^{44}}$

$\approx \dfrac{2388.90}{(1.00494)^{44}}$

$\approx \dfrac{2388.90}{1.24222}$

$\approx \$1923.09$

25. $a = 4, \; r = \dfrac{1}{2}$

The first five terms are

$4, \; 4\left(\dfrac{1}{2}\right), \; 4\left(\dfrac{1}{2}\right)^2, \; 4\left(\dfrac{1}{2}\right)^3, \; 4\left(\dfrac{1}{2}\right)^4$

or $\; 4, \; 2, \; 1, \; \dfrac{1}{2}, \; \dfrac{1}{4}.$

27. $a = -2, \; r = -2$

The first five terms are

$-2, \; -2(-2), \; -2(-2)^2, \; -2(-2)^3, \; -2(-2)^4$

or $\; -2, \; 4, \; -8, \; 16, \; -32.$

The fifth term is -32.

29. $a = 8000, \; r = -\dfrac{1}{2}, \; n = 5$

$S_n = \dfrac{a(r^n - 1)}{r - 1}$

$S_5 = \dfrac{8000\left[\left(-\dfrac{1}{2}\right)^5 - 1\right]}{-\dfrac{1}{2} - 1}$

$= \dfrac{8000\left(-\dfrac{33}{32}\right)}{-\dfrac{3}{2}}$

$= \dfrac{-8250}{-\dfrac{3}{2}}$

$= (-8250)\left(-\dfrac{2}{3}\right)$

$= 5500$

31. $s_{\overline{n}|i} = \dfrac{(1 + i)^n - 1}{i}$

$s_{\overline{20}|.05} = \dfrac{(1.05)^{20} - 1}{.05}$

$\approx \dfrac{1.6532977}{.05}$

$= 33.06595$

33. $R = 500, \; i = \dfrac{.06}{2} = .03,$

$n = 8 \cdot 2 = 16$

This is an ordinary annuity.

$S = R \cdot s_{\overline{n}|i}$

$S = R \cdot s_{\overline{16}|.03}$

$\approx 500(20.15688)$

$\approx \$10,078.44$

35. $R = 4000, \; i = \dfrac{.06}{4} = .015,$

$n = 7 \cdot 4 = 28$

This is an ordinary annuity.

$S = R \cdot s_{\overline{n}|i}$

$S = R \cdot s_{\overline{28}|.015}$

$\approx 4000(34.48148)$

$\approx \$137,925.91$

37. $R = 672, \; i = \dfrac{.08}{4} = .02,$

$n = 7 \cdot 4 = 28$

This is an annuity due, so we calculate $s_{\overline{n}|i}$ for one additional payment.

$R \cdot s_{\overline{29}|.02} \approx 672(38.79223)$

$\approx \$26,068.38$

Now subtract the amount of one payment to find the future value.

$$S = \$26,068.38 - \$672 = \$25,396.38$$

41. $S = 57,000$, $i = \dfrac{.06}{2} = .03$,

$n = \left(8\frac{1}{2}\right)(2) = 17$

Let R represent the amount of each paymemt.

$$S = R \cdot s_{\overline{n}|i}$$

$$57,000 = R \cdot s_{\overline{17}|}.03$$

$$R = \frac{57,000}{s_{\overline{17}|}.03}$$

$$\approx \frac{57,000}{21.76159}$$

$$\approx \$2619.29$$

43. $S = 1,056,788$, $i = \dfrac{.0812}{12} \approx .0067667$,

$n = \left(4\frac{1}{2}\right)(12) = 54$

Let R represent the amount of each payment.

$$S = R \cdot s_{\overline{n}|i}$$

$$1,056,788 = R \cdot s_{\overline{54}|}.0067667$$

$$R = \frac{1,056,788}{s_{\overline{54}|}.0067667}$$

$$\approx \frac{1,056,788}{64.92885}$$

$$\approx \$16,277.35$$

45. $R = 1500$, $i = \dfrac{.08}{4} = .02$,

$n = 7(4) = 28$

$$P = R \cdot a_{\overline{n}|i}$$

$$P = 1500 \cdot a_{\overline{28}|}.02$$

$$\approx 1500(21.28127)$$

$$\approx \$31,921.91$$

47. $R = 877.34$, $i = \dfrac{.094}{12} \approx .007833$,

$n = 17$

$$P = R \cdot a_{\overline{n}|i}$$

$$P = 877.34 \cdot a_{\overline{17}|}.007833$$

$$\approx 877.34(15.85875)$$

$$\approx \$13,913.48$$

49. $P = 80,000$, $i = .08$, $n = 9$

Let R represent the amount of each payment.

$$P = R \cdot a_{\overline{n}|i}$$

$$80,000 = R \cdot a_{\overline{9}|}.08$$

$$R = \frac{80,000}{a_{\overline{9}|}.08}$$

$$\approx \frac{80,000}{6.24689}$$

$$= \$12,806.38$$

51. $P = 32,000$, $i = \dfrac{.094}{4} = .0235$, $n = 17$

Let R represent the amount of each payment.

$$P = R \cdot a_{\overline{n}|i}$$

$$32,000 = R \cdot a_{\overline{17}|}.0235$$

$$R = \frac{32,000}{a_{\overline{17}|}.0235}$$

$$\approx \frac{32,000}{13.88246}$$

$$\approx \$2305.07$$

53. $P = 56,890$, $i = \frac{.1074}{12} = .00895$,

$n = 25 \cdot 12 = 300$

Let R represent the amount of each
payment.

$$P = R \cdot a_{\overline{n}|i}$$

$$56,890 = R \cdot a_{\overline{300}|.00895}$$

$$R = \frac{56,890}{a_{\overline{300}|.00895}}$$

$$\approx \frac{56,890}{104.0178}$$

$$\approx \$546.93$$

55. $896.06 of the fifth payment of
$1022.64 is interest.

57. In the first 3 months of the loan,
the total amount of interest paid is

$$899.58 + 898.71 + 897.83$$
$$= \$2696.12.$$

59. Find the maturity value of the loan,
the amount Tom Wilson must pay his
mother.

$$A = P(1 + rt)$$

$$= 5800\left[1 + .10\left(\frac{10}{12}\right)\right]$$

$$\approx 5800(1.08333)$$

$$\approx 6283.33$$

The bank will apply its discount rate
to the 6283.33, to obtain

Amount of discount

$$= 6283.33(.1345)\left(\frac{3}{12}\right)$$

$$\approx 211.28.$$

Tom's mother will receive from the
bank

$$\$6283.33 - \$211.28 = \$6072.05,$$

which is enough to buy the furniture.

61. $I = Prt$

$$t = \frac{I}{Pr}$$

$$= \frac{3255}{28,000(.115)}$$

$$\approx 1.011$$

The loan is for about 1.011 yr;
convert this to months.

$$(1.011 \text{ yr})\left(\frac{12 \text{ mo}}{1 \text{ yr}}\right) \approx 12.13 \text{ mo}$$

The loan is for about 12.13 months.

63. $A = 7500$, $i = \frac{.10}{2} = .05$, $n = 3(2) = 6$

Let P represent the lump sum.

$$A = P(1 + i)^n$$

$$P = \frac{A}{(1 + i)^n}$$

$$= \frac{7500}{(1.05)^6}$$

$$\approx \frac{7500}{1.34}$$

$$\approx 5596.62$$

She should deposit about $5596.62
today.

65. $R = 5000$, $i = \frac{.10}{2} = .05$,

$n = \left(7\frac{1}{2}\right)(2) = 15$

This is an ordinary annuity.

$$S = R \cdot s_{\overline{n}|i}$$

$$S = 5000 \cdot s_{\overline{15}|}.05$$

$$\approx 5000(21.57856)$$

$$\approx \$107{,}892.82$$

The amount of interest earned is

$$\$107{,}892.82 - 15(\$5000)$$

$$= \$32{,}892.82.$$

67. $P = 48{,}000$, $i = .10$, $n = 7$

Let R represent the amount of each payment.

$$P = R \cdot a_{\overline{n}|i}$$

$$R = \frac{P}{a_{\overline{n}|i}}$$

$$= \frac{48{,}000}{a_{\overline{7}|}.10}$$

$$\approx \frac{48{,}000}{4.86842}$$

$$\approx 9859.46$$

The owner should deposit about $9859.46 at the end of each year.

69. $P = 115{,}700$, $i = \frac{.105}{12} = .00875$,

$n = 300$

Let R represent the amount of each payment.

$$R = \frac{P}{a_{\overline{n}|i}}$$

$$R = \frac{115{,}700}{a_{\overline{300}|}.00875}$$

$$\approx \frac{115{,}700}{105.91182}$$

$$\approx 1092.42$$

Each monthly payment will be about $1092.42. The total amount of interest will be

$$300(\$1092.42) - \$115{,}700$$

$$= \$212{,}026.$$

71. The death benefit grows to

$$10{,}000(1.05)^7 \approx 14{,}071.00.$$

This 14,071 is the present value of an annuity due with $P = 14{,}071$,

$i = \frac{.03}{12} = .0025$, and $n = 120$. Let x

represent the amount of each monthly payment.

$$P = R \cdot a_{\overline{n+1}|i} - R$$

$$14{,}071 = x \cdot a_{\overline{121}|}.0025 - x$$

$$14{,}071 = (a_{\overline{121}|}.0025 - 1)x$$

$$14{,}071 \approx (104.301 - 1)x$$

$$14{,}071 \approx 103.301x$$

$$135 \approx x$$

Each payment is about $135, which corresponds to choice (d).

Extended Application

Note: Since the data for Exercises 1–4 is given to the nearest dollar, answers should be rounded to the nearest dollar.

1. Cash flow

$$= -.52(6228 + 2976) + .52(26{,}251)$$

$$+ .48(10{,}778)$$

$$= -4786.08 + 13{,}650.52 + 5173.44$$

$$= \$14{,}038$$

2. Cash flow

   = −.52(6228 + 2976) + .52(26,251)

     + .48(1347)

   = −4786.08 + 13,650.52 + 646.56

   = $9511

3. Cash flow

   = −.52(6386 + 2870) + .52(26,251)

     + .48(0)

   = −4813.12 + 13,650.52 + 0

   = $8837

4. Cash flow

   = −.52(6228 + 2976) + .52(10,618)

     + .48(6736)

   = −4786.08 + 5521.36 + 3233.28

   = $3968

CHAPTER 9   TEST

1. Find the simple interest on $1252 at 5% for 11 months.

2. Using a 360 day year, find the simple interest on $12,000 at 6.25% for 170 days.

3. Find the present value for the following future amount:

   $14,500 in 9 months at 7.2% simple interest.

4. Find the compound amount if $7000 is deposited for 8 yr in an account paying 6% per year compounded quarterly.

5. Find the present value of $8000 at 8.1% compounded monthly for 2 yr.

6. Find the fifth term of the geometric sequence with a = 4.7, r = 2.

7. Ralph deposits $100 at the end of each month for 3 yr in an account paying 6% compounded monthly. He then uses the money from this account to buy a certificate of deposit which pays 7.5% compounded annually. If he redeems the certicicate 4 yr later, how much money will he receive?

8. What amount must be deposited at the end of each quarter at 8% compounded quarterly to have $10,000 in 10 yr?

9. Find the present value of an ordinary annuity with payments of $800 per month at 6% compounded monthly for 4 yr.

10. Ms Morroco borrows $12,000 at 8% compounded quarterly, to be paid off with equal quarterly payments over 2 yr.

    (a) What quarterly payment is needed to amortize this loan?

    (b) Prepare an amortization table for the first 4 payments.

CHAPTER 9   TEST ANSWERS

1. $57.38   2. $354.17   3. $13,757.12   4. $11,272.27

5. $6807.23   6. 75.2   7. $5253.21   8. $165.56

9. $34,064.25

10. (a) $1638.12

(b)

Payment Number	Amount of Payment	Interest for Period	Portion to Principal	Principal at End of Period
0	———	———	———	$12,000.00
1	$1638.12	$240.00	$1398.12	$10,601.88
2	$1638.12	$212.04	$1426.08	$9175.80
3	$1638.12	$183.52	$1454.60	$7721.19
4	$1638.12	$154.42	$1483.70	$6237.49

CHAPTER 10   THE DERIVATIVE

Section 10.1

1. By reading the graph, as x gets
    closer to 3 from the left or the
    right, f(x) gets closer to 3.

    $$\lim_{x \to 3} f(x) = 3$$

3. By reading the graph, as x gets
    closer to -2 from the left, f(x)
    approaches -1. As x gets closer to
    -2 from the right, f(x) approaches
    1/2. Since these two values of f(x)
    are not equal,

    $$\lim_{x \to -2} f(x)$$

    does not exist.

5. By reading the graph, as x ap-
    proaches 1 from the left or the
    right, g(x) approaches 1.

    $$\lim_{x \to 1} g(x) = 1$$

9. From the table, as x approaches 1
    from the left or the right, f(x)
    approaches 2.

    $$\lim_{x \to 1} f(x) = 2$$

11. $k(x) = \dfrac{x^3 - 2x - 4}{x - 2}$, find $\lim_{x \to 2} k(x)$.

x	1.9	1.99	1.999
k(x)	9.41	9.9401	9.9941

x	2.001	2.01	2.1
k(x)	10.006	10.0601	10.61

As x approaches 2 from the left or
the right, k(x) approaches 10.

$$\lim_{x \to 2} k(x) = 10$$

13. $h(x) = \dfrac{\sqrt{x} - 2}{x - 1}$, find $\lim_{x \to 1} h(x)$.

x	.9	.99	.999
h(x)	-29.486833	-299.49874	-2999.4999

x	1.001	1.01	1.1
h(x)	3000.49988	300.49876	30.488088

$$\lim_{x \to 1^-} = -\infty$$

$$\lim_{x \to 1^+} = \infty$$

Thus, $\lim_{x \to 1} h(x)$ does not exist.

In Exercises 15-23, $\lim_{x \to 4} f(x) = 16$ and
$\lim_{x \to 4} g(x) = 8$.

15. $\lim_{x \to 4} [f(x) - g(x)]$

    $= \lim_{x \to 4} f(x) - \lim_{x \to 4} g(x)$

    $= 16 - 8 = 8$

17. $\lim_{x \to 4} \dfrac{f(x)}{g(x)}$

    $= \dfrac{\lim_{x \to 4} f(x)}{\lim_{x \to 4} g(x)}$

    $= \dfrac{16}{8} = 2$

19. $\lim\limits_{x \to 4} \sqrt{f(x)}$

$\quad = \lim\limits_{x \to 4} [f(x)]^{1/2}$

$\quad = [\lim\limits_{x \to 4} f(x)]^{1/2}$

$\quad = 16^{1/2} = 4$

21. $\lim\limits_{x \to 4} [g(x)]^3$

$\quad = [\lim\limits_{x \to 4} g(x)]^3$

$\quad = 8^3 = 512$

23. $\lim\limits_{x \to 4} \dfrac{f(x) + g(x)}{2g(x)}$

$\quad = \dfrac{\lim\limits_{x \to 4} [f(x) + g(x)]}{\lim\limits_{x \to 4} 2g(x)}$

$\quad = \dfrac{\lim\limits_{x \to 4} f(x) + \lim\limits_{x \to 4} g(x)}{2 \lim\limits_{x \to 4} g(x)}$

$\quad = \dfrac{16 + 8}{2(8)} = \dfrac{24}{16} = \dfrac{3}{2}$

25. $\lim\limits_{x \to 3} \dfrac{x^2 - 9}{x - 3}$

$\quad = \lim\limits_{x \to 3} \dfrac{(x - 3)(x + 3)}{x - 3}$

$\quad = \lim\limits_{x \to 3} (x + 3)$

$\quad = \lim\limits_{x \to 3} x + \lim\limits_{x \to 3} 3$

$\quad = 3 + 3$

$\quad = 6$

27. $\lim\limits_{x \to -2} \dfrac{x^2 - x - 6}{x + 2}$

$\quad = \lim\limits_{x \to -2} \dfrac{(x - 3)(x + 2)}{x + 2}$

$\quad = \lim\limits_{x \to -2} (x - 3)$

$\quad = \lim\limits_{x \to -2} x + \lim\limits_{x \to -2} (-3)$

$\quad = -2 - 3$

$\quad = -5$

29. $\lim\limits_{x \to 0} \dfrac{x^3 - 4x^2 + 8x}{2x}$

$\quad = \lim\limits_{x \to 0} \dfrac{x(x^2 - 4x + 8)}{2x}$

$\quad = \lim\limits_{x \to 0} \dfrac{x^2 - 4x + 8}{2}$

$\quad = \dfrac{0^2 - 4(0) + 8}{2} = 4$

31. $\lim\limits_{x \to 0} \dfrac{\dfrac{1}{x + 3} - \dfrac{1}{3}}{x}$

$\quad = \lim\limits_{x \to 0} \left(\dfrac{1}{x + 3} - \dfrac{1}{3}\right)\left(\dfrac{1}{x}\right)$

$\quad = \lim\limits_{x \to 0} \left[\dfrac{3}{3(x + 3)} - \dfrac{x + 3}{3(x + 3)}\right]\left(\dfrac{1}{x}\right)$

$\quad = \lim\limits_{x \to 0} \dfrac{3 - x - 3}{3(x + 3)(x)}$

$\quad = \lim\limits_{x \to 0} \dfrac{-x}{3(x + 3)x}$

$\quad = \lim\limits_{x \to 0} \dfrac{-1}{3(x + 3)}$

$\quad = \dfrac{-1}{3(0 + 3)}$

$\quad = -\dfrac{1}{9}$

33. $\lim\limits_{x \to 25} \dfrac{\sqrt{x} - 5}{x - 25}$

$\quad = \lim\limits_{x \to 25} \dfrac{\sqrt{x} - 5}{x - 25} \cdot \dfrac{\sqrt{x} + 5}{\sqrt{x} + 5}$

$\quad = \lim\limits_{x \to 25} \dfrac{x - 25}{(x - 25)(\sqrt{x} + 5)}$

$\quad = \lim\limits_{x \to 25} \dfrac{1}{\sqrt{x} + 5}$

$\quad = \dfrac{1}{\sqrt{25} + 5}$

$\quad = \dfrac{1}{10}$

35. $\lim\limits_{h \to 0} \dfrac{(x + h)^2 - x^2}{h}$

$= \lim\limits_{h \to 0} \dfrac{x^2 + 2hx + h^2 - x^2}{h}$

$= \lim\limits_{h \to 0} \dfrac{2hx + h^2}{h}$

$= \lim\limits_{h \to 0} \dfrac{h(2x + h)}{h}$

$= \lim\limits_{h \to 0} (2x + h)$

$= 2x + 0$

$= 2x$

37. $F(x) = \dfrac{3x}{(x + 2)^3}$

(a)

x	-3	-2.1	-2.01
F(x)	9	6300	6,030,000

x	-1.99	-1.9	-1
F(x)	-5,970,000	-5700	-3

$\lim\limits_{x \to -2^-} F(x) = \infty$

$\lim\limits_{x \to -2^+} F(x) = -\infty$

Thus, $\lim\limits_{x \to -2} F(x)$ does not exist.

(b) The vertical asymptote occurs when $(x + 2)^3 = 0$, or $x = -2$.

(c) If $x = a$ is an asymptote for the graph of $f(x)$, then $\lim\limits_{x \to a} f(x)$ does not exist.

39. By reading the graph, the function is discontinuous at $x = -1$.
f(-1) does not exist, since there is no value shown on the graph for $x = -1$.

$\lim\limits_{x \to -1} f(x) = \dfrac{1}{2}$ because as x approaches -1 from the left or the right, f(x) approaches $\dfrac{1}{2}$.

41. By reading the graph, the function is discontinuous at $x = 1$.
f(1) = 2, since the heavy dot above $x = 1$ is at 2.
$\lim\limits_{x \to 1} f(x) = -2$ because as x approaches 1 from the left or the right, f(x) approaches -2.

43. By reading the graph, the function is discontinuous at $x = -5$ and $x = 0$.
f(-5) does not exist since $x = -5$ is a vertical asymptote.
f(0) does not exist since there is no value shown on the graph for $x = 0$.
$\lim\limits_{x \to -5} f(x)$ does not exist since as x approaches -5 from the left, f(x) becomes infinitely large and as x approaches -5 from the right, f(x) becomes infinitely small.
$\lim\limits_{x \to 0} f(x) = 0$ because as x approaches 0 from the left or the right, f(x) approaches 0.

45. $g(x) = \dfrac{1}{x(x - 2)}$;

$x = 0$, $x = 2$, $x = 4$
Since g is a rational function, it will not be continuous when $x(x - 2) = 0$. Therefore, g is not

continuous for x = 0 or x = 2, but
is continuous at all other points,
including x = 4.

47. $k(x) = \dfrac{5 + x}{2 + x}$;

x = 0, x = -2, x = -5
Since k is a rational function,
it will not be continuous when
2 + x = 0. Therefore, k is not
continuous for x = -2, but is
continuous at all other points,
including x = 0 and x = -5.

49. $g(x) = \dfrac{x^2 - 4}{x - 2}$;

x = 0, x = 2, x = -2
Since g is a rational function,
it will not be continuous when
x - 2 = 0. Therefore, g is not
continuous for x = 2, but is
continuous at all other points,
including x = 0 and x = -2.

51. $p(x) = x^2 - 4x + 11$;

x = 0, x = 2, x = -1
Since p is a polynomial function,
it is continuous for all real values
of x, including x = 0, x = 2, and
x = -1.

53. $p(x) = \dfrac{|x + 2|}{x + 2}$; x = -2, x = 0, x = 2

Since p is a rational function, it
is not continuous at x = -2, but
is continuous at all other points,
including x = 0 and x = 2.

55. (a) $\lim\limits_{t \to 12} G(t)$

As t approaches 12, the value of
G(t) for the corresponding point
on the graph approaches 3.
Thus, $\lim\limits_{t \to 12} G(t) = 3$.

   (b) $\lim\limits_{t \to 16} G(t)$

   $\lim\limits_{t \to 16^+} G(t) = 1.5$

   $\lim\limits_{t \to 16^-} G(t) = 2$

Since $\lim\limits_{t \to 16^+} G(t) \neq \lim\limits_{t \to 16^-} G(t)$,

$\lim\limits_{t \to 16} G(t)$ does not exist.

   (c) G(16) is the value of function
       G(t) when t = 16. This value
       occurs at the heavy dot on the
       graph.
       $$G(16) = 2$$

   (d) The tipping point occurs at the
       break in the graph or when
       t = 16 months.

57. In dollars,

C(x) = 4x if 0 < x ≤ 150
C(x) = 3x if 150 < x ≤ 400
C(x) = 2.5x if 400 < x.

   (a) C(130) = 4(130) = $520

   (b) C(150) = 4(150) = $600

   (c) C(210) = 3(210) = $630

   (d) C(400) = 3(400) = $1200

   (e) C(500) = 2.5(500) = $1250

   (f) C is discontinuous at x = 150
       and x = 400 because those rep-
       resent points of price change.

59. In dollars,

$C(t) = 30t$ if $0 < t \leq 5$

$C(t) = 30(5) = 150$ if $t = 6$ or $t = 7$

$C(t) = 150 + 30(t - 7)$ if $7 < t \leq 12$.

The average cost per day is

$$A(t) = \frac{C(t)}{t}.$$

(a) $A(4) = \dfrac{30(4)}{4} = \$30$

(b) $A(5) = \dfrac{30(5)}{5} = \$30$

(c) $A(6) = \dfrac{150}{6} = \$25$

(d) $A(7) = \dfrac{150}{7} \approx \$21.43$

(e) $A(8) = \dfrac{150 + 30(8 - 7)}{8}$

$\quad\quad = \dfrac{180}{8} = \$22.50$

(f) $\lim\limits_{x \to 5^-} A(t) = 30$ because as t

approaches 5 from the left, $A(t)$ approaches 30 (think of the graph for $t = 1, 2, \ldots, 5$).

(g) $\lim\limits_{x \to 5^+} A(t) = 25$ because as t

approaches 5 from the right, $A(t)$ approaches 25.

(h) A is discontinuous at $t = 5$, $t = 6$, $t = 7$, and so on, because the average cost will differ for each different rental length.

61. The function is discontinuous at $t = m$, since the break in the graph occurs at that point.

Section 10.2

1. $y = x^2 + 2x = f(x)$ between $x = 0$ and $x = 3$

Average rate of change

$= \dfrac{f(3) - f(0)}{3 - 0}$

$= \dfrac{15 - 0}{3}$

$= 5$

3. $y = 2x^3 - 4x^2 + 6x = f(x)$ between $x = -1$ and $x = 1$

Average rate of change

$= \dfrac{f(1) - f(-1)}{1 - (-1)}$

$= \dfrac{4 - (-12)}{2} = \dfrac{16}{2}$

$= 8$

5. $y = \sqrt{x} = f(x)$ between $x = 1$ and $x = 4$

Average rate of change

$= \dfrac{f(4) - f(1)}{4 - 1}$

$= \dfrac{2 - 1}{3}$

$= \dfrac{1}{3}$

7. $y = \dfrac{1}{x - 1} = f(x)$ between $x = -2$ and $x = 0$

Average rate of change

$= \dfrac{f(0) - f(-2)}{0 - (-2)}$

$= \dfrac{-1 - \left(-\dfrac{1}{3}\right)}{2} = \dfrac{-\dfrac{2}{3}}{2}$

$= -\dfrac{1}{3}$

For Exercises 9 and 11, $s(t) = t^2 + 5t + 2$.

9. $\lim\limits_{h \to 0} \dfrac{s(6 + h) - s(6)}{h}$

$= \lim\limits_{h \to 0} \dfrac{(6+h)^2 + 5(6+h) + 2 - [6^2 + 5(6) + 2]}{h}$

$= \lim\limits_{h \to 0} \dfrac{h^2 + 17h + 68 - 68}{h}$

$= \lim\limits_{h \to 0} \dfrac{h^2 + 17h}{h}$

$= \lim\limits_{h \to 0} \dfrac{h(h + 17)}{h}$

$= \lim\limits_{h \to 0} (h + 17)$

$= 17$

11. $\lim\limits_{h \to 0} \dfrac{s(10 + h) - s(10)}{h}$

$= \lim\limits_{h \to 0} \dfrac{(10+h)^2 + 5(10+h) + 2 - [100 + 50 + 2]}{h}$

$= \lim\limits_{h \to 0} \dfrac{h^2 + 25h + 102 - 102}{h}$

$= \lim\limits_{h \to 0} \dfrac{h(h + 25)}{h}$

$= \lim\limits_{h \to 0} (h + 25)$

$= 25$

13. $s(t) = t^3 + 2t + 9$

$\lim\limits_{h \to 0} \dfrac{s(4 + h) - s(4)}{h}$

$= \lim\limits_{h \to 0} \dfrac{(4+h)^3 + 2(4+h) + 9 - [4^3 + 2(4) + 9]}{h}$

$= \lim\limits_{h \to 0} \dfrac{4^3 + 3(4^2)h + 3(4h^2) + h^3 + 8 + 2h + 9 - (4^3 + 8 + 9)}{h}$

$= \lim\limits_{h \to 0} \dfrac{h^3 + 36h^2 + 48h + 2h}{h}$

$= \lim\limits_{h \to 0} \dfrac{h^3 + 36h^2 + 50h}{h}$

$= \lim\limits_{h \to 0} \dfrac{h(h^2 + 36h + 50)}{h}$

$= \lim\limits_{h \to 0} (h^2 + 36h + 50)$

$= 50$

15. $s(t) = -4t^2 - 6$ at $t = 2$

$\lim\limits_{h \to 0} \dfrac{s(2 + h) - s(2)}{h}$

$= \lim\limits_{h \to 0} \dfrac{-4(2 + h)^2 - 6 - [-4(2)^2 - 6]}{h}$

$= \lim\limits_{h \to 0} \dfrac{-4(4 + 4h + h^2) - 6 + 16 + 6}{h}$

$= \lim\limits_{h \to 0} \dfrac{-16h - 4h^2}{h}$

$= \lim\limits_{h \to 0} \dfrac{h(-16 - 4h)}{h}$

$= \lim\limits_{h \to 0} (-16 - 4h)$

$= -16$

17. $F(x) = x^2 + 2$ at $x = 0$

$\lim\limits_{h \to 0} \dfrac{F(0 + h) - F(0)}{h}$

$= \lim\limits_{h \to 0} \dfrac{(0 + h)^2 + 2 - [0^2 + 2]}{h}$

$= \lim\limits_{h \to 0} \dfrac{h^2}{h}$

$= \lim\limits_{h \to 0} h$

$= 0$

19. The instantaneous rate of change is positive when f is increasing.

21. (a) $s(4) = 10$

$s(1) = 1$

Average rate of change

$= \dfrac{s(4) - s(1)}{4 - 1}$

$= \dfrac{10 - 1}{4 - 1}$

$= \dfrac{9}{3} = 3$

(b) $s(7) = 10$

$s(4) = 10$

Average rate of change

$$= \frac{s(7) - s(4)}{7 - 4}$$

$$= \frac{10 - 10}{7 - 4}$$

$$= \frac{0}{3} = 0$$

(c) $s(12) = 1$

   $s(7) = 10$

Average rate of change

$$= \frac{s(12) - s(7)}{12 - 7}$$

$$= \frac{1 - 10}{12 - 7} = -\frac{9}{5}$$

(d) Sales increase during the first four years, then stay constant until year 7, then decrease.

(e) Many answers are possible; usually this trend occurs with high-tech products or products that mirror trends, such as Walkman radios.

23. (a) $P(1987) = 11$

   $P(1985) = 14$

Average rate of change

$$= \frac{P(1987) - P(1985)}{1987 - 1985}$$

$$= \frac{11 - 14}{2}$$

$$= -\frac{3}{2}$$

(b) $P(1988) = 10$

   $P(1987) = 11$

Average rate of change

$$= \frac{P(1988) - P(1987)}{1988 - 1987}$$

$$= \frac{10 - 11}{1}$$

$$= -1$$

(c) $P(1989) = 11$

   $P(1988) = 10$

Average rate of change

$$= \frac{P(1989) - P(1988)}{1989 - 1988}$$

$$= \frac{11 - 10}{1} = 1$$

(d) $P(1990) = 10.5$

   $P(1989) = 11$

Average rate of change

$$= \frac{P(1990) - P(1989)}{1990 - 1989}$$

$$= \frac{10.5 - 11}{1} = -\frac{1}{2}$$

(e) The profit margin increased when the average rate of change was positive, from 1988 to 1989.

(f) The profit margin stabilized after declining from 1985 to 1987.

25.

$$f'(x) = \lim_{h \to 0} (12x + 6h - 4)$$

h	1	.1	.01	.001	.0001
4 + h	5	4.1	4.01	4.001	4.0001
P(4 + h)	31	19.12	18.1102	18.011	18.0011
P(4)	18	18	18	18	18
P(4 + h) − P(4)	13	1.12	.110212	.0109959	.00110817
$\frac{P(4 + h) - P(4)}{h}$	13	11.2	11.0212	10.9959	11.0817

By the table, it appears that

$$\lim_{h \to 0} \frac{P(4 + h) - P(4)}{h} = 11.$$

By the rule for limits, we have

$$\lim_{h \to 0} \frac{P(4 + h) - P(4)}{h}$$

$$= \lim_{h \to 0} \frac{2(4+h)^2 - 5(4+h) + 6 - 18}{h}$$

$$= \lim_{h \to 0} \frac{2h^2 + 11h + 18 - 18}{h}$$

$$= \lim_{h \to 0} \frac{2h^2 + 11h}{h}$$

$$= \lim_{h \to 0} 2h + 11$$

$$= 11$$

27. $N(p) = 80 - 5p^2$, $1 \le p \le 4$

(a) Average rate of change of demand is

$$\frac{N(3) - N(2)}{3 - 2}$$

$$= \frac{35 - 60}{1}$$

$$= -25 \text{ boxes per dollar.}$$

(b) Instantaneous rate of change when p is 2 is

$$\lim_{h \to 0} \frac{N(2 + h) - N(2)}{h}$$

$$= \lim_{h \to 0} \frac{80 - 5(2+h)^2 - [80 - 5(2)^2]}{h}$$

$$= \lim_{h \to 0} \frac{80 - 20 - 20h - 5h^2 - (80 - 20)}{h}$$

$$= \lim_{h \to 0} \frac{-5h^2 - 20h}{h}$$

$$= -20 \text{ boxes per dollar.}$$

(c) Instantaneous rate of change when p is 3 is

$$\lim_{h \to 0} \frac{80 - 5(3 + h)^2 - [80 - 5(3)^2]}{h}$$

$$= \lim_{h \to 0} \frac{80 - 45 - 30h - 5h^2 - 80 + 45}{h}$$

$$= \lim_{h \to 0} \frac{-30h - 5h^2}{h}$$

$$= -30 \text{ boxes per dollar.}$$

(d) As the price increases, the demand decreases; this is an expected change since a higher price usually reduces demand.

29. $C(1988) = 36$, $C(1987) = 25$

(a) $\dfrac{C(1988) - C(1987)}{1988 - 1987}$

$$= \frac{36 - 25}{1}$$

$$= 11$$

The average rate of change was $11 million.

$C(1989) = 35$, $C(1988) = 36$

$$\frac{C(1989) - C(1988)}{1989 - 1988}$$

$$= \frac{35 - 36}{1}$$

$$= -1$$

The average rate of change was -$1 million.

(b) $\dfrac{Cr(1988) - Cr(1987)}{1988 - 1987}$

$$= \frac{1 - 2}{1}$$

$$= -1$$

The average rate of change was -$1 million.

$$\frac{Cr(1989) - Cr(1988)}{1989 - 1988}$$

$$= \frac{10 - 1}{1}$$

$$= 9$$

The average rate of change was $9 million.

(c) Civil penalties were increasing from 1987 to 1988 and decreasing from 1988 to 1989. Criminal penalties decreased slightly from 1987 to 1988, then increased from 1988 to 1989. This indicates that criminal penalties began to replace civil penalties in 1988.

(d) $\frac{C(1989) - C(1981)}{1989 - 1981}$

$$= \frac{35 - 6.5}{8}$$

$$= 3.5$$

The average rate of change increased about $3.5 million. The general trend was upward. Penalties for polluting are being more strictly imposed.

31. (a) cocaine

$$\frac{f(1989) - f(1988)}{1989 - 1988}$$

$$= \frac{22.5 - 23}{1}$$

$$= -.5$$

The average rate of change was a decrease of -$.5 billion.

$$\frac{f(1990) - f(1989)}{1990 - 1989}$$

$$= \frac{17.5 - 22.5}{1}$$

$$= -5$$

The average rate of change was a decrease of -$5 billion.

(b) heroin

$$\frac{f(1989) - f(1988)}{1989 - 1988}$$

$$= \frac{15.5 - 16}{1}$$

$$= -\$.5$$

The average rate of change was a decrease of -$.5 billion.

$$\frac{f(1990) - f(1989)}{1990 - 1989}$$

$$= \frac{12.3 - 15.5}{1}$$

$$= -3.2$$

The average rate of change was a decrease of -$3.2 billion.

(c) marijuana

$$\frac{f(1989) - f(1988)}{1989 - 1988}$$

$$= \frac{10 - 11.2}{1}$$

$$= -1.2$$

The average rate of change was a decrease of -$1.2 billion.

$$\frac{f(1990) - f(1989)}{1990 - 1989}$$

$$= \frac{8.8 - 10}{1}$$

$$= -1.2$$

The average rate of change was a decrease of -$1.2 billion.

(d) The greatest decrease in spending was for cocaine from 1989 – 1990.

33. Let v = the average velocity of the car.

(a) From 0 sec to 2 sec

$$v = \frac{s(2) - s(0)}{2 - 0}$$

$$= \frac{10 - 0}{2 - 0} = 5 \text{ ft per sec.}$$

(b) From 2 sec to 6 sec

$$v = \frac{s(6) - s(2)}{6 - 2}$$

$$= \frac{14 - 10}{6 - 2} = 1 \text{ ft per sec.}$$

(c) From 6 sec to 10 sec

$$v = \frac{s(10) - s(6)}{10 - 6}$$

$$= \frac{18 - 14}{10 - 6} = 1 \text{ ft per sec.}$$

(d) From 10 sec to 12 sec

$$v = \frac{s(12) - s(10)}{12 - 10}$$

$$= \frac{30 - 18}{12 - 10} = 6 \text{ ft per sec.}$$

(e) From 12 sec to 18 sec

$$v = \frac{s(18) - s(12)}{18 - 12}$$

$$= \frac{36 - 30}{18 - 12} = 1 \text{ ft per sec.}$$

(f) The car is slowing down from 2 to 6 sec and 12 to 18 sec. The car is speeding up from 0 to 2 sec and 10 to 12 sec. It is maintaining a constant speed from 6 to 10 sec.

Section 10.3

1. $f(x) = -4x^2 + 11x; \; x = -2$

Step 1 $f(x + h)$

$$= -4(x + h)^2 + 11(x + h)$$

$$= -4(x^2 + 2xh + h^2)$$

$$\qquad + 11x + 11h$$

$$= -4x^2 - 8xh - 4h^2 + 11x + 11h$$

Step 2 $f(x + h) - f(x)$

$$= -4x^2 - 8xh - 4h^2 + 11x + 11h$$

$$\qquad - (-4x^2 + 11x)$$

$$= -8xh - 4h^2 + 11h$$

$$= h(-8x - 4h + 11)$$

Step 3 $\dfrac{f(x + h) - f(x)}{h}$

$$= \frac{h(-8x - 4h + 11)}{h}$$

$$= -8x - 4h + 11$$

Step 4 $f'(x) = \lim\limits_{h \to 0} \dfrac{f(x + h) - f(x)}{h}$

$$= \lim\limits_{h \to 0} (-8x - 4h + 11)$$

$$= -8x + 11$$

$f'(-2) = -8(-2) + 11 = 27$ is the slope of the tangent line at $x = -2$.

3. $f(x) = \dfrac{-2}{x}; \; x = 4$

$$f(x + h) = \frac{-2}{x + h}$$

$$f(x + h) - f(x) = \frac{-2}{x + h} - \frac{-2}{x}$$

$$= \frac{-2x + 2(x + h)}{x(x + h)}$$

$$= \frac{-2x + 2x + 2h}{x(x + h)}$$

$$= \frac{2h}{x(x + h)}$$

$$\frac{f(x + h) - f(x)}{h} = \frac{2h}{hx(x + h)}$$

$$= \frac{2}{x(x + h)}$$

$$= \frac{2}{x^2 + xh}$$

$$f'(x) = \lim_{h \to 0} \frac{f(x + h) - f(x)}{h}$$

$$= \lim_{h \to 0} \frac{2}{x^2 + xh}$$

$$= \frac{2}{x^2}$$

$f'(4) = \frac{2}{4^2} = \frac{2}{16} = \frac{1}{8}$ is the slope of

the tangent line at x = 4.

5. $f(x) = \sqrt{x}$; x = 16

Steps 1–3 are combined:

$$\frac{f(x + h) - f(x)}{h}$$

$$= \frac{\sqrt{x + h} - \sqrt{x}}{h}$$

$$= \frac{\sqrt{x + h} - \sqrt{x}}{h} \cdot \frac{\sqrt{x + h} + \sqrt{x}}{\sqrt{x + h} + \sqrt{x}}$$

$$= \frac{x + h - x}{h(\sqrt{x + h} + \sqrt{x})}$$

$$= \frac{1}{\sqrt{x + h} + \sqrt{x}}$$

$$f'(x) = \lim_{h \to 0} \frac{f(x + h) - f(x)}{h}$$

$$= \lim_{h \to 0} \frac{1}{\sqrt{x + h} + \sqrt{x}}$$

$$= \frac{1}{2\sqrt{x}}$$

$f'(16) = \frac{1}{2\sqrt{16}} = \frac{1}{8}$ is the slope of

the tangent line at x = 16.

7. $f(x) = x^2 + 2x$; x = 3

$$\frac{f(x + h) - f(x)}{h}$$

$$= \frac{[(x + h)^2 + 2(x + h)] - (x^2 + 2x)}{h}$$

$$= \frac{(x^2 + 2hx + h^2 + 2x + 2h) - (x^2 + 2x)}{h}$$

$$= \frac{2hx + h^2 + 2h}{h} = 2x + h + 2$$

$$f'(x) = \lim_{h \to 0} (2x + h + 2) = 2x + 2$$

$f'(3) = 2(3) + 2 = 8$ is the slope

of the tangent line at x = 3.

Use m = 8 and (3, 15) in the point-

slope form.

$$y - 15 = 8(x - 3)$$
$$y = 8x - 9$$

9. $f(x) = \frac{5}{x}$; x = 2

$$\frac{f(x + h) - f(x)}{h}$$

$$= \frac{\dfrac{5}{x + h} - \dfrac{5}{x}}{h}$$

$$= \frac{\dfrac{5x - 5(x + h)}{(x + h)x}}{h}$$

$$= \frac{5x - 5x - 5h}{h(x + h)(x)}$$

$$= \frac{-5h}{h(x + h)x}$$

$$= \frac{-5}{(x + h)x}$$

$$f'(x) = \lim_{h \to 0} \frac{-5}{(x + h)(x)} = \frac{-5}{x^2}$$

$f'(2) = \frac{-5}{2^2} = -\frac{5}{4}$ is the slope of the

tangent line at x = 2.

Now use $m = -\frac{5}{4}$ and $\left(2, \frac{5}{2}\right)$ in the point-slope form.

$$y - \frac{5}{2} = -\frac{5}{4}(x - 2)$$

$$y - \frac{5}{2} = -\frac{5}{4}x + \frac{10}{4}$$

$$y = -\frac{5}{4}x + 5$$

$$5x + 4y = 20$$

11. $f(x) = 4\sqrt{x}$; $x = 9$

$$\frac{f(x + h) - f(x)}{h}$$

$$= \frac{4\sqrt{x + h} - 4\sqrt{x}}{h} \cdot \frac{4\sqrt{x + h} + 4\sqrt{x}}{4\sqrt{x + h} + 4\sqrt{x}}$$

$$= \frac{16(x + h) - 16x}{h(4\sqrt{x + h} + 4\sqrt{x})}$$

$$f'(x) = \lim_{h \to 0} \frac{16(x + h) - 16x}{h(4\sqrt{x + h} + 4\sqrt{x})}$$

$$= \lim_{h \to 0} \frac{16h}{h(4\sqrt{x + h} + 4\sqrt{x})}$$

$$= \lim_{h \to 0} \frac{4}{(\sqrt{x + h} + \sqrt{x})} = \frac{4}{2\sqrt{x}}$$

$$= \frac{2}{\sqrt{x}}$$

$f'(9) = \frac{2}{\sqrt{9}} = \frac{2}{3}$ is the slope of the tangent line at $x = 9$.

Use $m = \frac{2}{3}$ and $(9, 12)$ in the point-slope form.

$$y - 12 = \frac{2}{3}(x - 9)$$

$$y = \frac{2}{3}x + 6$$

$$3y = 2x + 18$$

For Exercises 13–17, choose any two convenient points on each of the tangent lines.

13. Using the points $(5, 3)$ and $(6, 5)$, we have

$$m = \frac{5 - 3}{6 - 5} = \frac{2}{1}$$

$$= 2.$$

15. Using the points $(-2, 2)$ and $(3, 3)$, we have

$$m = \frac{3 - 2}{3 - (-2)}$$

$$= \frac{1}{5}.$$

17. Using the points $(-3, -3)$ and $(0, -3)$, we have

$$m = \frac{-3 - (-3)}{0 - 3} = \frac{0}{-3}$$

$$= 0$$

19. (a) $f(x) = 5$ is a horizontal line and has slope 0; the derivative is 0.

(b) $f(x) = x$ has slope 1; the derivative is 1.

(c) $f(x) = -x$ has slope of -1; the derivative is -1.

(d) $x = 3$ is vertical and has undefined slope; the derivative does not exist.

(e) $y = mx + b$ has slope m; the derivative is m.

21. $f(x) = \dfrac{x^2 - 1}{x + 2}$ is not differentiable

when $x + 2 = 0$ or $x = -2$ because the
function is undefined and a vertical
asymptote occurs there.

23. $f(x) = 6x^2 - 4x$

$\dfrac{f(x + h) - f(x)}{h}$

$= \dfrac{6(x + h)^2 - 4(x + h) - (6x^2 - 4x)}{h}$

$= \dfrac{12xh + 6h^2 - 4h}{h}$

$= 12x + 6h - 4$

$f'(x) = \lim_{h \to 0} (12x + 6h - 4)$

$= 12x - 4$

$f'(2) = 12(2) - 4 = 24 - 4 = 20$

$f'(0) = 12(0) - 4 = 0 - 4 = -4$

$f'(-3) = 12(-3) - 4 = -36 - 4 = -40$

25. $f(x) = -9x - 5$

$\dfrac{f(x + h) - f(x)}{h}$

$= \dfrac{-9(x + h) - 5 - [-9x - 5]}{h}$

$= \dfrac{-9x - 9h - 5 + 9x + 5}{h} = \dfrac{-9h}{h} = -9$

$f'(x) = \lim_{h \to 0} -9 = -9$

$f'(-2) = -9$

$f'(0) = -9$

$f'(-3) = -9$

27. $f(x) = \dfrac{6}{x}$

$\dfrac{f(x + h) - f(x)}{h} = \dfrac{\dfrac{6}{x + h} - \dfrac{6}{x}}{h}$

$= \dfrac{6x - 6(x + h)}{hx(x + h)}$

$= \dfrac{-6}{x(x + h)}$

$f'(x) = \lim_{h \to 0} \dfrac{-6}{x(x + h)}$

$= \dfrac{-6}{x^2}$

$f'(2) = \dfrac{-6}{(2)^2} = \dfrac{-6}{4} = -\dfrac{3}{2}$

$f'(0) = \dfrac{6}{0}$ does not exist.

$f'(-3) = \dfrac{-6}{(-3)^2} = \dfrac{-6}{9} = -\dfrac{2}{3}$

29. $f(x) = -3\sqrt{x}$

$\dfrac{f(x + h) - f(x)}{h}$

$= \dfrac{-3\sqrt{x + h} + 3\sqrt{x}}{h}$

$= \dfrac{-3\sqrt{x + h} + 3\sqrt{x}}{h} \cdot \dfrac{-3\sqrt{x + h} - 3\sqrt{x}}{-3\sqrt{x + h} - 3\sqrt{x}}$

$= \dfrac{9(x + h) - 9x}{h(-3\sqrt{x + h} - 3\sqrt{x})}$

$= \dfrac{9}{-3\sqrt{x + h} - 3\sqrt{x}}$

$= \dfrac{3}{-\sqrt{x + h} - \sqrt{x}}$

$f'(x) = \lim_{h \to 0} \dfrac{3}{-\sqrt{x + h} - \sqrt{x}}$

$= \dfrac{3}{-\sqrt{x} - \sqrt{x}} = -\dfrac{3}{2\sqrt{x}}$

$$f'(2) = -\frac{3}{2\sqrt{2}}$$

$$f'(0) = \frac{-3}{2\sqrt{0}} \text{ is undefined.}$$

$f'(0)$ does not exist.

$$f'(-3) = \frac{-3}{2\sqrt{-3}} \text{ is not a real number.}$$

$f'(-3)$ does not exist.

31. No derivative exists at $x = -6$ because the function is not defined at $x = -6$.

33. This function is continuous for all x. So $f(x)$ has derivatives for all values of x.

35. For $x = -5$ and $x = 0$, the function $f(x)$ is not defined. For $x = -3$ and $x = 2$ the graph of $f(x)$ has sharp points. For $x = 4$, the tangent to the graph is vertical. Therefore, no derivative exists for $x = -5$, $x = -3$, $x = 0$, $x = 2$, or $x = 4$.

37. (a) The rate of change of $f(x)$ is positive when $f(x)$ is increasing, that is, on (a, 0) and (b, c).

(b) The rate of change of $f(x)$ is negative when $f(x)$ is decreasing, that is, on (0, b).

(c) The rate of change is zero when the tangent to the graph is horizontal, that is, at $x = 0$ and $x = b$.

39. The zeros of graph (b) correspond to the turning points of graph (a), the points where the derivative is zero. Graph (a) gives the distance while graph (b) gives the velocity.

41. (a) At the first point, the slope of the tangent line is -1. The debt is decreasing at the rate of 1% per month. At the second point, the slope is positive and steeper, about 2. The debt is increasing at the rate of 2% per month. At the third point, the tangent line is horizontal, with zero slope, so the debt is not changing at this point.

(b) From the graph, during 1990 the debt level decreased from February to March, from April to May, and from August to September. The function is decreasing over these intervals.

(c) Since the overall direction of the curve is increasing, the average rate of change is also increasing.

43. $R(p) = 20p - \dfrac{p^2}{500}$

(a) $R'(p) = 20 - \dfrac{1}{250}x$

At $p = 1000$,

$$R'(1000) = 20 - \frac{1}{250}(1000)$$

$$= \$16 \text{ per table.}$$

(b) The marginal revenue for the 1001st table is approximately $R'(1000)$. From (a) this is about \$16.

(c) The actual revenue is

$R(1001) - R(1000)$

$$= 20(1001) - \frac{1000^2}{500}$$

$$- \left[20(1000 - \frac{1000^2}{500}\right]$$

$$= 18,015.998 - 18,000$$

$$= \$15.998 \quad \text{or} \quad \$16.$$

(d) The marginal revenue gives a good approximation of the actual revenue from the sale of the 1001st table.

45. $B(t) = 1000 + 50t - 5t^2$
    $B'(t) = 50 - 10t$

    (a) $B'(3) = 50 - 10(3) = 20$

    (b) $B'(5) = 50 - 10(5) = 0$

    (c) $B'(6) = 50 - 10(6) = -10$

    (d) $B'(t) = 50 - 10t$

    The population starts to decline when $B'(t) = 0$.

    $$50 - 10t = 0$$
    $$50 = 10t$$
    $$5 = t$$

    So at 5 hr, population begins to decline.

47. (a) From the graph, V_{mp} is just about at the turning point of the curve. Thus, the slope of the tangent line is approximately zero. The power expenditure is not changing.

    (b) From the graph, the slope of the tangent line at V_{mr} is approximately .1. The power expended is increasing .1 unit per unit increase in speed.

    (c) The slope of the tangent line at V_{opt} is a bit greater than that at V_{mr}, about .12. The power expended increases .12 units for each unit increase in speed.

    (d) The power level first decreases to V_{mp}, then increases at greater rates.

49. (a) From Exercise 46, the slope of the tangent at the first point is 1000, the slope of the tangent at the second point is 700, and the slope of the tangent at the third point is 250.
    Thus,

t	2	10	13
$f'(t)$	100	700	250

    (b) See the answer graph in the textbook.

Section 10.4

1. $y = 10x^3 - 9x^2 + 6x$

$y' = 10(3x^{3-1}) - 9(2x^{2-1}) + 6x^{1-1}$

$\quad = 30x^2 - 18x + 6$

3. $y = x^4 - 5x^3 + \dfrac{y^2}{9} + 5$

$y' = 4x^{4-1} - 5(3x^{3-1}) + \dfrac{1}{9}(2x^{2-1}) + 0$

$\quad = 4x^3 - 15x^2 + \dfrac{2}{9}x$

5. $f(x) = 6x^{1.5} - 4x^{.5}$

$f'(x) = 6(1.5x^{1.5-1}) - 4(.5x^{.5-1})$

$\quad = 9x^{.5} - 2x^{-.5}$ or $9x^{.5} - \dfrac{2}{x^{.5}}$

7. $y = -15x^{3.2} + 2x^{1.9}$

$y' = -15(3.2x^{3.2-1}) + 2(1.9x^{1.9-1})$

$\quad = -48x^{2.2} + 3.8x^{.9}$

9. $y = 8\sqrt{x} + 6x^{3/4}$

$\quad = 8x^{1/2} + 6x^{3/4}$

$y' = 8\left(\dfrac{1}{2}x^{1/2-1}\right) + 6\left(\dfrac{3}{4}x^{3/4-1}\right)$

$\quad = 4x^{-1/2} + \dfrac{9}{2}x^{-1/4}$

$\quad$ or $\dfrac{4}{x^{1/2}} + \dfrac{9}{2x^{1/4}}$

11. $g(x) = 6x^{-5} - x^{-1}$

$g'(x) = 6(-5)x^{-5-1} - (-1)x^{-1-1}$

$\quad = -30x^{-6} + x^{-2}$ or $\dfrac{-30}{x^6} + \dfrac{1}{x^2}$

13. $y = x^{-5} - x^{-2} + 5x^{-1}$

$y' = -5x^{-5-1} - (-2x^{-2-1}) + 5(-1x^{-1-1})$

$\quad = -5x^{-6} + 2x^{-3} - 5x^{-2}$

or $\dfrac{-5}{x^6} + \dfrac{2}{x^3} - \dfrac{5}{x^2}$

15. $f(t) = \dfrac{4}{t} + \dfrac{2}{t^3}$

$\quad = 4t^{-1} + 2t^{-3}$

$f'(t) = 4(-1t^{-1-1}) + 2(-3t^{-3-1})$

$\quad = -4t^{-2} - 6t^{-4}$

or $\dfrac{-4}{t^2} - \dfrac{6}{t^4}$

17. $y = \dfrac{3}{x^6} + \dfrac{1}{x^5} - \dfrac{7}{x^2}$

$\quad = 3x^{-6} + x^{-5} - 7x^{-2}$

$y' = 3(-6x^{-7}) + (-5x^{-6}) - 7(-2x^{-3})$

$\quad = -18x^{-7} - 5x^{-6} + 14x^{-3}$

or $\dfrac{-18}{x^7} - \dfrac{5}{x^6} + \dfrac{14}{x^3}$

19. $h(x) = x^{-1/2} - 14x^{-3/2}$

$h'(x) = -\dfrac{1}{2}x^{-3/2} - 14\left(-\dfrac{3}{2}x^{-5/2}\right)$

$\quad = \dfrac{-x^{-3/2}}{2} + 21x^{-5/2}$

$\quad$ or $\dfrac{-1}{2x^{3/2}} + \dfrac{21}{x^{5/2}}$

21. $y = \dfrac{-2}{\sqrt[3]{x}}$

$\quad = \dfrac{-2}{x^{1/3}} = -2x^{-1/3}$

$y' = -2\left(-\dfrac{1}{3}x^{-4/3}\right)$

$\quad = \dfrac{2x^{-4/3}}{3}$

$\quad$ or $\dfrac{2}{3x^{4/3}}$

23. $y = 8x^{-5} - 9x^{-4}$

$\dfrac{dy}{dx} = 8(-5x^{-6}) - 9(-4x^{-5})$

$\qquad = -40x^{-6} + 36x^{-5}$

or $\dfrac{-40}{x^6} + \dfrac{36}{x^5}$

25. $D_x\left[9x^{-1/2} + \dfrac{2}{x^{3/2}}\right]$

$\qquad = D_x[9x^{-1/2} + 2x^{-3/2}]$

$\qquad = 9\left(-\dfrac{1}{2}x^{-3/2}\right) + 2\left(-\dfrac{3}{2}x^{-5/2}\right)$

$\qquad = -\dfrac{9}{2}x^{-3/2} - 3x^{-5/2}$

or $\dfrac{-9}{2x^{3/2}} - \dfrac{3}{x^{5/2}}$

27. $f(x) = \dfrac{x^2}{6} - 4x$

$\qquad = \dfrac{1}{6}x^2 - 4x$

$f'(x) = \dfrac{1}{6}(2x) - 4$

$\qquad = \dfrac{1}{3}x - 4$

$f'(-2) = \dfrac{1}{3}(-2) - 4$

$\qquad = -\dfrac{2}{3} - 4 = -\dfrac{14}{3}$

29. $D_x\left(\dfrac{6}{x}\right) = D_x(6x^{-1})$

$\qquad = 6(-1x^{-2})$

$\qquad = -6x^{-2}$

$\qquad = -\dfrac{6}{x^2}$

Choice (c) equals $D_x\left(\dfrac{6}{x}\right)$.

33. $y = -2x^5 - 7x^3 + 8x^2$

$y' = -2(5x^4) - 7(3x^2) + 8(2x)$

$\qquad = -10x^4 - 21x^2 + 16x$

$y'(1) = -10(1)^4 - 21(1)^2 + 16(1)$

$\qquad = -10 - 21 + 16$

$\qquad = -15$ is the slope of the tangent line at $x = 1$.

Use $(1, -1)$ to obtain the equation.

$y - (-1) = -15(x - 1)$

$y + 1 = -15x + 15$

$15x + y = 14$

35. $y = -x^{-3} + x^{-2}$

$y' = -3(-x^{-4}) + (-2x^{-3})$

$\qquad = 3x^{-4} - 2x^{-3}$

$\qquad = \dfrac{3}{x^4} - \dfrac{2}{x^3}$

$y'(1) = \dfrac{3}{(1)^4} - \dfrac{2}{(1)^3} = \dfrac{3}{1} - \dfrac{2}{1}$

$\qquad = 3 - 2$

$\qquad = 1$ is the slope of the tangent line at $x = 1$.

37. $f(x) = 6x^2 + 4x - 9$

$f'(x) = 12x + 4$

Let $f'(x) = -2$ to find the point where the slope of the tangent line is -2.

$12x + 4 = -2$

$12x = -6$

$x = -\dfrac{6}{12} = -\dfrac{1}{2}$

Find the y-coordinate.

$f(x) = 6x^2 + 4x - 9$

$$f\left(-\frac{1}{2}\right) = 6\left(-\frac{1}{2}\right)^2 + 4\left(-\frac{1}{2}\right) - 9$$

$$= \frac{6}{4} - 2 - 9$$

$$= \frac{6}{4} - \frac{44}{4}$$

$$= -\frac{38}{4} = -\frac{19}{2}$$

The slope of the tangent line at $\left(-\frac{1}{2}, -\frac{19}{2}\right)$ is -2.

41. $C(x) = 3000 - 20x + .03x^2$

$$R(x) = xp = x\left(\frac{5000 - x}{100}\right)$$

$$= x\left(50 - \frac{x}{100}\right)$$

(from Exercise 40)

The profit equation is found as follows.

$P = R - C$

$$= x\left(50 - \frac{x}{100}\right) - [3000 - 20x + .03x^2]$$

$$= 50x - \frac{x^2}{100} - 3000 + 20x - .03x^2$$

$$= 50x + 20x - \frac{x^2}{100} - .03x^2 - 3000$$

$$= 70x - .01x^2 - .03x^2 - 3000$$

Therefore, $P = 70x - .04x^2 - 3000$.

Now, the marginal profit is

$P'(x) = 70 - .04(2x) - 0$

$\quad\quad = 70 - .08x.$

(a) $P'(500) = 70 - .08(500)$

$\quad\quad\quad\quad = 70 - 40$

$\quad\quad\quad\quad = 30$

(b) $P'(815) = 70 - .08(815)$

$\quad\quad\quad\quad = 70 - 65.2$

$\quad\quad\quad\quad = 4.8$

(c) $P'(1000) = 70 - .08(1000)$

$\quad\quad\quad\quad\quad = 70 - 80$

$\quad\quad\quad\quad\quad = -10$

43. $S(t) = 100 - 100t^{-1}$

$S'(t) = -100(-1t^{-2})$

$\quad\quad = 100t^{-2}$

$\quad\quad = \frac{100}{t^2}$

(a) $S'(1) = \frac{100}{(1)^2} = \frac{100}{1} = 100$

(b) $S'(10) = \frac{100}{(10)^2} = \frac{100}{100} = 1$

45. $C(x) = 2x;\ R(x) = 6x - \frac{x^2}{1000}$

(a) $C'(x) = 2$

(b) $R'(x) = 6 - \frac{2x}{1000} = 6 - \frac{x}{500}$

(c) $P(x) = R(x) - C(x)$

$$= \left(6x - \frac{x^2}{1000}\right) - 2x$$

$$= 4x - \frac{x^2}{1000}$$

$$P'(x) = 4 - \frac{2x}{1000} = 4 - \frac{x}{500}$$

(d) $P'(x) = 4 - \frac{x}{500} = 0$

$$4 = \frac{x}{500}$$

$$x = 2000$$

(e) Using part (d), we see that marginal profit is 0 when $x = 2000$ units.

The profit for 2000 units produced is

$$P(x) = 4(2000) - \frac{(2000)^2}{1000}$$

$$= 8000 - 4000$$

$$= \$4000.$$

47. $V = \pi r^2 h$

    Since $h = 80$ mm,

    $V = \pi r^2 (80) = 80\pi r^2$.

    (a) $V'(r) = 80\pi(2r)$

    $= 160\pi r$

    (b) $V'(4) = 160\pi(4) = 640\pi$ cu mm

    (c) $V'(6) = 160\pi(6) = 960\pi$ cu mm

    (d) $V'(8) = 160\pi(8) = 1280\pi$ cu mm

    (e) As the radius increases, the volume increases.

49. $P(t) = \dfrac{100}{t}$

    (a) $P(1) = \dfrac{100}{1} = 100$

    (b) $P(100) = \dfrac{100}{100} = 1$

    (c) $P(t) = \dfrac{100}{t} = 100t^{-1}$

    $P'(t) = 100(-1t^{-2})$

    $= -100t^{-2}$

    $= \dfrac{-100}{t^2}$

    $P'(100) = \dfrac{-100}{(100)^2}$

    $= \dfrac{-1}{100}$

    $= -.01$

    The percent of acid is decreasing at the rate of .01 per day after 100 days.

51. $s(t) = 25t^2 - 9t + 8$

    (a) $v(t) = s'(t) = 25(2t) - 9 + 0$

    $= 50t - 9$

    (b) $v(0) = 50(0) - 9 = -9$

    $v(5) = 50(5) - 9 = 241$

    $v(10) = 50(10) - 9 = 491$

53. $s(t) = -2t^3 + 4t^2 - 1$

    (a) $v(t) = s'(t) = -2(3t^2) + 4(2t) - 0$

    $= -6t^2 + 8t$

    (b) $v(0) = -6(0)^2 + 8(0) = 0$

    $v(5) = -6(5)^2 + 8(5)$

    $= -6(25) + 40 = -110$

    $v(10) = -6(10)^2 + 8(10)$

    $= -6(100) + 80 = -520$

55. $s(t) = -16t^2 + 64t$

    (a) $v(t) = s'(t) = -16(2t) + 64$

    $= -32t + 64$

    $v(2) = -32(2) + 64$

    $= -64 + 64$

    $= 0$ ft per sec

    $v(3) = -32(3) + 64$

    $= -96 + 64$

    $= -32$ ft per sec

    (b) As the ball travels upward, its speed decreases because of the force of gravity until, at maximum height, its speed is 0 ft per sec.

    Since $v = 0$ at maximum height,

    $0 = -32t + 64$

    $-64 = -32t$

    $2 = t$

    It takes 2 seconds to reach maximum height.

    (c) $s(2) = -16(2)^2 + 64(2)$

    $= -16(4) + 128$

    $= -64 + 128$

    $= 64$

    It will go 64 ft high.

Section 10.5

1. $y = (2x - 5)(x + 4)$

$y' = (2x - 5)(1) + (x + 4)(2)$

$\quad = 2x - 5 + 2x + 8$

$\quad = 4x + 3$

3. $y = (3x^2 + 2)(2x - 1)$

$y' = (3x^2 + 2)(2) + (2x - 1)(6x)$

$\quad = 6x^2 + 4 + 12x^2 - 6x$

$\quad = 18x^2 - 6x + 4$

5. $y = (2t^2 - 6t)(t + 2)$

$y' = (2t^2 - 6t)(1) + (4t - 6)(t + 2)$

$\quad = 2t^2 - 6t + 4t^2 + 2t - 12$

$\quad = 6t^2 - 4t - 12$

7. $y = (2x^2 - 4x)(5x^2 + 4)$

$y' = (2x^2 - 4x)(10x)$

$\qquad + (4x - 4)(5x^2 + 4)$

$\quad = 20x^3 - 40x^2 + 20x^3 + 16x$

$\qquad - 20x^2 - 16$

$\quad = 40x^3 - 60x^2 + 16x - 16$

9. $y = (7x - 6)^2 = (7x - 6)(7x - 6)$

$y' = (7x - 6)(7) + (7)(7x - 6)$

$\quad = 49x - 42 + 49x - 42$

$\quad = 98x - 84$

11. $g(t) = (3t^2 + 2)^2$

$\qquad = (3t^2 + 2)(3t^2 + 2)$

$g'(t) = (3t^2 + 2)(6t) + (6t)(3t^2 + 2)$

$\qquad = 18t^3 + 12t + 18t^3 + 12t$

$\qquad = 36t^3 + 24t$

13. $y = (2x - 3)(\sqrt{x} - 1)$

$\quad = (2x - 3)(x^{1/2} - 1)$

$y' = (2x - 3)\left(\frac{1}{2}x^{-1/2}\right) + 2(x^{1/2} - 1)$

$\quad = x^{1/2} - \frac{3}{2}x^{-1/2} + 2x^{1/2} - 2$

$\quad = 3x^{1/2} - \frac{3x^{-1/2}}{2} - 2$

or $3x^{1/2} - \frac{3}{2x^{1/2}} - 2$

15. $g(x) = (-3\sqrt{x} + 6)(4\sqrt{x} - 2)$

$\qquad = (-3x^{1/2} + 6)(4x^{1/2} - 2)$

$g'(x) = (-3x^{1/2} + 6)(2x^{-1/2})$

$\qquad + \left(-\frac{3}{2}x^{-1/2}\right)(4x^{1/2} - 2)$

$\qquad = -6x^0 + 12x^{-1/2} - 6x^0 + 3x^{-1/2}$

$\qquad = -6 + 12x^{-1/2} - 6 + 3x^{-1/2}$

$\qquad = -12 + 15x^{-1/2}$

or $-12 + \frac{15}{x^{1/2}}$

17. $f(x) = \frac{6x - 11}{8x + 1}$

$f'(x) = \frac{(8x + 1)(6) - (6x - 11)(8)}{(8x + 1)^2}$

$\qquad = \frac{48x + 6 - 48x + 88}{(8x + 1)^2}$

$\qquad = \frac{94}{(8x + 1)^2}$

19. $y = \frac{-4}{2x - 11}$

$y' = \frac{(2x - 11)(0) + 4(2)}{(2x - 11)^2}$

$\quad = \frac{8}{(2x - 11)^2}$

21. $y = \dfrac{9 - 7t}{1 - t}$

$y' = \dfrac{(1 - t)(-7) - (9 - 7t)(-1)}{(1 - t)^2}$

$\quad = \dfrac{-7 + 7t + 9 - 7t}{(1 - t)^2}$

$\quad = \dfrac{2}{(1 - t)^2}$

23. $y = \dfrac{x^2 - 4x}{x + 3}$

$y' = \dfrac{(x + 3)(2x - 4) - (x^2 - 4x)(1)}{(x + 3)^2}$

$\quad = \dfrac{2x^2 + 6x - 4x - 12 - x^2 + 4x}{(x + 3)^2}$

$\quad = \dfrac{x^2 + 6x - 12}{(x + 3)^2}$

25. $y = \dfrac{-x^2 + 6x}{4x^2 + 1}$

$y' = \dfrac{(4x^2 + 1)(-2x + 6) - (-x^2 + 6x)(8x)}{(4x^2 + 1)^2}$

$\quad = \dfrac{-8x^3 + 24x^2 - 2x + 6 + 8x^3 - 48x^2}{(4x^2 + 1)^2}$

$\quad = \dfrac{-24x^2 - 2x + 6}{(4x^2 + 1)^2}$

27. $k(x) = \dfrac{x^2 + 7x - 2}{x - 2}$

$k'(x) = \dfrac{(x - 2)(2x + 7) - (x^2 + 7x - 2)(1)}{(x - 2)^2}$

$\quad = \dfrac{2x^2 + 7x - 4x - 14 - x^2 - 7x + 2}{(x - 2)^2}$

$\quad = \dfrac{x^2 - 4x - 12}{(x - 2)^2}$

29. $r(t) = \dfrac{\sqrt{t}}{2t + 3} = \dfrac{t^{1/2}}{2t + 3}$

$r'(t) = \dfrac{(2t + 3)\left(\frac{1}{2}t^{-1/2}\right) - (t^{1/2})(2)}{(2t + 3)^2}$

$\quad = \dfrac{t^{1/2} + \frac{3}{2}t^{-1/2} - 2t^{1/2}}{(2t + 3)^2}$

$= \dfrac{-t^{1/2} + \dfrac{3}{2t^{1/2}}}{(2t + 3)^2}$

$= \dfrac{-\sqrt{t} + \dfrac{3}{2\sqrt{t}}}{(2t + 3)^2}$

or $\dfrac{-2t + 3}{2\sqrt{t}(2t + 3)^2}$

31. The two terms in the numerator are reversed. The correct work follows.

$D_x\left(\dfrac{2x + 5}{x^2 - 1}\right)$

$= \dfrac{(x^2 - 1)(2) - (2x + 5)(2x)}{(x^2 - 1)^2}$

$= \dfrac{2x^2 - 2 - 4x^2 - 10x}{(x^2 - 1)^2}$

$= \dfrac{-2x^2 - 10x - 2}{(x^2 - 1)^2}$

33. $f(x) = x/(x - 2)$, at $(3, 3)$

$m = f'(x) = \dfrac{(x - 2)(1) - x(1)}{(x - 2)^2}$

$\quad = -\dfrac{2}{(x - 2)^2}$

At $(3, 3)$,

$m = -\dfrac{2}{(3 - 2)^2} = -2$

Use the point-slope form.

$y - 3 = -2(x - 3)$

$y = -2x + 9$

35. $C(x) = \dfrac{3x + 2}{x + 4}$

$\overline{C}(x) = \dfrac{C(x)}{x} = \dfrac{3x + 2}{x^2 + 4x}$

(a) $\overline{C}(10) = \dfrac{3(10) + 2}{10^2 + 4(10)} = \dfrac{32}{140}$

$\approx .2286$ hundreds of dollars

or $22.86 per unit

(b) $\bar{C}(20) = \dfrac{3(20) + 2}{(20)^2 + 4(20)} = \dfrac{62}{480}$

$\approx .1292$ hundreds of dollars

or $12.92 per unit

(c) $\bar{C}(x) = \dfrac{3x + 2}{x^2 + 4x}$ per unit

(d) $\bar{C}'(x)$

$= \dfrac{(x^2 + 4x)(3) - (3x + 2)(2x + 4)}{(x^2 + 4x)^2}$

$= \dfrac{3x^2 + 12x - 6x^2 - 12x - 4x - 8}{(x^2 + 4x)^2}$

$= \dfrac{-3x^2 - 4x - 8}{(x^2 + 4x)^2}$

37. $G(x) = \dfrac{1}{200}\left(\dfrac{800 + x^2}{x}\right)$

(a) $G'(x)$

$= \dfrac{1}{200}\left[\dfrac{x(2x) - (800 + x^2)1}{x^2}\right]$

$= \dfrac{1}{200}\left(\dfrac{2x^2 - 800 - x^2}{x^2}\right)$

$= \dfrac{x^2 - 800}{200x^2}$

$G'(20) = \dfrac{(20)^2 - 800}{200(20)^2}$

$= -\dfrac{1}{200} < 0$

Tell the driver to go faster.

(b) $G'(40) = \dfrac{(40)^2 - 800}{200(40)^2}$

$= \dfrac{1}{400} > 0$

Tell the driver to go slower.

39. $s(x) = \dfrac{x}{m + nx}$;   m and n constants

(a) $s'(x) = \dfrac{(m + nx)1 - x(n)}{(m + nx)^2}$

$= \dfrac{m + nx - nx}{(m + nx)^2}$

$= \dfrac{m}{(m + nx)^2}$

(b) x = 50, m = 10, n = 3

$s'(50) = \dfrac{m}{(m + 50n)^2}$

$= \dfrac{10}{[10 + 50(3)]^2}$

$= \dfrac{1}{2560}$

$\approx .000391$ mm per ml

41. $f(t) = \dfrac{90t}{99t - 90}$

$f'(t) = \dfrac{(99t - 90)(90) - (90t)(99)}{(99t - 90)^2}$

$= \dfrac{-8100}{(99t - 90)^2}$

(a) $f'(1) = \dfrac{-8100}{(99 - 90)^2}$

$= \dfrac{-8100}{9^2}$

$= \dfrac{-8100}{81} = -100$

$f'(10) = \dfrac{-8100}{[99(10) - 90]^2}$

$= \dfrac{-8100}{(900)^2}$

$= \dfrac{-8100}{810,000}$

$= -\dfrac{1}{100}$ or $-.01$

Section 10.6

For Exercises 1-5, $f(x) = 4x^2 - 2x$ and $g(x) = 8x + 1$.

1. $g(2) = 8(2) + 1 = 17$

   $f[g(2)] = f[17] = 4(17)^2 - 2(17)$

   $\qquad = 1156 - 34 = 1122$

3. $f(2) = 4(2)^2 - 2(2)$

   $\qquad = 16 - 4 = 12$

   $g[f(2)] = g[12] = 8(12) + 1$

   $\qquad = 96 + 1 = 97$

5. $f[g(k)] = 4(8x + 1)^2 - 2(8k + 1)$

   $\qquad = 4(64k^2 + 16k + 1)$

   $\qquad\quad - 16k - 2$

   $\qquad = 256k^2 + 48k + 2$

7. $f(x) = \dfrac{x}{8} + 12$; $g(x) = 3x - 1$

   $f[g(x)] = \dfrac{3x - 1}{8} + 12$

   $\qquad = \dfrac{3x - 1}{8} + \dfrac{96}{8}$

   $\qquad = \dfrac{3x + 95}{8}$

   $g[f(x)] = 3[\dfrac{x}{8} + 12] - 1$

   $\qquad = \dfrac{3x}{8} + 36 - 1$

   $\qquad = \dfrac{3x}{8} + 35$

   $\qquad = \dfrac{3x}{8} + \dfrac{280}{8} = \dfrac{3x + 280}{8}$

9. $f(x) = \dfrac{1}{x}$; $g(x) = x^2$

   $f[g(x)] = \dfrac{1}{x^2}$

   $g[f(x)] = \left(\dfrac{1}{x}\right)^2$

   $\qquad = \dfrac{1}{x^2}$

11. $f(x) = \sqrt{x + 2}$; $g(x) = 8x^2 - 6$

   $f[g(x)] = \sqrt{(8x^2 - 6) + 2}$

   $\qquad = \sqrt{8x^2 - 4}$

   $g[f(x)] = 8(\sqrt{x + 2})^2 - 6$

   $\qquad = 8x + 16 - 6$

   $\qquad = 8x + 10$

13. $f(x) = \sqrt{x + 1}$; $g(x) = \dfrac{-1}{x}$

   $f[g(x)] = \sqrt{\dfrac{-1}{x} + 1}$

   $\qquad = \sqrt{\dfrac{x - 1}{x}}$

   $g[f(x)] = \dfrac{-1}{\sqrt{x + 1}}$

17. $y = (5 - x)^{2/5}$

   If $f(x) = x^{2/5}$ and

   $\qquad g(x) = 5 - x$,

   then $y = f[g(x)] = (5 - x)^{2/5}$.

19. $y = -\sqrt{13 + 7x}$

   If $f(x) = -\sqrt{x}$ and

   $\qquad g(x) = 13 + 7x$,

   then $y = f[g(x)] = -\sqrt{13 + 7x}$.

21. $y = (x^2 + 5x)^{1/3} - 2(x^2 + 5x)^{2/3} + 7$

If $f(x) = x^{1/3} - 2x^{2/3} + 7$ and

$g(x) = x^2 + 5x$,

then $y = f[g(x)] = (x^2 + 5x)^{1/3} - 2(x^2 + 5x)^{2/3} + 7$.

23. $y = (2x^3 + 9x)^5$

Let $f(x) = x^5$ and $g(x) = 2x^3 + 9x$. Then $(2x^3 + 9x)^5 = f[g(x)]$.

$D_x(2x^3 + 9x)^5 = f'[g(x)] \cdot g'(x)$

$\quad f'(x) = 5x^4$

$f'[g(x)] = 5[g(x)]^4$

$\qquad\quad = 5(2x^3 + 9x)^4$

$g'(x) = 6x^2 + 9$

$D_x(2x^3 + 9x)^5 = 5(2x^3 + 9x)^4(6x^2 + 9)$

25. $f(x) = -8(3x^4 + 2)^3$

Use the generalized power rule with $y = 3x^4 + 2$, $n = 3$ and $u' = 12x^3$.

$f'(x) = -8[3(3x^4 + 2)^{3-1} \cdot 12x^3]$

$\qquad = -8[36x^3(3x^4 + 2)^2]$

$\qquad = -288x^3(3x^4 + 2)^2$

27. $s(t) = 12(2t^4 + 5)^{3/2}$

Use the generalized power rule with $u = 2t^4 + 5$, $n = 3/2$ and $u' = 8t^3$.

$s'(t) = 12[\frac{3}{2}(2t^4 + 5)^{1/2} \cdot 8t^3]$

$\qquad = 12[12t^3(2t^4 + 5)^{1/2}]$

$\qquad = 144t^3(2t^4 + 5)^{1/2}$

29. $f(t) = 8\sqrt{4t^2 + 7}$

$\qquad = 8(4t^2 + 7)^{1/2}$

Use the generalized power rule with $y = 4t^2 + 7$, $n = 1/2$, and $u' = 8t$.

$$f'(t) = 8\left[\frac{1}{2}(4t^2 + 7)^{-1/2} \cdot 8t\right]$$

$$= 8\left[4t(4t^2 + 7)^{-1/2}\right]$$

$$= 32t(4t^2 + 7)^{-1/2}$$

$$= \frac{32t}{(4t^2 + 7)^{1/2}}$$

$$= \frac{32t}{\sqrt{4t^2 + 7}}$$

31. $r(t) = 4t(2t^5 + 3)^2$

Use the product rule and the power rule.

$$r'(t) = 4t[2(2t^5 + 3) \cdot 10t^4] + (2t^5 + 3)^2 \cdot 4$$

$$= 80t^5(2t^5 + 3) + 4(2t^5 + 3)^2$$

$$= 4(2t^5 + 3)[20t^5 + (2t^5 + 3)]$$

$$= 4(2t^5 + 3)(22t^5 + 3)$$

33. $y = (x^3 + 2)(x^2 - 1)^2$

Use the product rule and the power rule.

$$y' = (x^3 + 2)[2(x^2 - 1) \cdot 2x] + (x^2 - 1)^2(3x^2)$$

$$= (x^3 + 2)[4x(x^2 - 1)] + 3x^2(x^2 - 1)^2$$

$$= (x^2 - 1)[4x(x^3 + 2) + 3x^2(x^2 - 1)]$$

$$= (x^2 - 1)(4x^4 + 8x + 3x^4 - 3x^2)$$

$$= (x^2 - 1)(7x^4 - 3x^2 + 8x)$$

35. $y = (5x^6 + x)^2\sqrt{2x}$

$$= (5x^6 + x)^2(2x)^{1/2}$$

$$y' = (5x^6 + x)^2\left[\frac{1}{2}(2x)^{-1/2} \cdot 2\right] + (2x)^{1/2}[2(5x^6 + x)(30x^5 + 1)]$$

$$= (5x^6 + x)^2(2x)^{-1/2} + (2x)^{1/2} \cdot (60x^5 + 2)(5x^6 + x)$$

$$= (5x^6 + x)(2x)^{-1/2}[(5x^6 + x) + (2x)^1(60x^5 + 2)]$$

$$= (5x^6 + x)(2x)^{-1/2}(5x^6 + x + 12x^6 + 4x)$$

$$= \frac{(5x^6 + x)(125x^6 + 5x)}{(2x)^{1/2}}$$

$$= \frac{(5x^6 + x)(125x^6 + 5x)}{\sqrt{2x}}$$

37. $y = \dfrac{1}{(3x^2 - 4)^5} = (3x^2 - 4)^{-5}$

$y' = -5(3x^2 - 4)^{-6} \cdot 6x$

$= -30x(3x^2 - 4)^{-6}$

$= \dfrac{-30x}{(3x^2 - 4)^6}$

39. $p(t) = \dfrac{(2t + 3)^3}{4t^2 - 1}$

$p'(t) = \dfrac{(4t^2 - 1)[3(2t + 3)^2 \cdot 2] - (2t + 3)^3(8t)}{(4t^2 - 1)^2}$

$= \dfrac{6(4t^2 - 1)(2t + 3)^2 - 8t(2t + 3)^3}{(4t^2 - 1)^2}$

$= \dfrac{(2t + 3)^2[6(4t^2 - 1) - 8t(2t + 3)]}{(4t^2 - 1)^2}$

$= \dfrac{(2t + 3)^2[24t^2 - 6 - 16t^2 - 24t]}{(4t^2 - 1)^2}$

$= \dfrac{(2t + 3)^2[8t^2 - 24t - 6]}{(4t^2 - 1)^2}$

$= \dfrac{2(2t + 3)^2(4t^2 - 12t - 3)}{(4t^2 - 1)^2}$

41. $y = \dfrac{x^2 + 4x}{(3x^3 + 2)^4}$

$y' = \dfrac{(3x^3 + 2)^4(2x + 4) - (x^2 + 4x)[4(3x^3 + 2)^3 \cdot 9x^2]}{[(3x^3 + 2)^4]^2}$

$= \dfrac{(3x^3 + 2)^4(2x + 4) - 36x^2(x^2 + 4x)(3x^3 + 2)^3}{(3x^3 + 2)^8}$

$= \dfrac{2(3x^3 + 2)^3[(3x^3 + 2)(x + 2) - 18x^2(x^2 + 4x)]}{(3x^3 + 2)^8}$

$= \dfrac{2(3x^4 + 6x^3 + 2x + 4 - 18x^4 - 72x^3)}{(3x^3 + 2)^5}$

$= \dfrac{-30x^4 - 132x^3 + 4x + 8}{(3x^3 + 2)^5}$

43. (a) $D_x(f[g(x)])$ at $x = 1$

$= f'[g(1)] \cdot g'(1)$

$= f'(2) \cdot \left(\dfrac{2}{7}\right)$

$= -7\left(\dfrac{2}{7}\right)$

$= -2$

(b) $D_x(f[g(x)])$ at $x = 2$

$\quad = f'[g(2)] \cdot g'(2)$

$\quad = f'(3) \cdot \left(\frac{3}{7}\right)$

$\quad = -8\left(\frac{3}{7}\right)$

$\quad = -\frac{24}{7}$

45. $D(p) = \frac{-p^2}{100} + 500$; $p(c) = 2c - 10$

The demand in terms of the cost is

$D(c) = D[p(c)]$.

$\quad = \frac{-(2c - 10)^2}{100} + 500$

$\quad = \frac{-4(c - 5)^2}{100} + 500$

$\quad = \frac{-c^2 + 10c - 25}{25} + 500$

$\quad = \frac{-c^2 + 10c - 25 + 12,500}{25}$

$\quad = \frac{-c^2 + 10c + 12,475}{25}$

47. $A = 1500\left(1 + \frac{r}{36,500}\right)^{1825}$

dA/dr is the rate of change of A with respect to r.

$\frac{dA}{dr} = 1500(1825)\left(1 + \frac{r}{36,500}\right)^{1824}\left(\frac{1}{36,500}\right)$

$\quad = 75\left(1 + \frac{r}{36,500}\right)^{1824}$

For $r = 6\%$,

$\frac{dA}{dr} = 75\left(1 + \frac{6}{36,500}\right)^{1824} = \$101.22$.

For $r = 8\%$,

$\frac{dA}{dr} = 75\left(1 + \frac{8}{36,500}\right)^{1824} = \$111.86$.

For $r = 9\%$,

$\frac{dA}{dr} = 75\left(1 + \frac{9}{36,500}\right)^{1824} = \$117.59$.

49. $V = \frac{6000}{1 + .3t + .1t^2}$

The rate of change of the value is

$V'(t)$

$\quad = \frac{(1 + .3t + .1t^2)(0) - 6000(.3 + .2t)}{(1 + .3t + .1t^2)^2}$

$\quad = \frac{-6000(.3 + .2t)}{(1 + .3t + .1t^2)^2}$.

(a) 2 years after purchase the rate of decrease in the value is

$\quad = \frac{-6000[.3 + .2(2)]}{[1 + .3(2) + .1(2)^2]^2}$

$\quad = \frac{-6000(.3 + .4)}{(1 + .6 + .4)^2}$

$\quad = \frac{-4200}{4}$

$\quad = -\$1050$.

(b) 4 years after purchase:

$\quad = \frac{-6000[.3 + .2(4)]}{[1 + .3(4) + .1(4)^2]^2}$

$\quad = \frac{-6600}{14.44}$

$\quad = -\$457.06$.

51. The demand function is $p = 300/x^{1/3}$; $x = 8n$.

The marginal revenue product is

$\frac{dR}{dn} = \left(p + x\frac{dp}{dx}\right)\frac{dx}{dn}$

$\quad = [300x^{-1/3} + 8n(-100x^{-4/3})](8)$.

When $n = 8$,

$\frac{dR}{dn} = \left[\frac{300}{2} - 64\left(\frac{100}{256}\right)\right](8)$

$\quad = 125(8)$

$\quad = \$400$ per additional worker.

53. $P(x) = 2x^2 + 1$; $x = f(a) = 3a + 2$

$P[f(a)] = 2(3a + 2)^2 + 1$
$= 2(9a^2 + 12a + 4) + 1$
$= 18a^2 + 24a + 9$

55. $r(t) = 2t$; $A(r) = \pi r^2$

$A[r(t)] = \pi(2t)^2$
$= 4\pi t^2$

$A = 4\pi t^2$ gives the area of the pollution in terms of the time since the pollutants were first emitted.

57. $C(t) = \frac{1}{2}(2t + 1)^{-1/2}$

$C'(t) = \frac{1}{2}(-\frac{1}{2})(2t + 1)^{-3/2}(2)$

$= -\frac{1}{2}(2t + 1)^{-3/2}$

(a) $C'(0) = -\frac{1}{2}[2(0) + 1]^{-3/2}$

$= -\frac{1}{2} = -.5$

(b) $C'(4) = -\frac{1}{2}[2(4) + 1]^{-3/2}$

$= -\frac{1}{2}(9)^{-3/2}$

$= \frac{-1}{2} \cdot \frac{1}{(\sqrt{9})^3}$

$= -\frac{1}{54} \approx -.02$

(c) $C'(7.5) = -\frac{1}{2}[2(7.5) + 1]^{-3/2}$

$= -\frac{1}{2}(16)^{-3/2}$

$= -\frac{1}{2}\left(\frac{1}{(\sqrt{16})^3}\right)$

$= -\frac{1}{128} \approx -.008$

(d) C is always decreasing because

$C' = -\frac{1}{2}(2t + 1)^{-3/2}$ is always

negative for $t \geq 0$.

(The amount of calcium in the bloodstream will continue to decrease over time.)

Chapter 10 Review Exercises

3. As x approaches -3 from the left or the right, f(x) approaches 4.

$\lim\limits_{x \to -3} f(x) = 4$

5. Since there is a vertical asymptote at $x = 4$, the limit does not exist.

7. $\lim\limits_{x \to -1} (2x^2 + 3x + 5)$

$= 2(-1)^2 + 3(-1) + 5$
$= 4$

9. $\lim\limits_{x \to 6} \frac{2x + 5}{x - 3}$

$= \frac{2(6) + 5}{6 + (-3)}$

$= \frac{17}{3}$

11. $\lim\limits_{x \to 4} \frac{x^2 - 16}{x - 4}$

$= \lim\limits_{x \to 4} \frac{(x - 4)(x + 4)}{x - 4}$

$= \lim\limits_{x \to 4} (x + 4)$

$= 4 + 4$

$= 8$

13. $\lim\limits_{x \to -4} \dfrac{2x^2 + 3x - 20}{x + 4}$

$= \lim\limits_{x \to -4} \dfrac{(2x - 5)(x + 4)}{x - 4}$

$= \lim\limits_{x \to -4} (2x - 5)$

$= 2(-4) - 5$

$= -13$

15. $\lim\limits_{x \to 9} \dfrac{\sqrt{x} - 3}{x - 9}$

$= \lim\limits_{x \to 9} \dfrac{\sqrt{x} - 3}{x - 9} \cdot \dfrac{\sqrt{x} + 3}{\sqrt{x} + 3}$

$= \lim\limits_{x \to 9} \dfrac{x - 9}{(x - 9)(\sqrt{x} + 3)}$

$= \lim\limits_{x \to 9} \dfrac{1}{\sqrt{x} + 3}$

$= \dfrac{1}{(\sqrt{9}) + (3)}$

$= \dfrac{1}{6}$

17. As shown on the graph, $f(x)$ is discontinuous at x_2 and x_4.

19. $f(x) = \dfrac{-5}{3x(2x - 1)}$;

$x = -5,\ 0,\ -1/3,\ 1/2$

Since f is a rational function, it is discontinuous when $3x(2x - 1) = 0$, for $x = 0$ and $x = 1/2$. It is continuous at $x = -5$ and $x = -1/3$.

21. $f(x) = \dfrac{x - 6}{x + 5}$; $x = 6,\ -5,\ 0$

Since f is a rational function, it is discontinuous when $x + 5 = 0$, or when $x = -5$. It is continuous at $x = 6$ and $x = 0$.

23. $f(x) = x^2 + 3x - 4$; $x = 1,\ -4,\ 0$

Since $f(x)$ is a polynomial function, it is continuous at all points, including $x = 1$, $x = -4$, and $x = 0$.

25. (a) At $x = 0$, $f(0) = 0$.

At $x = 4$, $f(4) = 1$.

Average rate of change

$= \dfrac{f(0) - f(4)}{0 - 4} = \dfrac{0 - 1}{0 - 4} = \dfrac{1}{4}$

(b) At $x = 8$, $f(8) = 4$.

At $x = 2$, $f(2) = 4$.

Average rate of change

$= \dfrac{f(8) - f(2)}{8 - 2} = \dfrac{4 - 4}{8 - 2} = 0$

(c) At $x = 2$, $f(2) = 4$.

At $x = 4$, $f(4) = 1$.

Average rate of change

$= \dfrac{f(2) - f(4)}{2 - 4} = \dfrac{4 - 1}{2 - 4} = -\dfrac{3}{2}$

27. $y = -2x^3 - x^2 + 5 = f(x)$

Average rate of change

$= \dfrac{f(6) - f(-2)}{6 - (-2)}$

$f(6) = -2(6)^3 - (6)^2 + 5 = -463$

$f(-2) = -2(-2)^3 - (-2)^2 + 5 = 17$

Average rate of change

$= \dfrac{-463 - 17}{6 + 2} = \dfrac{-480}{8} = -60$

$y' = -6x^2 - 2x$

Instantaneous rate of change at $x = -2$:

$-6(-2)^2 - 2(-2) = -6(4) + 4 = -20$

29. $y = \dfrac{x + 4}{x - 1} = f(x)$

$f(5) = \dfrac{5 + 4}{5 - 1} = \dfrac{9}{4}$

$f(2) = \dfrac{2 + 4}{2 - 1} = 6$

Average rate of change

$= \dfrac{\dfrac{9}{4} - 6}{5 - 2} = \dfrac{\dfrac{-15}{4}}{3} = -\dfrac{5}{4}$

$y' = \dfrac{(x - 1)(1) - (x + 4)(1)}{(x - 1)^2}$

$= \dfrac{x - 1 - x - 4}{(x - 1)^2} = \dfrac{-5}{(x - 1)^2}$

Instantaneous rate of change at
$x = 2$:

$$\dfrac{-5}{(2 - 1)^2} = \dfrac{-5}{1} = -5$$

31. $y = 5x^2 + 6x$

$y' = \lim\limits_{h \to 0} \dfrac{f(x + h) - f(x)}{h}$

$= \lim\limits_{h \to 0} \dfrac{[5(x + h)^2 + 6(x + h)] - [5x^2 + 6x]}{h}$

$= \lim\limits_{h \to 0} \dfrac{5(x^2 + 2xh + h^2) + 6x + 6h - 5x^2 - 6x}{h}$

$= \lim\limits_{h \to 0} \dfrac{5x^2 + 10xh + 5h^2 + 6x + 6h - 5x^2 - 6x}{h}$

$= \lim\limits_{h \to 0} \dfrac{10xh + 5h^2 + 6h}{h}$

$= \lim\limits_{h \to 0} \dfrac{h(10x + 5h + 6)}{h}$

$= \lim\limits_{h \to 0} (10x + 5h + 6)$

$= 10x + 6$

33. $y = x^2 - 6x$, tangent at $x = 2$

$y' = 2x - 6$

Slope $= y'(2) = 2(2) - 6 = -2$

Use $(2, -8)$ and point—slope form.

$$y - (-8) = -2(x - 2)$$
$$y + 8 = -2x + 4$$
$$y + 2x = -4$$

35. $y = \dfrac{3}{x - 1}$, tangent at $x = -1$

$y = \dfrac{3}{x - 1} = 3(x - 1)^{-1}$

$y' = 3(-1)(x - 1)^{-2}(1)$

$= -3(x - 1)^{-2}$

Slope $= y'(-1) = -3(-1 - 1)^{-2} = -\dfrac{3}{4}$

Use $(-1, -3/2)$ and point—slope form.

$$y - \left(\dfrac{-3}{2}\right) = -\dfrac{3}{4}[x - 1(-1)]$$
$$y + \dfrac{3}{2} = -\dfrac{3}{4}(x + 1)$$
$$y = -\dfrac{3}{4}x - \dfrac{9}{4}$$
$$3x + 4y = -9$$

37. $y = \dfrac{3}{x^2 - 1}$

$= 3(x^2 - 1)^{-1}$

$y' = 3[-1(x^2 - 1)^{-2}2x]$

$= -6x(x^2 - 1)^{-2}$

$= \dfrac{-6x}{(x^2 - 1)^2}$

slope $= y'(2) = \dfrac{-6(2)}{(4 - 1)^2}$

$= \dfrac{-12}{9}$

$= -\dfrac{4}{3}$

Use $(2, 1)$ and point—slope form.

$$y - 1 = -\dfrac{4}{3}(x - 2)$$
$$3y - 3 = -4x + 8$$
$$4x + 3y = 11$$

39. $y = -\sqrt{8x + 1}$

$\quad = -(8x + 1)^{1/2}$

$y' = -\left[\dfrac{1}{2}(8x + 1)^{-1/2} \cdot 8\right]$

$\quad = -4(8x + 1)^{-1/2}$

$\quad = \dfrac{-4}{(8x + 1)^{1/2}}$

slope $= y'(3) = \dfrac{-4}{(8(3) + 1)^{1/2}}$

$\quad\quad\quad\quad\quad = -\dfrac{4}{5}$

Use $(3, -5)$ and point—slope form.

$y - (-5) = -\dfrac{4}{5}(x - 3)$

$5y + 25 = -4x + 12$

$4x + 5y = -13$

43. $y = x^3 - 4x^2$

$y' = 3x^2 - 4(2x) = 3x^2 - 8x$

45. $y = -3x^{-2}$

$y' = (-3)(-2)x^{-3}$

$\quad = 6x^{-3}$ or $\dfrac{6}{x^3}$

47. $f(x) = 6x^{-1} - 2\sqrt{x} = 6x^{-1} - 2(x)^{1/2}$

$f'(x) = 6(-x^{-2}) - 2\left(\dfrac{1}{2}x^{-1/2}\right)$

$\quad = -6x^{-2} - x^{-1/2}$

$\quad$ or $\dfrac{-6}{x^2} - \dfrac{1}{x^{1/2}}$

49. $y = (-5t + 4)(t^3 - 2t^2)$

$y' = (-5t + 4)(3t^2 - 4t) + (t^3 - 2t^2)(-5)$

$\quad = -15t^3 + 20t^2 + 12t^2 - 16t - 5t^3 + 10t^2$

$\quad = -20t^3 + 42t^2 - 16t$

51. $p(t) = 8t^{3/4}(7t - 2)$

$p'(t) = 8t^{3/4}(7) + (7t - 2)\left[8\left(\dfrac{3}{4}\right)t^{-1/4}\right]$

$\quad = 56t^{3/4} + (7t - 2)(6t^{-1/4})$

$\quad = 56t^{3/4} + 42t^{3/4} - 12t^{-1/4}$

$\quad = 98t^{3/4} - 12t^{-1/4}$

or $98t^{3/4} - \dfrac{12}{t^{1/4}}$

53. $y = 15x^{-3/5}\left(6 - \dfrac{x}{3}\right)$

$y' = 15x^{-3/5}\left(-\dfrac{1}{3}\right) + \left(6 - \dfrac{x}{3}\right)\left[15\left(-\dfrac{3}{5}x^{-8/5}\right)\right]$

$\quad = -5x^{-3/5} - 9x^{-8/5}\left(6 - \dfrac{x}{3}\right)$

$\quad = -5x^{-3/5} - 54x^{-8/5} + 3x^{-3/5}$

$\quad = -2x^{-3/5} - 54x^{-8/5}$

or $-\dfrac{2}{x^{3/5}} - \dfrac{54}{x^{8/5}}$

55. $r(x) = \dfrac{-8}{2x + 1} = -8(2x + 1)^{-1}$

$r'(x) = -8[(-1)(2x + 1)^{-2}(2)]$

$\quad = 16(2x + 1)^{-2}$

$\quad = \dfrac{16}{(2x + 1)^2}$

57. $y = \dfrac{2x^3 - 5x^2}{x + 2}$

$y' = \dfrac{(x + 2)(6x^2 - 10x) - (2x^3 - 5x^2)(1)}{(x + 2)^2}$

$\quad = \dfrac{6x^3 + 12x^2 - 10x^2 - 20x - 2x^3 + 5x^2}{(x + 2)^2}$

$\quad = \dfrac{4x^3 + 7x^2 - 20x}{(x + 2)^2}$

59. $k(x) = (5x - 1)^6$

$k'(x) = 6(5x - 1)^5(5)$

$\quad = 30(5x - 1)^5$

61. $y = -3\sqrt{8t - 1} = -3(8t - 1)^{1/2}$

$y' = -3\left[\frac{1}{2}(8t - 1)^{-1/2}(8)\right]$

$= -12(8t - 1)^{-1/2}$

or $\dfrac{-12}{(8t - 1)^{1/2}}$

63. $y = 4x^2(3x - 2)^5$

$y' = (4x^2)[5(3x - 2)^4(3)] + (3x - 2)^5(8x)$

$= 60x^2(3x - 2)^4 + 8x(3x - 2)^5$

$= 4x(3x - 2)^4[15x + 2(3x - 2)]$

$= 4x(3x - 2)^4(15x + 6x - 4)$

$= 4x(3x - 2)^4(21x - 4)$

65. $s(t) = \dfrac{t^3 - 2t}{(4t - 3)^4}$

$s'(t)$

$= \dfrac{(4t - 3)^4(3t^2 - 2) - (t^3 - 2t)4(4t - 3)^3(4)}{[(4t - 3)^4]^2}$

$= \dfrac{(4t - 3)^4(3t^2 - 2) - 16(t^3 - 2t)(4t - 3)^3}{(4t - 3)^8}$

$= \dfrac{(4t - 3)^3[(4t - 3)(3t^2 - 2) - 16(t^3 - 2t)]}{(4t - 3)^8}$

$= \dfrac{(4t - 3)^3(12t^3 - 9t^2 - 8t + 6 - 16t^3 + 32t)}{(4t - 3)^8}$

$= \dfrac{-4t^3 - 9t^2 + 24t + 6}{(4t - 3)^5}$

67. $D_x\left[\dfrac{2x + \sqrt{x}}{1 - x}\right]$

$= D_x\left[\dfrac{2x + x^{1/2}}{1 - x}\right]$

$= \dfrac{(1 - x)\left(2 + \frac{1}{2}x^{-1/2}\right) - (2x + x^{1/2})(-1)}{(1 - x)^2}$

$= \dfrac{2 + \frac{1}{2}x^{-1/2} - 2x - \frac{1}{2}x^{1/2} + 2x + x^{1/2}}{(1 - x)^2}$

$= \dfrac{2 + \frac{1}{2}x^{-1/2} + \frac{1}{2}x^{1/2}}{(1 - x)^2}$

$= \dfrac{\frac{1}{2}x^{-1/2}[4x^{1/2} + x + 1]}{(1 - x)^2}$

$= \dfrac{4x^{1/2} + x + 1}{2x^{1/2}(1 - x)^2}$

69. $y = \dfrac{\sqrt{x - 1}}{x} = \dfrac{(x - 1)^{1/2}}{x}$

$\dfrac{dy}{dx} = \dfrac{x\left[\frac{1}{2}(x - 1)^{-1/2}\right] - (x - 1)^{1/2}(1)}{x^2}$

$= \dfrac{x(x - 1)^{-1/2} - 2(x - 1)^{1/2}}{2x^2}$

$= \dfrac{(x - 1)^{-1/2}[x - 2(x - 1)]}{2x^2}$

$= \dfrac{(x - 1)^{-1/2}(x - 2x + 2)}{2x^2}$

$= \dfrac{2 - x}{2x^2(x - 1)^{1/2}}$

71. $f'(-2)$ if $f(t) = \dfrac{2 - 3t}{\sqrt{2 + t}}$

When $t = -2$, $f(t)$ is not defined.
Therefore, $f'(-2)$ does not exist.

73. (d) $F(x) = \dfrac{x^2}{1500}$ can be differentiated
using the power rule.

75. $C(x) = \sqrt{3x + 2}$

$\overline{C}(x) = \dfrac{C(x)}{x} = \dfrac{\sqrt{3x + 2}}{x} = \dfrac{(3x + 2)^{1/2}}{x}$

$\overline{C}'(x) = \dfrac{x\left[\frac{1}{2}(3x + 2)^{-1/2}(3)\right] - (3x + 2)^{1/2}(1)}{x^2}$

$= \dfrac{\frac{3}{2}x(3x + 2)^{-1/2} - (3x + 2)^{1/2}}{x^2}$

$= \dfrac{3x(3x + 2)^{-1/2} - 2(3x + 2)^{1/2}}{2x^2}$

$= \dfrac{(3x + 2)^{-1/2}[3x - 2(3x + 2)]}{2x^2}$

$= \dfrac{3x - 6x - 4}{2x^2(3x + 2)^{1/2}} = \dfrac{-3x - 4}{2x^2(3x + 2)^{1/2}}$

77. $C(x) = (4x + 3)^4$

$$\overline{C}(x) = \frac{C(x)}{x} = \frac{(4x + 3)^4}{x}$$

$$\overline{C}'(x) = \frac{x[4(4x + 3)^3(4)] - (4x + 3)^4(1)}{x^2}$$

$$= \frac{16x(4x + 3)^3 - (4x + 3)^4}{x^2}$$

$$= \frac{(4x + 3)^3[16x - (4x + 3)]}{x^2}$$

$$= \frac{(4x + 3)^3(12x - 3)}{x^2}$$

79. $P(x) = \dfrac{x^2}{x - 1}$

$$P'(x) = \frac{(x - 1)2x - (x^2)(1)}{(x - 1)^2}$$

$$= \frac{2x^2 - 2x - x^2}{(x - 1)^2}$$

$$= \frac{x^2 - 2x}{(x - 1)^2}$$

(a) $P'(4) = \dfrac{(4)^2 - 2(4)}{(4 - 1)^2}$

$$= \frac{16 - 8}{9}$$

$$= \frac{8}{9}$$

In dollars, this is $\frac{8}{9}(100) = \$88.89$, which represents the approximate increase in profit from selling the fifth unit.

(b) $P'(12) = \dfrac{(12)^2 - 2(12)}{(12 - 1)^2}$

$$= \frac{144 - 24}{121}$$

$$= \frac{120}{121}$$

In dollars, this is $\frac{120}{121}(100) =$ $\$99.17$, which represents the approximate increase in profit from selling the thirteenth unit.

(c) $P'(20) = \dfrac{(20)^2 - 2(20)}{(20 - 1)^2}$

$$= \frac{400 - 40}{361}$$

$$= \frac{360}{361}$$

In dollars, this is $\frac{360}{361}(100) =$ $\$99.72$, which represents the approximate increase in profit from selling the twenty-first unit.

(d) As the number of units sold increases, the marginal profit increases.

81. $R(x) = 5000 + 16x - 3x^2$

(a) $R'(x) = 16 - 6x$

(b) Since x is in hundreds of dollars, $1000 corresponds to x = 10.

$$R'(10) = 16 - 6(10)$$
$$= 16 - 60 = -44$$

So an increase of $100 on advertising when advertising expenditures are $1000 will result in the revenue decreasing by $44.

83. $P(x) = 15x + 25x^2$

(a) $P(6) = 15(6) + 25(6)^2$
$$= 90 + 900 = 990$$
$$P(7) = 15(7) + 25(7)^2$$
$$= 105 + 1225 = 1330$$

Average rate of change

$$= \frac{P(7) - P(6)}{7 - 6}$$

$$= \frac{1330 - 990}{1}$$

$$= 340 \text{ cents} \quad \text{or} \quad \$3.40$$

(b) $P(6) = 990$

$P(6.5) = 15(6.5) + 25(6.5)^2$

$= 97.5 + 1056.25$

$= 1153.75$

Average rate of change

$= \dfrac{P(6.5) - P(6)}{6.5 - 6}$

$= \dfrac{1153.75 - 990}{.5}$

$= 327.5$ cents   or $3.28

(c) $P(6) = 990$

$P(6.1) = 15(6.1) + 25(6.1)^2$

$= 91.5 + 930.25$

$= 1021.75$

Average rate of change

$= \dfrac{P(6.1) - P(6)}{6.1 - 6}$

$= \dfrac{1021.75 - 990}{.1}$

$= 317.5$ cents   or $3.18

(d) $P'(x) = 15 + 50x$

$P'(6) = 15 + 50(6)$

$= 15 + 300$

$= 315$ cents   or $3.15

(e) $P'(20) = 15 + 50(20)$

$= 1015$ cents   or $10.15

(f) $P'(30) = 15 + 50(30)$

$= 1515$ cents   or $15.15

(g) The domain of x is $[0, \infty)$ since pounds cannot be measured with negative numbers.

(h) Since $P'(x) = 15 + 50x$ gives the marginal profit, and $x \geq 0$, $P'(x)$ can never be negative.

(i) $\overline{P}(x) = \dfrac{P(x)}{x}$

$= \dfrac{15x + 25x^2}{x}$

$= 15 + 25x$

(j) $\overline{P}'(x) = 25$

(k) The marginal average profit cannot change since $\overline{P}'(x)$ is constant. The profit per pound never changes.

For Exercises 85 and 87, use a computer or graphing calculator. Solutions will vary according to the program that is used. Answers are given.

85. (a) The graph shows that there are no values of x where the derivative is positive.

(b) The graph shows that there are no values of x where the derivative is zero.

(c) The graph shows that the derivative is negative for all x-values, that is, for x in the interval $(-\infty, \infty)$.

(d) Since the derivative is always negative, the graph of g(x) is always decreasing.

87. (a) The graph shows that the derivative is positive where x-values are in the interval $(-1, 1)$.

(b) $G'(x) = 0$ where $x = -1$.

(c) The graph shows that the derivative is negative for $(-\infty, -1)$ and $(1, \infty)$.

(d) The derivative is 0 when G(x)
    is at a low point. It is
    positive where G(x) is increasing
    and negative where G(x) is
    decreasing.

CHAPTER 10   TEST

1. Find each of the following limits, if it exists.

    (a) $\lim\limits_{x \to 1} f(x)$ (b) $\lim\limits_{x \to -1} f(x)$

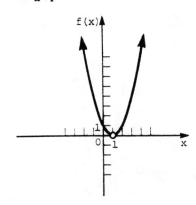
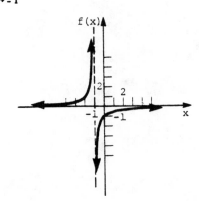

    (c) $\lim\limits_{x \to 2} f(x)$ (d) $\lim\limits_{x \to -2} f(x)$

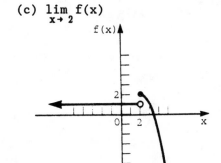
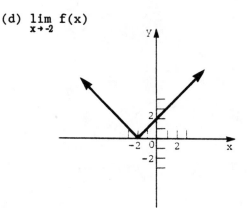

Find each of the following limits, if it exists.

2. $\lim\limits_{x \to 2} \left(\dfrac{1}{x} + 1\right)(3x - 2)$ 3. $\lim\limits_{x \to 0} \dfrac{3x + 5}{4x}$

4. $\lim\limits_{x \to 2} \dfrac{x^2 - 5x + 6}{x - 2}$ 5. $\lim\limits_{x \to 1} \dfrac{x - 1}{\sqrt{x} - 1}$

6. Use the graph to find the average rate of change of f on the given intervals.

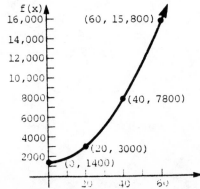

(a) From x = 0 to x = 20   (b) From x = 20 to x = 60

(c) From x = 0 to x = 40   (d) From x = 0 to x = 60

7. Find the average rate of change of $f(x) = x^3 - 5x$ from x = 1 to x = 5.

Use the definition of the derivative to find the derivative of each function.

8. $y = x^3 - 5x^2$ 9. $y = \dfrac{3}{x}$

10. Suppose the total profit in thousands of dollars from selling x units is given by

$$P(x) = 2x^2 - 6x + 9.$$

(a) Find the average rate of change of profit as x increases from 2 to 4.

(b) Find the marginal profit when 10 units are sold.

(c) Find the average rate of change of profit when sales are increased from 100 to 200 units.

Find the derivative of each function.

11. $y = 2x^4 - 3x^3 + 4x^2 - x + 1$ 12. $f(x) = x^{3/4} - x^{-2/3} + x^{-1}$

13. $f(x) = -2x^{-2} - 3\sqrt{x}$

14. Find the slope of the tangent line to $y = -2x + \dfrac{1}{x} + \sqrt{x}$ at x = 1.

Find the equation of the tangent line.

Find the derivative of each function.

15. $f(x) = (2x^2 - 5x)(3x^2 + 1)$ 16. $y = \dfrac{x^2 + x}{x^3 - 1}$

17. Management has determined that the cost in thousands of dollars for producing x units is given by the equation

$$C(x) = \frac{3x^2}{x^2 + 1} = 200.$$

Find and interpret the marginal cost when x = 3.

18. At the beginning of an experiment, a culture is determined to have
     2×10^4 bacteria. Thereafter, the number of bacteria observed at
     time t (in hours) is given by the equation

$$B(t) = 10^4(2 - 3\sqrt{t} + 2t + t^2).$$

How fast is the population growing at the end of 4 hours?

19. Find each of the following.

(a) $\dfrac{dy}{dt}$ if $y = \dfrac{\sqrt{x - 2}}{x + 1}$ (b) $D_x [\sqrt{(3x^2 - 1)^3}]$

(c) $f'(2)$ if $f(x) = \dfrac{2t - 1}{\sqrt{t + 2}}$

Find the derivative of each function.

20. $y = -2x\sqrt{3x - 1}$ 21. $y = (5x^3 - 2x)^5$

22. Is $f(x) = \dfrac{x - 2}{x(3 - x)(x + 4)}$ continuous at the given values of x?

(a) $x = 2$ (b) $x = 0$ (c) $x = 5$ (d) $x = 3$

23. Use the graph to answer the following questions.

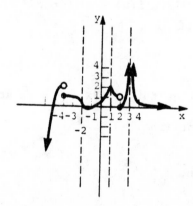

(a) On which of the following intervals is the graph continuous?

$(-5, -2),\ (-3, 2),\ (1, 4)$

(b) Where is the function discontinuous?

24. Explain the difference between the average cost function and the marginal cost function.

25. Under what circumstances could $\lim\limits_{x \to a} f(x)$ fail to exist?

CHAPTER 10   TEST ANSWERS

1. (a) 0   (b) Does not exist   (c) Does not exist   (d) 0   2. 6

3. Does not exist   4. −1   5. 2   6. (a) 80   (b) 320   (c) 160

(d) 240   7. 26   8. $y' = 3x^2 - 10x$ 9. $y' = -3/x^2$

10. (a) $6000 per unit   (b) $34,000 per unit   (c) $594,000 per unit

11. $y' = 8x^3 - 9x^2 + 8x - 1$ 12. $f'(x) = \dfrac{3}{4x^{1/4}} + \dfrac{2}{3x^{5/3}} - \dfrac{1}{x^2}$ 13. $f'(x) = \dfrac{4}{x^3} - \dfrac{3}{2\sqrt{x}}$

14. $-\dfrac{5}{2}$; $5x + 2y = 5$ 15. $f'(x) = 24x^3 - 45x^2 + 4x - 5$ 16. $y' = \dfrac{-x^4 - 2x^3 - 2x - 1}{(x^3 - 1)^2}$

17. $C'(3) = .18$; after three units have been produced, the cost to produce one more unit will be approximately $.18(1000)$ or $180.

18. 9.25×10^4 bacteria per hour   19. (a) $\dfrac{5 - x}{2\sqrt{x} - 2(x + 1)^2}$ (b) $9x\sqrt{3x^2 - 1}$

(c) $\dfrac{13}{24}$ 20. $y' = \dfrac{2 - 9x}{\sqrt{3x - 1}}$ 21. $y' = 5(5x^3 - 2x)(15x^2 - 2)$

22. (a) Yes (b) No (c) Yes (d) No 23. (a) $(-3, 2)$ (b) At $x = -4$, $x = 2$, and $x = 3$

24. The average cost function gives the cost per item when n items are produced. The marginal cost function gives the cost of the (n + 1)th item when n items are produced.

25. The limit may not exist if $x = a$ is a vertical asymptote or if the limit from the left does not equal the limit from the right.

CHAPTER 11 CURVE SKETCHING

Section 11.1

1. By reading the graph, $f(x)$ is

   (a) increasing on $(1, \infty)$ and

   (b) decreasing on $(-\infty, 1)$.

3. By reading the graph, $g(x)$ is

   (a) increasing on $(-\infty, -2)$ and

   (b) decreasing on $(-2, \infty)$.

5. By reading the graph, $h(x)$ is

   (a) increasing on $(-\infty, -4)$ and
   $(-2, \infty)$ and

   (b) decreasing on $(-4, -2)$.

7. By reading the graph, $f(x)$ is

   (a) increasing on $(-7, -4)$ and
   $(-2, \infty)$ and

   (b) decreasing on $(-\infty, -7)$ and
   $(-4, -2)$.

9. $f(x) = x^2 + 12x - 6$

   $f'(x) = 2x + 12$

   $f'(x)$ is zero when

   $$2x(x + 6) = 0$$
   $$x = -6.$$

   Since $f(x)$ is polynomial function, there are no values where f fails to exist. -6 is the only critical number.

   -6 determines two intervals on a number line.

Testing a point in each interval (any point will do other than the one where $x = -6$), we find the following.

   (a) $f'(0) = 12 > 0$, so $f(x)$ is increasing on $(-6, \infty)$.

   (b) $f'(-8) = -4 < 0$, so $f(x)$ is decreasing on $(-\infty, -6)$.

11. $y = 2 + 3.6x - 1.2x^2$

    $y' = 3.6 - 2.4x$

    y' is zero when

    $$3.6 - 2.4x = 0$$
    $$x = \frac{3}{2}.$$

    Test a point in each interval.

    (a) When $x = 1$, $y' = 1.2 > 0$, so y is increasing on $(-\infty, 3/2)$.

    (b) When $x = 2$, $y' = -1.2 < 0$, so y is decreasing on $(3/2, \infty)$.

13. $f(x) = \frac{2}{3}x^3 - x^2 - 24x - 4$

    $f'(x) = 2x^2 - 2x - 24$
    $\qquad = 2(x^2 - x - 12)$
    $\qquad = 2(x + 3)(x - 4)$

    $f'(x)$ is zero when $x = -3$ or $x = 4$.

Test a point in each interval.

$f'(-4) = 16 > 0$

$f'(0) = -24 < 0$

$f'(5) = 16 > 0$

(a) $f'(x)$ is increasing on $(-\infty, -3)$ and $(4, \infty)$.

(b) $f(x)$ is decreasing on $(-3, 4)$.

15. $f(x) = 4x^3 - 15x^2 - 72x + 5$

$f'(x) = 12x^2 - 30x - 72$

$= 6(2x^2 - 5x - 12)$

$= 6(2x + 3)(x - 4)$

$f'(x)$ is zero when $x = -\dfrac{3}{2}$ or $x = 4$.

$f'(-2) = 36 > 0$

$f'(0) = -72 < 0$

$f'(5) = 78 > 0$

(a) $f(x)$ is increasing on $(-\infty, -3/2)$ and $(4, \infty)$.

(b) $f(x)$ is decreasing on $(-3/2, 4)$.

17. $y = -3x + 6$

$y' = -3 < 0$

(a) Since y' is always negative, the function is increasing on no interval.

(b) y' is always negative so the function is decreasing everywhere, or on the interval $(-\infty, \infty)$.

19. $f(x) = \dfrac{x + 2}{x + 1}$

$f'(x) = \dfrac{(x + 1)(1) - (x + 2)1}{(x + 1)^2}$

$= \dfrac{-1}{(x + 1)^2}$

Since $f(x)$ is a rational function, it fails to exist when $x + 1 = 0$, that is, when $x = -1$. Since the derivative is never zero, there are no other critical points.

$f'(-2) = -1 < 0$

$f'(0) = -1 < 0$

(a) $f(x)$ is increasing on no interval.

(b) $f(x)$ is decreasing everywhere that it is defined, on $(-\infty, -1)$ and on $(-1, \infty)$.

21. $y = |x + 4|$

$y = |x + 4|$ is equivalent to

$y = \begin{cases} x + 4 & \text{if } x \geq -4 \\ -x - 4 & \text{if } x < -4. \end{cases}$

Thus, we can think of y as a polynomial function over the entire real number line. (The student should test a few values of x to be sure that these functions are equivalent.) Now we can take the derivative:

$y' = \begin{cases} 1 & \text{if } x \geq -4 \\ -1 & \text{if } x < -4. \end{cases}$

(a) y is increasing on $(-4, \infty)$.

(b) y is decreasing on $(-\infty, -4)$.

23. $f(x) = -\sqrt{x - 1} = -(x - 1)^{1/2}$

Note that $f(x)$ is defined only for $x \geq 1$ and exists for all such values.

$f'(x) = -\frac{1}{2}(x - 1)^{-1/2}$

$= \frac{-1}{2\sqrt{x - 1}} < 0$ for all x.

Since $f'(x)$ is never zero, there are no critical points.

(a) $f(x)$ is increasing on no interval.

(b) $f(x)$ is decreasing on $(1, \infty)$.

25. $y = \sqrt{x^2 + 1} = (x^2 + 1)^{1/2}$

$y' = \frac{1}{2}(x^2 + 1)^{-1/2}(2x)$

$= x(x^2 + 1)^{-1/2}$

$= \frac{x}{\sqrt{x^2 + 1}}$

$y' = 0$ when $x = 0$.

Since y does not fail to exist for any x, and since $y' = 0$ when $x = 0$, $x = 0$ is the only critical point.

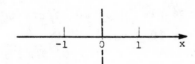

$y'(1) = \frac{1}{\sqrt{2}} > 0$

$y'(-1) = \frac{-1}{\sqrt{2}} < 0$

(a) y is increasing on $(0, \infty)$.

(b) y is decreasing on $(-\infty, 0)$.

27. $f(x) = x^{2/3}$

$f'(x) = \frac{2}{3}x^{-1/3} = \frac{2}{3x^{1/3}}$

$f'(x)$ is never zero, but fails to exist when $x = 0$.

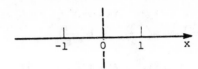

$f'(-1) = -\frac{2}{3} < 0$

$f'(1) = \frac{2}{3} > 0$

(a) $f(x)$ is increasing on $(0, \infty)$.

(b) $f(x)$ is decreasing on $(-\infty, 0)$.

29. $f(x) = ax^2 + bx + c$

$f'(x) = 2ax + b$

Let $f'(x) = 0$ to find the critical number.

$$2ax + b = 0$$
$$2ax = -b$$
$$x = \frac{-b}{2a}$$

Choose a value in the interval $(-\infty, -b/2a)$. Since $a > 0$,

$$\frac{-b}{2a} - \frac{1}{2a} = \frac{-b - 1}{2a} < \frac{-b}{2a}.$$

$f'\left(\frac{-b - 1}{2a}\right) = 2a\left(\frac{-b - 1}{2a}\right) + b$

$= -1 < 0$

Choose a value in the interval $(-b/2a, \infty)$. Since $a > 0$,

$$\frac{-b}{2a} + \frac{1}{2a} = \frac{-b + 1}{2a} > \frac{-b}{2a}.$$

$f'\left(\frac{-b + 1}{2a}\right) = 1 > 0.$

f(x) is increasing on $(-b/2a, \infty)$ and decreasing on $(-\infty, -b/2a)$.

This tells us that the curve opens upward and $x = -b/2a$ is the x-coordinate of the vertex.

$$f\left(\frac{-b}{2a}\right) = a\left(\frac{-b}{2a}\right)^2 + b\left(\frac{-b}{2a}\right) + c$$

$$= \frac{ab^2}{4a^2} - \frac{b^2}{2a} + c$$

$$= \frac{b^2}{4a} - \frac{2b^2}{4a} + \frac{4ac}{4a}$$

$$= \frac{4ac - b^2}{4a}$$

The vertex is $\left(-\dfrac{b}{2a}, \dfrac{4ac - b^2}{4a}\right)$.

31. $C(x) = x^3 - 2x^2 + 8x + 50$

$C'(x) = 3x^2 - 4x + 8$

Since $C(x)$ is a polynomial function, the only critical points are found by solving $C'(x) = 0$.

$3x^2 - 4x + 8 = 0$

$$x = \frac{4 \pm \sqrt{16 - 96}}{6} = \frac{4 \pm \sqrt{-80}}{6}$$

Since $\sqrt{-80}$ is not a real number, there are no critical points. The function either always increases or always decreases.

Test any point, say $x = 0$.

$$C'(0) = 8 > 0$$

(a) $C(x)$ is decreasing nowhere.

(b) $C(x)$ is increasing everywhere.

33. $C(x) = 4.8x - .0004x^2,\ 0 \le x \le 2250$

$R(x) = 8.4x - .002x^2,\ 0 \le x \le 2250$

$P(x) = R(x) - C(x)$

$\quad\quad = 8.4x - .002x^2 - (4.8x - .0004x^2)$

$\quad\quad = 3.6x - .0016x^2$

$P'(x) = 3.6 - .0032x$

$P'(x) = 0$ when

$$x = \frac{3.6}{.0032}$$

$$= 1125.$$

$P'(0) = 3.6 > 0$

$P'(1200) = -.24 < 0$

P is increasing on $[0, 1125)$.

35. $P(t) = 2 + 50t - \dfrac{5}{2}t^2$

$P'(t) = 50 - 5t$

$\quad\quad = 5(10 - t)$

$P'(t)$ is zero when $t = 10$.

$P'(9) = 50 - 5(9) = 5 > 0$

$P'(11) = 50 - 5(11) = -5 < 0$

P(t) is increasing on $(-\infty, 10)$ and decreasing on $(10, \infty)$, so the number of people infected starts to decline after 10 days.

37. $K(x) = \dfrac{4x}{3x^2 + 27}$ for $x \ge 0$

$K'(x) = \dfrac{(3x^2 + 27)4 - 4x(6x)}{(3x^2 + 27)^2}$

$\quad\quad = \dfrac{108 - 12x^2}{(3x^2 + 27)^2}$

$K'(x)$ is zero when

$$108 - 12x^2 = 0$$

$$12(9 - x^2) = 0$$

$$(3 - x)(3 + x) = 0$$

$$x = 3 \quad \text{or} \quad x = -3.$$

However, x ≥ 0, so x = 3 is the only critical point.

$K'(0) = \dfrac{108}{27^2} > 0$

$K'(4) = \dfrac{-84}{(48 + 27)^2} < 0$

(a) K(x) is increasing on (0, 3).
    (Note: x must be at least 0.)

(b) K(x) is decreasing on (3, ∞).

39. As shown on the graph,

   (a) horsepower increases with engine speed on (1000, 6100)

   (b) horsepower decreases with engine speed on (6100, 6500)

   (c) torque increases with engine speed on (1000, 3000) and (3600, 4200)

   (d) torque decreases with engine speed on (3000, 3600) and (4200, 6500).

Section 11.2

1. As shown on the graph, the relative minimum of -4 occurs when x = 1.

3. As shown on the graph, the relative maximum of 3 occurs when x = -2.

5. As shown on the graph, the relative maximum of 3 occurs when x = -4 and the relative minimum of 1 occurs when x = -2.

7. As shown on the graph, the relative maximum of 3 occurs when x = -4; the relative minimum of -2 occurs when x = -7 and x = -2.

9. f(x) = x² + 12x - 8
    f'(x) = 2x + 12
          = 2(x + 6)

    f'(x) is zero when x = -6.

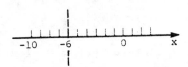

    Test f'(x) at -10 and 0.

    f'(-10) = -8 < 0
     f'(0) = 12 > 0

    Thus, we see that f(x) is decreasing on (-∞, -6) and increasing on (-6, ∞) so f(-6) is a relative minimum.

    f(-6) = (-6)² + 12(-6) - 8
          = 36 - 72 - 8
          = -44

    Relative minimum of -44 at -6

11. f(x) = 4 - 3x - .5x²
     f'(x) = -3 - x

     f'(x) is zero when x = -3.

f'(-4) = 1 > 0

f'(0) = -3 < 0

f(x) is increasing on $(-\infty, -3)$ and decreasing on $(-3, \infty)$. f(-3) is a relative maximum.

$$f(-3) = 4 - 3(-3) - .5(-3)^2$$
$$= 4 + 9 - .5(9)$$
$$= 8.5$$

Relative maximum of 8.5 at -3

13. $f(x) = x^3 + 6x^2 + 9x - 8$

$$f'(x) = 3x^2 + 12x + 9$$
$$= 3(x^2 + 4x + 3)$$
$$= 3(x + 3)(x + 1)$$

f'(x) is zero when x = -1 and x = -3.

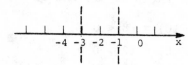

f'(-4) = 9 > 0

f'(-2) = -3 < 0

f'(0) = 9 > 0

Thus, f(x) is increasing on $(-\infty, -3)$, decreasing on $(-3, -1)$ and increasing on $(-1, \infty)$.
f(x) has a relative maximum at -3 and a relative minimum at -1.

$$f(-3) = -8$$
$$f(-1) = -12$$

Relative maximum of -8 at -3; relative minimum of -12 at -1

15. $f(x) = -\dfrac{4}{3}x^3 - \dfrac{21}{2}x^2 - 5x + 8$

$$f'(x) = -4x^2 - 21x - 5$$
$$= (-4x - 1)(x + 5)$$

f'(x) is zero when x = -5, or $x = -\dfrac{1}{4}$.

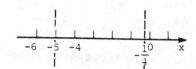

f'(-6) = -23 < 0

f'(-4) = 15 > 0

f'(0) = -5 < 0

f(x) is decreasing on $(-\infty, -5)$, increasing on $(-5, -1/4)$, and decreasing on $(-1/4, \infty)$. f(x) has a relative minimum at -5 and a relative maximum at -1/4.

$$f(-5) = -\dfrac{377}{6}$$
$$f\left(-\dfrac{1}{4}\right) = \dfrac{827}{96}$$

Relative maximum of 827/96 at -1/4; relative minimum of -377/6 at -5

17. $f(x) = 2x^3 - 21x^2 + 60x + 5$

$$f'(x) = 6x^2 - 42x + 60$$
$$= 6(x^2 - 7x + 10)$$
$$= 6(x - 5)(x - 2)$$

f'(x) is zero when x = 5 or x = 2.

f'(0) = 60 > 0

f'(3) = -12 < 0

f'(6) = 24 > 0

f(x) is increasing on $(-\infty, 2)$ and $(5, \infty)$; f(x) is decreasing on $(2, 5)$.

f(2) = 57

f(5) = 30

Relative maximum of 57 at 2;
relative minimum of 30 at 5

19. $f(x) = x^4 - 18x^2 - 4$

$f'(x) = 4x^3 - 36x$

$= 4x(x^2 - 9)$

$= 4x(x + 3)(x - 3)$

f'(x) is zero when x = 0 or x = -3
or x = 3.

$f'(-4) = 4(-4)^3 - 36(-4) = -112 < 0$

$f'(-1) = -4 + 36 = 32 > 0$

$f'(1) = 4 - 36 = -32 < 0$

$f'(4) = 4(4)^3 - 36(4) = 112 > 0$

f(x) is decreasing on (-∞, -3) and
(0, 4); f(x) is increasing on
(-3, 0) and (3, ∞).

f(-3) = -85

f(0) = -4

f(3) = -85

Relative maximum of -4 at 0;
relative minimum of -85 at 3
and -3

21. $f(x) = -(8 - 5x)^{2/3}$

$f'(x) = -\frac{2}{3}(8 - 5x)^{-1/3}(-5)$

$= \frac{10}{3(8 - 5x)^{1/3}}$

f'(x) is never zero but fails to
exist if 8 - 5x = 0.   8/5 is a
critical number.

$f'(0) = \frac{10}{3(8)^{1/3}} = \frac{5}{3} > 0$

$f'(2) = \frac{10}{3(8 - 10)^{1/3}} \approx -2.6 < 0$

f(x) is increasing on (-∞, 8/5) and
decreasing on (8/5, ∞).

$f\left(\frac{8}{5}\right) = -\left[8 - 5\left(\frac{8}{5}\right)\right]^{2/3} = 0$

Relative maximum of 0 at 8/5

23. $f(x) = 2x + 3x^{2/3}$

$f'(x) = 2 + 2x^{-1/3}$

$= 2 + \frac{2}{\sqrt[3]{x}}$

Critical numbers:

$2 + \frac{2}{\sqrt[3]{x}} = 0$ $\sqrt[3]{x} = 0$

$\frac{2}{\sqrt[3]{x}} = -2$ $x = 0$

$\frac{1}{\sqrt[3]{x}} = -1$

$x = (-1)^3$

$x = -1$

$f'(-2) = 2 + \frac{2}{\sqrt[3]{-2}} \approx .41 > 0$

$f'\left(-\frac{1}{2}\right) = 2 + \frac{2}{\sqrt[3]{-\frac{1}{2}}}$

$= 2 + \frac{2\sqrt[3]{2}}{-1} \approx -.52 < 0$

$f'(1) = 2 + \dfrac{2}{\sqrt[3]{1}} = 4 > 0$

$f(x)$ is increasing on $(-\infty, -1)$ and $(0, \infty)$.

$f(x)$ is decreasing on $(-1, 0)$.

$f(-1) = 2(-1) + 3(-1)^{2/3} = 1$

$f(0) = 0$

Relative maximum of 1 at -1; relative minimum of 0 at 0

25. $f(x) = x - \dfrac{1}{x}$

$f'(x) = 1 + \dfrac{1}{x^2}$ is never zero, but fails to exist at $x = 0$.

Since $f(x)$ also fails to exist at $x = 0$, there are no critical numbers and no relative extrema.

27. $f(x) = \dfrac{x^2}{x^2 + 1}$

$f'(x) = \dfrac{(x^2 + 1)2x - x^2(2x)}{(x^2 + 1)^2}$

$= \dfrac{2x}{(x^2 + 1)^2}$

$f'(x)$ is zero when $x = 0$.

$f'(-1) = -\dfrac{1}{2} < 0$

$f'(1) = \dfrac{1}{2} > 0$

$f(x)$ is decreasing on $(-\infty, 0)$ and increasing on $(0, \infty)$.

$f(0) = 0$

Relative minimum of 0 at 0

29. $f(x) = \dfrac{x^2 - 2x + 1}{x - 3}$

$f'(x) = \dfrac{(x - 3)(2x - 2) - (x^2 - 2x + 1)(1)}{(x - 3)^2}$

$= \dfrac{x^2 - 6x + 5}{(x - 3)^2}$

Find critical numbers:

$$x^2 - 6x + 5 = 0$$
$$(x - 5)(x - 1) = 0$$
$$x = 5 \quad \text{or} \quad x = 1$$

Note that $f(x)$ and $f'(x)$ do not exist at $x = 3$.

$f'(0) = \dfrac{5}{9} > 0$

$f'(2) = -3 < 0$

$f'(6) = \dfrac{5}{9} > 0$

$f(x)$ is increasing on $(-\infty, 1)$ and $(5, \infty)$.

$f(x)$ is decreasing on $(1, 5)$

$f(1) = 0$

$f(5) = 8$

Relative maximum of 0 at 1; relative minimum of 8 at 5

31. $y = -2x^2 + 8x - 1$

$y' = -4x + 8$

$= 4(2 - x)$

The vertex occurs when $y' = 0$ or $x = 2$.

$$-2(2)^2 + 8(2) - 1 = 7$$

The vertex is at $(2, 7)$.

33. $y = 2x^2 - 5x + 2$

$y' = 4x - 5 = 0$ when

$$x = \frac{5}{4}$$

$$2\left(\frac{5}{4}\right)^2 - 5\left(\frac{5}{4}\right) + 2 = -\frac{9}{8}$$

The vertex is at $\left(\frac{5}{4}, -\frac{9}{8}\right)$.

35. $C(x) = 75 + 10x$; $p = 70 - 2x$

$P(x) = R(x) - C(x)$

$\quad = px - C(x)$

$\quad = (70 - 2x)x - (75 + 10x)$

$\quad = 60x - 2x^2 - 75$

$P'(x) = 60 - 4x$

$P'(x) = 0$ when $x = 15$.

(a) $p = 70 - 2x$

$\quad = 70 - 2(15)$

$\quad = 40$, the price per unit that produces maximum profit.

(b) As shown above, the maximum profit occurs when 15 units are sold.

(c) $P(x) = 60x - 2x^2 - 75$

$P(15) = 60(15) - 2(15)^2 - 75$

$\quad = 375$

The maximum profit is 375.

37. $C(x) = -13x^2 + 300x$;

$p = -x^2 - x + 360$

$P(x) = R(x) - C(x)$

$\quad = px - C(x)$

$\quad = (-x^2 - x + 360)x$

$\quad\quad - (-13x^2 + 300x)$

$\quad = -x^3 + 12x^2 + 60x$

$P'(x) = -3x^2 + 24x + 60$

$\quad\quad = -3(x^2 - 8x - 20)$

$\quad\quad = -3(x + 2)(x - 10)$

$P'(x) = 0$ when $x = -2$ or $x = 10$. Disregard the negative value.

(a) $p = -x^2 - x + 360$

$\quad = -(10)^2 - 10 + 360$

$\quad = 250$, the price per unit that produces maximum profit.

(b) As shown above, the maximum profit occurs when 10 units are sold.

(c) $P(x) = -x^3 + 12x^3 + 60x$

$P(10) = -1000 + 1200 + 600$

$\quad = 800$

The maximum profit is 800.

39. $p = D(q) = \frac{1}{3}q^2 - \frac{25}{2}q + 100$,

$0 \le q \le 16$

$R = pq$

$\quad = \left(\frac{1}{3}q^2 - \frac{25}{2}q + 100\right)q$

$\quad = \frac{1}{3}q^3 - \frac{25}{2}q^2 + 100q$

$R' = q^2 - 25q + 100$

Set $R' = 0$.

$0 = q^2 - 25q + 100$

$\quad = (q - 20)(q - 5)$

$q = 20$ or $q = 5$

Since $0 \le q \le 16$, test only $q = 5$ for maximum.

$R'(4) = 16 - 100 + 100$

$\quad\quad = 16 > 0$

$R'(6) = 36 - 150 + 100$

$\quad\quad = -14 < 0$

So q = 5 maximizes R.

$$p = \frac{1}{3}q^2 - \frac{25}{2}q + 100$$

$$= \frac{1}{3}(25) - \frac{1}{2}(25) + 100$$

$$= \frac{275}{6}$$

41. $a(t) = .008t^3 - .27t^2 + 2.02t + 7$

t = the number of hours after 12 noon

$a'(t) = 3(.008t^2) - 2(.27t) + 2.02$

Set a' = 0 and use the quadratic formula to solve for t.

$.024t^2 - .54t + 2.02 = 0$

$.012t^2 - .27t + 1.01 = 0$

$$t = \frac{.27 \pm \sqrt{(-.27)^2 - 4(.012)(1.01)}}{2(.012)}$$

$t \approx 17.76$ or $t \approx 4.74$

Test for maximum or minimum:

$a'(17) = .012(17)^2 - .27(17) + 1.01$

$= 3.468 - 4.59 + 1.01 < 0$

$a'(18) = .012(18)^2 - .27(18) + 1.01$

$= 3.888 - 4.59 + 1.01 > 0$

Minimum activity occurs when t = 17.76.

$a'(4) = .012(4)^2 - .27(4) + 1.01$

$= .192 - 1.08 + 1.01 > 0$

$a'(5) = .012(5)^2 - .27(5) + 1.01 < 0$

$= .3 - 1.35 + 1.01 < 0$

Maximum activity occurs when t = 4.74.

When t = 4.74, the time is 4.74 hours past noon.

Since .74 hours = .74(60)

= 44 minutes,

the time is 4:44 P.M.

When t = 17.76, the time is 17.76 hours past noon or

17.76 - 12 = 5.76 hours past midnight.

Since .76 hours = .76(60)

= 46 minutes,

the time is 5:46 A.M.

Activity level is highest at 4:44 P.M. and lowest at 5:46 A.M.

43. $R(t) = \frac{20t}{t^2 + 100}$

$R'(t) = \frac{20(t^2 + 100) - 20t(2t)}{(t^2 + 100)^2}$

$= \frac{2000 - 20t^2}{(t^2 + 100)^2}$

R'(t) = 0 when $2000 - 20t^2 = 0$.

$-20t^2 = -2000$

$t^2 = 100$

$t = \pm 10$

Disregard the negative value.

Verify that t = 10 gives a maximum rating:

$R'(9) = .0116 > 0$

$R'(11) = -.0086 < 0$

The film should be 10 minutes long.

Exercises 45 and 47 should be solved by computer or graphing calculator methods. The solutions will vary according to the method or computer program that is used.

45. $f(x) = x^5 - x^4 + 4x^3 - 30x^2 + 5x + 6$

The answer is that there is a relative maximum of 6.2 at .085 and a relative minimum of −57.7 at 2.2.

47. $f(x) = .001x^6 - .02x^5 - x^4 - 4x^3 + 12x^2 - 9x - 17$

The answer is that there are relative maxima of 280 at -5.1 and of -18.96 at .89 and a relative minimum of -19.08 at .56.

Section 11.3

1. As shown on the graph, the absolute maximum occurs at x_3; there is no absolute minimum. (There is no functional value that is less than all others.)

3. As shown on the graph, there are no absolute extrema.

5. As shown on the graph, the absolute minimum occurs at x_1; there is no absolute maximum.

7. As shown on the graph, the absolute maximum occurs at x_1; the absolute minimum occurs at x_2.

9. $f(x) = x^2 + 6x + 2$; [-4, 0]

Find critical number(s):

$$f'(x) = 2x + 6 = 0$$
$$2(x + 3) = 0$$
$$x = -3$$

Evaluate the function at -3 and at the endpoints.

x	f(x)	
-4	-6	
-3	-7	Absolute minimum
0	0	Absolute maximum

11. $f(x) = 5 - 8x - 4x^2$; [-5, 1]

Find critical number(s).

$$f'(x) = -8 + 8x = 0$$
$$8(x - 1) = 0$$
$$x = 1$$

Here, an endpoint is also a critical number.

x	f(x)	
-5	-55	Absolute minimum
1	-7	Absolute maximum

13. $f(x) = x^3 - 3x^2 - 24x + 5$; [-3, 6]

Find critical numbers:

$$f'(x) = 3x^2 - 6x - 24 = 0$$
$$3(x^2 - 2x - 8) = 0$$
$$3(x + 2)(x - 4) = 0$$
$$x = -2 \text{ or } 4$$

x	f(x)	
-3	23	
-2	32	Absolute maximum
4	-75	Absolute minimum
6	-31	

15. $f(x) = \frac{1}{3}x^3 - \frac{1}{2}x^2 - 6x + 3$; [-4, 4]

Find critical numbers:

$f'(x) = x^2 - x - 6 = 0$

$\qquad (x + 2)(x - 3) = 0$

$\qquad x = -2 \text{ or } 3$

x	f(x)	
-4	$-\frac{7}{3} \approx -2.3$	
-2	$\frac{31}{3} \approx 10.3$	Absolute maximum
3	$-\frac{21}{2} \approx -10.5$	Absolute minimum
4	$-\frac{23}{3} \approx -7.7$	

17. $f(x) = x^4 - 32x^2 - 7$; [-5, 6]

$f'(x) = 4x^3 - 64x = 0$

$\qquad 4x(x^2 - 16) = 0$

$4x(x - 4)(x + 4) = 0$

$\qquad x = 0, 4, \text{ or } -4$

x	f(x)	
-5	-182	
-4	-263	Absolute minimum
0	-7	
4	-263	Absolute minimum
6	137	Absolute maximum

19. $f(x) = \frac{1}{1 + x}$; [0, 2]

$f'(x) = \frac{-1}{(1 + x)^2}$

$f'(x)$ is never zero. Although $f'(x)$ does not exist if x = -1, -1 is not in the given interval.

x	f(x)	
0	1	Absolute maximum
2	$\frac{1}{3}$	Absolute minimum

21. $f(x) = \frac{8 + x}{8 - x}$; [4, 6]

$f'(x) = \frac{(8 - x)(1) - (8 + x)(-1)}{(8 - x)^2}$

$\qquad = \frac{16}{(8 - x)^2}$

$f'(x)$ is never zero. Although $f'(x)$ fails to exist if x = 8, 8 is not in the given interval.

x	f(x)	
4	3	Absolute minimum
6	7	Absolute maximum

23. $f(x) = \frac{x}{x^2 + 2}$; [0, 4]

$f'(x) = \frac{(x^2 + 2)1 - x(2x)}{(x^2 + 2)^2}$

$\qquad = \frac{-x^2 + 2}{(x^2 + 2)^2} = 0$

$\qquad\qquad -x^2 + 2 = 0$

$\qquad\qquad\qquad x^2 = 2$

$x = \sqrt{2}$ or $x = -\sqrt{2}$ but $-\sqrt{2}$ is not in [0, 4].

$f'(x)$ is defined for all x.

x	f(x)	
0	0	Absolute minimum
$\sqrt{2}$	$\frac{\sqrt{2}}{4} \approx .35$	Absolute maximum
4	$\frac{2}{9} \approx .22$	

25. $f(x) = (x^2 + 18)^{2/3}$; [-3, 3]

$f'(x) = \frac{2}{3}(x^2 + 18)^{-1/3}(2x)$

$\qquad = \frac{4x}{3(x^2 + 18)^{1/3}}$

This derivative always exists, and is 0 when

$$\frac{4x}{3(x^2 + 18)^{1/3}} = 0,$$

$$4x = 0$$

$$x = 0.$$

x	f(x)	
0	$18^{2/3} \approx 6.87$	Absolute minimum
-3	9	Absolute maximum
3	9	Absolute maximum

27. $f(x) = (x + 1)(x + 2)^2$; $[-4, 0]$

$f'(x) = (x + 1)(2)(x + 2)(1)$
$\qquad + (x + 2)^2(1)$
$\qquad = 2(x + 1)(x + 2) + (x + 2)^2$
$\qquad = [2(x + 1) + x + 2](x + 2) = 0$
$\qquad\qquad (3x + 4)(x + 2) = 0$

$$x = -\frac{4}{3} \quad \text{or} \quad -2$$

x	f(x)	
-4	-12	Absolute minimum
-2	0	
$-\frac{4}{3}$	$-\frac{4}{9}$	
0	4	Absolute maximum

29. $f(x) = \dfrac{1}{\sqrt{x^2 + 1}}$; $[-1, 1]$

$f'(x)$

$= \dfrac{(x^2 + 1)^{1/2}(0) - 1\left(\frac{1}{2}\right)(x^2 + 1)^{-1/2}(2x)}{x^2 + 1}$

$= \dfrac{-x(x^2 + 1)^{-1/2}}{x^2 + 1} = 0$

$-x(x^2 + 1)^{-3/2} = 0$

$\dfrac{-x}{(\sqrt{x^2 + 1})^3} = 0$

$$x = 0$$

$f'(x)$ is defined for all x.

x	f(x)	
-1	$1/\sqrt{2} \approx .71$	Absolute minimum
0	1	Absolute maximum
1	$1/\sqrt{2} \approx .71$	Absolute minimum

31. $C(x) = x^2 + 200x + 100$

Since x is given in hundreds, the domain is $[1, 12]$.

Average cost per unit is

$$\overline{C}(x) = \frac{C(x)}{x} = \frac{x^2 + 200x + 100}{x}$$

$$= x + 200 + \frac{100}{x}.$$

$$\overline{C}'(x) = 1 - \frac{100}{x^2} = 0 \text{ when}$$

$$1 = \frac{100}{x^2}$$

$$x^2 = 100$$

$$x = 10.$$

Test for minimum.

$$\overline{C}'(5) = 1 - \frac{100}{25} = -3 < 0$$

$$\overline{C}'(15) = 1 - \frac{100}{225} = 1 - \frac{4}{9} = \frac{5}{9} > 0$$

The average cost per unit is as small as possible when x = 10, or 1000 manuals are produced. If the student produces 1000 manuals, his cost will be

$$C(10) = 10^2 + 200(10) + 100$$
$$= \$2200.$$

Thus, in order to make a profit, for each copy the student will have to charge at least

$$\frac{2200}{1000} = 2.2,$$

or more than $2.20.

33. $C(x) = 81x^2 + 17x + 324$

(a) $1 \leq x \leq 10$

$$\bar{C}(x) = \frac{C(x)}{x} = \frac{81x^2 + 17x + 324}{x}$$

$$= 81x + 17 + \frac{324}{x}$$

$$\bar{C}'(x) = 81 - \frac{324}{x^2} = 0 \text{ when}$$

$$81 = \frac{324}{x^2}$$

$$x^2 = 4$$

$$x = 2.$$

Test for minimum.

$\bar{C}'(1) = -243 < 0$

$\bar{C}'(3) = 45 > 0$

$\bar{C}(x)$ is a minimum when $x = 2$.

$$\bar{C}(2) = 81(2) + 17 + \frac{324}{2} = 341$$

The minimum for $1 \leq x \leq 10$ is 341.

(b) $10 \leq x \leq 20$

There are no critical values in this interval. Test the end-points.

$\bar{C}(10) = 859.4$

$\bar{C}(20) = 1620 + 17 + 16.2$

$= 1653.2$

The minimum for $10 \leq x \leq 20$ is 859.4.

35. $f(x) = \frac{x^2 + 36}{2x}, \ 1 \leq x \leq 12$

$$f'(x) = \frac{2x(2x) - (x^2 + 36)(2)}{(2x)^2}$$

$$= \frac{4x^2 - 2x^2 - 72}{4x^2}$$

$$= \frac{2x^2 - 72}{4x^2}$$

$$= \frac{2(x^2 - 36)}{4x^2}$$

$$= \frac{(x + 6)(x - 6)}{2x^2}$$

$f'(x) = 0$ when $x = 6$ and when $x = -6$. Only 6 is in the interval $1 \leq x \leq -12$.

Test for relative maximum or minimum.

$$f'(5) = \frac{(11)(-1)}{50} < 0$$

$$f'(7) = \frac{(13)(1)}{98} > 0$$

So the minimum is at $x = 6$, or at 6 months. Since $f(6) = 6$, the minimum percent is 6%.

37. $M(x) = -\frac{1}{45}x^2 + 2x - 20, \ 30 \leq x \leq 65$

$$M'(x) = -\frac{1}{45}(2x) + 2$$

$$= -\frac{2x}{45} + 2$$

When $M'(x) = 0$,

$$-\frac{2x}{45} + 2 = 0$$

$$2 = \frac{2x}{45}$$

$$45 = x.$$

x	M(x)
30	20
45	25
65	$145/9 \approx 16.1$

The absolute maximum miles per gallon is 25 and the absolute minimum miles per gallon is about 16.1.

39. Total area $A(x) = \pi\left(\dfrac{x}{2\pi}\right)^2 + \left(\dfrac{12 - x}{4}\right)^2$

$$= \dfrac{x^2}{4\pi} + \dfrac{(12 - x)^2}{16}$$

$A'(x) = \dfrac{x}{2\pi} - \dfrac{12 - x}{8} = 0$

$$\dfrac{4x - \pi(12 - x)}{8\pi} = 0$$

$$x = \dfrac{12\pi}{4 + \pi} \approx 5.28$$

x	Area
0	9
5.28	5.04
12	11.46

The total area is minimized when the piece used to form the circle is $\dfrac{12\pi}{4 + \pi}$ feet, or about 5.28 feet.

Exercises 41 and 43 should be solved by computer or graphing calculator methods. The solutions will vary according to the method or computer program that is used.

41. $f(x) = \dfrac{x^3 + 2x + 5}{x^4 + 3x^3 + 10}$; $[-3, 0]$

The answer is the absolute maximum is at 0 and the absolute minimum is at -2.4

43. $f(x) = x^{4/5}$ at $x^2 - 4\sqrt{x}$; $[0, 4]$

The answer is the absolute maximum is at 0 and the absolute minimum is at .74.

Section 11.4

1. $f(x) = 3x^3 - 4x + 5$
   $f'(x) = 9x^2 - 4$
   $f''(x) = 18x$

   $f''(0) = 18(0) = 0$
   $f''(2) = 18(2) = 36$
   $f''(-3) = 18(-3) = -54$

3. $f(x) = 3x^4 - 5x^3 + 2x^2$
   $f'(x) = 12x^3 - 15x^2 + 4x$
   $f''(x) = 36x^2 - 30x + 4$

   $f''(0) = 36(0)^2 - 30(0) + 4 = 4$
   $f''(2) = 36(2)^2 - 30(2) + 4 = 88$
   $f''(-3) = 36(-3)^2 - 30(-3) + 4 = 418$

5. $f(x) = 3x^2 - 4x + 8$
   $f'(x) = 6x - 4$
   $f''(x) = 6$

   $f''(0) = 6$
   $f''(2) = 6$
   $f''(-3) = 6$

7. $f(x) = (x + 4)^3$
   $f'(x) = 3(x + 4)^2(1)$
   $\quad\; = 3(x + 4)^2$
   $f''(x) = 6(x + 4)(1)$
   $\quad\quad\; = 6(x + 4)$

$f''(0) = 6(0 + 4) = 24$

$f''(2) = 6(2 + 4) = 36$

$f''(-3) = 6(-3 + 4) = 6$

9. $f(x) = \dfrac{2x + 1}{x - 2}$

$f'(x) = \dfrac{(x - 2)(2) - (2x + 1)(1)}{(x - 2)^2}$

$= \dfrac{-5}{(x - 2)^2}$

$f''(x) = \dfrac{(x - 2)^2(0) - (-5)2(x - 2)(1)}{(x - 2)^4}$

$= \dfrac{10}{(x - 2)^3}$

$f''(0) = \dfrac{10}{(0 - 2)^3} = -\dfrac{5}{4}$

$f''(2)$ does not exist.

$f''(-3) = \dfrac{10}{(-3 - 2)^3} = -\dfrac{2}{25}$

11. $f(x) = \dfrac{x^2}{1 + x}$

$f'(x) = \dfrac{(1 + x)(2x) - x^2(1)}{(1 + x)^2}$

$= \dfrac{2x + x^2}{(1 + x)^2}$

$f''(x) = \dfrac{(1+x)^2(2+2x) - (2x+x^2)(2)(1+x)}{(1 + x)^4}$

$= \dfrac{(1+x)(2 + 2x) - (2x + x^2)(2)}{(1 + x)^3}$

$= \dfrac{2}{(1 + x)^3}$

$f''(0) = 2$

$f''(2) = \dfrac{2}{27}$

$f''(-3) = -\dfrac{1}{4}$

13. $f(x) = \sqrt{x + 4} = (x + 4)^{1/2}$

$f'(x) = \dfrac{1}{2}(x + 4)^{-1/2}$

$f''(x) = \dfrac{1}{2}\left(-\dfrac{1}{2}\right)(x + 4)^{-3/2}$

$= -\dfrac{1}{4}(x + 4)^{-3/2}$

$= \dfrac{-1}{4(x + 4)^{3/2}}$

$f''(0) = \dfrac{-1}{4(0 + 4)^{3/2}} = \dfrac{-1}{4(4^{3/2})}$

$= \dfrac{-1}{4(8)}$

$= -\dfrac{1}{32}$

$f''(2) = \dfrac{-1}{4(2 + 4)^{3/2}}$

$= \dfrac{-1}{4(6^{3/2})} \approx -.0170$

$f''(-3) = \dfrac{-1}{4(-3 + 4)^{3/2}} = \dfrac{-1}{4(1^{3/2})}$

$= -\dfrac{1}{4}$

15. $f(x) = 5x^{3/5}$

$f'(x) = 3x^{-2/5}$

$f''(x) = -\dfrac{6}{5}x^{-7/5}$ or $\dfrac{-6}{5x^{7/5}}$

$f''(0)$ does not exist.

$f''(2) = -\dfrac{6}{5}(2^{-7/5}) = \dfrac{-6}{5(2^{7/5})}$

$\approx -.4547$

$f''(-3) = -\dfrac{6}{5}(-3)^{-7/5} = \dfrac{-6}{5(-3)^{7/5}}$

$\approx .2578$

17. $f(x) = -x^4 + 2x^2 + 8$

$f'(x) = -4x^3 + 4x$

$f''(x) = -12x^2 + 4$

$f'''(x) = -24x$

$f^{(4)}(x) = -24$

19. $f(x) = 4x^5 + 6x^4 - x^2 + 2$

$f'(x) = 20x^4 + 24x^3 - 2x$

$f''(x) = 80x^3 + 72x^2 - 2$

$f'''(x) = 240x^2 + 144x$

$f^{(4)}(x) = 480x + 144$

21. $f(x) = \dfrac{x - 1}{x + 2}$

$f'(x) = \dfrac{(x + 2) - (x - 1)}{(x + 2)^2}$

$= \dfrac{3}{(x + 2)^2}$

$f''(x) = \dfrac{-3(2)(x + 2)}{(x + 2)^4}$

$= \dfrac{-6}{(x + 2)^3}$

$f'''(x) = \dfrac{(-6)(-3)(x + 2)^2}{(x + 2)^6}$

$= \dfrac{18}{(x + 2)^4} \quad \text{or} \quad 18(x + 2)^{-4}$

$f^{(4)}(x) = \dfrac{-18(4)(x + 2)^3}{(x + 2)^8}$

$= \dfrac{-72}{(x + 2)^5}$

or $-72(x + 2)^{-5}$

23. $f(x) = \dfrac{3x}{x - 2}$

$f'(x) = \dfrac{(x - 2)(3) - 3x(1)}{(x - 2)^2}$

$= \dfrac{-6}{(x - 2)^2}$

$f''(x) = \dfrac{-6(-2)(x - 2)}{(x - 2)^4}$

$= \dfrac{12}{(x - 2)^3}$

$f'''(x) = \dfrac{-12(3)(x - 2)^2}{(x - 2)^6}$

$= -36(x - 2)^{-4}$

or $\dfrac{-36}{(x - 2)^4}$

$f^{(4)}(x) = \dfrac{-36(-4)(x - 2)^3}{(x - 2)^8}$

$= 144(x - 2)^{-5}$

or $\dfrac{144}{(x - 2)^5}$

25. Concave upward on $(2, \infty)$

Concave downward on $(-\infty, 2)$

Point of inflection at $(2, 3)$

27. Concave upward on $(-\infty, -1)$ and $(8, \infty)$.

Concave downward on $(-1, 8)$

Points of inflection at $(-1, 7)$ and $(8, 6)$

29. Concave upward on $(2, \infty)$

Concave downward on $(-\infty, 2)$

No points of inflection

31. $f(x) = x^2 + 10x - 9$

$f'(x) = 2x + 10$

$f''(x) = 2 > 0$ for all x.

Always concave upward

No points of inflection

33. $f(x) = x^3 + 3x^2 - 45x - 3$

$f'(x) = 3x^2 + 6x - 45$

$f''(x) = 6x + 6$

$f''(x) = 6x + 6 > 0$ when

$6(x + 1) > 0$

$x > -1.$

Concave upward on $(-1, \infty)$

$f''(x) = 6x + 6 < 0$ when

$\qquad 6(x + 1) < 0$

$\qquad\qquad x < -1.$

Concave downward on $(-\infty, -1)$

$f''(x) = 6x + 6 = 0$

$\qquad 6(x + 1) = 0$

$\qquad\qquad x = -1$

$f(1) = 44$

Point of inflection at $(-1, 44)$

35. $f(x) = -2x^3 + 9x^2 + 168x - 3$

$f'(x) = -6x^2 + 18x + 168$

$f''(x) = -12x + 18$

$f''(x) = -12x + 18 > 0$ when

$\qquad -6(2x - 3) > 0$

$\qquad\qquad 2x - 3 < 0$

$\qquad\qquad\qquad x < \dfrac{3}{2}.$

Concave upward on $(-\infty, 3/2)$

$f''(x) = -12x + 18 < 0$ when

$\qquad -6(2x - 3) < 0$

$\qquad\qquad 2x - 3 > 0$

$\qquad\qquad\qquad x > \dfrac{3}{2}.$

Concave downward on $(3/2, \infty)$

$f''(x) = -12x + 18 = 0$ when

$\qquad -6(2x + 3) = 0$

$\qquad\qquad 2x + 3 = 0$

$\qquad\qquad\qquad x = \dfrac{3}{2}.$

$\qquad f\left(\dfrac{3}{2}\right) = \dfrac{525}{2}$

Point of inflection at $(3/2, 525/2)$

37. $f(x) = \dfrac{3}{x - 5}$

$f'(x) = \dfrac{-3}{(x - 5)^2}$

$f''(x) = \dfrac{-3(-2)(x - 5)}{(x - 5)^4}$

$\qquad = \dfrac{6}{(x - 5)^3}$

$f''(x) = \dfrac{6}{(x - 5)^3} > 0$ when

$\qquad (x - 5)^3 > 0$

$\qquad\qquad x - 5 > 0$

$\qquad\qquad\qquad x > 5.$

Concave upward on $(5, \infty)$

$f''(x) = \dfrac{6}{(x - 5)^3} < 0$ when

$\qquad (x - 5)^3 < 0$

$\qquad\qquad x - 5 < 0$

$\qquad\qquad\qquad x < 5.$

Concave downward on $(-\infty, 5)$

$f''(x) \neq 0$ for any value of x; it does not exist when x = 5. There is a change of concavity there, but no point of inflection since f(5) does not exist.

39. $f(x) = x(x + 5)^2$

$f'(x) = x(2)(x + 5) + (x + 5)^2$

$\qquad = (x + 5)(2x + x + 5)$

$\qquad = (x + 5)(3x + 5)$

$f''(x) = (x + 5)(3) + (3x + 5)$

$\qquad = 3x + 15 + 3x + 5$

$\qquad = 6x + 20$

$f''(x) = 6x + 20 > 0$ when

$\qquad 2(3x + 10) > 0$

$\qquad\qquad 3x > -10$

$\qquad\qquad\qquad x > -\dfrac{10}{3}.$

Concave upward on $(-10/3, \infty)$

$f''(x) = 6x + 20 < 0$ when

$$2(3x + 10) < 0$$
$$3x < -10$$
$$x < -\frac{10}{3}.$$

Concave downward on $(-\infty, -10/3)$

$$f\left(-\frac{10}{3}\right) = -\frac{10}{3}\left(-\frac{10}{3} + 5\right)^2$$

$$= \frac{-10}{3}\left(\frac{-10 + 15}{3}\right)^2$$

$$= -\frac{10}{3} \cdot \frac{25}{9}$$

$$= -\frac{250}{27}$$

Point of inflection at

$$(-10/3, -250/27)$$

41. $f(x) = -x^2 - 10x - 25$

$f'(x) = -2x - 10$

$$= -2(x + 5) = 0$$

Critical number:

$$x = -5$$

$f''(x) = -2 < 0$ for all x.

The curve is concave downward, which means a relative maximum occurs at $x = -5$.

43. $f(x) = 3x^3 - 3x^2 + 1$

$f'(x) = 9x^2 - 6x$

$$= 3x(3x - 2) = 0$$

Critical numbers: $x = 0$ or $x = \frac{2}{3}$

$f''(x) = 18x - 6$

$f''(0) = -6 < 0$, which means that a relative maximum occurs at $x = 0$.

$f''\left(\frac{2}{3}\right) = 6 > 0$, which means that a

relative minimum occurs at $x = 2/3$.

45. $f(x) = (x + 3)^4$

$f'(x) = 4(x + 3)^3$

$$= 4(x + 3)^3 = 0$$

Critical number: $x = -3$

$f''(x) = 12(x + 3)^2$

$f''(-3) = 12(-3 + 3)^2 = 0$

The second derivative test fails. Use the first derivative test.

$f'(-4) = 4(-4 + 3)^3$

$$= 4(-1)^3 = -4 < 0$$

This indicates that f is decreasing on $(-\infty, -3)$.

$f'(0) = 4(0 + 3)^3 > 0$

This indicates that f is increasing on $(-3, \infty)$.

A relative minimum occurs at -3.

47. (a) The left side of the graph changes from concave upward to concave downward at the point of inflection between car phones and CD players. The rate of growth of sales begins to decline at the point of inflection.

(b) Food processors are closest to the right-hand point of inflection. This inflection point indicates that the rate of decline of sales is beginning to slow.

49. $R(x) = \frac{4}{27}(-x^3 + 66x^2 + 1050x - 400)$;

$0 \le x \le 25$

$R'(x) = \frac{4}{27}(-3x^2 + 132x + 1050)$

$R''(x) = \frac{4}{27}(-6x + 132)$

A point of diminishing returns occurs at a point of inflection, or where $R''(x) = 0$.

$$\frac{4}{27}(-6x + 132) = 0$$

$$-6x + 132 = 0$$

$$6x = 132$$

$$x = 22$$

Test $R''(x)$ to determine whether concavity changes at $x = 22$.

$R''(20) = \frac{4}{27}(-6 \cdot 20 + 132) = \frac{16}{9} > 0$

$R''(24) = \frac{4}{27}(-6 \cdot 24 + 132) = -\frac{16}{9} < 0$

$R(x)$ is concave up on $(0, 22)$ and concave down on $(22, 25)$.

$R(22)$

$= \frac{4}{27}[-22^3 + 66(22)^2 + 1050(22) - 400]$

≈ 6517.9

The point of diminishing returns is

$(22, 6517.9)$.

51. Let $D(q)$ represent the demand function.

The revenue function, $R(q)$, is

$R(q) = qD(q)$.

The marginal revenue is given by

$R'(q) = qD'(q) + D(q)(1)$

$\quad\quad = qD'(q) + D(q)$.

$R''(q) = qD''(q) + D'(q)(1) + D'(q)$

$\quad\quad = qD''(q) + 2D'(q)$

gives the rate of decline of marginal revenue.

$D'(q)$ gives the rate of decline of price.

If marginal revenue declines more quickly than price,

$\quad qD''(q) + 2D'(q) - D'(q) < 0$

or $qD''(q) + D'(q) < 0$.

53. (a) $R(t) = t^2(t - 18) + 96t + 1000$;

$\quad\quad 0 < t < 8$

$\quad\quad\quad = t^3 - 18t^2 + 96t + 1000$

$R'(t) = 3t^2 - 36t + 96$

Set $R'(t) = 0$.

$\quad\quad 3t^2 - 36t + 96 = 0$

$\quad\quad\quad t^2 - 12t + 32 = 0$

$\quad\quad\quad (t - 8)(t - 4) = 0$

$\quad\quad\quad t = 8$ or $t = 4$

8 is not in the domain of $R(t)$.

$R'' = 6t - 36$

$R''(4) = -12 < 0$ implies that $R(t)$ is maximized at $t = 4$ hr.

(b) $R(4) = 16(-14) + 96(4) + 1000$

$\quad\quad\quad = -224 + 384 + 1000$

$\quad\quad\quad = 1160$

The maximum population is 1160 million.

55. (a) $K(x) = \dfrac{4x}{3x^2 + 27}$; $x > 0$

$$K'(x) = \frac{(3x^2 + 27)(4) - 4x(6x)}{(3x^2 + 27)^2}$$

$$= \frac{-12x^2 + 108}{(3x^2 + 27)^2}$$

Set $K'(x) = 0$.

$$-12(x^2 - 9) = 0$$
$$x = -3 \quad \text{or} \quad x = 3$$

-3 is not in the domain of $K(x)$.

$$K''(x) = \frac{(3x^2 + 27)^2(-24x) - (-12x^2 + 108)(2)(3x^2 + 27)(6x)}{(3x^2 + 27)^4}$$

$$= \frac{(3x^2 + 27)^2(-24x) - (-2x(3x^2 + 27)(-12x^2 + 108)}{(3x^2 + 27)^4}$$

$K''(3) = -\dfrac{72}{54} < 0$ implies that $K(x)$ is maximized at $x = 3$ hours.

(b) $K(3) = \dfrac{4(3)}{3(3)^2 + 27}$

$$= \frac{2}{9}$$

The maximum concentration is 2/9%.

57. $s(t) = 8t^2 + 4t$

$v(t) = s'(t) = 16t + 4$

$a(t) = v'(t) = s''(t) = 16$

$v(0) = 4$ cm/sec

$v(4) = 68$ cm/sec

$a(0) = 16$ cm/sec^2

$a(4) = 16$ cm/sec^2

59. $s(t) = -5t^3 - 8t^2 + 6t - 3$

$v(t) = s'(t) = -15t^2 - 16t + 6$

$a(t) = v'(t) = s''(t) = -30t - 16$

$v(0) = 6$ cm/sec

$v(4) = -298$ cm/sec

$a(0) = -16$ cm/sec^2

$a(4) = -136$ cm/sec^2

61. $s(t) = \dfrac{-2}{3t + 4} = -2(3t + 4)^{-1}$

$v(t) = s'(t) = 2(3t + 4)^{-2}(3)$

$\qquad = 6(3t + 4)^{-2}$

$\qquad = \dfrac{6}{(3t + 4)^2}$

$a(t) = v'(t) = s''(t)$

$\quad = -12(3t + 4)^{-3}(3)$

$\quad = -36(3t + 4)^{-3}$

$\quad = \dfrac{-36}{(3t + 4)^3}$

$v(0) = \dfrac{6}{(0 + 4)^2} = \dfrac{3}{8}$ cm/sec

$v(4) = \dfrac{6}{(12 + 4)^2} = \dfrac{3}{128}$ cm/sec

$a(0) = \dfrac{-36}{(0 + 4)^3} = -\dfrac{9}{16}$ cm/sec^2

$a(4) = \dfrac{-36}{(12 + 4)^3} = -\dfrac{9}{1024}$ cm/sec^2

63. $s(t) = -16t^2$

$v(t) = s'(t) = -32t$

(a) $v(3) = -32(3) = -96$ ft/sec

(b) $v(5) = -32(5) = -160$ ft/sec

(c) $v(8) = -32(8) = -256$ ft/sec

(d) $a(t) = v'(t) = s''(t)$

$\quad = -32$ ft/sec^2

Exercises 65 and 67 should be solved by computer or graphing calculator methods. The solutions will vary according to the method or computer program that is used.

65. $f(x) = .25x^4 - 2x^3 + 3.5x^2 + 4x - 1$; $(-5, 5)$ in steps of .5

The answers are

(a) increasing on $(0, 2)$ and $(4, 5)$; decreasing on $(-5, -.5)$ and $(2.5, 3.5)$

(b) minima between $-.5$ and 0, and between 3.5 and 4; a maximum between 2 and 2.5

(c) concave upward on $(-5, .5)$ and $(3.5, 5)$; concave downward on $(.5, 3.5)$

(d) inflection points between .5 and 1, and between 3 and 3.5.

67. $f(x) = 3.1x^4 - 4.3x^3 + 5.82$; $(-1, 2)$ in steps of .2

The answers are

(a) decreasing on $(-1, 1)$; increasing on $(1.2, 2)$

(b) minimum between 1 and 1.2

(c) concave upward on $(-1, 0)$ and $(.8, 2)$; concave downward on $(0, .8)$

(d) inflection points between .6 and .8, and at 0.

Section 11.5

1. As x increases without bound ($x \to \infty$), the graph approaches the line $y = 3$. Thus $\lim\limits_{x \to \infty} f(x)$ exists and has the value 3.

3. $\lim\limits_{x \to \infty} \dfrac{3x}{5x - 1}$

$= \lim\limits_{x \to \infty} \dfrac{\dfrac{3x}{x}}{\dfrac{5x - 1}{x}}$

$= \lim\limits_{x \to \infty} \dfrac{3}{5 - \dfrac{1}{x}}$

$$= \frac{\lim_{x \to \infty} 3}{\lim_{x \to \infty} 5 + \lim_{x \to \infty} \left(-\frac{1}{x}\right)}$$

$$= \frac{3}{5 + 0} = \frac{3}{5}$$

5. $\lim_{x \to \infty} \dfrac{2x + 3}{4x - 7}$

$$= \lim_{x \to \infty} \frac{2 + \dfrac{3}{x}}{4 - \dfrac{7}{x}}$$

$$= \frac{\lim_{x \to \infty} 2 + \lim_{x \to \infty} \dfrac{3}{x}}{\lim_{x \to \infty} 4 + \lim_{x \to \infty} \left(-\dfrac{7}{x}\right)}$$

$$= \frac{2 + 0}{4 + 0}$$

7. $\lim_{x \to \infty} \dfrac{x^2 + 2x}{2x^2 - 2x + 1}$

$$= \lim_{x \to \infty} \frac{1 + \dfrac{2}{x}}{2 - \dfrac{2}{x} + \dfrac{1}{x^2}}$$

$$= \frac{\lim_{x \to \infty} \left(1 + \dfrac{2}{x}\right)}{\lim_{x \to \infty} \left(2 - \dfrac{2}{x} + \dfrac{1}{x^2}\right)}$$

$$= \frac{\lim_{x \to \infty} 1 + \lim_{x \to \infty} \dfrac{2}{x}}{\lim_{x \to \infty} 2 + \lim_{x \to \infty} \left(-\dfrac{2}{x}\right) + \lim_{x \to \infty} \dfrac{1}{x^2}}$$

$$= \frac{1 + 0}{2 + 0 + 0}$$

$$= \frac{1}{2}$$

9. $\lim_{x \to \infty} \dfrac{3x^3 + 2x - 1}{2x^4 - 3x^3 - 2}$

$$= \lim_{x \to \infty} \frac{\dfrac{3}{x} + \dfrac{2}{x^3} - \dfrac{1}{x^4}}{2 - \dfrac{3}{x} - \dfrac{2}{x^4}}$$

$$= \frac{\lim_{x \to \infty} \dfrac{3}{x} + \lim_{x \to \infty} \dfrac{2}{x^3} + \lim_{x \to \infty} \left(\dfrac{-1}{x^4}\right)}{\lim_{x \to \infty} 2 + \lim_{x \to \infty} \left(\dfrac{-3}{x}\right) + \lim_{x \to \infty} \left(\dfrac{-2}{x^4}\right)}$$

$$= \frac{0 + 0 + 0}{2 + 0 + 0}$$

$$= 0$$

11. $\lim_{x \to \infty} (\sqrt{x^2 + 4} - x)$

$$= \lim_{x \to \infty} (\sqrt{x^2 + 4} - x) \cdot \frac{(\sqrt{x^3 + 4} + x)}{(\sqrt{x^2 + 4} + x)}$$

$$= \lim_{x \to \infty} \frac{(x^2 + 4) - x^2}{\sqrt{x^2 + 4} + x}$$

$$= \lim_{x \to \infty} \frac{4}{\sqrt{x^2 + 4} + x}$$

$$= \lim_{x \to \infty} \frac{\dfrac{4}{x}}{\sqrt{\dfrac{x^2 + 4}{x^2}} + \dfrac{x}{x}}$$

$$= \lim_{x \to \infty} \frac{\dfrac{4}{x}}{\sqrt{1 + \dfrac{4}{x^2}} + 1}$$

$$= \frac{\lim_{x \to \infty} \dfrac{4}{x}}{\lim_{x \to \infty} \left(\sqrt{1 + \dfrac{4}{x^2}} + 1\right)}$$

$$= \frac{0}{\sqrt{\lim_{x \to \infty} \left(1 + \dfrac{4}{x^2}\right)} + \lim_{x \to \infty} 1}$$

$$= \frac{0}{1 + 1} = \frac{0}{2}$$

$$= 0$$

For Exercises 13–33, see the answer graphs in the back of the textbook.

13. $f(x) = -2x^3 - 9x^2 + 108x - 10$

$f'(x) = -6x^2 - 18x + 108$

$= -6(x^2 + 3x - 18) = 0$

$(x + 6)(x - 3) = 0$

Critical numbers: $x = -6$ and $x = 3$

Critical points: $(-6, -550)$ and $(3, 179)$

$f''(x) = -12x - 18$

$f''(-6) = 54 > 0$

$f''(3) = -54 < 0$

Relative maximum at 3, relative minimum at -6

Increasing on $(-6, 3)$

Decreasing on $(-\infty, -6)$ and $(3, \infty)$

$f''(x) = -12x - 18 = 0$

$-6(2x + 3) = 0$

$x = -\dfrac{3}{2}$

Point of inflection at $(-1.5, -185.5)$

Concave up on $(-\infty, -1.5)$

Concave down on $(-1.5, \infty)$

15. $f(x) = 2x^3 + \dfrac{7}{2}x^2 - 5x + 3$

$f'(x) = 6x^2 + 7x - 5 = 0$

$(2x - 1)(3x + 5) = 0$

Critical numbers: $x = \dfrac{1}{2}$ and $x = -\dfrac{5}{3}$

Critical points: $1/2, 1.625)$ and $(-5/3, 11.80)$

$f''(x) = 12x + 7$

$f''\left(\dfrac{1}{2}\right) = 13 > 0$

$f''\left(-\dfrac{5}{3}\right) = -13 < 0$

Relative maximum at $-5/3$, relative minimum at $1/2$

Increasing on $(-\infty, -5/3)$ and $(1/2, \infty)$

Decreasing on $(-5/3, 1/2)$

$f''(x) = 12x + 7 = 0$

$x = -\dfrac{7}{12}$

Point of inflection at $(-7/12, 6.71)$

Concave up on $(-7/12, \infty)$

Concave down on $(-\infty, -7/12)$

17. $f(x) = x^4 - 18x^2 + 5$

$f'(x) = 4x^3 - 36x = 0$

$4x(x^2 - 9) = 0$

$4(x)(x - 3)(x + 3) = 0$

Critical numbers $x = 0$, $x = 3$, and $x = -3$

Critical points: $(0, 5)$, $(3, -76)$ and $(-3, -76)$

$f''(x) = 12x^2 - 36$

$f''(-3) = 72 > 0$

$f''(0) = -36 < 0$

$f''(3) = 72 > 0$

Relative maximum at 0, relative minimum at -3 and 3

Increasing on $(-3, 0)$ and $(3, \infty)$

Decreasing on $(-\infty, -3)$ and $(0, 3)$

$f''(x) = 12x^2 - 36 = 0$

$12(x^2 - 3) = 0$

$x = \pm\sqrt{3}$

Points of inflection at $(\sqrt{3}, -40)$, $(-\sqrt{3}, -40)$

Concave up on $(-\infty, -\sqrt{3})$ and $(\sqrt{3}, \infty)$

Concave down on $(-\sqrt{3}, \sqrt{3})$

x-intercepts: $0 = x^4 - 18x^2 + 5$

Let $y = x^2$.

$0 = y^2 - 18y + 5$

Use the quadratic formula

$y = 9 \pm 2\sqrt{19} \approx 17.72$ or $.28$

$x = \pm 4.2$ or $\pm .53$

y-intercept: $y = 0^4 - 18(0)^2 + 5 = 5$

19. $f(x) = x^4 - 2x^3$

$f'(x) = 4x^3 - 6x^2 = 0$

$2x^2(2x - 3) = 0$

Critical numbers: $x = 0$ and $x = 3/2$

Critical point: $(0, 0)$ and

$(3/2, -27/16)$

$f''(x) = 12x^2 - 12x$

$f''(0) = 0$

$f''\left(\frac{3}{2}\right) = 9 > 0$

Relative minimum at $3/2$

No relative extremum at 0

Increasing on $(3/2, \infty)$

Decreasing on $(-\infty, 3/2)$

$f''(x) = 12x^2 - 12x = 0$

$12x(x - 1) = 0$

$x = 0, x = 1$

Points of inflection at $(0, 0)$ and $(1, -1)$

Concave up on $(-\infty, 0)$ and $(1, \infty)$

Concave down on $(0, 1)$

x-intercepts: $0 = x^4 - 2x^3$

$0 = x^3(x - 2)$

$x = 0, x = 2$

y-intercept: $y = 0^4 - 2(0)^3 = 0$

21. $f(x) = x + \dfrac{2}{x} = x + 2x^{-1}$

Vertical asymptote at $x = 0$

$f'(x) = 1 - 2x^{-2} = 1 - \dfrac{2}{x^2}$

$1 - \dfrac{2}{x^2} = 0$

$\dfrac{x^2 - 2}{x^2} = 0$

$x = \pm\sqrt{2}$

Critical numbers: $x = \sqrt{2}$ and $x = -\sqrt{2}$

Critical points: $(\sqrt{2}, 2\sqrt{2})$ and $(-\sqrt{2}, -2\sqrt{2})$

$f''(x) = 4x^{-3} = \dfrac{4}{x^3}$

$f''(\sqrt{2}) = \dfrac{2}{\sqrt{2}} > 0$

$f''(-\sqrt{2}) = -\dfrac{2}{\sqrt{2}} < 0$

Relative maximum at $-\sqrt{2}$

Relative minimum at $\sqrt{2}$

Increasing on $(-\infty, -\sqrt{2})$ and $(\sqrt{2}, \infty)$

Decreasing on $(-\sqrt{2}, 0)$ and $(0, \sqrt{2})$

(Recall that $f(x)$ does not exist at $x = 0$.)

$f''(x) = \dfrac{4}{x^3}$ is never zero.

There are no points of inflection.

Concave up on $(0, \infty)$

Concave down on $(-\infty, 0)$

$f(x)$ is never zero, so there are no x-intercepts.

$f(x)$ does not exist for $x = 0$ so there is no y-intercept.

$y = x$ is an oblique asymptote.

23. $f(x) = \dfrac{x^2 + 25}{x}$

Vertical asymptote at x = 0

$f'(x) = \dfrac{x(2x) - (x^2 + 25)}{x^2}$

$= \dfrac{x^2 - 25}{x^2}$

$f'(x) = \dfrac{(x - 5)(x + 5)}{x^2} = 0$

Critical numbers: x = 5 and x = -5
Critical points: (5, 10) and
(-5, -10)

$f''(x) = \dfrac{x^2(2x) - 2x(x^2 - 25)}{x^4}$

$= \dfrac{50}{x^3}$

$f''(-5) = -\dfrac{2}{5} < 0$

$f''(5) = \dfrac{2}{5} > 0$

Relative maximum at -5,
relative minimum at 5
Increasing on (-∞, -5) and (5, ∞)
Decreasing on (-5, 0) and (0, 5)
(Recall that f(x) does not exist at
x = 0.)

$f''(x) = \dfrac{50}{x^3}$ is never zero.

There are no points of inflection.
Concave up on (0, ∞)
Concave down on (-∞, 0)
f(x) is never zero, so there are no
x-intercepts.
f(x) doe not exist at x = 0 so there
is no y-intercept.
y = x is an oblique asymptote.

25. $f(x) = \dfrac{x - 1}{x + 1}$

Vertical asymptote at x = -1
Horizontal asymptote at y = 1

$f'(x) = \dfrac{(x + 1) - (x - 1)}{(x + 1)^2}$

$= \dfrac{2}{(x + 1)^2}$

f'(x) is never zero.
f'(x) fails to exist for x = -1.

$f''(x) = \dfrac{(x + 1)^2(0) - 2(2)(x + 1)}{(x + 1)^4}$

$= \dfrac{x - 3}{(x + 1)^3}$

f"(x) fails to exist for x = -1.
No critical values, no maximum or
minimum
No points of inflection

$f''(-2) = 5 > 0$

$f''(0) = -8 < 0$

Concave up on (-∞, -1)
Concave down on (-1, ∞)

x-intercept: $\dfrac{x - 1}{x + 1} = 0$

$x = 1$

y-intercept: $y = \dfrac{0 - 1}{0 + 1} = -1$

27. $f(x) = \dfrac{x}{x^2 + 1}$

Horizontal asymptote at y = 0

$f'(x) = \dfrac{(x^2 + 1)(1) - x(2x)}{(x^2 + 1)^2}$

$= \dfrac{1 - x^2}{(x^2 + 1)^2} = 0$

$1 - x^2 = 0$

Critical numbers: $x = 1$ and $x = -1$

Critical points: $(1, 1/2)$ and

$(-1, -1/2)$

$f''(x)$

$$= \frac{(x^2 + 1)^2(-2x) - (1 - x^2)(2)(x^2 + 1)(2x)}{(x^2 + 1)^3}$$

$$= \frac{-2x^3 - 2x - 4x + 4x^3}{(x^2 + 1)^3}$$

$$= \frac{2x^3 - 6x}{(x^2 + 1)^3}$$

$$f''(1) = -\frac{1}{2} < 0$$

$$f''(-1) = \frac{1}{2} > 0$$

Relative maximum at 1

Relative minimum at -1

Increasing on $(-1, 1)$

Decreasing on $(-\infty, -1)$ and $(1, \infty)$

$$f''(x) = \frac{2x^3 - 6x}{(x^2 + 1)^3} = 0$$

$$2x^3 - 6x = 0$$

$$2x(x^2 - 3) = 0$$

$$x = 0, \ x = \pm\sqrt{3}$$

Points of inflection at $(0, 0)$,

$\left(\sqrt{3}, \frac{\sqrt{3}}{4}\right)$ and $\left(-\sqrt{3}, -\frac{\sqrt{3}}{4}\right)$

Concave up on $(-\sqrt{3}, 0)$ and $(\sqrt{3}, \infty)$

Concave down on $(-\infty, -\sqrt{3})$ and $(0, \sqrt{3})$

x-intercept: $0 = \frac{x}{x^2 + 1}$

$$0 = x$$

y-intercept: $y = \frac{0}{0^2 + 1} = 0$

29. $f(x) = \frac{1}{x^2 - 4}$

Vertical asymptotes at $x = 2$ and

$x = -2$

Horizontal asymptote at $y = 0$

$$f'(x) = \frac{(x^2 - 4)(0) - 1(2x)}{(x^2 - 4)^2}$$

$$= \frac{-2x}{(x^2 - 4)^2} = 0$$

Critical number: $x = 0$

Critical point: $\left(0, -\frac{1}{4}\right)$

$$f''(x) = \frac{(x^2 - 4)^2(-2) - (-2x)(2)(x^2 - 4)(2x)}{(x^2 - 4)^4}$$

$$= \frac{6x^2 + 8}{(x^2 - 4)^3}$$

$$f''(0) = -\frac{1}{8} < 0$$

Relative maximum at 0

Increasing on $(-\infty, -2)$ and $(-2, 0)$

Decreasing on $(0, 2)$ and $(2, \infty)$

(Recall that $f(x)$ does not exist at

$x = 2$ and $x = -2$.)

$f''(x) = \frac{6x^2 + 8}{(x^2 - 4)^3}$ can never be zero.

There are no points of inflection.

$$f''(-3) = \frac{62}{125} > 0$$

$$f''(3) = \frac{62}{125} > 0$$

Concave up on $(-\infty, -2)$ and $(2, \infty)$

Concave down on $(-2, 2)$

$f(x)$ is never zero so there is no

x-intercept.

y-intercept: $y = \frac{1}{0^2 - 4} = -\frac{1}{4}$

Graphs for Exercises 31 and 33 may vary.
Examples are given in the answers at the
back of the textbook.

31.   (a) indicates that there can be no
      asymptotes, sharp "corners", holes,
      or jumps. The graph must be one
      smooth curve.

      (b) and (c) indicate relative maxima
      at -3 and 4 and a relative minimum at
      1.

      (d) and (e) are consistant with (g).

      (f) indicates turning points at the
      critical numbers -3 and 4.

33.   (a) indicates that the curve may not
      contain breaks.

      (b) and (c) indicate relative minima
      at -6 and 3 and a relative maximum at
      1.

      (d) and (e), when combined with (b)
      and (c) show that concavity does not
      change between relative
            extrema.

      (f) gives the y-intercept.

35. $\lim\limits_{x \to \infty} \bar{c}(x)$

      $= \lim\limits_{x \to \infty} \dfrac{15,000 + 64}{x}$

      $= \lim\limits_{x \to \infty} \dfrac{\dfrac{15,000}{x} + 6}{1}$

      $= \dfrac{\lim\limits_{x \to \infty} \dfrac{15,000}{x} + \lim\limits_{x \to \infty} 6}{\lim\limits_{x \to \infty} 1}$

      $= \dfrac{0 + 6}{1}$

      $= 6$

The average cost approaches 6 as the
number of tapes produced becomes very
large.

37. $A(h) = \dfrac{.17h}{h^2 + 2}$

      $\lim\limits_{h \to \infty} A(h) = \lim\limits_{h \to \infty} \dfrac{.17h}{\left(h + \dfrac{2}{h}\right)h}$

      $= \lim\limits_{h \to \infty} \dfrac{.17}{h + \dfrac{2}{h}}$

      $= 0$

The concentration of the drug in the
bloodstream approaches 0 as the number of hours after injection increases, which means that the drug in
the bloodstream eventually
dissipates.

Exercises 39–43 should be solved by computer or graphing calculator methods.
The solutions will vary according to the
method or computer program that is used.

39. $\lim\limits_{x \to \infty} \dfrac{\sqrt{9x^2 + 5}}{2x}$

      (a) The answer is 1.5.

41. $\lim\limits_{x \to -\infty} \dfrac{\sqrt{36x^2 + 2x + 7}}{3x}$

      (a) The answer is -2.

43. $\lim\limits_{x \to \infty} \dfrac{(1 + 5x^{1/3} + 2x^{5/3})^3}{x^5}$

      (a) The answer is 8.

Chapter 11   Review Exercises

5. $f(x) = x^2 - 5x + 3$

$f'(x) = 2x - 5 = 0$

Critical number: $x = \dfrac{5}{2}$

$f''(x) = 2 > 0$ for all x.

$f(x)$ is a minimum at $x = 5/2$.

$f(x)$ is increasing on $(5/2, \infty)$

and decreasing on $(-\infty, 5/2)$.

7. $f(x) = -x^3 - 5x^2 + 8x - 6$

$f'(x) = -3x^2 - 10x + 8$

$3x^2 + 10x - 8 = 0$

$(3x - 2)(x + 4) = 0$

Critical numbers: $x = \dfrac{2}{3}$ or $x = -4$

$f''(x) = -6x - 10$

$f''\left(\dfrac{2}{3}\right) = -14 < 0$

$f(x)$ is a maximum at $x = 2/3$.

$f''(-4) = 14 > 0$

$f(x)$ is a minimum at $x = -4$.

$f(x)$ is increasing on $(-4, 2/3)$;

decreasing on $(-\infty, -4)$ and on

$(2/3, \infty)$.

9. $f(x) = \dfrac{6}{x - 4}$

$f'(x) = \dfrac{-6}{(x - 4)^2} < 0$ for all x, but

not defined for $x = 4$.

$f(x)$ is never increasing, it is

decreasing on $(-\infty, 4)$ and $(4, \infty)$.

11. $f(x) = -x^2 + 4x - 8$

$f'(x) = -2x + 4 = 0$

Critical number: $x = 2$

$f''(x) = -2 < 0$ for all x, so f(2) is

a relative maximum.

$$f(2) = -4$$

Relative maximum of -4 at 2

13. $f(x) = 2x^2 - 8x + 1$

$f'(x) = 4x - 8 = 0$

Critical number: $x = 2$

$f''(x) = 4 > 0$ for all x, $f(2)$ is

a relative minimum.

$$f(2) = -7$$

Relative minimum of -7 at 2

15. $f(x) = 2x^3 + 3x^2 - 36x + 20$

$f'(x) = 6x^2 + 6x - 36 = 0$

$6(x^2 + x - 6) = 0$

$(x + 3)(x - 2) = 0$

Critical numbers: $x = -3$, $x = 2$

$f''(x) = 12x + 6$

$f''(-3) = -30 < 0$, so a maximum occurs

at $x = -3$.

$f''(2) = 30 > 0$, so a minimum occurs

at $x = 2$.

$f(-3) = 101$

$f(2) = -24$

Relative maximum of 101 at -3

Relative minimum of -24 at 2

17. $f(x) = 3x^4 - 5x^2 - 11x$

$f'(x) = 12x^3 - 10x - 11$

$f''(x) = 36x^2 - 10$

$f''(1) = 36(1)^2 - 10 = 26$

$f''(-3) = 36(-3)^2 - 10 = 314$

$f''(1) = \dfrac{1}{(1 + 1)^{3/2}}$

$= \dfrac{1}{2^{3/2}} \approx .354$

$f''(-3) = \dfrac{1}{(9 + 1)^{3/2}}$

$= \dfrac{1}{10^{3/2}} \approx .032$

19. $f(x) = \dfrac{5x - 1}{2x + 3}$

$f'(x) = \dfrac{(2x + 3)5 - (5x - 1)2}{(2x + 3)^2}$

$= \dfrac{17}{(2x + 3)^2}$

$f''(x) = \dfrac{(2x + 3)^2(0) - 17(2)(2x + 3)(2)}{(2x + 3)^4}$

$= \dfrac{-68}{(2x + 3)^3}$ or $-68(2x + 3)^{-3}$

$f''(1) = \dfrac{-68}{[2(1) + 3]^3} = -\dfrac{68}{125}$

$f''(-3) = \dfrac{-68}{[2(-3) + 3]^3}$

$= \dfrac{-68}{-27} = \dfrac{68}{27}$

21. $f(t) = \sqrt{t^2 + 1} = (t^2 + 1)^{1/2}$

$f'(t) = \dfrac{1}{2}(t^2 + 1)^{-1/2}(2t)$

$= t(t^2 + 1)^{-1/2}$

$f''(t) = (t^2 + 1)^{-1/2}(1)$

$\qquad + t[\left(-\dfrac{1}{2}\right)(t^2 + 1)^{-3/2}(2t)]$

$= (t^2 + 1)^{-1/2} - t^2(t^2 + 1)^{-3/2}$

$= \dfrac{1}{(t^2 + 1)^{1/2}} - \dfrac{t^2}{(t^2 + 1)^{3/2}}$

$= \dfrac{t^2 + 1 - t^2}{(t^2 + 1)^{3/2}}$

$= (t^2 + 1)^{-3/2}$

or $\dfrac{1}{(t^2 + 1)^{3/2}}$

23. $f(x) \; -x^2 + 5x + 1; \; [1, 4]$

$f'(x) = -2x + 5 = 0$ when

$x = \dfrac{5}{2}$

$f(1) = 5$

$f\left(\dfrac{5}{2}\right) = \dfrac{29}{4}$

$f(4) = 5$

Absolute maximum of 29/4 at 5/2;
absolute minimum of 5 at 1 and 4

25. $f(x) = x^3 + 2x^2 - 15x + 3; \; [-4, 2]$

$f'(x) = 3x^2 + 4x - 15 = 0$ when

$(3x - 5)(x + 3) = 0$

$x = \dfrac{5}{3}$ or $x = -3$

$f(-4) = 31$

$f(-3) = 39$

$f\left(\dfrac{5}{3}\right) = -\dfrac{319}{27}$

$f(2) = -11$

Absolute maximum of 39 at -3;
absolute minimum of -319/27 at 5/3

27. $\lim\limits_{x \to \infty} g(x)$ does not exist because $g(x)$
increases without bound as x gets
smaller and smaller.

29. $\lim\limits_{x\to\infty}\dfrac{x^2+5}{5x^2-1}$

$$=\lim\limits_{x\to\infty}\frac{\dfrac{x^2}{x^2}-\dfrac{5}{x^2}}{\dfrac{5x^2}{x^2}-\dfrac{1}{x^2}}$$

$$=\frac{\lim\limits_{x\to\infty}1-\lim\limits_{x\to\infty}\dfrac{5}{x^2}}{\lim\limits_{x\to\infty}5-\lim\limits_{x\to\infty}\dfrac{1}{x^2}}$$

$$=\frac{1-0}{5-0}=\frac{1}{5}$$

31. $\lim\limits_{x\to-\infty}\left(\dfrac{3}{4}+\dfrac{2}{x}-\dfrac{5}{x^2}\right)$

$$=\lim\limits_{x\to-\infty}\frac{3}{4}+\lim\limits_{x\to-\infty}\frac{2}{x}+\lim\limits_{x\to-\infty}\frac{-5}{x^2}$$

$$=\frac{3}{4}+0+0$$

$$=\frac{3}{4}$$

For Exercises 33–45, see the answer graphs in the back of the textbook.

33. $f(x)=-2x^3-\dfrac{1}{2}x^2+x-3$

$f'(x)=-6x^2-x+1=0$

$(3x-1)(2x+1)=0$

Critical numbers: $x=1/3$ and $x=-1/2$

Critical points: $(1/3,\,-2.80)$ and $(-1/2,\,-3.375)$

$f''(x)=-12x-1$

$f''\left(\dfrac{1}{3}\right)=-5<0$

$f''\left(-\dfrac{1}{2}\right)=5>0$

Relative maximum at $1/3$

Relative minimum at $-1/2$

Increasing on $(-1/2,\,1/3)$

Decreasing on $(-\infty,\,-1/2)$ and $(1/3,\,\infty)$

$f''(x)=-12x-1=0$

$$x=-\frac{1}{12}$$

Point of inflection at $(-1/12,\,-3.09)$

Concave up on $(-\infty,\,-1/12)$

Concave down on $(-1/12,\,\infty)$

y–intercept:

$$y=-2(0)^3-\frac{1}{2}(0)^2+(0)-3=-3$$

35. $f(x)=x^4-\dfrac{4}{3}x^3-4x^2+1$

$f'(x)=4x^3-4x^2-8x=0$

$4x(x^2-x-2)=0$

$4x(x-2)(x+1)=0$

Critical numbers: $x=0$, $x=2$, or $x=-1$

Critical points: $(0,1)$, $(2,\,-29/3)$ and $(-1,\,-2/3)$

$f''(x)=12x^2-8x-8$

$\qquad\quad=4(3x^2-2x-2)$

$f''(-1)=12>0$

$f''(0)=-8<0$

$f''(2)=8>0$

Relative maximum at 0

Relative minima at -1 and 2

Increasing on $(-1,\,0)$ and $(2,\,\infty)$

Decreasing on $(-\infty,\,-1)$ and $(0,\,2)$

$f''(x)=4(3x^2-2x-2)=0$

$$x=\frac{2\pm\sqrt{4-(-24)}}{6}$$

$$=\frac{1\pm\sqrt{7}}{3}$$

Points of inflection at

$\left(\dfrac{1 + \sqrt{7}}{3}, -3.47\right)$ and $\left(\dfrac{1 - \sqrt{7}}{3}, .11\right)$

Concave up on $\left(-\infty, \dfrac{1 - \sqrt{7}}{3}\right)$ and

$\left(\dfrac{1 + \sqrt{7}}{3}, \infty\right)$

Concave down on $\left(\dfrac{1 - \sqrt{7}}{3}, \dfrac{1 + \sqrt{7}}{3}\right)$

y-intercept:

$y = (0)^4 - \dfrac{4}{3}(0)^3 - 4(0)^2 + 1 = 1$

37. $f(x) = \dfrac{x - 1}{2x + 1}$

Vertical asymptote at $x = -\dfrac{1}{2}$

Horizontal asymptote at $y = \dfrac{1}{2}$

$f'(x) = \dfrac{(2x + 1) - (x - 1)(2)}{(2x + 1)^2}$

$= \dfrac{3}{(2x + 1)^2}$

f' is never zero.

$f(x)$ has no extrema.

$f''(x) = \dfrac{-12}{(2x + 1)^3}$

$f''(0) = -12 < 0$

$f''(-1) = 12 > 0$

Concave up on $\left(-\infty, -\dfrac{1}{2}\right)$

Concave down on $\left(-\dfrac{1}{2}, \infty\right)$

x-intercept: $\dfrac{x - 1}{2x + 1} = 0$

$x = 1$

y-intercept: $y = \dfrac{0 - 1}{2(0) + 1} = -1$

39. $f(x) = -4x^3 - x^2 + 4x + 5$

$f'(x) = -12x^2 - 2x + 4$

$= -2(6x^2 + x - 2) = 0$

$(3x + 2)(2x - 1) = 0$

Critical numbers: $x = -2/3$ and
$x = 1/2$

Critical points: $(-2/3, 3.07)$ and
$(1/2, 6.25)$

$f''(x) = -24x - 2$

$= -2(12x + 1)$

$f''\left(-\dfrac{2}{3}\right) = 14 > 0$

$f''\left(\dfrac{1}{2}\right) = -14 < 0$

Relative maximum at $1/2$
Relative minimum at $-2/3$
Increasing on $(-2/3, 1/2)$
Decreasing on $(-\infty, -2/3)$ and $(1/2, \infty)$

$f''(x) = -2(12x + 1) = 0$

$x = -\dfrac{1}{12}$

Point of inflection at $\left(-\dfrac{1}{12}, 4.66\right)$

Concave up on $\left(-\infty, -\dfrac{1}{12}\right)$

Concave down on $\left(-\dfrac{1}{12}, \infty\right)$

y-intercept:

$y = -4(0)^3 - (0)^2 + 4(0) + 5 = 5$

41. $f(x) = x^4 + 2x^2$

$f'(x) = 4x^3 + 4x$

$= 4x(x^2 + 1) = 0$

Critical number: $x = 0$
Critical point: $(0, 0)$

$f''(x) = 12x^2 + 4$

$f''(0) = 4 > 0$

Relative minimum at 0

Increasing on $(0, \infty)$

Decreasing on $(-\infty, 0)$

$f''(x) = 12x^2 + 4$

$\qquad 4(3x^2 + 1) \neq 0$ for any x

No points of inflection

$f''(-1) = 16 > 0$

$f''(1) = 16 > 0$

Concave up on $(-\infty, \infty)$

x-intercept and y-intercept at $(0, 0)$

43. $f(x) = \dfrac{x^2 + 4}{x}$

Vertical asymptote at $x = 0$

Oblique asymptote $y = x$

$f'(x) = \dfrac{x(2x) - (x^2 + 4)}{x^2}$

$\qquad = \dfrac{x^2 - 4}{x^2} = 0$

Critical numbers: $x = -2$ and $x = 2$

Critical points: $(-2, -4)$ and $(2, 4)$

$f''(x) = \dfrac{8}{x^3}$

$f''(-2) = -1 < 0$

$f''(2) = 1 > 0$

Relative maximum at -2

Relative minimum at 2

Increasing on $(-\infty, -2)$ and $(2, \infty)$

Decreasing on $(-2, 0)$ and $(0, 2)$

$f''(x) = \dfrac{8}{x^3} > 0$ for all x.

No inflection points

Concave up on $(0, \infty)$

Concave down on $(-\infty, 0)$

No x- or y-intercepts

45. $f(x) = \dfrac{2x}{3 - x}$

Vertical asymptote at $x = 3$

Horizontal asymptote at $y = -2$

$f'(x) = \dfrac{(3 - x)2 - (2x)(-1)}{(3 - x)^2}$

$\qquad = \dfrac{6}{(3 - x)^2}$

$f'(x)$ is never zero.

No critical values, no relative extrema

$f'(0) = \dfrac{2}{3} > 0$

$f'(4) = 6 > 0$

Increasing on $(-\infty, 3)$ and $(3, \infty)$

$f''(x) = \dfrac{12}{(3 - x)^3}$

$f''(x)$ is never zero; no inflection points.

$f''(0) = \dfrac{12}{27} > 0$

$f''(4) = -12 < 0$

Concave up on $(-\infty, 3)$

Concave down on $(3, \infty)$

x-intercept and y-intercept at $(0, 0)$

CHAPTER 11   TEST

Find the largest open intervals where each of the following functions is (a) increasing or (b) decreasing.

1.

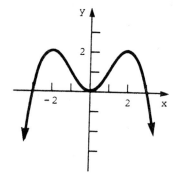

2. $f(x) = 3x^3 - 3x^2 - 3x + 5$

3. $f(x) = \dfrac{x - 2}{x + 3}$

4.   For a certain product, the demand equation is $p = 10 - .05x$. The cost function for this product is $C = 5x - .75$. Over what interval(s) is the profit increasing?

5.   What are critical numbers? Why are they significant?

Find the locations and values of all relative maxima and minima.

6. $f(x) = 4x^3 - \dfrac{9}{2}x^2 - 3x - 1$ 7. $f(x) = 4x^{2/3}$

8.   A manufacturer estimates that the cost in dollars per unit for a production run of x thousand units is given by $C = 3x^2 - 60x + 320$. How many thousand units should be produced during each run to minimize the cost per unit, and what is the minimum cost per unit?

9.   Use the derivative to find the vertex of the parabola $y = -2x^2 + 6x + 9$.

10.   What conditions must a function, $f(x)$, meet for the Extreme Value Theorem to hold?

Find the locations of any absolute extrema for the functions with graphs as follows.

11.

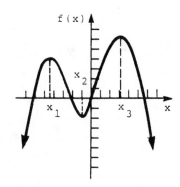

12.

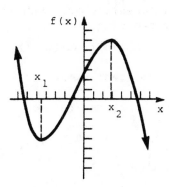

Find the locations of all absolute extrema for the functions defined as follow, with the specified domwins.

13. $f(x) = x^3 - 12x$; $[0, 4]$

14. $f(x) = \dfrac{x^2 + 4}{x}$; $[-3, -1]$

15. Why is it important to check the endpoints of the domain when checking for absolute extrema?

Find the second derivative of each function, then find f"(−3).

16. $f(x) = -5x^3 + 3x^2 - x + 1$

17. $f(t) = \sqrt{t^2 - 5}$

18. Find the largest open intervals where $f(x) = x^3 - 3x^2 - 9x - 1$ is concave upward or concave downward. Find the location of any points of inflection. Graph the function.

19. The function $s(t) = t^3 - 9t^2 + 15t + 25$ gives the displacement in centimeters at time t (in seconds) of a particle moving along a line. Find the velocity and acceleration functions. Then find the velocity and acceleration at t = 0 and t = 2.

20. What is the difference between velocity and acceleration?

Evaluate each limit that exists.

21. $\lim\limits_{x \to \infty} \dfrac{2x}{7x + 1}$

22. $\lim\limits_{x \to -\infty} \dfrac{3x^2 + 2x}{4x^3 - x}$

23. Graph $f(x) = \frac{2}{3}x^3 - \frac{5}{2}x^2 - 3x + 1$. Give critical points, regions where the function is increasing or decreasing, points of inflection, regions where the function is concave up or concave down.

24. Sketch a graph of a single function that has all the properties listed.

(a) continuous for all real numbers

(b) increasing on $(-3, 2)$ and $(4, \infty)$

(c) decreasing on $(-\infty, -3)$ and $(2, 4)$

(d) concave upward on $(-\infty, 0)$

(e) concave downward on $(0, 4)$ and $(4, \infty)$

(f) differentiable everywhere except $x = 4$

(g) $f'(-3) = f'(2) = 0$

(h) inflection point at $(0, 0)$

25. The cost for manufacturing a certain computer component is

$$C(x) = 12{,}000 + 11x$$

where x is the number of components manufactured. The average cost per component, denoted by $\overline{C}(x)$, is found by dividing $C(x)$ by x. Find and interpret $\lim\limits_{x \to \infty} \overline{C}(x)$.

CHAPTER 11 TEST ANSWERS

1. (a) $(-\infty, -2)$ and $(0, 2)$ (b) $(-2, 0)$ and $(2, \infty)$ 2. (a) $(-\infty, -1/3)$ and

$(1, \infty)$ (b) $(-1/3, 1)$ 3. (a) $(-\infty, -3)$ and $(-3, \infty)$ (b) never decreasing

4. $(0, 50)$

5. Critical numbers are x–values for which $f'(x) = 0$ or $f'(x)$ does not exist. They tell where the derivative may change signs, and are used to tell where a function is increasing or decreasing.

6. Relative maximum of $-19/32$ at $1/4$; relative minimum of $-9/2$ at 1

7. Relative minimum of 0 at 0 8. 10 thousand units; $20 per unit

9. $(3/2, 27/2)$ 10. $f(x)$ must be continuous on a closed interval.

11. Absolute maximum at x_3 12. No absolute extrema

13. Absolute maximum of 16 at 4; absolute minimum of -16 at 2

14. Absolute maximum of -4 at -2; absolute minimum of -5 at -1

15. The smallest or largest value in a closed interval may occur at the endpoints.

16. $f''(x) = -30x + 6$; $f''(-3) = 96$ 17. $f''(t) = \dfrac{-5}{(t^2 - 5)^{3/2}}$; $f''(-3) = -\dfrac{5}{8}$

18. Concave upward on $(1, \infty)$; concave downward on $(-\infty, 1)$; point of inflection $(1, -12)$

19. $v(t) = 3t^2 - 18t + 15$; $a(t) = 6t - 18$; $v(0) = 15$ cm/sec; $a(0) = -18$ cm/sec^2; $v(2) = -9$ cm/sec; $a(2) = -6$ cm/sec^2

20. Velocity gives the rate of change of position relative to time. Accelation gives the rate of change of velocity.

21. $2/7$ 22. 0

$f(x) = x^3 - 3x^2 - 9x - 1$

23. Critical points: $\left(-\frac{1}{2}, \frac{43}{24}\right)$ and $\left(3, -\frac{25}{2}\right)$; increasing on $\left(-\infty, -\frac{1}{2}\right)$ and $(3, \infty)$;

decreasing on $\left(-\frac{1}{2}, 3\right)$; point of inflection at $\left(\frac{5}{4}, -\frac{257}{48}\right)$; concave up on $\left(\frac{5}{4}, \infty\right)$;

concave down on $\left(-\infty, \frac{5}{4}\right)$

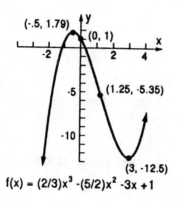

(-.5, 1.79)   (0, 1)   (1.25, -5.35)   (3, -12.5)

$f(x) = (2/3)x^3 - (5/2)x^2 - 3x + 1$

24.

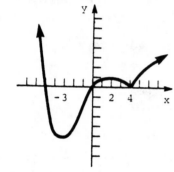

25. 11; the average cost approaches 11 as the number of components manufactured becomes very large.

CHAPTER 12 APPLICATIONS OF THE DERIVATIVE

Section 12.1

1. $x + y = 100$, $P = xy$

   (a) $y = 100 - x$

   (b) $P = xy = x(100 - x)$
   $$= 100x - x^2$$

   (c) Since $y = 100 - x$ and x and y are nonnegative numbers, $x \geq 0$ and $100 - x \geq 0$ or $x \leq 100$. The domain of P is $[0, 100]$

   (d) $P' = 100 - 2x$
   $$100 - 2x = 0$$
   $$2(50 - x) = 0$$
   $$x = 50$$

   (e)

x	P
0	0
50	2500
100	0

   (f) From the chart, the maximum value of P is 2500; this occurs when $x = 50$ and $y = 50$.

3. $x + y = 200$
   $P = x^2 + y^2$

   (a) $y = 200 - x$

   (b) $P = x^2 + (200 - x)^2$
   $$= x^2 + 40,000 - 400x + x^2$$
   $$= 2x^2 - 400x + 40,000$$

   (c) Since $y = 200 - x$ and x and y are nonnegative numbers, the domain of P is $[0, 200]$.

   (d) $P' = 4x - 400$
   $$4x - 400 = 0$$
   $$4(x - 100) = 0$$
   $$x = 100$$

   (e)

x	P
0	40,000
100	20,000
200	40,000

   (f) The minimum value of P is 20,000; this occurs when $x = 100$ and $y = 100$.

5. $x + y = 150$

   Maximize x^2y.

   (a) $y = 150 - x$

   (b) Let $P = x^2y = x^2(150 - x)$
   $$= 150x^2 - x^3$$

   (c) Since $y = 150 - x$ and x and y are nonnegative, the domain of P is $[0, 150]$.

   (d) $P' = 300x - 3x^2$
   $$300x - 3x^2 = 0$$
   $$3x(100 - x) = 0$$
   $$x = 0 \quad \text{or} \quad x = 100$$

   (e)

x	P
0	0
100	50,000
150	0

   (f) The maximum value of x^2y occurs when $x = 100$ and $y = 50$. The maximum value is 500,000.

7. x − y = 10

    Minimize xy.

    (a) y = x − 10

    (b) Let P = xy = x(x − 10)

                    = x² − 10x

    (c) Since y = x − 10 and x and y are
        nonnegative numbers, the domain
        of P is [10, ∞).

    (d) P′ = 2x − 10

        2x − 10 = 0

        2(x − 5) = 0

             x = 5

        Note that 5 is not in the
        domain.

    (e)
x	P
10	0

    (f) P′(10) = 10 > 0 implies that
        P(10) is a minimum. The minimum
        value of xy occurs when x = 10
        and y = 0. The minimum value is
        0.

9. Let x = the width
    and y = the length.

    (a) The perimeter is

        P = 2x + y

          = 1200.

    So, y = 1200 − 2x.

    (b) Area = xy = x(1200 − 2x)

        A(x) = 1200x − 2x²

    (c) A′ = 1200 − 4x

            1200 − 4x = 0

                1200 = 4x

                 300 = x

A″ = −4, which implies that
x = 300 m leads to the maximum area.

(d) If x = 300,

    y = 1200 − 2(300) = 600.

    The maximum area is

    (300)(600) = 180,000 m².

11. Let x = the width of the rectangle
    and y = the total length of the
            rectangle.

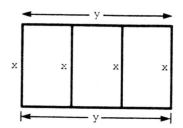

An equation for the fencing is

        3600 = 4x + 2y

          2y = 3600 − 4x

           y = 1800 − 2x.

Area = xy = x(1800 − 2x)

    A(x) = 1800x − 2x²

A′ = 1800 − 4x

1800 − 4x = 0

    1800 = 4x

     450 = x

A″ = −4, which implies that x = 450
is the location of a maximum.

If x = 450, y = 1800 − 2(450) = 900.

The maximum area is

    (450)(900) = 405,000 m².

13. Let $8 - x =$ the distance the hunter will travel on the river.

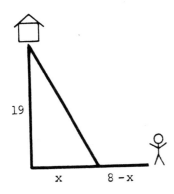

Then $\sqrt{19^2 + x^2} =$ the distance he will travel on land.

Since the rate on the river is 5 mph, the rate on land is 2 mph, and $t = d/r$,

$$\frac{8 - x}{5} = \text{the time on the river}$$

$$\frac{\sqrt{361 + x^2}}{2} = \text{the time on land.}$$

The total time is

$$T(x) = \frac{8 - x}{5} + \frac{\sqrt{361 + x^2}}{2}$$

$$= \frac{8}{5} - \frac{1}{5}x + \frac{1}{2}(361 + x^2)^{1/2}.$$

$$T'(x) = -\frac{1}{5} + \frac{1}{4} \cdot 2x(361 + x^2)^{-1/2}$$

$$-\frac{1}{5} + \frac{x}{2(361 + x^2)^{1/2}} = 0$$

$$\frac{1}{5} = \frac{x}{2(361 + x^2)^{1/2}}$$

$$2(361 + x^2)^{1/2} = 5x$$

$$4(361 + x^2) = 25x^2$$

$$1444 + 4x^2 = 25x^2$$

$$1444 = 21x^2$$

$$\frac{38}{\sqrt{21}} = x$$

$$8.29 \approx x$$

8.29 is not possible, since the cabin is only 8 miles west. Check the endpoints.

y	T(x)
0	17.5
8	10.3

$T(x)$ is minimized when $x = 8$.

The distance along the river is given by $8 - x$, so the hunter should travel $8 - 8 = 0$ miles along the river. He should complete the entire trip on land.

15. $p(x) = 100 - \dfrac{x}{10}$

(a) Revenue from sale of x thousand candy bars:

$$R(x) = 1000xp$$

$$= 1000x\left(100 - \frac{x}{10}\right)$$

$$= 100,000x - 100x^2$$

(b) $R'(x) = 100,000 - 200x$

$$100,000 - 200x = 0$$

$$100,000 = 200x$$

$$500 = x$$

The maximum revenue occurs when 500 thousand bars are sold.

(c) $R(500) = 100,000(500)$

$$- 100(500)^2$$

$$= 25,000,000$$

The maximum revenue is 25,000,000 cents.

17. $G(x) = \frac{1}{32}\left(\frac{64}{x} + \frac{x}{50}\right)$

The cost of fuel on the 400-mile trip is \$1.60 per gallon.
The total cost of fuel is

$C(x) = $ (Cost per gallon)
$\times\ G(x) \times$ (miles traveled)

$= 1.60\left(\frac{1}{32}\right)\left(\frac{64}{x} + \frac{x}{50}\right)\ (400)$

$= 20\left(\frac{64}{x} + \frac{x}{50}\right)$

$= \frac{1280}{x} + \frac{2}{5}x$ dollars.

(a) $C'(x) = \frac{-1280}{x^2} + \frac{2}{5} = 0$

$x = \sqrt{3200}$

≈ 56.6 mi/hr

$C''(x) = \frac{-1280(-2)}{x^3} > 0$ for all

$x > 0$ which implies $C(x)$ is
minimized when $x \approx 56.6$ mph.

(b) $C(10\sqrt{32}) = \frac{1280}{10\sqrt{32}} + \frac{2}{5}(10\sqrt{32})$

$\approx 22.61 + 22.63$

$= 45.24$

The minimum total cost is
\$45.24.

19. Let $x = $ the length at \$3 per foot
$y = $ the width at \$6 per foot

$xy = 20,000$

$y = \frac{20,000}{x}$

Perimeter $= 2x + 2y = 2x + \frac{40,000}{x}$

Cost $= 2x(\$3) + \frac{40,000}{x}(\$6)$

$= 6x + \frac{240,000}{x}$

Minimize cost.

$C' = 6 - \frac{240,000}{x^2}$

$6 - \frac{240,000}{x^2} = 0$

$6 = \frac{240,000}{x^2}$

$6x^2 = 240,000$

$x^2 = 40,000$

$x = 200$

$y = \frac{20,000}{200} = 100$

400 ft at \$3 per foot will cost
\$1200.   200 ft at \$6 per foot will
cost \$1200.   The entire cost will
be \$2400.

21. Profit is 5 dollars per seat for
$60 \le x \le 80$ seats.
Profit (in dollars) is

$5 - .05(x - 80)$

per seat for $x > 80$ seats.
We expect that the number of seats
which makes the total profit a maximum will be greater than 80 because
after 80 the profit is still increasing, though at a slower rate.
(Thus we know the function is concave down and its one extrema will
be a maximum.)

(a) The total profit for x seats is

$P(x) = [5 - .05(x - 80)]x$

$= (5 - .05x + 4)x$

$= (9 - .05x)x$

$= 9x - .05x^2.$

$$P'(x) = 9 - .10x$$
$$9 - .10x = 0$$
$$9 = .10x$$
$$x = 90$$

90 seats will produce maximum profit.

(b) $P(90) = 9(90) - .05(90)^2$

$$= 810 - .05(8100)^2$$

$$= 405$$

The maximum profit is \$405.

23. Let x = a side of the base
    h = the height of the box.

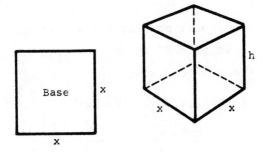

An equation for the volume of the box is

$$V = x^2 h.$$
$$32 = x^2 h$$
$$h = \frac{32}{x^2}.$$

The box is open at the top so the area of the surface material m(x) in square inches is the area of the base plus the area of the four sides.

$$m(x) = x^2 + 4xh$$

$$= x^2 + 4x\left(\frac{32}{x^2}\right)$$

$$= x^2 + \frac{128}{x}$$

$$m'(x) = 2x - \frac{128}{x^2}$$

$$\frac{2x^3 - 128}{x^2} = 0$$

$$2x^3 - 128 = 0$$

$$2(x^3 - 64) = 0$$

$$x = 4$$

$m'(x) = 2 + \dfrac{256}{x^3} > 0$ since x > 0.

So, x = 4 minimizes the surface material.

If $x = 4$, $h = \dfrac{32}{x^2} = \dfrac{32}{16} = 2$.

The dimensions that will minimize the surface material are 4 in by 4 in by 2 in.

25. Let   x = the width.
    Then 2x = the length
           h = the height.

An equation for volume is

$$V = (2x)(x)h = 2x^2 h$$
$$36 = 2x^2 h.$$

So, $h = \dfrac{18}{x^2}$.

The surface area S(x) is the sum of the areas of the base and the four sides.

$$S(x) = (2x)(x) + 2xh + 2(2x)h$$

$$= 2x^2 + 6xh$$

$$= 2x^2 + 6x\left(\frac{18}{x^2}\right)$$

$$= 2x^2 + \frac{108}{x}$$

$$S'(x) = 4x - \frac{108}{x^2}$$

$$\frac{4x^3 - 108}{x^2} = 0$$

$$4(x^3 - 27) = 0$$

$$x = 3$$

$$S''(x) = 4 + \frac{108(2)}{x^3}$$

$$= 4 + \frac{216}{x^3} > 0 \text{ and } x > 0.$$

So $x = 3$ minimizes the surface material.

If $x = 3$, $h = \frac{18}{x^2} = \frac{18}{9} = 2$.

The dimensions are 3 ft by 6 ft by 2 ft.

27. 120 centimeters of ribbon are available; it will cover 4 heights and 8 radii.

$$4h + 8r = 120$$

$$h + 2r = 30$$

$$h = 30 - 2r$$

$$V = \pi r^2 h$$

$$V = \pi r^2 (30 - 2r)$$

$$= 30\pi r^2 - 2\pi r^3$$

Maximize volume.

$$V' = 60\pi r - 6\pi r^2$$

$$60\pi r - 6\pi r^2 = 0$$

$$6\pi r(10 - r) = 0$$

$$r = 0 \quad \text{or} \quad r = 10$$

r	V
0	0
10	1000π

$V'' = 6\pi - 12\pi r < 0$ for $r = 10$, which implies that $r = 10$ gives maximum volume.

When $r = 10$, $h = 30 - 2(10) = 10$. The volume is maximized when the radius and height are both 10 cm.

29. $V = \pi r^2 h$

$16 = \pi r^2 h$,

so

$$h = \frac{16}{\pi r^2}.$$

The total cost is the sum of the cost of the top and bottom and the cost of the sides, or

$$C = \$2(2)(\pi r^2) + \$1(2\pi rh)$$

$$= 4(\pi r^2) + 1(2\pi r)\left(\frac{16}{\pi r^2}\right)$$

$$= 4\pi r^2 + \frac{32}{r}.$$

Minimize cost.

$$C' = 8\pi r - \frac{32}{r^2}$$

$$8\pi r - \frac{32}{r^2} = 0$$

$$8\pi r^3 = 32$$

$$\pi r^3 = 4$$

$$r = \sqrt[3]{\frac{4}{\pi}}$$

$$\approx 1.08$$

$$h = \frac{16}{\pi(1.08)^2} \approx 4.34$$

The radius should be 1.08 ft and the height should be 4.34 ft. If these rounded values for the height and radius are used, the cost is

$2(2)(\pi r^2) + \$1(2\pi rh)$

$= 4\pi(1.08)^2 + 2\pi(1.08)(4.34)$

$= \$44.11.$

31. Let x = the width of printed material

and y = the length of printed material.

Then, the area of the printed material is

$$xy = 36$$

$$y = \frac{36}{x}.$$

Also, x + 2 = the width of a page

and   y + 3 = the length of a page.

The area of a page is

$A = (x + 2)(y + 3)$

$= xy + 2y + 3x + 6$

$= 36 + 2\left(\frac{36}{x}\right) + 3x + 6$

$= 42 + \frac{72}{x} + 3x.$

$A' = -\frac{72}{x^2} + 3 = 0$

$x^2 = 24$

$x = \sqrt{24}$

$= 2\sqrt{6}$

$A'' = \frac{216}{x^3} > 0$ when $x = 2\sqrt{6}$, which implies that A is minimized when $x = 2\sqrt{6}$.

$$y = \frac{36}{x} = \frac{36}{2\sqrt{6}} = \frac{18}{\sqrt{6}} = \frac{18\sqrt{6}}{6} = 3\sqrt{6}$$

The width of a page is

$x + 2 = 2\sqrt{6} + 2$

≈ 6.9 in.

The length of a page is

$y + 3 = 3\sqrt{6} + 3$

≈ 10.3 in.

33. Distance on shore: 7 - x miles
Cost on shore: $400 per mile
Distance underwater: $\sqrt{x^2 + 36}$
Cost underwater: $500 per mile
Find the distance from A, that is, (7 - x), to minimize cost, C(x).

$C(x) = (7 - x)(400) + (\sqrt{x^2 + 36})(500)$

$= 2800 - 400x + 500(x^2 + 36)^{1/2}$

$C'(x) = -400 + 500\left(\frac{1}{2}\right)(x^2 + 36)^{-1/2}(2x)$

$= -400 + \frac{500x}{\sqrt{x^2 + 36}}$

If $C'(x) = 0$,

$\frac{500x}{\sqrt{x^2 + 36}} = 400$

$\frac{5x}{4} = \sqrt{x^2 + 36}$

$\frac{25}{16}x^2 = x^2 + 36$

$\frac{9}{16}x^2 = 36$

$x^2 = \frac{36 \cdot 16}{9}$

$x = \frac{6 \cdot 4}{3} = 8.$

(Discard the negative solution.) x = 8 is impossible since Point A is only 7 miles from point C.

Check the endpoints.

x	C(x)
0	5800
7	4610

The cost is minimized when x = 7. $7 - x = 7 - 7 = 0$, so the company should angle the cable at Point A.

35. $H(S) = f(S) - S$

$f(S) = -.1S^2 + 11S$

$H(S) = -.1S^2 + 11S - S$

$\quad\quad = -.1S^2 + 10S$

$H'(S) = -.2S + 10$

If $H'(S) = -.2S + 10 = 0$,

$\quad\quad .2S = 10$

$\quad\quad\quad S = 50.$

The number of creatures needed to sustain the population is $S_0 = 50$ thousand.

$H''(S) = -.2 < 0$,

so $H(S)$ is a maximum at $S_0 = 50$.

$H(S_0) = -.1(50)^2 + 10(50)$

$\quad\quad\quad = -250 + 500$

$\quad\quad\quad = 250$

The maximum sustainable harvest is 250 thousand.

37. $H(S) = f(S) - S$

$f(S) = 15\sqrt{S}$

$H(S) = 15S^{1/2} - S$

$H'(S) = \frac{1}{2} \cdot 15S^{-1/2} - 1$

If $H'(S) = 0$,

$\quad\quad 7.5 = \sqrt{S}$

$\quad\quad\quad S = 56.25.$

The number of creatures needed to sustain the population is $S_0 = 56.25$ thousand.

$H''(S) = -\frac{1}{4} \cdot 15S^{-3/2} < 0$,

so H is maximum at $S_0 = 56.25$.

$H(S_0) = 15\sqrt{56.25} - 56.25$

$\quad\quad\quad = 56.25$

The maximum sustainable harvest is 56.25 thousand.

39. $H(S) = f(S) - S$

$f(S) = \dfrac{25S}{S + 2}$

$H'(S) = \dfrac{(S + 2)(25) - 25S}{(S + 2)^2} - 1$

$\quad\quad = \dfrac{25S + 50 - 25S - (S + 2)^2}{(S + 2)^2}$

$\quad\quad = \dfrac{50 - (S^2 + 4S + 4)}{(S + 2)^2}$

$\quad\quad = \dfrac{-S^2 - 4S + 46}{(S + 2)^2}$

$H'(S) = 0$

$S^2 + 4S - 46 = 0$

$S = \dfrac{-4 \pm \sqrt{16 + 184}}{2}$

$\quad = 5.071$

(Discard the negative solution.) The number of creatures needed to sustain the population is $S_0 = 5.071$ thousand.

$H'' = \dfrac{(S+2)^2(-2S-4)-(-S^2-4S+46)(2S+4)}{(S + 2)^4} < 0$,

so H is a maximum at $S_0 = 5.017$.

$$H(S_0) = \frac{25(5.017)}{7.017} - 5.017$$

$$= 12.86$$

The maximum sustainable harvest is 12.86 thousand.

41. Let x = distance from P to A.

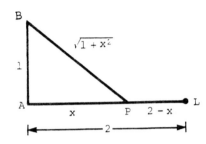

Energy used over land: 1 unit per mile

Energy used over water: 4/3 units per mile

Distance over land: (2 - x) mi

Distance over water: $\sqrt{1 + x^2}$ mi

Find the location of P to minimize energy used.

$$E(x) = 1(2 - x) + \frac{4}{3}\sqrt{1 + x^2}$$

$$E'(x) = -1 + \frac{4}{3}\left(\frac{1}{2}\right)(1 + x^2)^{-1/2}(2x)$$

If E'(x) = 0,

$$\frac{4}{3}x(1 + x^2)^{-1/2} = 1$$

$$\frac{4x}{3(1 + x^2)^{1/2}} = 1$$

$$\frac{4}{3}x = (1 + x^2)^{1/2}$$

$$\frac{16}{9}x^2 = 1 + x^2$$

$$\frac{7}{9}x^2 = 1$$

$$x^2 = \frac{9}{7}$$

$$x = \frac{3}{\sqrt{7}}$$

$$= \frac{3\sqrt{7}}{7}.$$

Point P is $3\sqrt{7}/7 \approx 1.134$ mi from Point A.

43. (a) Distance traveled by salmon = d

Net speed, or rate of salmon =

v - v_0

Time for journey = $\frac{\text{distance}}{\text{rate}}$

$$= \frac{d}{v - v_0} = t$$

(b) Let k_1 = constant of proportionality in expression for energy expended per hour, or

$$k_1 v^\alpha.$$

Then,

$$\begin{pmatrix} \text{total energy} \\ \text{expended over} \\ \text{journey} \end{pmatrix} = \begin{pmatrix} \text{energy} \\ \text{expended} \\ \text{per hour} \end{pmatrix} \times \begin{pmatrix} \text{time} \\ \text{for} \\ \text{journey} \end{pmatrix}$$

$$T = (k_1 v^\alpha)\left(\frac{d}{v - v_0}\right)$$

$$= k_1 d\left(\frac{v^\alpha}{v - v_0}\right).$$

Since d is constant for each journey, let k = $k_1 d$.

Then,

$$T = k\left(\frac{v^\alpha}{v - v_0}\right).$$

(c)

$$T' = \frac{k\alpha v^{\alpha-1}(v - v_0) - (1)(kv^\alpha)}{(v - v_0)^2} = 0$$

$$k\alpha v^{\alpha-1}(v - v_0) - (kv^\alpha) = 0$$

$$\alpha kv^\alpha v^{-1}(v - v_0) - kv^\alpha = 0$$

$$kv^\alpha[\alpha v^{-1}(v - v_0) - 1] = 0$$

$$kv^\alpha = 0 \quad \text{or} \quad \alpha v^{-1}(v - v_0) - 1 = 0$$

$$v = 0 \quad \text{or} \quad \frac{\alpha(v - v_0)}{v} = 1$$
(impossible)

$$\alpha(v - v_0) = v$$

$$\alpha v - \alpha v_0 = v$$

$$\alpha v - v = \alpha v_0$$

$$\alpha(\alpha - 1) = \alpha v_0$$

$$v = \frac{\alpha v_0}{\alpha - 1} \quad \text{(minimize time)}$$

(d)

$$t = \frac{d}{v - v_0}$$

$$40 = \frac{20}{v - 2}$$

$$40(v - 2) = 20$$

$$v = 2.5 \text{ mph}$$

Now, using $v = \frac{\alpha v_0}{\alpha - 1}$, we have

$$2.5 = \frac{\alpha(2)}{\alpha - 1}$$

$$2.5(\alpha - 1) = 2\alpha$$

$$2.5\alpha - 2.5 = 2\alpha$$

$$.5\alpha = 2.5$$

$$\alpha = 5.$$

Assume that the current of the river stays constant and that salmon swim at constant velocity.

Exercises 45–51 should be solved by computer methods. The solutions will vary according to the computer programs that are used.

45. The answer is the radius is 5.206 cm, the height is 11.75 cm.

47. The answer is the radius is 5.242 cm, the height is 11.58 cm

49. $C(x) = .01x^3 + .05x^2 + .2x + 28$

The answer is $A(x) = .01x^2 + .05x + .2 + 28/x$, $x = 10.41$.

51. $C(x) = 30 + 42x^{1/2} + .2x^{3/2} + .03x^{5/2}$

The answer is $A(x) = 30/x + 42x^{-1/2} + .2x^{1/2} + .03x^{3/2}$, $x = 23.49$.

Section 12.2

1. $x = \sqrt{\dfrac{kM}{2f}}$ *Equation 3*

$$T(x) = \left(f + \frac{gM}{x}\right)x + \frac{kM}{2x}$$

$$= f \cdot x + gM + \frac{kM}{2x}$$

$$T'(x) = f - \frac{kM}{2}x^{-2}$$

When $T'(x) = 0$,

$$f = \frac{kM}{2}x^{-2}$$

$$x^2 = \frac{kM}{2f}$$

$$x = \sqrt{\frac{kM}{2f}}.$$

$$T'' = -\frac{kM}{2}(-2x^{-3})$$

$$= \frac{kM}{x^3}$$

Since $x > 0$, $T'' > 0$ and therefore

$x = \sqrt{\dfrac{kM}{2f}}$ is minimum.

3. Use equation (3) with

   M = 100,000 units produced annually

   k = \$1 cost to store one unit for one year

   f = \$500 cost to set up, fixed cost

   x = the number of batches per year.

   $$x = \sqrt{\frac{kM}{2f}} = \sqrt{\frac{(1)(100,000)}{2(500)}}$$

   $$= \sqrt{100} = 10$$

5. From Exercise 3, the number of batches is 10 and the number of units is 100,000.

   The number of units per batch is
   $\dfrac{100,000}{10} = 10,000.$

7. Use equation (3) with

   f = the fixed cost for one order

   = \$60

   M = the total number of units in one year

   = 100,000

   k = the cost of storing one unit for one year

   = \$.50

   x = the number of units annually

$$x = \sqrt{\frac{kM}{2f}} = \sqrt{\frac{(.50)(100,000)}{2(60)}}$$

≈ 20 (The number of units must be an integer.)

The optimum number of copies per order is

$$\frac{M}{x} = \frac{100,000}{20} = 5000.$$

9. In the inventory problem, the manufacturing cost is

   $$f + \frac{gM}{x}.$$

   The total manufacturing cost is

   $$\left(f + \frac{gM}{x}\right)x.$$

   Total storage cost is $\left(\dfrac{M}{x}\right)k$, where $\dfrac{M}{x}$ is the size of each batch and k is the cost of storing one unit. The total cost is

   $$T(x) = \left(f + \frac{gM}{x}\right)x + \frac{kM}{x}$$

   $$= fx + gM + \frac{kM}{x}$$

   $$T'(x) = f - kMx^{-2}$$

   $$f - kMx^{-2} = 0$$

   $$f = \frac{kM}{x^2}$$

   $$x^2 f = kM$$

   $$x^2 = \frac{kM}{f}$$

   $$x = \sqrt{\frac{kM}{f}}$$

   $T''(x) = kMx^{-3} > 0$, so $x = \sqrt{\dfrac{kM}{f}}$

   minimizes the total cost.

11. In the inventory problem, total manufacturing cost is $\left(f + \frac{gM}{x}\right)x$.

In this case, storage cost is

$$\left(\frac{M}{2x}\right)k_1 + \left(\frac{M}{x}\right)k_2.$$

The total cost is

$$T(x) = \left(f + \frac{gM}{x}\right)x + \left(\frac{M}{2x}\right)k_1 + \left(\frac{M}{x}\right)k_2$$

$$= fx + gM + \frac{Mk_1}{2x} + \frac{Mk_2}{x}$$

$$= fx + gM + \frac{Mk_1 + 2Mk_2}{2x}$$

$$= fx + gM + \frac{(k_1 + 2k_2)M}{2x}$$

$$T'(x) = f - \frac{(k_1 + 2k_2)M}{2x^2}$$

$$f - \frac{(k_1 + 2k_2)M}{2x^2} = 0$$

$$2fx^2 = (k_1 + 2k_2)M$$

$$x^2 = \frac{(k_1 + 2k_2)M}{2f}$$

$$x = \sqrt{\frac{(k_1 + 2k_2)M}{2f}}$$

$$T''(x) = \frac{(k_1 + k_2)M}{2x^3} > 0, \text{ so}$$

$$x = \sqrt{\frac{(k_1 + 2k_2)M}{2f}} \text{ minimizes the}$$

total cost.

13. $q = 25,000 - 50p$

(a) $\dfrac{dq}{dp} = -50$

$$E = -\frac{p}{q} \cdot \frac{dq}{dp}$$

$$= -\frac{p}{25,000 - 50p}(-50)$$

$$= \frac{p}{500 - p}$$

(b) $R = pq$

$$\frac{dR}{dp} = q(1 - E)$$

When R is maximum, $q(1 - E) = 0$. Since $q = 0$ means no revenue, set $1 - E = 0$.

$$E = 1$$

From (a),

$$\frac{p}{500 - p} = 1$$

$$p = 500 - p$$

$$p = 250.$$

$$q = 25,000 - 50p$$

$$= 25,000 - 50(250)$$

$$= 12,500$$

Total revenue is maximized if $q = 12,500$.

15. $q = \dfrac{3000}{p}$

$$\frac{dq}{dp} = -\frac{3000}{p^2}$$

(a) $E = -\dfrac{p}{q} \cdot \dfrac{dq}{dp}$

$$= -\frac{p}{\frac{3000}{p}}\left(-\frac{3000}{p^2}\right)$$

$$= \left(-\frac{p^2}{3000}\right)\left(-\frac{3000}{p^2}\right)$$

$$= 1$$

(b) Since E is a constant, the demand always has unit elasticity.

$$\frac{dR}{dp} = q(1 - E)$$

$$= q(1 - 1)$$

$$= 0$$

So no value of q maximizes total revenue.

17. $q = 300 - 2p$

$\dfrac{dq}{dp} = -2$

$E = -\dfrac{p}{q} \cdot \dfrac{dq}{dp}$

$E = -\dfrac{p(-2)}{300 - 2p}$

$ = \dfrac{2p}{300 - 2p}$

(a) When $p = \$100$,

$E = \dfrac{200}{300 - 200}$

$ = 2 > 1.$

$E = 2$; elastic; a percentage increase in price will result in a greater percentage decrease in demand.

(b) When $p = \$50$,

$E = \dfrac{100}{300 - 100}$

$ = \dfrac{1}{2} < 1.$

$E = 1/2$; inelastic; a percentage change in price will result in a smaller percentage change in demand.

19. $p = (25 - q)^{1/2}$

$p^2 = 25 - q$

$q = 25 - p^2$

$\dfrac{dq}{dp} = -2p$

$\phantom{\dfrac{dq}{dp}} = -2(25 - q)^{1/2}$

$E = -\dfrac{p}{q}\dfrac{dq}{dp}$

$ = -\dfrac{\sqrt{25 - q}}{q}(-2\sqrt{25 - q})$

$ = \dfrac{2(25 - q)}{q}$

(a) When $q = 16$,

$E = \dfrac{2(9)}{16}$

$ = \dfrac{9}{8}.$

(b) $R = pq$

$ = p(25 - p^2)$

$ = 25p - p^3$

$R' = 25 - 3p^2$

If $R' = 0$,

$3p^2 = 25$

$p = \dfrac{5}{\sqrt{3}}$

$ = \dfrac{5\sqrt{3}}{3}.$

$q = 25 - p^2$

$ = 25 - \dfrac{25}{3}$

$ = \dfrac{50}{3}$

Total revenue is maximized when $q = 50/3$, $p = 5\sqrt{3}/3$.

(c) We have concluded that total revenue is maximized at the price where demand has unit elasticity. Thus, when $q = 50/3$ (or R is maximized), $E = 1$.

21. $E = -\dfrac{p}{q} \cdot \dfrac{dq}{dp}$

Since $p \neq 0$, $E = 0$ when $\dfrac{dq}{dp} = 0$. The derivative is zero, which implies that the demand function has a horizontal tangent line at the value of p where $E = 0$.

$\dfrac{dy}{dx}$

Section 12.3

1. $4x^2 + 3y^2 = 6$

$$\frac{d}{dx}(4x^2 + 3y^2) = \frac{d}{dx}(6)$$

$$\frac{d}{dx}(4x^2) + \frac{d}{dx}(3y^2) = \frac{d}{dx}(6)$$

$$8x + 3 \cdot 2y\frac{dy}{dx} = 0$$

$$6y\frac{dy}{dx} = -8x$$

$$\frac{dy}{dx} = -\frac{4x}{3y}$$

3. $2xy + y^2 = 8$

$$\frac{d}{dx}(2xy + y^2) = \frac{d}{dx}(8)$$

$$\frac{d}{dx}(2xy) + \frac{d}{dx}(y^2) = \frac{d}{dx}(8)$$

$$2x\frac{dy}{dx} + y(2) + 2y\frac{dy}{dx} = 0$$

$$(2x + 2y)\frac{dy}{dx} = -2y$$

$$\frac{dy}{dx} = \frac{-2y}{2x + 2y}$$

$$\frac{dy}{dx} = \frac{-y}{x + y}$$

5. $6xy^2 - 8y + 1 = 0$

$$\frac{d}{dx}(6xy^2 - 8y + 1) = \frac{d}{dx}(0)$$

$$6x\frac{d}{dx}(y^2) + y^2\frac{d}{dx}(6x) - 8\frac{dy}{dx} = 0$$

$$6x(2)y\frac{dy}{dx} + y^2(6) - 8\frac{dy}{dx} = 0$$

$$12xy\frac{dy}{dx} + 6y^2 - 8\frac{dy}{dx} = 0$$

$$(12xy - 8)\frac{dy}{dx} = -6y^2$$

$$\frac{dy}{dx} = \frac{-6y^2}{12xy - 8}$$

$$\frac{dy}{dx} = \frac{-3y^2}{6xy - 4}$$

7. $6x^2 + 8xy + y^2 = 6$

$$\frac{d}{dx}(6x^2 + 8xy + y^2) = \frac{d}{dx}(6)$$

$$12x + \frac{d}{dx}(8xy) + \frac{d}{dx}(y^2) = 0$$

$$12x + 8x\frac{dy}{dx} + y\frac{d}{dx}(8x) + 2y\frac{dy}{dx} = 0$$

$$12x + 8x\frac{dy}{dx} + 8y + 2y\frac{dy}{dx} = 0$$

$$(8x + 2y)\frac{dy}{dx} = -12x - 8y$$

$$\frac{dy}{dx} = \frac{-12x - 8y}{8x + 2y}$$

$$\frac{dy}{dx} = \frac{-6x - 4y}{4x + y}$$

9. $x^3 = y^2 + 4$

$$\frac{d}{dx}(x^3) = \frac{d}{dx}(y^2 + 4)$$

$$3x^2 = \frac{d}{dx}(y^2) + \frac{d}{dx}(4)$$

$$3x^2 = 2y\frac{dy}{dx} + 0$$

$$\frac{3x^2}{2y} = \frac{dy}{dx}$$

11. $\dfrac{1}{x} - \dfrac{1}{y} = 2$

$$\frac{d}{dx}\left(\frac{1}{x} - \frac{1}{y}\right) = \frac{d}{dx}(2)$$

$$-\frac{1}{x^2} - \frac{d}{dx}\left(\frac{1}{y}\right) = 0$$

$$-\frac{1}{x^2} + \frac{1}{y^2} \cdot \frac{dy}{dx} = 0$$

$$\frac{1}{y^2} \cdot \frac{dy}{dx} = \frac{1}{x^2}$$

$$\frac{dy}{dx} = \frac{y^2}{x^2}$$

13. $3x^2 = \dfrac{2 - y}{2 + y}$

$$\frac{d}{dx}(3x^2) = \frac{d}{dx}\left(\frac{2 - y}{2 + y}\right)$$

$$6x = \frac{(2 + y)\frac{d}{dx}(2 - y) - (2 - y)\frac{d}{dx}(2 + y)}{(2 + y)^2}$$

$$6x = \frac{(2 + y)\left(-\frac{dy}{dx}\right) - (2 - y)\frac{dy}{dx}}{(2 + y)^2}$$

$$6x = \frac{-4\frac{dy}{dx}}{(2 + y)^2}$$

$$6x(2 + y)^2 = -4\frac{dy}{dx}$$

$$\frac{-3x(2 + y)^2}{2} = \frac{dy}{dx}$$

15. $x^2 y + y^3 = 4$

$$\frac{d}{dx}(x^2 y + y^3) = \frac{d}{dx}(4)$$

$$\frac{d}{dx}(x^2 y) + \frac{d}{dx}(y^3) = 0$$

$$x^2\frac{dy}{dx} + y(2x) + 3y^2\frac{dy}{dx} = 0$$

$$(x^2 + 3y^2)\frac{dy}{dx} = -2xy$$

$$\frac{dy}{dx} = \frac{-2xy}{x^2 + 3y^2}$$

17. $\sqrt{x} + \sqrt{y} = 4$

$$\frac{d}{dx}(x^{1/2} + y^{1/2}) = \frac{d}{dx}4$$

$$\frac{1}{2}x^{-1/2} + \frac{1}{2}y^{-1/2}\frac{dy}{dx} = 0$$

$$\frac{1}{2}y^{-1/2}\frac{dy}{dx} = 0 - \frac{1}{2}x^{-1/2}$$

$$\frac{dy}{dx} = 2y^{1/2}\left(-\frac{1}{2}x^{-1/2}\right)$$

$$\frac{dy}{dx} = \frac{-y^{1/2}}{x^{1/2}}$$

19. $\sqrt{xy} + y = 1$

$$\frac{d}{dx}(\sqrt{xy} + y) = \frac{d}{dx}(1)$$

$$\frac{d}{dx}(x^{1/2} y^{1/2}) + \frac{d}{dx}(y) = 0$$

$$x^{1/2}\frac{d}{dx}(y^{1/2}) + y^{1/2}\frac{d}{dx}(x^{1/2}) + \frac{dy}{dx} = 0$$

$$x^{1/2}\cdot\frac{1}{2}y^{-1/2}\frac{dy}{dx} + \frac{1}{2}x^{-1/2}y^{1/2} + \frac{dy}{dx} = 0$$

$$\left(\frac{1}{2}x^{1/2}y^{-1/2} + 1\right)\frac{dy}{dx} = -\frac{1}{2}x^{-1/2}y^{1/2}$$

$$\frac{dy}{dx} = \frac{-\frac{1}{2}x^{-1/2}y^{1/2}}{\frac{1}{2}x^{1/2}y^{-1/2} + 1}$$

$$= \frac{-\frac{1}{2}x^{-1/2}y^{1/2}}{\frac{1}{2}(x^{1/2}y^{-1/2} + 2)}$$

$$= \frac{-x^{-1/2}y^{1/2}}{x^{1/2}y^{-1/2} + 2}$$

21. $x^4 y^3 + 4x^{3/2} = 6y^{3/2} + 5$

$$\frac{d}{dx}(x^4 y^3 + 4x^{3/2}) = \frac{d}{dx}(6y^{3/2} + 5)$$

$$\frac{d}{dx}(x^4 y^3) + \frac{d}{dx}(4x^{3/2}) = \frac{d}{dx}(6y^{3/2}) + \frac{d}{dx}(5)$$

$$4x^3 y^3 + x^4 \cdot 3y^2\frac{dy}{dx} + 6x^{1/2} = 9y^{1/2}\frac{dy}{dx} + 0$$

$$4x^3 y^3 + 6x^{1/2} = 9y^{1/2}\frac{dy}{dx} - 3x^4 y^2\frac{dy}{dx}$$

$$4x^3 y^3 + 6x^{1/2} = (9y^{1/2} - 3x^4 y^2)\frac{dy}{dx}$$

$$\frac{4x^3 y^3 + 6x^{1/2}}{9y^{1/2} - 3x^4 y^2} = \frac{dy}{dx}$$

23. $(x^2 + y^3)^4 = x + 2y + 4$

$$\frac{d}{dx}(x^2 + y^2)^4 = \frac{d}{dx}(x + 2y + 4)$$

$$4(x^2 + y^3)^3\frac{d}{dx}(x^2 + y^3) = \frac{d}{dx}(x) + \frac{d}{dx}(2y)$$

$$+ \frac{d}{dx}(4)$$

$$4(x^2 + y^3)^3\left(2x + 3y^2\frac{dy}{dx}\right) = 1 + 2\frac{dy}{dx} + 0$$

$$8x(x^2 + y^3)^3 + 12y^2(x^2 + y^3)^3 \frac{dy}{dx}$$

$$= 1 + 2\frac{dy}{dx}$$

$$8x(x^2 + y^3)^3 - 1 = 2\frac{dy}{dx} - 12y^2(x^2 + y^3)^3 \frac{dy}{dx}$$

$$8x(x^2 + y^3)^3 - 1 = [2 - 12y^2(x^2 + y^3)^3]\frac{dy}{dx}$$

$$\frac{8x(x^2 + y^3)^3 - 1}{2 - 12y^2(x^2 + y^3)^3} = \frac{dy}{dx}$$

25. $x^2 + y^2 = 25$; tangent at $(-3, 4)$

$$\frac{d}{dx}(x^2 + y^2) = \frac{d}{dx}(25)$$

$$2x + 2y\frac{dy}{dx} = 0$$

$$2y\frac{dy}{dx} = -2x$$

$$\frac{dy}{dx} = -\frac{x}{y}$$

$$m = -\frac{x}{y} = -\frac{-3}{4} = \frac{3}{4}$$

$$y - y_1 = m(x - x_1)$$

$$y - 4 = \frac{3}{4}[x - (-3)]$$

$$4y - 16 = 3x + 9$$

$$4y = 3x + 25$$

27. $x^2y^2 = 1$; tangent at $(-1, 1)$

$$\frac{d}{dx}(x^2y^2) = \frac{d}{dx}(1)$$

$$x^2\frac{d}{dx}(y^2) + y^2\frac{d}{dx}(x^2) = 0$$

$$x^2(2y)\frac{dy}{dx} + y^2(2x) = 0$$

$$2x^2y\frac{dy}{dx} = -2xy^2$$

$$\frac{dy}{dx} = \frac{-2xy^2}{2x^2y} = -\frac{y}{x}$$

$$m = -\frac{y}{x} = -\frac{1}{-1} = 1$$

$$y - 1 = 1[x - (-1)]$$

$$y = x + 1 + 1$$

$$y = x + 2$$

29. $x^2 + \sqrt{y} = 7$; tangent at $(2, 9)$

$$\frac{d}{dx}(x^2 + \sqrt{y}) = \frac{d}{dx}(7)$$

$$2x + \frac{1}{2}y^{-1/2}\frac{dy}{dx} = 0$$

$$\frac{1}{2}y^{-1/2}\frac{dy}{dx} = -2x$$

$$\frac{dy}{dx} = -2x(2y^{1/2})$$

$$\frac{dy}{dx} = -4xy^{1/2}$$

$$m = -4xy^{1/2} = -4(2)(9^{1/2}) = -24$$

$$y - 9 = -24(x - 2)$$

$$24x + y = 57$$

31. $y + \frac{\sqrt{x}}{y} = 3$; tangent at $(4, 2)$

$$\frac{d}{dx}\left(y + \frac{\sqrt{x}}{y}\right) = \frac{d}{dx}(3)$$

$$\frac{dy}{dx} + \frac{d}{dx}\left(\frac{\sqrt{x}}{y}\right) = 0$$

$$\frac{dy}{dx} + \frac{y\left(\frac{1}{2}\right)x^{-1/2} - \sqrt{x}\frac{dy}{dx}}{y^2} = 0$$

$$\frac{dy}{dx} = \frac{-\frac{1}{2}yx^{-1/2} + \sqrt{x}\frac{dy}{dx}}{y^2}$$

$$y^2\frac{dy}{dx} = -\frac{1}{2}yx^{-1/2} + \sqrt{x}\frac{dy}{dx}$$

$$(y^2 - \sqrt{x})\frac{dy}{dx} = -\frac{1}{2}yx^{-1/2}$$

$$\frac{dy}{dx} = \frac{-y}{2x^{1/2}(y^2 - \sqrt{x})}$$

$$m = \frac{-y}{2x^{1/2}(y^2 - \sqrt{x})}$$

$$= \frac{-2}{2(2)(4 - 2)}$$

$$= -\frac{1}{4}$$

$$y - 2 = -\frac{1}{4}(x - 4)$$

$$y = -\frac{1}{4}x + 3$$

$$x + 4y = 12$$

33. $x^2 + y^2 = 100$

(a) Lines are tangent at points where x = 6. By substituting x = 6 in the equation, we find the points are (6, 8) and (6, -8).

$$\frac{d}{dx}(x^2 + y^2) = \frac{d}{dx}(100)$$

$$2x + 2y\frac{dy}{dx} = 0$$

$$2y\frac{dy}{dx} = -2x$$

$$dy = -\frac{x}{y}$$

$$m_1 = -\frac{x}{y} = -\frac{6}{8} = -\frac{3}{4}$$

$$m_2 = -\frac{x}{y} = -\frac{6}{-8} = \frac{3}{4}$$

First tangent:

$$y - 8 = -\frac{3}{4}(x - 6)$$

$$y = -\frac{3}{4}x + \frac{50}{4}$$

$$3x + 4y = 50$$

Second tangent:

$$y - (-8) = \frac{3}{4}(x - 6)$$

$$y + 8 = \frac{3}{4}x - \frac{18}{4}$$

$$y = \frac{3}{4}x - \frac{50}{4}$$

$$-3x + 4y = -50$$

(b) See the answer graph in the back of the textbook.

35. $y^5 - y - x^2 = -1$; tangent at (1, 1)

$$\frac{d}{dx}(y^5 - y - x^2) = \frac{dy}{dx}(-1)$$

$$5y^4\frac{dy}{dx} - \frac{dy}{dx} - 2x = 0$$

$$(5y^4 - 1)\frac{dy}{dx} = 2x$$

$$\frac{dy}{dx} = \frac{2x}{5y^4 - 1}$$

$$m = \frac{2x}{5y^4 - 1}$$

$$= \frac{2(1)}{5(1)^4 - 1}$$

$$= \frac{2}{4}$$

$$= \frac{1}{2}$$

$$y - 1 = \frac{1}{2}(x - 1)$$

$$y = \frac{1}{2}x + \frac{1}{2}$$

$$2y = x + 1$$

37. $x^2 + y^2 + 1 = 0$

$$\frac{d}{dx}(x^2 + y^2 + 1) = \frac{d}{dx}(0)$$

$$2x + 2y\frac{dy}{dx} = 0$$

$$\frac{dy}{dx} = \frac{-2x}{2y}$$

$$= -\frac{x}{y}$$

If x and y are real numbers, x^2 and y^2 are nonnegative. 1 + a non-negative number cannot equal zero so there is no function y = f(x) that satisfies $x^2 + y^2 + 1 = 0$.

39. $\sqrt{u} + \sqrt{2v + 1} = 5$

$$\frac{dv}{du}(\sqrt{u} + \sqrt{2v + 1}) = \frac{dv}{du}(5)$$

$$\frac{1}{2}u^{-1/2} + \frac{1}{2}(2v + 1)^{-1/2}(2)\frac{dv}{du} = 0$$

$$(2v + 1)^{-1/2}\frac{dv}{du} = -\frac{1}{2}u^{-1/2}$$

$$\frac{dv}{du} = -\frac{(2v + 1)^{1/2}}{2u^{1/2}}$$

41. $C^2 = x^2 + 100\sqrt{x} + 50$

(a) $2C\dfrac{dC}{dx} = 2x + \dfrac{1}{2}(100)x^{-1/2}$

$$\frac{dC}{dx} = \frac{2x + 50x^{-1/2}}{2C}$$

$$\frac{dC}{dx} = \frac{x + 25x^{-1/2}}{C} \cdot \frac{x^{1/2}}{x^{1/2}}$$

$$\frac{dC}{dx} = \frac{x^{3/2} + 25}{Cx^{1/2}}$$

When x = 5, the approximate increase in cost of an additional unit is

$$\frac{(5)^{3/2} + 25}{(5^2 + 100\sqrt{5} + 50)^{1/2}(5)^{1/2}}$$

$$= \frac{36.18}{(17.28)\sqrt{5}}$$

$$= .94.$$

(b) $900(x - 5)^2 + 25R^2 = 22,500$

$$R^2 = 900 - 36(x - 5)^2$$

$$2R\frac{dR}{dx} = -72(x - 5)$$

$$\frac{dR}{dx} = \frac{-36(x - 5)}{R} = \frac{180 - 36x}{R}$$

When x = 5, the approximate change in revenue for a unit increase in sales is

$$\frac{180 - 36(5)}{R} = \frac{0}{R} = 0.$$

43. $2s^2 + \sqrt{st} - 4 = 3t$

$$4s\frac{ds}{dt} + \frac{1}{2}(st)^{-1/2}\left(s + t\frac{ds}{dt}\right) = 3$$

$$4s\frac{ds}{dt} + \frac{s + t\dfrac{ds}{dt}}{2\sqrt{st}} = 3$$

$$\frac{8s(\sqrt{st})\dfrac{ds}{dt} + s + t\dfrac{ds}{dt}}{2\sqrt{st}} = 3$$

$$\frac{(8s\sqrt{st} + t)\dfrac{ds}{dt} + s}{2\sqrt{st}} = 3$$

$$(8s\sqrt{st} + t)\frac{ds}{dt} = 6\sqrt{st} - s$$

$$\frac{ds}{dt} = \frac{-s + 6\sqrt{st}}{8s\sqrt{st} + t}$$

Section 12.4

1. $y = 9x^2 + 2x;\ \dfrac{dx}{dt} = 4,\ x = 6$

$$\frac{dy}{dt} = 18x\frac{dx}{dt} + 2\frac{dx}{dt}$$

$$= 18(6)(4) + 2(4) = 440$$

3. $y^2 - 5x^2 = -1;\ -\dfrac{dx}{dt} = -3,\ x = 1,$

$y = 2$

$$2y\frac{dy}{dt} - 10x\frac{dx}{dt} = 0$$

$$y\frac{dy}{dt} = 5x\frac{dx}{dt}$$

$$2\frac{dy}{dt} = 5(-3)$$

$$\frac{dy}{dt} = -\frac{15}{2}$$

5. $xy - 5x + 2y^3 = -70;\ \dfrac{dx}{dt} = -5,$

$x = 2,\ y = -3$

$x\dfrac{dy}{dt} + y\dfrac{dx}{dt} - 5\dfrac{dx}{dt} + 6y^2\,\dfrac{dy}{dt} = 0$

$(x + 6y^2)\dfrac{dy}{dt} + (y - 5)\dfrac{dx}{dt} = 0$

$(x + 6y^2)\dfrac{dy}{dt} = (5 - y)\dfrac{dy}{dt}$

$\dfrac{dy}{dt} = \dfrac{(5 - y)\dfrac{dx}{dt}}{x + 6y^2}$

$\phantom{\dfrac{dy}{dt}} = \dfrac{[5 - (-3)](-5)}{2 + 6(-3)^2}$

$\phantom{\dfrac{dy}{dt}} = \dfrac{-40}{56} = -\dfrac{5}{7}$

7. $\dfrac{x^2 + y}{x - y} = 9;\ \dfrac{dx}{dt} = 2,\ x = 4,\ y = 2$

$\dfrac{(x - y)\left(2x\dfrac{dx}{dt} + \dfrac{dy}{dt}\right) - (x^2 + y)\left(\dfrac{dx}{dt} - \dfrac{dy}{dt}\right)}{(x - y)^2} = 0$

$\dfrac{2x(x - y)\dfrac{dx}{dt} + (x - y)\dfrac{dy}{dt} - (x^2 + y)\dfrac{dx}{dt} + (x^2 + y)\dfrac{dy}{dt}}{(x - y)^2} = 0$

$[2x(x - y) - (x^2 + y)]\dfrac{dx}{dt} + [(x - y) + (x^2 + y)]\dfrac{dy}{dt} = 0$

$\dfrac{dy}{dt} = \dfrac{[(x^2 + y) - 2x(x - y)]\dfrac{dx}{dt}}{(x - y) + (x^2 + y)}$

$\dfrac{dy}{dt} = \dfrac{(-x^2 + y + 2xy)\dfrac{dx}{dt}}{x + x^2}$

$\phantom{\dfrac{dy}{dt}} = \dfrac{[-(4)^2 + 2 + 2(4)(2)](2)}{4 + 4^2}$

$\phantom{\dfrac{dy}{dt}} = \dfrac{4}{20} = \dfrac{1}{5}$

9. $C = .1x^2 + 10{,}000$; $x = 100$, $\dfrac{dx}{dt} = 10$

$$\frac{dC}{dt} = .1(2x)\frac{dx}{dt}$$

$$= .1(200)(10)$$

$$= 200$$

The cost is changing at a rate of $200 per month.

11. $R = 50x - .4x^2$; $C = 5x + 15$; $x = 40$;
$\dfrac{dx}{dt} = 10$

(a) $\dfrac{dR}{dt} = 50\dfrac{dx}{dt} - .8x\dfrac{dx}{dt}$

$$= 50(10) - .8(40)(10)$$

$$= 500 - 320$$

$$= 180$$

Revenue is increasing at a rate of $180 per day.

(b) $\dfrac{dC}{dt} = 5\dfrac{dx}{dt}$

$$= 5(10) = 50$$

Cost is increasing at a rate of $50 per day.

(c) Profit = Revenue - Cost, or

$$P = R - C$$

$$\frac{dP}{dt} = \frac{dR}{dt} - \frac{dC}{dt}$$

$$= 180 - 50$$

$$= 130$$

Profit is increasing at a rate of $130 per day.

13. $p = \dfrac{8000}{q}$; $p = 3.50$,

$$\frac{dp}{dt} = .15p$$

because price is increasing at 15% as well.

$$pq = 8000$$

$$p\frac{dq}{dt} + q\frac{dp}{dt} = 0$$

$$\frac{dq}{dt} = \frac{-q\dfrac{dp}{dt}}{p}$$

$$= \frac{-\left(\dfrac{8000}{3.50}\right)(.15)(3.50)}{3.50}$$

$$= -343$$

Demand is decreasing at a rate of approximately 343 units per unit time.

15. $V = k(R^2 - r^2)$; $r = 2$ mm, $k = 3$,
$\dfrac{dr}{dt} = .02$ mm per minute, and R is constant.

$$V = k(R^2 - r^2)$$

$$V = 3(R^2 - r^2)$$

$$\frac{dV}{dt} = 3\left(0 - 2\frac{dr}{dt}\right)$$

$$= -6r\frac{dr}{dt}$$

$$= -6(2)(.02)$$

$$= -.24 \text{ mm/min}$$

17. $V = k(R^2 - r^2)$; $k = 555.6$,
$R = .02$ mm, $\dfrac{dR}{dt} = .003$ mm per minute, r is constant.

$V = k(R^2 - r^2)$

$V = 555.6(R^2 - r^2)$

$\dfrac{dV}{dt} = 555.6\left(2R\dfrac{dR}{dt} - 0\right)$

$\quad = 555.6(2)(.02)(.003)$

$\quad = .067 \text{ mm/min}$

19. $T(x) = \dfrac{2 + x}{2 + x^2}$, $x = 4$, $\dfrac{dx}{dt} = 4$

Find $\dfrac{dT}{dt}$.

$\dfrac{dT}{dt} = \dfrac{(2 + x^2)\dfrac{d}{dt}(2 + x) - (2 + x)\dfrac{d}{dt}(2 + x^2)}{(2 + x^2)^2}$

$\quad = \dfrac{(2 + x^2)\dfrac{dx}{dt} - (2 + x)(2x)\dfrac{dx}{dt}}{(2 + x^2)^2}$

$\quad = \dfrac{(2 + x^2)\dfrac{dx}{dt} - (4x + 2x^2)\dfrac{dx}{dt}}{(2 + x^2)^2}$

$\quad = \dfrac{(2 - 4x - x^2)\dfrac{dx}{dt}}{(2 + x^2)^2}$

$\quad = \dfrac{(2 - 16 - 16)4}{(2 + 16)^2}$

$\quad = -\dfrac{120}{324}$

$\quad = -.370$

21. Let x = the distance of the base of the ladder from the base of the building

y = the distance up the side of the building to the top of the ladder.

Find $\dfrac{dy}{dt}$ when x = 7 ft and $\dfrac{dx}{dt} = 4$ ft/min.

Since $y = \sqrt{25^2 - x^2}$, when x = 7,

$\quad y = 24 \text{ ft.}$

$x^2 + y^2 = 25^2$

$\dfrac{d}{dt}(x^2 + y^2) = \dfrac{d}{dt}(25)^2$

$2x\dfrac{dx}{dt} + 2y\dfrac{dy}{dt} = 0$

$\quad 2y\dfrac{dy}{dt} = -2x\dfrac{dx}{dt}$

$\quad \dfrac{dy}{dt} = \dfrac{-2x}{2y}\dfrac{dx}{dt} = -\dfrac{x}{y}\dfrac{dx}{dt}$

$\quad\quad = -\dfrac{7}{24}(4)$

$\quad\quad = -\dfrac{7}{6}$

The ladder is sliding down the building at the rate of 7/6 ft/min.

23. Let r = the radius of the circle formed by the ripple.

Find $\dfrac{dA}{dt}$ = when r = 4 ft and $\dfrac{dr}{dt} = 2$ ft/min.

$\quad A = \pi r^2$

$\dfrac{dA}{dt} = 2\pi r\dfrac{dr}{dt}$

$\quad = 2\pi(4)(2)$

$\quad = 16\pi \text{ ft}^2/\text{min}$

25. Let r = the radius of the base of the conical pile.

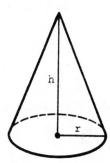

Find $\dfrac{dV}{dt}$ when r = 5 in, $\dfrac{dr}{dt}$ = 1 in/min

h = 2r for all t.

$V = \dfrac{\pi}{3}r^2 h$

$V = \dfrac{\pi}{3}r^2(2r)$

$\quad = \dfrac{2\pi}{3}r^3$

$\dfrac{dV}{dt} = \dfrac{3 \cdot 2\pi r^3}{3} \dfrac{dr}{dt}$

$\dfrac{dV}{dt} = 2\pi(5^2)(1)$

$\quad = 50\pi$ in³/min

V = (xh)(16)

$\quad = \left(\dfrac{h}{2}h\right)16$

$\quad = 8h^2$

$\dfrac{dV}{dt} = 16h\dfrac{dh}{dt}$

$\dfrac{1}{16h}\dfrac{dV}{dt} = \dfrac{dh}{dt}$

$\dfrac{1}{16(4)}(4) = \dfrac{dh}{dt}$

$\dfrac{dh}{dt} = \dfrac{1}{16}$ ft/min

27. Let x = one-half the width of the
   triangular cross section

   h = the height of the water

   V = the volume of the water.

   $\dfrac{dV}{dt}$ = 4 cu ft per min

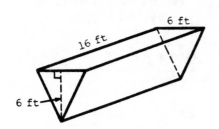
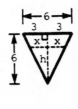

Find $\dfrac{dh}{dt}$ when h = 4.

$V = \begin{bmatrix} \text{Area of} \\ \text{triangular} \\ \text{cross section} \end{bmatrix}\Big(\text{length}\Big)$

Area of triangular cross section

$\quad = \dfrac{1}{2}(\text{base})(\text{altitude})$

$\quad = \dfrac{1}{2}(2x)(h)$

$\quad = xh$

By similar triangles,

$\quad \dfrac{3}{x} = \dfrac{6}{h}$, so x = $\dfrac{h}{2}$.

29. Let x = the horizontal length

   r = the rope length.

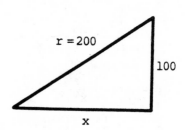

$\dfrac{dx}{dt}$ = 50 ft per min

$200^2 = x^2 + 100^2$

$173.2 = x$

$r^2 = x^2 + 100$

$2r\dfrac{dr}{dt} = 2x\dfrac{dx}{dt} + 0$

$r\dfrac{dr}{dt} = x\dfrac{dx}{dt}$

$\dfrac{dr}{dt} = \dfrac{x\dfrac{dx}{dt}}{r}$

$\dfrac{dr}{dt} = \dfrac{173.2(50)}{200}$

$\quad = 43.3$ ft/min

Section 12.5

1. $y = 6x^2$

   $dy = 2(6x)dx$

   $= 12x\ dx$

3. $y = 7x^2 - 9x + 6$

   $dy = [2(7x) - 9)dx$

   $= (14x - 9)dx$

5. $y = 2\sqrt{x}$

   $dy = 2\left(\frac{1}{2}\right)x^{-1/2}\ dx$

   $= x^{-1/2}\ dx$

7. $y = \dfrac{8x - 2}{x - 3}$

   $dy = \dfrac{(x - 3)(8) - (8x - 2)}{(x - 3)^2}\ dx$

   $= \dfrac{-22}{(x - 3)^2}\ dx$

9. $y = x^2\left(x - \dfrac{1}{x} + 2\right)$

   $dy = \left[2x\left(x - \dfrac{1}{x} + 2\right) + x^2\left(1 + \dfrac{1}{x^2}\right)\right]dx$

   $= (2x^2 - 2 + 4x + x^2 + 1)dx$

   $= (3x^2 + 4x - 1)dx$

11. $y = \left(2 - \dfrac{3}{x}\right)\left(1 + \dfrac{1}{x}\right)$

   $dy = \left[\left(2 - \dfrac{3}{x}\right)\left(-\dfrac{1}{x^2}\right) + \left(1 + \dfrac{1}{x}\right)\left(\dfrac{3}{x^2}\right)\right]dx$

   $= \left[\dfrac{-2}{x^2} + \dfrac{3}{x^3} + \dfrac{3}{x^2} + \dfrac{3}{x^3}\right]dx$

   $= \left(\dfrac{1}{x^2} + \dfrac{6}{x^3}\right)dx$ or $(x^{-2} + 6x^{-3})dx$

13. $y = 2x^2 - 5x;\ x = -2,\ \Delta x = .2$

   $dy = (4x - 5)dx$

   $\approx (4x - 5)\Delta x$

   $= [4(-2) - 5](.2)$

   $= -2.6$

15. $y = x^3 - 2x^2 + 3,\ x = 1,\ \Delta x = -.1$

   $dy = (3x^2 - 4x)dx$

   $\approx (3x^2 - 4x)\Delta x$

   $= [3(1^2) - 4(1)](-.1)$

   $= .1$

17. $y = \sqrt{3x},\ x = 1,\ \Delta x = .15$

   $dy = \sqrt{3}\left(\dfrac{1}{2}x^{-1/2}\right)dx$

   $\approx \dfrac{\sqrt{3}}{2\sqrt{x}}\Delta x$

   $= \dfrac{1.73}{2}(.15)$

   $= .130$

19. $y = \dfrac{2x - 5}{x + 1};\ x = 2,\ \Delta x = -.03$

   $dy = \dfrac{(x + 1)(2) - (2x - 5)(1)}{(x + 1)^2}\ dx$

   $= \dfrac{7}{(x + 1)^2}\ dx$

   $\approx \dfrac{7}{(x + 1)^2}\ \Delta x$

   $= \dfrac{7}{(2 + 1)^2}(-.03)$

   $= -.023$

21. $y = -6\left(2 - \dfrac{1}{x^2}\right),\ x = -1,\ \Delta x = .02$

   $dy = -6\left(\dfrac{2}{x^3}\right)dx$

   $\approx -\dfrac{12}{x^3}\ \Delta x$

   $= \dfrac{-12}{(-1)^3}(.02)$

   $= .24$

23. $y = \dfrac{1 + x}{\sqrt{x}}$, $x = 9$, $\Delta x = -.03$

$dy = \dfrac{\sqrt{x}(1) - (1 + x)\left(\frac{1}{2}\right)(x^{-1/2})}{x}\, dx$

$\approx \dfrac{2\sqrt{x} - (1 + x)(x^{-1/2})}{2x}\, \Delta x$

$= \dfrac{2\sqrt{9} - (1 + 9)(9^{-1/2})}{2(9)}(-.03)$

$= -.00444$

25. Let d = the demand in thousands of
pounds

x = the price in dollars.

$d(x) = -5x^3 - 2x^2 + 1500$

(a) $x = 2$, $\Delta x = .50$

$dd = (-15x^2 - 4x)dx$

$\approx (-15x^2 - 4x)\Delta x$

$= [-15(4) - 4(2)](.50)$

$= -34$ thousand pounds

(b) $x = 6$, $\Delta x = .30$

$dd \approx [-15(36) - 4(6)](.30)$

$= -169.2$ thousand pounds

27. $H(r) = \dfrac{300}{1 + .03r^2}$

$r = 10$, $\Delta r = .5$

$dH = \dfrac{-300(.06r)}{(1 + .03r^2)^2}\, dr$

$\approx \dfrac{-300(.06)r}{(1 + .03r^2)^2}\, \Delta r$

$= \dfrac{-300(.06)(10)}{(1 + 3)^2}(.5)$

$= \dfrac{90}{16}$

$= -5.625$ housing starts

29. $P(x) = -390 + 24x + 5x^2 - \frac{1}{3}x^3$

$x = 1000$, $\Delta x = 1$

$dP = (24 + 10x - x^2)dx$

$\approx (24 + 10x - x^2)\Delta x$

$= (24 + 10,000 - 100,000) \cdot 1$

$\approx -\$990,000$

31. Let x = the number of beach balls

V = the volume of x beach balls.

Then $\dfrac{dV}{dr} \approx$ the volume of material in
beach balls since they are
hollow.

$V = \frac{4}{3}\pi r^3 x$

$r = 6$ in, $x = 5000$, $\Delta r = .03$ in

$dV = \frac{4}{3}\pi(3r^2x + r^3)\Delta r$

$= \frac{4}{3}\pi(3 \cdot 36 \cdot 5000 + 36)(.03)$

$= 21,600\pi$

$21,600\pi$ in³ of material would be
needed.

33. $P(x) = \dfrac{25x}{8 + x^2}$

$dP = \dfrac{(8 + x^2)(25) - 25x(2x)}{(8 + x^2)^2}\, dx$

$\approx \dfrac{(8 + x^2)(25) - 25x)(2x)}{(8 + x^2)^2}\, \Delta x$

(a) $x = 2$, $\Delta x = .5$

$dP = \dfrac{[(8 + 4)(25) - (25(2)(4)](.5)}{(8 + 4)^2}$

$= .347$ million

(b) $x = 3$, $\Delta x = .25$

$dP \approx \Delta P = [(8 + 9)(25) - 25(3)(6)](.25)$

$= -.022$ million

35. r changes from 14 mm to 16 mm so

$\Delta r = 2.$

$V = \frac{4}{3}\pi r^3$

$dV = \frac{4}{3}(3)\pi r^2 \; dr$

$\Delta V = 4\pi r^2 \Delta r$

$\quad = 4\pi(14)^2(2)$

$\quad = 1568\pi \; \text{mm}^3$

37. r increases from 20 mm to 22 mm so

$\Delta r = 2.$

$A = \pi r^2$

$dA = 2\pi r \; dr$

$\Delta A \approx 2\pi r \Delta r$

$\quad = 2\pi(20)(2)$

$\quad = 80\pi \; \text{mm}^2$

39. r = 3 cm, $\Delta r = -.2$ cm

$V = \frac{4}{3}\pi r^3$

$dV = 4\pi r^2 dr$

$\Delta V \approx 4\pi r^2 \Delta r$

$\quad = 4\pi(9)(-.2)$

$\quad = -7.2\pi \; \text{cm}^3$

41. r = 4.87 in, $\Delta r = \pm.040$

$A = \pi r^2$

$dA = 2\pi r \; dr$

$\Delta A \approx 2\pi r \Delta r$

$\quad = 2\pi(4.87)(\pm.040)$

$\quad = \pm1.224 \; \text{in}^2$

43. h = 7.284 in, r = 1.09 ± .007 in

$V = \frac{1}{3}\pi r^2 h$

$dV = \frac{2}{3}\pi r h \; dr$

$\Delta V \approx \frac{2}{3}\pi r h \; \Delta r$

$\quad = \frac{2}{3}\pi(1.09)(7.284)(.007)$

$\quad = \pm.116 \; \text{in}^3$

Chapter 12 Review Exercises

5. $x^2 y^3 + 4xy = 2$

$\frac{d}{dx}(x^2 y^3 + 4xy) = \frac{d}{dx}(2)$

$2xy^3 + 3y^2\left(\frac{dy}{dx}\right)x^2 + 4y + 4x\frac{dy}{dx} = 0$

$(3x^2 y^2 + 4x)\frac{dy}{dx} = -2xy^3 - 4y$

$\frac{dy}{dx} = \frac{-2xy^3 - 4y}{3x^2 y^2 + 4x}$

7. $9\sqrt{x} + 4y^3 = \frac{2}{x}$

$\frac{d}{dx}(9\sqrt{x} + 4y^3) = \frac{d}{dx}\left(\frac{2}{x}\right)$

$\frac{9}{2}x^{-1/2} + 12y^2 \frac{dy}{dx} = \frac{-2}{x^2}$

$12y^2 \frac{dy}{dx} = \frac{-2}{x^2} - \frac{9x^{-1/2}}{2}$

$12y^2 \frac{dy}{dx} = \frac{-4 - 9x^{3/2}}{2x^2}$

$\frac{dy}{dx} = \frac{-4 - 9x^{3/2}}{24x^2 y^2}$

9. $\dfrac{x + 2y}{x - 3y} = y^{1/2}$

$x + 2y = y^{1/2}(x - 3y)$

$\dfrac{d}{dx}(x + 2y) = \dfrac{d}{dx}[y^{1/2}(x - 3y)]$

$1 + 2\dfrac{dy}{dx} = y^{1/2}\left(1 - 3\dfrac{dy}{dx}\right)$

$\qquad\qquad + \dfrac{1}{2}(x - 3y)y^{-1/2}\,\dfrac{dy}{dx}$

$1 + 2\dfrac{dy}{dx} = y^{1/2} - 3y^{1/2}\,\dfrac{dy}{dx} + \dfrac{1}{2}xy^{-1/2}\,\dfrac{dy}{dx}$

$\qquad\qquad - \dfrac{3}{2}y^{1/2}\,\dfrac{dy}{dx}$

$\left(2 + 3y^{1/2} - \dfrac{1}{2}xy^{-1/2} + \dfrac{3}{2}y^{1/2}\right)\dfrac{dy}{dx} = y^{1/2} - 1$

$\dfrac{2y^{1/2}\left(2 + \dfrac{9}{2}y^{1/2} - \dfrac{1}{2}xy^{-1/2}\right)}{2y^{1/2}}\,\dfrac{dy}{dx} = y^{1/2} - 1$

$\left(\dfrac{4y^{1/2} + 9y - x}{2y^{1/2}}\right)\dfrac{dy}{dx} = y^{1/2} - 1$

$\dfrac{dy}{dx} = \dfrac{2y - 2y^{1/2}}{4y^{1/2} + 9y - x}$

11. $(4y^2 - 3x)^{2/3} = 6x$

$\dfrac{d}{dx}[(4y^2 - 3x)^{2/3}] = \dfrac{d}{dx}(6x)$

$\dfrac{2}{3}(4y^2 - 3x)^{-1/3}\left(8y\,\dfrac{dy}{dx} - 3\right) = 6$

$8y\dfrac{dy}{dx} - 3 = 6\left(\dfrac{3}{2}\right)(4y^2 - 3x)^{1/3}$

$\dfrac{dy}{dx} = \dfrac{9(4y^2 - 3x)^{1/3} + 3}{8y}$

13. $\sqrt{2x} - 4yx = -22$, tangent line at $(2, 3)$

$\dfrac{d}{dx}(\sqrt{2x} - 4yx) = \dfrac{d}{dx}(-22)$

$\dfrac{1}{2}(2x)^{-1/2}(2) - 4y - 4x\,\dfrac{dy}{dx} = 0$

$(2x)^{-1/2} - 4y = 4x\dfrac{dy}{dx}$

$\dfrac{dy}{dx} = \dfrac{(2x)^{-1/2} - 4y}{4x} = \dfrac{\dfrac{1}{\sqrt{2x}} - 4y}{4x}$

At $(2, 3)$,

$m = \dfrac{\dfrac{1}{\sqrt{2 \cdot 2}} - 4 \cdot 3}{4 \cdot 2} = \dfrac{\dfrac{1}{2} - 12}{8}$

$\quad = -\dfrac{23}{16}$

The equation of the tangent line is

$y - y_1 = m(x - x_1)$

$y - 3 = -\dfrac{23}{16}(x - 2)$

$23x + 16y = 94.$

15. $y = 8x^3 - 7x^2,\ \dfrac{dx}{dt} = 4,\ x = 2$

$\dfrac{dy}{dt} = \dfrac{d}{dt}(8x^3 - 7x^2)$

$\quad = 24x^2\,\dfrac{dx}{dt} - 14x\,\dfrac{dx}{dt}$

$\quad = 24(2)^2(4) - 14(2)(4)$

$\quad = 272$

17. $y = \dfrac{1 + \sqrt{x}}{1 - \sqrt{x}},\ \dfrac{dx}{dt} = -4,\ x = 4$

$\dfrac{dy}{dt} = \dfrac{d}{dt}\left[\dfrac{1 + \sqrt{x}}{1 - \sqrt{x}}\right]$

$\quad = \dfrac{(1-\sqrt{x})\left(\dfrac{1}{2}x^{-1/2}\dfrac{dx}{dt}\right) - (1 + \sqrt{x})\left(-\dfrac{1}{2}\right)(x^{-1/2}\dfrac{dx}{dt})}{(1 - \sqrt{x})^2}$

$\quad = \dfrac{(1 - 2)\left(\dfrac{1}{2 \cdot 2}\right)(-4) - (1 + 2)\left(\dfrac{-1}{2 \cdot 2}\right)(-4)}{(1 - 2)^2}$

$\quad = \dfrac{1 - 3}{1}$

$\quad = -2$

21. $y = 4(x^2 - 1)^3$

$dy = 12(x^2 - 1)^2(2x)dx$

$\quad = 24x(x^2 - 1)^2dx$

23. $y = \sqrt{9 + x^3} = (9 + x^3)^{1/2}$

$dy = \frac{1}{2}(9 + x^3)^{-1/2}(3x^2)dx$

$= \frac{3x^2}{2}(9 + x^3)^{-1/2} dx$

25. $y = \frac{3x - 7}{2x + 1}$; $x = 2$, $\Delta x = .003$

$dy = \frac{(3)(2x + 1) - (2)(3x - 7)}{(2x + 1)^2} dx$

$dy = \frac{17}{(2x + 1)^2} dx$

$\approx \frac{17}{(2x + 1)^2}\Delta x$

$= \frac{17}{(2[2] + 1)^2}(.003)$

$= .00204$

29. $x = 2 + y$

$P = xy^2 = (2 + y)y^2$

$= 2y^2 + y^3$

$P' = 4y + 3y^2 = 0$

$y(4 + 3y) = 0$

$y = 0$ or $4 + 3y = 0$

$y = -\frac{4}{3}$

$P'' = 4 + 6y$

If $y = 0$, $P'' = 4$, which implies that $y = 0$ is location of minimum.

If $y = -\frac{4}{3}$, $P'' = -4$, which implies that $y = -\frac{4}{3}$ is location of maximum.

$x = 2 + y = 2 + 0 = 2$

So xy^2 is minimized when $x = 2$ and $y = 0$.

31. (a) $P(x) = -x^3 + 10x^2 - 12x$

$P'(x) = -3x^2 + 20x - 12 = 0$

$3x^2 - 20x + 12 = 0$

$(3x - 2)(x - 6) = 0$

$3x - 2 = 0$ or $x - 6 = 0$

$x = \frac{2}{3}$ or $x = 6$

$P''(x) = -6x + 20$

$P''\left(\frac{2}{3}\right) = 16,$

which implies that $x = 2/3$ is location of minimum.

$P''(6) = -16,$

which implies that $x = 6$ is location of maximum.

Thus, 600 boxes will produce a maximum profit.

(b) Maximum profit

$= P(6) = -(6)^3 + 10(6)^2 - 12(6)$

$P(6) = 72$

The maximum profit is $720.

33. $V = \pi r^2 h = 40$, so $h = \frac{40}{\pi r^2}$

$A = 2\pi r^2 + 2\pi rh$

$= 2\pi r^2 + 2\pi r\left(\frac{40}{\pi r^2}\right)$

$= 2\pi r^2 + \frac{80}{r}$

Cost $= C(r) = 4(2\pi r^2) + 3\left(\frac{80}{r}\right)$

$= 8\pi r^2 + \frac{240}{r}$

$C'(r) = 16\pi r - \frac{240}{r^2}$

$16\pi r - \frac{240}{r^2} = 240$

$16\pi r^3 = 240$

$r^3 = \frac{15}{\pi}$

$r \approx 1.684$

$C''(r) = 16\pi + \dfrac{240}{r^3} > 0$, so

$r = 1.684$ minimizes cost.

$h = \dfrac{40}{\pi r^2} = \dfrac{40}{\pi(1.684)^2} = 4.490$

The radius should be 1.684 in and the height should be 4.490 in.

35. $M = 980{,}000$ cases sold per year

$k = \$2$, cost to store 1 case for 1 yr

$f = \$20$, fixed cost for order

$x =$ the number of cases per order

$x = \sqrt{\dfrac{kM}{2f}}$

$\quad = \sqrt{\dfrac{2(980{,}000)}{2(20)}}$

$\quad = \sqrt{49{,}000}$

$\quad \approx 221$

Each order will consist of

$\dfrac{M}{x} = \dfrac{980{,}000}{221} = 4434$ cases.

37. $M = 240{,}000$ cases per year

$k = \$2$, cost to store 1 case for 1 yr

$f = \$15$, fixed cost for 1 batch

$x =$ the number of batches that should be produced annually

$x = \sqrt{\dfrac{kM}{2f}}$

$\quad = \sqrt{\dfrac{2(240{,}000)}{2(15)}}$

$\quad = \sqrt{16{,}000} \approx 126$ batches

39. $q = \dfrac{A}{p^k}$

$\dfrac{dq}{dp} = -k\dfrac{A}{p^{k+1}}$

$E = -\dfrac{p}{q} \cdot \dfrac{dq}{dp}$

$\quad = -\dfrac{p}{\dfrac{A}{p^k}}\left(-k\dfrac{A}{p^{k+1}}\right)$

$\quad = \left(-\dfrac{p^{k+1}}{A}\right)\left(-k\dfrac{A}{p^{k+1}}\right)$

$\quad = k$

The demand is elastic when $k > 1$ and inelastic when $k < 1$.

41. Let $x =$ the distance from the base of the ladder to the building

$y =$ the height on the building at the top of the ladder.

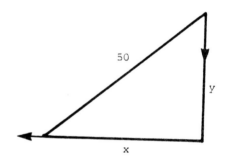

$\dfrac{dy}{dt} = 2$ ft/min;

$(50)^2 = x^2 + y^2$

$0 = 2x\dfrac{dx}{dt} + 2y\dfrac{dy}{dt}$

$\dfrac{dx}{dt} = -\dfrac{y}{x}\dfrac{dy}{dt}$

When $x = 30$, $y = \sqrt{2500 - (30)^2} = 40$.

So $\dfrac{dx}{dt} = -\dfrac{40}{30}(2) = -\dfrac{80}{30} = -\dfrac{8}{3}$.

The base of the ladder is slipping away from the building at a rate of 8/3 ft/min.

43. Let x = one-half the width of the triangular cross section

h = the height of the water

V = the volume of the water.

$\frac{dV}{dt} = 3.5$ ft³/min

Find $\frac{dV}{dt}$ when $h = \frac{1}{3}$.

$$V = \begin{bmatrix} \text{Area of} \\ \text{triangular} \\ \text{side} \end{bmatrix} \begin{bmatrix} \text{length} \end{bmatrix}$$

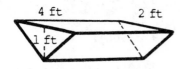

Area of triangular cross section

$= \frac{1}{2}$(base)(altitude)

$= \frac{1}{2}(2x)(h)$

$= xh$

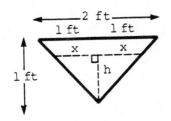

By similar triangles, $\frac{1}{x} = \frac{1}{h}$,

so x = h.

$V = (xh)(4)$

$\quad = h^2 \cdot 4$

$\quad = 4h^2$

$\frac{dV}{dt} = 8h\frac{dh}{dt}$

$\frac{1}{8h}\frac{dV}{dt} = \frac{dh}{dt}$

$\frac{1}{8\left(\frac{1}{3}\right)}(3.5) = \frac{dh}{dt}$

$\frac{dh}{dt} = \frac{21}{16} = 1.3125$ ft/min

45. A = s²; s = 9.2", Δs = ±.04

ds = 2sds

ΔA ≈ 2sΔs

$\quad = 2(9.2)(\pm.04)$

$\quad = \pm.736$ in²

47. $-12x + x^3 + y + y^2 = 4$

$\frac{d}{dx}(-12x + x^3 + y + y^2) = \frac{d}{dx}(4)$

$-12 + 3x^2 + \frac{dy}{dx} + 2y\frac{dy}{dx} = 0$

$(1 + 2y)\frac{dy}{dx} = 12 - 3x^2$

$\frac{dy}{dx} = \frac{12 - 3x^2}{1 + 2y}$

(a) If $\frac{dy}{dx} = 0$, $12 - 3x^2 = 0$

$12 = 3x^2$

$\pm 2 = x$

x = 2: $-24 + 8 + y + y^2 = 4$

$y + y^2 = 20$

$y^2 + y - 20 = 0$

$(y + 5)(y - 4) = 0$

$y = -5,\ y = 4$

(2, -5) and (2, 4) are critical points.

x = -2:

$24 - 8 + y + y^2 = 4$

$y + y^2 = -12$

$y^2 + y + 12 = 0$

$y = \frac{-1 \pm \sqrt{1^2 - 48}}{2}$

This leads to imaginary roots.
x = -2 does not produce critical
points.

(b)

x	y_1	y_2
1.9	-4.99	3.99
2	-5	4
2.1	-4.99	3.99

The point (2, -5) is a relative
minimum.
The point (2, 4) is a relative
maximum.

Extended Application

1. $Z(m) = \dfrac{C_1}{m} + DtC_2 + DC_2\left(\dfrac{m-1}{2}\right)$

$Z'(m) = -\dfrac{C_1}{m^2} + 0 + \dfrac{DC_3}{2}$

$= -\dfrac{C_1}{m^2} + \dfrac{DC_3}{2}$

2. $Z'(m) = 0$ when

$\dfrac{DC_3}{2} = \dfrac{C_1}{m^2}$

$m^2 = \dfrac{2C_1}{DC_3}$

$m = \sqrt{\dfrac{2C_1}{DC_3}}$

3. $D = 3$, $t = 12$, $C_1 = \$15,000$,
$C_3 = 900$

$m = \sqrt{\dfrac{2(15,000)}{3(900)}}$

$= \sqrt{\dfrac{30,000}{2700}}$

$= \sqrt{\dfrac{100}{9}}$

≈ 3.33

4. $3 < 3.33 < 4$
$m^+ = 4$ and $m^- = 3$

5. $Z(m) = \dfrac{C_1 + mDtC_2}{m} + \dfrac{DC_3\left[\dfrac{m(m-1)}{2}\right]}{m}$

$= \dfrac{C_1}{m} + DtC_2 + DC_3\left(\dfrac{m-1}{2}\right)$

$C_1 = 15,000$; $D = 3$, $t = 12$,
$C_2 = 100$; $C_3 = 900$

$Z(m^+) = Z(4)$

$= \dfrac{15,000}{4} + 3(12)(100)$

$+ 3(900)\left(\dfrac{4-1}{2}\right)$

$= 3750 + 3600 + 4050$

$= \$11,400$

$Z(m^-) = Z(3)$

$= \dfrac{15,000}{3} + 3600$

$+ 3\dfrac{(900)(3-1)}{2}$

$= 5000 + 3600 + 2700$

$= \$11,300$

6. Since $Z(3) < Z(4)$, the optimal time
interval is 3 months.

$N = mD$
$= 3 \cdot 3$
$= 9$

There should be 9 trainees per batch

CHAPTER 12 TEST

1. Find two nonnegative numbers x and y such that x + y = 50 and P = xy is maximized.

2. A travel agency offers a tour to the Bahamas for 12 people at $800 each. For each person more than the original 12, up to a total of 30 people, who signs up for the cruise, the fare is reduced $20. What tour group size produces the greatest revenue for the travel agency?

3. $320 is available for fencing for a rectangular garden. The fencing for the two sides parallel to the back of the house costs $6 per linear foot, while the fencing for the other sides costs $2 per linear foot. Find the dimensions that will maximize the area of the garden.

4. A small beauty supply store sells 600 hair dryers each year. Each hair dryer costs the store $5. It costs the store $3 a year to store one hair dryer for one year. The fixed cost of placing the order is $36. Find the number of orders that should be placed each year. Find the number of dryers that should be ordered each time.

5. A demand function is given by

$$q = 100 - 2p$$

where q is the number of units produced and p is the price in dollars.

    (a) Find E.   (b) Find the price that maximizes the total revenue.

6. What does elasticity of demand measure? What does it mean for demand to be elastic? What does it mean for demand to be inelastic?

7. Find $\frac{dy}{dx}$ for $x^3 - x^2y + y^2 = 0$.

8. Find an equation for the tangent line to the graph of $2x^2 - y^2 = 2$ at $(3, 4)$.

9. Suppose $3x^2 + 4y^2 + 6 = 0$.

    Use implicit differentiation to find dy/dx. Explain why your result is meaningless.

10. Find $\dfrac{dy}{dt}$ if $y = 3x^3 - 2x^2$, $\dfrac{dx}{dt} = -2$, and $x = 3$.

11. Find $\dfrac{dy}{dt}$ if $2xy - y^2 = x$, $x = -1$, $y = 3$, and $\dfrac{dx}{dt} = 2$.

12. When solving related rates problems, how should you interpret a negative derivative?

13. A ladder 25 ft long is leaning against a vertical wall. If the bottom of the ladder is pulled horizontally away from the wall at 3 ft/sec, how fast is the top of ladder sliding down the wall when the bottom is 15 ft from the wall?

14. A real estate developer estimates that his monthly sales are given by

$$S = 30y - xy - \frac{y^2}{3000},$$

    where y is the average cost of a new house in the development and x percent is the current interest rate for home mortages. If the current rate is 11% and is rising at a rate of 1/4% per month and the current average price of a new home is $90,000 and is increasing at a rate of $500 per month, how fast are his expected sales changing?

Given the revenue and cost functions $R = 30q - .2q^2$ and $C = 12q + 20$, where q is the daily production (and sales), answer Problems 15–17 when 50 units are produced and the rate of change of production is 8 units per day.

15. Find the rate of change of revenue with respect to time.

16. Find the rate of change of cost with respect to time.

17. Find the rate of change of profit with respect to time.

18. If $y = \dfrac{2x - 3}{x^2 - 7}$, find dy.

19. Find the value of dy if $y = \sqrt{5 - x^2}$, $x = 1$, $\Delta x = .01$.

20. The radius of a sphere is claimed to be 5 cm with a possible error of .01 cm. Estimate the possible error in the volume of the sphere.

CHAPTER 12   TEST ANSWERS

1. 25 and 25

2. 26 people

3. $13\frac{1}{3}$ ft on each side parallel to the back, 40 ft on each of the other sides

4. 5 orders, 120 dryers per order

5. (a) $E = \dfrac{2p}{100 - 2p}$ (b) $25

6. The instantaneous responsiveness of demand to price; the relative change in demand is greater than the relative change in price; the relative change in demand is less than the relative change in price.

7. $\dfrac{dy}{dx} = \dfrac{2xy - 3x^2}{2y - x^2}$

8. $3x - 2y = 1$

9. $-\dfrac{3x}{4y}$; no function exists such that $3x^2 + 4y^2 + 6 = 0$

10. -138

11. 1.25

12. A decrease in rate or travel in a negative direction

13. $-\dfrac{9}{4}$ ft/sec

14. Decreasing $43,000/mo

15. Increasing at a rate of $80/day

16. Increasing at a rate of $96/day

17. Decreasing at a rate of $16/day

18. $dy = \dfrac{-2x^2 + 6x - 14}{(x^2 - 7)^2} dx$

19. $-.005$

20. 3.14 cm³

CHAPTER 13 EXPONENTIAL AND LOGARITHMIC
FUNCTIONS

Section 13.1

For Exercises 1–15, see the graphs in the
answer section at the back of your textbook.

1. $y = 3^x$

x	-2	-1	0	1	2
y	1/9	1/3	1	3	9

3. $y = 3^{-x/2}$

x	-4	-2	0	2	4
y	9	3	1	1/3	1/9

5. $y = \left(\frac{1}{4}\right)^x + 1$

x	-2	-1	0	1	2
y	17	5	2	5/4	17/16

7. $y = -3^{2x-1}$

x	-1	-1/2	0	1/2	1
2x - 1	-3	-2	-1	0	1
y	-1/27	-1/9	-1/3	-1	-3

9. $y = e^{-x^2+1}$

x	-2	-1	0	1	2
y	.05	1	e ≈ 2.72	1	.05

11. $y = 10 - 5e^{-x}$

x	-2	-1	0	1	2	3	4
y	-26.9	-3.6	5	8.2	9.3	9.8	9.9

13. $y = x \cdot 2^x$

x	-3	-2	-1	-.5	0	1	2
y	-.38	-.5	-.5	-.35	0	2	8

15. $y = \dfrac{e^x - e^{-x}}{2}$

x	-3	-2	-1	0	1	2
y	-10.02	-3.6	-1.2	0	1.2	3.6

17. To solve the equation $4^x = 6$ using
the method described in this
section, 4 and 6 must be written as
powers of the same base. This is
not easily done.

19. $4^x = 64$

$4^x = 4^3$

$x = 3$

21. $e^x = \dfrac{1}{e^2}$

$e^x = e^{-2}$

$x = -2$

23. $25^x = 125^{x-2}$

$(5^2)^x = (5^3)^{x-2}$

$5^{2x} = 5^{3x-6}$

$2x = 3x - 6$

$6 = x$

25. $16^{x+2} = 64^{2x-1}$

$(2^4)^{x+2} = (2^6)^{2x-1}$

$2^{4x+8} = 2^{12x-6}$

$4x + 8 = 12x - 6$

$14 = 8x$

$\frac{7}{4} = x$

27. $e^{-x} = (e^2)^{x+3}$

$e^{-x} = e^{2x+6}$

$-x = 2x + 6$

$-3x = 6$

$x = -2$

29. $5^{-|x|} = \frac{1}{25}$

$5^{-|x|} = 5^{-2}$

$-|x| = -2$

$|x| = 2$

$x = 2 \quad \text{or} \quad x = -2$

31. $5^{x^2+x} = 1$

$5^{x^2+x} = 5^0$

$x^2 + x = 0$

$x(x + 1) = 0$

$x = 0 \quad \text{or} \quad x + 1 = 0$

$x = 0 \quad \text{or} \qquad x = -1$

33. $9^{x+4} = 3^{x^2}$

$(3^2)^{x+4} = 3^{x^2}$

$3^{2x+8} = 3^{x^2}$

$2x + 8 = x^2$

$0 = x^2 - 2x - 8$

$0 = (x + 2)(x - 4)$

$x = -2 \quad \text{or} \quad x = 4$

35. $f(x) = e^x$

(a) $\dfrac{f(1) - f(0)}{1 - 0} = \dfrac{e^1 - e^0}{1} = 1.718$

(b) $\dfrac{f(.1) - f(0)}{.1 - 0} = \dfrac{e^{.1} - e^0}{.1} = 1.052$

(c) $\dfrac{f(.01) - f(0)}{.01 - 0} = \dfrac{e^{.01} - e^0}{.01}$

$= 1.005$

(d) $\dfrac{f(.001) - f(0)}{.001 - 0} = \dfrac{e^{.001} - e^0}{.001}$

$= 1.001$

(e) The slope of the graph of $f(x)$ at $x = 0$ is about 1.

39. $A = P\left(1 + \dfrac{r}{m}\right)^{tm}$, $P = 26,000$, $r = .12$, $t = 3$

(a) annually, $m = 1$

$A = 26,000\left(1 + \dfrac{.12}{1}\right)^{3 \cdot 1}$

$= 26,000(1.12)^3$

$= 36,528.13$

Interest $= \$36,528.13 - \$26,000$

$= \$10,528.13$

(b) semiannually, $m = 2$

$A = 26,000\left(1 + \dfrac{.12}{2}\right)^{3 \cdot 2}$

$= 26,000(1.06)^6$

$= 36,881.50$

Interest $= \$36,881.50 - \$26,000$

$= \$10,881.50$

(c) quarterly, m = 4

$$A = 26{,}000\left(1 + \frac{.12}{4}\right)^{3 \cdot 4}$$

$$= 26{,}000(1.03)^{12}$$

$$= 37{,}069.78$$

Interest = $37,069.78 − $26,000

$$= \$11{,}069.78$$

(d) monthly, m = 12

$$A = 26{,}000\left(1 + \frac{.12}{12}\right)^{3 \cdot 12}$$

$$= 26{,}000(1.01)^{36}$$

$$= 37{,}199.99$$

Interest = $37,199.99 − $26,000

$$= \$11{,}199.99$$

41. $A = P\left(1 + \frac{r}{m}\right)^{tm}$, P = 17,500, r = .10,

m = 2, t = 6

$$A = 17{,}500\left(1 + \frac{.10}{2}\right)^{2 \cdot 6}$$

$$= 17{,}500(1.05)^{12}$$

$$= 31{,}427.49$$

She will owe $31,427.49.

43. $A = P\left(1 + \frac{r}{m}\right)^{tm}$, P = 5000, A = 8000,

t = 4

(a) m = 1

$$8000 = 5000\left(1 + \frac{r}{1}\right)^{4 \cdot 1}$$

$$\frac{8}{5} = (1 + r)^4$$

$$\left(\frac{8}{5}\right)^{1/4} - 1 = r$$

$$.125 = r$$

The interest rate is 12.5%.

(b) m = 4

$$8000 = 5000\left(1 + \frac{r}{4}\right)^{4 \cdot 4}$$

$$\frac{8}{5} = \left(1 + \frac{r}{4}\right)^{16}$$

$$\left(\frac{8}{5}\right)^{1/16} - 1 = \frac{r}{4}$$

$$4\left[\left(\frac{8}{5}\right)^{1/16} - 1\right] = r$$

$$.119 = r$$

The interest rate is 11.9%.

45. $p(t) = 250 - 120(2.8)^{-.5t}$

(a) $p(2) = 250 - 120(2.8)^{-.5(2)}$

$$\approx 207$$

(b) $p(4) = 240 - 120(2.8)^{-.5(4)}$

$$\approx 235$$

(c) $p(10) = 250 - 120(2.8)^{-.5(10)}$

$$\approx 249$$

(d) See the answer graph in the back of your textbook.

(e) The number of symbols typed per minute gets very close to 250.

(f) The limit appears to be 250 symbols per minute.

47. $A(t) = 2600e^{.018t}$

(a) 1970: t = 20

$$A(20) = 2600e^{.018(20)}$$

$$= 2600e^{.36}$$

$$= 3727 \text{ million}$$

This is very close to the actual population of 3700 million.

(b) 1990: t = 40

$$A(40) = 2600e^{.018(40)}$$
$$= 2600e^{.72}$$
$$= 5341 \text{ million}$$

(c) 2000: t = 50

$$A(50) = 2600e^{.018(50)}$$
$$= 2600e^{.9}$$
$$= 6395 \text{ million}$$

49. $y = 6 \cdot 2^t$

(a) t = 0

$$y = 6 \cdot 2^0$$
$$= 6$$

6 animals were introduced originally.

(b) $24 = 6 \cdot 2^t$
$$4 = 2^t$$
$$2^2 = 2^t$$
$$2 = t$$

It will take 2 yr.

(c) $384 = 6 \cdot 2^t$
$$64 = 2^t$$
$$2^6 = 2^t$$
$$6 = t$$

It will take 6 yr.

51. $Q(t) = 1000(5^{-.3t})$

(a) $Q(6) = 1000[5^{-.3(6)}]$
$$= 1000(5^{-1.8})$$
$$= 55 \text{ g}$$

(b) $8 = 1000(5^{-.3t})$
$$\frac{1}{125} = 5^{-.3t}$$
$$5^{-3} = 5^{-.3t}$$
$$-3 = -.3t$$
$$10 = t$$

It will take 10 mo.

53. This exercise should be solved by a graphing calculator or computer methods. The solution will vary according to the method that is used. See the answer graph at the back of the textbook.

55. This exercise should be solved by a graphing calculator or computer methods. The solution will vary according to the method that is used. See the answer graph at the back of the textbook.

Section 13.2

1. $2^3 = 8$

Since $a^y = x$ means $y = \log_a x$, the equation in logarithmic form is

$$\log_2 8 = 3.$$

3. $3^4 = 81$
$$\log_3 81 = 4$$

5. $\left(\frac{1}{3}\right)^{-2} = 9$
$$\log_{1/3} 9 = -2$$

7. $\log_2 128 = 7$

Since $y = \log_a x$ means $a^y = x$, the equation in exponential form is

$$2^7 = 128.$$

9. $\log_{25} \dfrac{1}{25} = -1$

$$25^{-1} = \dfrac{1}{25}$$

11. $\log 10,000 = 4$

$\log_{10} 10,000 = 4$

$$10^4 = 10,000$$

When no base is written, $\log_{10}$ is understood.

13. Let $\log_5 25 = x$.

Then, $5^x = 25$.

$$5^x = 5^2$$
$$x = 2$$

Thus, $\log_5 25 = 2$.

15. $\log_4 64 = x$

$$4^x = 64$$
$$4^x = 4^3$$
$$x = 3$$

17. $\log_2 \dfrac{1}{4} = x$

$$2^x = \dfrac{1}{4}$$
$$2^x = 2^{-2}$$
$$x = -2$$

19. $\log_2 \sqrt[3]{1/4} = x$

$$2^x = \left(\dfrac{1}{4}\right)^{1/3}$$
$$2^x = \left(\dfrac{1}{2^2}\right)^{1/3}$$
$$2^x = 2^{-2/3}$$
$$x = -\dfrac{2}{3}$$

21. $\ln e = x$

Recall that $\ln$ is $\log_e$.

$$e^x = e$$
$$x = 1$$

23. $\ln e^{5/3} = x$

$$e^x = e^{5/3}$$
$$x = \dfrac{5}{3}$$

25. The logarithm to the base 3 of 4 is written $\log_3 4$. The subscript denotes the base.

For Exercises 27–31, see the answer graphs in the textbook.

27. $y = \log_4 x$

$4^y = x$

To find plotting points, use values of y to find x.

x	1/16	1/4	1	4	16
y	-2	-1	0	1	2

29. $y = \log_{1/3}(x - 1)$

$\left(\frac{1}{3}\right)^y = x - 1$

$\left(\frac{1}{3}\right)^y + 1 = x$

x	10	4	2	4/3	10/9
y	-2	-1	0	1	2

31. $y = \ln x^2$

Use a calculator to find plotting points.

x	-2	-1	0	1	2
y	1.39	0	does not exist	0	1.39

x cannot be zero because there is no value of y such that $e^y = 0$.

33. $\log_9 7m = \log_9 7 + \log_9 m$

35. $\log_3 \frac{3p}{5k}$

$= \log_3 3p - \log_3 5k$

$= (\log_3 3 + \log_3 p) - (\log_3 5 + \log_3 k)$

$= 1 + \log_3 p - \log_3 5 - \log_3 k$

37. $\log_3 \frac{5\sqrt{2}}{\sqrt[4]{7}}$

$= \log_3 5\sqrt{2} - \log_3 \sqrt[4]{7}$

$= \log_3 5 \cdot 2^{1/2} - \log_3 7^{1/4}$

$= \log_3 5 + \log_3 2^{1/2} - \log_3 7^{1/4}$

$= \log_3 5 + \frac{1}{2} \log_3 2 - \frac{1}{4} \log_3 7$

For Exercises 39–41, $\log_b 2 = a$ and $\log_b 3 = c$.

39. $\log_b 8 = \log_b 2^3$

$= 3 \log_b 2$

$= 3a$

41. $\log_b 72b = \log_b 72 + \log_b b$

$= \log_b 72 + 1$

$= \log_b 2^3 \cdot 3^2 + 1$

$= \log_b 2^3 + \log_b 3^2 + 1$

$= 3 \log_b 2 + 2 \log_b 3 + 1$

$= 3a + 2c + 1$

43. $\log_5 20 = \frac{\ln 20}{\ln 5}$

$\approx \frac{3}{1.61}$

≈ 1.86

45. $\log_{1.2} 5.5 = \frac{\ln 5.5}{\ln 1.2}$

≈ 9.35

47. $\log_3 1.1^{-2.4} = -2.4(\log_3 1.1)$

$= -2.4\left(\frac{\ln 1.1}{\ln 3}\right)$

$\approx -.21$

49. $\log_x 25 = -2$

$x^{-2} = 25$

$(x^{-2})^{-1/2} = 25^{-1/2}$

$x = \frac{1}{5}$

51. $\log_8 4 = z$

$8^z = 4$

$(2^3)^z = 2^2$

$2^{3z} = 2^2$

$3z = 2$

$z = \dfrac{2}{3}$

53. $\log_r 7 = \dfrac{1}{2}$

$r^{1/2} = 7$

$(r^{1/2})^2 = 7^2$

$r = 49$

55. $\log_5 (9x - 4) = 1$

$5^1 = 9x - 4$

$9 = 9x$

$1 = x$

57. $\log_9 m - \log_9 (m - 4) = -2$

$\log_9 \dfrac{m}{m - 4} = -2$

$9^{-2} = \dfrac{m}{m - 4}$

$\dfrac{1}{81} = \dfrac{m}{m - 4}$

$m - 4 = 81m$

$-4 = 80m$

$-.05 = m$

This value is not possible since $\log_9 (-.05)$ does not exist. Thus, there is no solution to the original equation.

59. $4^x = 12$

$x \log 4 = \log 12$

$x = \dfrac{\log 12}{\log 4}$

≈ 1.79

61. $e^{2y} = 12$

$\ln e^{2y} = \ln 12$

$2y \ln e = \ln 12$

$2y(1) = \ln 12$

$y = \dfrac{\ln 12}{2}$

≈ 1.24

63. $10e^{3z-7} = 5$

$\ln 10e^{3z-7} = \ln 5$

$\ln 10 + \ln e^{3z-7} = \ln 5$

$\ln 10 + (3z - 7) \ln e = \ln 5$

$3z - 7 = \ln 5 - \ln 10$

$3z = \ln 5 - \ln 10 + 7$

$z = \dfrac{\ln 5 - \ln 10 + 7}{3}$

≈ 2.10

65. Prove: $\log_a x^r = r \log_a x.$

Let $s = \log_a x.$

The equivalent exponential equation is $a^s = x.$

Now $\log_a x^r = \log_a (a^s)^r$

$= \log_a a^{rs}$

So, $\log_a x^r = rs$

or $\log_a x^r = r \log_a x.$

67. $P = 15,000, \ r = .06, \ m = 2$

$A = P\left(1 + \dfrac{r}{m}\right)^{mt}$

$= 15,000\left(1 + \dfrac{.06}{2}\right)^{2t}$

(a) $A = 2P$

Notice that the doubling time is the same for all values of P.

$$2P = P(1 + .03)^{2t}$$

$$2 = 1.03^{2t}$$

$$\ln 2 = \ln 1.03^{2t}$$

$$= 2t \ln 1.03$$

$$t = \frac{\ln 2}{2 \ln 1.03}$$

$$\approx 11.7 \text{ yr}$$

(b) $A = 3P$

$$3P = P(1 + .03)^{2t}$$

$$3 = 1.03^{2t}$$

$$\ln 3 = 1.03^{2t}$$

$$= 2t \ln 1.03$$

$$t = \frac{\ln 3}{2 \ln 1.03}$$

$$t \approx 18.6 \text{ yr}$$

(c) Since $.05 \le .06 \le .12$, the rule of 72 approximates the doubling time in (a).

$$\frac{72}{100r} = \frac{72}{100(.06)} = 12 \text{ yr}$$

69. $H = \dfrac{-1}{\ln 2} [P_1 \ln P_1 + P_2 \ln P_2$

$$+ P_3 \ln P_3 + P_4 \ln P_4]$$

Let $P_1 = .521$, $P_2 = .324$, $P_3 = .081$, $P_4 = .074$.

$$H = \frac{-1}{\ln 2} [.521 \ln .521$$

$$+ .324 \ln .324$$

$$+ .081 \ln .081$$

$$+ .074 \ln .074$$

$$= 1.589$$

71. From the graph, a body mass of .3 kg (on horizontal axis) corresponds to oxygen consumption of 4.3 ml/min. A body mass of .7 kg has oxygen consumption of 7.8 ml/min.

73. Decibal rating: $10 \log \dfrac{I}{I_0}$

(a) Intensity, $I = 115I_0$

$$10 \log \left(\frac{115I_0}{I_0} \right)$$

$$= 10 \cdot \log 115$$

$$\approx 21$$

(b) $I = 9{,}500{,}000I_0$

$$10 \log \left(\frac{9.5 \times 10^6 I_0}{I_0} \right)$$

$$= 10 \log 9.5 \times 10^6$$

$$\approx 70$$

(c) $I = 1{,}200{,}000{,}000I_0$

$$10 \log \left(\frac{1.2 \times 10^9 I_0}{I_0} \right)$$

$$= 10 \log 1.2 \times 10^9$$

$$\approx 91$$

(d) $I = 895{,}000{,}000{,}000I_0$

$$10 \log \left(\frac{8.95 \times 10^{11} I_0}{I_0} \right)$$

$$= 10 \log 8.95 \times 10^{11}$$

$$\approx 120$$

(e) $I = 109{,}000{,}000{,}000{,}000I_0$

$$10 \log \left(\frac{1.09 \times 10^{14} I_0}{I_0} \right)$$

$$= 10 \log 1.09 \times 10^{14}$$

$$\approx 140$$

(f) Let I_1 = the intensity of 96 decibel sound

I_2 = the intensity of 98 decibel sound.

$$x_1 = \frac{I_1}{I_0} \text{ and } x_2 = \frac{I_2}{I_0}$$

Then $I_1 = x_1 I_0$ and $I_2 = x_2 I_0$.

$$96 = 10 \log x_1$$

$$9.6 = \log x_1$$

$$10^{9.6} = x_1$$

$$4{,}000{,}000{,}000 \approx x_1$$

$I_1 \approx 4,000,000,000 I_0$

$$98 = 10 \log x_2$$
$$9.8 = \log x_2$$
$$10^{9.8} = x_2$$
$$6,300,000,000 \approx x_2$$

$I_2 \approx 6,300,000,000 I_0$

The difference is about

$2,300,000,000 I_0$.

75. $pH = -\log [H^+]$

(a) For pure water:

$$7 = -\log [H^+]$$
$$-7 = \log [H^+]$$
$$10^{-7} = [H^+]$$
$$.0000001 = [H^+]$$

For acid rain:

$$4 = -\log [H^+]$$
$$-4 = \log [H^+]$$
$$10^{-4} = [H^+]$$
$$.0001 = [H^+]$$

$$\frac{.0001}{.0000001} = 1000$$

The acid rain has a hydrogen ion concentration 1000 times greater than pure water.

(b) For laundry solution:

$$11 = -\log [H^+]$$
$$10^{-11} = [H^+]$$

For black coffee:

$$5 = -\log [H^+]$$
$$10^{-5} = [H^+]$$

$$\frac{10^{-5}}{10^{-11}} = 10^6 \quad \text{or} \quad 1,000,000$$

The black coffee has a hydrogen ion concentration 1,000,000 times greater than laundry solution.

Exercises 77 and 79 should be solved by graphing calculator or computer methods. The solutions will vary according to the method that is used. See the answer graphs in the back of the textbook.

Section 13.3

1. $r = 5\%$ compounded monthly
    $m = 12$

$$r_E = \left(1 + \frac{r}{m}\right)^m - 1$$
$$= \left(1 + \frac{.05}{12}\right)^{12} - 1$$
$$\approx .0512$$
$$= 5.12\%$$

3. $r = 10\%$ compounded semiannually
    $m = 2$

$$r_E = \left(1 + \frac{.10}{2}\right)^2 - 1$$
$$= .1025$$
$$= 10.25\%$$

5. $r = 11\%$ compounded continuously

$$r_E = e^r - 1$$
$$= e^{.11} - 1$$
$$\approx .1163$$
$$= 11.63\%$$

7. A = $2000, r = 6%, m = 2, t = 11

$$P = A\left(1 + \frac{r}{m}\right)^{-tm}$$

$$= 2000\left(1 + \frac{.06}{2}\right)^{-11(2)}$$

$$\approx \$1043.79$$

9. A = $10,000, r = 10%, m = 4, t = 8

$$P = A\left(1 + \frac{r}{m}\right)^{-tm}$$

$$= 10,000\left(1 + \frac{.10}{4}\right)^{-8(4)}$$

$$\approx \$4537.71$$

11. A = $7300, r = 11% compounded continuously, t = 3

$$A = Pe^{rt}$$

$$P = \frac{A}{e^{rt}}$$

$$= \frac{7300}{e^{.11(3)}}$$

$$\approx \$5248.14$$

17. $A(t) = A_0 e^{kt}$, t = T, $A(T) = \frac{1}{2} A_0$

$$\frac{1}{2} A_0 = A_0 e^{kT}$$

$$\frac{1}{2} = e^{kT}$$

$$\ln \frac{1}{2} = \ln e^{kT}$$

$$\ln 1 - \ln 2 = kT \ln e$$

Since ln 1 = 0 and ln e = 1,

$$0 - \ln 2 = kT$$

$$-\frac{\ln 2}{T} = k.$$

19. $A = Pe^{rt}$

(a) r = 3%

$$A = 10e^{.03(3)}$$

$$= \$10.94$$

(b) r = 4%

$$A = 10e^{.04(3)}$$

$$= \$11.27$$

(c) r = 5%

$$A = 10e^{.05(3)}$$

$$= \$11.62$$

21. P = $60,000

(a) r = 10% compounded quarterly:

$$A = P\left(1 + \frac{r}{m}\right)^{tm}$$

$$= 60,000\left(1 + \frac{.10}{4}\right)^{5 \cdot 4}$$

$$\approx \$98,316.99$$

r = 9.75% compounded continously:

$$A = Pe^{rt}$$

$$= 60,000\, e^{.0975(5)}$$

$$\approx \$97,694.43$$

Linda will earn more money at 10% compounded quarterly.

(b) She will earn $622.56 more.

(c) r = 10%, m = 4:

$$r_E = \left(1 + \frac{r}{m}\right)^m - 1$$

$$= \left(1 + \frac{.10}{4}\right)^4 - 1$$

$$\approx .1038$$

$$= 10.38\%$$

r = 9.75% compounded continously:

$$r_E = e^r - 1$$
$$= e^{.0975} - 1$$
$$\approx .1024$$
$$= 10.24\%$$

(d) A = $80,000

$$A = Pe^{rt}$$
$$80,000 = 60,000e^{.0975t}$$
$$\frac{4}{3} = e^{.0975t}$$
$$\ln \frac{4}{3} = \ln e^{.0975t}$$
$$\ln 4 - \ln 3 = .0975t$$
$$\frac{\ln 4 - \ln 3}{.0975} = t$$
$$2.95 \approx t$$

$60,000 will grow to $80,000 in about 2.95 years.

23. r = 7.2%, m = 4

$$r_E = \left(1 + \frac{.072}{4}\right)^4 - 1$$
$$\approx .0740$$
$$= 7.40\%$$

25. A = $20,000, t = 4, r = 8%, m = 1

$$A = P\left(1 + \frac{r}{m}\right)^{mt}$$
$$20,000 = P\left(1 + \frac{.08}{1}\right)^{1 \cdot 4}$$
$$\frac{20,000}{(1.08)^4} = P$$
$$\$14,700.60 = P$$

27. A = $20,000, t = 5

(a) r = .08, m = 4

$$A = P\left(1 + \frac{r}{m}\right)^{mt}$$
$$20,000 = P\left(1 + \frac{.08}{4}\right)^{5(4)}$$
$$\frac{20,000}{(1.02)^{20}} = P$$
$$\$13,459.43 = P$$

(b) Interest = 20,000 - 13,459.43
$$= \$6540.57$$

(c) P = $10,000

$$A = 10,000\left(1 + \frac{.08}{4}\right)^{5(4)}$$
$$= \$14,859.47$$

The amount needed will be

$20,000 - $14,859.47 = $5140.53

29. $S(x) = 1000 - 800e^{-x}$

(a) $S(0) = 1000 - 800e^0$
$$= 1000 - 800$$
$$= 200$$

(b) $S(x) = 500$

$$500 = 1000 - 800e^{-x}$$
$$-500 = -800e^{-x}$$
$$\frac{5}{8} = e^{-x}$$
$$\ln \frac{5}{8} = \ln e^{-x}$$
$$-\ln \frac{5}{8} = x$$
$$.47 \approx x$$

Sales reach 500 in about 1/2 yr.

(c) Since $800e^{-x}$ will never actually be zero, $S(x) = 1000 - 800e^{-x}$ will never be 1000.

(d) $\lim\limits_{x \to \infty} (1000 - 800e^{-x})$

$= \lim\limits_{x \to \infty} 1000 - \lim\limits_{x \to \infty} 800e^{-x}$

$= 1000 - 0 = 1000$

31. $y = y_0 e^{kt}$

(a) $y = 125$, $t = 2$

$125 = 100e^{k(2)}$

$1.25 = e^{2k}$

$\ln 1.25 = 2k$

$.11 = k$

The equation is

$y = 100e^{.11t}$

(b) $y = 500$

$500 = 100e^{.11t}$

$5 = e^{.11t}$

$\ln 5 = .11t$

$14.6 = t$

There will be 500 lice in about 15 mo.

33. $y = y_0 e^{kt}$

(a) $y = 20,000$, $y_0 = 50,000$, $t = 9$

$20,000 = 50,000e^{9k}$

$.4 = e^{9k}$

$\ln .4 = 9k$

$-.102 = k$

The equation is

$y = 50,000e^{-.102t}$.

(b) $\frac{1}{2}(50,000) = 25,000$

$25,000 = 50,000e^{-.102t}$

$.5 = e^{-.102t}$

$\ln .5 = -.102t$

$6.8 = t$

Half the bacteria remain after about 6.8 hours.

35. $G(t)$

$= \dfrac{(2500)(1000)}{1000 + (2500 - 1000)e^{-(.0004)(2500)t}}$

$= \dfrac{2,500,000}{1000 + 1500\,e^{-t}}$

(a) $G(.2) = \dfrac{2,500,000}{1000 + 1500e^{-.2}}$

≈ 1100

(b) $G(1) = \dfrac{2,500,000}{1000 + 1500e^{-1}}$

≈ 1600

(c) $G(3) = \dfrac{2,500,000}{1000 + 1500e^{-3}}$

≈ 2300

(d) $2000 = \dfrac{2,500,000}{1000 + 1500e^{-t}}$

$2000(1000 + 1550e^{-t}) = 2,500,000$

$2,000,000 + 3,000,000e^{-t} = 2,500,000$

$3,000,000e^{-t} = 500,000$

$e^{-t} = .167$

$-t = \ln (.167)$

$t \approx 1.8$ **decades**

37. $y = y_0 e^{kt}$

Let $y = .37y_0$ and $t = 5$; that is the number of survivors in 5 yr is 37% of the total number.

$.37y_0 = y_0 e^{k(5)}$

$.37 = e^{5k}$

$\ln .37 = 5k$

$\dfrac{\ln .37}{5} = k$

$-.1989 = k$

39. $P(x) = 500 - 500e^{-x}$

(a) $P_0 = P(0) = 500 - 500e^0$

$\qquad = 0$

(b) $P(2) = 500 - 500e^{-2}$

$\qquad \approx 432$

(c) $P(5) = 500 - 500e^{-5}$

$\qquad \approx 497$

(d) $P = 400$

$\qquad 400 = 500 - 500e^{-x}$

$\qquad -100 = -500e^{-x}$

$\qquad e^{-x} = \dfrac{1}{5}$

$\qquad e^x = 5$

$\qquad x = \ln 5 \approx 1.6$

400 items per day are produced after about 1.6 days.

(e) $\lim\limits_{x \to \infty} (500 - 500e^{-x})$

$\qquad = \lim\limits_{x \to \infty} 500 - \lim\limits_{x \to \infty} 500e^{-x}$

$\qquad = 500 - 0$

$\qquad = 500$

(f) See the answer graph in your textbook.

41. $\qquad A(t) = A_0 e^{kt}$

$\qquad .60\, A_0 = A_0 e^{(-\ln 2/5600)t}$

$\qquad .60 = e^{(-\ln/5600)t}$

$\qquad \ln .60 = -\dfrac{\ln 2}{5600}t$

$\qquad \dfrac{5600(\ln .60)}{-\ln 2} = t$

$\qquad\qquad 4127 \approx t$

The sample was about 4100 yr old.

43. $\qquad \dfrac{1}{2}A_0 = A_0 e^{-.00043t}$

$\qquad \dfrac{1}{2} = e^{-.00043t}$

$\qquad \ln \dfrac{1}{2} = -.00043t$

$\qquad \ln 1 - \ln 2 = -.00043t$

$\qquad \dfrac{0 - \ln 2}{-.00043} = t$

$\qquad\qquad 1612 \approx t$

The half-life of radium 226 is about 1600 yr.

45. $y = 40e^{-.004t}$

(a) $t = 180$

$\qquad y = 40e^{-.004(180)}$

$\qquad = 40e^{-.72}$

$\qquad \approx 19.5$ watts

(b) $\qquad 20 = 40e^{-.004t}$

$\qquad \dfrac{1}{2} = e^{-.004t}$

$\qquad \ln \dfrac{1}{2} = -.004t$

$\qquad \dfrac{\ln 1 - \ln 2}{-.004} = t$

$\qquad\qquad 173 \approx t$

(c) The power will never be completely gone. $\lim\limits_{t \to \infty} 40e^{-.004t} = 0$, but it will never actually reach zero.

47. (a) Let t = the number of degrees Celsius.

$y = y_0 \cdot e^{kt}$, $y_0 = 10$ when $t = 0°$.

To find k, let $y = 11$, when $t = 10°$.

$$11 = 10e^{10k}$$

$$e^{10k} = \frac{11}{10}$$

$$10k = \ln 1.1$$

$$k = \frac{\ln 1.1}{10}$$

$$\approx .0095$$

$$y = 10 \, e^{.0095t}$$

(b) Let $y = 15$,

Solve for t.

$$15 = 10e^{.0095t}$$

$$\ln 1.5 = .0095t$$

$$t = \frac{\ln 1.5}{.0095}$$

$$\approx 42.7° \text{ C}$$

49. $f(t) = T_0 + Ce^{-kt}$

$$25 = 20 + 100e^{-.1t}$$

$$5 = 100e^{-.1t}$$

$$e^{-.1t} = .05$$

$$-.1t = \ln .05$$

$$t = \frac{\ln .05}{-.1}$$

$$\approx 30$$

It will take about 30 min.

Section 13.4

1. $y = \ln (8x)$

$$y' = \frac{d}{dx}(\ln 8x)$$

$$= \frac{d}{dx}(\ln 8 + \ln x)$$

$$= \frac{d}{dx}(\ln 8) + \frac{d}{dx}(\ln x)$$

$$= 0 + \frac{1}{x}$$

$$= \frac{1}{x}$$

3. $y = \ln (3 - x)$

$$g(x) = 3 - x$$

$$g'(x) = -1$$

$$y' = \frac{g'(x)}{g(x)}$$

$$= \frac{-1}{3 - x} \text{ or } \frac{1}{x - 3}$$

5. $y = \ln |2x^2 - 7x|$

$$g(x) = 2x^2 - 7x$$

$$g'(x) = 4x - 7$$

$$y' = \frac{4x - 7}{2x^2 - 7x}$$

7. $y = \sqrt{x + 5}$

$$g(x) = \sqrt{x + 5}$$

$$= (x + 5)^{1/2}$$

$$g'(x) = \frac{1}{2}(x + 5)^{-1/2}$$

$$y' = \frac{\frac{1}{2}(x + 5)^{-1/2}}{(x + 5)^{1/2}}$$

$$= \frac{1}{2(x + 5)}$$

9. $y = \ln (x^4 + 5x^2)^{3/2}$

$$= \frac{3}{2} \ln (x^4 + 5x^2)$$

$$y' = \frac{3}{2}D_x [\ln (x^4 + 5x^2)]$$

$$g(x) = x^4 + 5x^2$$

$$g'(x) = 4x^3 + 10x$$

$$y' = \frac{3}{2}\left(\frac{4x^3 + 10x}{x^4 + 5x^2}\right)$$

$$= \frac{3}{2}\left[\frac{2x(2x^2 + 5)}{x^2(x^2 + 5)}\right]$$

$$= \frac{3(2x^2 + 5)}{x(x^2 + 5)}$$

11. $y = -3x \ln (x + 2)$

Use the product rule.

$y' = -3x \left[\frac{d}{dx} \ln (x + 2) \right]$

$\quad + \ln (x + 2) \left[\frac{d}{dx} (-3x) \right]$

$\quad = -3x \left(\frac{1}{x + 2} \right) + \left[\ln (x + 2) \right] (-3)$

$\quad = -\frac{3x}{x + 2} - 3 \ln (x + 2)$

13. $y = x^2 \ln |x|$

Use the product rule.

$y' = x^2 \left(\frac{1}{x} \right) + 2x \ln |x|$

$\quad = x + 2x \ln |x|$

15. $y = \frac{2 \ln (x + 3)}{x^2}$

Use the quotient rule.

$y' = \frac{x^2 \left(\frac{2}{x + 3} \right) - 2 \ln (x + 3) \cdot 2x}{(x^2)^2}$

$\quad = \frac{\frac{2x^2}{x + 3} - 4x \ln (x + 3)}{x^4}$

$\quad = \frac{2x^2 - 4x(x + 3) \ln (x + 3)}{x^4 (x + 3)}$

$\quad = \frac{x[2x - 4(x + 3) \ln (x + 3)]}{x^4 (x + 3)}$

$\quad = \frac{2x - 4(x + 3) \ln (x + 3)}{x^3 (x + 3)}$

17. $y = \frac{\ln x}{4x + 7}$

Use the quotient rule.

$y' = \frac{(4x + 7) \left(\frac{1}{x} \right) - (\ln x)(4)}{(4x + 7)^2}$

$\quad = \frac{\frac{4x + 7}{x} - 4 \ln x}{(4x + 7)^2}$

$\quad = \frac{4x + 7 - 4x \ln x}{x(4x + 7)^2}$

19. $y = \frac{3x^2}{\ln x}$

$y' = \frac{(\ln x)(6x) - 3x^2 \left(\frac{1}{x} \right)}{(\ln x)^2}$

$\quad = \frac{6x \ln x - 3x}{(\ln x)^2}$

21. $y = (\ln |x + 1|)^4$

$y' = 4(\ln |x + 1|)^3 \left(\frac{1}{x + 1} \right)$

$\quad = \frac{4(\ln |x + 1|)^3}{x + 1}$

23. $\quad y = \ln |\ln x|$

$g(x) = \ln x$

$g'(x) = \frac{1}{x}$

$\quad y' = \frac{g'(x)}{g(x)}$

$\quad = \frac{\frac{1}{x}}{\ln x}$

$\quad = \frac{1}{x \ln x}$

27. $y = \log (6x)$

$\quad = \frac{\ln 6x}{\ln 10}$

$\quad = \frac{1}{\ln 10} (\ln 6x)$

$$y' = \frac{1}{\ln 10}\left(\frac{1}{x}\right)$$

$$= \frac{1}{x \ln 10}$$

29. $y = \log |1 - x| = \frac{\ln |1 - x|}{\ln 10}$

$$y' = \frac{1}{\ln 10} \cdot \frac{d}{dx}[\ln |1 - x|]$$

$$= \frac{1}{\ln 10} \cdot \frac{-1}{1 - x}$$

$$= -\frac{1}{(\ln 10)(1 - x)}$$

or $\frac{1}{(\ln 10)(x - 1)}$

31. $y = \log_5 \sqrt{5x + 2}$

$$= \log_5 (5x + 2)^{1/2}$$

$$= \frac{\ln (5x + 2)^{1/2}}{\ln 5}$$

$$y' = \frac{1}{\ln 5} \cdot \frac{\frac{1}{2}(5x + 2)^{-1/2} (5)}{(5x + 2)^{1/2}}$$

$$= \frac{5}{2 \ln 5 (5x + 2)}$$

33. $y = \log_3 (x^2 + 2x)^{3/2}$

$$= \frac{\ln (x^2 + 2x)^{3/2}}{\ln 3}$$

$$= \frac{1}{\ln 3} \cdot \frac{3}{2} \ln (x^2 + 2x)$$

$$y' = \frac{3}{2 \ln 3} \cdot \frac{2x + 2}{x^2 + 2x}$$

$$= \frac{3}{2 \ln 3} \cdot \frac{2(x + 1)}{x^2 + 2x}$$

$$= \frac{3(x + 1)}{(\ln 3)(x^2 + 2x)}$$

For Exercises 35–39, see the answer graphs in the textbook.

35. $y = x \ln x, \ x > 0$

To find any maxima or minima, find the first derivative.

$$y = x \cdot \ln x$$

$$y' = x\left(\frac{1}{x}\right) + \ln x$$

$$= 1 + \ln x$$

Set the derivative equal to 0.

$$1 + \ln x = 0$$

$$\ln x = -1$$

$$e^{-1} = x$$

$$x = \frac{1}{e} \approx .3679$$

$$y = \frac{1}{e} \ln \left(\frac{1}{e}\right)$$

$$= \frac{1}{e} \ln e^{-1}$$

$$= -\frac{1}{e} \approx -.3679$$

Find the second derivative to determine whether y is a maximum or minimum at $x = 1/e$.

$$y'' = 0 + \frac{1}{x} = \frac{1}{x} > 0 \text{ since } x > 0.$$

There is a minimum value of

$$y = -\frac{1}{e} \approx -.3679 \text{ at}$$

$$x = \frac{1}{e} \approx .3679.$$

37. $y = x \ln |x|$

Find the first derivative.

$$y' = x\left(\frac{1}{x}\right) + \ln |x| = 1 + \ln |x|$$

Set $y' = 0$.

$$\ln |x| = -1$$
$$e^{-1} = |x|$$

If $x > 0$, $x = 1/e \approx .3679$.
If $x < 0$, $x = -1/e \approx -.3679$.

$$y'' = \frac{1}{x} > 0 \text{ if } x > 0,$$

which indicates a minimum.
At $x = 1/e \approx .3679$, there is a minimum:

$$y = \frac{1}{e} \ln \frac{1}{e} = \frac{1}{e} \ln e^{-1}$$

$$= -\frac{1}{e} \approx -.3679.$$

$$y'' = \frac{1}{x} < 0 \text{ if } x < 0,$$

which indicates a maximum.
At $x = -1/e \approx -.3679$, there is a maximum.

$$y = -\frac{1}{e} \ln \left| -\frac{1}{e} \right| = -\frac{1}{e} \ln e^{-1}$$

$$= \frac{1}{e} \approx .3679.$$

39. $y = \dfrac{\ln x}{x}$, $x > 0$

$x \le 0$ is not in the domain of the function.

$$y' = \frac{x\left(\frac{1}{x}\right) - \ln x}{x^2}$$

$$= \frac{1 - \ln x}{x^2}$$

Set $y' = 0$.

$$\ln x = 1$$
$$x = e$$

$$y'' = \frac{x^2(0) - (\ln x)2x}{x^4} < 0$$

$$= -\frac{2 \ln x}{x^3} < 0$$

At $x = e$,

$$y'' = -2e^{-3} < 0.$$

Therefore, at $x = e \approx 2.718$, there is a maximum of

$$y = \frac{\ln e}{e}$$

$$= \frac{1}{e} \approx .3679.$$

41. $f(x) = \ln x$

The slope of the tangent line is

$$f'(x) = \frac{1}{x}.$$

$$\lim_{x \to \infty} \left(\frac{1}{x}\right) = 0$$

$$\lim_{x \to \infty} \left(\frac{1}{x}\right) = \infty$$

As $x \to \infty$ the slope approaches 0; as $x \to \infty$, the slope becomes infinitely large.

43. $\dfrac{d}{dx} \ln |ax|$ Note: a is a constant.

$$= \frac{d}{dx}(\ln |a| + \ln |x|)$$

$$= \frac{d}{dx} \ln |a| + \frac{d}{dx} \ln |x|$$

$$= 0 + \frac{d}{dx} \ln |x|$$

$$= \frac{d}{dx} \ln |x|$$

Therefore,

$$\frac{d}{dx} \ln |ax| = \frac{d}{dx} \ln |x|.$$

45. $p = 100 + \dfrac{50}{\ln x}$

(a) $R = px$

$$R = 100x + \frac{50x}{\ln x}$$

The marginal revenue is

$$\frac{dR}{dx} = 100 + \frac{(\ln x)(50) - 50x\left(\frac{1}{x}\right)}{(\ln x)^2}$$

$$= 100 + \frac{50(\ln x - 1)}{(\ln x)^2}.$$

(b) The revenue from one more unit is dR/dx for x = 8.

$$100 + \frac{50(\ln 8 - 1)}{(\ln 8)^2} = \$112.48$$

(c) The manager can use the information from (b) to decide if it is reasonable to sell additional items. If the revenue does not exceed the cost, there will be no profit.

47. p = 100 - 10 ln x, 1 < x ≤ 20,000
x = 6n, n = number of employees

(a) The revenue function is

R(x) = px
$$= (100 - 10 \ln x)x$$
$$= 100x - 10x \ln x,$$
$$1 < x < 20,000.$$

(b) The marginal revenue product function is

$$\frac{dR}{dn} = \frac{dR}{dx} \cdot \frac{dx}{dn}$$

$$= \left[100 - 10\left(x \cdot \frac{1}{x} + \ln x\right)\right]6$$

$$= (90 - 10 \ln x)6$$

$$= 60(9 - \ln x)$$

$$= 60(9 - \ln 6n).$$

(c) When x = 20, the marginal revenue product is

$$60(9 - \ln 20) \approx 360.$$

Hiring an additional worker will produce an increase of about $360.

49. P(t) = (t + 100) ln (t + 2)

$$P'(t) = (t + 100)\left(\frac{1}{t + 2}\right) + \ln (t + 2)$$

$$P'(2) = (102)\left(\frac{1}{4}\right) + \ln (4)$$

$$\approx 26.9$$

$$P'(8) = (108)\left(\frac{1}{10}\right) + \ln (10)$$

$$\approx 13.1$$

51. This exercise should be solved by a graphing calculator or by computer methods. The solution will vary according to the method that is used. See the answer graph at the back of the textbook. The minimum of about .85 should occur at about .7.

Section 13.5

1. $y = e^{4x}$
Let g(x) = 4x,
with g'(x) = 4.
       $y' = 4e^{4x}$

3. $y = -8e^{2x}$
$y' = -8(2e^{2x})$
$= -16e^{2x}$

5. $y = -16e^{x+1}$
g(x) = x + 1
g'(x) = 1
$y' = -16[(1)e^{x+1}]$
$= -16e^{x+1}$

7. $\quad y = e^{x^2}$

$\quad\quad g(x) = x^2$

$\quad\quad g'(x) = 2x$

$\quad\quad y' = 2xe^{x^2}$

9. $\quad y = 3e^{2x^2}$

$\quad\quad g(x) = 2x^2$

$\quad\quad g'(x) = 4x$

$\quad\quad y' = 3(4xe^{2x^2})$

$\quad\quad\quad = 12xe^{2x^2}$

11. $\quad y = 4e^{2x^2-4}$

$\quad\quad g(x) = 2x^2 - 4$

$\quad\quad g'(x) = 4x$

$\quad\quad y' = 4[(4x)e^{2x^2-4}]$

$\quad\quad\quad = 16xe^{2x^2-4}$

13. $\quad y = xe^x$

Use the product rule.

$\quad y' = xe^x \quad e^x$

$\quad\quad = e^x(x + 1)$

15. $\quad y = (x - 3)^2 e^{2x}$

Use the product rule.

$\quad y' = (x - 3)^2(2)e^{2x} + e^{2x}(2)(x - 3)$

$\quad\quad = 2(x - 3)^2 e^{2x} + 2(x - 3)e^{2x}$

$\quad\quad = 2(x - 3)e^{2x}[(x - 3) + 1]$

$\quad\quad = 2(x - 3)(x - 2)e^{2x}$

17. $\quad y = e^{x^2} \ln x, \quad x > 0$

$\quad y' = e^{x^2}\left(\frac{1}{x}\right) + (\ln x)(2x)e^{x^2}$

$\quad\quad = \frac{e^{x^2}}{x} + 2xe^{x^2} \ln x$

19. $\quad y = \frac{e^x}{\ln x}, \quad x > 0$

Use the quotient rule.

$\quad y' = \frac{(\ln x)e^x - e^x\left(\frac{1}{x}\right)}{(\ln x)^2} \cdot \frac{x}{x}$

$\quad\quad = \frac{xe^x \ln x - e^x}{x(\ln x)^2}$

21. $\quad y = \frac{x^2}{e^x}$

Use the quotient rule.

$\quad y' = \frac{e^x(2x) - x^2 e^x}{(e^x)^2}$

$\quad\quad = \frac{xe^x(2 - x)}{e^{2x}}$

$\quad\quad = \frac{x(2 - x)}{e^x}$

23. $\quad y = \frac{e^x + e^{-x}}{x}$

$\quad y' = \frac{x(e^x - e^{-x}) - (e^x + e^{-x})}{x^2}$

25. $\quad y = \frac{5000}{1 + 10e^{.4x}}$

$\quad y' = \frac{(1 + 10e^{.4x}) \cdot 0 - 5000[0 + 10(.4)e^{.4x}]}{(1 + 10e^{.4x})^2}$

$\quad\quad = \frac{-20,000e^{.4x}}{(1 + 10e^{.4x})^2}$

27. $\quad y = \frac{10,000}{9 + 4e^{-.2x}}$

$\quad y' = \frac{(9 + 4e^{-.2x}) \cdot 0 - 10,000[0 + 4(.-2)e^{-.2x}]}{(9 + 4e^{-.2x})^2}$

$\quad\quad = \frac{8000e^{-.2x}}{(9 + 4e^{-.2x})^2}$

29. $y = (2x + e^{-x^2})^2$

Use the chain rule.

$y' = 2(2x + e^{-x^2})(2 - 2xe^{-x^2})$

31. $y = 8^{5x}$

$= e^{\ln 8^{5x}}$

$= e^{5x \ln 8}$

$y' = 5(\ln 8)e^{5x \ln 8}$

$= 5(\ln 8)8^{5x}$

33. $y = 3 \cdot 4^{x^2+2} = 3 \cdot e^{\ln 4^{x^2+2}} = 3 \cdot e^{(x^2+2)\ln 4}$

$y' = 3(\ln 4)(2x)e^{(x^2+2)\ln 4} = 6x(\ln 4)e^{(x^2+2)\ln 4} = 6x(\ln 4)4^{x^2+2}$

35. $y = 2 \cdot 3^{\sqrt{x}} = 2 \cdot e^{\ln 3^{\sqrt{x}}} = 2 \cdot e^{\sqrt{x} \ln 3}$

$y' = 2 \cdot \ln 3 \, \frac{1}{2}x^{-1/2} \, e^{\sqrt{x} \ln 3} = \frac{\ln 3e^{\sqrt{x} \ln 3}}{\sqrt{x}} = \frac{(\ln 3)(3^{\sqrt{x}})}{\sqrt{x}}$

37. $S'(x) = \dfrac{3,000,000e^{-.3x}}{(1 + 100e^{-.3x})^2} = (3,000,000e^{-.3x})(1 + 100e^{-.3x})^{-2}$

$S''(x) = (3,000,000e^{-.3x})[-2(1 + 100e^{-.3x})^{-3}(-30e^{-.3x})]$

$\qquad + (1 + 100e^{-.3x})^{-2}(-900,000e^{-.3x})$

$\qquad = (3,000,000e^{-.3x})(60e^{-.3x})(1 + 100e^{-.3x})^{-3} + (1 + 100e^{-.3x})^{-2}(-900,000e^{-.3x})$

$\qquad = (1 + 100e^{-.3x})^{-3}[180,000,000e^{-.6x} + (1 + 100e^{-.3x})(-900,000e^{-.3x})]$

$\qquad = (1 + 100e^{-.3x})^{-3}(90,000,000e^{-.6x} - 900,000e^{-.3x})$

$\qquad = (1 + 100e^{-.3x})(900,000e^{-.3x})(100e^{-.3x} - 1)$

$S''(x) = 0$ when $100e^{-.3x} - 1 = 0$.

$$100e^{-.3x} = 1$$

$$e^{-.3x} = .01$$

$$-.3x = \ln .01$$

$$x = \frac{\ln .01}{-.3} \approx 15.4$$

$$S(15.4) = \frac{100,000}{1 + 100e^{-.3(15.4)}}$$

$$\approx 50,370.7$$

The inflection point is approximately (15.4, 50,000).

For Exercises 39–43, see the answer graphs in your textbook.

39. $y = -xe^x$

$y' = -xe^x + e^x(-1)$

$\quad = -xe^x - e^x$

Set $y' = 0$ for critical points.

$\quad 0 = -xe^x - e^x$

$\quad 0 = e^x(-x - 1)$

$e^x = 0$ has no solution.

$-x - 1 = 0$

$\quad -x = 1$

$\quad\quad x = -1$

Check second derivative for concavity.

$\quad y'' = -xe^x + e^x(-1) - e^x$

$\quad\quad = -xe^x - e^x - e^x$

$\quad\quad = -xe^x - 2e^x$

At $x = -1$,

$\quad y'' = e^{-1} - 2e^{-1}$

$\quad\quad = -e^{-1} < 0$

So y is concave down at $x = -1$ where y is a maximum.

$\quad y = -(-1)e^{-1}$

$\quad\quad = \dfrac{1}{e} \approx .3679$

Maximum at $1/e \approx .3679$ at $x = -1$
Set $y'' = 0$ to find inflection points.

$\quad -xe^x - 2e^x = 0$

$\quad -e^x(x + 2) = 0$

$\quad\quad\quad x = -2$

$y = -(-2)e^{-2} = 2e^{-2}$

Inflection point at $(-2, 2e^{-2})$

41. $y = x^2 e^{-x}$

$y' = x^2 e^{-x}(-1) + e^{-x}(2x)$

$\quad = -x^2 e^{-x} + 2xe^{-x}$

$\quad = xe^{-x}(-x + 2)$

Critical points:

$\quad 0 = xe^{-x}(-x + 2)$

$\quad x = 0 \quad$ or $\quad -x + 2 = 0$

$\quad\quad\quad\quad\quad\quad\quad x = 2$

Check second derivative for concavity.

$y'' = -x^2(-e^{-x}) + (-2x)(e^{-x})$

$\quad\quad + 2x(-e^{-x}) + 2e^{-x}$

$\quad = x^2 e^{-x} - 2xe^{-x} - 2xe^{-x} + 2e^{-x}$

$\quad = x^2 e^{-x} - 4xe^{-x} + 2e^{-x}$

If $x = 0$, $y'' = 0 - 0 + 2 = 2 > 0$,
y is concave up, minimum, and $y = 0$.

If $x = 2$, $y'' = 2^2 e^{-2} - 4(2)e^{-2} + 2e^{-2}$

$\quad\quad = 4e^{-2} - 8e^{-2} + 2e^{-2}$

$\quad\quad = -2e^{-2} < 0$,

y is concave down, maximum, and

$\quad y = 2^2 e^{-2} = 4e^{-2} \approx .54.$

Minimum of 0 at $x = 0$, maximum of $4/e^2 \approx .54$ at $x = 2$
Set $y'' = 0$ to find inflection points.

$x^2 e^{-x} - 4xe^{-x} + 2e^{-x} = 0$

$\quad e^{-x}(x^2 - 4x + 2) = 0$

$\quad\quad x = \dfrac{4 \pm \sqrt{16 - 8}}{2}$

$\quad\quad\quad \approx 3.4 \quad$ or $\quad .6$

$\quad\quad y = (3.4)^2 e^{-3.4} \approx .38$

$\quad\quad y = (.6)^2 e^{-.6} \approx .19$

Inflection points at $(3.4, .38)$ and $(.6, .19)$

43. $y = e^x + e^{-x}$

$y' = e^x + (-e^{-x})$

Critical points:

$0 = e^x - e^{-x}$

$\dfrac{1}{e^x} = e^x$

$1 = e^{2x}$

$x = 0$

Check second derivative for concavity.

$y'' = e^x + e^{-x}$

At $x = 0$, $y'' = e^0 + e^{-0}$

$\qquad\qquad = 1 + 1 = 2 > 0,$

y is concave up, minimum, and

$y = e^0 + e^{-0} = 1 + 1 = 2.$

Minimum of 2 at $x = 0$

Set $y'' = 0$ to find inflection points.

$e^x + e^{-x} = 0$ has no solution.

No inflection point

45. $f(x) = e^x$

$\qquad y = e^x$

$f'(x) = e^x$

Similarly,

$\quad f''(x) = e^x,$

$\quad f'''(x) = e^x$ and

$f^{(n)}(x) = e^x.$

47. $y = e^{ax}$ where a is any constant.

$\ln y = \ln (e^{ax})$

$\ln y = ax$

$\dfrac{1}{y}\dfrac{dy}{dx} = a$

$\dfrac{dy}{dx} = ay$

$\dfrac{dy}{dx} = a\, e^{ax}$

That is, $\dfrac{d}{dx}\, e^{ax} = ae^{ax}$ for any constant a.

49. $P(x) = e^{-.02x}$

(a) $P(1) = e^{-.02(1)} = e^{-.02} \approx .98$

(b) $P(10) = e^{-.02(10)} = e^{-.2} \approx .82$

(c) $P(100) = e^{-.02(100)} = e^{-2} \approx .14$

(d) $P'(t) = -.02e^{-.02x}$

$P'(100) = -.02e^{-.02(100)}$

$\qquad\qquad = -.02e^{-2}$

$\qquad\qquad \approx -.0027$

This is the rate of change in the proportion that are wearable when $x = 100$, or after 100 days.

(e) As the number of days increases, the proportion of wearable shoes decreases. This is reasonable because shoes wear out overtime.

51.   (a) $p = 400e^{-.2x}$

$E = -\dfrac{p}{x} \cdot \dfrac{dx}{dp}$

$\quad = -\dfrac{400e^{-.2x}}{x} \cdot \dfrac{1}{400(-.2e^{-.2x})}$

$\quad = \dfrac{-400e^{-.2x}}{400x(-.2)e^{-.2x}}$

$\quad = \dfrac{1}{.2x} = \dfrac{5}{x}$

(b) $R = xp$

$R = x(400e^{-.2x})$

$\quad = 400xe^{-.2x}$

$R'(x) = 400e^{-.2x} + (400x)(-.2)e^{-.2x}$

$\qquad = 400e^{-.2x} - 80xe^{-.2x}$

$\qquad = 80e^{-.2x}(5 - x)$

$\qquad 80e^{-.2x}(5 - x) = 0$ when $x = 5$

R′(4) > 0 and R′(6) < 0, so there is a relative maximum when x = 5.

53. $A(t) = 500e^{-.25t}$

  $A'(t) = 500(-.25)e^{-.25t}$

  $= -125e^{-.25t}$

  (a) $A'(4) = -125e^{-.25(4)}$

  $= -125e^{-1}$

  ≈ -46.0

  (b) $A'(6) = -125e^{-.25(6)}$

  $= -125e^{-1.5}$

  ≈ -27.9

  (c) $A'(10) = -125e^{-.25(10)}$

  $= -125e^{-2.5}$

  ≈ -10.3

  (d) $\lim_{t \to \infty} A'(t) = \lim_{t \to \infty} -125e^{-.25t} = 0$

  As the number of years increase, the rate of change approaches zero.

  (e) A′(t) will never equal zero, although as t increases, it approaches zero. Powers of e are always positive.

55. This exercise should be solved by a graphing calculator or computer methods. The solution will vary according to the method that is used. See the answer graph at the back of the textbook. There are no relative extrema. An inflection point occurs at (1, 0).

Chapter 13   Review Exercises

3. $2^{3x} = \frac{1}{8}$

  $2^{3x} = 2^{-3}$

  $3x = -3$

  $x = -1$

5. $9^{2y-1} = 27^y$

  $3^{2(2y-1)} = 3^{3y}$

  $2(2y - 1) = 3y$

  $4y - 2 = 3y$

  $y = 2$

For Exercises 7–13, see the answer graphs in your textbook.

7. $y = 5^x$

x	-2	-1	0	1	2
y	1/25	1/5	1	5	25

9. $y = \left(\frac{1}{5}\right)^{2x-3}$

x	0	1	2
y	125	5	1/5

11. $y = \log_2 (x - 1)$

  $2^y = x - 1$

  $x = 1 + 2^y$

x	2	3	5	9
y	0	1	2	3

13. $y = -\log_3 x$

$-y = \log_3 x$

$3^{-y} = x$

x	1/3	1	3	9
y	1	0	-1	-2

15. $2^6 = 64$

$\log_2 64 = 6$

17. $e^{.09} = 1.09417$

$\ln 1.09417 = .09$

19. $\log_2 32 = 5$

$2^5 = 32$

21. $\ln 82.9 = 4.41763$

$e^{4.41763} = 82.9$

23. $\log_3 81 = x$

$3^x = 81$

$3^x = 3^4$

$x = 4$

25. $\log_{32} 16 = x$

$32^x = 16$

$2^{5x} = 2^4$

$5x = 4$

$x = \dfrac{4}{5}$

27. $\log_{100} 1000 = x$

$100^x = 1000$

$(10^2)^x = 10^3$

$2x = 3$

$x = \dfrac{3}{2}$

29. $\log_5 3k + \log_5 7k^3$

$= \log_5 3k(7k^3)$

$= \log_5 (21k^4)$

31. $2 \log_2 x - 3 \log_2 m$

$= \log_2 x^2 - \log_2 m^3$

$= \log_2 \left(\dfrac{x^2}{m^3}\right)$

33. $8^P = 19$

$\ln 8^P = \ln 19$

$p \ln 8 = \ln 19$

$p = \dfrac{\ln 19}{\ln 8}$

≈ 1.416

35. $2^{1-m} = 7$

$\ln 2^{1-m} = \ln 7$

$(1 - m) \ln 2 = \ln 7$

$1 - m = \dfrac{\ln 7}{\ln 2}$

$-m = \dfrac{\ln 7}{\ln 2} - 1$

$m = 1 - \dfrac{\ln 7}{\ln 2}$

≈ -1.807

37. $e^{-5-2x} = 5$

$\ln e^{-5-2x} = \ln 5$

$(-5 - 2x) \ln e = \ln 5$

$(-5 - 2x) \cdot 1 = \ln 5$

$-2x = \ln 5 + 5$

$x = \dfrac{\ln 5 + 5}{-2}$

≈ -3.305

39.
$$\left(1 + \frac{m}{3}\right)^5 = 10$$

$$\left[\left(1 + \frac{m}{3}\right)^5\right]^{1/5} = 10^{1/5}$$

$$1 + \frac{m}{3} = 10^{1/5}$$

$$\frac{m}{3} = 10^{1/5} - 1$$

$$m = 3(10^{1/5} - 1)$$

$$\approx 1.7547$$

41. $y = -6e^{2x}$
$$y' = -6(2e^{2x}) = -12\ e^{2x}$$

43. $y = e^{-2x^3}$
$$g(x) = -2x^3$$
$$g'(x) = -6x^2$$
$$y' = -6x^2\ e^{-2x^3}$$

45. $y = 5x \cdot e^{2x}$

Use the product rule.

$$y' = 5x(2e^{2x}) + e^{2x}(5)$$
$$= 10xe^{2x} + 5e^{2x}$$
$$= 5e^{2x}(2x + 1)$$

47. $y = \ln(2 + x^2)$
$$g(x) = 2 + x^2$$
$$g'(x) = 2x$$
$$y' = \frac{2x}{2 + x^2}$$

49. $y = \frac{\ln|3x|}{x - 3}$

$$y' = \frac{(x - 3)\left(\frac{1}{3x}\right)(3) - (\ln|3x|)1}{(x - 3)^2}$$

$$= \frac{\frac{x - 3}{x} - \ln|3x|}{(x - 3)^2} \cdot \frac{x}{x}$$

$$= \frac{x - 3 - x\ln|3x|}{x(x - 3)^2}$$

51. $y = \frac{xe^x}{\ln(x^2 - 1)}$

$$y' = \frac{\ln(x^2 - 1)[xe^x + e^x] - xe^x\left(\frac{1}{x^2 - 1}\right)(2x)}{[\ln(x^2 - 1)]^2}$$

$$= \frac{e^x(x + 1)\ln(x^2 - 1) - \frac{2x^2 e^x}{x^2 - 1}}{[\ln(x^2 - 1)]^2} \cdot \frac{x^2 - 1}{x^2 - 1}$$

$$= \frac{e^x(x + 1)(x^2 - 1)\ln(x^2 - 1) - 2x^2 e^x}{(x^2 - 1)[\ln(x^2 - 1)]^2}$$

53. $y = (x^2 + e^x)^2$

Use the chain rule.

$$y' = 2(x^2 + e^x)(2x + e^x)$$

55. $y = x \cdot e^x$

Find the first derivative and set it equal to 0.

$$y' = xe^x + e^x = 0$$
$$e^x(x + 1) = 0$$
$$x = -1$$

To determine minimum or maximum, check the second derivative.

$$y'' = xe^x + e^x + e^x = xe^x + 2e^x$$

At $x = -1$, $y'' = -1e^{-1} + 2e^{-1}$

$$= -\frac{1}{e} + \frac{2}{e} = \frac{1}{e} > 0$$

y is concave up, minimum.

$$y = -1e^{-1}$$
$$= -\frac{1}{e} \approx -.368$$

Relative minimum of $-e^{-1} \approx -.368$ at $x = -1$

Set $y'' = 0$ to find inflection points.

$$xe^x + 2e^x = 0$$

$$e^x(x + 2) = 0$$

$$x = -2$$

$$y = -2e^{-2} \approx .27$$

Inflection point at $(-2, -.27)$

See the answer graph in your textbook.

57. $y = \dfrac{e^x}{x - 1}$

Find the first derivative and set it equal to zero.

$$y' = \frac{(x - 1)e^x - e^x}{(x - 1)^2} = 0$$

$$\frac{e^x(x - 1 - 1)}{(x - 1)^2} = 0$$

$$\frac{e^x(x - 2)}{(x - 1)^2} = 0$$

$$x - 2 = 0$$

$$x = 2$$

To determine minimum or maximum, check the second derivative.

$$y'' = \frac{(x-1)^2[e^x+(x-2)e^x]-e^x(x-2)(2)(x-1)}{(x - 1)^4}$$

$$= \frac{e^x(x^2 - 4x + 5)}{(x - 1)^3}$$

At $x = 2$,

$$y'' = \frac{1^2(e^2) - 0}{1^4} = e^2 > 0$$

y is concave up, minimum at $x = 2$.

$$y = \frac{e^2}{1} = e^2 \approx 7.39$$

Relative minimum at $e^2 \approx 7.39$ at $x = 2$

Set $y'' = 0$ to find inflection points.

$$\frac{e^x(x^2 - 4x + 5)}{(x - 1)^3} = 0$$

$$x^2 - 4x + 5 = 0$$

$$x = \frac{4 \pm \sqrt{16 - 20}}{2}$$

This quadratic equation has no real roots. There are no inflection points. See the answer graph in your textbook.

59. $P = \$6902$, $r = 12\%$, $t = 8$, $m = 2$

$$A = P\left(1 + \frac{r}{m}\right)^{tm}$$

$$A = 6902\left(1 + \frac{.12}{2}\right)^{8(2)}$$

$$= 6902(1.06)^{16}$$

$$= \$17,533.51$$

Interest $= A - P$

$$= \$17,533.51 - \$6902$$

$$= \$10,631.51$$

For Exercises 61–65, use $A = Pe^{rt}$.

61. $P = \$12,104$, $r = 8\%$, $t = 2$

$$A = 12,104e^{.08(2)}$$

$$= \$14,204.18$$

63. $P = \$12,104$, $r = 8\%$, $t = 7$

$$A = 12,104e^{.08(7)}$$

$$= \$21,190.14$$

65. $P = \$12,000$, $r = .05$, $t = 8$

$$A = 12,000e^{.05(8)}$$

$$= 12,000e^{.40}$$

$$= \$17,901.90$$

67. $r = 9\%$, $m = 12$

$$r_E = \left(1 + \frac{r}{m}\right)^m - 1$$

$$= \left(1 + \frac{.09}{12}\right)^{12} - 1$$

$$= .0938 = 9.38\%$$

69. $r = 9\%$ compounded continuously

$$r_E = e^r - 1$$

$$= e^{.09} - 1$$

$$= .0942 = 9.42\%$$

71. $A = \$2000$, $r = 6\%$, $t = 5$, $m = 1$

$$P = A\left(1 + \frac{r}{m}\right)^{-tm}$$

$$= 2000\left(1 + \frac{.06}{1}\right)^{-5(1)}$$

$$= 2000(1.06)^{-5}$$

$$= \$1494.52$$

73. $A = \$43,200$, $r = 8\%$, $t = 4$, $m = 4$

$$P = 43,200\left(1 + \frac{.08}{4}\right)^{-4(4)}$$

$$= 43,200(1.02)^{-16}$$

$$= \$31,468.86$$

75. (a) $P(x) = 100 - 100e^{-.8x}$

$$\lim_{x \to \infty} P(x) = \lim_{x \to \infty} \left(100 - 100e^{-.8x}\right)$$

$$= \lim_{x \to \infty} 100 - \lim_{x \to \infty} 100e^{-.8x}$$

$$= 100 - 0$$

$$= 100$$

(b) An experienced worker, who has had many days on the job should produce 100 items per day.

(c)
$$50 = 100 - 100e^{-.8x}$$

$$-50 = -100e^{-.8x}$$

$$\frac{1}{2} = e^{-.8x}$$

$$\ln\left(\frac{1}{2}\right) = -.8x$$

$$\frac{-\ln 2}{-.8} = x$$

$$1 \approx x$$

A new employee will produce 50 items after 1 day on the job.

77. $P = \$1$, $r = .08$

$$A = Pe^{rt}, \quad A = 3(1)$$

$$3 = 1e^{.08t}$$

$$\ln 3 = .08t$$

$$\frac{\ln 3}{.08} = t$$

$$13.7 = t$$

It would take about 13.7 yr.

79. $P = A\left(1 + \frac{r}{m}\right)^{-tm}$

$$P = 25,000\left(1 + \frac{.06}{12}\right)^{-3(12)}$$

$$= 25,000(1.005)^{-36}$$

$$= \$20,891.12$$

81. $I(x) \geq 1$

$$I(x) = 10e^{-.3x}$$

$$10e^{-.3x} \geq 1$$

$$e^{-.3x} \geq .1$$

$$-.3x \geq \ln .1$$

$$x \leq \frac{\ln .1}{-.3} \approx 7.7$$

The greatest depth is about 7.7 m.

83. $g(t) = \frac{c}{a} + \left(g_0 - \frac{c}{a}\right)e^{-at}$

  (a) $g'(t) = -a\left(g_0 - \frac{c}{a}\right)e^{-at}$

  $= (-ag_0 + c)e^{-at}$

  $g_0 = .08$, $c = .1$ and $a = 1.3$

  $g'(t) = [(-1.3)(.08) + .1]e^{-at}$

  $= -.004e^{-1.3t}$

  $g'(t)$ cannot equal zero. The maximum amount occurs when the glucose is first infused. The maximum amount is $g_0 = .08$ g.

  (b) $g(t) = \frac{c}{a} + \left(g_0 - \frac{c}{a}\right)e^{-at}$

  $.1 = \frac{.1}{1.3} + \left(.08 - \frac{.1}{1.3}\right)e^{-1.3t}$

  $.1 - \frac{.1}{1.3} = \left(.08 - \frac{.1}{1.3}\right)e^{-1.3t}$

  $7.5 = e^{-1.3t}$

  $\ln 7.5 = -1.3t$

  $-\frac{1}{1.3} \ln 7.5 = t$

  $t = -1.55$

  Since time is never negative, the amount of glucose is never .1 g. (As shown in part (a), the maximum amount is .08 g.)

  (c) If the glucose is continuously infused into the bloodstream, it continuously decreases at a slower and slower rate over time.

85. $t = (1.26 \times 10^9)\dfrac{\ln\left[1 + 8.33\left(\frac{A}{K}\right)\right]}{\ln 2}$

  (a) $A = 0$, $K > 0$

  $t = (1.26 \times 10^9)\dfrac{\ln[1 + 8.33(0)]}{\ln 2}$

  $= (1.26 \times 10^9)(0) = 0$ years

  (b) $t = (1.26 \times 10^9)\dfrac{\ln[1 + 8.33(.212)]}{\ln 2}$

  $= (1.26 \times 10^9)\dfrac{\ln 2.76596}{\ln 2}$

  $= 1,849,403,169$

  or about 1.85×10^9 years.

  (c) $t = (1.26 \times 10^9)\dfrac{\ln(1 + 8.33r)}{\ln 2}$

  $\dfrac{dt}{dr} = \dfrac{(1.26 \times 10^9)}{\ln 2} \cdot \dfrac{1}{1 + 8.33r}(8.33)$

  $= \dfrac{10.4958 \times 10^9}{\ln 2(1 + 8.33r)}$

  (d) As r increases, t increases, but at a slower and slower rate. As r decreases, t decreases at a faster and faster rate.

87. $L(t) = 9 + 2e^{.15t}$

  where $t = 0$ corresponds to 1982.

  (a) $L(0) = 9 + 2e^{.15(0)}$

  $= 9 + 2$

  $= 11$

  (b) $L(4) = 9 + 2e^{.15(4)}$

  ≈ 12.6

  (c) $L(10) = 9 + 2e^{.15(10)}$

  ≈ 18.0

  (d) See the answer graph in the textbook.

89. $f(x) = a^x$; $a > 0$, $a \neq 1$.

  (a) The domain is $(-\infty, \infty)$.

  (b) The range is $(0, \infty)$.

  (c) The y-intercept is 1.

  (d) There are no discontinuities.

  (e) $y = 0$ is an asymptote.

(f) f(x) is increasing when a is greater than 1.

(g) f(x) is decreasing when a is between 0 and 1.

Extended Application

1. (a) P = 2000, i = .1, n = 20, t = .37

$$A = P[(1 + i)^n(1 - t) + t]$$
$$= 2000[(1 + .1)^{20}(1 - .37) + .37]$$
$$= \$9216.65$$
$$M = \frac{9216.65}{2000}$$
$$= 4.6$$

(b) $A = 2000[1 + (1 - t)i]^n$
$$= 2000[1 + .63(.1)]^{20}$$
$$= 2000(1.063)^{20}$$
$$= \$6787.27$$
$$m = \frac{6787.27}{2000}$$
$$= 3.4$$

(c) $\frac{M}{m} = \frac{4.6}{3.4}$
$$= 1.35$$

Investment (a) will yield approximately 35% more after-tax dollars than investment (b).

2. (a) P = 2000, i = .1, n = 10, t = .37

$$A = P[(1 + i)^n(1 - t) + t]$$
$$= 2000[(1 + .1)^{10}(1 - .37) + .37]$$
$$= \$4008.12$$
$$M = \frac{4008.12}{2000}$$
$$= 2.0$$

(b) $A = 2000[1 + (1 - t)i]^n$
$$= 2000[1 + (1 - .37).1]^{10}$$
$$= \$3684.37$$
$$M = \frac{3684.37}{2.0}$$
$$= 1.8$$

(c) $\frac{M}{m} = \frac{2.0}{1.8}$
$$= 1.11$$

Investment (a) will yield approximately 11% more after-tax dollars than investment (b).

3. M is an increasing function of t. The longer you leave your money in the account, the greater yield in after-tax dollars.

4. As n increases, the ratio $\frac{M}{m}$ increases for given values of i and t; that is, m is an increasing function of n.

5. The multiplier function is an increasing function of i. The advantage of the IRA over a regular account widens as the interest rate i increases and is particularly dramatic for high income tax rates.

CHAPTER 13   TEST

1. Solve $8^{2y-1} = 4^{y+1}$

2. Graph the function $y = 4^{x-1}$.

3. $315 is deposited in an account paying 6% compounded quarterly for 3 yr. Find the following.

   (a) The amount in the account after 3 yr.

   (b) The amount of interest earned by this deposit

4. (a) Explain the concept of effective rate.

   (b) Find the effective rate for the situation in Exercise 3.

5. Use exponents to write the equation $\ln 42.8 = 3.75654$.

6. Evaluate $\log_8 16$.

7. Evaluate $\ln 543$.

Use properties of logarithms to simplify the following.

8. $\log_2 3k + \log_2 4k^2$

9. $4 \log_3 r - 3 \log_3 m$

Solve each equation. Round to the nearest thousandth.

10. $2^{x+1} = 10$

11. $\left(1 + \dfrac{2m}{3}\right)^5 = 8$

12. Find the interest rate needed for $5000 to grow to $10,000 in 8 yr with continuous compounding.

13. How long will it take for $1 to triple at an average rate of 7% compounded continuously?

14. Suppose sales of a certain item are given by $S(x) = 2500 - 1500e^{-2x}$ where x represents the number of years that the item has been on the market and $S(x)$ represents sales in thousands. Find the limit on sales.

15. Find the derivative of $y = \dfrac{\ln |3x - 1|}{x - 2}$.

16. Find all relative maxima or minima of $y = 2x \ln x$, $x > 0$. Sketch the graph of the function.

Find the derivative of each function.

17. $y = 3x^2 e^{2x}$

18. $y = (e^{2x} - \ln |x|)^3$

19. Find all relative maxima or minima of $y = 4xe^{-x}$ and sketch the graph of the function.

20. The concentration of a certain drug in the bloodstream at time t in minutes is given by

$$c(t) = e^{-t} - e^{-3t}.$$

Find the concentration at each of the following times.

(a) $t = 0$ (b) $t = 1$ (c) $t = 2$

(d) Find the maximum concentration and when it occurs.

21. The population of Smalltown has grown exponentially from 14,000 in 1990 to 16,500 in 1993. At this rate, in what year will the population reach 17,400?

22. Potassium 42 decays exponentially. A sample which contained 1000 g 5 hr ago has decreased to 758 g at present.

(a) Write an exponential equation to express the amount, y, present after t hr.

(b) What is the half-life of potassium 42?

23. What is meant by the half-life of a substance?

24. Find the present value of $15,000 at 6% compounded quarterly for 4 yr.

25. Mr. Jones needs $20,000 for a down payment on a house in 5 yr. How much must he deposit now at 5.8% compounded quarterly in order to have $20,000 in 5 yr?

CHAPTER 13   TEST ANSWERS

1. 5/4

2.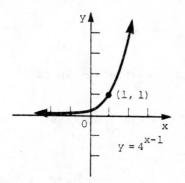
$y = 4^{x-1}$

3. **(a)** $376.62   **(b)** $61.62

4. **(a)** The effective interest rate is the actual annual rate yielded when compounding occurs.

    (b) 6.14%

5. $e^{3.75654} = 42.8$

6. 4/3

7. 6.29711

8. $\log_2 12k^3$

9. $\log_3 \dfrac{r^4}{m^3}$

10. 2.322

11. .774

12. 8.66%

13. About 15.7 yr

14. 2,500,000

15. $\dfrac{3x - 6 - (3x - 1)\ln|3x - 1|}{(3x - 1)(x - 2)^2}$

16. Relative minimum of $-2/e$ at $x = 1/e$

17. $6x^2e^{2x} + 6xe^{2x}$

18. $\dfrac{3(e^{2x} - \ln|x|)^2(2xe^{2x} - 1)}{x}$

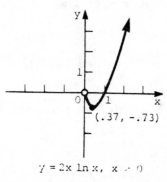
$y = 2x \ln x, \; x > 0$
$(.37, -.73)$

19. Relative maximum of $4/e$ at $x = 1$

20. **(a)** 0   **(b)** .318   **(c)** .133

    (d) Maximum of .38 at $t = .55$

21. 1994

22. **(a)** $y = 1000e^{-.0554t}$

    (b) About 12.5 hr

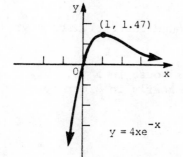
$(1, 1.47)$
$y = 4xe^{-x}$

23. The half-life is the amount of time needed for half the quantity to decay.

24. $11,820.47

25. $14,996.47

CHAPTER 14   INTEGRATION

Section 14.1

1. $\int 4x \ dx = 4 \int x \ dx$

$= 4 \cdot \dfrac{1}{1+1} x^{1+1} + C$

$= \dfrac{4x^2}{2} + C$

$= 2x^2 + C$

3. $\int 5t^2 \ dt = 5 \int t^2 \ dt$

$= 5 \cdot \dfrac{1}{2+1} t^{2+1} + C$

$= \dfrac{5t^3}{3} + C$

5. $\int 6 \ dk = 6 \int 1 \ dk = 6 \int k^0 \ dy$

$= 6 \cdot \dfrac{1}{1} k^{0+1} + C$

$= 6k + C$

7. $\int (2z + 3) \ dz$

$= 2 \int z \ dz + 3 \int z^0 \ dz$

$= 2 \cdot \dfrac{1}{1+1} z^{1+1} + 3 \cdot \dfrac{1}{0+1} z^{0+1} + C$

$= z^2 + 3z + C$

9. $\int (x^2 + 6x) \, dx = \int x^2 \ dx + 6 \int x \ dx$

$= \dfrac{x^3}{3} + \dfrac{6x^2}{2} + C$

$= \dfrac{x^3}{3} + 3x^2 + C$

11. $\int (t^2 - 4t + 5) \, dt$

$= \int t^2 \ dt - 4 \int t \ dt + 5 \int t^0 \ dt$

$= \dfrac{t^3}{3} - \dfrac{4t^2}{2} + 5t + C$

$= \dfrac{t^3}{3} - 2t^2 + 5t + C$

13. $\int (4z^3 + 3z^2 + 2z - 6) \, dz$

$= 4 \int z^3 \ dz + 3 \int z^2 \ dz + 2 \int z \ dz$

$- 6 \int z^0 \ dz$

$= \dfrac{4z^4}{4} + \dfrac{3z^3}{3} + \dfrac{2z^2}{2} - 6z + C$

$= z^4 + z^3 + z^2 - 6z + C$

15. $\int 5\sqrt{z} \ dz = 5 \int z^{1/2} \ dz$

$= \dfrac{5z^{3/2}}{3/2} + C$

$= 5 \left(\dfrac{2}{3}\right) z^{3/2} + C$

$= \dfrac{10z^{3/2}}{3} + C$

17. $\int (u^{1/2} + u^{3/2}) \, du$

$= \int u^{1/2} \ du + \int u^{3/2} \ du$

$= \dfrac{u^{3/2}}{3/2} + \dfrac{u^{5/2}}{5/2} + C$

$= \dfrac{2u^{3/2}}{3} + \dfrac{2u^{5/2}}{5} + C$

19. $\int (15x\sqrt{x} + 2\sqrt{x}) \, dx$

$= 15 \int x(x^{1/2}) \ dx + 2 \int x^{1/2} \ dx$

$= 15 \int x^{3/2} \ dx + 2 \int x^{1/2} \ dx$

$$= \frac{15x^{5/2}}{5/2} + \frac{2x^{3/2}}{3/2} + C$$

$$= 15\left(\frac{2}{5}\right)x^{5/2} + 2\left(\frac{2}{3}\right)x^{3/2} + C$$

$$= 6x^{5/2} + \frac{4x^{3/2}}{3} + C$$

21. $\int (10u^{3/2} - 14u^{5/2})\,du$

$$= 10\int u^{3/2}\,du - 14\int u^{5/2}\,du$$

$$= \frac{10u^{5/2}}{5/2} - \frac{14u^{7/2}}{7/2} + C$$

$$= 10\left(\frac{2}{5}\right)u^{5/2} - 14\left(\frac{2}{7}\right)u^{7/2} + C$$

$$= 4u^{5/2} - 4u^{7/2} + C$$

23. $\int \left(\frac{1}{z^2}\right)dz = \int z^{-2}\,dz$

$$= \frac{z^{-2+1}}{-2+1} + C$$

$$= \frac{z^{-1}}{-1} + C$$

$$= -\frac{1}{z} + C$$

25. $\int \left(\frac{1}{y^3} - \frac{1}{\sqrt{y}}\right)dy$

$$= \int y^{-3}\,dy - \int y^{-1/2}\,dy$$

$$= \frac{y^{-2}}{-2} - \frac{y^{1/2}}{1/2} + C$$

$$= \frac{-1}{2y^2} - 2y^{1/2} + C$$

27. $\int (-9t^{-2} - 2t^{-1})\,dt$

$$= -9\int t^{-2}\,dt - 2\int t^{-1}\,dt$$

$$= \frac{-9t^{-1}}{-1} - 2\int \frac{dt}{t}$$

$$= \frac{9}{t} - 2\ln|t| + C$$

29. $\int e^{2t}\,dt = \frac{e^{2t}}{2} + C$

31. $\int 3e^{-.2x}\,dx = 3\int e^{-.2x}\,dx$

$$= 3\left(\frac{1}{-.2}\right)e^{-.2x} + C$$

$$= \frac{3(e^{-.2x})}{-.2} + C$$

$$= -15e^{-.2x} + C$$

33. $\int \left(\frac{3}{x} + 4e^{-.5x}\right)dx$

$$= 3\int \frac{dx}{x} + 4\int e^{-.5x}\,dx$$

$$= 3\ln|x| + \frac{4e^{-.5x}}{-.5} + C$$

$$= 3\ln|x| - 8e^{-.5x} + C$$

35. $\int \frac{1+2t^3}{t}\,dt = \int \left(\frac{1}{t} + 2t^2\right)dt$

$$= \int \frac{1}{t}\,dt + 2\int t^2\,dt$$

$$= \ln|t| + \frac{2t^3}{3} + C$$

37. $\int (e^{2u} + 4u)\,du = \frac{e^{2u}}{2} + \frac{4u^2}{2} + C$

$$= \frac{e^{2u}}{2} + 2u^2 + C$$

39. $\int (x+1)^2\,dx = \int (x^2 + 2x + 1)\,dx$

$$= \frac{x^3}{3} + \frac{2x^2}{2} + x + C$$

$$= \frac{x^3}{3} + x^2 + x + C$$

41. $\displaystyle\int \frac{\sqrt{x}+1}{\sqrt[3]{x}}\, dx = \int \left(\frac{\sqrt{x}}{\sqrt[3]{x}} + \frac{1}{\sqrt[3]{x}}\right) dx$

$\displaystyle = \int (x^{(1/2\,-\,1/3\,)} + x^{-1/3})\, dx$

$\displaystyle = \int x^{1/6}\, dx + \int x^{-1/3}\, dx$

$\displaystyle = \frac{x^{7/6}}{7/6} + \frac{x^{2/3}}{2/3} + C$

$\displaystyle = \frac{6x^{7/6}}{7} + \frac{3x^{2/3}}{2} + C$

43. Find $f(x)$ such that $f'(x) = x^{2/3}$, and $(1,\ 3/5)$ is on the curve.

$\displaystyle\int x^{2/3} = \frac{x^{5/3}}{5/3} + C$

$\displaystyle f(x) = \frac{3x^{5/3}}{5} + C$

Since $(1,\ 3/5)$ is on the curve,

$\displaystyle f(1) = \frac{3}{5}.$

$\displaystyle f(1) = \frac{3(1)^{5/3}}{5} + C = \frac{3}{5}$

$\displaystyle \frac{3}{5} + C = \frac{3}{5}$

$\displaystyle C = 0$

Thus,

$\displaystyle f(x) = \frac{3x^{5/3}}{5}.$

45. $C'(x) = 4x - 5$, fixed cost is \$8.

$\displaystyle C(x) = \int (4x - 5)\, dx$

$\displaystyle = \frac{4x^2}{2} - 5x + k$

$\displaystyle = 2x^2 - 5x + k$

$C(0) = 2(0)^2 - 5(0) + k = k$

Since $C(0) = 8$, $k = 8$.

$C(x) = 2x^2 - 5x + 8$

47. $C'(x) = .2x^2 + 5x$, fixed cost is \$10.

$\displaystyle C(x) = \int (.2x^2 + 5x)\, dx$

$\displaystyle = \frac{.2x^3}{3} + \frac{5x^2}{2} + k$

$\displaystyle = \frac{.2x^3}{3} + \frac{5x^2}{2} + k$

$\displaystyle C(0) = \frac{.2(0)^3}{3} + \frac{5(0)^2}{2} + k = k$

Since $C(0) = 10$, $k = 10$.

$\displaystyle C(x) = \frac{.2x^3}{3} + \frac{5x^2}{2} + 10$

49. $C'(x) = .03e^{.01x}$, fixed cost is \$8.

$\displaystyle C(x) = \int .03e^{.01x}\, dx$

$\displaystyle = .03 \int e^{.01x}\, dx$

$\displaystyle = .03\left(\frac{1}{.01}e^{.01x}\right) + k$

$\displaystyle = 3e^{.01x} + k$

$C(0) = 3e^{.01(0)} + k = 3(1) + k$

$= 3 + k$

Since $C(0) = 8$, $3 + k = 8$, and $k = 5$.

So, $C(x) = 3e^{.01x} + 5$.

51. $C'(x) = x^{2/3} + 2$, 8 units costs \$58, so

$\displaystyle C(x) = \int (x^{2/3} + 2)\, dx$

$\displaystyle = \frac{3x^{5/3}}{5} + 2x + k$

$$C(8) = \frac{3(8)^{5/3}}{5} + 2(8) + k$$

$$= \frac{3(32)}{5} + 16 + k$$

Since $C(8) = 58$,

$$58 - 16 - \frac{96}{5} = k$$

$$\frac{114}{5} = k.$$

So, $C(x) = \frac{3x^{5/3}}{5} + 2x + \frac{114}{5}$.

53. $C'(x) = x + \frac{1}{x^2}$, 2 units cost \$5.50,

so

$$C(2) = 5.50.$$

$$C(x) = \int \left(x + \frac{1}{x^2} \right) dx$$

$$= \int (x + x^{-2}) dx$$

$$= \frac{x^2}{2} + \frac{x^{-1}}{-1} + k$$

$$C(x) = \frac{x^2}{2} - \frac{1}{x} + k$$

$$C(2) = \frac{(2)^2}{2} - \frac{1}{2} + k$$

$$= 2 - \frac{1}{2} + k$$

Since $C(2) = 5.50$,

$$5.50 - 1.5 = k$$

$$4 = k.$$

So, $C(x) = \frac{x^2}{2} - \frac{1}{x} + 4$.

55. $C'(x) = 5x - \frac{1}{x}$, 10 units cost

\$94.20, so

$$C(10) = 94.20.$$

$$C(x) = \int \left(5x - \frac{1}{x} \right) dx$$

$$= \frac{5x^2}{2} - \ln x + k$$

$$C(10) = \frac{5(10)^2}{2} - \ln (10) + k$$

$$= 250 - 2.30 + k$$

Since $C(10) = 94.20$,

$$94.20 = 247.70 + k$$

$$-153.50 = k$$

So, $C(x) = \frac{5x^2}{2} - \ln x - 153.50$.

57. $P'(x) = 2x + 20$, profit is -50 when
0 hamburgers are sold.

$$P(x) = \int (2x + 20) dx$$

$$= \frac{2x^2}{2} + 20x + k$$

$$= x^2 + 20x + k$$

$$P(0) = 0^2 + 20(0) + k$$

Since $P(0) = -50$,

$$k = -50.$$

$$P(x) = x^2 + 20x - 50$$

59. **(a)** $f'(t) = .01e^{-.01t}$

$$f(t) = \int .01e^{-.01t} \, dt$$

$$= -\frac{.01e^{-.01t}}{.01} + k$$

$$= -e^{-.01t} + k$$

(b) $f(0) = -e^{-.01(0)} + k$

$$= -e^0 + k$$

$$= -1 + k$$

Since $f(0) = 0$,

$$0 = -1 + k$$

$$k = 1.$$

$$f(t) = -e^{-.01t} + 1$$

$$f(10) = -e^{-.01(10)} + 1$$
$$= -e^{-.1} + 1$$
$$= -.905 + 1$$
$$= .095 \text{ unit}$$

61. $a(t) = t^2 + 1$

$$v(t) = \int (t^2 + 1)dt$$

$$= \frac{t^3}{3} + t + C$$

$$v(0) = \frac{0^3}{3} + 0 + C$$

Since $v(0) = 6$,

$$C = 6.$$

$$v(t) = \frac{t^3}{3} + t + 6$$

63. $a(t) = -32$

$$v(t) = \int -32 \, dt = -32t + C_1$$

$$v(0) = -32(0) + C_1$$

Since $v(0) = 0$,

$$C_1 = 0.$$

$$v(t) = -32t$$

$$s(t) = \int -32t \, dt$$

$$= \frac{-32t^2}{2} + C_2$$

$$= -16t^2 + C_2$$

At $t = 0$, the plane is at 6400 ft.
That is, $s(0) = 6400$.

$$s(0) = -16(0)^2 + C_2$$
$$6400 = 0 + C_2$$
$$C_2 = 6400$$
$$s(t) = -16t^2 + 6400$$

When the object hits the ground,
$S(t) = 0$.

$$-16t^2 + 6400 = 0$$
$$-16t^2 = -6400$$
$$t^2 = 400$$
$$t = \pm 20$$

Discard -20 since time must be positive in this problem.
The object hits the ground in 20 sec.

65. $a(t) = \frac{15}{2}\sqrt{t} = 3e^{-t}$

$$v(t) = \int \left(\frac{15}{2}\sqrt{t} + 3e^{-t}\right)dt$$

$$= \int \left(\frac{15}{2}t^{1/2} + 3e^{-t}\right)dt$$

$$= \frac{15}{2}\left(\frac{t^{3/2}}{3/2}\right) + 3\left(\frac{1}{-1}e^{-t}\right) + C_1$$

$$= 5t^{3/2} - 3e^{-t} + C_1$$

$$v(0) = 5(0)^{3/2} - 3e^{-0} + C_1 = -3 + C_1$$

Since $v(0) = -3$, $C_1 = 0$.

$$v(t) = 5t^{3/2} - 3e^{-t}$$

$$s(t) = \int (5t^{3/2} - 3e^{-t})dt$$

$$= 5\left(\frac{t^{5/2}}{5/2}\right) - 3\left(-\frac{1}{1}e^{-t}\right) + C_2$$

$$= 2t^{5/2} + 3e^{-t} + C_2$$

$$s(0) = 2(0)^{5/2} + 3e^{-0} + C_2 = 3 + C_2$$

Since $s(0) = 4$, $C_2 = 1$.

So $s(t) = 2t^{5/2} + 3e^{-t} + 1$.

Section 14.2

1. $\int 4(2x + 3)^4 \, dx = 2 \int 2(2x + 3)^4 \, dx$

Let $u = 2x + 3$, so that

   $du = 2 \, dx$.

$$= 2 \int u^4 \, du$$

$$= \frac{2 \cdot u^5}{5} + C$$

$$= \frac{2(2x + 3)^5}{5} + C$$

3. $\int \frac{2 \, dm}{(2m + 1)^3} = \int 2(2m + 1)^{-3} \, dm$

Let $u = 2m + 1$, so that

   $du = 2 \, dm$.

$$= \int u^{-3} \, du$$

$$= \frac{u^{-2}}{-2} + C$$

$$= \frac{-(2m + 1)^{-2}}{2} + C$$

5. $\int \frac{2x + 2}{(x^2 + 2x - 4)^4} \, dx$

$$= \int (2x + 2)(x^2 + 2x - 4)^{-4} \, dx$$

Let $u = x^2 + 2x - 4$, so that

   $du = (2x + 2) \, dx$.

$$= \int u^{-4} \, du$$

$$= \frac{u^{-3}}{-3} + C$$

$$= \frac{-(x^2 + 2x - 4)^{-3}}{3} + C$$

7. $\int z\sqrt{z^2 - 5} \, dz = \int z(z^2 - 5)^{1/2} \, dz$

$$= \frac{1}{2} \int 2z(z^2 - 5)^{1/2} \, dz$$

Let $u = z^2 - 5$, so that

   $du = 2z \, dz$.

$$= \frac{1}{2} \int u^{1/2} \, du$$

$$= \frac{1}{2} \cdot \frac{u^{3/2}}{3/2} + C$$

$$= \frac{1}{2}\left(\frac{2}{3}\right)u^{3/2} + C$$

$$= \frac{(z^2 - 5)^{3/2}}{3} + C$$

9. $\int (-4e^{2p}) \, dp = -2 \int 2e^{2p} \, dp$

Let $u = 2p$, so that

   $du = 2 \, dp$.

$$= -2 \int e^u \, du$$

$$= -2e^u + C$$

$$= -2e^{2p} + C$$

11. $\int 3x^2 \, e^{2x^3} \, dx = \frac{1}{2} \int 2 \cdot 3x^2 \, e^{2x^3} \, dx$

Let $u = 2x^3$, so that

   $du = 6x^2 \, dx$.

$$= \frac{1}{2} \int e^u \, du$$

$$= \frac{1}{2}e^u + C$$

$$= \frac{e^{2x^3}}{2} + C$$

13. $\int (1 - t)e^{2t - t^2} \, dt$

$$= \frac{1}{2} \int 2(1 - t)e^{2t - t^2} \, dt$$
$$\qquad\qquad \textit{Multiply by } 1/2 \cdot 2$$

Let $u = 2t - t^2$, so that

   $du = (2 - 2t) \, dt$

$$= 2(1 - t) \, dt.$$

$$= \frac{1}{2} \int e^u \, du$$

$$= \frac{e^u}{2} + C$$

$$= \frac{e^{2t-t^2}}{2} + C$$

15. $\int \frac{e^{1/z}}{z^2} \, dz = -\int e^{1/z} \cdot \frac{-1}{z^2} \, dz$

Let $u = 1/z$, so that

$$du = \frac{-1}{z^2} \, dx.$$

$$= -\int e^u \, du$$

$$= -e^u + C$$

$$= -e^{1/z} + C$$

17. $\int \frac{-8}{1 + 3x} \, dx = -8 \int \frac{1}{1 + 3x} \, dx$

$$= -8\left(\frac{1}{3}\right) \int \frac{3}{1 + 3x} \, dx$$

Let $u = 1 + 3x$, so that

$$du = 3 \, dx.$$

$$= \frac{-8}{3} \int \frac{du}{u}$$

$$= \frac{-8}{3} \ln |u| + C$$

$$= \frac{-8 \ln |1 + 3x|}{3} + C$$

19. $\int \frac{dt}{2t + 1} = \frac{1}{2} \int \frac{2 \, dt}{2t + 1}$

Let $u = 2t + 1$, so that

$$du = 2 \, dt.$$

$$= \frac{1}{2} \int \frac{du}{u}$$

$$= \frac{1}{2} \ln |u| + C$$

$$= \frac{\ln |2t + 1|}{2} + C$$

21. $\int \frac{v \, dv}{(3v^2 + 2)^4} = \frac{1}{6} \int \frac{6v \, dv}{(3v^2 + 2)^4}$

Let $u = 3v^2 + 2$, so that

$$du = 6v \, dv.$$

$$= \frac{1}{6} \int \frac{du}{u^4}$$

$$= \frac{1}{6} \int u^{-4} \, du$$

$$= \left(\frac{1}{6}\right)\frac{u^{-3}}{-3} + C$$

$$= -\frac{1}{18}(3v^2 + 2)^{-3} + C$$

$$= \frac{-(3v^2 + 2)^{-3}}{18} + C$$

23. $\int \frac{x - 1}{(2x^2 - 4x)^2} \, dx = \frac{1}{4} \int \frac{4(x - 1) \, dx}{(2x^2 - 4x)^2}$

Let $u = 2x^2 - 4x$, so that

$$du = (4x - 4) \, dx$$
$$= 4(x - 1) \, dx.$$

$$= \frac{1}{4} \int \frac{du}{u^2}$$

$$= \frac{1}{4} \int u^{-2} \, du$$

$$= \frac{1}{4}\left(\frac{u^{-1}}{-1}\right) + C$$

$$= \frac{-(2x^2 - 4x)^{-1}}{4} + C$$

25. $\int \left(\frac{1}{r} + r\right)\left(1 - \frac{1}{r^2}\right) dr = \int u \, du$

Let $u = 1/r + r$, so that

$$du = \left(-\frac{1}{r^2} + 1\right) dr$$

$$= \left(1 - \frac{1}{r^2}\right) dr.$$

$$= \frac{u^2}{2} + C$$

$$= \frac{1}{2}\left(\frac{1}{r} + r\right)^2 + C$$

$$= \frac{\left(\frac{1}{r} + r\right)^2}{2} + C$$

27. $\displaystyle\int \frac{x^2 + 1}{(x^3 + 3x)^{2/3}}\, dx$

$\displaystyle = \frac{1}{3}\int \frac{3(x^2 + 1)\, dx}{(x^3 + 3x)^{2/3}}$

Let $u = x^3 + 3x$, so that

$\quad du = (3x^2 + 3)\, dx$.

$\quad = 3(x^2 + 1)\, dx$

$\displaystyle = \frac{1}{3}\int \frac{du}{u^{2/3}}$

$\displaystyle = \frac{1}{3}\int u^{-2/3}\, du$

$\displaystyle = \frac{1}{3}\left(\frac{u^{1/3}}{1/3}\right) + C$

$\displaystyle = u^{1/3} + C$

$\displaystyle = (x^3 + 3x)^{1/3} + C$

29. $\displaystyle\int p(p + 1)^5\, dp$

Let $u = p + 1$, so that

$\quad du = dp$; also, $p = u - 1$.

$\displaystyle = \int (u - 1)u^5\, du$

$\displaystyle = \int (u^6 - u^5)\, du$

$\displaystyle = \frac{u^7}{u} - \frac{u^6}{6} + C$

$\displaystyle = \frac{(p + 1)^7}{7} - \frac{(p + 1)^6}{6} + C$

31. $\displaystyle\int t\sqrt{5t - 1}\, dt$

$\displaystyle = \frac{1}{5}\int 5t(5t - 1)^{1/2}\, dt$

Let $u = 5t - 1$, so that

$\quad du = 5\, dt$; also,

$\quad t = \dfrac{u + 1}{5}$.

$\displaystyle = \frac{1}{5}\int \left(\frac{u + 1}{5}\right)u^{1/2}\, du$

$\displaystyle = \frac{1}{25}\int (u^{3/2} + u^{1/2})\, du$

$\displaystyle = \frac{1}{25}\left(\frac{u^{5/2}}{5/2} + \frac{u^{3/2}}{3/2}\right) = C$

$\displaystyle = \frac{1}{25}\left[\frac{2}{5}(5t - 1)^{5/2} + \frac{2}{3}(5t - 1)^{3/2}\right] + C$

$\displaystyle = \frac{2(5t - 1)^{5/2}}{125} + \frac{2(5t - 1)^{3/2}}{75} + C$

33. $\displaystyle\int \frac{u}{\sqrt{u - 1}}\, du$

$\displaystyle = \int u(u - 1)^{-1/2}\, du$

Let $w = u - 1$, so that

$\quad dw = du$ and

$\quad u = w + 1$.

$\displaystyle = \int (w + 1)w^{-1/2}\, dw$

$\displaystyle = \int (w^{1/2} + w^{-1/2})\, dw$

$\displaystyle = \frac{w^{3/2}}{3/2} + \frac{w^{1/2}}{1/2} + C$

$\displaystyle = \frac{2(u - 1)^{3/2}}{3} + 2(u - 1)^{1/2} + C$

35. $\displaystyle\int (\sqrt{x^2 + 12x})(x + 6)\, dx$

$\displaystyle = \int (x^2 + 12x)^{1/2}(x + 6)\, dx$

Let $x^2 + 12x = u$, so that

$\quad (2x + 12)\, dx = du$

$\quad 2(x + 6)\, dx = du$.

$\displaystyle = \frac{1}{2}\int u^{1/2}\, du$

$\displaystyle = \frac{1}{2}\left(\frac{2}{3}\right)u^{3/2} + C$

$\displaystyle = \frac{(x^2 + 12x)^{3/2}}{3} + C$

37. $\displaystyle\int \frac{t}{t^2 + 2}\, dt$

Let $t^2 + 2 = u$, so that

$\qquad 2t\, dt = du.$

$\displaystyle = \frac{1}{2} \int \frac{du}{u}$

$\displaystyle = \frac{1}{2} \ln |u| + C$

$\displaystyle = \frac{\ln (t^2 + 2)}{2} + C$

39. $\displaystyle\int ze^{2z^2}\, dz$

Let $2z^2 = u$, so that

$\qquad 4z\, dz = du.$

$\displaystyle = \frac{1}{4} \int e^u\, du$

$\displaystyle = \frac{1}{4} e^u + C$

$\displaystyle = \frac{e^{2z^2}}{4} + C$

41. $\displaystyle\int \frac{(1 + \ln x)^2}{x}\, dx$

Let $1 + \ln x = u$, so that

$\qquad \dfrac{1}{x}\, dx = du,$

$\displaystyle = \int u^2\, du$

$\displaystyle = \frac{1}{3} u^3 + C$

$\displaystyle = \frac{(1 + \ln x)^3}{3} + C$

43. $\displaystyle\int x^{3/2} \sqrt{x^{5/2} + 4}\, dx$

Let $x^{5/2} + 4 = u$, so that

$\qquad \dfrac{5}{2} x^{3/2}\, dx = du$

$\qquad x^{3/2}\, dx = \dfrac{2}{5}\, du.$

$\displaystyle = \frac{2}{5} \int \sqrt{u}\, du$

$\displaystyle = \frac{2}{5} \int u^{1/2}\, du$

$\displaystyle = \frac{2}{5} \left(\frac{u^{3/2}}{3/2} \right) + C$

$\displaystyle = \frac{4}{15} u^{3/2} + C$

$\displaystyle = \frac{4}{15} (x^{5/2} + 4)^{3/2} + C$

45. **(a)** $R'(x) = 2x(x^2 + 50)^2$

$\qquad R(x) = \displaystyle\int 2x(x^2 + 50)^2\, dx$

Let $u = x^2 + 50$, so that

$\qquad du = 2x\, dx.$

$\qquad R = \displaystyle\int u^2\, du$

$\qquad = \dfrac{u^3}{3} + C$

$\qquad = \dfrac{1}{3}(x^2 + 50)^3 + C$

$\qquad R(3) = \dfrac{1}{3}(9 + 50)^3 + C$

Since $R(3) = \$206{,}379$,

$\dfrac{1}{3}(59)^3 + C = 206{,}379$

$\dfrac{205{,}379}{3} + C = 206{,}379$

$\qquad C = 137{,}919.33.$

$R(x) = \dfrac{(x^2 + 50)^3}{3} + 137{,}919.33$

(b)

$R(x) = \dfrac{(x^2 + 50)^3}{3} + 137{,}919.33 \geq 450{,}000$

$\dfrac{(x^2 + 50)^3}{3} \geq 312{,}080.67$

$(x^2 + 50)^3 \geq 936{,}242.01$

$x^2 + 50 \geq 97.83$

$x^2 \geq 47.83$

$x \geq 6.92$

At least 7 planes must be sold.

47. **(a)** $p'(x) = xe^{-x^2}$

Let $-x^2 = u$, so that

$-2x\, dx = du$

$x\, dx = -\dfrac{du}{2}$

$p = -\dfrac{1}{2} \int e^u\, du$

$= -\dfrac{1}{2}e^u$

$= -\dfrac{e^{-x^2}}{2} + C$

$p(3) = -\dfrac{e^{-9}}{2} + C$

Since $10,000 = .01$ millions and $p(3) = .01$,

$-\dfrac{e^{-9}}{2} + C = .01$

$C = .01 + \dfrac{e^{-9}}{2}$

$= .01006$

$\approx .01$.

$p(x) = \dfrac{-e^{-x^2}}{2} + .01$

(b) $\displaystyle\lim_{x \to \infty} (x) = \lim_{x \to \infty} \left(\dfrac{-e^{-x^2}}{2} + .01\right)$

$= \displaystyle\lim_{x \to \infty} \left(-\dfrac{1}{2e^{x^2}} + .01\right)$

$= .01$

Since profit is expressed in millions of dollars, the profit approaches $.01(1,000,000) = \$10,000$.

49. **(a)** $D'(x) = \dfrac{2}{x + 9}$

$D(x) = \displaystyle\int D'(x)\, dx$

$= \displaystyle\int \dfrac{2}{x + 9}\, dx = 2\int \dfrac{dx}{x + 9}$

$D(x) = 2 \ln |x + 9| + C$

If $x = 1$, $D(x) = 2.5$.

So

$2.5 = 2 \ln |1 + 9| + C$

$2.5 = 2 \ln 10 + C$

$2.5 - 2 \ln 10 = C$

$-2.11 \approx C$.

Thus

$D(x) = 2 \ln |x + 9| - 2.11$.

(b)

$D(x)$

$= 2 \ln |x + 9| - 2.11 = 3$

$2 \ln |x + 9| = 5.11$

$\ln |x + 9| = 2.555$

$|x + 9| = e^{2.555}$

$|x + 9| = 12.87$

$x + 9 = \pm 12.87$

$x = -9 \pm 12.87$

$x = 3.87 \quad \text{or} \quad x = -21.87$

Disregard negative solution. Thus, about 3.9 mg of the drug is necessary.

Section 14.3

1. $\displaystyle\sum_{i=1}^{3} 3i = 3(1) + 3(2) + 3(3) = 18$

3. $\displaystyle\sum_{i=1}^{5} (2i + 7)$

$= [2(1) + 7] + [2(2) + 7] + [2(3) + 7]$

$\quad + [2(4) + 7] + [2(5) + 7]$

$= 9 + 11 + 13 + 15 + 17$

$= 65$

5. $\displaystyle\sum_{i=1}^{4} x_i = x_1 + x_2 + x_3 + x_4$

$\qquad\qquad = -5 + 8 + 7 + 10$

$\qquad\qquad = 20$

7. $f(x) = x - 3$ and $x_1 = 4$, $x_2 = 6$,

$\quad x_3 = 7$

$\quad \displaystyle\sum_{i=1}^{3} f(x_i)$

$\qquad = \displaystyle\sum_{i=1}^{3} (x_i - 3)$

$\qquad = (x_1 - 3) + (x_2 - 3) + (x_3 - 3)$

$\qquad = (4 - 3) + (6 - 3) + (7 - 3)$

$\qquad = 1 + 3 + 4$

$\qquad = 8$

9. $f(x) = 2x + 1$, $x_1 = 0$, $x_2 = 2$,

$\quad x_3 = 4$, $x_4 = 6$, and $\Delta x = 2$

$\quad$ **(a)** $\displaystyle\sum_{i=1}^{4} f(x_i)\Delta x$

$\qquad = f(x_1)\Delta x + f(x_2)\Delta x + f(x_3)\Delta x$

$\qquad\quad + f(x_4)\Delta x$

$\qquad = f(0)(2) + f(2)(2) + f(4)(2)$

$\qquad\quad + f(6)(2)$

$\qquad = [2(0) + 1](2) + [2(2) + 1](2)$

$\qquad\quad + [2(4) + 1](2) + [2(6) + 1](2)$

$\qquad = 2 + 5(2) + 9(2) + 13(2)$

$\qquad = 56$

(b)

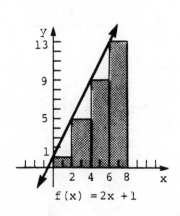

$f(x) = 2x + 1$

The sum of these rectangles approxi-

mates $\displaystyle\int_0^8 (2x + 1)\,dx$.

11. $f(x) = 3x + 2$ from $x = 1$ to $x = 5$

$\quad$ For $n = 2$ rectangles:

$$\Delta x = \frac{5 - 1}{2} = 2$$

i	x_i	$f(x_i)$
1	1	5
2	3	11

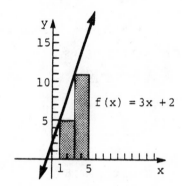

$f(x) = 3x + 2$

$A = \displaystyle\sum_{i=1}^{2} f(x_i)\Delta x = f(x_1)\Delta x + f(x_2)\Delta x$

$\qquad\qquad\qquad = 5(2) + 11(2)$

$\qquad\qquad\qquad = 32$

For $n = 4$ rectangles:

$$\Delta x = \frac{5 - 1}{4} = 2$$

i	x_i	$f(x_i)$
1	1	5
2	2	8
3	3	11
4	4	14

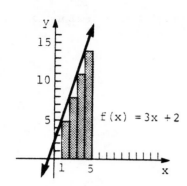

$f(x) = 3x + 2$

$$A = \sum_{i=1}^{4} f(x_i)\Delta x$$

$$= 5(1) + 8(1) + 11(1) + 14(1)$$

$$= 38$$

13. $f(x) = x + 5$ from $x = 2$ to $x = 4$

For $n = 2$ rectangles:

$$\Delta x = \frac{4 - 2}{2} = 1$$

i	x_i	$f(x_i)$
1	2	7
2	3	8

$$A = \sum_{i=1}^{2} f(x_i)\Delta x = 7(1) + 8(1)$$
$$= 15$$

For $n = 4$ rectangles:

$$\Delta x = \frac{4 - 2}{4} = \frac{1}{2}$$

i	x_i	$f(x_i)$
1	2	7
2	5/2	15/2
3	3	8
4	7/2	17/2

$$A = \sum_{i=1}^{4} f(x_i)\Delta x$$

$$= 7\left(\frac{1}{2}\right) + \frac{15}{2}\left(\frac{1}{2}\right) + 8\left(\frac{1}{2}\right) + \frac{17}{2}\left(\frac{1}{2}\right)$$

$$= \frac{31}{2}$$

15. $f(x) = x^2$ from $x = 1$ to 5

For $n = 2$ rectangles:

$$\Delta x = \frac{5 - 1}{2} = 2$$

i	x_i	$f(x_i)$
1	1	1
2	3	9

$$A = \sum_{i=1}^{2} f(x_i)\Delta x = 1(2) + 9(2)$$
$$= 20$$

For $n = 4$ rectangles:

$$\Delta x = \frac{5 - 1}{4} = 1$$

i	x_i	$f(x_i)$
1	1	1
2	2	4
3	3	9
4	4	16

$$A = \sum_{i=1}^{4} f(x_i)\Delta x$$

$$= 1(1) + 4(1) + 9(1) + 16(1)$$

$$= 30$$

17. $f(x) = x^2 + 2$ from $x = -2$ to $x = 2$

For $n = 2$ rectangles:

$$\Delta x = \frac{2 - (-2)}{2} = 2$$

i	x_i	$f(x_i)$
1	-2	6
2	0	2

$$A = \sum_{i=1}^{2} f(x_i)\Delta x = 6(2) + 2(2)$$
$$= 16$$

For n = 4 rectangles:

$$\Delta x = \frac{2 - (-2)}{4} = 1$$

i	x_i	$f(x_i)$
1	-2	6
2	-1	3
3	0	2
4	1	3

$$A = \sum_{i=1}^{4} f(x_i)\Delta x = 6 + 3 + 2 + 3$$
$$= 14$$

19. $f(x) = e^x - 1$ from x = 0 to x = 4

For n = 2 rectangles:

$$\Delta x = \frac{4 - 0}{2} = 2$$

i	x_i	$f(x_i)$
1	0	0
2	2	6.39

$$A = \sum_{i=1}^{2} f(x_i)\Delta x = 0(2) + 6.39(2)$$
$$= 12.8$$

For n = 4 rectangles:

$$\Delta x = \frac{4 - 0}{4} = 1$$

i	x_i	$f(x_i)$
1	0	0
2	1	1.72
3	2	6.39
4	3	19.09

$$A = \sum_{i=1}^{4} f(x_i)\Delta x$$
$$= 0 + 1.72 + 6.39 + 10.09$$
$$= 27.2$$

21. $f(x) = \frac{1}{x}$ from x = 1 to 5

For n = 2 rectangles:

$$\Delta x = \frac{5 - 1}{2} = 2$$

i	x_i	$f(x_i)$
1	1	1
2	3	.333

$$A = \sum_{i=1}^{2} f(x_i)\Delta x = 1(2) + .333(2)$$
$$= 2.67$$

For n = 4 rectangles:

$$\Delta x = \frac{5 - 1}{4} = 1$$

i	x_i	$f(x_i)$
1	1	1
2	2	.5
3	3	.333
4	4	.25

$$A = \sum_{i=1}^{4} f(x_i)\Delta x = 1 + .5 + .333 + .25$$
$$= 2.08$$

23. $f(x) = \frac{x}{2}$ between x = 0 and x = 4

(a) For n = 4 rectangles:

$$\Delta x = \frac{4 - 0}{4} = 1$$

i	x_i	$f(x_i)$
1	0	0
2	1	.5
3	2	1
4	3	1.5

$$A = \sum_{i=1}^{4} f(x_i)\Delta x = 0 + .5 + 1 + 1.5$$
$$= 3$$

(b) For n = 8 rectangles:

$$\Delta x = \frac{4 - 0}{8} = .5$$

i	x_i	$f(x_i)$
1	0	0
2	.5	.25
3	1	.5
4	1.5	.75
5	2	1
6	2.5	1.25
7	3	1.5
8	3.5	1.75

$$A = \sum_{i=1}^{8} f(x_i)\Delta x$$

$$= 0(.5) + .25(.5) + .5(.5)$$
$$+ .75(.5) + 1(.5) + 1.25(.5)$$
$$+ 1.5(.5) + 1.75$$

$$= 3.5$$

(c)

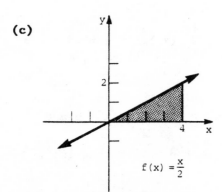

$$f(x) = \frac{x}{2}$$

$$\int_0^4 f(x)\,dx = \int_0^4 \frac{x}{2}\,dx$$

$$= \frac{1}{2}(\text{base})(\text{height})$$

$$= \frac{1}{2}(4)(2)$$

$$= 4$$

25. $\int_0^3 2x\,dx$

Graph y = 2x.

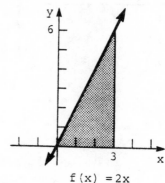

f(x) = 2x

$\int_0^3 2x\,dx$ is the area of a triangle

with base = 3 − 0 = 3 and altitude 6.

$$\text{Area} = \frac{1}{2}(\text{base})(\text{altitude})$$

$$= \frac{1}{2}(3)(6)$$

$$= 9$$

27. $\int_{-3}^3 \sqrt{9 - x^2}\,dx$

Graph $y = \sqrt{9 - x^2}$.

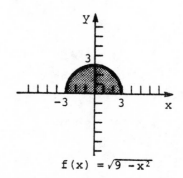

$f(x) = \sqrt{9 - x^2}$

$\int_{-3}^3 \sqrt{9 - x^2}\,dx$ is the area of a semi-

circle with radius 3 centered at the
origin.

Area $= \frac{1}{2}\pi r^2$

$\qquad = \frac{1}{2}\pi(3)^2$

$\qquad = \frac{9}{2}\pi$

29. $\displaystyle\int_{1}^{3}(5-x)\,dx$

Graph $y = 5 - x$.

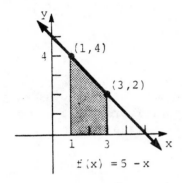

(1,4)

(3,2)

$f(x) = 5 - x$

$\displaystyle\int_{1}^{3}(5-x)\,dx$ is the area of a

trapezoid with bases of length 4
and 2 and height of length 2.

Area $= \frac{1}{2}(\text{height})(\text{base}_1 + \text{base}_2)$

$\qquad = \frac{1}{2}(2)(4+2)$

$\qquad = 6$

For Exercises 31–39, readings on the
graphs and answers may vary.

31. (a) Read values for estimated production per day, current policy
base, on the graph for every 2 yr
from 1990 to 2008. These are the
left sides of rectangles with width
$\Delta x = 2$.

$\Sigma = 8.5(2) + 8(2) + 7.7(2) + 7.5(2)$

$\qquad + 7.2(2) + 7.0(2) + 6.7(2)$

$\qquad + 6.5(2) + 6.3(2) + 6.1(2)$

$\qquad \approx 143$ million barrels

This sum must be multiplied by 365
since there are 365 days in a year.
The total production is $365(143) \approx$
52 billion barrels.

(b) Read values for estimated production, with strategy, on the graph
for every 2 yr from 1990 to 2008.
These are the left sides of rectangles with width $\Delta x = 2$.

$\Sigma = 8.5(2) + 8(2) + 7.7(2) + 7.7(2)$

$\qquad + 8.4(2) + 9(2) + 9.5(2)$

$\qquad + 9.8(2) + 10.2(2) + 10(2)$

$\qquad \approx 177$

Total production is $177(365) \approx 64$
billion barrels.

33. (a) Read the value for United States
for every year from $x = 1987$ to
$x = 1990$. These are the left sides
of rectangles with width $\Delta x = 1$.

$\Sigma = 9.9(1) + 10.2(1) + 10.5(1)$

$\qquad + 10.8(1)$

$\qquad \approx 41.4$

This sum must be multiplied by 2000
hr/yr, so $41.4(2000) \approx \$83,000$.

(b) Read the value for Canada for
every year from $x = 1987$ to $x =$
1990. Thee are the left sides of
rectangles with width $\Delta x = 1$.

$\Sigma = 9.5(1) + 10.8(1) + 11.8(1)$

$\qquad + 12.2(1)$

$\qquad \approx 44.3$

This sum must be multiplied by 2000 hr/yr, so 44.3(2000) ≈ $89,000.

35. Read the value of the function for every minute from x = 0 to x = 19. These are the left sides of rectangles with width Δx = 1.

$$\sum_{i=1}^{20} f(x_i)\Delta x$$

≈ 0(1) + 2.5(1) + 2.8(1) + 3(1)

+ 3(1) + 3(1) + 3(1) + 3(1)

+ 3(1) + 3(1) + 3(1) + 1(1)

+ .8(1) + .6(1) + .4(1) + .3(1)

+ .2(1) + .2(1) + .2(1) + .2(1)

≈ 33.2

The total volume of oxygen inhaled is 33.2 liters.

37. Read the value for the speed every 3 sec from x = 3 to x = 27. These are the right sides of rectangles with width Δx = 3. Then read the speed for x = 28, which is the right side of a rectangle with width Δx = 1.

$$\sum_{i=1}^{9} f(x_i)\Delta x$$

≈ 30(3) + 45(3) + 60(3) + 68(3)

+ 76(3) + 85(3) + 90(3) + 94(3)

= 1742

$\frac{1742}{3600}$(5280) ≈ 2600

The BMW 733i traveled about 2600 ft.

39. **(a)** Read the value for a plain glass window facing south for every 2 hr from 6 to 6. These are the heights of rectangles with width Δx = 2.

Σ = 10(2) + 28(2) + 80(2) + 108(2)

+ 80(2) + 28(2) + 10(2)

≈ 690 BTUs

(b) Read the value for a window with Shadescreen facing south for every 2 hr from 6 to 6. These are the heights of rectangles with width Δx = 2.

Σ = 4(2) + 10(2) + 20(2) + 22(2)

+ 20(2) + 10(2) + 4(2)

≈ 180 BTUs

For Exercises 41 and 43 use computer methods. Answers will vary depending upon the computer program that is used. The answers are given.

41. $f(x) = x^2 e^{-x}$; [-1, 3]

The area is 1.91837.

43. $f(x) = \dfrac{e^x - e^{-x}}{2}$; [0, 4]

The area is 25.7659.

Section 14.4

1. $\displaystyle\int_{-2}^{4} (-1)\ dp = -1 \int_{-2}^{4} dp$

$$= -1 \cdot p \Big|_{-2}^{4}$$

$$= -1[4 - (-2)]$$

$$= -6$$

3. $\displaystyle\int_{-1}^{2} (3t - 1)\,dt$

$$= 3\int_{-1}^{2} t\,dt - \int_{-1}^{2} dt$$

$$= \frac{3}{2}t^2 \Big|_{-1}^{2} - t\Big|_{-1}^{2}$$

$$= \frac{3}{2}[2^2 - (-1)^2] - [2 - (-1)]$$

$$= \frac{3}{2}(4 - 1) - (2 + 1)$$

$$= \frac{9}{2} - 3 = \frac{9 - 6}{2}$$

$$= \frac{3}{2}$$

$\displaystyle\int_{0}^{2} 3\sqrt{4u + 1}\,du$

$$= \frac{3}{4}\int_{0}^{2} \sqrt{4u + 1}\,(4\,du)$$

$$= \frac{3}{4}\int_{1}^{9} x^{1/2}\,dx$$

$$= \frac{3}{4}\cdot\frac{x^{3/2}}{3/2}\Big|_{1}^{9}$$

$$= \frac{3}{4}\cdot\frac{2}{3}(9^{2/3} - 1^{3/2})$$

$$= \frac{1}{2}(27 - 1) = \frac{26}{2}$$

$$= 13$$

5. $\displaystyle\int_{0}^{2} (5x^2 - 4x + 2)\,dx$

$$= 5\int_{0}^{2} x^2\,dx - 4\int_{0}^{2} x\,dx + 2\int_{0}^{2} dx$$

$$= \frac{5x^3}{3}\Big|_{0}^{2} - 2x^2\Big|_{0}^{2} + 2x\Big|_{0}^{2}$$

$$= \frac{5}{3}(2^3 - 0^3) - 2(2^2 - 0^2) + 2(2 - 0)$$

$$= \frac{5}{3}(8) - 2(4) + 2(2)$$

$$= \frac{40 - 24 + 12}{3}$$

$$= \frac{28}{3}$$

7. $\displaystyle\int_{0}^{2} 3\sqrt{4u + 1}\,du$

Let $4u + 1 = x$, so that

$\qquad 4\,du = dx$.

When $u = 0$, $x = 4(0) + 1 = 1$.

When $u = 2$, $x = 4(2) + 1 = 9$.

9. $\displaystyle\int_{0}^{1} 2(t^{1/2} - t)\,dt$

$$= 2\int_{0}^{1} t^{1/2} - 2\int_{0}^{1} t\,dt$$

$$= 2\frac{t^{3/2}}{3/2}\Big|_{0}^{1} - 2\frac{t^2}{2}\Big|_{0}^{1}$$

$$= \frac{4}{3}(1^{3/2} - 0^{3/2}) - (1^2 - 0^2)$$

$$= \frac{4}{3} - 1 = \frac{1}{3}$$

11. $\displaystyle\int_{1}^{4} (5y\sqrt{y} + 3\sqrt{y})\,dy$

$$= 5\int_{1}^{4} y^{3/2}\,dy + 3\int_{1}^{4} y^{1/2}\,dy$$

$$= 5\left(\frac{y^{5/2}}{5/2}\right)\Big|_{1}^{4} + 3\left(\frac{y^{3/2}}{3/2}\right)\Big|_{1}^{4}$$

$$= 2y^{5/2}\Big|_{1}^{4} + 2y^{3/2}\Big|_{1}^{4}$$

$$= 2(4^{5/2} - 1) + 2(4^{3/2} - 1)$$
$$= 2(32 - 1) + 2(8 - 1)$$
$$= 62 + 14$$
$$= 76$$

$$= \frac{-5}{5} - \frac{-5}{1} + \frac{-1}{2(25)} - \frac{-1}{2(1)}$$
$$= \frac{-5}{5} + \frac{5}{1} - \frac{1}{50} + \frac{1}{2}$$
$$= \frac{-50 + 250 - 1 + 25}{50} = \frac{224}{50}$$
$$= \frac{112}{25}$$

13. $\displaystyle\int_4^6 \frac{2}{(x - 3)^2}\, dx$

Let $x - 3 = u$, so that
$$dx = du$$
$$x = u + 3.$$

When $x = 6$, $u = 6 - 3 = 3$.
When $x = 4$, $u = 4 - 3 = 1$.

$$\int_4^6 \frac{2}{(x - 3)^2} = 2 \int_4^6 (x - 3)^{-2}\, dx$$
$$= 2 \int_1^3 u^{-2}\, du$$
$$= 2 \cdot \frac{u^{-1}}{-1}\Big|_1^3 = -2 \cdot u^{-1}\Big|_1^3$$
$$= -2\left(\frac{1}{3} - 1\right) = -2\left(-\frac{2}{3}\right)$$
$$= \frac{4}{3}$$

15. $\displaystyle\int_1^5 (5n^{-2} + n^{-3})\, dn$

$$= \int_1^5 5n^{-2}\, dn + \int_1^5 n^{-3}\, dn$$
$$= \frac{5n^{-1}}{-1}\Big|_1^5 + \frac{n^{-2}}{-2}\Big|_1^5$$
$$= \frac{-5}{n}\Big|_1^5 + \frac{-1}{2n^2}\Big|_1^5$$

17. $\displaystyle\int_2^3 \left(2e^{-.1A} + \frac{3}{A}\right) dA$

$$= 2 \int_2^3 e^{-.1A}\, dA + 3 \int_2^3 \frac{1}{A}\, dA$$
$$= 2\frac{e^{-.1A}}{-.1}\Big|_2^3 + 3 \ln |A|\,\Big|_2^3$$
$$= -20e^{-.1A}\Big|_2^3 + 3 \ln |A|\,\Big|_2^3$$
$$= 20e^{-.2} - 20e^{-.3} + 3 \ln 3$$
$$\quad - 3 \ln 2$$
$$\approx 2.775$$

19. $\displaystyle\int_1^2 \left(e^{5u} - \frac{1}{u^2}\right) du$

$$= \int_1^2 e^{5u}\, du - \int_1^2 \frac{1}{u^2}\, du$$
$$= \frac{e^{5u}}{5}\Big|_1^2 + \frac{1}{u}\Big|_1^2$$
$$= \frac{e^{10}}{5} - \frac{e^5}{5} + \frac{1}{2} - 1$$
$$= \frac{e^{10}}{5} - \frac{e^5}{5} - \frac{1}{2}$$
$$\approx 4375.1$$

21. $\displaystyle\int_{-1}^{0} y(2y^2 - 3)^5 \, dy$

Let $u = 2y^2 - 3$, so that

$\qquad du = 4y \, dy$ and $\dfrac{1}{4} \, du = y \, dy.$

When $y = -1$, $u = 2(-1)^2 - 3 = -1.$
When $y = 0$, $u = 2(0)^2 - 3 = -3.$

$\dfrac{1}{4}\displaystyle\int_{-1}^{-3} u^5 \, du = \dfrac{1}{4} \cdot \dfrac{u^6}{6}\Big|_{-1}^{-3}$

$\qquad\qquad = \dfrac{1}{24}u^6\Big|_{-1}^{-3}$

$\qquad\qquad = \dfrac{1}{24}(-3)^6 - \dfrac{1}{24}(-1)^6$

$\qquad\qquad = \dfrac{729}{24} - \dfrac{1}{24}$

$\qquad\qquad = \dfrac{728}{24}$

$\qquad\qquad = \dfrac{91}{3}$

23. $\displaystyle\int_{1}^{64} \dfrac{\sqrt{z} - 2}{\sqrt[3]{z}} \, dz$

$= \displaystyle\int_{1}^{64} \left(\dfrac{z^{1/2}}{z^{1/3}} - 2z^{-1/3}\right) dz$

$= \displaystyle\int_{1}^{64} z^{1/6} \, dz - 2\displaystyle\int_{1}^{64} z^{-1/3} \, dz$

$= \dfrac{z^{7/6}}{7/6}\Big|_{1}^{64} - 2\dfrac{z^{2/3}}{2/3}\Big|_{1}^{64}$

$= \dfrac{6z^{7/6}}{7}\Big|_{1}^{64} - 3z^{2/3}\Big|_{1}^{64}$

$= \dfrac{6(64)^{7/6}}{7} - \dfrac{6(1)^{7/6}}{7}$
$\qquad - 3(64^{2/3} - 1^{2/3})$

$= \dfrac{6(128)}{7} - \dfrac{6}{7} - 3(16 - 1)$

$= \dfrac{768 - 6 - 315}{7}$

$= \dfrac{447}{7} \approx 63.857$

25. $\displaystyle\int_{1}^{2} \dfrac{\ln x}{x} \, dx$

Let $u = \ln x$, so that

$\qquad du = \dfrac{1}{x} \, dx.$

When $x = 1$, $u = \ln 1 = 0.$
When $x = 2$, $u = \ln 2.$

$\displaystyle\int_{0}^{\ln 2} u \, du = \dfrac{u^2}{2}\Big|_{0}^{\ln 2}$

$\qquad\qquad = \dfrac{(\ln 2)^2}{2} - 0$

$\qquad\qquad \approx .24023$

27. $\displaystyle\int_{0}^{8} x^{1/3} \sqrt{x^{4/3} + 9} \, dx$

Let $u = x^{4/3} + 9$, so that

$\qquad du = \dfrac{4}{3}x^{1/3} \, dx$ and $\dfrac{3}{4} \, du = x^{1/3} \, dx.$

When $x = 0$, $u = 0^{4/3} + 9 = 9.$
When $x = 8$, $u = 8^{4/3} + 9 = 25.$

$\dfrac{3}{4}\displaystyle\int_{9}^{25} \sqrt{u} \, du = \dfrac{3}{4}\displaystyle\int_{9}^{25} u^{1/2} \, du$

$\qquad\qquad = \dfrac{3}{4} \cdot \dfrac{u^{3/2}}{3/2}\Big|_{9}^{25}$

$\qquad\qquad = \dfrac{1}{2}u^{3/2}\Big|_{9}^{25}$

$\qquad\qquad = \dfrac{1}{2}(25)^{3/2} - \dfrac{1}{2}(9)^{3/2}$

$\qquad\qquad = \dfrac{125}{2} - \dfrac{27}{2}$

$\qquad\qquad = 49$

29. $\displaystyle\int_0^1 \frac{e^t}{(3 + e^t)^2} \, dt$

Let $u = 3 + e^t$, so that

$\quad du = e^t \, dt.$

When $t = 0$, $u = 3 + e^0 = 4$.
When $t = 1$, $u = 3 + e$.

$\displaystyle\int_4^{3+e} \frac{1}{u^2} \, du = \int_4^{3+e} u^{-2} \, du$

$\displaystyle\qquad = \frac{u^{-1}}{-1}\Bigg|_4^{3+e}$

$\displaystyle\qquad = \frac{-1}{u}\Bigg|_4^{3+e}$

$\displaystyle\qquad = -\frac{1}{3 + e} + \frac{1}{4}$

$\displaystyle\qquad \approx .075122$

31. $\displaystyle\int_1^{49} \frac{(1 + \sqrt{x})^{4/3}}{\sqrt{x}} \, dx$

Let $u = 1 + \sqrt{x}$, so that

$\quad du = \frac{1}{2}x^{-1/2} \, dx$ and $2 \, du = \frac{1}{\sqrt{x}} \, dx$.

When $x = 1$, $u = 1 + \sqrt{1} = 2$.
When $x = 49$, $u = 1 + \sqrt{49} = 8$.

$\displaystyle 2\int_2^8 u^{4/3} \, du = 2 \cdot \frac{u^{7/3}}{7/3}\Bigg|_2^8$

$\displaystyle\qquad = \frac{6}{7}u^{7/3}\Bigg|_2^8$

$\displaystyle\qquad = \frac{6}{7}(8)^{7/3} - \frac{6}{7}(2)^{7/3}$

$\displaystyle\qquad = \frac{6}{7}(128 - 2^{7/3})$

$\displaystyle\qquad \approx 105.39$

33. $f(x) = 2x + 3$; $[8, 10]$

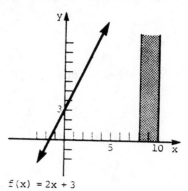

$f(x) = 2x + 3$

Graph does not cross x-axis in given interval $[8, 10]$.

$\displaystyle\int_8^{10} (2x + 3) \, dx$

$\displaystyle\qquad = \left(\frac{2x^2}{2} + 3x\right)\Bigg|_8^{10}$

$\displaystyle\qquad = (10^2 + 30) - (8^2 + 24)$

$\displaystyle\qquad = 42$

35. $f(x) = 2 - 2x^2$; $[0, 5]$

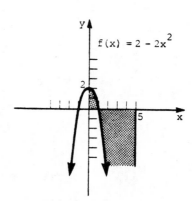

$f(x) = 2 - 2x^2$

Find the points where the graph crosses the x-axis by solving

$\qquad 2 - 2x^2 = 0$

$\qquad\qquad 2x^2 = 2$

$\qquad\qquad\ x^2 = 1$

$\qquad\qquad\quad x = \pm 1.$

The only solution in the interval $[0, 5]$ is 1.

The total area is

$$\int_0^1 (2 - 2x^2)\,dx + \left| \int_1^5 (2 - 2x^2)\,dx \right|$$

$$= \left.\left(2x - \frac{2x^3}{3}\right)\right|_0^1 + \left| \left.\left(2x - \frac{2x^3}{3}\right)\right|_1^5 \right|$$

$$= 2 - \frac{2}{3} + \left| 10 - \frac{2(5^3)}{3} - 2 + \frac{2}{3} \right|$$

$$= \frac{4}{3} + \left| \frac{-224}{3} \right|$$

$$= \frac{228}{3}$$

$$= 76.$$

37. $f(x) = x^2 + 4x - 5;\ [-6,\ 3]$

$f(x) = (x + 5)(x - 1)$

$f(x) = 0$ when $x = -5$ or $x = 1$ which means the graph crosses the x-axis at $x = -5$ and $x = 1$.

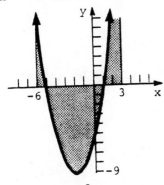

$$f(x) = x^2 + 4x - 5$$

The total area is

$$\int_{-6}^{-5} (x^2 + 4x - 5)\,dx + \left| \int_{-5}^{1} (x^2 + 4x - 5)\,dx \right|$$

$$+ \int_{1}^{3} (x^2 + 4x - 5)\,dx.$$

$$= \left.\frac{x^3}{3} + 2x^2 - 5x\right|_{-6}^{-5} + \left| \left.\frac{x^3}{3} + 2x^2 - 5x\right|_{-5}^{1} \right|$$

$$+ \left.\frac{x^3}{3} + 2x^2 - 5x\right|_{1}^{3}$$

$$= \left(\frac{100}{3} - 30\right) + \left| \left(-\frac{8}{3} - \frac{100}{3}\right| + \left(12 + \frac{8}{3}\right)\right.$$

$$= \frac{10}{3} + 36 + \frac{44}{3}$$

$$= 54$$

39. $f(x) = x^3;\ [-1,\ 3]$

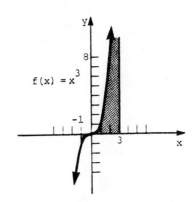

The solution

$$x^3 = 0$$
$$x = 0$$

indicates that the graph crosses the x-axis at 0 in the given interval $[-1,\ 3]$.

The total area is

$$\left| \int_{-1}^{0} x^3\,dx \right| + \int_{0}^{3} x^3\,dx$$

$$= \left| \left.\frac{x^4}{4}\right|_{-1}^{0} \right| + \left.\frac{x^4}{4}\right|_{0}^{3}$$

$$= \left| \left(0 - \frac{1}{4}\right) \right| + \left(\frac{3^4}{4} - 0\right)$$

$$= \frac{1}{4} + \frac{81}{4} = \frac{82}{4}$$

$$= \frac{41}{2}$$

41. $f(x) = e^x - 1$; $[-1, 2]$

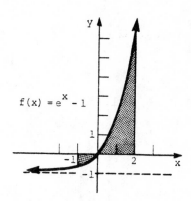

Solve

$$e^x - 1 = 0.$$
$$e^x = 1$$
$$x \ln e = \ln 1$$
$$x = 0$$

The graph crosses the x-axis at 0 in the given interval $[-1, 2]$. The total area is

$$\left| \int_{-1}^{0} (e^x - 1)\,dx \right| + \int_{0}^{2} (e^x - 1)\,dx$$

$$= \left| (e^x - x) \Big|_{-1}^{0} \right| + (e^x - x) \Big|_{0}^{2}$$

$$= \left| (1 - 0) - (e^{-1} + 1) \right|$$
$$\quad + (e^2 - 2) - (1 - 0)$$

$$= \left| 1 - e^{-1} - 1 \right| + e^2 - 2 - 1$$

$$= \frac{1}{e} + e^2 - 3$$

$$\approx 4.757$$

43. $f(x) = \frac{1}{x}$; $[1, e]$

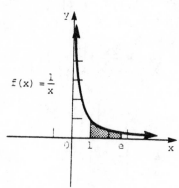

$\frac{1}{x} = 0$ has no solution so the graph does not cross the x-axis in the given interval $[1, e]$.

$$\int_{1}^{e} \frac{1}{x}\,dx = \ln x \Big|_{1}^{e}$$

$$= \ln e - \ln 1$$
$$= 1$$

45. $y = 4 - x^2$; $[0, 3]$

From the graph, we see that the total area is

$$\int_{0}^{2} (4 - x^2)\,dx + \left| \int_{2}^{3} (4 - x^2)\,dx \right|$$

$$= \left(4x - \frac{x^3}{3} \right) \Big|_{0}^{2} + \left| \left(4x - \frac{x^3}{3} \right) \Big|_{2}^{3} \right|$$

$$= \left[\left(8 - \frac{8}{3} \right) - 0 \right]$$
$$\quad + \left| \left[(12 - 9) - \left(8 - \frac{8}{3} \right) \right] \right|$$

$$= \frac{16}{3} + \left| 3 - \frac{16}{3} \right|$$

$$= \frac{16}{3} + \frac{7}{3}$$

$$= \frac{23}{3}$$

47. $y = e^2 - e^x$; $[1, 3]$

From the graph, we see that the total area is

$$\int_1^2 (e^2 - e^x)\,dx + \left|\int_2^3 (e^2 - e^x)\,dx\right|$$

$$= (e^2x - e^x)\Big|_1^2 + \left|(e^2x - e^x)\Big|_2^3\right|$$

$$= (2e^2 - e^2) - (e^2 - e)$$
$$\quad + \left|(3e^2 - e^3) - (2e^2 - e^2)\right|$$

$$= e^2 - e^2 + e$$
$$\quad + \left|3e^2 - e^3 - 2e^2 + e^2\right|$$

$$= e + \left|2e^2 - e^3\right|$$

$$= e + e^3 - 2e^2 \qquad e^3 > 2e^2$$

$$\approx 8.026$$

49. Let $F'(x) = f(x)$, and hence $F(x)$ is an antiderivative of $f(x)$.

$$\int_a^a f(x)\,dx = F(x)\Big|_a^a$$

$$= F(a) - F(a)$$

$$= 0$$

51. Let $F'(x) = f(x)$, and hence $F(x)$ is an antiderivative of $f(x)$.

$$\int_a^c f(x)\,dx + \int_c^b f(x)\,dx$$

$$= F(x)\Big|_a^c + F(x)\Big|_c^b$$

$$= [F(c) - F(a)] + [F(b) - F(a)]$$

$$= -F(a) + F(b)$$

$$= F(b) - F(a)$$

$$= f(x)\Big|_a^b$$

$$= \int_a^b f(x)\,dx$$

Therefore,

$$\int_a^b f(x)\,dx = \int_a^c f(x)\,dx + \int_c^b f(x)\,dx.$$

53. $E'(x) = 4x + 2$ is the rate of expenditure per day

(a) The total expenditure in hundreds of dollars in 10 days is

$$\int_0^{10} (4x + 2)\,dx$$

$$= \left(\frac{4x^2}{2} + 2x\right)\Big|_0^{10}$$

$$= 2(100) + 20 - 0$$

$$= 220.$$

Therefore, since $220(100) = 22{,}000$, the total expenditure is $22,000.

(b) From the tenth to the twentieth day:

$$\int_{10}^{20} (4x + 2)\,dx$$

$$= \left(\frac{4x^2}{2} + 2x\right)\Big|_{10}^{20}$$

$$= [2(400) + 40] - [2(100) + 20]$$

$$= 620$$

That is, $62,000 is spent.

(c) Let a = the number of days on the job.

The total expenditure in a days is

$$\int_0^a (4x + 2)dx$$

$$= (2x^2 + 2x)\Big|_0^a$$

$$= 2a^2 + 2a.$$

If no more than $5000, or 50(100) is spent, $50 \geq 2a^2 + 2a$.

Solve $50 = 2a^2 + 2a$ by the quadratic formula.

$$2a^2 + 2a - 50 = 0$$

$$a^2 + a - 25 = 0$$

$$a = \frac{-1 \pm \sqrt{1 - 4(-25)}}{2}$$

$$a = \frac{-1 + \sqrt{101}}{2} \quad \text{or} \quad a = \frac{-1 - \sqrt{101}}{2}$$

$$\approx 4.5 \qquad\qquad = -5.5$$

To spend less than $5000, the company must complete the job within 4.5 days.

55. $P'(t) = (3t + 3)(t^2 + 2t + 2)^{1/3}$

(a) $\int_0^3 3(t + 1)(t^2 + 2t + 2)^{1/3}\ dt$

Let $u = t^2 + 2t + 2$, so that

$du = (2t + 2)dt$ and $\frac{1}{2}du = (t + 1)dt$.

When $t = 0$, $u = 0^2 + 2 \cdot 0 + 2 = 2$.
When $t = 3$, $u = 3^2 + 2 \cdot 3 + 2 = 17$.

$$\frac{3}{2}\int_2^{17} u^{1/3}\ du = \frac{3}{2} \cdot \frac{u^{4/3}}{4/3}\Big|_2^{17}$$

$$= \frac{9}{8}u^{4/3}\Big|_2^{17}$$

$$= \frac{9}{8}(17)^{4/3} - \frac{9}{8}(2)^{4/3}$$

$$\approx 46.341$$

Total profits for the first 3 yr were

$$\frac{9000}{8}(17^{4/3} - 2^{4/3}) \approx \$46,341.$$

(b) $\int_3^4 3(t + 1)(t^2 + 2t + 2)^{1/3}\ dt$

Let $u = t^2 + 2t + 2$, so that
$du = (2t + 2)dt = 2(t + 1)dt$ and $\frac{3}{2}du = 3(t + 1)dt$.

When $t = 3$, $u = 3^2 + 2 \cdot 3 + 2 = 17$.
When $t = 4$, $u = 4^2 + 2 \cdot 4 + 2 = 26$.

$$\frac{3}{2}\int_{17}^{26} u^{1/3}\ du = \frac{9}{8}u^{4/3}\Big|_{17}^{26}$$

$$= \frac{9}{8}(26)^{4/3} - \frac{9}{8}(17)^{4/3}$$

$$\approx 37.477$$

Profit in the fourth year was

$$\frac{9000}{8}(26^{4/3} - 17^{4/3}) \approx \$37,477.$$

(c) $\lim\limits_{t \to \infty} P'(t)$

$$= \lim_{t \to \infty} (3t + 3)(t^2 + 2t + 2)^{1/3}$$

$$= \infty$$

The annual profits increase slowly without bound.

57. $P'(t) = 140t^{5/2}$

The total concentration of polutants in 4 yr is

$$\int_0^4 140t^{5/2}\,dt = \frac{140t^{7/2}}{7/2}\Big|_0^4$$

$$= 40t^{7/2}\Big|_0^4$$

$$= 40(4^{7/2}) - 0$$

$$= 5120.$$

Since $5120 > 4850$, the factory cannot operate for 4 yr without killing all the fish.

59. Growth rate is $.2 + 4t^{-4}$ ft/yr.

(a) Total growth in the second year is

$$\int_1^2 (.2 + 4t^{-4})\,dt$$

$$= \left(.2t + \frac{4t^{-3}}{-3}\right)\Big|_1^2$$

$$= \left[.2(2) - \frac{4}{3}(2)^{-3}\right] - \left[.2 - \frac{4}{3}\right]$$

$$= .2 - \frac{1}{6} + \frac{4}{3}$$

$$\approx 1.37 \text{ ft.}$$

(b) Total growth in the third year is

$$\int_2^3 (.2 + 4t^{-4})\,dt$$

$$= \left(.2t - \frac{4}{3}t^{-3}\right)\Big|_2^3$$

$$= \left[.2(3) - \frac{4}{81}\right] - \left[.2(2) - \frac{1}{6}\right]$$

$$= .2 - \frac{4}{81} + \frac{1}{6} \approx .32 \text{ ft.}$$

61. $R'(t) = \frac{5}{t} + \frac{2}{t^2}$

(a) Total reaction from $t = 1$ to $t = 12$ is

$$\int_1^{12} \left(\frac{5}{t} + 2t^{-2}\right)dt$$

$$= (5\ln t - 2t^{-1})\Big|_1^{12}$$

$$= \left(5\ln 12 - \frac{1}{6}\right) - (5\ln 1 - 2)$$

$$\approx 14.26.$$

(b) Total reaction from $t = 12$ to $t = 24$ is

$$\int_{12}^{24} \left(\frac{5}{t} + 2t^{-2}\right)dt$$

$$= \left(5\ln t - \frac{2}{t}\right)\Big|_{12}^{24}$$

$$= \left(5\ln 24 - \frac{1}{12}\right) - \left(5\ln 12 - \frac{1}{6}\right)$$

$$\approx 3.55.$$

63. $c'(t) = ke^{rt}$

(a) $c(t) = 1.2e^{.04t}$

(b) $\int_0^{10} 1.2e^{.04t}\,dt$

(c) $= \frac{1.2e^{.04t}}{.04}\Big|_0^{10}$

$$= 30e^{.04t}\Big|_0^{10}$$

$$= 30e^{.4} - 30$$

$$\approx 14.75 \text{ billion}$$

(d) $\displaystyle\int_0^T 1.2e^{.04t}\ dt = 30e^{.04t}\Big|_0^T$

$$= 30e^{.04T} - 30$$

Solve

$$20 = 30e^{.04T} - 30.$$

$$50 = 30e^{.04T}$$

$$\frac{5}{3} = e^{.04T}$$

$$\ln\frac{5}{3} = .04T\ \ln e$$

$$T = \frac{\ln\frac{5}{3}}{.04}$$

$$\approx 12.8\ \text{yr}$$

(e) $\displaystyle\int_0^T 1/2e^{.02t}\ dt = 60e^{.02t}\Big|_0^T$

$$= 60e^{.02T} - 60$$

Solve

$$20 = 60e^{.02T} - 60.$$

$$80 = 60e^{.02T}$$

$$\frac{4}{3} = e^{.02T}$$

$$\ln\frac{4}{3} = .02T\ \ln e$$

$$T = \frac{\ln\frac{4}{3}}{.02}$$

$$\approx 14.4\ \text{yr}$$

65. $C'(t) = 72e^{.014t}$

$$\int_0^T 72e^{.014t}\ dt = \frac{72e^{.014t}}{.014}\Big|_0^T$$

$$= \frac{72e^{.014T}}{.014} - \frac{72e^0}{.014}$$

$$= \frac{72}{.014}(e^{.014T} - 1)$$

$$\approx 5142.9(e^{.014T} - 1)$$

Section 14.5

1. $x = -2$, $x = 1$, $y = x^2 + 4$, $y = 0$

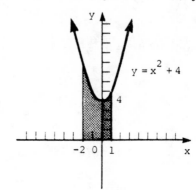

$$\int_{-2}^1 [(x^2 + 4) - 0]\,dx$$

$$= \left(\frac{x^3}{3} + 4x\right)\Big|_{-2}^1$$

$$= \left(\frac{1}{3} + 4\right) - \left(-\frac{8}{3} - 8\right)$$

$$= 3 + 12$$

$$= 15$$

3. $x = -3$, $x = 1$, $y = x + 1$, $y = 0$

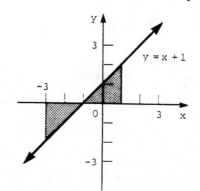

The region is composed of two separate regions because $y = x + 1$ intersects $y = 0$ at $x = -1$.
Let $f(x) = x + 1$, $g(x) = 0$.
In interval $[-3, -1]$, $g(x) \geq f(x)$.
In interval $[-1, 1]$, $f(x) \geq g(x)$.

$$\int_{-3}^{-1} [0 - (x + 1)]\,dx + \int_{-1}^{1} [(x + 1) - 0]\,dx$$

$$= \left(\frac{-x^2}{2} - x\right)\Big|_{-3}^{-1} + \left(\frac{x^2}{2} + x\right)\Big|_{-1}^{1}$$

$$= \left(-\frac{1}{2} + 1\right) - \left(-\frac{9}{2} + 3\right)$$

$$\quad + \left(\frac{1}{2} + 1\right) - \left(\frac{1}{2} - 1\right)$$

$$= 4$$

$$= \left(\frac{x^3}{3} - 3x - x^2\right)\Big|_{-2}^{-1} + \left(x^2 - \frac{x^3}{3} + 3x\right)\Big|_{-1}^{1}$$

$$= -\frac{1}{3} + 3 - 1 - \left(-\frac{8}{3} + 6 - 4\right) + 1 - \frac{1}{3} + 3$$

$$\quad - \left(1 + \frac{1}{3} - 3\right)$$

$$= \frac{5}{3} + 6 = \frac{23}{3}$$

5. x = -2, x = 1, y = 2x, y = x² - 3

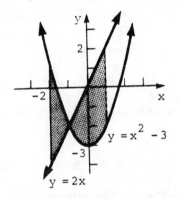

Find the intersections of y = 2x and y = x² - 3 by substituting for y.

$$2x = x^2 - 3$$
$$0 = x^2 - 2x - 3$$
$$0 = (x - 3)(x + 1)$$

The only intersection in [-2, 1] is at x = -1.
In interval [-2, -1], (x² - 3) ≥ 2x.
In interval [-1, 1], 2x ≥ (x² - 3).

$$\int_{-2}^{-1} [(x^2 - 3) - (2x)]\,dx$$

$$+ \int_{-1}^{1} [(2x) - (x^2 - 3)]\,dx$$

$$= \int_{-2}^{-1} (x^2 - 3 - 2x)\,dx + \int_{-1}^{1} (2x - x^2 + 3)\,dx$$

7. y = x² - 30
   y = 10 - 3x

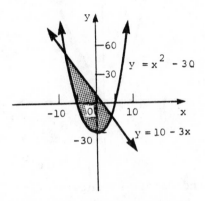

Find points of intersection:

$$x^2 - 30 = 10 - 3x$$
$$x^2 + 3x - 40 = 0$$
$$(x + 8)(x - 5) = 0$$
$$x = -8 \quad \text{or} \quad x = 5$$

Let f(x) = 10 - 3x and g(x) = x² - 30. The area between the curves is given by

$$\int_{-8}^{5} [f(x) - g(x)]\,dx$$

$$= \int_{-8}^{5} [(10 - 3x) - (x^2 - 30)]\,dx$$

$$= \int_{-8}^{5} (-x^2 - 3x + 40)\,dx$$

$$= \left(\frac{-x^3}{3} - \frac{3x^2}{2} + 40x\right)\Big|_{-8}^{5}$$

$$= \frac{-5^3}{3} - \frac{3(5)^2}{2} + 40(5)$$

$$- \left[\frac{-(-8)^3}{3} - \frac{3(-8)^2}{2} + 40(-8)\right]$$

$$= \frac{-125}{3} - \frac{75}{2} + 200 - \frac{512}{3}$$

$$+ \frac{192}{2} + 320$$

$$\approx 366.1667.$$

9. $y = x^2$, $y = 2x$

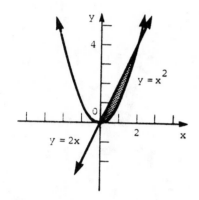

Find points of intersection:

$$x^2 = 2x$$
$$x^2 - 2x = 0$$
$$x(x - 2) = 0$$
$$x = 0 \quad \text{or} \quad x = 2$$

Let $f(x) = 2x$ and $g(x) = x^2$.
The area between the curves is given by

$$\int_0^2 [f(x) - g(x)]\,dx$$

$$= \int_0^2 (2x - x^2)\,dx$$

$$= \left(\frac{2x^2}{2} - \frac{x^3}{3}\right)\Bigg|_0^2$$

$$= 4 - \frac{8}{3} = \frac{4}{3}.$$

11. $x = 1$, $x = 6$, $y = \frac{1}{x}$, $y = -1$

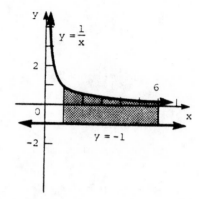

Find the points of intersection.

$$-1 = \frac{1}{x}$$

$$x = -1$$

The curves do not intersect in the interval $[1, 6]$.

If $f(x) = \frac{1}{x}$ and $g(x) = -1$, the area between the curves is

$$\int_1^6 [f(x) - g(x)]\,dx$$

$$= \int_1^6 \left[\frac{1}{x} - (-1)\right]dx$$

$$= \int_1^6 \left(\frac{1}{x} + 1\right)dx$$

$$= (\ln|x| + x)\Bigg|_1^6$$

$$= \ln 6 + 6 - \ln 1 - 1$$
$$= 5 + \ln 6$$
$$\approx 6.792.$$

13. $x = -1$, $x = 1$, $y = e^x$, $y = 3 - e^x$.

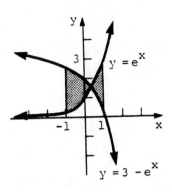

To find the point of intersection, set $e^x = 3 - e^x$ and solve for x.

$$e^x = 3 - e^x$$

$$2e^x = 3$$

$$e^x = \frac{3}{2}$$

$$\ln e^x = \ln \frac{3}{2}$$

$$x \ln e = \ln \frac{3}{2}$$

$$x = \ln \frac{3}{2}$$

The area of the region between the curves from $x = -1$ to $x = 1$ is

$$\int_{-1}^{\ln 3/2} [(3 - e^x) - e^x]dx$$

$$+ \int_{\ln 3/2}^{1} [e^x - (3 - e^x)]dx$$

$$= \int_{-1}^{\ln 3/2} (3 - 2e^x)dx + \int_{\ln 3/2}^{1} (2e^x - 3)dx$$

$$= (3x - 2e^x)\Big|_{-1}^{\ln 3/2} + (2e^x - 3x)\Big|_{\ln 3/2}^{1}$$

$$= \left[\left(3 \ln \frac{3}{2} - 2e^{\ln 3/2}\right) - (3(-1) - 2e^{-1})\right]$$

$$+ \left[2e^1 - 3(1) - \left(2e^{\ln 3/2} - 3 \ln \frac{3}{2}\right)\right]$$

$$= \left[\left(3 \ln \frac{3}{2} - 3\right) - \left(-3 - \frac{2}{e}\right)\right]$$

$$+ \left[2e - 3 - \left(3 - 3 \ln \frac{3}{2}\right)\right]$$

$$= 6 \ln \frac{3}{2} + \frac{2}{e} + 2e - 6$$

$$\approx 2.6051.$$

15. $x = 1$, $x = 2$, $y = e^x$, $y = \frac{1}{x}$

Sketch the graph of $y = e^x$ and $y = \frac{1}{x}$.

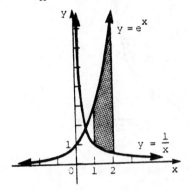

Let $f(x) = e^x$ and $g(x) = \frac{1}{x}$.

For all x in $[1, 2]$ $f(x) \geq g(x)$. The area between the curves is

$$\int_{1}^{2} [f(x) - g(x)]dx$$

$$= \int_{1}^{2} \left(e^x - \frac{1}{x}\right)dx$$

$$= (e^x - \ln |x|)\Big|_{1}^{2}$$

$$= e^2 - \ln 2 - e + \ln 1$$

$$= e^2 - e - \ln 2$$

$$\approx 3.978.$$

17. $y = x^3 - x^2 + x + 1$, $y = 2x^2 - x + 1$

Find the points of intersection.

$$x^3 - x^2 + x + 1 = 2x^2 - x + 1$$
$$x^3 - 3x^2 + 2x = 0$$
$$x(x^2 - 3x + 2) = 0$$
$$x(x - 2)(x - 1) = 0$$

So the points of intersection are at $x = 0$, $x = 1$, and $x = 2$.

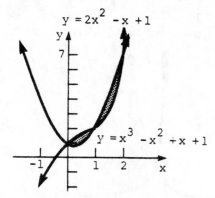

Area between the curves is

$$\int_0^1 [(x^3 - x^2 + x + 1) - (2x^2 - x + 1)]\,dx$$

$$+ \int_1^2 [(2x^2 - x + 1) - (x^3 - x^2 + x + 1)]\,dx$$

$$= \int_0^1 (x^3 - 3x^2 + 2x)\,dx + \int_1^2 (-x^3 + 3x^2 - 2x)\,dx$$

$$= \left(\frac{x^4}{4} - x^3 + x^2\right)\Big|_0^1 + \left(\frac{-x^4}{4} + x^3 - x^2\right)\Big|_1^2$$

$$= \left[\left(\frac{1}{4} - 1 + 1\right) - (0)\right]$$

$$+ \left[(-4 + 8 - 4) - \left(-\frac{1}{4} + 1 - 1\right)\right]$$

$$= \frac{1}{4} + \frac{1}{4}$$

$$= \frac{1}{2}.$$

19. $y = x^4 + \ln (x + 10)$,
$y = x^3 + \ln (x + 10)$

Find the points of intersection.

$$x^4 + \ln (x + 10) = x^3 + \ln (x + 10)$$
$$x^4 - x^3 = 0$$
$$x^3(x - 1) = 0$$
$$x = 0 \quad \text{or} \quad x = 1$$

The points of intersection are at $x = 0$ and $x = 1$.

The area between the curves is

$$\int_0^1 [(x^3 + \ln (x + 10)) - (x^4 + \ln (x + 10))]\,dx$$

$$= \int_0^1 (x^3 - x^4)\,dx$$

$$= \left(\frac{x^4}{4} - \frac{x^5}{5}\right)\Big|_0^1$$

$$= \left(\frac{1}{4} - \frac{1}{5}\right) - (0)$$

$$= \frac{1}{20}.$$

21. $y = x^{4/3}$, $y = 2x^{1/3}$

Find the points of intersection.

$$x^{4/3} = 2x^{1/3}$$
$$x^{4/3} - 2x^{1/3} = 0$$
$$x^{1/3}(x - 2) = 0$$
$$x = 0 \quad \text{or} \quad x = 2$$

The points of intersection are at $x = 0$ and $x = 2$.

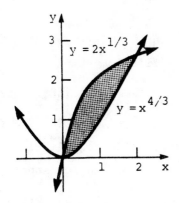

The area between the curves is

$$\int_0^2 (2x^{1/3} - x^{4/3})\,dx$$

$$= 2\frac{x^{4/3}}{4/3} - \frac{x^{7/3}}{7/3}\Big|_0^2$$

$$= \frac{3}{2}x^{4/3} - \frac{3}{7}x^{7/3}\Big|_0^2$$

$$= \left[\frac{3}{2}(2)^{4/3} - \frac{3}{7}(2)^{7/3}\right] - 0$$

$$= \frac{3(2^{4/3})}{2} - \frac{3(2^{7/3})}{7}$$

$$\approx 1.6199.$$

23.   (a) It is profitable to use the machine until $S(x) = C(x)$.

$$150 - x^2 = x^2 + \frac{11}{4}x$$

$$2x^2 + \frac{11}{4}x - 150 = 0$$

$$8x^2 + 11x - 600 = 0$$

$$x = \frac{-11 \pm \sqrt{121 - 4(8)(-600)}}{16}$$

$$= \frac{-11 \pm 139}{16} \quad \text{\textit{Discard negative solution}}$$

$$= 8 \text{ yr}$$

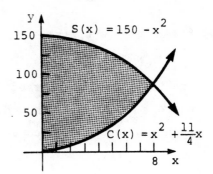

(b) Since $150 - x^2 > x^2 + \frac{11}{4}x$, in the interval $[0, 8]$, the net total savings are in the first year.

$$\int_0^1 \left[(150 - x^2) - \left(x^2 + \frac{11}{4}x\right)\right]dx$$

$$= \int_0^1 \left(-2x^2 - \frac{11}{4}x + 150\right)dx$$

$$= \left(\frac{-2x^3}{3} - \frac{11x^2}{8} + 150x\right)\Big|_0^1$$

$$= -\frac{2}{3} - \frac{11}{8} + 150$$

$$\approx \$148.$$

(c) The net total savings over the entire period of use are

$$\int_0^8 \left[(150 - x^2) - \left(x^2 + \frac{11}{4}x\right)\right]dx$$

$$= \left(\frac{-2x^3}{3} - \frac{11x^2}{8} + 150x\right)\Big|_0^8$$

$$= \frac{-2(8^3)}{3} - \frac{11(8^2)}{8} + 150(8)$$

$$= \frac{-1024}{3} - \frac{704}{8} + 1200$$

$$\approx \$771.$$

25.   **(a)** $E(x) = e^{.1x}$ and $I(x) = 98.8 - e^{.1x}$

To find the point of intersection, where profit will be maximized, set the functions equal to each other and solve for x.

$$e^{.1x} = 98.8 - e^{.1x}$$
$$2e^{.1x} = 98.8$$
$$e^{.1x} = 49.4$$
$$.1x = \ln 49.4$$
$$x = \frac{\ln 49.4}{.1}$$
$$x \approx 39 \text{ days}$$

(b) The total income for 39 days is

$$\int_0^{39} (98.8 - e^{.1x}) dx$$

$$= \left. \left(98.8x - \frac{e^{.1x}}{.1}\right) \right|_0^{39}$$

$$= \left. (98.8x - 10e^{.1x}) \right|_0^{39}$$

$$= [98.8(39) - 10e^{3.9}] - (0 - 10)$$
$$= \$3369.18$$

(c) The total expenditures for 39 days is

$$\int_0^{39} e^{.1x} dx = \left. \frac{e^{.1x}}{.1} \right|_0^{39}$$

$$= \left. 10e^{.1x} \right|_0^{39}$$

$$= 10e^{3.9} - 10$$
$$= \$484.02$$

(d) Profit = Income - Expense
$$= 3369.18 - 484.02$$
$$= \$2885.16$$

27. $S(q) = q^{5/2} + 2q^{3/2} + 50$, $q = 16$ is the equilibrium quantity.

$$\text{Producers' surplus} = \int_0^{q_0} [p_0 - S(q)] dq$$

where p_0 is the equilibrium price and q_0 is equilibrium supply.

$$p_0 = S(16) = (16)^{5/2} + 2(16)^{3/2} + 50$$
$$= 1202$$

Therefore, the producers' surplus is

$$\int_0^{16} [1202 - (q^{5/2} + 2q^{3/2} + 50] dq$$

$$= \int_0^{16} (1152 - q^{5/2} - 2q^{3/2}) dq$$

$$= \left. \left(1152q - \frac{2}{7}q^{7/2} - \frac{4}{5}q^{5/2}\right) \right|_0^{16}$$

$$= 1152(16) - \frac{2}{7}(16)^{7/2} - \frac{4}{5}(16)^{5/2}$$

$$= 18,432 - \frac{32,768}{7} - \frac{4096}{5}$$

$$= \$12,931.66.$$

29. $D(q) = \dfrac{100}{(3q + 1)^2}$, $q = 3$ is the equilibrium quantity.

$$\text{Consumers' surplus} = \int_0^{q_0} [D(q) - p_0] dq$$

$$p_0 = D(3) = 1$$

Therefore, the consumers' surplus is

$$\int_0^3 \left[\frac{100}{(3q + 1)^2} - 1 \right] dq$$

$$\int_0^3 \frac{100}{(3q + 1)^2} dq - \int 1 \, dq.$$

Let u = 3q + 1 so that

$$du = 3 \ dq \text{ and } \frac{1}{3} \ du = dq.$$

$$= \frac{1}{3} \int_1^{10} \frac{100}{u^2} \ du - \int_0^3 1 \ dq$$

$$= \frac{100}{3} \int_1^{10} u^{-2} \ du - \int_0^3 1 \ dq$$

$$= \frac{100}{3} \cdot \frac{u^{-1}}{-1} \Big|_1^{10} - q \Big|_0^3$$

$$= -\frac{100}{3u} \Big|_1^{10} - 3$$

$$= \frac{-100}{30} + \frac{100}{3} - 3$$

$$= 27$$

31. $S(q) = q^2 + 10q$

$D(q) = 900 - 20q - q^2$

(a) The graphs of the supply and demand functions are parabolas with vertices at (−5, −25) and (−10, 1900), respectively. See the answer graph at the back of the textbook.

(b) The graphs intersect at the point where the y-coordinates are equal.

$$q^2 + 10q = 900 - 20q - q^2$$
$$2q^2 + 30q - 900 = 0$$
$$q^2 + 15q - 450 = 0$$
$$(q + 30)(q - 15) = 0$$
$$q = -30 \quad \text{or} \quad q = 15.$$

Disregard the negative solution. The supply and demand functions are in equilibrium when q = 15.

$$S(15) = 15^2 + 10(15) = 375$$

The point is (15, 375).

(c) The consumers' surplus is

$$\int_0^{q_0} [(D(q) - p_0)] \ dq$$

$p_0 = D(15) = 375$

$$\int_0^{15} [(900 - 20q - q^2) - 375] \ dq$$

$$\int_0^{15} (525 - 20q - q^2) \ dq$$

$$= \left(525q - 10q^2 - \frac{1}{3}q^3\right) \Big|_0^{15}$$

$$= \left[525(15) - 10(15)^2 - \frac{1}{3}(15)^3\right] - 0$$

$$= 4500.$$

(d) The producers' surplus is

$$\int_0^{q_0} [p_0 - S(q)] \ dq$$

$p_0 = S(15) = 375$

$$\int_0^{15} [375 - (q^2 + 10q)] \ dq$$

$$\int_0^{15} (375 - q^2 - 10q) \ dq$$

$$= 375q - \frac{1}{3}q^3 - 5q^2 \Big|_0^{15}$$

$$= \left[375(15) - \frac{1}{3}(15)^3 - 5(15)^2\right] - 0$$

$$= 3375.$$

33. $I(x) = .9x^2 + .1x$

(a) $I(.1) = .9(.1)^2 + .1(.1)$
$$= .019$$

The lower 10% of income producers earn 1.9% of total income of the population.

(b) $I(.4) = .9(.4)^2 + .1(.4) = .184$

The lower 40% of income producers earn 18.4% of income of the population.

(c) $I(.6) = .9(.6)^2 + .1(.6) = .384$

The lower 60% of income producers earn 38.4% of total income of the population.

(d) $I(.9) = .9(.9)^2 + .1(.9) = .819$

The lower 90% of income producers earn 81.9% of total income of the population.

(e) See the answer graph in your textbook.

(f) To find points of intersection, solve

$$x = .9x^2 + .1x$$
$$.9x^2 - .9x = 0$$
$$.9x(x - 1) = 0$$
$$x = 0 \quad \text{or} \quad x = 1$$

The area between the curves is given by

$$\int_0^1 [x - (.9x^2 + .1x)]\,dx$$

$$= \int_0^1 (.9x - .9x^2)\,dx$$

$$= \left(\frac{.9x^2}{2} - \frac{.9x^3}{3}\right)\Bigg|_0^1$$

$$= \frac{.9}{2} - \frac{.9}{3}$$

$$= .15.$$

For Exercises 35 and 37, use computer methods. The solutions will vary depending on the computer program used. The answers are given.

35. $y = \ln x$ and $y = xe^x$; $[1, 4]$
    The area is 161.2.

37. $y = \sqrt{9 - x^2}$ and $y = \sqrt{x + 1}$; $[-1, 3]$
    The area is 5.516.

Chapter 14   Review Exercises

5. $\int 6\,dx = 6x + C$

7. $\int (2x + 3)\,dx = \dfrac{2x^2}{2} + 3x + C$
$$= x^2 + 3x + C$$

9. $\int (x^2 - 3x + 2)\,dx$
$$= \frac{x^3}{3} - \frac{3x^2}{2} + 2x + C$$

11. $\int 3\sqrt{x}\,dx = 3\int x^{1/2}\,dx$
$$= \frac{3x^{3/2}}{3/2} + C$$
$$= 2x^{3/2} + C$$

13. $\int (x^{1/2} + 3x^{-2/3})\,dx$
$$= \frac{x^{3/2}}{3/2} + \frac{3x^{1/3}}{1/3} + C$$
$$= \frac{2x^{3/2}}{3} + 9x^{1/3} + C$$

15. $\displaystyle\int \frac{-4}{x^3}\, dx = \int -4x^{-3}\, dx$

$$= \frac{-4x^{-2}}{-2} + C$$

$$= 2x^{-2} + C$$

17. $\displaystyle\int -3e^{2x}\, dx = \frac{-3e^{2x}}{2} + C$

19. $\displaystyle\int \frac{2}{x-1}\, dx = 2\int \frac{dx}{x-1}$

Let $u = x - 1$, so that

$\quad du = dx.$

$$= 2\int \frac{du}{u}$$

$$= 2\ln |u| + C$$

$$= 2\ln |x-1| + C$$

21. $\displaystyle\int xe^{3x^2}\, dx = \frac{1}{6}\int 6xe^{3x^2}\, dx$

Let $u = 3x^2$, so that

$\quad du = 6x\, dx.$

$$= \frac{1}{6}\int e^u\, du$$

$$= \frac{1}{6}e^u + C$$

$$= \frac{e^{3x^2}}{6} + C$$

23. $\displaystyle\int \frac{3x}{x^2-1}\, dx = 3\left(\frac{1}{2}\right)\int \frac{2x\, dx}{x^2-1}$

Let $u = x^2 - 1$, so that

$\quad du = 2x\, dx.$

$$= \frac{3}{2}\int \frac{du}{u}$$

$$= \frac{3}{2}\ln |u| + C$$

$$= \frac{3\ln |x^2-1|}{2} + C$$

25. $\displaystyle\int \frac{x^2\, dx}{(x^3+5)^4} = \frac{1}{3}\int \frac{3x^2\, dx}{(x^3+5)^4}$

Let $u = x^3 + 5$, so that

$\quad du = 3x^2\, dx.$

$$= \frac{1}{3}\int \frac{du}{u^4}$$

$$= \frac{1}{3}\int u^{-4}\, du$$

$$= \frac{1}{3}\left(\frac{u^{-3}}{-3}\right) + C$$

$$= \frac{-(x^3+5)^{-3}}{6} + C$$

27. $\displaystyle\int \frac{4x-5}{2x^2-5x}\, dx$

Let $u = 2x^2 - 5x$, so that

$\quad du = (4x-5)dx.$

$$= \int \frac{du}{u}$$

$$= \ln |u| + C$$

$$= \ln |2x^2-5x| + C$$

29. $\displaystyle\int \frac{x^3}{e^{3x^4}}\, dx = \int x^3 e^{-3x^4}$

$$= -\frac{1}{12}\int -12x^3 e^{-3x^4}\, dx$$

Let $u = -3x^4$, so that

$\quad du = -12x^3\, dx.$

$$= -\frac{1}{12}\int e^u\, du$$

$$= -\frac{1}{12}e^u + C$$

$$= \frac{-e^{-3x^4}}{12} + C$$

31. $\int -2e^{-5x}\ dx = \frac{-2}{-5}\int -5e^{-5x}\ dx$

Let $u = 5x$, so that
$du = -5\ dx$.

$$= \frac{2}{5}\int e^u\ du$$

$$= \frac{2}{5}e^u + C$$

$$= \frac{2e^{-5x}}{5} + C$$

33. $\sum_{i=1}^{4} (i^2 - i)$

$= (1^2 - 1) + (2^2 - 2) + (3^2 - 3)$

$+ (4^2 - 4)$

$= 0 + 2 + 6 + 12$

$= 20$

35. $f(x) = 2x + 3$, from $x = 0$ to $x = 4$

$$\Delta x = \frac{4 - 0}{4} = 1$$

i	x_i	$f(x_i)$
1	0	3
2	1	5
3	2	7
4	3	9

$A = \sum_{i=1}^{4} f(x_i)\Delta x$

$= 3(1) + 5(1) + 7(1) + 9(1)$

$= 24$

39. $\int_{1}^{6} (2x^2 + x)\,dx$

$$= \left(\frac{2x^3}{3} + \frac{x^2}{2}\right)\Big|_{1}^{6}$$

$$= \left[\frac{2(6)^3}{3} + \frac{(6)^2}{2}\right] - \left[\frac{2(1)^3}{3} + \frac{(1)^2}{2}\right]$$

$$= 144 + 18 - \frac{2}{3} - \frac{1}{2}$$

$$= 162 - \frac{2}{3} - \frac{1}{2}$$

$$= \frac{965}{6}$$

$$\approx 160.83$$

41. $\int_{2}^{3} (5x^{-2} + x^{-4})\,dx$

$$= \left(\frac{5x^{-1}}{-1} + \frac{x^{-3}}{-3}\right)\Big|_{2}^{3}$$

$$= \left(\frac{-5}{x} - \frac{1}{3x^3}\right)\Big|_{2}^{3}$$

$$= \left(\frac{-5}{3} - \frac{1}{81}\right) - \left(\frac{-5}{2} - \frac{1}{24}\right)$$

$$= \frac{-5}{3} - \frac{1}{81} + \frac{5}{2} + \frac{1}{24}$$

$$= \frac{-1080 - 8 + 1620 + 27}{648}$$

$$= \frac{559}{648}$$

$$\approx .863$$

43. $\int_{1}^{6} 8x^{-1}\ dx = \int_{1}^{6} \frac{8}{x}\ dx$

$$= 8(\ln x)\Big|_{1}^{6}$$

$$= 8(\ln 6 - \ln 1)$$

$$= 8\ln 6$$

$$\approx 14.334$$

45. $\displaystyle\int_{1}^{6} \frac{5}{2}e^{4x}\ dx = \frac{1}{4} \cdot \frac{5}{2} \int_{1}^{6} 4e^{4x}\ dx$

$$= \frac{5e^{4x}}{8}\Big|_{1}^{6}$$

$$= \frac{5(e^{24} - e^{4})}{8}$$

$$\approx 1.656 \times 10^{10}$$

47. $\displaystyle\int_{0}^{1} x\sqrt{5x^2 + 4}\ dx$

Let $u = 5x^2 + 4$, so that

$du = 10x\ dx$ and $\dfrac{1}{10}\ du = x\ dx$.

When $x = 0$, $u = 5(0^2) + 4 = 4$.

When $x = 1$, $u = 5(1^2) + 4 = 9$.

$$= \frac{1}{10} \int_{4}^{9} \sqrt{u}\ du = \frac{1}{10} \int_{4}^{9} u^{1/2}\ du$$

$$= \frac{1}{10} \cdot \frac{u^{3/2}}{3/2}\Big|_{4}^{9} = \frac{1}{15}u^{3/2}\Big|_{4}^{9}$$

$$= \frac{1}{15}(9)^{3/2} - \frac{1}{15}(4)^{3/2}$$

$$= \frac{27}{15} - \frac{8}{15}$$

$$= \frac{19}{15}$$

49. $f(x) = x(x + 2)^6;\ [-2,\ 0]$

$$\text{Area} = \int_{-2}^{0} [0 - x(x + 2)^6]\,dx$$

$$= -\int_{-2}^{0} x(x + 2)^6\ dx$$

Let $u = x + 2$, so that

$du = dx$ and $x = u - 2$.

When $x = -2$, $u = 0$.

When $x = 0$, $u = 2$.

$$= -\int_{0}^{2} (u - 2)u^6\ du$$

$$= -\int_{0}^{2} u^7\ du + 2\int_{0}^{2} u^6\ du$$

$$= -\frac{u^8}{8}\Big|_{0}^{2} + 2 \cdot \frac{u^7}{7}\Big|_{0}^{2}$$

$$= -\frac{256}{8} + \frac{256}{7}$$

$$= -32 + \frac{256}{7}$$

$$= \frac{32}{7}$$

51. $f(x) = 1 + e^{-x};\ [0,\ 4]$

$$\int_{0}^{4} (1 + e^{-x})dx = (x - e^{-x})\Big|_{0}^{4}$$

$$= (4 - e^{-4}) - (0 - e^{0})$$

$$= 5 - e^{-4}$$

$$\approx 4.982$$

53. $f(x) = x^2 - 4x,\ g(x) = x - 6$

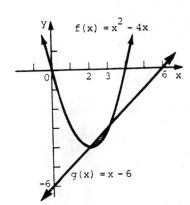

Points of intersection:

$$x^2 - 4x = x - 6$$
$$x^2 - 5x + 6 = 0$$
$$(x - 3)(x - 2) = 0$$
$$x = 2 \quad \text{or} \quad x = 3$$

Since $g(x) \geq f(x)$ in the interval [2, 3], the area between the graphs is

$$\int_2^3 [g(x) - f(x)]dx$$

$$= \int_2^3 [(x - 6) - (x^2 - 4x)]dx$$

$$= \int_2^3 (-x^2 + 5x - 6)dx$$

$$= \left(\frac{-x^3}{3} + \frac{5x^2}{2} - 6x\right)\Big|_2^3$$

$$= \frac{-27}{3} + \frac{5(9)}{2} - 6(3) - \frac{-8}{3}$$

$$\quad - \frac{5(4)}{2} + 6(2)$$

$$= -\frac{19}{3} + \frac{25}{2} - 6$$

$$= \frac{1}{6}$$

55. $f(x) = 5 - x^2$, $g(x) = x^2 - 3$, $x = 0$, $x = 4$

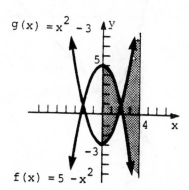

Find the points of intersection.

$$5 - x^2 = x^2 - 3$$
$$8 = 2x^2$$
$$4 = x^2$$
$$\pm 2 = x$$

So the curves intersect at $x = 2$, -2.

Thus, the area

$$= \int_0^2 [(5 - x^2) - (x^2 - 3)]dx$$

$$\quad + \int_2^4 [(x^2 - 3) - (5 - x^2)]dx$$

$$= \int_0^2 (-2x^2 + 8)dx + \int_2^4 (2x^2 - 8)dx$$

$$= \left(\frac{-2x^3}{3} + 8x\right)\Big|_0^2 + \left(\frac{2x^3}{3} - 8x\right)\Big|_2^4$$

$$= \frac{-16}{3} + 16 + \left(\frac{128}{3} - 32\right) - \left(\frac{16}{3} - 16\right)$$

$$= \frac{32}{3} + \frac{128}{3} - 32 - \frac{16}{3} + 16$$

$$= 32.$$

57. $C'(x) = 10 - 2x$; fixed cost is \$4

$$C(x) = \int (10 - 2x)dx$$

$$= 10x - \frac{2x^2}{2} + C$$

$$= 10x - x^2 + C$$
$$C(0) = 10(0) - 0^2 + C$$

Since $C(0) = 4$,

$$C = 4.$$
$$C(x) = 10x - x^2 + 4$$

59.
$$C(x) = \int 3(2x - 1)^{1/2}\, dx$$
$$= \frac{3}{2} \int 2(2x - 1)^{1/2}\, dx$$

Let $u = 2x - 1$, so that

$du = 2\, dx$.

$$= \frac{3}{2} \int u^{1/2}\, du$$
$$= \frac{3}{2}\left(\frac{u^{3/2}}{3/2}\right) + C$$
$$= (2x - 1)^{3/2} + C$$
$$C(13) = [2(13) - 1]^{3/2} + C$$

Since $C(13) = 270$,

$$270 = 25^{3/2} + C$$
$$270 = 125 + C$$
$$C = 145.$$
$$C(x) = (2x - 1)^{3/2} + 145$$

61. Read values for the rate of investment income accumulation for every 2 year from year 1 to yr 9. These are the heights of rectangles with width $\Delta x = 2$.

Total accumulated income
$$= 11{,}000(2) + 9000(2) + 12{,}000(2)$$
$$+ 10{,}000(2) = 6000(2)$$
$$\approx \$96{,}000$$

63. $S' = \sqrt{x} + 2$
$$S(x) = \int_0^9 (x^{1/2} + 2)\, dx$$
$$= \left(\frac{x^{3/2}}{3/2} + 2x\right)\Big|_0^9$$
$$= \frac{2}{3}(9)^{3/2} + 18$$
$$= 36$$

Total sales are 36(1000), or 36,000 units.

65. $S'(x) = 225 - x^2$, $C'(x) = x^2 + 25x + 150$
$$S'(x) = C'(x)$$
$$225 - x^2 = x^2 + 25x + 150$$
$$2x^2 + 25x - 75 = 0$$
$$(2x - 5)(x + 15) = 0$$
$$x = \frac{5}{2} = 2.5 \text{ yr}$$

$$\int_0^{2.5} [225 - x^2) - (x^2 + 25x + 150)]dx$$
$$= \int_0^{2.5} (-2x^2 - 25x + 75)dx$$
$$= \left(\frac{-2x^3}{3} - \frac{25x^2}{2} + 75x\right)\Big|_0^{2.5}$$
$$= \frac{-2(2.5^3)}{3} - \frac{25(2.5^2)}{2} + 75(2.5)$$
$$\approx 98.95833$$

Net savings are 98,958.33, or about \$99,000.

67. The total number of infected people over the first four months is
$$\int_0^4 \frac{100t}{t^2 + 1}\, dt, \text{ where } t \text{ is time in}$$
months.

Let $u = t^2 + 1$, so that

$du = 2t\, dt$ and $50\, du = 100t\, dt$

$$= 50 \int_{1}^{17} \frac{1}{u} \, du = 50 \ln |u| \Big|_{1}^{17}$$

$$= 50 \ln 17 - 50 \ln |1|$$

$$= 50 \ln 17$$

$$\approx 141.66$$

Approximately 142 people are infected.

69. For each month, subtract the average temperature from 65° if it falls below 65° F, then multiply this number times the number of days in the month. The sum is the total number of heating degree days. Readings may vary, but the sum is approximately 4500 degree-days. (The actual value is 4868 degree-days.)

3.
$$15,000,000 = \frac{63,000}{.06}(e^{.06t_1} - 1)$$

$$\frac{15,000,000(.06)}{63,000} + 1 = e^{.06t_1}$$

$$15.286 \approx e^{.06t_1}$$

$$\ln 15.286 \approx .06t_1 \ln e$$

$$t_1 \approx \frac{\ln 15.286}{.06}$$

$$\approx 45.4$$

or about 45.4 yr

4.
$$2,000,000 = \frac{2200}{.04}(e^{.04t_1} - 1)$$

$$\frac{2,000,000(.04)}{2200} + 1 = e^{.04t_1}$$

$$37.36 = e^{.04t_1}$$

$$\ln 37.36 = .04t_1$$

$$t_1 \approx 90.5$$

or about 90 yr

Extended Application

1. $2,000,000 \div 19,600 \approx 102$ yr

2.
$$2,000,000 = \frac{19,600}{.02}(e^{.02t_1} - 1)$$

$$\frac{2,000,000(.02)}{19,600} = e^{.02t_1} - 1$$

$$2.0408 + 1 = e^{.02t_1}$$

$$3.0408 = e^{.02t_1}$$

$$\ln 3.0408 = .02t_1$$

$$t_1 = \frac{\ln 3.0408}{.02}$$

$$t_1 \approx 55.6,$$

or about 55.6 yr

CHAPTER 14   TEST

1. Explain what an antiderivative of a function f(x) is.

Find each indefinite integral.

2. $\int (3x^3 - 5x^2 + x + 1)\,dx$ 3. $\int \left(\frac{5}{x} + e^{\cdot\,5x}\right) dx$

4. $\int \dfrac{2x^3 - 3x^2}{\sqrt{x}}\,dx$

5. Find the cost function if $C'(x) = 200 + 2x^{-1/4}$ and 16 units cost \$4000.

6. A ball is thrown upward at time t = 0 with initial velocity of 64 ft/sec from a height of 100 ft. Assume that a(t) = −32 ft/sec². Find v(t) and s(t).

Use substitution to find each indefinite integral.

7. $\int 6x^2 (3x^3 - 5)^8 \, dx$ 8. $\int \dfrac{6x + 5}{3x^2 + 5x}\,dx$

9. $\int \sqrt[3]{2x^2 - 8x}\,(x - 2)\,dx$ 10. $\int 4x^3 e^{-x^4}\,dx$

11. A city's population is predicted to grow at a rate of

$$P'(x) = \frac{400e^{10t}}{1 + e^{10t}}$$

people per year where t is the time in years from the present. Find the total population 3 yr from now if P(0) = 100,000.

12. Evaluate $\displaystyle\sum_{i=1}^{4} \frac{2}{i^2}$.

13. Approximate the area under the graph of $f(x) = x^2 + x$ and above the x-axis from x = 0 to x = 2 using 4 rectangles. Let the height of each rectangle be the function value of the left side.

14. Approximate the value of $\int_1^5 x^2 \, dx$ by using the formula for the area

of a trapezoid, $A = \frac{1}{2}(B + b)h$, where B and b are the lengths of the

parallel sides and h is the distance between them.

Evaluate the following definite integrals.

15. $\int_1^5 (4x^3 - 5x) \, dx$

16. $\int_1^5 \left(\frac{3}{x^2} + \frac{2}{x}\right) dx$

17. $\int_1^2 \sqrt{3r - 2} \, dr$

18. $\int_0^1 4xe^{x^2 + 1} \, dx$

19. $\int_{-2}^1 3x(x^2 - 4)^5 \, dx$

20. Find the area of the region between the x-axis and $f(x) = e^{x/2}$ on the interval $[0, 2]$.

21. The rate at which a substance grows is given by $R(x) = 500e^{.5x}$, where x is the time in days. What is the total accumulated growth after 4 days?

22. Find the area of the region enclosed by $f(x) = -x + 4$, $g(x) = -x^2 + 6x - 6$, $x = 2$, and $x = 4$.

23. Find the area of the region enclosed by $f(x) = 5x$ and $g(x) = x^3 - 4x$.

24. A company has determined that the use of a new process would produce a savings rate (in thousands of dollars) of

$$S(x) = 2x + 7,$$

where x is the number of years the process is used. However, the use of this process also creates additional costs (in thousands of dollars) according to the rate-of-cost function

$$C(x) = x^2 + 2x + 3.$$

(a) For how many years does the new process save the company money?

(b) Find the net savings in thousands of dollars over this period.

25. Suppose that the supply function of a commodity is p = .05q + 5 and the demand function is p = 12 - .02q.

    (a) Find the producers' surplus.

    (b) Find the consumers' surplus.

CHAPTER 14   TEST ANSWERS

1. An antiderivative of $f(x)$ is $F(x)$ such that $F'(x) = f(x)$.

2. $\frac{3}{4}x^4 - \frac{5}{3}x^3 + \frac{x^2}{2} + x + C$

3. $5 \ln |x| + 2e^{.5x} + C$

4. $\frac{4}{7}x^{7/2} - \frac{6}{5}x^{5/2} + C$

5. $C(x) = 200x + \frac{8}{x}x^{3/4} + 778.67$

6. $v(t) = -32t + 64$; $s(t) = -16t^2 + 64t + 100$

7. $\frac{2}{27}(3x^3 - 5)^9 + C$

8. $\ln |3x^2 + 5x| + C$

9. $\frac{3}{16}(2x^2 - 8x)^{4/3} + C$

10. $-e^{-x^4} + C$

11. 101,172

12. $2\frac{61}{72}$ or 2.85

13. 3.25   14. 52   15. 564   16. 5.62   17. $\frac{14}{9}$ or 1.56

18. $2e^2 - 2e$ or 9.34   19. 182.25   20. $2e - 2$ or 3.44

21. 6389   22. 3.33   23. 40.5

24. (a) 2 yr   (b) $5.33 thousand   25. (a) 250   (b) 100

CHAPTER 15   FURTHER TECHNIQUES AND
APPLICATIONS OF INTEGRATION

Section 15.1

1. $\int xe^x \, dx$

Let $dv = e^x \, dx$ and $u = x$.

Then $v = \int e^x \, dx$ and $du = dx$.

$$v = e^x + C$$

Use the formula

$$\int u \, dv = uv - \int v \, du.$$

$$\int xe^x \, dx = xe^x - \int e^x \, dx$$

$$= xe^x - e^x + C$$

3. $\int (5x - 9)e^{-3x} \, dx$

Let $dv = e^{-3x} \, dx$ and $u = 5x - 9$.

Then $v = \int e^{-3x} \, dx$ and $du = 5 \, dx$.

$$v = \frac{e^{-3x}}{-3}$$

$$\int (5x - 9)e^{-3x} \, dx$$

$$= \frac{-(5x - 9)e^{-3x}}{3} + \int \frac{5}{3} e^{-3x} \, dx$$

$$= \frac{-5xe^{-3x}}{3} + 3e^{-3x} + \frac{5e^{-3x}}{-9} + C$$

$$= \frac{-5xe^{-3x}}{3} - \frac{5e^{-3x}}{9} + 3e^{-3x} + C$$

or $\frac{-5xe^{-3x}}{3} + \frac{22e^{-3x}}{9} + C$

5. $\int_0^1 \frac{2x + 1}{e^x} \, dx$

$$= \int_0^1 (2x + 1)e^{-x} \, dx$$

Let $dv = e^{-x} \, dx$ and $u = 2x + 1$.

Then $v = \int e^{-x} \, dx$ and $du = 2 \, dx$

$$v = -e^{-x}$$

$$\int \frac{2x + 1}{e^x} \, dx$$

$$= -(2x + 1)e^{-x} + \int 2e^{-x} \, dx$$

$$= -(2x + 1)e^{-x} - 2e^{-x}$$

$$\int_0^1 \frac{2x + 1}{e^x} \, dx$$

$$= \left[-(2x + 1)e^{-x} - 2e^{-x}\right]\Big|_0^1$$

$$= \left[-(3)e^{-1} - 2e^{-1}\right] - (-1 - 2)$$

$$= -5e^{-1} + 3$$

$$\approx 1.1606$$

7. $\int_1^4 \ln 2x \, dx$

Let $dv = dx$ and $u = \ln 2x$.

Then $v = x$ and $du = \frac{1}{x}$.

$$\int \ln 2x \, dx = x \ln 2x - \int dx$$

$$\int_1^4 \ln 2x \, dx$$

$$= (x \ln 2x - x)\Big|_1^4$$

$$= (4 \ln 8 - 4) - (\ln 2 - 1)$$

$$= 4 \ln 2^3 - 4 - \ln 2 + 1$$

$$= 12 \ln 2 - \ln 2 - 3$$

$$= 11 \ln 2 - 3$$

$$\approx 4.6246$$

9. $\int x \ln x \, dx$

Let $dv = x \, dx$ and $u = \ln x$.

Then $v = \dfrac{x^2}{2}$ and $du = \dfrac{1}{x} \, dx$.

$\int x \ln x \, dx = \dfrac{x^2}{2} \ln x - \int \dfrac{x}{2} \, dx$

$= \dfrac{x^2 \ln x}{2} - \dfrac{x^2}{4} + C$

11. The area is $\displaystyle\int_{2}^{4} (x - 2)e^x \, dx$.

Let $dv = e^x \, dx$ and $u = x - 2$.

Then $v = e^x$ and $du = dx$.

$\int (x - 2)e^x \, dx$

$= (x - 2)e^x \int e^x \, dx$

$\displaystyle\int_{2}^{4} (x - 2)e^x \, dx$

$= \left[(x - 2)e^x - e^x \right] \Big|_{2}^{4}$

$= (2e^4 - e^4) - (0 - e^2)$

$= e^4 + e^2$

≈ 61.9872

13. $\int x^2 e^{2x} \, dx$

Let $u = x^2$ and $dv = e^{2x} \, dx$.

Use column integration.

D	I
x^2	e^{2x}
$2x$	$e^{2x}/2$
2	$e^{2x}/4$
0	$e^{2x}/8$

$\int x^2 e^{2x} \, dx$

$= x^2(e^{2x}/2) - 2x(e^{2x}/4)$
$\quad + 2e^{2x}/8 + C$

$= \dfrac{x^2 e^{2x}}{2} - \dfrac{xe^{2x}}{2} + \dfrac{e^{2x}}{4} + C$

15. $\displaystyle\int_{0}^{5} x\sqrt[3]{x^2 + 2} \, dx$

$= \displaystyle\int_{0}^{5} x(x^2 + 2)^{1/3} \, dx$

$= \dfrac{1}{2} \displaystyle\int_{0}^{5} 2x(x^2 + 2)^{1/3} \, dx$

Let $u = x^2 + 2$. Then $du = 2x \, dx$.

If $x = 5$, $u = 27$

$\quad x = 0$, $u = 2$.

$= \dfrac{1}{2} \displaystyle\int_{2}^{27} u^{1/3} \, du$

$= \dfrac{1}{2}\left(\dfrac{u^{4/3}}{1}\right)\left(\dfrac{3}{4}\right) \Big|_{2}^{27}$

$= \dfrac{3}{8}(27)^{4/3} - \dfrac{3}{8}(2)^{4/3}$

$= \dfrac{243}{8} - \dfrac{3(2^{4/3})}{8}$

$= \dfrac{243}{8} - \dfrac{3\sqrt[3]{2}}{4} \approx 29.4301$

17. $\int (8x + 7) \ln (5x) \, dx$

Let $dv = (8x + 7)dx$ and $u = \ln 5x$.

Then $v = \dfrac{8x^2}{2} + 7x$ and $du = \dfrac{1}{5x}(5)dx$

$= 4x^2 + 7x \qquad\qquad = \dfrac{1}{x} \, dx$.

$$\int (8x + 7) \ln (5x)dx$$

$$= (4x^2 + 7x) \ln 5x - \int (4x + 7)dx$$

$$= (4x^2 + 7x) \ln 5x - \left(\frac{4x^2}{2} + 7x\right) + C$$

$$= 4x^2 \ln (5x) + 7x \ln (5x)$$
$$- 2x^2 - 7x + C$$

19. $\int x^2\sqrt{x + 2}\ dx$

Let $u = x^2$ and $dv = (x + 2)^{1/2}\ dx$.
Use column integration.

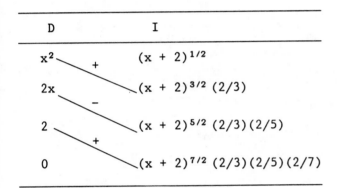

$$\int x^2\sqrt{x + 2}\ dx$$

$$= x^2(x + 2)^{3/2}(2/3)$$
$$- 2x(x + 2)^{5/2}(2/3)(2/5)$$
$$+ 2(x + 2)^{7/2}(2/3)(2/5)(2/7) + C$$
$$= \frac{2x^2(x + 2)^{3/2}}{3} - \frac{8x(x + 2)^{5/2}}{15}$$
$$+ \frac{16(x + 2)^{7/2}}{105} + C$$

21. $\int_0^1 \frac{x^3\ dx}{\sqrt{3 + x^2}}$

$$= \int_0^1 x^3(3x + x^2)^{-1/2}\ dx$$

Let $dv = x(3 + x^2)^{-1/2}\ dx$ and $u = x^2$.

Then $v = \dfrac{2(3 + x^2)^{1/2}}{2}$ and $du = 2x\ dx$.

$$v = (3 + x^2)^{1/2}$$

$$\int \frac{x^3\ dx}{\sqrt{3 + x^2}}$$

$$= x^2(3 + x^2)^{1/2} - \int 2x(3 + x^2)^{1/2}\ dx$$

$$= x^2(3 + x^2)^{1/2} - \frac{2}{3}(3 + x^2)^{3/2}$$

$$\int_0^1 \frac{x^3\ dx}{\sqrt{3 + x^2}}$$

$$= \left[x^2(3 + x^2)^{1/2} - \frac{2}{3}(3 + x^2)^{3/2}\right]\Bigg|_0^1$$

$$= 4^{1/2} - \frac{2}{3}(4^{3/2}) - 0 + \frac{2}{3}(3^{3/2})$$

$$= 2 - \frac{2}{3}(8) + \frac{2}{3}(3^{3/2})$$

$$\approx .13077$$

23. $\int \dfrac{-4}{\sqrt{x^2 + 36}}\ dx$

$$= -4 \int \frac{dx}{\sqrt{x^2 + 36}}$$

If $a = 6$, this integral matches
entry 5 in the table.

$$= -4 \ln \left|x + \sqrt{x^2 + 36}\right| + C$$

25. $\int \dfrac{6}{x^2 - 9}\ dx$

$$= 6 \int \frac{1}{x^2 - 9}\ dx$$

Use entry 8 from table with $a = 3$

$$= 6\left[\frac{1}{2(3)} \ln \left|\frac{x - 3}{x + 3}\right|\right] + C$$

$$= \ln \left|\frac{x - 3}{x + 3}\right| + C$$

27. $\int \dfrac{-4}{x\sqrt{9-x^2}}\, dx$

$= -4 \int \dfrac{dx}{x\sqrt{9-x^2}}$

Use entry 9 from table with a = 3.

$= -4\left(-\dfrac{1}{3}\ln\left|\dfrac{3+\sqrt{9-x^2}}{x}\right|\right) + C$

$= \dfrac{4}{3}\ln\left|\dfrac{3+\sqrt{9-x^2}}{x}\right| + C$

Let u = 2x.

Then du = 2 dx.

$= 2 \int \dfrac{du}{u^2 - 1}$

Use entry 8 from table with a = 1.

$= 2 \cdot \dfrac{1}{2}\ln\left|\dfrac{u-1}{u+1}\right| + C$

Substitute 2x for u.

$= \ln\left|\dfrac{2x-1}{2x+1}\right| + C$

29. $\int \dfrac{-2x}{3x+1}\, dx$

$= -2 \int \dfrac{x\, dx}{3x+1}$

Use entry 11 from table with a = 3, b = 1.

$= -2\left(\dfrac{x}{3} - \dfrac{1}{9}\ln|3x+1|\right) + C$

$= -\dfrac{2x}{3} + \dfrac{2}{9}\ln|3x+1| + C$

35. $\int \dfrac{3}{x\sqrt{1-9x^2}}\, dx$

$= 3 \int \dfrac{3\, dx}{3x\sqrt{1-9x^2}}$

Let   u = 3x.

Then du = 3 dx.

$= 3 \int \dfrac{du}{u\sqrt{1-u^2}}$

Use entry 9 from table with a = 1.

$= -3\ln\left|\dfrac{1+\sqrt{1-u^2}}{u}\right| + C$

$= -3\ln\left|\dfrac{1+\sqrt{1-9x^2}}{3x}\right| + C$

31. $\int \dfrac{2}{3x(3x-5)}\, dx$

$= \dfrac{2}{3} \int \dfrac{1}{x(3x-5)}\, dx$

Use entry 13 from table with a = 3, b = -5.

$= \dfrac{2}{3}\left(-\dfrac{1}{5}\ln\left|\dfrac{x}{3x-5}\right|\right) + C$

$= -\dfrac{2}{15}\ln\left|\dfrac{x}{3x-5}\right| + C$

37. $\int \dfrac{4x}{2x+3}\, dx$

$= 4 \int \dfrac{x}{2x+3}\, dx$

Use entry 11 from table with a = 2, b = 3.

$= 4\left(\dfrac{x}{2} - \dfrac{3}{4}\ln|2x+3|\right) + C$

$= 2x - 3\ln|2x+3| + C$

33. $\int \dfrac{4}{4x^2-1}\, dx$

$= 4 \int \dfrac{dx}{4x^2-1}$

$= 2 \int \dfrac{2\, dx}{4x^2-1}$

39. $\int \dfrac{-x}{(5x - 1)^2} \, dx$

$= -\int \dfrac{x \, dx}{(5x - 1)^2}$

Use entry 12 from table with a = 5, b = -1.

$= -\left[\dfrac{-1}{25(5x - 1)} + \dfrac{1}{25} \ln |5x - 1|\right] + C$

$= \dfrac{1}{25(5x - 1)} - \dfrac{\ln |5x - 1|}{25} + C$

41. $\int x^n \cdot \ln |x| \, dx$

Let $u = \ln |x|$ and $dv = x^n dx$.

Use column integration.

D	I
$\ln |x|$	x^n
$\dfrac{1}{x}$	$\dfrac{1}{n + 1}x^{n + 1}$

$\int x^n \cdot \ln |x| \, dx$

$= \dfrac{1}{n + 1}x^{n + 1} \ln |x| - \int \left[\dfrac{1}{x} \cdot \dfrac{1}{n + 1}x^{n + 1}\right] dx$

$= \dfrac{1}{n + 1}x^{n + 1} \ln |x| - \int \dfrac{1}{n + 1}x^n \, dx$

$= \dfrac{1}{n + 1}x^{n + 1} \ln |x| - \dfrac{1}{(n + 1)^2}x^{n + 1} + C$

$= x^{n + 1}\left[\dfrac{\ln |x|}{n + 1} - \dfrac{1}{(n + 1)^2}\right] + C$

Notice that $n \neq -1$.

43. $\int x\sqrt{x + 1} \, dx$

(a) Let $u = x$ and $dv = \sqrt{x + 1} \, dx$. Use column integration.

D	I
x	$\sqrt{x + 1}$
1	$\left(\dfrac{2}{3}\right)(x + 1)^{3/2}$
0	$\left(\dfrac{4}{15}\right)(x + 1)^{5/2}$

$\int x\sqrt{x + 1} \, dx$

$= \left(\dfrac{2}{3}\right)x(x + 1)^{3/2} - \left(\dfrac{4}{15}\right)(x + 1)^{5/2} + C$

(b) Let $u = x + 1$; then $u - 1 = x$ and $du = dx$.

$\int x\sqrt{x + 1} \, dx$

$= \int (u - 1)u^{1/2} \, du$

$= \int (u^{3/2} - u^{1/2}) \, du$

$= \left(\dfrac{2}{5}\right)u^{5/2} - \left(\dfrac{2}{3}\right)u^{3/2} + C$

$= \left(\dfrac{2}{5}\right)(x + 1)^{5/2} - \left(\dfrac{2}{3}\right)(x + 1)^{3/2} + C$

45. $r(x) = \displaystyle\int_1^6 2x^2 e^{-x} \, dx$

Let $dv = e^{-x} dx$ and $u = 2x^2$.

Then $v = -e^{-x}$ and $du = 4x \, dx$.

Use column integration.

D	I
$2x^2$	e^{-x}
$4x$	$-e^{-x}$
4	e^{-x}
0	$-e^{-x}$

$\int 2x^2 e^{-x} \, dx$

$= 2x^2(-e^{-x}) - 4x(e^{-x}) + 4(-e^{-x})$

$= -2x^2 e^{-x} - 4xe^{-x} - 4e^{-x}$

$= -2e^{-x}(x^2 + 2x + 2)$

$\int_1^6 2x^2 e^{-x} dx$

$= -2e^{-x}(x^2 + 2x + 2)\Big|_1^6$

$= -100e^{-6} + 10e^{-1}$

≈ 3.431

47. $h'(x) = \sqrt{x^2 + 16}$

$h(x) = \int_0^7 \sqrt{x^2 + 16} \, dx$

If $a = 4$, this matches entry 15 in the table.

$\int_0^7 \sqrt{x^2 + 16} \, dx$

$= \left(\frac{x}{2}\sqrt{x^2 + 16} + 8 \ln \left|x + \sqrt{x^2 + 16}\right|\right)\Big|_0^7$

$= \frac{7\sqrt{65}}{2} + 8 \ln (7 + \sqrt{65}) - 8 \ln 4$

≈ 38.8

Section 15.2

1. $\int_0^2 x^2 \, dx$

$n = 4$, $b = 2$, $a = 0$, $f(x) = x^2$

i	x_i	$f(x_i)$
0	0	0
1	1/2	.25
2	1	1
3	3/2	2.25
4	2	4

(a) Trapezoidal rule

$\int_0^2 x^2 \, dx$

$\approx \frac{2 - 0}{4}\left[\frac{1}{2}(0) + .25 + 1 + 2.25 + \frac{1}{2}(4)\right]$

$= .5(5.5) = 2.7500$

(b) Simpson's rule

$\int_0^2 x^2 \, dx$

$\approx \frac{2}{3(4)}[0 + 4(.25) + 2(1) + 4(2.25) + 4]$

$= \frac{2}{12}(16) \approx 2.6667$

(c) Exact value

$\int_0^2 x^2 \, dx = \frac{x^3}{3}\Big|_0^2 = \frac{8}{3} \approx 2.6667$

3. $\int_{-1}^{3} \dfrac{1}{4-x}$

n = 4, b = 3, a = -1, $f(x) = \dfrac{1}{4-x}$

i	x_i	$f(x_i)$
0	-1	.2
1	0	.25
2	1	.3333
3	2	.5
4	3	1

(a) Trapezoidal rule

$\int_{-1}^{3} \dfrac{1}{4-x}\, dx$

$\approx \dfrac{3-(-1)}{4}$

$\bullet \left[\dfrac{1}{2}(.2) + .25 + .3333 + .5 + \dfrac{1}{2}(1) \right]$

≈ 1.6833

(b) Simpson's rule

$\int_{-1}^{3} \dfrac{1}{4-x}\, dx$

$\approx \dfrac{3-(-1)}{12}$

$\bullet\, [.2 + 4(.25) + 2(.3333) + 4(.5) + 1]$

≈ 1.6222

(c) Exact value

$\int_{-1}^{3} \dfrac{1}{4-x}\, dx = -\int_{-1}^{3} \dfrac{-dx}{4-x}$

$\qquad = -(\ln |4-x|) \Big|_{-1}^{3}$

$\qquad = -\ln 1 + \ln 5$

$\qquad = \ln 5 \approx 1.6094$

5. $\int_{-2}^{2} (2x^2 + 1)\, dx$

n = 4, b = 2, a = -2, $f(x) = 2x^2 + 1$

i	x	f(x)
0	-2	9
1	-1	3
2	0	1
3	1	3
4	2	9

(a) Trapezoidal rule

$\int_{-2}^{2} (2x^2 + 1)\, dx$

$\approx \dfrac{2-(-2)}{4}$

$\bullet \left[\dfrac{1}{2}(9) + 3 + 1 + 3 + \dfrac{1}{2}(9) \right]$

≈ 16

(b) Simpson's rule

$\int_{-2}^{2} (2x^2 + 1)\, dx$

$\approx \dfrac{2-(-2)}{12}$

$\bullet\, [9 + 4(3) + 2(1) + 4(3) + 9]$

≈ 14.6667

(c) Exact value

$\int_{-2}^{2} (2x^2 + 1)\, dx$

$= \left(\dfrac{2x^3}{3} + x \right) \Big|_{-2}^{2}$

$= \left(\dfrac{16}{3} + 2 \right) - \left(-\dfrac{16}{3} - 2 \right)$

$= \dfrac{32}{3} + 4$

$= \dfrac{44}{3}$

≈ 14.6667

7. $\displaystyle\int_1^5 \frac{1}{x^2}\, dx$

$n = 4,\ b = 5,\ a = 1,\ f(x) = \dfrac{1}{x^2}$

i	x_i	$f(x_i)$
0	1	1
1	2	.25
2	3	.1111
3	4	.0625
4	5	.04

(a) Trapezoidal rule

$\displaystyle\int_1^5 \frac{1}{x^2}\, dx$

$\approx \dfrac{5-1}{4}$

$\cdot \left[\dfrac{1}{2}(1) + .25 + .1111 + .0625 + \dfrac{1}{2}(.04)\right]$

$\approx .9436$

(b) Simpson's rule

$\displaystyle\int_1^5 \frac{1}{x^2}\, dx$

$\approx \dfrac{5-1}{12}$

$\cdot [1 + 4(.25) + 2(.1111)$

$+ 4(.0625) + .04)]$

$\approx .8374$

(c) Exact value

$\displaystyle\int_1^5 x^{-2}\, dx = -x^{-1}\Big|_1^5$

$= -\dfrac{1}{5} + 1$

$= \dfrac{4}{5} = .8$

9. $\displaystyle\int_0^4 \sqrt{x^2 + 1}\, dx$

$n = 4,\ b = 4,\ a = 0,\ f(x) = \sqrt{x^2 + 1}$

i	x_i	$f(x_i)$
0	0	1
1	1	1.4142
2	2	2.2361
3	3	3.1623
4	4	4.1231

(a) Trapezoidal rule

$\displaystyle\int_0^4 \sqrt{x^2 + 1}\, dx$

$\approx \dfrac{4-0}{4}\left[\dfrac{1}{2}(1) + 1.4142 + 2.2361 + 3.1623\right.$

$\left.+ \dfrac{1}{2}(4.1231)\right]$

≈ 9.3741

(b) Simpson's rule

$\displaystyle\int_0^4 \sqrt{x^2 + 1}\, dx$

$\approx \dfrac{4-0}{12}[1 + 4(1.4142) + 2(2.2361)$

$+ 4(3.1623) + 4.1231]$

≈ 9.3004

(c) Exact value

$\displaystyle\int_0^4 \sqrt{x^2 + 1}\, dx$

If $a = 1$, this integral matches entry 15 in the table.

$$\int_0^4 \sqrt{x^2 + 1}\ dx$$

$$= \left(\frac{x}{2}\sqrt{x^2 + 1} + \frac{1}{2}\ln |x + \sqrt{x^2 + 1}|\right)\Big|_0^4$$

$$= 2\sqrt{17} + \frac{1}{2}\ln (4 + \sqrt{17}) - \frac{1}{2}\ln 1$$

$$= 2\sqrt{17} + \frac{1}{2}\ln (4 + \sqrt{17})$$

$$\approx 9.2936$$

11. $y = \sqrt{4 - x^2}$

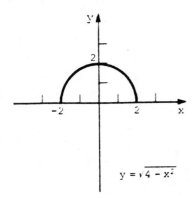

$y = \sqrt{4 - x^2}$

$n = 8$, $b = 2$, $a = -2$, $f(x) = \sqrt{4 - x^2}$

i	x_i	y
0	-2.0	0
1	-1.5	1.32289
2	-1.0	1.73205
3	-.5	1.93649
4	0	2
5	.5	1.93649
6	1.0	1.73205
7	1.5	1.32289
8	2.0	0

(a) Trapezoidal rule

$$\int_{-2}^2 \sqrt{4 - x^2}\ dx$$

$$\approx \frac{2 - (-2)}{8}$$

$$\cdot \left[\frac{1}{2}(0) + 1.32289 + 1.73205 + \cdots\right.$$

$$\left. + \frac{1}{2}(0)\right]$$

$$\approx 5.9914$$

(b) Simpson's rule

$$\int_{-2}^2 \sqrt{4 - x^2}\ dx$$

$$\approx \frac{2 - (-2)}{8}$$

$$\cdot [0 + 4(1.32289) + 2(1.73205)$$

$$+ 4(1.93649) + 2(2) + 4(1.93649)$$

$$+ 2(1.73205) + 4(1.32289) + 0]$$

$$\approx 6.1672$$

(c) Area of semicircle $= \frac{1}{2}\pi r^2$

$$= \frac{1}{2}\pi(2)^2$$

$$\approx 6.2832$$

Simpson's rule is more accurate.

13. **(a)** See the answer graph in your textbook.

(b) $A = \dfrac{7 - 1}{6}\left[\frac{1}{2}(.4) + .6 + .9 + 1.1\right.$

$$\left. + 1.3 + 1.4 + \frac{1}{2}(1.6)\right]$$

$$= 6.3$$

(c) $A = \dfrac{7 - 1}{3(6)}[.4 + 4(.6) + 2(.9)$

$$+ 4(1.1) + 2(1.3) + 4(1.4)$$

$$+ 1.6]$$

$$\approx 6.27$$

15. $y = e^{-t^2} + \dfrac{1}{t}$

The total reaction is

$$\int_1^9 \left(e^{-t^2} + \frac{1}{t}\right) dt.$$

$n = 8$, $b = 9$, $a = 1$, $f(t) = e^{-t^2} + \dfrac{1}{t}$

i	x_i	$f(x_i)$
0	1	1.3679
1	2	.5183
2	3	.3335
3	4	.2500
4	5	.2000
5	6	.1667
6	7	.1429
7	8	.1250
8	9	.1111

(a) Trapezoidal rule

$$\int_1^9 \left(e^{-t^2} + \frac{1}{t}\right) dt$$

$$\approx \frac{9-1}{8}\left[\frac{1}{2}(1.3679) + .5183 + .3335\right.$$

$$\left. + \cdots + \frac{1}{2}(.1111)\right]$$

$$\approx 2.4759$$

(b) Simpson's rule

$$\int_1^9 \left(e^{-t^2} + \frac{1}{t}\right) dt$$

$$\approx \frac{9-1}{3(8)}[1.3679 + 4(.5183) + 2(.3334)$$

$$+ 4(.2500) + 2(.2000) + 4(.1667)$$

$$+ 2(.1429) + 4(.1250) + .1111]$$

$$\approx 2.3572$$

17. Note that heights may differ depending on the readings of the graph. Thus, answers may vary.

$n = 10$, $b = 20$, $a = 0$

i	x_i	y
0	0	0
1	2	5
2	4	3
3	6	2
4	8	1.5
5	10	1.2
6	12	1
7	14	.5
8	16	.3
9	18	.2
10	20	.2

Area under curve for Formulation A

$$= \frac{20-0}{10}\left[\frac{1}{2}(0) + 5 + 3 + 2 + 1.5 + 1.2\right.$$

$$\left. + 1 + .5 + .3 + .2 + \frac{1}{2}(.2)\right]$$

$$= 2(14.8)$$

$$\approx 30 \text{ mcg/ml}$$

This represents the total amount of drug available to the patient.

19. As in Exercise 17, readings on the graph may vary, so answers may vary. The area both under the curve for Formulation A and above the minimum effective concentration line is on the interval $[1/2, 6]$.

Area under curve for Formulation A on [1/2, 1], with n = 1,

$$= \frac{1 - \frac{1}{2}}{1} \left[\frac{1}{2}(2 + 6) \right]$$

$$= \frac{1}{2}(4) = 2$$

Area under curve for Formulation A on [1, 6], with n = 5,

$$= \frac{6 - 1}{5} \left[\frac{1}{2}(6) + 5 + 4 + 3 + 2.4 + \frac{1}{2}(2) \right]$$

$$= 18.4$$

Area under minimum effective concentration line on [1/2, 6]

$$= 5.5(2)$$

$$= 11.0$$

Area under the curve for Formulation A and above minimum effective concentration line

$$= 2 + 18.4 - 11.0$$

$$\approx 9 \text{ mcg/ml}$$

This represents the total effective amount of drug available to the patient.

21. **(a)** See the answer graph in textbook.

(b) $\frac{(7 - 1)}{6} \left[\frac{1}{2}(4) + 7 + 11 + 9 + 15 \right.$

$\left. + 16 + \frac{1}{2}(23) \right]$

$= 71.5$

(c) $\frac{(7 - 1)}{3(6)} [4 + 4(7) + 2(11) + 4(9)$

$+ 2(15) + 4(16) + 23]$

$= 69.0$

Exercises 23–31 should be solved by computer methods. The solutions will vary according to the computer program that is used. The answers are given.

23. $\int_{4}^{8} \ln(x^2 - 10) \, dx$

Trapezoidal rule: 12.6027
Simpson's rule: 12.6029

25. $\int_{-2}^{2} \sqrt{9 - 2x^2} \, dx$

Trapezoidal rule: 9.83271
Simpson's rule: 9.83377

27. $\int_{1}^{5} (2x^2 + 3x - 1)^{2/5} \, dx$

Trapezoidal rule: 14.5192
Simpson's rule: 14.5193

29. Trapezoidal rule: 3979.24
Simpson's rule: 3979.24

31. **(a)** Trapezoidal rule: .682673
Simpson's rule: .682689

(b) Trapezoidal rule: .954471
Simpson's rule: .954500

(c) Trapezoidal rule: .997292
Simpson's rule: .997300

33. As n changes from 4 to 8, for example, the error changes from .020703 to .005200.

$$.020703a = .005200$$

$$a \approx \frac{1}{4}$$

Similar results would be obtained using other values for n.

The error is multiplied by 1/4.

35. As n changes form 4 to 8, the error changes from .0005208 to .0000326.

$$.0005208a = .0000326$$

$$a \approx \frac{1}{16}$$

Similar results would be obtained using other values for n.

The error is multiplied by 1/16.

Section 15.3

1. $f(x) = x$, $y = 0$, $x = 0$, $x = 2$

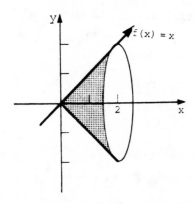

$$V = \pi \int_0^2 x^2 \ dx = \frac{\pi x^3}{2}\bigg|_0^2$$

$$= \frac{\pi(8)}{3} - 0$$

$$= \frac{8\pi}{3}$$

3. $f(x) = 2x + 1$, $y = 0$, $x = 4$, $x = 4$

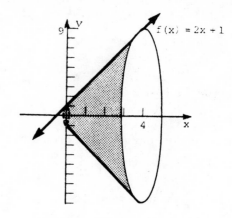

$$V = \pi \int_0^4 (2x + 1)^2 \ dx$$

Let $u = 2x + 1$.

Then $du = 2 \ dx$.

If $x = 4$, $u = 9$.

If $x = 0$, $u = 1$.

$$V = \frac{1}{2}\pi \int_0^4 2(2x + 1)^2 \ dx$$

$$= \frac{1}{2}\pi \int_1^9 u^2 \ du$$

$$= \frac{\pi}{2}\left(\frac{u^3}{3}\right)\bigg|_1^9$$

$$= \frac{729\pi}{6} - \frac{\pi}{6}$$

$$= \frac{728\pi}{6}$$

$$= \frac{364\pi}{3}$$

5. $f(x) = \frac{1}{3}x + 2$, $y = 0$, $x = 1$, $x = 3$

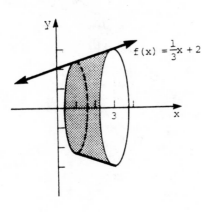

$$V = \pi \int_1^2 (\sqrt{x})^2 \, dx = \pi \int_1^2 x \, dx$$

$$= \left. \frac{\pi x^2}{2} \right|_1^2$$

$$= 2\pi - \frac{\pi}{2}$$

$$= \frac{3\pi}{2}$$

9. $f(x) = \sqrt{2x + 1}$, $y = 0$, $x = 1$, $x = 4$

$$V = \pi \int_1^3 \left(\frac{1}{3}x + 2\right)^2 \, dx$$

$$= 3\pi \int_1^3 \frac{1}{3}\left(\frac{1}{3}x + 2\right)^2 \, dx$$

$$= 3\pi \left. \frac{\left(\frac{1}{3}x + 2\right)^3}{3} \right|_1^3$$

$$= \pi \left.\left(\frac{1}{3}x + 2\right)^3 \right|_1^3$$

$$= 27\pi - \frac{343\pi}{27}$$

$$= \frac{386\pi}{27}$$

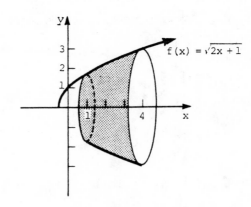

$$V = \pi \int_1^4 (\sqrt{2x + 1})^2 \, dx$$

$$= \pi \int_1^4 (2x + 1) \, dx$$

$$= \pi \left.\left(\frac{2x^2}{2} + x\right) \right|_1^4$$

$$= \pi[(16 + 4) - 2]$$

$$= 18\pi$$

7. $f(x) = \sqrt{x}$, $y = 0$, $x = 1$, $x = 2$

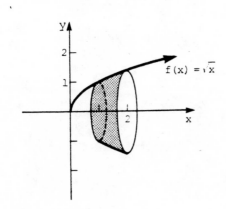

11. $f(x) = e^x$; $y = 0$, $x = 0$, $x = 2$

$$V = \pi \int_0^2 e^{2x} \, dx = \frac{\pi e^{2x}}{2} \Big|_0^2$$

$$= \frac{\pi e^4}{2} - \frac{\pi}{2}$$

$$= \frac{\pi}{2}(e^4 - 1)$$

$$\approx 84.19$$

13. $f(x) = \dfrac{1}{\sqrt{x}}$, $y = 0$, $x = 1$, $x = 4$

$$V = \pi \int_1^4 \frac{1}{(\sqrt{x})^2} \, dx$$

$$= \pi \int_1^4 \frac{1}{x} \, dx$$

$$= \pi \ln |x| \Big|_1^4$$

$$= \pi \ln 4 - \pi \ln 1$$

$$= \pi \ln 4 \approx 4.36$$

15. $f(x) = x^2$, $y = 0$, $x = 1$, $x = 5$

$$V = \pi \int_1^5 x^4 \, dx = \frac{\pi x^5}{5} \Big|_1^5$$

$$= 625\pi - \frac{\pi}{5}$$

$$= \frac{3124\pi}{5}$$

17. $f(x) = 1 - x^2$, $y = 0$

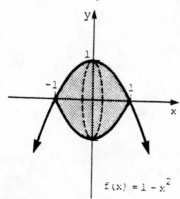

$f(x) = 1 - x^2$

Since $f(x) = 1 - x^2$ intersects $y = 0$ where

$$1 - x^2 = 0$$

$$x = \pm 1,$$

$$a = -1 \quad \text{and} \quad b = 1.$$

$$V = \pi \int_{-1}^1 (1 - x^2)^2 \, dx$$

$$= \pi \int_{-1}^1 (1 - 2x^2 + x^4) \, dx$$

$$= \pi \left(x - \frac{2x^3}{3} + \frac{x^5}{5} \right) \Big|_{-1}^1$$

$$= \pi \left(1 - \frac{2}{3} + \frac{1}{5} \right) - \pi \left(-1 + \frac{2}{3} - \frac{1}{5} \right)$$

$$= 2\pi - \frac{4\pi}{3} + \frac{2\pi}{5}$$

$$= \frac{16\pi}{15}$$

19. $f(x) = \sqrt{1 - x^2}$, $r = \sqrt{1} = 1$

$$V = \pi \int_{-1}^1 (\sqrt{1 - x^2})^2 \, dx$$

$$= \pi \int_{-1}^1 (1 - x^2) \, dx$$

$$= \pi \left(x - \frac{x^3}{3} \right) \Big|_{-1}^1$$

$$= \pi \left(1 - \frac{1}{3} \right) - \pi \left(-1 + \frac{1}{3} \right)$$

$$= 2\pi - \frac{2}{3}\pi$$

$$= \frac{4\pi}{3}$$

21. $f(x) = \sqrt{r^2 - x^2}$

$$V = \pi \int_{-r}^{r} (\sqrt{r^2 - x^2})^2 \, dx$$

$$= \pi \int_{-r}^{r} (r^2 - x^2) \, dx$$

$$= \pi\left(r^2 x - \frac{x^3}{3}\right)\Big|_{-r}^{r}$$

$$= \pi\left(r^3 - \frac{r^3}{3}\right) - \pi\left(-r^3 + \frac{r^3}{3}\right)$$

$$= 2r^3\pi - \frac{2r^3\pi}{3})$$

$$= \frac{4r^3\pi}{3}$$

23. $f(x) = r, \ x = 0, \ x = h$

Graph $f(x) = r$; then show solid of revolution formed by rotating about the x-axis the region bounded by $f(x), \ x = 0, \ x = h$.

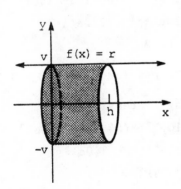

$$\int_{0}^{h} \pi r^2 \, dx = \pi r^2 x \Big|_{0}^{h}$$

$$= \pi r^2 h - 0$$

$$= \pi r^2 h$$

25. $f(x) = x^2 - 2; \ [0, 5]$

Average value

$$= \frac{1}{5 - 0}\int_{0}^{5} (x^2 - 2) \, dx$$

$$= \frac{1}{5}\left(\frac{x^3}{3} - 2x\right)\Big|_{0}^{5}$$

$$= \frac{1}{5}\left(\frac{125}{3} - 10\right)$$

$$= \frac{1}{5}\left(\frac{95}{3}\right)$$

$$= \frac{19}{3}$$

27. $f(x) = \sqrt{x + 1}; \ [3, 8]$

Average value

$$= \frac{1}{8 - 3}\int_{3}^{8} \sqrt{x + 1} \, dx$$

$$= \frac{1}{5}\int_{3}^{8} (x + 1)^{1/2} \, dx$$

$$= \frac{1}{5} \cdot \frac{2}{3}(x + 1)^{3/2}\Big|_{3}^{8}$$

$$= \frac{2}{15}(9^{3/2} - 4^{3/2})$$

$$= \frac{2}{15}(27 - 8)$$

$$= \frac{38}{15}$$

29. $f(x) = e^{x/5}; \ [0, 5]$

Average value

$$= \frac{1}{5 - 0}\int_{0}^{e} e^{x/5} \, dx$$

$$= \int_{0}^{5} \frac{1}{5}e^{x/5} \, dx$$

$$= e^{x/5}\Big|_{0}^{5} = e^1 - e^0$$

$$= e - 1 \approx 1.718$$

31. $f(x) = x^2e^{2x}$; $[0, 2]$

Average value

$$= \frac{1}{2 - 0} \int_0^2 x^2 e^{2x}\ dx$$

Let $u = x^2$ and $dv = e^{2x}dx$.

Use column integration.

D	I
x^2	e^{2x}
$2x$	$(1/2)e^{2x}$
2	$(1/4)e^{2x}$
0	$(1/8)e^{2x}$

$$\frac{1}{2 - 0} \int_0^2 x^2 e^{2x}\ dx$$

$$= \frac{1}{2}\left(\left[(x^2)\left(\frac{1}{2}\right)e^{2x} - (2x)\left(\frac{1}{4}\right)e^{2x}\right.\right.$$

$$\left.\left. + 2\left(\frac{1}{8}\right)e^{2x}\right]\ \bigg|_0^2\right)$$

$$= \frac{1}{2}\left(2e^4 - e^4 + \frac{1}{4}e^4 - \frac{1}{4}\right)$$

$$= \frac{(5e^4 - 1)}{8} \approx 33.999$$

33. $R(t) = te^{-.1t}$

During the nth hour is the interval $(n - 1, n)$.

Average intensity during nth hour

$$= \frac{1}{n - (n - 1)} \int_{n-1}^n te^{-.1t}dt$$

$$= \int_{n-1}^n te^{-.1t}dt$$

Let $u = t$ and $dv = e^{-.1t}dt$.

D	I
t	$e^{-.1t}$
1	$-10e^{-.1t}$
0	$100e^{-.1t}$

$$\int_{n-1}^n te^{-.1t}\ dt$$

$$= (-10te^{-.1t} - 100e^{-.1t})\ \bigg|_{n-1}^n$$

(a) Second hour, $n = 2$

Average intensity

$$= -10e^{-.2}(12) + 10e^{-.1}(11)$$

$$= 110e^{-.1} - 120e^{-.2}$$

$$\approx 1.2844$$

(b) Twelfth hour, $n = 12$

Average intensity

$$= -10e^{-1.2}(12 + 10) + 10e^{-1.1}(11 + 10)$$

$$= 210e^{-1.1} - 220e^{-1.2}$$

$$\approx 3.6402$$

(c) Twenty-fourth hour, $n = 24$

Average intensity

$$= -10e^{-2.4}(24 + 10) + 10e^{-2.3}(23 + 10)$$

$$= 330e^{-2.3} - 340e^{-2.4}$$

$$\approx 2.2413$$

35. $W(t) = -6t^2 + 10t + 80$

(a) $W(0) = -6(0)^2 + 10(0) + 80$

$\qquad W(0) = 80$

(b) $W'(t) = -12t + 10$

If $W'(t) = 0$, $t = \frac{5}{6}$.

$$W\left(\frac{5}{6}\right) = -6\left(\frac{5}{6}\right)^2 + 10\left(\frac{5}{6}\right) + 80$$

$$= -\frac{25}{6} + \frac{50}{6} + 80$$

$$= \frac{505}{6}$$

≈ 84 words per minute
when $t = 5/6$

(c) Average speed

$$= \frac{1}{5 - 0}\int_0^5 (-6t^2 + 10t + 80)dt$$

$$= \frac{1}{5}(-2t^3 + 5t^2 + 80t)\Big|_0^5$$

$$= \frac{1}{5}(-250 + 125 + 400)$$

$$= \frac{275}{5}$$

$$= 55 \text{ words per minute}$$

Section 15.4

1. $f(x) = 1000$

(a) $P = \int_0^{10} 1000e^{-.12x}\ dx$

$$= \frac{1000}{-.12}\ e^{-.12x}\Big|_0^{10}$$

$$= \frac{1000}{-.12}\ e^{-1.2} + \frac{1000}{.12}$$

$$= \$5823.38$$

Store the value for P without round-
ing in your calculator.

(b) $A = e^{.12(10)}\int_0^{10} 1000e^{-.12x}\ dx$

$$= e^{.12(10)}P$$

$$= \$19,334.31$$

3. $f(x) = 500$

(a) $P = \int_0^{10} 500e^{-.12x}\ dx$

$$= \frac{500}{.12}\ e^{-.12x}\Big|_0^{10}$$

$$= \frac{500}{-.12}e^{-1.2} + \frac{500}{.12}$$

$$= \$2911.69$$

(b) $A = e^{.12(10)}\int_0^{10} 500e^{-.12x}\ dx$

$$= e^{1.2}\ P$$

$$= \$9667.16$$

5. $f(x) = 400e^{.03x}$

(a) $\int_0^{10} 400e^{.03x} \cdot e^{-.12x}\ dx$

$$= \int_0^{10} 400e^{-.09}dx$$

$$= \frac{400}{-.09}\ e^{-.09x}\Big|_0^{10}$$

$$= \frac{400}{-.09}\ e^{-.9} + \frac{400}{.09}$$

$$= \$2637.47$$

(b) $e^{.12(10)}\int_0^{10} 400e^{-.09x}dx$

$$= e^{1.2}\left(\frac{400}{-.09}\ e^{-.9} + \frac{400}{.09}\right)$$

$$= \$8756.70$$

7. (a) $\displaystyle\int_0^{10} 5000e^{-.01x} \cdot e^{-.12x} \; dx$

$= 1000 \displaystyle\int_0^{10} 5000e^{-.13x} \; dx$

$= \dfrac{5000}{-.13} \; e^{-.13x} \Big|_0^{10}$

$= \dfrac{5000}{-.13} \; e^{-1.3} + \dfrac{5000}{.13}$

$= \$27,979.55$

(b) $e^{.12(10)} \displaystyle\int_0^{10} 5000e^{-.13x} \; dx$

$= e^{1.2}\left(\dfrac{5000}{-.13} \; e^{-1.3} + \dfrac{5000}{.13}\right)$

$= \$92,895.37$

9. $f(x) = .1x$

(a) $P = \displaystyle\int_0^{10} .1xe^{-.12x} \; dx$

Let $u = .1x$ and $dv = e^{-.12x}$.

Then $du = .1dx$ and $v = \dfrac{e^{-.12x}}{-.12}$.

$\displaystyle\int .1xe^{-.12x} \; dx$

$= .1x\dfrac{e^{-.12x}}{-.12} - \displaystyle\int \dfrac{.1}{-.12} \; e^{-.12x}dx$

$= \dfrac{.1xe^{-.12x}}{-.12} + \dfrac{.1}{.12}\left(\dfrac{e^{-.12x}}{-.12}\right)$

$\displaystyle\int_0^{10} .1xe^{-.12x}dx$

$= \left[\dfrac{.1xe^{-.12x}}{-.12} + \dfrac{.1}{.12}\left(\dfrac{e^{-.12x}}{-.12}\right)\right]\Big|_0^{10}$

$= \dfrac{1}{-.12} \; e^{-1.2} + \dfrac{.1e^{-1.2}}{-.12(.12)}$

$\qquad + \dfrac{.1}{(.12)^2}$

$= \$2.34$

(b) $A = e^{.12(10)} \displaystyle\int_0^{10} .1xe^{-.12x} \; dx$

$\approx \$7.78$

$= e^{.12(10)}P$

$= \$7.78$

11. $f(x) = .01x + 100$

(a) $P = \displaystyle\int_0^{10} (.01x + 100)e^{-.12x} \; dx$

$= \displaystyle\int_0^{10} .01xe^{-.12x} \; dx$

$\quad + \displaystyle\int_0^{10} 100e^{-.12x} \; dx$

Let $u = .01x$ and $dv = e^{-.12x}$.

Then $du = .01dx$ and $v = \dfrac{e^{-.12x}}{-.12}$.

$\displaystyle\int .01xe^{-.12x} \; dx$

$= \dfrac{.01xe^{-.12x}}{-.12} - \displaystyle\int \dfrac{.01e^{-.12x}}{-.12} \; dx$

$= \dfrac{.01xe^{-.12x}}{-.12} - \dfrac{.01e^{-.12x}}{(-.12)^2}$

$\displaystyle\int_0^{10} .01xe^{-.12x}dx + \displaystyle\int_0^{10} 100e^{-.12x} \; dx$

$= \left(\dfrac{.01xe^{-.12x}}{-.12} - \dfrac{.01e^{-.12x}}{(-.12)^2}\right)\Big|_0^{10}$

$\quad + \dfrac{100e^{-.12x}}{-.12}\Big|_0^{10}$

$= \dfrac{.1e^{-1.2}}{-.12} - \dfrac{.01e^{-1.2}}{(-.12)^2} - 0 + \dfrac{.01}{(-.12)^2}$

$\quad + \dfrac{100e^{-1.2}}{-.12} - \dfrac{100}{-.12}$

$P = \$582.57$

(b) $e^{.12(10)} \int_0^{10} (.01x + 100)e^{-.12x}dx$

$= e^{1.2}P$

$\approx \$1934.20$

13. $f(x) = 1000x - 100x^2$

(a) $P = \int_0^{10} (1000x - 100x^2)e^{-.12x} \, dx$

$= \int_0^{10} 1000xe^{-.12x} \, dx$

$- \int_0^{10} 100x^2e^{-.12x} \, dx$

Let $u = 1000x$ and $dv = e^{-.12x} \, dx$; then $du = 1000 \, dx$ and

$v = \dfrac{-1}{.12}e^{-.12x} \, dx.$

$\int 1000xe^{-.12x} \, dx$

$= \dfrac{1000xe^{-.12x}}{(-.12)} - \int \dfrac{1000e^{-.12x}}{-.12} \, dx$

$= \dfrac{1000xe^{-.12x}}{(-.12)} - \dfrac{1000e^{-.12x}}{(-.12)^2} + C$

For $\int 100x^2e^{-.12x} \, dx$, let $u = 100x^2$ and $dv = e^{-.12x} \, dx.$

Use column integration

D	I
$100x^2$	$e^{-.12x}$
$200x$	$\dfrac{1}{(-.12)}e^{-.12x}$
200	$\dfrac{1}{(-.12)^2}e^{-.12x}$
0	$\dfrac{1}{(-.12)^3}e^{-.12x}$

$\int 100x^2e^{-.12x} \, dx$

$= \left(\dfrac{100}{-.12}\right)x^2e^{-.12x} - \dfrac{200}{(-.12)^2}xe^{-.12x}$

$+ \dfrac{200}{(-.12)^3}e^{-.12x} + C$

$\int_0^{10} 1000xe^{-.12x} \, dx - \int_0^{10} 100x^2e^{-.12x} \, dx$

$= \left(\dfrac{1000xe^{-.12x}}{(-.12)} - \dfrac{1000e^{-.12x}}{(-.12)^2}\right.$

$- \dfrac{100}{(-.12)}x^2e^{-.12x} + \dfrac{200}{(-.12)^2}xe^{-.12x}$

$\left. - \dfrac{200}{(-.12)^3}e^{-.12x}\right)\Bigg|_0^{10}$

$= \dfrac{10,000e^{-1.2}}{(-.12)} - \dfrac{1000}{(-.12)^2}e^{-1.2} - \dfrac{10,000}{(-.12)}e^{-1.2}$

$+ \dfrac{2000}{(-.12)^2}e^{-1.2} - \dfrac{200}{(-.12)^3}e^{-1.2}$

$+ \dfrac{1000}{(-.12)^2} + \dfrac{200}{(-.12)^3}$

$= \dfrac{1000}{(-.12)^2}(e^{-1.2} + 1)$

$- \dfrac{200}{(-.12)^3}(e^{-1.2} - 1)$

$= \$9480.41$

(b)

$e^{.12(10)} \int_0^{10} (1000x - 100x^2)e^{-.12x} \, dx$

$= e^{1.2}(P)$

$= \$31,476.07$

15. $A = e^{.14(3)} \int_0^3 20,000e^{-.14x} \, dx$

$= e^{.42}\left(\dfrac{20,000}{-.14}e^{-.14x}\right)\Bigg|_0^3$

$= e^{.42}\left(\dfrac{60,000}{-.14}e^{-.42} + \dfrac{20,000}{.14}\right)$

$\approx \$74,565.94$

17. (a) Present value

$$= \int_0^8 5000e^{-.01x}e^{-.08x}\ dx$$

$$= \int_0^8 5000e^{-.09x}\ dx$$

$$= \left(\frac{5000}{-.09}\ e^{-.09x}\right)\Big|_0^8$$

$$= \frac{5000e^{-.72}}{-.09} + \frac{5000}{.09}$$

$$= \$28,513.76$$

(b) Final amount

$$= e^{.08(8)}\int_0^8 5000e^{-.01x}e^{-.08x}dx$$

$$= e^{.64}(28,513.76)$$

$$= \$54,075.80$$

19. $$P = \int_0^5 (1500 - 60x^2)e^{-.10x}\ dx$$

$$= \int_0^5 1500e^{-.10x}\ dx$$

$$\quad - \int_0^5 60x^2e^{-.10x}\ dx$$

$$= 1500\int_0^5 e^{-.10x}\ dx$$

$$\quad - 60\int_0^5 x^2e^{-.10x}\ dx$$

Second integral, by column integration.

D		I
x^2	$+$	$e^{-.10x}$
$2x$	$-$	$e^{-.10x}/-.10$
2	$+$	$e^{-.10x}/.01$
0		$e^{-.10x}/-.001$

$$-60\int x^2e^{-.10x}dx$$

$$= \frac{x^2e^{-.10x}}{-.10} - \frac{2xe^{-.10x}}{.01} + \frac{2e^{-.10x}}{-.001} + C$$

$$1500\int_0^5 e^{-.10x}dx - 60\int_0^5 x^2e^{-.10x}dx$$

$$= \frac{1500}{-.10}\ e^{-.10x}\Big|_0^5$$

$$\quad - 60\left(-\frac{x^2e^{-.10x}}{.10} - \frac{2xe^{-.10x}}{.01}\right.$$

$$\quad \left. - \frac{2e^{-.10x}}{.001}\right)\Big|_0^5$$

$$= -15,000e^{-.5} + 15,000$$

$$\quad + 60\left(\frac{25e^{-.5}}{.1} + \frac{10e^{-.5}}{.01} + \frac{2e^{-.5}}{.001}\right.$$

$$\quad \left. - 0 - \frac{2}{.001}\right)$$

$$= -15,000e^{-.5} + 15,000 + 15,000e^{-.5}$$

$$\quad + 60,000e^{-.5} + 120,000e^{-.5}$$

$$\quad - 120,000$$

$$= 180,000e^{-.5} - 105,000$$

$$= \$4175.52$$

Section 15.5

1. $\displaystyle\int_{2}^{\infty} \frac{1}{x^2}\, dx = \lim_{b\to\infty}\int_{2}^{b} x^{-2}\,dx$

$\displaystyle = \lim_{b\to\infty}\int (-x^{-1})\,\Big|_{2}^{b}$

$\displaystyle = \lim_{b\to\infty}\left(-\frac{1}{b} + \frac{1}{2}\right)$

$\displaystyle = \lim_{b\to\infty}\left(-\frac{1}{b}\right) + \lim_{b\to\infty}\frac{1}{2}$

As $b\to\infty$, $-\dfrac{1}{b}\to 0$. The integral is convergent.

$\displaystyle\int_{2}^{\infty} \frac{1}{x^2}\, dx = 0 + \frac{1}{2} = \frac{1}{2}$

3. $\displaystyle\int_{1}^{\infty} \frac{1}{\sqrt{x}}\, dx = \lim_{b\to\infty}\int_{1}^{b} x^{-1/2}\, dx$

$\displaystyle = \lim_{b\to\infty}(2x^{1/2})\,\Big|_{1}^{b}$

$\displaystyle = \lim_{b\to\infty}(2\sqrt{b} - 2)$

$\displaystyle = \lim_{b\to\infty} 2\sqrt{b} - \lim_{b\to\infty} 2$

As $b\to\infty$, $2\sqrt{b}\to\infty$.
The integral is divergent.

5. $\displaystyle\int_{-\infty}^{-1} \frac{2}{x^3}\, dx$

$\displaystyle = \int_{-\infty}^{-1} 2x^{-3}\,dx$

$\displaystyle = \lim_{a\to-\infty}\int_{a}^{-1} 2x^{-3}\, dx$

$\displaystyle = \lim_{a\to-\infty}\left(\frac{2x^{-2}}{-2}\right)\Big|_{a}^{-1}$

$\displaystyle = \lim_{a\to-\infty}\left(-1 + \frac{1}{a^2}\right)$

As $a\to-\infty$, $\dfrac{1}{a^2}\to 0$. The integral is convergent.

$\displaystyle\int_{-\infty}^{-1} \frac{2}{x^3}\, dx = -1 + 0 = -1$

7. $\displaystyle\int_{1}^{\infty} \frac{1}{x^{1.001}}\, dx$

$\displaystyle = \int_{1}^{\infty} x^{-1.001}\, dx$

$\displaystyle = \lim_{b\to\infty}\int_{1}^{b} x^{-1.001}\,dx$

$\displaystyle = \lim_{b\to\infty}\left(\frac{x^{-.001}}{-.001}\right)\Big|_{1}^{b}$

$\displaystyle = \lim_{b\to\infty}\left(-\frac{1}{(.001)b^{.001}} + \frac{1}{.001}\right)$

As $b\to\infty$, $-\dfrac{1}{.001 b^{.001}}\to 0$.
The integral is convergent.

$\displaystyle\int_{1}^{\infty} \frac{1}{x^{1.001}}\, dx = 0 + \frac{1}{.001} = 1000$

9. $\displaystyle\int_{-\infty}^{-1} x^{-2}\, dx = \lim_{a\to-\infty}\int_{a}^{-1} x^{-2}\, dx$

$\displaystyle = \lim_{a\to-\infty}(-x^{-1})\,\Big|_{a}^{-1}$

$\displaystyle = \lim_{a\to-\infty}\left(1 + \frac{1}{a}\right)$

$\displaystyle = 1 + 0 = 1$

The integral is convergent.

11. $\displaystyle\int_{-\infty}^{-1} x^{-8/3} \, dx = \lim_{a \to -\infty} \int_{a}^{-1} x^{-8/3} \, dx$

$\displaystyle = \lim_{a \to -\infty} \left(-\frac{3}{5} x^{-5/3} \right) \Big|_{a}^{-1}$

$\displaystyle = \lim_{a \to -\infty} \left(\frac{3}{5} + \frac{3}{5a^{5/3}} \right)$

$\displaystyle = \frac{3}{5} + 0 = \frac{3}{5}$

The integral is convergent.

13. $\displaystyle\int_{0}^{\infty} 4e^{-4x} \, dx = \lim_{b \to \infty} \int_{0}^{b} 4e^{-4x} \, dx$

$\displaystyle = \lim_{b \to \infty} \left(\frac{4e^{-4x}}{-4} \right) \Big|_{0}^{b}$

$\displaystyle = \lim_{b \to \infty} \left(-e^{-4b} + 1 \right)$

$\displaystyle = \lim_{b \to \infty} \left(-\frac{1}{e^{4b}} + 1 \right)$

$= 0 + 1 = 1$

The integral is convergent.

15. $\displaystyle\int_{-\infty}^{0} 4e^{x} \, dx = \lim_{a \to -\infty} \int_{a}^{0} 4e^{x} \, dx$

$\displaystyle = \lim_{a \to -\infty} (4e^{x}) \Big|_{a}^{0}$

$\displaystyle = \lim_{a \to -\infty} (4 - 4e^{a})$

Since a approaches $-\infty$, e^{a} is in the denominator of a fraction.

As $a \to -\infty$, $-4e^{a} \to 0$. The integral is convergent.

$\displaystyle\int_{-\infty}^{0} 4e^{x} \, dx = 4 - 0 = 4$

17. $\displaystyle\int_{-\infty}^{-1} \ln |x| \, dx = \lim_{a \to -\infty} \int_{a}^{-1} \ln |x| \, dx$

Let $u = \ln |x|$ and $dv = dx$.

Then $du = \dfrac{1}{x} \, dx$ and $v = x$.

$\displaystyle\int \ln |x| \, dx = x \ln |x| - \int \frac{x}{x} \, dx$

$\displaystyle = x \ln |x| - x + C$

$\displaystyle\int_{-\infty}^{-1} \ln |x| \, dx$

$\displaystyle = \lim_{a \to -\infty} (x \ln |x| - x) \Big|_{a}^{-1}$

$\displaystyle = \lim_{a \to -\infty} (-\ln 1 + 1 - a \ln |a| + a)$

$\displaystyle = \lim_{a \to -\infty} (1 + a - a \ln |a|)$

The integral is divergent, since as $a \to -\infty$,

$$(a - a \ln |a|) \to \infty.$$

19. $\displaystyle\int_{1}^{\infty} \frac{dx}{(x + 1)^{2}} = \lim_{b \to \infty} \int_{0}^{b} \frac{dx}{(x + 1)^{2}}$

(Ue substitution.)

$\displaystyle = \lim_{b \to \infty} -(x + 1)^{-1} \Big|_{0}^{b}$

$\displaystyle = \lim_{b \to \infty} \left(\frac{-1}{b + 1} + 1 \right)$

As $b \to \infty$, $-\dfrac{1}{b + 1} \to 0$. The integral is convergent.

$\displaystyle\int_{0}^{\infty} \frac{dx}{(x + 1)^{2}} = 0 + 1 = 1$

21. $\int_{-\infty}^{-1} \frac{2x - 1}{x^2 - x} \, dx$

$= \lim_{a \to -\infty} \int_{a}^{-1} \frac{2x - 1}{x^2 - x} \, dx$

(Use substitution.)

$= \lim_{a \to -\infty} \ln |x^2 - x| \Big|_{a}^{-1}$

$= \lim_{a \to -\infty} (\ln 2 - \ln |a^2 - a|)$

As $a \to -\infty$, $\ln |a^2 - a) \to \infty$. The integral is divergent.

23. $\int_{2}^{\infty} \frac{1}{x \ln x} \, dx$

$= \lim_{b \to \infty} \int_{2}^{b} \frac{1}{x \ln x} \, dx$

(Use substitution.)

$= \lim_{b \to \infty} \left[\ln (\ln x) \Big| \right]_{2}^{b}$

$= \lim_{b \to \infty} [\ln (\ln b) - \ln (\ln 2)]$

As $b \to \infty$, $\ln (\ln b) \to \infty$. The integral is divergent.

25. $\int_{0}^{\infty} x e^{2x} dx = \lim_{b \to \infty} \int_{0}^{b} x e^{2x} \, dx$

Let $dv = e^{2x} dx$ and $u = x$.

Then $v = \frac{1}{2} e^{2x}$ and $du = dx$.

$\int x e^{2x} dx = \frac{x}{2} e^{2x} - \int \frac{1}{2} e^{2x} dx$

$= \frac{x}{2} e^{2x} - \frac{1}{4} e^{2x}$

$= \frac{1}{4} (2x - 1) e^{2x}$

$\int_{0}^{\infty} x e^{2x} dx$

$= \lim_{b \to \infty} \left[\frac{1}{4} (2x - 1) e^{2x} \right] \Big|_{0}^{b}$

$= \lim_{b \to \infty} \left[\frac{1}{4} (2b - 1) e^{2b} - \frac{1}{4} (-1)(1) \right]$

$= \lim_{b \to \infty} \left[\frac{1}{4} (2b - 1) e^{2b} + \frac{1}{4} \right]$

As $b \to \infty$, $\frac{1}{4} (2b - 1) e^{2b} \to \infty$. The integral is divergent.

27. $\int_{-\infty}^{-1} \frac{2}{3x(2x - 7)} \, dx$

$= \lim_{a \to -\infty} \int_{a}^{-1} \left[\frac{2}{3} \cdot \frac{1}{x(2x - 7)} \right] \, dx$

Use entry 13 from the table.

$= \lim_{a \to -\infty} \left(\frac{2}{3} \left[-\frac{1}{7} \ln \left| \frac{x}{2x - 7} \right| \right] \Big|_{a}^{-1} \right)$

$= \lim_{a \to -\infty} \frac{2}{21} \left(-\ln \frac{1}{9} + \ln \left| \frac{a}{2a - 7} \right| \right)$

As $a \to -\infty$, $\ln \left| \frac{a}{2a - 7} \right| \to \ln \frac{1}{2}$.

$= \frac{2}{21} \left(\ln \frac{1}{2} - \ln \frac{1}{9} \right)$

$= \frac{2}{21} \left(\ln \frac{9}{2} \right)$

$= \frac{2 \ln 4.5}{21} \approx .143$

29. $\int_{1}^{\infty} \frac{4}{9x(x + 1)^2} \, dx$

$= \lim_{b \to \infty} \frac{4}{9} \int_{1}^{b} \frac{dx}{x(x + 1)^2}$

Use entry 14 from the table.

$$= \frac{4}{9} \lim_{b \to \infty} \left(\frac{1}{x + 1} + \ln \left| \frac{x}{x + 1} \right| \right) \Big|_1^b$$

$$= \frac{4}{9} \lim_{b \to \infty} \left(\frac{1}{b + 1} + \ln \frac{b}{b + 1} \right.$$

$$\left. - \frac{1}{2} - \ln \frac{1}{2} \right)$$

As $b \to \infty$, $\frac{1}{b + 1} \to 0$ and $\ln \frac{b}{b + 1} \to \ln 1 = 0$. The integral is convergent.

$$\int_1^\infty \frac{4}{9x(x + 1)^2}\, dx$$

$$= \frac{4}{9} \left(0 + 0 - \frac{1}{2} - \ln \frac{1}{2} \right)$$

$$= \frac{4}{9} \left(-\frac{1}{2} - \ln 1 + \ln 2 \right)$$

$$= \frac{4}{9} \left(\ln 2 - \frac{1}{2} \right) \approx .086$$

31. $\displaystyle \int_0^\infty \frac{1}{\sqrt{1 + x^2}}\, dx$

$$= \lim_{b \to \infty} \int_0^b \frac{1}{\sqrt{1 + x^2}}\, dx$$

Use entry 5 from the table.

$$= \lim_{b \to \infty} \left(\ln \left| x + \sqrt{x^2 + 1} \right| \Big|_0^b \right)$$

$$= \lim_{b \to \infty} \left(\ln \left| b + \sqrt{b^2 + 1} \right| + \ln 1 \right)$$

$$= \lim_{b \to \infty} \ln \left| b + \sqrt{b^2 + 1} \right|$$

As $b \to \infty$, $\ln \left| b + \sqrt{b^2 + 1} \right| \to \infty$. The integral is divergent.

33. $f(x) = \dfrac{1}{x - 1}$ for $(-\infty, 0]$

$$\int_{-\infty}^0 \frac{1}{x - 1}\, dx$$

$$= \lim_{a \to -\infty} \int_a^0 \frac{dx}{x - 1}$$

$$= \lim_{a \to -\infty} \left(\ln |x - 1| \right) \Big|_a^0$$

$$= \lim_{a \to -\infty} \left(\ln |-1| - \ln |a - 1| \right)$$

But $\displaystyle \lim_{a \to -\infty} \left(\ln |a - 1| \right) = \infty$.

The integral is divergent, so the area cannot be found.

35. $f(x) = \dfrac{1}{(x - 1)^2}$ for $(-\infty, 0]$

$$\int_{-\infty}^0 \frac{1}{(x - 1)^2}$$

$$= \lim_{a \to -\infty} \int_a^0 \frac{1}{(x - 1)^2}$$
(Use substitution.)

$$= \lim_{a \to -\infty} -(x - 1)^{-1} \Big|_a^0$$

$$= \lim_{a \to -\infty} \left(-\frac{1}{-1} + \frac{1}{a - 1} \right)$$

As $a \to -\infty$, $\dfrac{1}{a - 1} \to 0$. The integral is convergent.

$$= 1 + 0$$

$$= 1$$

Therefore, the area is 1.

37. $\displaystyle\int_{-\infty}^{\infty} xe^{-x^2}\,dx$

$$= \int_{-\infty}^{\infty} xe^{-x^2}\,dx + \int_{0}^{\infty} xe^{-x^2}\,dx$$

Let $u = -x^2$, so that

$du = -2x\,dx.$

$$= \lim_{a\to -\infty}\left(-\frac{1}{2}\int_{a}^{0} -2xe^{-x^2}\,dx\right.$$

$$+ \lim_{b\to\infty}\left(-\frac{1}{2}\int_{0}^{b} -2xe^{-x^2}\,dx\right)$$

$$= \lim_{a\to -\infty}\left(-\frac{1}{2}e^{-x^2}\right)\Big|_{a}^{0}$$

$$+ \lim_{b\to\infty}\left(-\frac{1}{2}e^{-x^2}\right)\Big|_{0}^{b}$$

$$= \lim_{a\to -\infty}\left(-\frac{1}{2} + \frac{1}{2e^{a^2}}\right)$$

$$+ \lim_{b\to\infty}\left(-\frac{1}{2e^{b^2}} + \frac{1}{2}\right)$$

$$= -\frac{1}{2} + \frac{1}{2}$$

$$= 0$$

The area is 0.

39. $\displaystyle\int_{1}^{\infty} \frac{1}{x^p}\,dx$

Case $p < 1$:

$$\int_{1}^{\infty} \frac{1}{x^p}\,dx$$

$$= \int_{1}^{\infty} x^{-p}\,dx$$

$$= \lim_{a\to\infty}\int_{1}^{a} x^{-p}\,dx$$

$$= \lim_{a\to\infty}\left[\frac{x^{-p+1}}{(-p+1)}\Big|_{1}^{a}\right]$$

$$= \lim_{a\to\infty}\left[\frac{1}{(-p+1)}\left(a^{-p+1} - 1\right)\right]$$

$$= \lim_{a\to\infty}\left[\frac{1}{(-p+1)}a^{1-p} - \frac{1}{(-p+1)}\right]$$

Since $p < 1$, $1 - p$ is positive and, as $a \to \infty$, $a^{1-p} \to \infty$. The integral diverges.

$p = 1$:

$$\int_{1}^{\infty} \frac{1}{x^p}\,dx = \int_{1}^{\infty} \frac{1}{x}\,dx$$

$$= \lim_{a\to\infty}\int_{1}^{a} \frac{1}{x}\,dx$$

$$= \lim_{a\to\infty}\left(\ln|x|\,\Big|_{1}^{a}\right)$$

$$= \lim_{a\to\infty}\left(\ln|a| - \ln 1\right)$$

$$= \lim_{a\to\infty}\ln|a|$$

As $a \to \infty$, $\ln|a| \to \infty$. The integral diverges.

Case $p > 1$:

$$\int_{1}^{\infty} \frac{1}{x^p}\,dx$$

$$= \lim_{a\to\infty}\int_{1}^{a} x^{-p}\,dx$$

$$= \lim_{a\to\infty}\left(\frac{x^{-p+1}}{-p+1}\Big|_{1}^{a}\right)$$

$$= \lim_{a\to\infty}\left[\frac{a^{-p+1}}{(-p+1)} - \frac{1}{(-p+1)}\right]$$

Since $p > 1$, $-p + 1 < 0$; thus as

$a \to \infty$, $\dfrac{a^{-p+1}}{(-p + 1)} \to 0$.

Hence, $\displaystyle\lim_{a \to \infty} \left[\dfrac{a^{-p+1}}{(-p + 1)} - \dfrac{1}{(-p + 1)} \right]$

$= 0 - \dfrac{1}{(-p + 1)}$

$= \dfrac{-1}{-p + 1}$

$= \dfrac{1}{p + 1}$

The integral converges.

41. Capital value is

$\displaystyle\int_0^\infty 60{,}000e^{-.08t}\, dt$

$= \displaystyle\lim_{b \to \infty} \int_0^b 60{,}000e^{-.08t}\, dt$

$= \displaystyle\lim_{b \to \infty} \left(\dfrac{60{,}000}{-.08} e^{-.08t} \right) \Bigg|_0^b$

$= -750{,}000 \displaystyle\lim_{b \to \infty} (e^{-.08b} - 1)$

$= -750{,}000 \displaystyle\lim_{b \to \infty} \left(\dfrac{1}{e^{.08b}} - 1 \right)$

As $b \to \infty$, $\dfrac{1}{e^{.08b}} \to 0$.

$= -750{,}000(-1)$

$= \$750{,}000$

43. (a) $\displaystyle\int_0^\infty 6000e^{-.08t}\, dt$

$= \displaystyle\lim_{b \to \infty} \int_0^b 6000e^{-.08t}\, dt$

$= \displaystyle\lim_{b \to \infty} \dfrac{6000e^{-.08t}}{-.08} \Bigg|_0^b$

$= \displaystyle\lim_{b \to \infty} \left(\dfrac{6000e^{-.08b}}{-.08} - \dfrac{6000}{-.08} \right)$

$= 0 + 75{,}000 = \$75{,}000$

(b) $\displaystyle\int_0^\infty 6000e^{-.1t}\, dt$

$= \displaystyle\lim_{b \to \infty} \int_0^b 6000e^{-.1t}\, dt$

$= \displaystyle\lim_{b \to \infty} \dfrac{6000e^{-.1t}}{-.1} \Bigg|_0^b$

$= \displaystyle\lim_{b \to \infty} \left(\dfrac{6000e^{-.1b}}{-.1} - \dfrac{6000}{-.1} \right)$

$= 0 + 60{,}000$

$= \$60{,}000$

45. $\displaystyle\int_0^\infty 3000e^{-.1t}\, dt$

$= \displaystyle\lim_{b \to \infty} \int_0^b 3000^{-.1t}\, dt$

$= \displaystyle\lim_{b \to \infty} \dfrac{3000e^{-.1b}}{-.1} \Bigg|_0^b$

$= \displaystyle\lim_{b \to \infty} \left(\dfrac{3000e^{-.1b}}{-.1} + \dfrac{3000}{.1} \right)$

$= 0 + 30{,}000$

$= \$30{,}000$

47. $\displaystyle\int_0^\infty 50e^{-.06t}\, dt$

$= 50 \displaystyle\lim_{b \to \infty} \int_0^b e^{-.06t}\, dt$

$= 50 \displaystyle\lim_{b \to \infty} \dfrac{e^{-.06t}}{-.06} \Bigg|_0^b$

$= \dfrac{50}{-.06} \displaystyle\lim_{b \to \infty} (e^{-.06b} - e^0)$

$= -\dfrac{50}{.06}(0 - 1)$

$= \dfrac{50}{.06}$

$= 833.33$

Section 15.6

1. $\dfrac{dy}{dx} = -2x + 3x^2$

$y = \displaystyle\int (-2x + 3x^2)\,dx$

$= \dfrac{-2x^2}{2} + \dfrac{3x^3}{3} + C$

$= -x^2 + x^3 + C$

3. $3x^2 - 2\dfrac{dy}{dx} = 0$

Solve for $\dfrac{dy}{dx}$.

$\dfrac{dy}{dx} = \dfrac{3x^3}{2}$

$y = \dfrac{3}{2}\displaystyle\int x^3\,dx$

$= \dfrac{3}{2}\left(\dfrac{x^4}{4}\right) + C$

$= \dfrac{3x^4}{8} + C$

5. $y\dfrac{dy}{dx} = x$

Separate the variables and take antiderivatives.

$\displaystyle\int y\,dy = \int x\,dx$

$\dfrac{y^2}{2} = \dfrac{x^2}{2} + K$

$y^2 = \dfrac{2}{2}x^2 + 2K$

$y^2 = x^2 + C$

7. $\dfrac{dy}{dx} = 2xy$

$\displaystyle\int \dfrac{dy}{y} = \int 2x\,dx$

$\ln|y| = \dfrac{2x^2}{2} + C$

$\ln|y| = x^2 + C$

$y = \pm e^{x^2} + C$

$e^{\ln|y|} = e^{x^2 + C}$

$y = \pm e^{x^2} \cdot e^C$

$y = ke^{x^2}$

9. $\dfrac{dy}{dx} = 3x^2 y - 2xy$

$\dfrac{dy}{dx} = y(3x^2 - 2x)$

$\displaystyle\int \dfrac{dy}{y} = \int (3x^2 - 2x)\,dx$

$\ln|y| = \dfrac{3x^3}{3} - \dfrac{2x^2}{2} + C$

$e^{\ln|y|} = e^{x^3 - x^2 + C}$

$y = \pm(e^{x^3 - x^2})e^C$

$y = ke^{(x^3 - x^2)}$

11. $\dfrac{dy}{dx} = \dfrac{y}{x}, \; x > 0$

$\displaystyle\int \dfrac{dy}{y} = \int \dfrac{dx}{x}$

$\ln|y| = \ln x + C$

$e^{\ln|y|} = e^{\ln x + C}$

$y = \pm e^{\ln x} \cdot e^C$

$y = Me^{\ln x}$

$y = Mx$

13.
$$\frac{dy}{dx} = y - 5$$

$$\int \frac{dy}{y - 5} = \int dx$$

$$\ln |y - 5| = x + C$$

$$e^{\ln |y-5|} = e^{x+C}$$

$$y - 5 = \pm e^x \cdot e^C$$

$$y - 5 = Me^x$$

$$y = Me^x + 5$$

15.
$$\frac{dy}{dx} = y^2 e^x$$

$$\int y^{-2} dy = \int e^x dx$$

$$-y^{-1} = e^x + C$$

$$-\frac{1}{y} = e^x + C$$

$$y = \frac{-1}{e^x + C}$$

17. $\frac{dy}{dx} + 2x = 3x^2$

$$\frac{dy}{dx} = 3x^2 - 2x$$

$$y = \frac{3x^3}{3} - \frac{2x^2}{2} + C$$

$$y = x^3 - x^2 + C$$

Since y = 2 when x = 0,

$$2 = 0 - 0 + C$$

$$C = 2.$$

Thus,

$$y = x^3 - x^2 + 2.$$

19. $2\frac{dy}{dx} = 4xe^{-x}$

$$\frac{dy}{dx} = 2xe^{-x}$$

Use tables for antiderivatives or integrate by parts.

$$y = 2(-x - 1)e^{-x} + C$$

Since y = 42 when x = 0,

$$42 = 2(0 - 1)(1) + C$$

$$42 = -2 + C$$

$$C = 44.$$

Thus,

$$y = -2xe^{-x} - 2e^{-x} + 44.$$

21. $\frac{dy}{dx} = \frac{x^2}{y}$; y = 3 when x = 0.

$$\int y \, dy = \int x^2 \, dx$$

$$\frac{y^2}{2} = \frac{x^3}{3} + C$$

$$y^2 = \frac{2}{3}x^3 + 2C$$

$$y^2 = \frac{2}{3}x^3 + k$$

Since y = 3 when x = 0,

$$9 = 0 + k$$

$$k = 0.$$

So $y^2 = \frac{2}{3}x^3 + 9.$

23. $(2x + 3)y = \frac{dy}{dx}$; y = 1 when x = 0.

$$\int (2x + 3) dx = \int \frac{dy}{y}$$

$$\frac{2x^2}{2} + 3x + C = \ln |y|$$

$$e^{x^2+3x+C} = e^{\ln|y|}$$

$$y = (e^{x^2+3x})(\pm e^C)$$

$$y = ke^{x^2+3x}$$

Since y = 1 when x = 0,

$$1 = ke^{0+0}$$

$$k = 1.$$

So $y = e^{x^2+3x}.$

25. $\dfrac{dy}{dx} = \dfrac{y^2}{x}$; $y = 5$ when $x = e$.

$$\int y^{-2}\, dy = \int \frac{dx}{x}$$

$$-y^{-1} = \ln |x| + C$$

$$-\frac{1}{y} = \ln |x| + C$$

$$y = \frac{-1}{\ln |x| + C}$$

Since $y = 5$ when $x = e$,

$$5 = \frac{-1}{\ln e + C}$$

$$5 = \frac{-1}{1 + C}$$

$$5 + 5C = -1$$

$$5C = -6$$

$$C = -\frac{6}{5}.$$

So $y = \dfrac{-1}{\ln |x| - \dfrac{6}{5}}$

$$= \frac{-5}{5 \ln |x| - 6}$$

27. $\dfrac{dy}{dx} = \dfrac{2x + 1}{y - 3}$; $y = 4$ when $x = 0$.

$$\int (y - 3)\, dy = \int (2x + 1)\, dx$$

$$\frac{y^2}{2} - 3y = \frac{2x^2}{2} + x + C$$

Since $y = 4$ when $x = 0$,

$$\frac{16}{2} - 12 = 0 + 0 + C$$

$$C = -4.$$

So $\dfrac{y^2}{2} - 3y = x^2 + x - 4.$

29. $\dfrac{dy}{dx} = (y - 1)^2 e^x$; $y = 2$ when $x = 0$.

$$\int (y - 1)^{-2}\, dy = \int e^x\, dx$$

$$-(y - 1)^{-1} = e^x + C$$

$$-\frac{1}{y - 1} = e^x + C$$

$$y - 1 = \frac{-1}{e^x + C}$$

$$y = 1 - \frac{1}{e^x + C}$$

Since $y = 2$ when $x = 0$,

$$2 = 1 - \frac{1}{1 + C}$$

$$1 = -\frac{1}{1 + C}$$

$$1 + C = -1$$

$$C = -2.$$

So $y = 1 - \dfrac{1}{e^x - 2}$

$$= \frac{e^x - 2 - 1}{e^x - 1}$$

$$= \frac{e^x - 3}{e^x - 2}.$$

31. $\dfrac{dy}{dx} = \dfrac{k}{N}(N - y)y$

(a) $\dfrac{N\, dy}{(N - y)y} = k\, dx$

Since $\dfrac{1}{y} + \dfrac{1}{N - y} = \dfrac{N}{(N - y)y}$,

$$\int \frac{dy}{y} + \int \frac{dy}{N - y} = kx$$

$$\ln \left|\frac{y}{N - y}\right| = kx + C$$

$$\frac{y}{N - y} = Ce^{kx}.$$

For $0 < y < N$, $Ce^{kx} > 0$.

For $0 < N < y$, $Ce^{kx} < 0$.

Solve for y.

$$y = \frac{Ce^{kx}N}{1 + Ce^{kx}} = \frac{N}{1 + C^{-1}e^{-kx}}$$

Let $b = C^{-1}$ for $0 < y < N$.

$$y = \frac{N}{1 + be^{-kx}}$$

Let $-b = C^{-1}$ for $0 < N < y$.

$$y = \frac{N}{1 + be^{-kx}}$$

(b) For $0 < y < N$; $t = 0$, $y = y_0$.

$$y_0 = \frac{N}{1 + be^0} = \frac{N}{1 + b}$$

Solve for b.

$$b = \frac{N - y_0}{y_0}$$

(c) For $0 < N < y$; $t = 0$, $y = y_0$.

$$y_0 = \frac{N}{1 - be^0} = \frac{N}{1 - b}$$

Solve for b.

$$b = \frac{y_0 - N}{y_0}$$

33. **(a)** $0 < y_0 < N$ implies that $y_0 > 0$, $N > 0$, and $N - y_0 > 0$.

Therefore, $b = \frac{N - y_0}{y_0} > 0$.

Also $e^{-kx} > 0$ for all x which implies that $1 + be^{-kx} > 1$.

(1) $y(x) = \frac{N}{1 + be^{-kx}} < N$ since $1 + be^{-kx} > 1$.

(2) $y(x) = \frac{N}{1 + be^{-kx}} > 0$ since $N > 0$ and $1 + be^{-kx} > 0$.

Combining statements (1) and (2), we have

$$0 < \frac{N}{1 + be^{-kx}} = y(x)$$

$$= \frac{N}{1 + be^{-kx}} < N$$

or $0 < y(x) < N$ for all x.

(b) $\lim\limits_{x \to \infty} \dfrac{N}{1 + be^{-kx}} = \dfrac{N}{1 + b(0)} = N$

$\lim\limits_{x \to -\infty} \dfrac{N}{1 + be^{-kx}} = 0$

Note that as $x \to -\infty$, $1 + be^{-kx}$ becomes infinitely large.

Therefore, the horizontal asymptotes are $y = N$ and $y = 0$.

(c) $y'(x) = \dfrac{(1 + be^{-kx})(0) - N(-kbe^{-kx})}{(1 + be^{-kx})^2}$

$= \dfrac{Nkbe^{-kx}}{(1 + be^{-kx})^2} > 0$ for all x.

Therefore, $y(x)$ is an increasing function.

(d) $y''(x) = \dfrac{(1 + be^{-kx})^2(-Nk^2be^{-kx}) - Nkbe^{-kx}[-2kbe^{-kx}(1 + be^{-kx})]}{(1 + be^{-kx})^4}$

$= \dfrac{-Nkbe^{-kx}(1 + be^{-kx})[k(1 + be^{-kx}) - 2kbe^{-kx}]}{(1 + be^{-kx})^4}$

$= \dfrac{-Nkbe^{-kx}(k - kbe^{-kx})}{(1 + be^{-kx})^3}$

$y''(x) = 0$ when

$k - kbe^{-kx} = 0$

$be^{-kx} = 1$

$e^{-kx} = \dfrac{1}{b}$

$-kx = \ln\left(\dfrac{1}{b}\right)$

$x = -\dfrac{\ln\left(\dfrac{1}{b}\right)}{k}$

$= \dfrac{\ln\left(\dfrac{1}{b}\right)^{-1}}{k}$

$= \dfrac{\ln b}{k}.$

When $x = \frac{\ln b}{k}$,

$$y = \frac{N}{1 + be^{-k\left(\frac{\ln b}{k}\right)}}$$

$$= \frac{N}{1 + be^{(-\ln b)}}$$

$$= \frac{N}{1 + be^{\ln(1/b)}}$$

$$= \frac{N}{1 + b\left(\frac{1}{b}\right)}$$

$$= \frac{N}{2}.$$

Therefore $\left(\frac{\ln b}{k}, \frac{N}{2}\right)$ is a point of inflection.

(e) To locate the maximum of $\frac{dy}{dx}$ we must consider, from part (d),

$$\frac{d}{dx}\left(\frac{dy}{dx}\right) = \frac{-Nkbe^{-kx}(k - kbe^{-kx})}{(1 + be^{-kx})^3}.$$

Since $y''(x) > 0$ for $x < \frac{\ln b}{k}$ and

$$y''(x) < 0 \text{ for } x > \frac{\ln b}{k},$$

we know that $x = \frac{\ln b}{k}$ locates a relative maximum of $\frac{dy}{dx}$.

35. $\frac{dy}{dx} = \frac{100}{32 - 4x}$

$$y = 100\left(-\frac{1}{4}\right) \ln |32 - 4x| + C$$

$$y = -25 \ln |32 - 4x| + C$$

Now, $y = 1000$ when $x = 0$.

$$1000 = -25 \ln |32| + C$$

$$C = 1000 + 25 \ln 32$$

$$C \approx 1086.64$$

Thus,

$$y = -25 \ln |32 - 4x| + 1086.64$$

(a) Let $x = 3$.

$$y = -25 \ln |32 - 12| + 1086.64$$

$$\approx \$1011.75$$

(b) Let $x = 5$.

$$y = -25 \ln |32 - 20| + 1086.64$$

$$\approx \$1024.52$$

(c) Advertising expenditures can never reach \$8000. If $x = 8$, the denominator becomes zero.

37. $\frac{dy}{dt} = -.06y$

See Example 4.

$$\int \frac{dy}{y} = \int -.06 \, dt$$

$$\ln |y| = -.06t + C$$

$$e^{\ln|y|} = e^{-.06t+C}$$

$$e^{\ln|y|} = e^{-.06t} \cdot e^C$$

$$|y| = e^{-.06t} \cdot e^C$$

$$y = Me^{-.06t}$$

Let $y = 1$ when $t = 0$.

Solve for M:

$$1 = Me^0$$

$$M = 1.$$

So $y = e^{-.06t}$.

If $y = .50$,

$$.50 = e^{-.06t}$$

$$\ln .5 = -.06t \ln e$$

$$t = \frac{-\ln .5}{.06}$$

$$\approx 11.6 \text{ yr.}$$

39. Let y = the number of firms that will be bankrupt.

$$\frac{dy}{dt} = .06(1500 - y)$$

See Example 5.

$$\int \frac{dy}{1500 - y} = .06 \int dt$$

$$-\ln|1500 - y| = .06t + C$$

$$\ln|1500 - y| = -.06t - C$$

$$|1500 - y| = e^{-.06t - C}$$

$$1500 - y = ke^{-.06t}$$

$$y = 1500 - ke^{-.06t}$$

Since $t = 0$, when $y = 100$.

$$100 = 1500 - k(1)$$

$$k = 1400.$$

$$y = 1500 - 1400e^{-.06t}$$

When $t = 2$ yr,

$$y = 1500 - 1400e^{-.06(2)}$$

$$y = 258.31.$$

About 260 firms will be bankrupt in 2 yr.

41. $E = -\frac{p}{q} \cdot \frac{dq}{dp}$

If $E = \frac{4p^2}{q^2}$,

$$\frac{4p^2}{q^2} = -\frac{p}{q} \cdot \frac{dq}{dp}$$

$$4p\,dp = -q\,dq.$$

$$\int 4p\,dp = -\int q\,dq$$

$$2p^2 = -\frac{1}{2}q^2 + C_1$$

Multiplying by 2, we have

$$4p^2 = -q^2 + 2C_2$$

$$q^2 = -4p^2 + 2C_2.$$

Let $C = 2C_2$. Then

$$q^2 = -4p^2 + C$$

$$q = \pm\sqrt{-4p^2 + C}.$$

Since q cannot be negative,

$$q = \sqrt{-4p^2 + C}.$$

43. $\frac{dy}{dt} = -.03y$

$$\int \frac{dy}{y} = -.03 \int dt$$

$$\ln|y| = -.03t + C$$

$$e^{\ln|y|} = e^{-.03t + C}$$

$$y = Me^{-.03t}$$

Since $y = 6$ when $t = 0$,

$$6 = Me^0$$

$$M = 6.$$

So $y = 6e^{-.03t}$.

If $t = 10$,

$$y = 6e^{-.03(10)}$$

$$\approx 4.4 \text{ cc.}$$

45. $\frac{dy}{dx} = .01(5000 - y)$

$$\int \frac{dy}{5000 - y} = \int .01\,dx$$

$$-\ln|5000 - y| = .01x + C$$

$$\ln|5000 - y| = -.01x + C$$

$$|5000 - y| = e^{-.01x - C}$$

$$|5000 - y| = e^{-.01x} \cdot e^C$$

$$5000 - y = Me^{-.01x}$$

$$y = 5000 - Me^{-.01x}$$

Since $y = 150$ when $x = 0$,

$$150 = 5000 - Me^0$$
$$M = 4850$$

So $y = 5000 - 4850e^{-.01x}$.

If $x = 5$,

$$y = 5000 - 4850e^{-.01(5)}$$
$$\approx 387.$$

(c) When $t = 5$,

$$y = e^{.8(5)}$$
$$= e^4$$
$$\approx 55.$$

(d) When $t = 10$,

$$y = e^{.8(10)}$$
$$= e^8$$
$$\approx 2981.$$

47.   **(a)** $\dfrac{dy}{dt} = ky$

$$\int \frac{dy}{y} = \int k \, dt$$
$$\ln |y| = kt + C$$
$$e^{\ln|y|} = e^{kt+C}$$
$$y = \pm(e^{kt})(e^C)$$
$$y = Me^{kt}$$

If $y = 1$ when $t = 0$ and $y = 5$ when $t = 2$, we have the system of equations:

$$1 = Me^{k(0)}$$
$$\frac{5 = Me^{2k}}{1 = M(1)}$$
$$M = 1$$

Substitute.

$$5 = (1)e^{2k}$$
$$e^{2k} = 5$$
$$2k \ln e = \ln 5$$
$$k = \frac{\ln 5}{2}$$
$$\approx .8$$

(b) If $k = .8$ and $M = 1$,

$$y = e^{.8t}.$$

When $t = 3$,

$$y = e^{.8(3)}$$
$$= e^{2.4}$$
$$\approx 11.$$

49.   **(a)** $\dfrac{dy}{dt} = -.05y$

(b) $$\int \frac{dy}{y} = -\int .05 \, dt$$
$$\ln |y| = -.05t + C$$
$$e^{\ln|y|} = e^{-.05+C}$$
$$y = \pm e^{-.05t} \cdot e^C$$
$$y = Me^{-.05t}$$

(c) Since $y = 90$ when $t = 0$.

$$90 = Me^0$$
$$M = 90.$$

So $y = 90e^{-.05t}$.

(d) At $t = 10$,

$$y = 90e^{-.05(10)}$$
$$\approx 55 \text{ g}.$$

Chapter 15 Review Exercises

5. $\int x(8-x)^{3/2}\,dx$

Let $u = x$ and $dv = (8-x)^{3/2}$.

Then $du = dx$ and $v = -\frac{2}{5}(8-x)^{5/2}$.

$\int x(8-x)^{3/2}\,dx$

$= -\frac{2}{5}x(8-x)^{5/2} + \int \frac{2}{5}(8-x)^{5/2}\,dx$

$= -\frac{2}{5}x(8-x)^{5/2} - \frac{2}{5}\left(\frac{2}{7}\right)(8-x)^{7/2} + C$

$= \frac{-2x}{5}(8-x)^{5/2} - \frac{4}{35}(8-x)^{7/2} + C$

7. $\int xe^x\,dx$

Let $u = x$ and $dv = e^x\,dx$.

Then $du = dx$ and $v = e^x$.

$\int xe^x\,dx = xe^x - \int e^x\,dx$

$= xe^x - e^x + C$

9. $\int \ln|2x+3|\,dx$

First, use substitution.

Let $a = 2x + 3$.

Then $da = 2\,dx$.

$\int \ln|2x+3|\,dx = \frac{1}{2}\int \ln|a|\,da$

Second, integrate by parts.

Let $u = \ln|a|$ and $dv = da$.

Then $du = \frac{1}{a}\,da$ and $v = a$.

$\frac{1}{2}\int \ln|a|\,da$

$= \frac{1}{2}(a\ln|a| - \int da)$

$= \frac{1}{2}(a\ln|a| - a) + C$

Finally, substitute $2x + 3$ for a.

$\int \ln|2x+3|\,dx$

$= \frac{1}{2}[(2x+3)\ln|2x+3| - (2x+3) + C]$

$= \frac{1}{2}(2x+3)[\ln|2x+3| - 1] + C$

11. $\int \frac{x}{9-4x^2}\,dx$

Use substitution.

Let $u = 9 - 4x^2$.

Then $du = -8x\,dx$.

$\int \frac{x}{9-4x^2}\,dx = -\frac{1}{8}\int \frac{-8x\,dx}{9-4x^2}$

$= -\frac{1}{8}\int \frac{du}{u}$

$= -\frac{1}{8}\ln|u| + C$

$= -\frac{1}{8}\ln|9-4x^2| + C$

13. $\int \frac{1}{9-4x^2}\,dx$

First use substitution.

Let $u = 2x$.

Then $du = 2\,dx$.

$\int \frac{1}{9-4x^2}\,dx$

$= \frac{1}{2}\int \frac{2}{9-4x^2}\,dx$

$= \frac{1}{2}\int \frac{1}{9-u^2}\,du$

Now use entry 7 in the table.

$$= \frac{1}{2} \left[\frac{1}{6} \ln \left| \frac{3 + u}{3 - u} \right| \right] + C$$

$$= \frac{1}{12} \ln \left| \frac{3 + u}{3 - u} \right| + C$$

$$= \frac{1}{12} \ln \left| \frac{3 + 2x}{3 - 2x} \right| + C$$

15. $\displaystyle\int_{1}^{2} \frac{1}{x\sqrt{9 - x^2}} \, dx$

(Use entry 9 in the table.)

$$= \left(-\frac{1}{3} \ln \left| \frac{3 + \sqrt{9 - x^2}}{x} \right| \right) \Big|_{1}^{2}$$

$$= -\frac{1}{3} \ln \left| \frac{3 + \sqrt{5}}{2} \right| + \frac{1}{3} \ln \left| \frac{3 + \sqrt{8}}{1} \right|$$

$$= \frac{1}{3} \left[\ln (3 + \sqrt{8}) - \ln \left(\frac{3 + \sqrt{5}}{2} \right) \right]$$

$$= \frac{1}{3} \ln \left[\frac{(3 + 2\sqrt{2})2}{3 + \sqrt{5}} \right]$$

$$= \frac{1}{3} \ln \left(\frac{6 + 4\sqrt{2}}{3 + \sqrt{5}} \right) \approx .26677$$

17. $\displaystyle\int_{1}^{e} x^3 \ln x \, dx$

Let $u = \ln x$ and $dv = x^3 \, dx$.
Use column integration.

D	I
$\ln x$	x^3
$\frac{1}{x}$	$\frac{1}{4}x^4$

$\displaystyle\int x^3 \ln x \, dx$

$$= \frac{1}{4}x^4 \ln x - \int \left(\frac{1}{4}x^4 \cdot \frac{1}{x} \right) dx$$

$$= \frac{1}{4}x^4 \ln x - \frac{1}{4} \int x^3 \, dx$$

$$= \frac{1}{4}x^4 \ln x - \frac{1}{16}x^4 + C$$

$\displaystyle\int_{1}^{e} x^3 \ln x \, dx$

$$= \left(\frac{1}{4}x^4 \ln x - \frac{1}{16}x^4 \right) \Big|_{1}^{e}$$

$$= \left(\frac{e^4}{4} \right)(1) - \frac{e^4}{16} - 0 + \frac{1}{16}$$

$$= \frac{e^4}{4} - \frac{e^4}{16} + \frac{1}{16}$$

$$= \frac{3e^4 + 1}{16}$$

$$\approx 10.300$$

19. $A = \displaystyle\int_{0}^{1} (3 + x^2)e^{2x} \, dx$

$$= \int_{0}^{1} 3e^{2x} \, dx + \int_{0}^{1} x^2 e^{2x} \, dx$$

$$\int 3e^{2x} \, dx = \frac{3}{2}e^{2x} + C$$

For the second integral, $\displaystyle\int x^2 e^{2x} \, dx$,
let $u = x^2$ and $dv = e^{2x} \, dx$
Use column integration.

D	I
x^2	e^{2x}
$2x$	$e^{2x}/2$
2	$e^{2x}/4$
0	$e^{2x}/8$

$$\int x^2 \, e^{2x} dx$$

$$= x^2 \frac{e^{2x}}{2} - 2x\frac{e^{2x}}{4} + 2\frac{e^{2x}}{8}$$

$$= \frac{x^2 e^{2x}}{2} - \frac{xe^{2x}}{2} + \frac{e^{2x}}{4}$$

$$A = \left(\frac{3}{2}e^{2x} + \frac{x^2 e^{2x}}{2} - \frac{xe^{2x}}{2} + \frac{e^{2x}}{4}\right)\Bigg|_0^1$$

$$= \frac{3}{2}e^2 + \frac{e^2}{2} - \frac{e^2}{2} + \frac{e^2}{4} - \left(\frac{3}{2} + \frac{1}{4}\right)$$

$$= \left(\frac{6 + 2 - 2 + 1}{4}\right)e^2 - \left(\frac{7}{4}\right)$$

$$= \frac{7}{4}(e^2 - 1)$$

$$\approx 11.181$$

21. $$\int_2^6 \frac{dx}{x^2 - 1}$$

Trapezoidal rule:

$$n = 4, \ b = 6, \ a = 2, \ f(x) = \frac{1}{x^2 - 1}$$

i	x_i	$f(x_i)$
0	2	.3333
1	3	.125
2	4	.0667
3	5	.0417
4	6	.0286

$$\int_2^6 \frac{dx}{x^2 - 1}$$

$$\approx \frac{6 - 2}{4}\left[\frac{1}{2}(.3333) + .125 + .0667 \right.$$

$$\left. + .0417 + \frac{1}{2}(.0286)\right]$$

$$\approx .4143$$

Exact value:

Use entry 8 in the table.

$$\int_2^6 \frac{dx}{x^2 - 1} = \frac{1}{2} \cdot \ln\left|\frac{x - 1}{x + 1}\right|\Bigg|_2^6$$

$$= \frac{1}{2} \ln \frac{5}{7} - \frac{1}{2} \ln \frac{1}{3}$$

$$= \frac{1}{2} \ln \frac{15}{7}$$

$$\approx .3811$$

23. $$\int_1^5 \ln x \, dx$$

Trapezoidal rule:

$$n = 4, \ b = 5, \ a = 1, \ f(x) = \ln x$$

i	x_i	$f(x_i)$
0	1	0
1	2	.6931
2	3	1.0986
3	4	1.3863
4	5	1.6094

$$\int_1^5 \ln x \, dx$$

$$\approx \frac{5 - 1}{4}\left[\frac{1}{2}(0) + .6931 + 1.0986 \right.$$

$$\left. + 1.3863 + \frac{1}{2}(1.6094)\right]$$

$$\approx 3.983$$

Exact value:

Use entry 4 in the table.

$$\int_1^5 \ln x \, dx$$

$$= x(\ln |x| - 1)\Bigg|_1^5$$

$$= 5(\ln 5 - 1) - (\ln 1 - 1)$$

$$= 5 \ln 5 - 5 - 0 + 1$$

$$= 5 \ln 5 - 4$$

$$\approx 4.047$$

25. $\displaystyle\int_{2}^{10} \frac{x\ dx}{x-1}$

$n = 4$, $b = 10$, $a = 2$, $f(x) = \dfrac{x}{x-1}$

i	x_i	$f(x_i)$
0	2	2
1	4	1.3333
2	6	1.2
3	8	1.1429
4	10	1.1111

$\displaystyle\int_{2}^{10} \frac{x}{x-1}\ dx$

$\approx \dfrac{10-2}{12}[2 + 4(1.3333) + 2(1.2)$

$\qquad + 4(1.1429) + 1.1111]$

≈ 10.28

27. $A = \displaystyle\int_{-1}^{1} \sqrt{1-x^2}\ dx$

$n = 6$, $b = 1$, $a = -1$, $f(x) = \sqrt{1-x^2}$

i	x_i	$f(x_i)$
0	-1	0
1	$-\dfrac{2}{3}$	.7454
2	$-\dfrac{1}{3}$	.9428
3	0	1
4	$\dfrac{1}{3}$	.9428
5	$\dfrac{2}{3}$	.7454
6	1	0

$A \approx \dfrac{1-(-1)}{6}\left[\dfrac{1}{2}(0) + .7454 + .9428\right.$

$\qquad + 1 + .9428 + .7454 + \left.\dfrac{1}{2}(0)\right]$

≈ 1.459

29. $f(x) = 2x - 1$; $y = 0$, $x = 3$

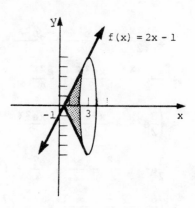

Since $f(x) = 2x - 1$ intersects $y = 0$ at $x = 1/2$, the integral has a lower bound $a = 1/2$.

$V = \pi \displaystyle\int_{1/2}^{3} (2x-1)^2\ dx$

$= \pi \displaystyle\int_{1/2}^{3} (4x^2 - 4x + 1)\ dx$

$= \pi\left(\dfrac{4x^3}{3} - \dfrac{4x^2}{2} + x\right)\Bigg|_{1/2}^{3}$

$= \pi\left(36 - 18 + 3 - \dfrac{1}{6} + \dfrac{1}{2} - \dfrac{1}{2}\right)$

$= \pi\left(21 - \dfrac{1}{6}\right)$

$= \dfrac{125}{6}\pi \approx 65.45$

31. $f(x) = e^{-x}$, $y = 0$, $x = -2$, $x = 1$

$V = \pi \displaystyle\int_{-2}^{1} e^{-2x}\ dx = \dfrac{\pi e^{-2x}}{-2}\Bigg|_{-2}^{1}$

$= \dfrac{\pi e^{-2}}{-2} + \dfrac{\pi e^4}{2}$

$= \dfrac{\pi(e^4 - e^{-2})}{2}$

≈ 85.55

33. $f(x) = 4 - x^2$, $y = 0$, $x = -1$, $x = 1$

$$V = \pi \int_{-1}^{1} (4 - x^2)^2 \, dx$$

$$= \pi \int_{-1}^{1} (16 - 8x^2 + x^4) \, dx$$

$$= \pi\left(16x - \frac{8x^3}{3} + \frac{x^5}{5}\right)\Big|_{-1}^{1}$$

$$= \pi\left(16 - \frac{8}{3} + \frac{1}{5} + 16 - \frac{8}{3} + \frac{1}{5}\right)$$

$$= \pi\left(32 - \frac{16}{3} + \frac{2}{5}\right)$$

$$= \frac{406\pi}{15} \approx 85.03$$

35. The frustum may be shown as follows.

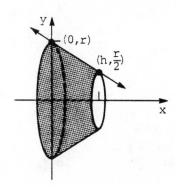

Use the two points given to find

$$f(x) = -\frac{r}{2h}x + r.$$

$$V = \pi \int_{0}^{h} \left(-\frac{r}{2h}x + r\right)^2 \, dx$$

$$= -\frac{2\pi h}{3r}\left(-\frac{r}{2h}x + r\right)^3 \Big|_{0}^{h}$$

$$= -\frac{2\pi h}{3r}\left[\left(-\frac{r}{2} + r\right)^3 - (0 + r)^3\right]$$

$$= -\frac{2\pi h}{3r}\left[\left(\frac{r}{2}\right)^3 - r^3\right]$$

$$-\frac{2\pi h}{3r}\left(\frac{r^3}{8} - r^3\right)$$

$$= -\frac{2\pi h}{3r}\left(-\frac{7r^3}{8}\right)$$

$$= \frac{7\pi r^2 h}{12}$$

37. $f(x) = \sqrt{x + 1}$

$$\frac{1}{b - a}\int_{a}^{b} f(x) \, dx$$

$$= \frac{1}{8 - 0}\int_{0}^{8} \sqrt{x + 1} \, dx$$

$$= \frac{1}{8}\int_{0}^{8} (x + 1)^{1/2} \, dx$$

$$= \frac{1}{8}\left(\frac{2}{3}\right)(x + 1)^{3/2}\Big|_{0}^{8}$$

$$= \frac{1}{12}(9)^{3/2} - \frac{1}{12}(1)$$

$$= \frac{27}{12} - \frac{1}{12} = \frac{26}{12}$$

$$= \frac{13}{6}$$

39. $\displaystyle\int_{1}^{\infty} x^{-1} \, dx$

$$= \lim_{b \to \infty}\int_{1}^{b} \frac{dx}{x}$$

$$= \lim_{b \to \infty} \ln x \Big|_{1}^{b}$$

$$= \lim_{b \to \infty} \ln b$$

As $b \to \infty$, $\ln b \to \infty$. The integral is divergent.

41. $\displaystyle\int_0^\infty \frac{dx}{(5x + 2)^2}$

$\displaystyle = \lim_{b\to\infty} \int_0^b (5x + 2)^{-2}\, dx$

$\displaystyle = \lim_{b\to\infty} \left[-\frac{1}{5}(5x + 2)^{-1}\right]\Big|_0^b$

$\displaystyle = \lim_{b\to\infty} \left[\frac{-1}{5(5x + 2)}\right]\Big|_0^b$

$\displaystyle = \lim_{b\to\infty} \left[\frac{-1}{5(5b + 2)} + \frac{1}{10}\right]$

$\displaystyle = \frac{1}{10}$

43. $\displaystyle\int_{-\infty}^0 \frac{x}{x^2 + 3}\, dx$

$\displaystyle = \lim_{a\to-\infty} \frac{1}{2} \int_a^0 \frac{2x\, dx}{x^2 + 3}$

$\displaystyle = \lim_{a\to-\infty} \frac{1}{2}(\ln |x^2 + 3|)\Big|_a^0$

$\displaystyle = \lim_{a\to-\infty} \frac{1}{2}(\ln 3 - \ln |a^2 + 3|)$

As $a \to -\infty$, $\frac{1}{2}(\ln 3 - \ln |a^2 + 3|) \to \infty$.

The integral is divergent.

45. $\displaystyle A = \int_{-\infty}^1 \frac{3}{(x - 2)^2}\, dx$

$\displaystyle = \lim_{a\to-\infty} \int_a^1 3(x - 2)^{-2} dx$

$\displaystyle = \lim_{a\to-\infty} \frac{3(x - 2)^{-1}}{-1}\Big|_a^1$

$\displaystyle = \lim_{a\to-\infty} \left(-\frac{3}{x - 2}\right)\Big|_a^1$

$\displaystyle = \lim_{a\to-\infty} \left(-\frac{3}{-1} + \frac{3}{a - 2}\right)$

$= 3$

51. $\dfrac{dy}{dx} = 2x^3 + 6x$

$dy = (2x^3 + 6x)dx$

$y = \dfrac{x^4}{2} + 3x^2 + C$

53. $\dfrac{dy}{dx} = 4e^x$

$dy = 4e^x\, dx$

$y = 4e^x + C$

55. $\dfrac{dy}{dx} = \dfrac{3x + 1}{y}$

$y\, dy = (3x + 1)dx$

$\dfrac{y^2}{2} = \dfrac{3x^2}{2} + x + C_1$

$y^2 = 3x^2 + 2x + C$

57. $\dfrac{dy}{dx} = \dfrac{2y + 1}{x}$

$\dfrac{dy}{2y + 1} = \dfrac{dx}{x}$

$\dfrac{1}{2}\left(\dfrac{2\, dy}{2y + 1}\right) = \dfrac{dx}{x}$

$\dfrac{1}{2} \ln |2y + 1| = \ln |x| + C_1$

$\ln |2y + 1|^{1/2} = \ln |x| + \ln k$

$\qquad\qquad Let\ ln\ k = C_1$

$\ln |2y + 1|^{1/2} = \ln k|x|$

$|2y + 1|^{1/2} = k\, |x|$

$2y + 1 = k^2 x^2$

$2y + 1 = Cx^2$

$2y = Cx^2 - 1$

$y = \dfrac{Cx^2 - 1}{2}$

59. $\dfrac{dy}{dx} = x^2 - 5x$; $y = 1$ when $x = 0$

$dy = (x^2 - 5x)\,dx$

$y = \dfrac{x^3}{3} - \dfrac{5x^2}{2} + C$

When $x = 0$, $y = 1$.

$1 = 0 - 0 + C$

$C = 1$

$y = \dfrac{x^3}{3} - \dfrac{5x^2}{2} + 1$

61. $\dfrac{dy}{dx} = 5(e^{-x} - 1)$; $y = 17$ when $x = 0$.

$dy = 5(e^{-x} - 1)\,dx$

$y = 5(-e^{-x} - x) + C$

$= -5e^{-x} - 5x + C$

$17 = -5e^0 - 0 + C$

$17 = -5 + C$

$22 = C$

$y = -5e^{-x} - 5x + 22$

63. $(5 - 2x)y = \dfrac{dy}{dx}$; $y = 2$ when $x = 0$.

$(5 - 2x)\,dx = \dfrac{dy}{y}$

$5x - x^2 + C = \ln |y|$

$e^{5x - x^2 + C} = y$

$e^{5x - x^2} \cdot e^C = y$

$Me^{5x - x^2} = y$

$Me^0 = 2$

$M = 2$

$y = 2e^{5x - x^2}$

65. $\dfrac{dy}{dx} = \dfrac{1 - 2x}{y + 3}$; $y = 16$ when $x = 0$.

$(y + 3)\,dy = (1 - 2x)\,dx$

$\dfrac{y^2}{2} + 3y = x - x^2 + C$

$\dfrac{16^2}{2} + 3(16) = 0 + C$

$176 = C$

$\dfrac{y^2}{2} + 3y = x - x^2 + 176$

$y^2 + 6y = 2x - 2x^2 + 352$

67. $R' = x(x - 50)^{1/2}$

$R = \displaystyle\int_{50}^{75} x(x - 50)^{1/2}\,dx$

Let $u = x$ and $dv = (x - 50)^{1/2}$.

Then $du = dx$ and $v = \dfrac{2}{3}(x - 50)^{3/2}$.

$\displaystyle\int x(x - 50)^{1/2}\,dx$

$= \dfrac{2}{3}x(x - 50)^{3/2} - \dfrac{2}{3}\displaystyle\int (x - 50)^{3/2}\,dx$

$= \dfrac{2}{3}x(x - 50)^{3/2} - \dfrac{2}{3} \cdot \dfrac{2}{5}(x - 50)^{5/2}$

$R = \left[\dfrac{2}{3}x(x - 50)^{3/2} - \dfrac{4}{15}(x - 50)^{5/2}\right]\Bigg|_{50}^{75}$

$= \dfrac{2}{3}(75)(25^{3/2}) - \dfrac{4}{15}(25^{5/2})$

$= 6250 - \dfrac{2500}{3}$

$= \dfrac{16{,}250}{3} \approx 5416.67$

69. Sales

$= \dfrac{7 - 1}{3(6)}[.7 + 4(1.2) + 2(1.5)$

$\quad + 4(1.9) + 2(2.2) + 4(2.4) + 2]$

$= \dfrac{1}{3}(32.10)$

$= 10.7$

Total sales are $10.7 million.

71. $f(x) = 25,000$; 12 yr; 10%

$$P = \int_0^{12} 25,000e^{-.10x}\,dx$$

$$= 25,000\left(\frac{e^{-.10x}}{-.10}\right)\Bigg|_0^{12}$$

$$= 250,000(-.3012 + 1)$$

$$= \$174,701.45$$

73. $f(x) = 30x$; 18 mo; 5%

$$P = \int_0^{1.5} 30xe^{-.05x}\,dx$$

Let $u = 30x$ and $dv = e^{-.05x}\,dx$.

Use column integration.

D	I
30x	$e^{-.05x}$
30	$-20e^{-.05x}$
0	$400e^{-.05x}$

$$\int_0^{1.5} 30xe^{-.05x}\,dx$$

$$= (-600xe^{-.05x} - 12,000e^{-.05x})\Bigg|_0^{1.5}$$

$$= -600(1.5)e^{-.075} - 12,000e^{-.075}$$
$$+ 0 + 12,000e^0$$

$$= -12,900e^{-.075} + 12,000$$

$$\approx \$32.11$$

75. $f(x) = 500e^{-.03x}$; 8 yr; 10% per year

$$e^{.10(8)}\int_0^8 500e^{-.03x}e^{-.10x}\,dx$$

$$= e^{.8}\int_0^8 500e^{-.13x}\,dx$$

$$= \frac{500e^{.8}e^{-.13x}}{-.13}\Bigg|_0^8$$

$$= -8559.77(e^{-1.04} - 1)$$

$$= \$5534.28$$

77. $f(x) = 1000 + 200x$; 10 yr; 9% per year

$$e^{(.09)(10)}\int_0^{10} (1000 + 200x)e^{-.09x}\,dx$$

$$= e^{.9}\left[\frac{1000}{-.09}e^{-.09x}\right.$$

$$\left. + \frac{200}{(.09)^2}(-.09x - 1)e^{-.09x}\right]\Bigg|_0^{10}$$

$$= e^{.9}\left[\frac{1000}{-.09}(e^{-.9} - 1)\right.$$

$$\left. + \frac{200}{(.09)^2}(-1.9e^{-.9} + 1)\right]$$

$$= \$30,035.17$$

79. $e^{.105(10)}\int_0^{10} 10,000e^{-.105x}\,dx$

$$= e^{1.05}\left(\frac{10,000e^{-.105x}}{-.105}\right)\Bigg|_0^{10}$$

$$= \frac{10,000e^{1.05}}{-.105}(e^{-1.05} - 1)$$

$$= -272,157.25(-.65006)$$

$$= \$176,919.15$$

81. (a) $\frac{dy}{dx} = 5e^{.2x}$

$dy = 5e^{.2x} \, dx$

$y = \frac{5}{.2}e^{.2x} + C$

$y = 25e^{.2x} + C$

When $x = 0$, $y = 0$.

$0 = 25e^0 + C$

$C = -25 \qquad e^0 = 1$

$y = 25e^{.2x} - 25$

When $x = 6$,

$y = 25e^{1.2} - 25$

$\approx 58.$

Sales are $5800.

(b) When $x = 12$,

$y = 25e^{2.4} - 25$

$\approx 251.$

Sales are $25,100.

83. $\frac{dA}{dt} = rA - D$; $t = 0$, $A = \$10,000$;

$r = .05$; $D = \$1000$

$\frac{dA}{dt} = .05A - 1000$

$\int \frac{dA}{.05A - 1000} = \int dt$

$\frac{1}{.05} \ln |.05A - 1000| = t + C_2$

$\ln |.05A - 1000| = .05t + C_1$

$.05A - 1000 = Ce^{.05t}$

$A = 20Ce^{.05t} + 20,000$

Since $t = 0$ when $A = 10,000$,

$10,000 = 20C(1) + 20,000$

$C = -500.$

$A = 20(-500)e^{.05t} + 20,000$

$A = 20,000 - 10,000e^{.05t}$

Find t for A = 0.

$0 = 20,000 - 10,000e^{.05t}$

$e^{.05t} = \frac{20,000}{10,000} = 2$

$\ln e^{.05t} = \ln 2$

$t = 20 \ln 2 = 13.86$

It will take about 13.9 yr.

85. $f(x) = 100e^{-.05x}$

The total amount of oil is

$\int_0^\infty 100e^{-.05x} \, dx$

$= \lim_{b \to \infty} \int_0^b 100e^{-.05x} \, dx$

$= \lim_{b \to \infty} \left(\frac{100e^{-.05x}}{-.05} \right) \Big|_0^b$

$= \lim_{b \to \infty} \left(\frac{-2000}{e^{.05b}} + 2000 \right)$

$= 0 + 2000$

$= 2000$ gal.

87. This exercise should be solved by computer or programmable calculator methods. The solutions will vary according to the method that is used.
The answers are given.

(a) .68270

(b) .95450

(c) .99730

(d) .99994

(e) The values approach 1 as n increases. The value of the integral over the interval $(-\infty, \infty)$ is 1.

Extended Application

1. $\displaystyle\int_0^w c\left(1 - \frac{t}{w}\right)\left(\frac{N}{m}\right)e^{-t/m}\ dt$

   $= \dfrac{Nc}{m}\left[\displaystyle\int_0^w \left(1 - \frac{t}{w}\right)e^{-t/m}\ dt\right]$

   $= \dfrac{Nc}{m}\left[\displaystyle\int_0^w e^{-t/m}\ dt - \int_0^w \frac{t}{w}e^{-t/m}\ dt\right]$

   Integrate $\displaystyle\int \frac{t}{w}e^{-t/m}\,dt$ by parts.

   Let $u = \dfrac{t}{w}$ and $dv = e^{-t/m}\,dt.$

   Use column integration.

D	I
$\dfrac{t}{w}$ $\quad+$	$e^{-t/m}$
$\dfrac{1}{w}$ $\quad-$	$-me^{-t/m}$
0	$m^2 e^{-t/m}$

   $\displaystyle\int \frac{t}{w}e^{-t/m}\ dt$

   $= -\dfrac{mt}{w}e^{-t/m} - \dfrac{m^2}{w}e^{-t/m} + C$

   $\dfrac{Nc}{m}\left[\displaystyle\int_0^w e^{-t/m}\,dt - \int_0^w \frac{t}{w}e^{-t/m}\,dt\right]$

   $= \dfrac{Nc}{m}\Big[-me^{-t/m}$

   $\quad - \left(-\dfrac{mt}{w}e^{-t/m} - \dfrac{m^2}{w}e^{-t/m}\right)\Big]\Big|_0^w$

   $= \dfrac{Nc}{m}\left(-me^{-t/m} + \dfrac{mt}{w}e^{-t/m} + \dfrac{m^2}{w}e^{-t/m}\right)\Big|_0^w$

   $= \dfrac{Nc}{m}\Big(-me^{-w/m} + me^{-w/m} + \dfrac{m^2}{w}e^{-w/m}$

   $\quad + m - 0 - \dfrac{m^2}{w}\Big)$

   $= \dfrac{Nc}{m}\left(\dfrac{m^2}{w}e^{-w/m} + m - \dfrac{m^2}{w}\right)$

   $= Nc\left(\dfrac{m}{w}e^{-w/m} + 1 - \dfrac{m}{w}\right)$

   $= Nc\left(1 - \dfrac{m}{w} + \dfrac{m}{w}e^{-w/m}\right)$

2. $r = 50\left(1 - \dfrac{48}{24} + \dfrac{48}{24}e^{-24/48}\right)$

   $= 50(1 - 2 + 2e^{-.5})$

   $= 10.65$

3. $r = 1000\left(1 - \dfrac{60}{24} + \dfrac{60}{24}e^{-24/60}\right)$

   $= 1000(1 - 2.5 + 2.5e^{-.4})$

   ≈ 175.8

4. $r = 1200\left(1 - \dfrac{30}{30} + \dfrac{30}{30}e^{-1}\right)$

   $= 1200(e^{-1})$

   ≈ 441.66

CHAPTER 15   TEST

Use integration by parts or column integration to find the integrals.

1. $\int x \ln |5x| \ dx$

2. $\int x\sqrt{2x - 1} \ dx$

3. $\int (2x - 1)e^x \ dx$

4. $\int \frac{x + 3}{(3x - 1)^4} \ dx$

5. Find the area between the x-axis and the graph of $f(x) = x\sqrt[3]{x - 3}$ from x = 4 to x = 11.

6. Given that the marginal profit in dollars earned from the sale of x computers is

$$P'(x) = xe^{.001x} - 100,$$

find the total profit from the sale of the first 500 computers.

7. Use the trapezoidal rule with n = 4 to approximate the value of

$$\int_0^2 \sqrt{1 + x^3} \ dx.$$

8. Use Simpson's rule with n = 6 to approximate the value of $\int_0^3 \frac{1}{x^2 + 1} \ dx.$

9. Find the area between the curve $y = e^{-x^2}$ and the x-axis from x = 0 to x = 3, using the trapezoidal rule with n = 6.

10. Find the area between the curve $y = \frac{x}{x^2 + 1}$ and the x-axis from x = 1 to x = 4, using Simpson's rule with n = 6.

Find the volume of the solid of revolution formed by rotating each of the following bounded regions about the x-axis.

11. $f(x) = 3x - 1$, y = 0, x = 4

12. $f(x) = \frac{1}{\sqrt{3x + 1}}$, y = 0, x = 0, x = 5

13. $f(x) = e^x$, y = 0, x = -1, x = 2

14.   Find the average value of $f(x) = x^3 - x^2$ from $a = -1$ to $b = 1$.

15.   The rate of depreciation t years after purchase of a certain machine is

$$D = 10,000(t - 7), \ 0 \le t \le 6.$$

What is the average depreciation over the first 3 yr?

16.   The rate of flow of an investment is $2000 per year for 10 yr. Find the present value if the annual interest rate is 11% compounded continuously.

17.   The rate of flow of an investment is given by

$$f(x) = 200e^{-.1x}$$

for 10 yr. Find the present value if the annual interest rate is 8% compounded continuously.

18.   The rate of flow of an investment is given by the function $f(x) = 1000$. Find the final amount at the end of 15 yr at an interest rate of 11% compounded continuously.

19.   An investment scheme is expected to produce a continuous flow of money, starting at $2000 and increasing exponentially at 4% per year for 8 yr. Find the present value at an interest rate of 12% compounded continuously.

Find the value of each interval that converges.

20. $\int_{2}^{\infty} \dfrac{14}{(x + 1)^2} \ dx$ 21. $\int_{-\infty}^{2} \dfrac{4}{x + 1} \ dx$

22. $\int_{1}^{\infty} e^{-3x} \ dx$

Find the area between the graph of the function and the x-axis over the given interval, if possible.

23. $f(x) = 8e^{-x}$ on $[1, \infty)$ 24. $f(x) = \dfrac{1}{(2x - 1)^2}$ on $[2, \infty)$

25.   Find the capital value of an asset that generates income at an annual rate of $4000 if the interest rate is 8% compounded continuously.

Find general solutions for the following differential equations.

26. $\dfrac{dy}{dx} = 3x^2 + 4x - 5$ 27. $\dfrac{dy}{dx} = 5e^{2x}$

28. $\dfrac{dy}{dx} = \dfrac{2x}{x^2 + 5}$

Find particular solutions of the following differential equations.

29. $\dfrac{dy}{dx} = 4x^3 + 2x^2 + 1$; $y = 4$ when $x = 1$

30. $\dfrac{dy}{dx} = \dfrac{1}{3x + 1}$; $y = 4$ when $x = 3$

31. $\dfrac{dy}{dx} = 4(e^{-x} - 1)$; $y = 5$ when $x = 0$

32. After use of an insecticide, the rate of decline of an insect population is

$$\frac{dy}{dt} = \frac{-8}{1 + 4t}$$

where t is the number of hours after the insecticide is applied. If there were 40 insects initially, how many are left after 20 hours?

Find general solutions for the following differential equations. (Some solutions may give y implicitly.)

33. $\dfrac{dy}{dx} = \dfrac{2x + 1}{y - 1}$ 34. $\dfrac{dy}{dx} = \dfrac{3y}{x + 1}$ 35. $\dfrac{dy}{dx} = \dfrac{e^x - x}{y + 1}$

Find particular solutions of the following differential equations.

36. $\sqrt{y}\,\dfrac{dy}{dx} = xy$; $y = 4$ when $x = 2$

37. $\dfrac{dy}{dx} = e^y \cdot x^3$; $y = 0$ when $x = 0$

CHAPTER 15   TEST ANSWERS

1. $\frac{1}{2}x^2 \ln |5x| - \frac{1}{4}x^2 + C$

2. $\frac{x}{3}(2x - 1)^{3/2} - \frac{1}{15}(2x - 1)^{5/2} + C$

3. $(2x - 3)e^x + C$

4. $-\frac{1}{9}\left[\frac{x + 3}{(3x - 1)^3}\right] - \frac{1}{54(3x - 1)^2} + C$ or $-\frac{9x + 17}{54(3x - 1)^3} + C$

5. 88.18   6. \$125,639.36   7. 3.28   8. 1.25   9. .89   10. 1.07

11. $\frac{1331}{9}\pi$ or 464.61   12. 2.90   13. 85.55   14. $-\frac{1}{3}$

15. 55,000 per year   16. \$12,129.62   17. \$927.45   18. \$38,245.27

19. \$11,817.69   20. $\frac{14}{3}$ 21. Divergent   22. .017

23. $\frac{8}{e}$ or 2.943   24. $\frac{1}{6}$ 25. \$50,000

26. $y = x^3 + 2x^2 - 5x + C$ 27. $y = \frac{5}{2}e^{2x} + C$ 28. $y = \ln (x^2 + 5) + C$

29. $y = x^4 + \frac{2}{3}x^3 + x + \frac{4}{3}$ 30. $y = \frac{1}{3} \ln |3x + 1| + 4 - \frac{1}{3} \ln 10$

31. $y = -4e^{-x} - 4x + 9$ 32. 31   33. $\frac{y^2}{2} - y = x^2 + x + C$

34. $y = C(x + 1)^3$ 35. $\frac{y^2}{2} + y = e^x - \frac{x^2}{2} + C$ 36. $y^{1/2} = \frac{x^2}{4} + 1$

37. $e^{-y} = 1 - \frac{x^4}{4}$ or $y = -\ln \left|1 - \frac{x^4}{4}\right|$

CHAPTER 16   MULTIVARIABLE CALCULUS

Section 16.1

1. $f(x, y) = 4x + 5y + 3$

    (a) $f(2, -1) = 4(2) + 5(-1) + 3$
    $$= 6$$

    (b) $f(-4, 1) = 4(-4) + 5(1) + 3$
    $$= -8$$

    (c) $f(-2, -3) = 4(-2) + 5(-3) + 3$
    $$= -20$$

    (d) $f(0, 8) = 4(0) + 5(8) + 3$
    $$= 43$$

3. $h(x, y) = = \sqrt{x^2 + 2y^2}$

    (a) $h(5, 3) = \sqrt{25 + 2(9)} = \sqrt{43}$

    (b) $h(2, 4) = \sqrt{4 + 32} = 6$

    (c) $h(-1, -3) = \sqrt{1 + 18} = \sqrt{19}$

    (d) $h(-3, -1) = \sqrt{9 + 2} = \sqrt{11}$

For Exercises 5–15, see the graphs in the answer section at the back of your textbook.

5. $x + y + z = 6$

    If $x = 0$ and $y = 0$, $z = 6$.
    If $x = 0$ and $z = 0$, $y = 6$.
    If $y = 0$ and $z = 0$, $x = 6$.

7. $2x + 3y + 4z = 12$

    If $x = 0$ and $y = 0$, $z = 3$.
    If $x = 0$ and $z = 0$, $y = 4$.
    If $y = 0$ and $z = 0$, $x = 6$.

9. $x + y = 4$

    If $x = 0$, $y = 4$.
    If $y = 0$, $x = 4$.

    There is no z-intercept.

11. $x = 2$

    The point $(2, 0, 0)$ is on the graph.
    There are no y- or z-intercepts.
    The plane is parallel to the yz-plane.

13. $3x + 2y + z = 18$

    For $z = 0$, $3x + 2y = 18$. Graph the line $3x + 2y = 18$ in the xy-plane.
    For $z = 2$, $3x + 2y = 16$. Graph the line $3x + 2y = 16$ in the plane $z = 2$.
    For $z = 4$, $3x + 2y = 14$. Graph the line $3x + 2y = 14$ in the plane $z = 4$.

15. $y^2 - x = -z$

    For $z = 0$, $x = y^2$. Graph $x = y^2$ in the xy-plane.
    For $z = 2$, $x = y^2 + 2$. Graph $x = y^2 + 2$ in the plane $z = 2$.
    For $z = 4$, $x = y^2 + 4$. Graph $x = y^2 + 4$ in the plane $z = 4$.

17. $P(x, y) = 100\left[\frac{3}{5}x^{-2/5} + \frac{2}{5}y^{-2/5}\right]^{-5}$

    (a) $P(32, 1)$
    $$= 100\left[\frac{3}{5}(32)^{-2/5} + \frac{2}{5}(1)^{-2/5}\right]^{-5}$$
    $$= 100\left[\frac{3}{5}\left(\frac{1}{4}\right) + \frac{2}{5}(1)\right]^{-5}$$

$$= 100\left[\frac{11}{20}\right]^{-5}$$

$$= 100\left(\frac{20}{11}\right)^{5}$$

$$\approx 1986.95$$

The production is approximately 1987 cameras.

(b) $P(1, 32)$

$$= 100\left[\frac{3}{5}(1)^{-2/5} + \frac{2}{5}(32)^{-2/5}\right]^{-5}$$

$$= 100\left[\frac{3}{5}(1) + \frac{2}{5}\left(\frac{1}{4}\right)\right]^{-5}$$

$$= 100\left(\frac{7}{10}\right)^{-5}$$

$$= 100\left(\frac{10}{7}\right)^{5}$$

$$\approx 595$$

The production is approximately 595 cameras.

(c) 32 work hours means that $x = 32$. 243 units of capital means that $y = 243$.

$P(32, 243)$

$$= 100\left[\frac{3}{5}(32)^{-2/5} + \frac{2}{5}(243)^{-2/5}\right]^{-5}$$

$$= 100\left[\frac{3}{5}\left(\frac{1}{4}\right) + \frac{2}{5}\left(\frac{1}{9}\right)\right]^{-5}$$

$$= 100\left(\frac{7}{36}\right)^{-5}$$

$$= 100\left(\frac{36}{7}\right)^{5}$$

$$\approx 359,767.81$$

The production is approximately 359,768 cameras.

19. $M = f(n, i, t)$

$$= \frac{(1 + i)^{n}(1 - t) + t}{[1 + (1 - t)i]^{n}}$$

Therefore,

$f(25, .06, .33)$

$$= \frac{(1 + .06)^{25}(1 - .33) + .33}{[1 + (1 - .33)(.06)]^{25}}$$

$$= \frac{(1.06)^{25}(.67) + .33}{[1 + (.67)(.06)]^{25}}$$

$$\approx 1.197$$

Since the value of $M \approx 1.197$, which is greater than 1, the IRA account grows faster.

21. $z = x^{.4}y^{.6}$ where $z = 500$

$$500 = x^{2/5} y^{3/5}$$

$$\frac{500}{x^{2/5}} = y^{3/5}$$

$$\left(\frac{500}{x^{2/5}}\right)^{5/3} = (y^{3/5})^{5/3}$$

$$y = \frac{(500)^{5/3}}{x^{2/3}}$$

$$y \approx \frac{31,498}{x^{2/3}}$$

See answer graph in the back of your textbook.

23. $C(x, y, z) = 200x + 100y + 50z$

25. $A = .202W^{.425}H^{.725}$

(a) $A = .202(72)^{.425}(1.78)^{.725}$

$$\approx 1.89 \text{ m}^2$$

(b) $A = .202(65)^{.425}(1.40)^{.725}$

$$\approx 1.52 \text{ m}^2$$

(c) $A = .202(70)^{.425}(1.60)^{.725}$

$$\approx 1.73 \text{ m}^2$$

(d) Answers vary.

27. Let the area be given by $g(L, W, H)$. Then,

$$g(L, W, H) = 2LW + 2WH + 2LH \text{ ft}^2$$

29. $z = x^2 + y^2$

The xy-trace is

$$z = x^2 + 0 = x^2.$$

The yz-trace is

$$z = 0 + y^2 = y^2.$$

Both are parabolas with vertices at the origin that open upward.
The xy-trace is

$$0 = x^2 + y^2.$$

This is a point, the origin.
The equation is represented by a paraboloid as shown in (c).

31. $x^2 - y^2 = z$

The xz-trace is

$$x^2 = z,$$

which is a parabola with vertex at the origin that opens upward.
The yz-trace is

$$-y^2 = z,$$

which is a parabola with vertex at the origin that opens downward.
The xy-trace is

$$x^2 - y^2 = 0$$
$$x^2 = y^2$$
$$x = y \quad \text{or} \quad x = -y,$$

which are two lines that intersect at the origin.

The equation is represented by a hyperbolic paraboloid, as shown in (e).

33. $\dfrac{x^2}{16} + \dfrac{y^2}{25} + \dfrac{z^2}{4} = 1$

xz-trace:

$$\dfrac{x^2}{16} + \dfrac{z^2}{4} = 1, \text{ an ellipse}$$

yz-trace:

$$\dfrac{y^2}{25} + \dfrac{z^2}{4} = 1, \text{ and ellipse}$$

xy-trace:

$$\dfrac{x^2}{16} + \dfrac{y^2}{25} = 1, \text{ an ellipse}$$

The graph is an ellipsoid as shown in (b).

35. **(a)** If $f(x, y) = 9x^2 - 3y^2$, then

$$\frac{f(x + h, y) - f(x, y)}{h}$$

$$= \frac{[9(x + h)^2 - 3y^2] - [9x^2 - 3y^2]}{h}$$

$$= \frac{9x^2 + 18xh + 9h^2 - 3y^2 - 9x^2 + 3y^2}{h}$$

$$= \frac{h(18x + 9h)}{h}$$

$$= 18x + 9h.$$

(b) If $f(x, y) = 9x^2 - 3y^2$, then

$$\frac{f(x, y + h) - f(x, y)}{h}$$

$$= \frac{[9x^2 - 3(y + h)^2] - [9x^2 - 3y^2]}{h}$$

$$= \frac{9x^2 - 3y^2 - 6yh - 3h^2 - 9x^2 + 3y^2}{h}$$

$$= \frac{-h(6y + 3h)}{h}$$

$$= -(6y + 3h)$$

$$= -6y - 3h.$$

Section 16.2

1. $z = f(x, y) = 12x^2 - 8xy + 3y^2$

   (a) $\dfrac{\partial z}{\partial x} = 24x - 8y$

   (b) $\dfrac{\partial z}{\partial y} = -8x + 6y$

   (c) $\dfrac{\partial f}{\partial x}(2, 3) = 24(2) - 8(3) = 24$

   (d) $f_y(1, -2) = -8(1) + 6(-2)$
   $$= -20$$

3. $f(x, y) = -2xy + 6y^3 + 2$

   $f_x = -2y$

   $f_y = -2x + 18y^2$

   $f_x(2, -1) = -2(-1) = 2$

   $f_y(-4, 3) = -2(-4) + 18(3)^2$
   $$= 8 + 18(9)$$
   $$= 170$$

5. $f(x, y) = 3x^3y^2$

   $f_x = 9x^2y^2$

   $f_y = 6x^3y$

   $f_x(2, -1) = 9(4)(1) = 36$

   $f_y(-4, 3) = 6(-64)3 = -1152$

7. $f(x, y) = e^{x+y}$

   $f_x = e^{x+y}$

   $f_y = e^{x+y}$

   $f_x(2, 1) = e^{2-1} = e^1 = e$

   $f_y(-4, 3) = e^{-4+3} = e^{-1} = \dfrac{1}{e}$

9. $f(x, y) = -5e^{3x-4y}$

   $f_x = -15e^{3x-4y}$

   $f_y = 20e^{3x-4y}$

   $f_x(2, -1) = -15e^{10}$

   $f_y(-4, 3) = 20e^{-12-12} = 20e^{-24}$

11. $f(x, y) = \dfrac{x^2 + y^3}{x^3 - y^2}$

    $f_x = \dfrac{2x(x^3 - y^2) - 3x^2(x^2 + y^3)}{(x^3 - y^2)^2}$

    $\quad = \dfrac{2x^4 - 2xy^2 - 3x^4 - 3x^2y^3}{(x^3 - y^2)^2}$

    $\quad = \dfrac{-x^4 - 2xy^2 - 3x^2y^3}{(x^3 - y^2)^2}$

    $f_y = \dfrac{3y^2(x^3 - y^2) - (-2y)(x^2 + y^3)}{(x^3 - y^2)^2}$

    $\quad = \dfrac{3x^3y^2 - 3y^4 + 2x^2y + 2y^4}{(x^3 - y^2)^2}$

    $\quad = \dfrac{3x^3y^2 - y^4 + 2x^2y}{(x^3 - y^2)^2}$

    $f_x(2, -1)$

    $\quad = \dfrac{-2^4 - 2(2)(-1)^2 - 3(2^2)(-1)^3}{[2^3 - (-1)^2]^2}$

    $\quad = -\dfrac{8}{49}$

    $f_y(-4, 3)$

    $\quad = \dfrac{3(-4)^3(3)^2 - 3^4 + 2(-4)^2(3)}{[(-4)^3 - 3^2]^2}$

    $\quad = -\dfrac{1713}{5329}$

13. $f(x, y) = \ln|1 + 3x^2y^3|$

    $f_x = \dfrac{1}{1 + 3x^2y^3} \cdot 6xy^3$

    $\quad = \dfrac{6xy^3}{1 + 3x^2y^3}$

    $f_y = \dfrac{9x^2y^2}{1 + 3x^2y^3}$

    $f_x(2, -1) = \dfrac{6(2)(-1)}{1 + 3(4)(-1)}$

    $\quad = \dfrac{-12}{1 - 12} = \dfrac{12}{11}$

    $f_y(-4, 3) = \dfrac{9(16)(9)}{1 + 3(16)(27)}$

    $\quad = \dfrac{1296}{1297}$

15. $f(x, y) = xe^{x^2y}$

$f_x = e^{x^2y} \cdot 1 + x(2xy)(e^{x^2y})$

$\quad = e^{x^2y}(1 + 2x^2y)$

$f_y = x^3 e^{x^2y}$

$f_x(2, -1) = e^{-4}(1 - 8) = -7e^{-4}$

$f_y(-4, 3) = -64e^{48}$

17. $f(x, y) = 6x^3y - 9y^2 + 2x$

$f_x = 18x^2y + 2$

$f_y = 6x^3 - 18y$

$f_{xx} = 36xy$

$f_{yy} = -18$

$f_{xy} = 18x^2 = f_{yx}$

19. $R(x, y) = 4x^2 - 5xy^3 + 12y^2x^2$

$R_x = 8x - 5y^3 + 24y^2x$

$R_y = -15xy^2 + 24yx^2$

$R_{xx} = 8 + 24y^2$

$R_{yy} = -30xy + 24x^2$

$R_{xy} = -15y^2 + 48xy = R_{yx}$

21. $r(x, y) = \dfrac{4x}{x + y}$

$r_x = \dfrac{4(x + y) - 4x}{(x + y)^2}$

$\quad = 4y(x + y)^{-2}$

$r_y = \dfrac{-4x}{(x + y)^2}$

$r_{xx} = -8y(x + y)^{-3} = \dfrac{-8y}{(x + y)^3}$

$r_{yy} = 8x(x + y)^{-3} = \dfrac{8x}{(x + y)^3}$

$r_{xy} = 4(x - y)(x + y)^{-3}$

$\quad = \dfrac{4x - 4y}{(x + y)^3} = r_{yx}$

23. $z = 4xe^y$

$z_x = 4e^y$

$z_y = 4xe^y$

$z_{xx} = 0$

$z_{yy} = 4xe^y$

$z_{xy} = 4e^y = z_{yx}$

25. $r = \ln |x + y|$

$r_x = \dfrac{1}{x + y}$

$r_y = \dfrac{1}{x + y}$

$r_{xx} = \dfrac{-1}{(x + y)^2}$

$r_{yy} = \dfrac{-1}{(x + y)^2}$

$r_{xy} = \dfrac{-1}{(x + y)^2} = r_{yx}$

27. $z = x \ln |xy|$

$z_x = \ln |xy| + 1$

$z_y = \dfrac{x}{y}$

$z_{xx} = \dfrac{1}{x}$

$z_{yy} = -xy^{-2} = \dfrac{-x}{y^2}$

$z_{xy} = \dfrac{1}{y} = z_{yx}$

29. $f(x, y) = 6x^2 + 6y^2 + 6xy + 36x - 5$

First, $f_x = 12x + 6y + 36$ and

$f_y = 12y + 6x$.

We must solve the system

$$12x + 6y + 36 = 0$$
$$12y + 6x = 0.$$

Multiply both sides of the first

equation by -2 and add.

$$-24x - 12y - 72 = 0$$
$$\underline{6x + 12y = 0}$$
$$-18x - 72 = 0$$
$$x = -4$$

Substitute into either equation to get $y = 2$.

31. $f(x, y) = 9xy - x^3 - y^3 - 6$

First, $f_x = 9y - 3x^2$ and
$f_y = 9x - 3y^2$.
We must solve the system

$$9y - 3x^2 = 0$$
$$9x - 3y^2 = 0.$$

From the first equation, $y = \frac{1}{3}x^2$.

Substitute into the second equation to get

$$9x - 3\left(\frac{1}{3}x^2\right)^2 = 0$$

$$9x - 3\left(\frac{1}{9}x^4\right) = 0$$

$$9x - \frac{1}{3}x^4 = 0.$$

Multiply by 3 to get

$$27x - x^4 = 0.$$

Now factor:

$$x(27 - x^3) = 0.$$

Set each factor equal to 0.

$$x = 0 \quad \text{or} \quad 27 - x^3 = 0$$
$$x = 3$$

Substitute into $y = \frac{x^2}{3}$.

$$y = 0 \quad \text{or} \quad y = 3$$

The solutions are $x = 0$, $y = 0$ and
$x = 3$, $y = 3$.

33. $f(x, y, z) = x^2 + yz + z^4$

$f_x = 2x$

$f_y = z$

$f_z = y + 4z^3$

$f_{yz} = 1$

35. $f(x, y, z) = \dfrac{6x - 5y}{4z + 5}$

$f_x = \dfrac{6}{4z + 5}$

$f_y = \dfrac{-5}{4z + 5}$

$f_z = \dfrac{-4(6x - 5y)}{(4z + 5)^2}$

$f_{yz} = \dfrac{20}{(4z + 5)^2}$

37. $f(x, y, z) = \ln \left| x^2 - 5xz^2 + y^4 \right|$

$f_x = \dfrac{2x - 5z^2}{x^2 - 5xz^2 + y^4}$

$f_y = \dfrac{4y^3}{x^2 - 5xz^2 + y^4}$

$f_z = \dfrac{-10xz}{x^2 - 5xz^2 + y^4}$

$f_{yz} = \dfrac{4y^3(10zx)}{(x^2 - 5xz^2 + y^4)^2}$

$\phantom{f_{yz}} = \dfrac{40xy^3z}{(x^2 - 5xz^2 + y^4)^2}$

39. $M(x, y) = 40x^2 + 30y^2 - 10xy + 30$

(a) $ M_y = 60y - 10x$
$ M_y(4, 2) = 120 - 40 = 80$

(b) $ M_x = 80x - 10y$
$ M_x(3, 6) = 240 - 60 = 180$

(c) $\dfrac{\partial M}{\partial x}(2, 5) = 80(2) - 10(5)$

$ = 110$

(d) $\dfrac{\partial M}{\partial y}(6, 7) = 60(7) - 10(6)$

$ = 360$

41. $f(p, i) = 132p - 2pi - .01p^2$

(a) $f(9400, 8)$

$= 132(9400) - 2(9400)(8)$

$\qquad - .01(9400)^2$

$= \$206,800$

The weekly sales are $\$206,800$.

(b) $f_p = 132 - 2i - .02p$, which represents the rate of change in weekly sales revenue per unit change in price when the interest rate remains constant.

$f_i = -2p$, which represents the rate of change in weekly sales revenue per unit change in interest rate when the list price remains constant.

(c) $p = 9400$ remains constant and i changes by 1 unit from 8 to 9.

$$f_i(p, i) = f_i(9400, 8)$$
$$= -2(9400)$$
$$= -18,800$$

Therefore, sales revenue declines by $\$18,800$.

43. $f(x, y) = \left(\frac{1}{3}x^{-1/3} + \frac{2}{3}y^{-1/3}\right)^{-3}$

(a) $f(27, 64)$

$= \left[\frac{1}{3}(27)^{-1/3} + \frac{2}{3}(64)^{-1/3}\right]^{-3}$

$= \left[\frac{1}{3}\left(\frac{1}{3}\right) + \frac{2}{3}\left(\frac{1}{4}\right)\right]^{-3}$

$= \left(\frac{1}{9} + \frac{1}{6}\right)^{-3}$

$= \left(\frac{5}{18}\right)^{-3}$

$= \left(\frac{18}{5}\right)^{3}$

$= 46.656$

The production is 46.656 hundred units.

(b) $f_x = -3\left(\frac{1}{3}x^{-1/3} + \frac{2}{3}y^{-1/3}\right)\left(-\frac{1}{9}x^{-4/3}\right)$

$f_x(27, 64)$

$= -3\left[\frac{1}{3}(27)^{-1/3} + \frac{2}{3}(64)^{-1/3}\right]^{-4}$

$\qquad \cdot \left(-\frac{1}{9}\right)(27)^{-4/3}$

$= -3\left(\frac{5}{18}\right)^{-4}\left(-\frac{1}{9}\right)\left(\frac{1}{81}\right)$

$= \frac{432}{625}$

$f_x(27, 64) = .6912$ hundred units, which represents the rate at which production is changing when labor changes by 1 unit from 27 to 28 and capital remains constant.

$f_y = -3\left(\frac{1}{3}x^{-1/3} + \frac{2}{3}y^{-1/3}\right)^{-4}\left(-\frac{2}{9}y^{-4/3}\right)$

$f_y(27, 64)$

$= -3\left[\frac{1}{3}(27)^{-1/3} + \frac{2}{3}(64)^{-1/3}\right]^{-4}$

$\qquad \cdot \left(-\frac{2}{9}\right)(64)^{-4/3}$

$= -3\left(\frac{5}{18}\right)^{-4}\left(-\frac{2}{9}\right)\left(\frac{1}{256}\right)$

$= \frac{2187}{5000}$

$= .4374$ hundred units,

which represents the rate at which production is changing when capital changes by 1 unit from 64 to 65 and labor remains constant.

(c) If labor increases by 1 unit then production would increase at the rate of

$f_x(x, y)$

$$= \frac{1}{3}x^{-4/3}\left(\frac{1}{3}x^{-1/3} + \frac{2}{3}y^{-1/3}\right)^{-4}.$$

(See part (b) of this solution.)

45. $z = x^{.4}y^{.6}$

The marginal productivity of labor is

$$\frac{\partial z}{\partial x} = .4x^{-.6}y^{.6} + x^{.4} \cdot 0$$

$$= .4x^{-.6}y^{.6}.$$

The marginal productivity of capital is

$$\frac{\partial z}{\partial y} = x^{.4}(.6y^{-.4}) + y^{.6} \cdot 0$$

$$= .6x^{.4}y^{-.4}.$$

47. $M(x, y) = 2xy + 10xy^2 + 30y^2 + 20$

(a) $\dfrac{\partial M}{\partial x} = 2y + 10y^2$

$\dfrac{\partial M}{\partial x}(20, 4) = 168$

(b) $\dfrac{\partial M}{\partial y} = 2x + 20yx + 60y$

$\dfrac{\partial M}{\partial y}(24, 10) = 5448$

(c) An increase in days since rain causes more of an increase in matings.

49. $A(W, H) = .202W^{.425}H^{.725}$

(a) $\dfrac{\partial A}{\partial W}(72, 1.8)$

$= .08585(72)^{-.575}(1.8)^{.725}$

$= .0112$

(b) $\dfrac{\partial A}{\partial H}(70, 1.6)$

$= .14645(70)^{.425}(1.6)^{-.275}$

$\approx .783$

51. $f(n, c) = \dfrac{1}{8}n^2 - \dfrac{1}{5}c + \dfrac{1937}{8}$

(a) $f(4, 1200)$

$= \dfrac{1}{8}(4)^2 - \dfrac{1}{5}(1200) + \dfrac{1937}{8}$

$= 2 - 240 + \dfrac{1937}{8}$

$= 4.125$ lb

(b) $\dfrac{\partial f}{\partial n} = \dfrac{1}{8}(2n) - \dfrac{1}{5}(0) + 0$

$= \dfrac{1}{4}n,$

which represents the rate of change of weight loss per unit change in number of workouts.

(c) $f_n(3, 1100) = \dfrac{1}{4}(3)$

$= \dfrac{3}{4}$ lb

represents an additional weight loss by adding the fourth workout.

53. $p = f(s, n, a) = .003a + .1(sn)^{1/2}$

(a) $f(8, 6, 450)$

$= .003(450) + .1[(8)(6)]^{1/2}$

$= 1.35 + .1(48)^{1/2}$

$= 2.0428$

$p \approx 2.04\%$

(b) $f(3, 3, 320)$

$= .003(320) + .1[(3)(3)]^{1/2}$

$= .96 + .1(9)^{1/2}$

$= 1.26$

$p = 1.26\%$

(c) $f_n = .003(0) + .1\left[\frac{1}{2}(sn)^{-1/2}(s)\right]$

$f_n(3, 3, 320) = .1\left(\frac{1}{2}\right)[(3)(3)]^{-1/2}(3)$

$$= (.1)\left(\frac{1}{2}\right)\left(\frac{1}{3}\right)(3)$$

$$= .05$$

$$p_n = .05\%$$

$f_a = .003$ for all ordered triples (s, n, a). Therefore, $p_a = .003\%$.

$p_n = .05\%$ is the rate of change of the probability for an additional semester of high school math.

$p_a = .003\%$ is the rate of change of the probability per unit of change in an SAT score.

Section 16.3

1. $f(x, y) = xy + x - y$

    $f_x = y + 1$, $f_y = x - 1$

    If $f_x = 0$, $y = -1$.

    If $f_y = 0$, $x = 1$.

    Therefore, $(1, -1)$ is the critical point.

    $$f_{xx} = 0,$$
    $$f_{yy} = 0,$$
    $$f_{xy} = 1.$$

    For $(1, -1)$,

    $$D = 0 \cdot 0 - 1^2 = -1 < 0.$$

    A saddle point is at $(1, -1)$.

3. $f(x, y) = x^2 - 2xy + 2y^2 + x - 5$

    $f_x = 2x - 2y + 1$, $f_y = -2x + 4y$

Solve the system $f_x = 0$, $f_y = 0$.

$$
\begin{array}{r}
2x - 2y + 1 = 0 \\
\underline{-2x + 4y \quad\;\; = 0} \\
2y + 1 = 0 \\
y = -\frac{1}{2}
\end{array}
$$

$$-2x + 4\left(-\frac{1}{2}\right) = 0$$

$$-2x = 2$$

$$x = -1$$

Therefore, $(-1, -1/2)$ is the critical point.

$$f_{xx} = 2,$$
$$f_{yy} = 4,$$
$$f_{xy} = -2.$$

For $\left(-1, -\frac{1}{2}\right)$,

$$D = 2 \cdot 4 - (-2)^2 = 4 > 0.$$

Since $f_{xx} = 2 > 0$, then a relative minimum is at $(-1, -1/2)$.

5. $f(x, y) = x^2 - xy + y^2 + 2x + 2y + 6$

    $f_x = 2x - y + 2$, $f_y = -x + 2y + 2$

Solve the system $f_x = 0$, $f_y = 0$.

$$
\begin{array}{r}
2x - y + 2 = 0 \\
\underline{-x + 2y + 2 = 0}
\end{array}
$$

$$
\begin{array}{r}
2x - y + 2 = 0 \\
\underline{-2x + 4y + 4 = 0} \\
3y + 6 = 0 \\
y = -2
\end{array}
$$

$$-x + 2(-2) + 2 = 0$$

$$x = -2$$

$(-2, -2)$ is the critical point.

$$f_{xx} = 2,$$
$$f_{yy} = 2,$$
$$f_{xy} = -1$$

For $(-2, -2)$

$$D = (2)(2) - (-1)^2 = 3 > 0.$$

Since $f_{xx} > 0$, a relative minimum is at $(-2, -2)$.

7. $f(x, y) = x^2 + 3xy + 3y^2 - 6x + 3y$
    $f_x = 2x + 3y - 6$, $f_y = 3x + 6y + 3$

Solve the system $f_x = 0$, $f_y = 0$.

$$2x + 3y - 6 = 0$$
$$3x + 6y + 3 = 0$$

$$-4x - 6y + 12 = 0$$
$$\underline{3x + 6y + 3 = 0}$$
$$-x + 15 = 0$$
$$x = 15$$

$$3(15) + 6y + 3 = 0$$
$$6y = -48$$
$$y = -8$$

$(15, -8)$ is the critical point.

$$f_{xx} = 2,$$
$$f_{yy} = 6,$$
$$f_{xy} = 3$$

For $(15, -8)$,

$$D = 2 \cdot 6 - 9 = 3 > 0.$$

Since $f_{xx} > 0$, a relative minimum is at $(15, -8)$.

9. $f(x, y) = 4xy - 10x^2 - 4y^2 + 8x + 8y + 9$
    $f_x = 4y - 20x + 8$, $f_y = 4x - 8y + 8$

$$4y - 20x + 8 = 0$$
$$4x - 8y + 8 = 0$$

$$4y - 20x + 8 = 0$$
$$\underline{-4y + 2x + 4 = 0}$$
$$-18x + 12 = 0$$

$$x = \frac{2}{3}$$

$$4y - 20\left(\frac{2}{3}\right) + 8 = 0$$

The critical point is $\left(\frac{2}{3}, \frac{4}{3}\right)$.

$$f_{xx} = -20,$$
$$f_{yy} = -8,$$
$$f_{xy} = 4$$

For $\left(\frac{2}{3}, \frac{4}{3}\right)$,

$$D = (-20)(-8) - 16 = 144 > 0.$$

Since $f_{xx} < 0$, a relative maximum is at $\left(\frac{2}{3}, \frac{4}{3}\right)$.

11. $f(x, y) = x^2 + xy - 2x - 2y + 2$
     $f_x = 2x + y - 2$, $f_y = x - 2$

$$2x + y - 2 = 0$$
$$x \qquad - 2 = 0$$
$$x = 2$$

$$2(2) + y - 2 = 0$$
$$y = -2$$

The critical point is $(2, -2)$.

$$f_{xx} = 2$$
$$f_{yy} = 0$$
$$f_{xy} = 1$$

For $(2, -2)$,

$$D = 2 \cdot 0 - 1^2 = -1 < 0.$$

A saddle point is at $(2, -2)$.

13. $f(x, y) = 2x^3 + 3y^2 - 12xy + 4$

$f_x = 6x^2 - 12y, \quad f_y = 6y - 12x$

$$6x^2 - 12y = 0$$
$$6y - 12x = 0$$

If $6y - 12x = 0$, $y = 2x$.
Substitute for y in the first equation.

$$6x^2 - 12(2x) = 0$$
$$6x(x - 4) = 0$$
$$x = 0 \quad \text{or} \quad x = 4$$

Then, $\quad y = 0 \quad \text{or} \quad y = 8$.

The critical points are $(0, 0)$ and $(4, 8)$.

$$f_{xx} = 12x,$$
$$f_{yy} = 6,$$
$$f_{xy} = -12$$

For $(0, 0)$,

$$D = 12(0)6 - (-12)^2$$
$$= -144 > 0.$$

A saddle point is at $(0, 0)$.
For $(4, 8)$,

$$D = 12(4)6 - (-12)^2$$
$$= 144 > 0.$$

Since $f_{xx} = 12(4) = 48 > 0$, a relative minimum is at $(4, 8)$.

15. $f(x, y) = x^2 + 4y^3 - 6xy - 1$

$f_x = 2x - 6y, \quad f_y = 12y^2 - 6x$

Solve $f_x = 0$ for x.

$$2x + 6y = 0$$
$$x = 3y$$

Substitute for x in $12y^2 - 6x = 0$.

$$12y^2 - 6(3y) = 0$$
$$6y(2y - 3) = 0$$
$$y = 0 \quad \text{or} \quad y = \frac{3}{2}$$

Then $\quad x = 0 \quad \text{or} \quad x = \frac{9}{2}$.

The critical points are $(0, 0)$ and $(9/2, 3/2)$.

$$f_{xx} = 2,$$
$$f_{yy} = 24y,$$
$$f_{xy} = -6$$

For $(0, 0)$,

$$D = 2 \cdot 24(0) - (-6)^2$$
$$= -36 < 0.$$

A saddle point is at $(0, 0)$.
For $(9/2, 3/2)$,

$$D = 2 \cdot 24\left(\frac{3}{2}\right) - (-6)^2$$
$$= 36 > 0.$$

Since $f_{xx} > 0$, a relative minimum is at $(9/2, 3/2)$.

17. $f(x, y) = e^{xy}$

$$f_x = ye^{xy}$$
$$f_y = xe^{xy}$$
$$ye^{xy} = 0$$
$$xe^{xy} = 0$$
$$x = y = 0$$

The critical point is $(0, 0)$.

$$f_{xx} = y^2e^{xy},$$
$$f_{yy} = x^2e^{xy},$$
$$f_{xy} = e^{xy} + xye^{xy}$$

For $(0, 0)$,

$$D = 0 \cdot 0 - (e^0)^2 = -1 < 0.$$

A saddle point is at $(0, 0)$.

21. $z = -3xy + x^3 - y^3 + \dfrac{1}{8}$

$f_x = -3y + 3x^2$, $f_y = -3x - 3y^2$

Solve the system $f_x = 0$, $f_y = 0$.

$$-3y + 3x^2 = 0$$
$$-3x - 3y^2 = 0$$
$$-y + x^2 = 0$$
$$-x - y^2 = 0$$

Solve the first equation for y, substitute into the second, and solve for x.

$$y = x^2$$
$$-x - x^4 = 0$$
$$x(1 + x^3) = 0$$
$$x = 0 \quad \text{or} \quad x = -1$$

Then $y = 0$ or $y = 1$.

The critical points are $(0, 0)$ and $(-1, 1)$.

$$f_{xx} = 6x$$
$$f_{yy} = -6y$$
$$f_{xy} = -3$$

For $(0, 0)$,

$$D = 0 \cdot 0 - (-3)^2 = -9 < 0.$$

A saddle point is at $(0, 0)$.

For $(-1, 1)$,

$$D = -6(-6) - (-3)^2 = 27 > 0.$$

$f_{xx} = 6(-1) = -6 < 0$.

$f(-1, 1)$

$\quad = -3(-1)(1) + (-1)^3 - 1^2 + \dfrac{1}{8}$

$\quad = 1\dfrac{1}{8}$

A relative maximum of $1\dfrac{1}{8}$ is at $(-1, 1)$.

The equation matches the graph in (a).

23. $z = y^4 - 2y^2 + x^2 - \dfrac{17}{16}$

$f_x = 2x$, $f_y = 4y^3 - 4y$

Solve the system $f_x = 0$, $f_y = 0$.

$$2x = 0 \quad (1)$$
$$4y^3 - 4y = 0 \quad (2)$$
$$4y(y^2 - 1) = 0$$
$$4y(y + 1)(y - 1) = 0$$

Equation 1 gives $x = 0$ and equation 2 gives $y = 0$, $y = -1$, or $y = 1$.

The critical points are $(0, 0)$, $(0, -1)$, and $(0, 1)$.

$$f_{xx} = 2,$$
$$f_{yy} = 12y^2 - 4,$$
$$f_{xy} = 0$$

For $(0, 0)$,

$$D = 2(12 \cdot 0^2 - 4) = -8 < 0.$$

A saddle point is at $(0, 0)$.

For $(0, -1)$,

$$D = 2[12(-1)^2 - 4] = 16 > 0.$$

$$f_{xx} = 2 > 0$$

$f(0, -1) = (-1)^4 - 2(-1)^2 + 0^2 - \dfrac{17}{16}$

$\quad = -2\dfrac{1}{16}$

A relative minimum of $-2\dfrac{1}{16}$ is at $(0, -1)$.

For (0, 1),

$$D = 2(12 \cdot 1^2 - 4) = 16 > 0$$

$$f_{xx} = 2 > 0$$

$$f(0, 1) = 1^4 - 2 \cdot 1^2 + 0^2 - \frac{17}{16}$$

$$= -2\frac{1}{16}$$

A relative minimum of $-2\frac{1}{16}$ is at (0, 1).

The equation matches the graph of (b).

25. $z = -x^4 + y^4 + 2x^2 - 2y^2 + \frac{1}{16}$

$f_x = -4x^3 + 4x$, $f_y = 4y^3 - 4y$

Solve $f_x = 0$, $f_y = 0$.

$$-4x^3 + 4x = 0 \quad (1)$$

$$4y^3 - 4y = 0 \quad (2)$$

$$-4x(x^2 - 1) = 0 \quad (1)$$

$$-4x(x + 1)(x - 1) = 0$$

$$4y(y^2 - 1) = 0 \quad (2)$$

$$4y(y + 1)(y - 1) = 0$$

Equation 1 gives x = 0, -1, or 1.
Equation 2 gives y = 0, -1, or 1.

Critical points are (0, 0), (0, -1), (0, 1), (-1, 0), (-1, -1), (-1, 1), (1, 0), (1, -1), (1, 1).

$$f_{xx} = -12x^2 + 4,$$

$$f_{yy} = 12y^2 - 4$$

$$f_{xy} = 0$$

For (0, 0),

$$D = 4(-4) - 0 = 16 < 0.$$

For (0, -1),

$$D = 4(8) - 0 = 32 > 0,$$

and $f_{xx} = 4 > 0$.

$$f(0, -1) = -\frac{15}{16}$$

For (0, 1),

$$D = 4(8) - 0 = 32 > 0,$$

and $f_{xx} = 4 > 0$.

$$f(0, 1) = -\frac{15}{16}$$

For (-1, 0),

$$D = -8(-4) - 0 = 32 > 0,$$

and $f_{xx} = -8 < 0$.

$$f(-1, 0) = 1\frac{1}{16}$$

For (-1, -1),

$$D = -8(8) - 0 = -64 < 0.$$

For (-1, 1),

$$D = -8(8) - 0 = -64 < 0.$$

For (1, 0),

$$D = -8(-4) = 32 > 0.$$

and $f_{xx} = -8 < 0$.

$$f(1, 0) = 1\frac{1}{16}$$

For (1, -1),

$$D = -8(8) - 0 = -64 < 0.$$

For (1, 1),

$$D = -8(8) - 0 = -64 < 0.$$

Saddle points are at (0, 0), (-1, -1), (-1, 1), (1, -1), and (1, 1).

Relative maximum of $1\frac{1}{16}$ is at

$(-1, 0)$ and $(1, 0)$.

Relative minimum of $-\frac{15}{16}$ is at

$(0, -1)$ and $(0, 1)$.

27. $f(x, y) = 1 - x^4 - y^4$

$f_x = -4x^3$, $f_y = -4y^3$

The system

$f_x = -4x^3 = 0$, $f_y = -4y^3 = 0$

gives the critical point $(0, 0)$.

$$f_{xx} = -12x^2,$$
$$f_{yy} = -12y^3,$$
$$f_{xy} = 0$$

For $(0, 0)$,

$$D = 0 \cdot 0 - 0^2 = 0.$$

Therefore, the test gives no information. Examine a graph of the function drawn by using level curves.

If $f(x, y) = 1$, then $x^4 + y^4 = 0$. The level curve is the point $(0, 0, 1)$.

If $f(x, y) = 0$, then $x^4 + y^4 = 1$. The level curve is the circle with center $(0, 0, 0)$ and radius 1.

If $(x, y) = -15$, then $x^4 + y^4 = 16$. The level curve is the curve with center $(0, 0, -15)$ and radius 2. The xz-trace is

$$z = 1 - x^4.$$

This curve has a maximum at $(0, 0, 1)$ and opens downward.

The yz-trace is

$$z = 1 - y^4.$$

This curve also has a maximum at $(0, 0, 1)$ and opens downward.

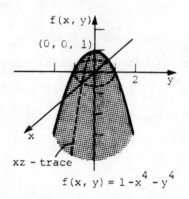

$f(x, y) = 1 - x^4 - y^4$

If $f(x, y) > 1$, then $x^4 + y^4 < 0$, which is impossible, so the function does not exist. Thus, the function has a relative maximum of 1 at $(0, 0)$.

29. $P(x, y) = 1000 + 24x - x^2 + 80y - y^2$

$$P_x = 24 - 2x$$
$$P_y = 80 - 2y$$
$$24 - 2x = 0$$
$$80 - 2y = 0$$
$$12 = x$$
$$40 = y$$

A critical point is $(12, 40)$.

$P_{xx} = -2$, $P_{yy} = -2$, $P_{xy} = 0$

For $(12, 40)$,

$$D = -2(-2) - 0^2 = 4 > 0.$$

Since $P_{xx} < 0$,

$$P(12, 40) = 1000 + 24(12) - (12)^2$$
$$+ 80(40) - (40)^2$$
$$= 2744$$

or \$274,400 is the maximum profit.

31. $C(x, y) = 2x^2 + 3y^2 - 2xy$
$$+ 2x - 126y + 3800$$

$$C_x = 4x - 2y + 2$$
$$C_y = 6y - 2x - 126$$

$$0 = 4x - 2y + 2$$
$$0 = 6y - 2x - 126$$

$$0 = 2x - y + 1$$
$$\underline{0 = -2x + 6y - 126} \quad \textit{Add}$$
$$0 = \qquad 5y - 125$$
$$y = 25$$
$$0 = 2x - 25 + 1$$

If $y = 25$, $x = 12$.

(12, 25) is a critical point.

$$C_{xx} = 4$$
$$C_{yy} = 6$$
$$C_{xy} = -2$$

For (12, 25).

$$D = 4 \cdot 6 - 4 = 20 > 0.$$

Since $C_{xx} > 0$, 12 units of electrical tape and 25 units of packing tape should be produced to yield a minimum cost.

$$C(12, 25)$$
$$= 2(12)^2 + 3(25)^2 - 2(12 \cdot 25)$$
$$+ 2(12) - 126(25) + 3800$$
$$= 2237$$

The minimum cost is \$2237.

Section 16.4

1. Maximize $f(x, y) = 2xy$
subject to $x + y = 12$.

1. Constraint $g(x, y) = x + y - 12$

2. Let $F(x, y, \lambda)$
$$= 2xy + \lambda(x + y - 12)$$

3. $F_x = 2y + \lambda$
$F_y = 2x + \lambda$
$F_\lambda = x + y - 12$

4. $2y + \lambda = 0$
$2x + \lambda = 0$
$x + y = 12$

5. $\lambda = -2y$ and $\lambda = -2x$
$$-2y = -2x$$
$$y = x$$

Substituting into the third equation gives

$$x + x = 12$$
$$x = 6.$$
Thus, $\qquad y = 6.$

Maximum is $f(6, 6) = 2 \cdot 6 \cdot 6 = 72$.

3. Maximize $f(x, y) = x^2y$,
subject to $2x + y = 4$.

1. $g(x, y) = 2x + y - 4$

2. $F(x, y, \lambda)$
$$= x^2y + \lambda(2x + y - 4)$$

3. $F_x = 2xy + 2\lambda$
$F_y = x^2 + \lambda$
$F_\lambda = 2x + y - 4$

4. $2xy + 2\lambda = 0$
$x^2 + \lambda = 0$
$2x + y = 4$

5. $xy = -\lambda$

   $x^2 = -\lambda$

$$xy = x^2$$
$$x(y - x) = 0$$
$$y = 0 \quad \text{or} \quad y = x$$

Substituting $x = 0$ into the third equation gives

$$2(0) + y = 4$$
$$y = 4.$$

Substituting $y = x$ into the third equation gives

$$2x + x = 4$$
$$x = \frac{4}{3}.$$

Thus, $y = \frac{4}{3}.$

$$f(0, 4) = 0^2(4) = 0$$

$$f\left(\frac{4}{3}, \frac{4}{3}\right) = \frac{16}{9} \cdot \frac{4}{3} = \frac{64}{27} > 0$$

Therefore, $f\left(\frac{4}{3}, \frac{4}{3}\right) = \frac{64}{27} \approx 2.4$ is a maximum.

5. Minimize $f(x, y) = x^2 + 2y^2 - xy$, subject to $x + y = 8$.

1. $g(x, y) = x + y - 8$

2. $F(x, y, \lambda)$
      $= x^2 + 2y^2 - xy + \lambda(x + y - 8)$

3. $F_x = 2x - y + \lambda$

   $F_y = 4y - x + \lambda$

   $F_\lambda = x + y - 8$

4. $2x - y + \lambda = 0$

   $4y - x + \lambda = 0$

   $x + y - 8 = 0$

5. Subtracting the second equation from the first equation to eliminate λ gives the new system of equations

$$x + y = 8$$
$$3x - 5y = 0$$

$$5x + 5y = 40$$
$$\underline{3x - 5y = 0}$$
$$8x = 40$$
$$x = 5$$

But $x + y = 8$, so $y = 3$.

$f(5, 3) = 25 + 18 - 15 = 28.$

$f(5, 3) = 28$ is a minimum.

7. Maximize $f(x, y) = x^2 - 10y^2$, subject to $x - y = 18$.

1. $g(x, y) = x - y - 18$

2. $F(x, y, \lambda)$
      $= x^2 - 10y^2 + \lambda(x - y - 18)$

3. $F_x = 2x + \lambda$

   $F_y = -20y - \lambda$

   $F_\lambda = x - y - 18$

4. $2x + \lambda = 0$

      $-20y - \lambda = 0$

      $x - y - 18 = 0$

5. Adding the first two equations to elimimate λ gives

$$2x - 20y = 0$$
$$x = 10y.$$

Substituting $x = 10y$ in the third equation gives

$$10y - y = 18$$
$$y = 2$$

Thus, $x = 20.$

$$f(20, 2) = 20^2 - 10(2)^2$$
$$= 400 - 40 = 360.$$

$f(20, 2) = 360$ is a relative maximum.

9. Maximize $f(x, y, z) = xyz^2$, subject to $x + y + z = 6$.

1. $g(x, y, z) = x + y + z - 6$

2. $F(x, y, \lambda)$
    $$= xyz^2 + \lambda(x + y + z - 6)$$

3. $F_x = yz^2 + \lambda$
    $F_y = xz^2 + \lambda$
    $F_z = 2zxy + \lambda$
    $F_\lambda = x + y + z - 6$

4. Setting F_x, F_y, F_z and F_λ equal to zero yields

$$yz^2 + \lambda = 0$$
$$xz^2 + \lambda = 0$$
$$2xyz + \lambda = 0$$
$$x + y + z - 6 = 0.$$

5. $\lambda = -yz^2$, $\lambda = -xz^2$, and $\lambda = -2xyz$

$$-yz^2 = -xz^2$$
$$z^2(x - y) = 0$$
$$x = y \quad \text{or} \quad z = 0$$
$$-yz^2 = -2xyz$$
$$2xyz - yz^2 = 0$$
$$yz(2x - z) = 0$$
$$y = 0, \ z = 0, \ \text{or} \ z = 2x$$

If $y = x$ and $z = 2x$, substituting into the fourth equation gives

$$x + x + 2x - 6 = 0$$
$$x = \frac{3}{2}.$$

Thus, $y = \frac{3}{2}$ and $z = 3$.

If $z = 0$, then $x + y = 6$ from the fourth equation.
If $y = 0$, then $x = 6$.
If $z = 2x$, then $x = 0$ or $y = 6$.
If $y = 2x$, then $x \pm y = 3$.
Possible extrema:

$(\frac{3}{2}, \frac{3}{2}, 3)$, $(6, 0, 0)$, $(0, 6, 0)$,

$(3, 3, 0)$

But $f(6, 0, 0) = f(0, 6, 0)$
$$= f(3, 3, 0) = 0$$
$$f\left(\frac{3}{2}, \frac{3}{2}, 3\right) = \frac{3}{2} \cdot \frac{3}{2} \cdot 9$$
$$= \frac{81}{4} > 0.$$

So $f\left(\frac{3}{2}, \frac{3}{2}, 3\right) = \frac{81}{4} = 20.25$

is a maximum.

11. Let $f(x, y) = xy^2$.

1. $g(x, y) = x + y - 18$

2. $F(x, y, \lambda)$
    $$= xy^2 + \lambda(x + y - 18)$$

3. $F_x = y^2 + \lambda$
    $F_y = 2yx + \lambda$
    $F_\lambda = x + y - 18$

4. $y^2 + \lambda = 0$
    $2yx + \lambda = 0$
    $x + y - 18 = 0$

5. $\lambda = -y^2$ and $\lambda = -2yx$

$$-y^2 = -2yx$$
$$y^2 - 2yx = 0$$
$$y(y - 2x) = 0.$$
$$y = 0 \quad \text{or} \quad y = 2x$$

If $y = 0$, from the third equation, $x = 18$.

If $y = 2x$, from the third equation,

$$x + 2x - 18 = 0$$
$$3x - 18 = 0$$
$$x = 6 \text{ and } y = 12.$$

$f(18, 0) = 18 \cdot 0 = 0$

$f(6, 12) = 6(144) = 864 > 0$

So $x = 6$, $y = 12$ will maximize

$f(x, y) = xy^2$.

13. Let x, y, and z be three numbers such that

$$x + y + z = 90$$
and $f(x, y, z) = xyz.$

1. $g(x, y, z) = x + y + z - 90$

2. $F(x, y, z)$

   $= xyz + \lambda(x + y + z - 90)$

3. $F_x = yz + \lambda$

   $F_y = xz + \lambda$

   $F_\lambda = xy + \lambda$

   $F_\lambda = x + y + z - 90$

4. $yz + \lambda = 0$ (1)

           $xz + \lambda = 0$ (2)

           $xy + \lambda = 0$ (3)

   $x + y + z - 90 = 0$ (4)

5. $\lambda = -yz$, $\lambda = -xz$, and $\lambda = -xy$

           $-yz = -xz$

          $yz - xz = 0$

          $(y - x)z = 0$

       $y - x = 0$ or $z = 0$

   $xz - xy = 0$

   $x(z - y) = 0$

       $x = 0$ or $z - y = 0$

Since $x = 0$ or $z = 0$ would not maximize $f(x, y, z) = xyz$, then $y - x = 0$ and $z - y = 0$ imply that $y = x = z$.

Substituting into the fourth equation gives

$$x + x + x - 90 = 0$$
$$x = 30.$$

$x = y = z = 30$ will maximize $f(x, y, z) = xyz.$

The numbers are 30, 30, and 30.

17. Let x be the width and y be the length of a field such that the cost in dollars to enclose the field is

$$6x + 6y + 4x + 4y = 1200$$
$$10x + 10y = 1200.$$

The area is

$f(x, y) = xy.$

1. $g(x, y) = 10x + 10y - 1200$

2. $F(x, y) = xy + \lambda(10x + 10y - 1200)$

3. $F_x = y + 10\lambda$

   $F_y = x + 10\lambda$

   $F_\lambda = 10x + 10y - 1200$

4. $y + 10\lambda = 0$

           $x + 10\lambda = 0$

   $10x + 10y - 1200 = 0$

5. $10\lambda = -y$ and $10\lambda = -x$

          $-y = -x$

           $y = x$

Substituting into the third equation gives

$$10x + 10x - 1200 = 0$$
$$20x - 1200 = 0$$
$$x = 60$$
$$y = 60.$$

The dimensions, 60 ft by 60 ft, will maximize the area.

19. $C(x, y) = 2x^2 + 6y^2 + 4xy + 10$, subject to $x + y = 10$

1. $g(x, y) = x + y - 10$

2. $F(x, y)$
$$= 2x^2 + 6y^2 + 4xy + 10$$
$$+ \lambda(x + y - 10)$$

3. $F_x = 4x + 4y + \lambda$
$F_y = 12y + 4x + \lambda$
$F_\lambda = x + y - 10$

4. $4x + 4y + \lambda = 0$
$12y + 4x + \lambda = 0$
$x + y - 10 = 0$

5. $\lambda = -4x + 4y$ and $\lambda = -12y - 4x$.
$$-4x + 4y = -12y - 4x$$
$$8y = 0$$
$$y = 0$$

Since $x + y - 10$, $x = 10$.
10 large kits and no small kits will maximize the cost.

21. $f(x, y) = 3x^{1/3} y^{2/3}$, subject to $80x + 150y = 40,000$

1. $g(x, y) = 80x + 150y - 40,000$

2. $F(x, y)$
$$= 3x^{1/3} y^{2/3} + \lambda(80x + 150y - 40,000)$$

3. and 4.

$F_x = x^{-2/3} + 80\lambda = 0$
$F_y = 2x^{1/3} y^{-1/3} + 150\lambda = 0$
$F_\lambda = 80x + 150y - 40,000 = 0$

5. $\dfrac{x^{-2/3} y^{2/3}}{80} = \dfrac{2x^{1/3} y^{-1/3}}{150}$

$$\frac{15y}{16} = x$$

Substitute into the third equation.

$$80\left(\frac{15y}{16}\right) + 150y - 40,000 = 0$$
$$y = 178 \text{ (rounded)}$$
$$= \frac{15(178)}{16}$$
$$x = 167$$

Use 167 units of labor and 178 units of capital to maximize production.

23. Let x and y be the dimensions of the field such that $2x + 2y = 200$, and the area is $f(x, y) = xy$.

1. $g(x, y) = 2x + 2y - 200$

2. $F(x, y)$
$$= xy + \lambda(2x + 2y - 200)$$

3. $F_x = y + 2\lambda$

$F_y = x + 2\lambda$
$F_\lambda = 2x + 2y - 200$

4. $y + 2\lambda = 0$

$x + 2\lambda = 0$

$2x + 2y - 200 = 0$

5. $2\lambda = -y$ and $2\lambda = -x$ so $x = y$.

$2x + 2x - 200 = 0$

$4x - 200 = 0$

$x = 50$

Thus, $y = 50$.

Dimensions of 50 m by 50 m will maximize the area.

25. Let x be the radius r of the circular base and y the height h of the can, such that the volume is

$$\pi x^2 y = 250\pi.$$

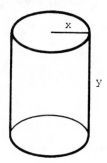

The surface is

$$f(x, y) = 2\pi xy + 2\pi x^2.$$

1. $g(x, y) = \pi x^2 y - 250\pi$

2. $F(x, y)$

$= 2\pi xy + 2\pi x^2 + \lambda(\pi x^2 y - 250\pi)$

3. $F_x = 2\pi y + 4\pi x + \lambda(2\pi xy)$

$F_y = 2\pi x + \lambda(\pi x^2)$

$F_\lambda = \pi x^2 y - 250\pi$

4. $2\pi y + 4\pi x + \lambda(2\pi xy) = 0$

$2\pi x + \lambda\pi x^2 = 0$

$\pi x^2 y - 250\pi = 0$

Simplifying these equations gives

$$y + 2x + \lambda xy = 0$$

$$2x + \lambda x^2 = 0$$

$$x^2 y - 250 = 0.$$

5. From the second equation

$$x(2 + \lambda x) = 0$$

$$x = 0 \quad \text{or} \quad \lambda = -\frac{2}{x}.$$

If x = 0, the volume will be 0, which is not possible.

Substituting $x = -\frac{2}{\lambda}$ into the first equation gives

$$y + 2\left(-\frac{2}{\lambda}\right) + \lambda\left(-\frac{2}{\lambda}\right)y = 0$$

$$y - \frac{4}{\lambda} - 2y = 0$$

$$-\frac{4}{\lambda} = y$$

$$\lambda = -\frac{4}{y}.$$

Since $\lambda = -2/x$, $y = 2x$.

Substituting into the third equation gives

$$x^2(2x) - 250 = 0$$

$$2x^3 - 250 = 0$$

$$x = 5$$

$$y = 10.$$

Since $g(1, 250) = 0$, and $f(1, 250) = 502\pi > f(5, 10) = 300\pi$, a can with radius of 5 in and height of 10 in will have minimum surface area.

27. Let x, y, and z be the dimensions of
the box such that the surface area
is

$$xy + 2yz + 2xz = 500$$

and the volume is

$$f(x, y, z) = xyz.$$

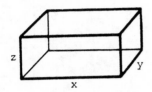

1. $g(x, y, z) - 500 = 0$

2. $F(x, y, z)$
    $= xyz + \lambda(xy + 2yz + 2xz - 500)$

3. and 4.
    $F_x = yz + \lambda(y + 2z) = 0$ (1)
    $F_y = xz + \lambda(x + 2z) = 0$ (2)
    $F_z = xy + \lambda(2y + 2x) = 0$ (3)
    $F_\lambda = xy + 2xz + 2yz - 500 = 0$ (4)

Multiplying Equation (1) by x,
Equation (2) by y, Equation (3)
by z gives

$$xyz + \lambda x(y + 2z) = 0$$
$$xyz + \lambda y(x + 2z) = 0$$
$$xyz + \lambda z(2y + 2x) = 0.$$

5. Subtracting the second equation from
the first equation gives

$$\lambda x(y + 2z) - \lambda y(x + 2z) = 0$$
$$2\lambda xz - 2\lambda yz = 0$$
$$\lambda z(x - y) = 0,$$
so $x = y.$

Subtracting the second equation from
the third equation gives

$$\lambda z(2y + 2x) - \lambda y(x + 2z) = 0$$
$$2\lambda xz - \lambda xy = 0$$
$$\lambda x(2z - y) = 0$$
so $z = \dfrac{y}{2}.$

Substituting into the fourth equation gives

$$y^2 + 2y\left(\frac{y}{2}\right) + 2y\left(\frac{y}{2}\right) - 500 = 0$$
$$3y^2 = 500$$
$$y = \sqrt{\frac{500}{3}}$$
$$\approx 12.9099$$
$$x \approx 12.9099$$
$$z \approx \frac{12.9099}{2}$$
$$\approx 6.4649$$

The dimensions are 12.91 m by
12.91 m by 6.45 m.

29. Let x, y, and z be the dimensions of
the box.
The surface area is

$$2xy + 2xz + 2yz.$$

We must minimize

$$f(x, y, z) = 2xy + 2xz + 2yz$$

subject to $xyz = 27$.

1. $g(x, y, z) = xyz - 27$

2. $F(x, y, z)$
    $= 2xy + 2xz + 2yz + \lambda(xyz - 27)$

3. $F_x = 2y + 2z + \lambda yz$
    $F_y = 2x + 2z + \lambda xz$
    $F_z = 2x + 2y + \lambda xy$
    $F_\lambda = xyz - 27$

4. $2y + 2z + \lambda yz = 0$

$2x + 2z + \lambda xz = 0$

$2x + 2y + \lambda xy = 0$

$xyz - 27 = 0$

5. The first and second equation gives

$\frac{2y + 2z}{yz} = -\lambda$ and $\frac{2x + 2z}{xz} = -\lambda$.

Thus,

$$\frac{2y + 2z}{yz} = \frac{2x + 2z}{xz}$$

$$2xyz + 2xz^2 = 2xyz + 2yz^2$$

$$2xz^2 - 2yz^2 = 0$$

$2z^2 = 0$ or $x - y = 0$

$z = 0$ (impossible) or $x = y$

The second and third equations give

$\frac{2x + 2z}{xz} = -\lambda$ and $\frac{2x + 2y}{xy} = -\lambda$.

Thus,

$$\frac{2x + 2z}{xz} = \frac{2x + 2y}{xy}$$

$$2x^2y + 2xyz = 2x^2z + 2xyz$$

$$2x^2y - 2x^2z = 0$$

$$2x^2(y - z) = 0$$

$2x^2 = 0$ or $y - z = 0$

$x = 0$ (impossible) or $y = z$

Therefore, $x = y = z$. Substituting into the fourth equation gives

$$x^3 - 27 = 0$$

$$x^3 = 27$$

$$x = 3$$

Thus, $y = 3$ and $z = 3$.

The dimensions that will minimize the surface area are 3 m by 3 m by 3 m.

Section 16.5

3. See the answer graph at the back of the textbook.

x	y	x^2	xy
4	3	16	12
5	7	25	35
8	17	64	136
12	28	144	336
14	35	196	490
Totals 43	90	445	1009

$n = 5$

$m = \frac{(43)90 - 5(1009)}{(43)^2 - 5(445)}$

$= \frac{-1175}{-376}$

$= 3.125$

$b = \frac{90 - 3.125(43)}{5}$

$= -8.875$

Therefore, the least squares equation is

$$y' = 3.125x - 8.875.$$

When $y' = 12$,

$$12 = 3.125x - 8.875$$

$$\frac{20.875}{3.125} = x$$

$$6.68 = x.$$

When $y' = 32$,

$$\frac{32 + 8.875}{3.125} = x$$

$$13.08 = x.$$

5. (a) See the answer graph at the back of the textbook.

(b)

x	y	x^2	xy
250	9.3	62,500	2325
300	10.8	90,000	3240
375	14.8	140,625	5550
425	14.0	180,625	5950
450	14.2	202,500	6390
475	15.3	225,625	7267.5
500	15.9	250,000	7950
575	19.1	330,625	10,982.5
600	19.2	360,000	11,520
650	21.0	422,500	13,650
Totals 4600	153.6	2,265,000	74,825

$$m = \frac{4600(153.6) - 10(74,825)}{(4600)^2 - 10(2,265,000)}$$

$$\approx .02798$$

$$\approx .03$$

$$b = \frac{153.6 - m(4600)}{10} \approx 2.49.$$

The least squares equation is

$$y' = .03x + 2.49$$

See graph.

(c) If x = 700, then $y' \approx 23.5$;
   if x = 750, then $y' \approx 25.0$

7. (a)

x	y	x^2	xy
369	299	136,161	110,331
379	280	143,641	106,120
482	393	232,324	189,426
493	388	243,049	191,284
496	385	246,016	190,960
567	423	321,489	239,841
521	374	277,441	194,854
482	357	232,324	172,074
391	106	152,881	41,446
417	332	173,889	138,444
Totals 4597	3337	2,153,215	1,574,780

$$m = \frac{4597(337) - 10(1,574,780)}{(4597)^2 - 10(2,153,215)} \approx 1.02$$

$$b = \frac{3337 - m(4597)}{10} \approx -135$$

Least squares equation is

$$y' = 1.02x - 135.$$

(b) If x = 500, then y' = 375, so we predict $375,000 in repair costs.

(c) There appears to be an approximately linear relationship.

9. (a) All answers are rounded to four significant digits.

$$y' = 49.2755 - 1.1924(42)$$
$$+ .1631(170.0)$$
$$\approx \$26,920$$

$$y' = 49.2755 - 1.1924(45)$$
$$+ .1631(170.0)$$
$$\approx \$23,340$$

$$y' = 49.2755 - 1.1924(48)$$
$$+ .1631(170.0)$$
$$\approx \$19,770$$

(b) $y = 49.2755 - 1.1924(42)$

$\qquad + .1631(185.0)$

$\qquad \approx \$29,370$

$y = 49.2755 - 1.1924(45)$

$\qquad + .1631(185.0)$

$\qquad \approx \$24,790$

$y = 49.2755 - 1.1924(48)$

$\qquad + .1631(185.0)$

$\qquad \approx \$22,210$

(c) \$42, which is obvious, since y is largest when the term being subtracted is smallest.

11. **(a)** and **(b)** See the answer graphs at the back of the textbook.

(c)

x	log, y	x^2	xy
0	3.00	0	0.00
1	3.21	1	3.21
2	3.43	4	6.86
3	3.65	9	10.95
4	3.87	16	15.48
5	4.09	25	20.45
Totals 15	21.25	55	56.95

$m = \dfrac{(15)(21.25) - 6(56.95)}{(15)^2 - 6(55)}$

$\quad = \dfrac{-22.95}{-105}$

$\quad \approx .22$

$b = \dfrac{21.25 - m(15)}{6}$

$\quad \approx 3.00$

$\qquad y' = .22x + 3.00$

(d) If x = 7 hr,

$\qquad y' = .22(7) + 3.00 = 4.54.$

The antilogarithm of 4.54 is the predicted number, or $10^{4.54} \approx 35,000$ bacteria.

13. $\Sigma x = 175,878$; $\Sigma y = 9989$; $\Sigma xy = 46,209,266$; $\Sigma x^2 = 872,066,218$; $\Sigma y^2 = 2,629,701$; and n = 38

(a)

$m = \dfrac{(175,878)(9989) - 38(46,209,266)}{(175,878)^2 - 38(872,066,218)}$

$\quad = \dfrac{893,234}{-2,205,445,400}$

$\quad \approx -.00041$

$b = \dfrac{9989 - m(175,878)}{38}$

$\quad \approx 265$

$\qquad y' = -.00041x + 265$

(b) If x = 2838, then

$\qquad y' = -.00041(2838) + 265 \approx 264.$

For Idaho, the average mathematics proficiency is 264.

If x = 7850, then

$\qquad y' = -.00041(7850) + 265 \approx 262.$

For Washington, D.C., the average mathematics proficiency is 262.

(c) The negative sign indicates that as expenditure per pupil increases, the mathematics proficiency decreases.

15.

	x	y	x^2	xy
	540	20	291,600	10,800
	510	16	260,100	8160
	490	10	240,100	4900
	560	8	313,600	4480
	470	12	220,900	5640
	600	11	360,000	6600
	540	10	291,600	5400
	580	8	336,400	4640
	680	15	462,400	10,200
	560	8	313,600	4480
	560	13	313,600	7280
	500	14	250,000	7000
	470	10	220,900	4700
	440	10	193,600	4400
	520	11	270,400	5720
	620	11	384,400	6820
	680	8	462,400	5440
	550	8	302,500	4400
	620	7	384,400	4340
Totals	10,490	210	5,872,500	115,400

(a) $m = \dfrac{(10,490)(210) - 19(115,400)}{(10,490)^2 - 19(5,872,500)}$

$= \dfrac{10,300}{-1,537,400}$

$\approx -.0067$

$b = \dfrac{210 - m(10,490)}{19}$

≈ 14.75

$y' = -.0067x + 14.75$

(b) If $x = 420$, then

$y' = -.0067(420) + 14.75 \approx 12.$

The predicted mathematics placement test score for a math SAT score of
420 is 12.

(c) If x = 620, then

$$y' = -.0067(620) + 14.75 \approx 11.$$

The predicted mathematics placement test score for a math SAT score of 620 is 11.

17.

x	y	x^2	xy
150	5000	22,500	750,000
175	5500	30,625	962,500
215	6000	46,225	1,290,000
250	6500	62,500	1,625,000
280	7000	78,400	1,960,000
310	7500	96,100	2,325,000
350	8000	122,500	2,800,000
370	8500	136,900	3,145,000
420	9000	176,400	3,780,000
450	9500	202,500	4,275,000
Totals 2970	72,500	974,650	22,912,500

$$m = \frac{(2970)(72,500) - 10(22,912,500)}{(2970)^2 - 10(974,650)}$$

$$= \frac{-13,800,000}{-925,600}$$

$$\approx 14.9$$

$$b = \frac{72,500 - m(2970)}{10}$$

$$\approx 2820$$

$$y' = 14.9x + 2820$$

(a) If x = 150, y' = 14.9(150) + 2820 ≈ 5060.

If x = 280, y' = 14.9(280) + 2820 ≈ 6990.

If x = 420, y' = 14.9(420) + 2820 ≈ 9080.

The following table compares the actual data with the predicted values.

Square feet	Actual BTUs	Predicted BTUs
150	5000	5060
280	7000	6990
420	9000	9080

The largest discrepancy is 80 BTUs for 420 ft². This is a good agreement.

(b) If $x = 230$, $y' = 14.9(230) +$
2820 $\approx$ 6250.

Since air conditioners are available
only with the given BTU choices,
6500 BTUs must be used.

Section 16.6

1. $z = 9x^4 - 5y^3$

$dz = 36x^3\ dx - 15y^2\ dy$

3. $z = \dfrac{x + y}{x - y}$

$f_x = \dfrac{(x - y)(1) - (x + y)(1)}{(x - y)^2}$

$\quad = \dfrac{-2y}{(x - y)^2}$

$f_y = \dfrac{(x - y)(1) - (x + y)(-1)}{(x - y)^2}$

$\quad = \dfrac{2x}{(x - y)^2}$

$dz = \dfrac{-2y}{(x - y)^2}\ dx + \dfrac{2x}{(x - y)^2}\ dy$

5. $z = 2\sqrt{xy} - \sqrt{x + y}$

$\quad = 2(xy)^{1/2} - (x + y)^{1/2}$

$f_x = 2y^{1/2}\left(\dfrac{1}{2}\right)(x^{-1/2}) - \dfrac{1}{2}(x + y)^{-1/2}(1)$

$\quad = y^{1/2}(x^{-1/2}) - \dfrac{1}{2}(x + y)^{-1/2}$

$f_y = 2x^{1/2}\left(\dfrac{1}{2}\right)(y^{-1/2})$

$\qquad - \dfrac{1}{2}(x + y)^{-1/2}(1)$

$\quad = x^{1/2}(y^{-1/2}) - \dfrac{1}{2}(x + y)^{-1/2}$

$dz = \left[\dfrac{y^{1/2}}{x^{1/2}} - \dfrac{1}{2(x + y)^{1/2}}\right] dx$

$\qquad + \left[\dfrac{x^{1/2}}{y^{1/2}} - \dfrac{1}{2(x + y)^{1/2}}\right] dy$

7. $z = (3x + 2)\sqrt{1 - 2y}$

$\quad = (3x + 2)(1 - 2y)^{1/2}$

$f_x = (1 - 2y)^{1/2}(3) + (3x + 2)(0)$

$\quad = 3(1 - 2y)^{1/2}$

$f_y = (3x + 2)\left(\dfrac{1}{2}\right)(1 - 2y)^{-1/2}(-2)$

$\qquad + (1 - 2y)^{1/2}(0)$

$\quad = -(3x + 2)(1 - 2y)^{-1/2}$

$dz = 3(1 - 2y)^{1/2}\ dx - \dfrac{3x + 2}{(1 - 2y)^{1/2}}\ dy$

9. $z = \ln(x^2 + 2y^4)$

$f_x = \dfrac{2x}{x^2 + 2y^4}$

$f_y = \dfrac{8y^3}{x^2 + 2y^4}$

$dz = \dfrac{2x}{x^2 + 2y^4}\ dx + \dfrac{8y^3}{x^2 + 2y^4}\ dy$

11. $z = xy^2 e^{x+y}$

$f_x = y^2(xe^{x+y} + e^{x+y})$

$\quad = y^2 e^{x+y}(x + 1)$

$f_y = x[y^2 e^{x+y} + e^{x+y}(2y)]$

$\quad = xye^{x+y}(y + 2)$

$dz = y^2 e^{x+y}(x + 1)dx$

$\qquad + xye^{x+y}(y + 2)dy$

13. $z = x^2 - y\ln x$

$f_x = 2x - \dfrac{y}{x}$

$f_y = -\ln x$

$dz = \left(2x - \dfrac{y}{x}\right)dx + (-\ln x\ dy)$

15. $w = x^4 yz^3$

$dw = 4x^3 yz^3\ dx + x^4 z^3\ dy$

$\qquad + 3x^4 yz^2\ dz$

17. $z = x^2 + 3xy + y^2$

$x = 4$, $y = -2$, $dx = .02$, $dy = -.03$

$f_x = 2x + 3y$

$f_y = 3x + 2y$

$dz = (2x + 3y)dx + (3x + 2y)dy$

Substitute the given values to get

$dz = [2(4) + 3(-2)](.02)$

$+ [3(4) + 2(-2)](-.03)$

$= .04 + (-.24) = -.2.$

19. $z = \dfrac{x - 4y}{x + 2y}$

$x = 0$, $y = 5$, $dx = -.03$,

$dy = .05$

$f_x = \dfrac{(x + 2y)(1) - (x - 4y)(1)}{(x + 2y)^2}$

$= \dfrac{6y}{(x + 2y)^2}$

$f_y = \dfrac{(x + 2y)(-4) - (x - 4y)(2)}{(x + 2y)^2}$

$= \dfrac{-6x}{(x + 2y)^2}$

$dz = \dfrac{6y}{(x + 2y)^2}\ dx - \dfrac{6x}{(x + 2y)^2}\ dy$

Substitute the given information.

$dz = \dfrac{30}{10^2}(-.03) - \dfrac{0}{10^2}(.05)$

$= \dfrac{3}{10}(-.03) = -.009$

21. $z = \ln(x^2 + y^2)$

$x = 2$, $y = 3$, $dx = .02$,

$dy = -.03$

$dz = \dfrac{2x\ dx}{x^2 + y^2} + \dfrac{2y\ dy}{x^2 + y^2}$

Substitute the given information.

$dz = \dfrac{2(2)(.02)}{2^2 + 3^2} + \dfrac{2(3)(-.03)}{2^2 + 3^2}$

$= -.00769$

23. $w = \dfrac{5x^2 + y^2}{z + 1}$

$x = -2$, $y = 1$, $z = 1$,

$dx = .02$, $dy = -.03$, $dz = .02$

$f_x = \dfrac{(z + 1)10x - (5x^2 + y^2)(0)}{(z + 1)^2}$

$= \dfrac{10x}{z + 1}$

$f_y = \dfrac{(z + 1)(2y) - (5x^2 + y^2)(0)}{(z + 1)^2}$

$= \dfrac{2y}{z + 1}$

$f_z = \dfrac{(z + 1)(0) - (5x^2 + y^2)(1)}{(z + 1)^2}$

$= \dfrac{-5x^2 - y^2}{(z + 1)^2}$

$dz = \dfrac{10x}{z + 1}\ dx + \dfrac{2y}{z + 1}\ dy$

$+ \dfrac{-5x^2 - y^2}{(z + 1)^2}\ dz$

Substitute the given information.

$= \dfrac{-20}{2}(.02) + \dfrac{2}{2}(-.03)$

$+ \dfrac{[-5(4) - 1](.02)}{(2)^2}$

$= -.2 - .03 - \dfrac{21}{4}(.02) = -.355$

25. The volume of the can is

$$V = \pi r^2 h$$

with $r = 2.5$ cm, $h = 14$ cm,

$dr = .08$ cm, $dh = .16$ cm.

$dV = 2\pi rh\ dr + \pi r^2\ dh$

$dV = 2\pi(2.5)(14)(.08) + \pi(2.5)^2(.16)$

$dV \approx 20.73$

Approximately 20.73 cm³ are needed.

27. The volume of the box is

$$V = LWH$$

with L = 10 in, W = 9 in, H = 14 in. Since .2 in is applied to each side and each dimension has a side at each end,

$$dL = dW = dH = 2(.2) = .4 \text{ in.}$$
$$dV = WH \, dL + LH \, dW + LW \, dH$$

Substitute.

$$dV = (9)(14)(.4) + (10)(14)(.4)$$
$$+ (10)(9)(.4)$$
$$dV = 142.4$$

Approximately 142.4 in³ are needed.

29. $z = x^{.65}y^{.35}$
$x = 50$, $y = 29$, $dx = 52 - 50 = 2$,
$dy = 27 - 29 = -2$

$f_x = y^{.35}(.65)(x^{-.35})$

$= .65\left(\dfrac{y}{x}\right)^{.35}$

$f_y = (x^{.65})(.35)(y^{-.65})$

$= .35\left(\dfrac{x}{y}\right)^{.65}$

$dz = .65\left(\dfrac{y}{x}\right)^{.35} dx + .35\left(\dfrac{x}{y}\right)^{.65} dy$

Substitute.

$dz = .65\left(\dfrac{29}{50}\right)^{.35}(2) + .35\left(\dfrac{50}{29}\right)^{.65}(-2)$

$dz = .0769$ units

31. The volume of the bone is

$$V = \pi r^2 h,$$

with h = 7 cm, r = 1.4 cm,

dr = .09 cm, dh = 2(.09) = 18 cm
$dV = 2\pi rh \, dr + \pi r^2 \, dh$
$dV = 2\pi(1.4)(7)(.09) + \pi(1.4)^2(.18)$
$dV = 6.65$

6.65 cm³ are used.

33. $C = \dfrac{b}{a - v} = b(a - v)^{-1}$, $a = 160$,

b = 200, v = 125, da = 145 − 160
= −15, db = 190 − 200 = −10,
v = 125, da = 145 − 160

$dC = -b(a - v)^{-2} da$

$\quad + \dfrac{1}{a - v} db + b(a - v)^{-2} dv$

$= \dfrac{-b}{(a - v)^2} da + \dfrac{1}{a - v} db$

$\quad + \dfrac{b}{(a - v)^2} dv$

$= \dfrac{-200}{(160 - 125)^2}(-15) + \dfrac{1}{160 - 125}(-10)$

$\quad + \dfrac{200}{(160 - 125)^2}(5)$

≈ 2.98 liters

35. The area is

$$A = \frac{1}{2}bh$$

with h = 42.6 cm, b = 23.4 cm,
dh = 1.2 cm, db = .9 cm.

$dA = \dfrac{1}{2}h \, db + \dfrac{1}{2}b \, dh$

Substitute.

$dA = \dfrac{1}{2}(42.6)(.9) + \dfrac{1}{2}(23.4)(1.2)$

$= 33.2$ cm²

Section 16.7

1. $\displaystyle\int_0^3 (x^3y + y)\,dx$

$\qquad = \left(\dfrac{x^4}{4}y + yx\right)\Big|_0^3$

$\qquad = \dfrac{81y}{4} + 3y$

$\qquad = \dfrac{93y}{4}$

3. $\displaystyle\int_4^8 \sqrt{6x + y}\ dx = \int_4^8 (6x + y)^{1/2}\,dx$

Let $u = 6x + y$.

Then $du = 6\ dx$.

When $x = 8$, $u = 48 + y$.

When $x = 4$, $u = 24 + y$.

$\displaystyle\int_{y+24}^{y+48} u^{1/2}\cdot\dfrac{1}{6}\ du$

$\qquad = \dfrac{1}{6}\cdot\dfrac{2}{3}u^{3/2}\Big|_{y+24}^{y+48}$

$\qquad = \dfrac{1}{9}u^{3/2}\Big|_{y+24}^{y+48}$

$\qquad = \dfrac{1}{9}[(48 + y)^{3/2} - (24 + y)^{3/2}]$

5. $\displaystyle\int_4^5 x\sqrt{x^2 + 3y}\ dy$

$\qquad = \displaystyle\int_4^5 x(x^2 + 3y)^{1/2}\,dy$

$\qquad = \dfrac{2x}{9}(x^2 + 3y)^{3/2}\Big|_4^5$

$\qquad = \dfrac{2x}{9}[(x^2 + 15)^{3/2} - (x^2 + 12)^{3/2}]$

7. $\displaystyle\int_4^9 \dfrac{3 + 5y}{\sqrt{x}}\ dx$

$\qquad = (3 + 5y)\displaystyle\int_4^9 x^{-1/2}\,dx$

$\qquad = (3 + 5y)2x^{1/2}\Big|_4^9$

$\qquad = (3 + 5y)2[\sqrt{9} - \sqrt{4}]$

$\qquad = 6 + 10y$

9. $\displaystyle\int_{-1}^1 e^{x+4y}\,dy$

Let $u = x + 4y$.

Then $du = 4\ dy$.

When $y = 1$, $u = x + 4$.

When $y = -1$, $u = x - 4$.

$\dfrac{1}{4}\displaystyle\int_{x-4}^{x+4} e^u\ du = \dfrac{1}{4}e^u\Big|_{x-4}^{x+4}$

$\qquad\qquad\qquad = \dfrac{1}{4}(e^{x+4} - e^{x-4})$

$\qquad\qquad\qquad = \dfrac{1}{4}e^{x+4} - \dfrac{1}{4}e^{x-4}$

11. $\displaystyle\int_0^5 xe^{x^2+9y}\ dx$

Let $u = x^2 + 9y$.

Then $du = 2x\ dx$.

When $x = 5$, $u = 25 + 9y$.

When $x = 0$, $u = 9y$.

$\dfrac{1}{2}\displaystyle\int_{9y}^{25+9y} e^u\ du = \dfrac{1}{2}(e^{25+9y} - e^{9y})$

$\qquad\qquad\qquad = \dfrac{1}{2}e^{25+9y} - \dfrac{1}{2}e^{9y}$

13. $\int_{1}^{2}\left[\int_{0}^{3}(x^3y + y)dx\right]dy$

From Exercise 1,

$\int_{0}^{3}(x^3y + y)dx = \dfrac{93y}{4}.$

$\int_{1}^{2}\left[\int_{0}^{3}(x^3y + y)dx\right]dy$

$= \int_{1}^{2}\dfrac{93y}{4}\,dy$

$= \dfrac{93}{8}y^2\Big|_{1}^{2}$

$= \dfrac{93}{8}(4 - 1)$

$= \dfrac{279}{8}$

15. $\int_{0}^{1}\left[\int_{3}^{6}x\sqrt{x^2 + 3y}\,dx\right]dy$

From Exercise 6,

$\int_{3}^{6}x\sqrt{x^2 + 3y}\,dx$

$= \dfrac{1}{3}[(36 + 3y)^{3/2} - (9 + 3y)^3]$

$\int_{0}^{1}\left[\int_{3}^{6}x\sqrt{x^2 + 3y}\,dx\right]dy$

$= \int_{0}^{1}\dfrac{1}{3}[(36 + 3y)^{3/2} - (9 + 3y)^{3/2}]dy$

Let $u = 36 + 3y$.
Then $du = 3\,dy$.
When $x = 0$, $u = 36$.
When $x = 1$, $w = 39$.

Let $z = 9 + 3y$.
Then $dz = 3\,dy$.
When $y = 0$, $z = 9$.
When $y = 1$, $z = 12$.

$\dfrac{1}{9}\left[\int_{36}^{39}u^{3/2}\,du - \int_{9}^{12}z^{3/2}\,dz\right]$

$= \dfrac{1}{9}\cdot\dfrac{2}{5}[(39)^{5/2} - (36)^{5/2}$

$\quad - (12)^{9/2} + (9)^{5/2})]$

$= \dfrac{2}{45}[(39)^{5/2} - (12)^{5/2} - 6^5 + 3^5]$

$= \dfrac{2}{45}(39^{5/2} - 12^{5/2} - 7533)$

17. $\int_{1}^{2}\left[\int_{4}^{9}\dfrac{3 + 5y}{\sqrt{x}}\,dx\right]dy$

From Exercise 7,

$\int_{4}^{9}\dfrac{3 + 5y}{\sqrt{x}}\,dx = 6 + 10y$

$\int_{1}^{2}\left[\int_{4}^{9}\dfrac{3 + 5y}{\sqrt{x}}\,dx\right]dy$

$= \int_{1}^{2}(6 + 10y)dy$

$= 6y\Big|_{1}^{2} + 5y^2\Big|_{1}^{2}$

$= 6(2 - 1) + 5(4 - 1)$

$= 6 + 15 = 21$

19. $\int_{1}^{2}\int_{1}^{2}\dfrac{dx\,dy}{xy}$

$= \int_{1}^{2}\left(\dfrac{\ln\,|x|}{y}\right)\Big|_{1}^{2}\,dy$

$= \int_{1}^{2}\dfrac{1}{y}(\ln 2 - \ln 1)dy$

$$= \ln 2 \int_1^2 \frac{1}{y}\,dy$$

$$= \ln 2(\ln |y|) \Big|_1^2$$

$$= \ln 2(\ln 2 - \ln 1)$$

$$= (\ln 2)^2$$

$$= \int_1^5 \left(\frac{4}{2} + 6y^2 - 0\right)dy$$

$$= (2y + 2y^3) \Big|_1^5$$

$$= 2 \cdot 5 + 2 \cdot 125 - (2 \cdot 1 + 2 \cdot 1)$$

$$= 256$$

21. $\displaystyle\int_2^4\int_3^5 \left(\frac{x}{y} + \frac{y}{3}\right)dx\,dy$

$$= \int_2^4 \left(\frac{x^2}{2y} + \frac{yx}{3}\right)\Big|_3^5 dy$$

$$= \int_2^4 \left[\frac{25}{2y} + \frac{5y}{3} - \left(\frac{9}{2y} + \frac{3y}{3}\right)\right]dy$$

$$= \int_2^4 \left(\frac{16}{2y} + \frac{2y}{3}\right)dy$$

$$= \left(8 \ln |y| + \frac{y^2}{3}\right)\Big|_2^4$$

$$= 8(\ln 4 - \ln 2) + \frac{16}{3} - \frac{4}{3}$$

$$= 8 \ln \frac{4}{2} + \frac{12}{3}$$

$$= 8 \ln 2 + 4$$

23. $\displaystyle\int\!\!\int_R (x + 3y^2)dx\,dy; \ 0 \le x \le 2,$

$1 \le y \le 5$

$$\int\!\!\int_R (x + 3y^2)dx\,dy$$

$$= \int_1^5\int_0^2 (x + 3y^2)dx\,dy$$

$$= \int_1^5 \left(\frac{x^2}{2} + 3y^2x\right)\Big|_0^2 dy$$

25. $\displaystyle\int\!\!\int_R \sqrt{x + y}\,dy\,dx; \ 1 \le x \le 3,$

$0 \le y \le 1$

$$\int\!\!\int_R \sqrt{x + y}\,dy\,dx$$

$$= \int_1^3\int_0^1 (x + y)^{1/2}\,dy\,dx$$

$$= \int_1^3 \left[\frac{2}{3}(x + y)^{3/2}\right]\Big|_0^1 dx$$

$$= \int_1^3 \frac{2}{3}[(x + 1)^{3/2} - x^{3/2}]dx$$

$$= \frac{2}{3} \cdot \frac{2}{5}[(x + 1)^{5/2} - x^{5/2}]\Big|_1^3$$

$$= \frac{4}{15}(4^{5/2} - 3^{5/2} - 2^{5/2} + 1^{5/2})$$

$$= \frac{4}{15}(32 - 3^{5/2} - 2^{5/2} + 1)$$

$$= \frac{4}{15}(33 - 3^{5/2} - 2^{5/2})$$

27. $\displaystyle\iint_R \frac{2}{(x + y)^2}\, dy\, dx;\ 2 \le x \le 3,$

$1 \le y \le 5$

$\displaystyle\iint_R \frac{2}{(x + y)^2}\, dy\, dx$

$\displaystyle= \int_2^3 \int_1^5 2(x + y)^{-2}\, dy\, dx$

$\displaystyle= \int_2^3 -2(x + y)^{-1}\Big|_1^5\, dy$

$\displaystyle= -2\int_2^3 \left[\frac{1}{(5 + y)} - \frac{1}{(1 + y)}\right] dy$

$\displaystyle= -2\left(\ln|5 + y| - \ln|1 + y|\right)\Big|_2^3$

$\displaystyle= -2\left(\ln\left|\frac{5 + y}{1 + y}\right|\right)\Big|_2^3$

$\displaystyle= -2\left(\ln 2 - \ln\frac{7}{3}\right)$

$\displaystyle= -2\,\ln\frac{6}{7}\quad \text{or}\quad 2\,\ln\frac{7}{6}$

29. $\displaystyle\iint_R y e^{(x + y^2)}\, dx\, dy;\ 2 \le x \le 3,$

$0 \le y \le 2$

$\displaystyle\iint_R y e^{(x + y^2)}\, dx\, dy$

$\displaystyle= \int_0^2 \int_2^3 y e^{x + y^2}\, dx\, dy$

$\displaystyle= \int_0^2 y e^{x + y^2}\Big|_2^3\, dy$

$\displaystyle= \int_0^2 \left(y e^{3 + y^2} - y e^{2 + y^2}\right) dy$

$\displaystyle= e^3 \int_0^2 y e^{y^2}\, dy - e^2 \int_0^2 y e^{y^2}\, dy$

$\displaystyle= \frac{e^3}{2}\left(e^{y^2}\right)\Big|_0^2 - \frac{e^2}{2}\left(e^{y^2}\right)\Big|_0^2$

$\displaystyle= \frac{e^3}{2}\left(e^4 - e^0\right) - \frac{e^2}{2}\left(e^4 - e^0\right)$

$\displaystyle= \frac{1}{2}\left(e^7 - e^6 - e^3 + e^2\right)$

31. $z = 6x + 2y + 5;\ -1 \le x \le 1,$

$0 \le y \le 3$

$\displaystyle V = \int_{-1}^1 \int_0^3 (6x + 2y + 5)\, dy\, dx$

$\displaystyle= \int_{-1}^1 (6xy + 5y + y^2)\Big|_0^3\, dx$

$\displaystyle= \int_{-1}^1 (18x + 15 + 9)\, dx$

$\displaystyle= (9x^2 + 24x)\Big|_{-1}^1$

$\displaystyle= 9 + 24 - (9 - 24) = 48$

33. $z = x^2;\ 0 \le x \le 1,\ 0 \le y \le 4$

$\displaystyle V = \int_0^1 \int_0^4 x^2\, dy\, dx$

$\displaystyle= \int_0^1 (x^2 y)\Big|_0^4\, dx$

$\displaystyle= \int_0^1 4x^2\, dx$

$\displaystyle= \frac{4}{3}x^3\Big|_0^1 = \frac{4}{3}$

35. $z = x\sqrt{x^2 + y}$; $0 \le x \le 1$, $0 \le y \le 1$

$$V = \int_0^1 \int_0^1 x\sqrt{x^2 + y} \; dx \; dy$$

Let $u = x^2 + y$.

Then du = 2x dx.

When x = 0, u = y.

When x = 1, u = 1 + y.

$$= \int_0^1 \left[\int_y^{1+y} u^{1/2} \, du \right] dy$$

$$= \int_0^1 \frac{1}{2}\left(\frac{2}{3}u^{3/2}\right)\Big|_y^{1+y} dy$$

$$= \int_0^1 \frac{1}{3}[(1 + y)^{3/2} - y^{3/2}] dy$$

$$= \frac{1}{3} \cdot \frac{2}{5}[(1 + y)^{5/2} - y^{5/2}]\Big|_0^1$$

$$= \frac{2}{15}(2^{5/2} - 1 - 1)$$

$$= \frac{2}{15}(2^{5/2} - 2)$$

37. $z = \dfrac{xy}{(x^2 + y^2)^2}$; $1 \le x \le 2$, $1 \le y \le 4$

$$V = \int_1^2 \int_1^4 \frac{xy}{(x^2 + y^2)^2} \; dy \; dx$$

$$= \int_1^2 \left[\int_1^4 xy(x^2 + y^2)^{-2} dy \right] dx$$

$$= \int_1^2 \left[\int_1^4 \frac{1}{2}x(x^2 + y^2)^{-2}(2y) dy \right] dx$$

$$= \int_1^2 \left[-\frac{1}{2}x(x^2 + y^2)^{-1} \right]\Big|_1^4 dx$$

$$= \int_1^2 \left[-\frac{1}{2}x(x^2 + 16)^{-1} + \frac{1}{2}x(x^2 + 1)^{-1} \right] dx$$

$$= -\frac{1}{2} \int_1^2 \frac{1}{2}(x^2 + 16)^{-1}(2x)(dx$$

$$+ \frac{1}{2} \int_1^2 \frac{1}{2}(x^2 + 1)^{-1}(2x) dx$$

$$= -\frac{1}{2} \cdot \frac{1}{2} \ln |x^2 + 16|\Big|_1^2$$

$$+ \frac{1}{2} \cdot \frac{1}{2} \ln |x^2 + 1|\Big|_1^2$$

$$= -\frac{1}{4} \cdot \ln 20 + \frac{1}{4} \ln 17$$

$$+ \frac{1}{4} \ln 5 - \frac{1}{4} \ln 2$$

$$= \frac{1}{4}(-\ln 20 + \ln 17 + \ln 5 - \ln 2)$$

$$= \frac{1}{4} \ln \frac{(17)(5)}{(20)(2)}$$

$$= \frac{1}{4} \ln \frac{17}{8}$$

39. $\displaystyle\iint_R xe^{xy} \; dx \; dy$; $0 \le x \le 2$; $0 \le y \le 1$

$$\iint_R xe^{xy} \; dx \; dy$$

$$= \int_0^2 \int_0^1 xe^{xy} \; dy \; dx$$

$$= \int_0^2 \frac{x}{x}e^{xy}\Big|_0^1 dx$$

$$= \int_0^2 (e^x - e^0) dx$$

$$= (e^x - x)\Big|_0^2$$

$$= e^2 - 2 - e^0 + 0$$

$$= e^2 - 3$$

41. $\int_2^4 \int_2^{x^2} (x^2 + y^2)\,dy\,dx$

$$= \int_2^4 (x^2 y + \frac{y^3}{3}) \Big|_2^{x^2} dx$$

$$= \int_2^4 (x^4 + \frac{x^6}{3} - 2x^2 - \frac{8}{3})\,dx$$

$$= (\frac{x^5}{5} + \frac{x^7}{21} - \frac{2}{3}x^3 - \frac{8}{3}x) \Big|_2^4$$

$$= \frac{1024}{5} + \frac{16,384}{21} - \frac{2}{3}(64) - \frac{8}{3}(4)$$

$$- (\frac{32}{5} + \frac{128}{21} - \frac{16}{3} - \frac{16}{3})$$

$$= \frac{1024}{5} - \frac{32}{5} + \frac{16,384 - 128}{21}$$

$$- \frac{128}{3} - \frac{32}{3} - (\frac{-32}{3})$$

$$= \frac{992}{5} + \frac{16,256}{21} - \frac{128}{3}$$

$$= \frac{20,832}{105} + \frac{81,280}{105} - \frac{4480}{105}$$

$$= \frac{97,632}{105} \approx 929.83$$

43. $\int_0^4 \int_0^x \sqrt{xy}\,dy\,dx$

$$= \int_0^4 \left[\frac{2(xy)^{3/2}}{3x}\right]\Big|_0^x dx$$

$$= \frac{2}{3} \int_0^4 \left[\frac{(\sqrt{x^2})^3}{x} - \frac{0}{x}\right] dx$$

$$= \frac{2}{3} \int_0^4 x^2\,dx = \frac{2}{3} \cdot \frac{x^3}{3}\Big|_0^4 = \frac{2}{9}(64)$$

$$= \frac{128}{9}$$

45. $\int_1^2 \int_y^{3y} \frac{1}{x}\,dx\,dy$

$$= \int_1^2 (\ln|x|)\Big|_y^{3y} dy$$

$$= \int_1^2 (\ln|3y| - \ln|y|)\,dy$$

$$= \int_1^2 \ln\left|\frac{3y}{y}\right| dy$$

$$= (\ln 3)(y)\Big|_1^2 = \ln 3$$

47. $\int_0^4 \int_1^{e^x} \frac{x}{y}\,dy\,dx$

$$= \int_0^4 (x \ln|y|)\Big|_1^{e^x} dx$$

$$= \int_0^4 (x \ln e^x - x \ln 1)\,dx$$

$$= \int_0^4 x^2\,dx = \frac{x^3}{3}\Big|_0^4 = \frac{64}{3}$$

49. $\iint_R (4x + 7y)\,dy\,dx; \; 1 \le x \le 3;$

$0 \le y \le x + 1$

$$\iint_R (4x + 7y)\,dy\,dx$$

$$= \int_1^3 \int_0^{x+1} (4x + 7y)\,dy\,dx$$

$$= \int_1^3 (4xy + \frac{7y^2}{2})\Big|_0^{x+1} dx$$

$$= \int_1^3 [4x^2 + 4x + \frac{7}{2}(x^2 + 2x + 1)]\,dx$$

$$= \int_1^3 \left(\frac{15}{2}x^2 + 11x + \frac{7}{2}\right)dx$$

$$= \left(\frac{15}{2} \cdot \frac{x^3}{3} + \frac{11x^2}{2} + \frac{7x}{2}\right)\Big|_1^3$$

$$= \frac{5}{2} \cdot 27 + \frac{11(9)}{2} + \frac{7(3)}{2}$$

$$\quad - \left(\frac{5}{2} + \frac{11}{2} + \frac{7}{2}\right)$$

$$= \frac{130}{2} + \frac{88}{2} + \frac{14}{2}$$

$$= \frac{232}{2} = 116$$

51. $\displaystyle\iint_R (4 - 4x^2)\,dy\,dx;\ 0 \le x \le 1,$

$0 \le y \le 2 - 2x$

$$\iint_R (4 - 4x^2)\,dy\,dx$$

$$= \int_0^2 \int_0^{2-2x} 4(1 - x^2)\,dy\,dx$$

$$= \int_0^1 [4(1 - x^2)y]\Big|_0^{2(1-x)}\,dx$$

$$= \int_0^1 4(1 - x^2)(2)(1 - x)\,dx$$

$$= 8 \int_0^1 (1 - x - x^2 + x^3)\,dx$$

$$= 8\left(x - \frac{x^2}{2} - \frac{x^3}{3} + \frac{x^4}{4}\right)\Big|_0^1$$

$$= 8\left(1 - \frac{1}{2} - \frac{1}{3} + \frac{1}{4}\right)$$

$$= 8\left(\frac{1}{2} - \frac{1}{12}\right)$$

$$= 8 \cdot \frac{5}{12} = \frac{10}{3}$$

53. $\displaystyle\iint_R e^{x/y^2}\,dx\,dy;\ 1 \le y \le 2;$

$0 \le x \le y^2$

$$\iint_R e^{x/y^2}\,dx\,dy$$

$$= \int_1^2 \int_0^{y^2} e^{x/y^2}\,dx\,dy$$

$$= \int_1^2 [y^2 e^{x/y^2}]\Big|_0^{y^2}\,dy$$

$$= \int_1^2 (y^2 e^{y^2/y^2} - y^2 e^0)\,dy$$

$$= \int_1^2 (ey^2 - y^2)\,dy$$

$$= (e - 1)\frac{y^3}{3}\Big|_1^2$$

$$= (e - 1)\left(\frac{8}{3} - \frac{1}{3}\right)$$

$$= \frac{7(e - 1)}{3}$$

55. $\displaystyle\iint_R x^3 y\,dx\,dy;\ R$ bounded by $y = x^2,$

$y = 2x$

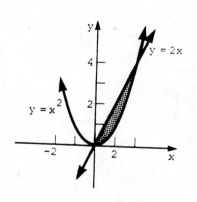

The points of intersection can be
determined by solving the following
system for x.

$$y = x^2$$
$$\underline{y = 2x}$$

$$x^2 = 2x$$
$$x(x - 2) = 0$$
$$x = 0 \quad \text{or} \quad x = 2$$

Therefore,

$$\iint_R x^3 y \; dx \; dy$$

$$= \int_0^2 \int_{x^2}^{2x} x^3 y \; dy \; dx$$

$$= \int_0^2 \left(x^3 \; \frac{y^2}{2}\right)\Bigg|_{x^2}^{2x} dx$$

$$= \int_0^2 \left[x^3 \; \frac{(4x^2)}{2} - x^3 \; \frac{(x^4)}{2}\right] dx$$

$$= \int_0^2 \left(2x^5 - \frac{x^7}{2}\right) dx$$

$$= \left(\frac{1}{3}x^6 - \frac{1}{16}x^8\right)\Bigg|_0^2 = \frac{1}{3} \cdot 2^6 - \frac{1}{16} \cdot 2^8$$

$$= \frac{64}{3} - 16 = \frac{16}{3}.$$

57. $\iint_R \dfrac{dy \; dx}{y}$; R bounded by y = x,

$y = \dfrac{1}{x}$, x = 2

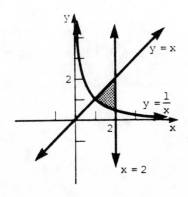

y = x and y = $\dfrac{1}{x}$
intersect at (1, 1).

$$\int_1^2 \int_{1/x}^x \frac{dy}{y} \; dx$$

$$= \int_1^2 \ln y \Bigg|_{1/x}^x dx$$

$$= \int_1^2 \left(\ln x - \ln \frac{1}{x}\right) dx$$

$$= \int_1^2 2 \ln x \; dx$$

$$= 2(x \ln x - x)\Bigg|_1^2$$

$$= 2[2 \ln 2 - 2) - (\ln 1 - 1)]$$
$$= 4 \ln 2 - 2$$

59. $f(x, y) = x^2 + y^2$; $0 \le x \le 2$,
$0 \le y \le 3$

The area of region R is

$$A = (2 - 0)(3 - 0) = 6.$$

The average value of

$$f(x, y) = x^2 + y^2 \text{ is}$$

$$\frac{1}{A} \iint_R (x^2 + y^2) \, dy \; dx$$

$$= \frac{1}{6} \int_0^2 \int_0^3 (x^2 + y^2) \, dy \; dx$$

$$= \frac{1}{6} \int_0^2 \left(x^2 y + \frac{y^3}{3}\right)\Bigg|_0^3 dx$$

$$= \frac{1}{6} \int_0^2 (3x^2 + 9) \, dx$$

$$= \frac{1}{6}(x^3 + 9x)\Bigg|_0^2$$

$$= \frac{1}{6}(8 + 18 - 0)$$

$$= \frac{1}{6} \cdot 26$$

$$= \frac{13}{3}.$$

61. $f(x, y)^{2x+y}$; $1 \le x \le 2$, $2 \le y \le 3$

The area of region R is

$$A = (2 - 1)(3 - 2) = 1.$$

The average value is

$$\frac{1}{A} \iint\limits_{R} e^{2x+y} \, dy \, dx$$

$$= \iint\limits_{R} e^{2x+y} \, dy \, dx$$

$$= \int_{1}^{2} \int_{2}^{3} e^{2x+y} \, dy \, dx$$

$$= \int_{1}^{2} (e^{2x+y}) \Big|_{2}^{3} \, dx$$

$$= \int_{1}^{2} (e^{2x+3} - e^{2x+2}) \, dx$$

$$= \frac{1}{2}(e^{2x+3} - e^{2x+2}) \Big|_{1}^{2}$$

$$= \frac{1}{2}(e^{4+3} - e^{2+3} - e^{4+2} + e^{4})$$

$$= \frac{e^{7} - e^{6} - e^{5} + e^{4}}{2}.$$

63. $C(x, y) = \frac{1}{9}x^2 + 2x + y^2 + 5y + 100$

Area $= (75 - 48)(60 - 20)$

$\quad = 27(40)$

$\quad = 1080$

The average cost is

$$\frac{1}{1080} \iint\limits_{R} \left(\frac{1}{9}x^2 + 2x + y^2 + 5y + 100\right) dx \, dy$$

$$= \frac{1}{1080} \int_{48}^{75} \int_{20}^{60} \left[\frac{1}{9}x^2 + 2x + 100\right.$$

$$\left. + y^2 + 5y\right] dy \, dx$$

$$= \frac{1}{1080} \int_{48}^{75} \left[\left(\frac{1}{9}x^2 + 2x + 100\right)y + \frac{y^3}{3}\right.$$

$$\left. + \frac{5}{2}y^2\right]\Big|_{20}^{60} dx$$

$$= \frac{1}{1080} \int_{48}^{75} \left[\left(\frac{1}{9}x^2 + 2x + 100\right)(60 - 20)\right.$$

$$\left. + \frac{1}{3}(60^3 - 20^3) + \frac{5}{2}(60^2 - 20^2)\right] dx$$

$$= \frac{40}{1080}\left(\frac{1}{27}x^3 + x^2 + 100x\right)\Big|_{48}^{75}$$

$$+ \frac{1}{1080}\left[\frac{208,000}{3} + \frac{5}{2}(3200)\right]x\Big|_{48}^{75}$$

$$= \frac{40}{1080}$$

$$\cdot \left[\frac{1}{27}(75^3 - 48^3) + 75^2 - 48^2 + 100(75 - 48)\right]$$

$$+ \frac{1}{1080}\left(\frac{208,000}{3} + 8000\right)(75 - 48)$$

$$= \frac{40}{1080}\left(\frac{311,283}{27} + 3321 + 2700\right)$$

$$+ \frac{1}{1080}\left(\frac{232,000}{3}\right)27$$

$$= \frac{40}{1080}(11,529 + 6021) + \frac{23,200}{12}$$

$$\approx \$2583$$

65. $P(x, y)$

$\quad = -(x - 100)^2 - (y - 50)^2 + 2000$

$\quad \text{Area} = (150 - 100)(80 - 40)$

$\quad\quad = (50)(40)$

$\quad\quad = 2000$

The average weekly profit is

$\frac{1}{2000} \int\int_R [-(x - 100)^2 - (y - 50)^2 + 2000]\,dy\,dx$

$= \frac{1}{2000} \int_{100}^{150} \int_{40}^{80} [-(x - 100)^2 - (y - 50)^2$

$\quad + 2000]\,dy\,dx$

$= \frac{1}{2000} \int_{100}^{150} \left[-(x - 100)^2 y - \frac{(y - 50)^3}{3}\right.$

$\quad \left. + 2000y\right]\Big|_{40}^{80} dx \Big|$

$= \frac{1}{2000} \int_{100}^{150} \left[-(x - 100)^2(80 - 40)\right.$

$\quad - \frac{(80 - 50)^3}{3} + \frac{(40 - 50)^3}{3}$

$\quad \left. + 2000(80 - 40)\right] dx$

$= \frac{1}{2000} \int_{100}^{150} \left[-40(x - 100)^2 - \frac{30^3}{3}\right.$

$\quad \left. + \frac{(-10)^3}{3} + 2000(40)\right] dx$

$= \frac{1}{2000} \int_{100}^{150} \left[-40(x - 100)^2 - \frac{28,000}{3}\right.$

$\quad \left. + 80,000\right] dx$

$= \frac{1}{2000}$

$\quad \cdot \left[\frac{-40(x - 100)^3}{3} - \frac{28,000}{3}x + 80,000x\right]\Big|_{100}^{150}$

$= \frac{1}{2000}\left[-\frac{40}{3}(150 - 100)^3 + \frac{40}{3}(100 - 100)^3\right.$

$\quad \left. - \frac{28,000}{3}(150 - 100) + 80,000(150 - 1000)\right]$

$= \frac{1}{2000}\left[-\frac{40}{3}(50)^3 + \frac{40}{3} \cdot 0\right.$

$\quad \left. - \frac{28,000}{3}(50) + 80,000(50)\right]$

$= \frac{1}{2000}\left(\frac{-5,000,000 - 1,400,000 + 12,000,000}{3}\right)$

$= \frac{1}{2000}\left(\frac{5,600,000}{3}\right) = \$933.33$

Chapter 16 Review Exercises

1. $f(x, y) = -4x^2 + 6xy - 3$

$\quad f(-1, 2) = -4(-1)^2 + 6(-1)(2) - 3$

$\quad\quad = -19$

$\quad f(6, -3) = -4(6)^2 + 6(6)(-3) - 3$

$\quad\quad = -4(36) + (-108) - 3$

$\quad\quad = -255$

3. $f(x, y) = \frac{x - 3y}{x + 4y}$

$\quad f(-1, 2) = \frac{-1 - 6}{-1 + 8} = \frac{-7}{7} = -1$

$\quad f(6, -3) = \frac{6 - 3(-3)}{6 + 4(-3)} = \frac{6 + 9}{6 - 12}$

$\quad\quad = \frac{15}{-6} = -\frac{5}{2}$

For Exercises 5–9, see the answer graphs at the back of the textbook.

5. The plane $x + y + z = 4$ intersects the axes at $(4, 0, 0)$, $(0, 4, 0)$, and $(0, 0, 4)$.

7. The plane $5x + 2y = 10$ intersects the x- and y-axes at $(2, 0, 0)$ and $(0, 5, 0)$. Note that there is no z-intercept since $x = y = 0$ is not a solution of the equation of the plane.

9. $x = 3$

The plane is parallel to the yz-plane. It intersects the x-axis at $(3, 0, 0)$.

11. Let $z = f(x, y) = -5x^2 + 7xy - y^2$

   (a) $\dfrac{\partial z}{\partial x} = -10x + 7y$

   (b) $\dfrac{\partial z}{\partial y} = 7x - 2y$

   $$\left(\dfrac{\partial z}{\partial y}\right)(-1, 4) = 7(-1) - 2(4)$$
   $$= -15$$

   (c) $f_{xy}(x, y) = 7$
   $f_{xy}(2, -1) = 7$

13. $f(x, y) = 9x^3y^2 - 5x$
   $f_x = 27x^2y^2 - 5$
   $f_y = 18x^3y$

15. $f(x, y) = \sqrt{4x^2 + y^2}$
   $f_x = \dfrac{1}{2}(4x^2 + y^2)^{-1/2}(8x)$

   $\quad = \dfrac{4x}{(4x^2 + y^2)^{1/2}}$

   $f_y = \dfrac{1}{2}(4x^2 + y^2)^{-1/2}(2y)$

   $\quad = \dfrac{y}{(4x^2 + y^2)^{1/2}}$

17. $f(x, y) = x^2 \cdot e^{2y}$
   $f_x = 2xe^{2y}$
   $f_y = 2x^2e^{2y}$

19. $f(x, y) = \ln|2x^2 + y^2|$
   $f_x = \dfrac{1}{2x^2 + y^2} \cdot 4x$

   $\quad = \dfrac{4x}{2x^2 + y^2}$

   $f_y = \dfrac{1}{2x^2 + y^2} \cdot 2y$

   $\quad = \dfrac{2y}{2x^2 + y^2}$

21. $f(x, y) = 4x^3y^2 - 8xy$
   $f_x = 12x^2y^2 - 8y$
   $f_{xx} = 24xy^2$
   $f_{xy} = 24x^2y - 8$

23. $f(x, y) = \dfrac{2x}{x - 2y}$

   $f_x = \dfrac{2(x - 2y) - (2x)}{(x - 2y)^2}$

   $\quad = \dfrac{-4y}{(x - 2y)^2}$

   $\quad = -4y(x - 2y)^{-2}$

   $f_{xx} = -4y(-2(x - 2y)^{-3})$

   $\quad = 8y(x - 2y)^{-3}$

   $\quad = \dfrac{8y}{(x - 2y)^3}$

Use the quotient rule on

   $$f_x = \dfrac{-4y}{(x - 2y)^2}$$

to get

$$f_{xy} = \frac{-4(x-2y)^2 + 4y[2(x-2y)(-2)]}{(x-2y)^4}$$

$$= \frac{-4(x-2y)[(x-2y)+4y]}{(x-2y)^4}$$

$$= \frac{-4(x+2y)}{(x-2y)^3}$$

$$= \frac{-4x-8y}{(x-2y)^3}.$$

25. $f(x, y) = x^2 e^y$

$$f_x = 2xe^y$$

$$f_{xx} = 2e^y$$

$$f_{xy} = 2xe^y$$

27. $f(x, y) = \ln|2 - x^2y|$

$$f_x = \frac{1}{2 - x^2y} \cdot -2xy$$

$$= \frac{2xy}{x^2y - 2}$$

$$f_{xx} = \frac{(x^2y - 2)2y - 2xy(2xy)}{(x^2y - 2)^2}$$

$$= \frac{2y[(x^2y - 2) - 2x^2y]}{(x^2y - 2)^2}$$

$$= \frac{2y(-x^2y - 2)}{(x^2y - 2)^2}$$

$$= \frac{-2x^2y^2 - 4y}{(2 - x^2y)^2}$$

$$f_{xy} = \frac{2x(x^2y - 2) - x^2(2xy)}{(x^2y - 2)^2}$$

$$= \frac{2x[(x^2y - 2) - x^2y]}{(x^2y - 2)^2}$$

$$= \frac{2x(-2)}{(x^2y - 2)^2}$$

$$= \frac{-4x}{(2 - x^2y)^2}$$

29. $z = x^2 + 2y^2 - 4y$

$$z_x = 2x$$

$$z_y = 4y - 4$$

Setting z_x and z_y equal to zero simultaneously implies $x = 0$ and $y = 1$.

$$z_{xx} = 2, \ z_{yy} = 4, \ z_{xy} = 0$$

For $(0, 1)$,

$$D = 2 \cdot 4 - 0 = 8 > 0.$$

Since

$$z_{xx}(0, 1) = 2 > 0,$$

then z has a relative minimum at $(0, 1)$.

31. $f(x, y) = x^2 + 5xy - 10x + 3y^2 - 12y$

$$f_x = 2x + 5y - 10$$

$$f_y = 5x + 6y - 12$$

Setting f_x and f_y equal to zero and solving yields

$$2x + 5y = 10$$
$$5x + 6y = 12$$

$$-10x - 25y = -50$$
$$\underline{10x + 12y = \ \ 24}$$
$$-13y = -26$$
$$y = 2$$

$$2x + 10 = 10$$
$$x = 0$$

$$f_{xx} = 2, \ f_{yy} = 6, \ f_{xy} = 5,$$

For $(0, 2)$,

$$D = 2 \cdot 6 - 5^2 = -13 < 0.$$

Therefore f has a saddle point at $(0, 2)$.

33. $z = \frac{1}{2}x^2 + \frac{1}{2}y^2 + 2xy - 5x - 7y + 10$

$z_x = x + 2y - 5$

$z_y = y + 2x - 7$

Setting $z_x = z_y = 0$ and solving yields

$$x + 2y = 5$$
$$2x + y = 7.$$
$$-2x - 4y = -10$$
$$\underline{2x + y = \quad 7}$$
$$-3y = -3$$
$$y = 1, \ x = 3.$$

$z_{xx} = 1, \ z_{yy} = 1, \ z_{xy} = 2$

For (3, 1),

$$D = 1 \cdot 1 - 4 = -3 < 0.$$

Therefore z has a saddle point at (3, 1).

35. $z = x^3 + y^2 + 2xy - 4x - 3y - 2$

$z_x = 3x^2 + 2y - 4$

$z_y = 2y + 2x - 3$

Setting $z_x = z_y = 0$ yields

$$3x^2 + 2y - 4 = 0 \quad (1)$$
$$2y + 2x - 3 = 0. \quad (2)$$

Solving for 2y in Equation (2) gives $2y = -2x + 3$. Substitute into Equation (1).

$$3x^2 + (-2x) + 3 - 4 = 0$$
$$3x^2 - 2x - 1 = 0$$
$$(3x + 1)(x - 1) = 0$$
$$x = \frac{-1}{3} \quad \text{or} \quad x = 1$$
$$y = \frac{11}{6} \quad \text{or} \quad y = \frac{1}{2}$$

$z_{xx} = 6x, \ z_{yy} = 2, \ z_{xy} = 2$

For (-1/3, 11/6),

$$D = 6\left(\frac{-1}{3}\right)(2) - 4$$
$$= -4 - 4 = -8 < 0, \text{ so}$$

so z has a saddle point at (-1/3, 11/6).

For (1, 1/2),

$$D = 6(1)(2) - 4 = 8 > 0.$$

$z_{xx}\left(1, \frac{1}{2}\right) = 6 > 0$, so

z has a relative minimum at (1, 1/2).

37. $f(x, y) = x^2 y; \ x + y = 4$

1. $g(x) = x + y - 4$

2. $F(x, y) = x^2 y + \lambda(x + y - 4)$

3. $F_x = 2xy + \lambda$

   $F_y = x^2 + \lambda$

   $F_\lambda = x + y - 4$

4. $2xy + \lambda = 0$

      $x^2 + \lambda = 0$

   $x + y - 4 = 0$

5. $\lambda = -2xy$

        $\lambda = -x^2$

     $-2xy = -x^2$

$$2xy - x^2 = 0$$
$$x(2y - x) = 0$$
$$x = 0 \quad \text{or} \quad 2y = x$$

Substituting into the third equation gives

$$y = 4 \quad \text{or} \quad y = \frac{4}{3}.$$

If $y = \frac{4}{3}, \ x = \frac{8}{3}.$

The critical points are $(0, 4)$ and $\left(\frac{8}{3}, \frac{4}{3}\right)$.

$$f(0, 4) = 0$$

$$f\left(\frac{8}{3}, \frac{4}{3}\right) = \frac{64}{9} \cdot \frac{4}{3}$$

$$= \frac{256}{27}$$

Therefore, f has a minimum of 0 at $(0, 4)$ and a maximum of $256/27$ at $(8/3, 4/3)$.

39. Let x and y be the numbers such that $x + y = 80$ and $f(x, y) = x^2 y$.

1. $g(x) = x + y - 80$

2. $F(x, y) = x^2 y + \lambda(x + y - 80)$

3. $F_x = 2xy + \lambda$

$F_y = x^2 + \lambda$

$F_\lambda = x + y - 80$

4. $2xy + \lambda = 0$

$x^2 + \lambda = 0$

$x + y - 80 = 0$

5. $\lambda = -2xy$

$\lambda = -x^2$

$-2xy = -x^2$

$2xy - x^2 = 0$

$x(2y - x) = 0$

$x = 0$ or $x = 2y$

Substituting into the third equation gives

$y = 80$ or $2y + y - 80 = 0$

$3y = 80$

$y = \frac{80}{3}$

and $x = \frac{160}{3}$.

$f(0, 80) = 0 \cdot 80^2 = 0$

$$f\left(\frac{160}{3}, \frac{80}{3}\right) = \frac{(160)^2}{9} \frac{(80)}{3}$$

$$= \frac{2{,}048{,}000}{27} > f(0, 80)$$

f has maximum at $(160/3, 80/3)$. Therefore, if $x = 160/3$ and $y = 80/3$, then $x^2 y$ is maximized.

41. Let x be the length of each of the square faces of the box and y be the length of the box.

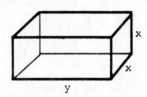

Since the volume must be 125, the constraint is $125 = x^2 y$.

$f(x, y) = 2x^2 + 4xy$ is the surface area of the box.

1. $g(x) = x^2 y - 125$

2. $F(x, y) = 2x^2 + 4xy + \lambda(x^2 y - 125)$

3. $F_x = 4x + 4y + 2xy\lambda$

$F_y = 4x + x^2 \lambda$

$F_\lambda = x^2 y - 125$

4. $4x + 4y + 2xy\lambda = 0$ (1)

$4x + x^2 \lambda = 0$ (2)

$x^2 y - 125 = 0$ (3)

5. Factoring Equation (2) gives

$$x(4 + x\lambda) = 0$$

$$x = 0 \quad \text{or} \quad 4 + x\lambda = 0.$$

Since $x = 0$ is not a solution of Equation (3), then

$$4 + x\lambda = 0$$

$$\lambda = \frac{-4}{x}.$$

Substituting into Equation (1) gives

$$4x + 4y + 2xy\left(\frac{-4}{x}\right) = 0$$

or $4x + 4y - 8y = 0$

$$x = y.$$

Substituting $x = y$ into Equation (3) gives

$$x^2y - 125 = 0$$
$$y^3 = 125$$
$$y = 5.$$

Therefore $x = y = 5$. The dimensions are 5 in by 5 in by 5 in.

43. $z = 7x^3y - 4y^3$

$dz = 21x^2y\,dx + (7x^3 - 12y^2)dy$

45. $z = x^2ye^{x-y}$

$f_x = x^2ye^{x-y} + e^{x-y} \cdot y(2x)$

$\quad = xye^{x-y}(x + 2)$

$f_y = x^2ye^{x-y}(-1) + x^2e^{x-y}$

$\quad = x^2e^{x-y}(1 - y)$

$dz = xye^{x-y}(2 + x)dx + x^2e^{x-y}(1 - y)dy$

47. $w = x^5 + y^4 - z^3$

$dw = 5x^4dx + 4y^3dy - 3z^2dz$

49. $z = 2x^2 - 4y^2 + 6xy;\ x = 2,\ y = -3,$

$dx = .01,\ dy = .05$

$f_x = 4x + 6y$

$f_y = -8y + 6x$

$dz = (4x + 6y)dx + (-8y + 6x)dy$

Substitute.

$dz = (4(2) + 6(-3))(.01)$

$\quad + (-8(-3) + 6(2))(.05)$

$\quad = -.1 + 1.8 = 1.7$

51. $\displaystyle\int_0^4 (x^2y^2 + 5x)\,dx$

$$= \left(\frac{x^3}{3}y^2 + \frac{5}{2}x^2\right)\Bigg|_0^4$$

$$= \frac{64}{3}y^2 + \frac{5(4)^2}{2}$$

$$= \frac{64y^2}{3} + 40$$

53. $\displaystyle\int_2^5 \sqrt{6x + 3y}\ dx$

Let $u = 6x + 3y$

$\quad du = 6\ dx$

When $x = 2,\ u = 12 + 3y.$

When $x = 5,\ u = 30 + 3y.$

$$= \frac{1}{6}\int_{12+3y}^{30+3y} u^{1/2}\ du$$

$$= \frac{1}{6}\left(\frac{2}{3}u^{3/2}\right)\Bigg|_{12+3y}^{30+3y}$$

$$= \frac{1}{9}[(30 + 3y)^{3/2} - (12 + 3y)^{3/2}]$$

55. $\displaystyle\int_4^9 \frac{6y - 8}{\sqrt{x}}\ dx$

$$= (6y - 8)(2\sqrt{x})\Bigg|_4^9$$

$$= 2(3y - 4)(2)(\sqrt{9} - \sqrt{4})$$

$$= 4(3y - 4)$$

$$= 12y - 16$$

57. $\displaystyle\int_0^5 \frac{6x}{\sqrt{4x^2 + 2y^2}}\ dx$

Let $u = 4x^2 + 2y^2$

$\quad du = 8x\ dx$

When x = 0, u = $2y^2$.

When x = 5, u = $100 + 2y^2$.

$$= \frac{3}{4}\int_{2y^2}^{100+2y^2} u^{-1/2}\, du$$

$$= \frac{3}{4}(2u^{1/2})\Big|_{2y^2}^{100+2y^2}$$

$$= \frac{3}{4}\cdot 2[(100 + 2y^2)^{1/2} - (2y^2)^{1/2}]$$

$$= \frac{3}{2}[(100 + 2y^2)^{1/2} - (2y^2)^{1/2}]$$

59. $\int_0^2\left[\int_0^4 (x^2y^2 + 5x)\,dx\right]dy$

See Exercise 51.

$$= \int_0^2 \left(\frac{64}{3}y^2 + 40\right)dy$$

$$= \left(\frac{64}{3}y^3 + 40y\right)\Big|_0^2$$

$$= \frac{64}{9}(8) + 40(2)$$

$$= \frac{512}{9} + \frac{720}{9}$$

$$= \frac{1232}{9}$$

61. $\int_3^4\left[\int_2^5 \sqrt{6x + 3y}\ dx\right]dy$

See Exercise 53.

$$= \int_3^4 \frac{1}{9}\left[(30 + 3y)^{3/2} - (12 + 3y)^{3/2}\right]dy$$

$$= \frac{1}{3}\cdot\frac{1}{9}\cdot\frac{2}{5}$$

$$\cdot\,[(30 + 3y)^{5/2} - (12 + 3y)^{5/2}]\Big|_3^4$$

$$= \frac{2}{135}[(42)^{5/2} - (24)^{5/2} - (39)^{5/2}$$

$$+ (21)^{5/2}]$$

63. $\int_2^4\int_2^4 \frac{dx\ dy}{y}$

$$= \int_2^4 \left(\frac{1}{y}x\right)\Big|_2^4 dy$$

$$= \int_2^4 \left[\frac{1}{y}(4 - 2)\right]dy$$

$$= 2\,\ln\,|y|\,\Big|_2^4$$

$$= 2\,\ln\,\left|\frac{4}{2}\right| = 2\,\ln\,2\ \text{or}\ \ln 4$$

65. $\int\int_R (x^2 + y^2)\,dx\ dy;\ 0 \le x \le 2,$

$0 \le y \le 3$

$$\int\int_R (x^2 + y^2)\,dx\ dy$$

$$= \int_0^3\int_0^2 (x^2 + y^2)\,dx\ dy$$

$$= \int_0^3 \left(\frac{x^3}{3} + y^2x\right)\Big|_0^2 dy$$

$$= \int_0^3 \left(\frac{8}{3} + 2y^2\right)dy$$

$$= \left(\frac{8}{3}y + \frac{2}{3}y^3\right)\Big|_0^3$$

$$= 8 + 18 = 26$$

67. $\displaystyle\iint_R \sqrt{y + x}\ dx\ dy;\ 0 \le x \le 7,$

$1 \le y \le 9$

$\displaystyle\iint_R \sqrt{y + x}\ dx\ dy$

$\displaystyle = \int_0^7 \int_1^9 \sqrt{y + x}\ dy\ dx$

$\displaystyle = \int_0^7 \left[\frac{2}{3}(y + x)^{3/2} \right]\Big|_1^9\ dx$

$\displaystyle = \int_0^7 \frac{2}{3}[(9 + x)^{3/2} - (1 + x)^{3/2}]\,dx$

$\displaystyle = \frac{2}{3} \cdot \frac{2}{5}[(9 + x)^{5/2} - (1 + x)^{5/2}]\Big|_0^7$

$\displaystyle = \frac{4}{15}[(16)^{5/2} - (8)^{5/2} - (9)^{5/2}$

$\displaystyle \quad + (1)^{5/2}]$

$\displaystyle = \frac{4}{15}[4^5 - (2\sqrt{2})^5 - 3^5 + 1]$

$\displaystyle = \frac{4}{15}(1024 - 32(4\sqrt{2}) - 243 + 1)$

$\displaystyle = \frac{4}{15}(782 - 128\sqrt{2})$

$\displaystyle = = \frac{4}{15}(782 - 8^{5/2})$

69. $z = x + 9y + 8;\ 1 \le x \le 6,$

$0 \le y \le 8$

$\displaystyle V = \iint_R (x + 9y + 8)\,dx\ dy$

$\displaystyle = \int_0^8 \int_1^6 (x + 9y + 8)\,dx\ dy$

$\displaystyle = \int_0^8 \left[\frac{x^2}{2} + (9y + 8)x\right]\Big|_1^6\ dy$

$\displaystyle = \int_0^8 \left[\frac{36 - 1}{2} + (9y + 8)(6 - 1)\right]dy$

$\displaystyle = \left[\frac{35}{2}y + 5\left(\frac{9}{2}y^2 + 8y\right)\right]\Big|_0^8$

$\displaystyle = \frac{35}{2}(8) + 5\left[\left(\frac{9}{2}\right)(64) + 8 \cdot 8\right]$

$\displaystyle = \frac{35(8)}{2} + 5\left(\frac{9(64)}{2} + 8 \cdot 8\right)$

$= 140 + 5(352)$

$= 1900$

71. $\displaystyle\int_0^1 \int_0^{2x} xy\ dy\ dx$

$\displaystyle = \int_0^1 \left(\frac{xy^2}{2}\right)\Big|_0^{2x}\ dx$

$\displaystyle = \int_0^1 \frac{x}{2}(4x^2 - 0)\,dx$

$\displaystyle = \int_0^1 2x^3\ dx$

$\displaystyle = \left(\frac{1}{2}x^4\right)\Big|_0^1 = \frac{1}{2}$

73. $\displaystyle\int_0^1 \int_{x^2}^x x^3 y\ dy\ dx$

$\displaystyle \int_0^1 \left(\frac{x^3}{2}y^2\right)\Big|_{x^2}^x\ dx$

$\displaystyle = \int_0^1 \frac{x^3}{2}(x^2 - x^4)\,dx$

$\displaystyle = \frac{1}{2}\int_0^1 (x^5 - x^7)\,dx$

$\displaystyle = \frac{1}{2}\left(\frac{x^6}{6} - \frac{x^8}{8}\right)\Big|_0^1$

$$= \frac{1}{2}\left(\frac{1}{6} - \frac{1}{8}\right)$$

$$= \frac{1}{2} \cdot \frac{1}{24}$$

$$= \frac{1}{48}$$

75. $\displaystyle\iint_R (2x + 3y)\,dx\,dy; \quad 0 \le y \le 1,$

$y \le x \le 2 - y$

$\displaystyle\int_0^1 \int_y^{2-y} (2x + 3y)\,dx\,dy$

$= \displaystyle\int_0^1 (x^2 + 3xy)\Big|_y^{2-y}\,dy$

$= \displaystyle\int_0^1 [(2-y)^2 - y^2 + 3y(2-y-y)]\,dy$

$= \displaystyle\int_0^1 (4 - 4y + y^2 - y^2 + 6y - 6y^2)\,dy$

$= \displaystyle\int_0^1 (4 + 2y - 6y^2)\,dy$

$= (4y + y^2 - 2y^3)\Big|_0^1$

$= 4 + 1 - 2$

$= 3$

77. $C(x, y) = 2x^2 + 4y^2 - 3xy + \sqrt{x}$

(a) $C(10, 5)$

$= 2(10)^2 + 4(5)^2 - 3(10)(5) + \sqrt{10}$

$= 200 + 100 - 150 + \sqrt{10}$

$= \$(150 + \sqrt{10})$

(b) $C(15, 10)$

$= 2(15)^2 + 4(10)^2 - 3(15)(10)$

$\quad + \sqrt{15}$

$= 450 + 400 - 450 + \sqrt{15}$

$= \$(400 + \sqrt{15})$

(c) $C(20, 20)$

$= 2(20)^2 + 4(20)^2 - 3(20)(20)$

$\quad + \sqrt{20}$

$= 800 + 1600 - 1200 + \sqrt{20}$

$= \$(1200 + 2\sqrt{5})$

79. $z = x^{.6}y^{.4}$

(a) The marginal productivity of labor is

$$\frac{\partial z}{\partial x} = .6x^{-.4}y^{.4}$$

$$= \frac{.6y^{.4}}{x^{.4}}$$

(b) The marginal productivity of capital is

$$\frac{\partial z}{\partial y} = .4x^{.6}y^{-.6}$$

$$= \frac{.4x^{.6}}{y^{.6}}$$

81. (a)

x	y	x^2	xy
3	4	9	12
5	11	25	55
7	20	49	140
8	23	64	184
Totals 23	58	147	391

$$m = \frac{23(58) - 4(391)}{(23)^2 - 4(147)} \approx 3.90$$

$$b = \frac{58 - m(23)}{4} \approx -7.92$$

Least squares equation is

$$y' = 3.90x - 7.92.$$

(b) If x = 6,

$$y' = 3.90(6) - 7.92 = 15.48$$

(in ten-thousands).

The predicted earnings are $154,800.

83. $P(x, y) = \dfrac{x}{x + 5y} + \dfrac{y + x}{y}$

$$= x(x + 5y)^{-1} + \dfrac{y + x}{y}$$

x = 75, y = 50, dx = -3, dy = 2

$$dP = \left[\dfrac{(x + 5y)\cdot 1 - x\cdot 1}{(x + 5y)^2} + \dfrac{1}{y}\right]dx$$

$$+ \left[-5x(x + 5y)^{-2} + \dfrac{y\cdot 1 - (y + x)}{y^2}\right]dy$$

$$= \left[\dfrac{5y}{(x + 5y)^2} + \dfrac{1}{y}\right]dx$$

$$+ \left[\dfrac{-5x}{(x + 5y)^2} - \dfrac{x}{y^2}\right]dy$$

$$= \left[\dfrac{5\cdot 50}{(75 + 5\cdot 50)^2} + \dfrac{1}{50}\right](-3)$$

$$+ \left[\dfrac{-5(75)}{(75 + 5\cdot 50)^2} - \dfrac{75}{50^2}\right](2)$$

$$\approx -.1342$$

Expect profits to decrease by $13.42.

85. $V = \dfrac{4}{3}\pi r^3$, r = 2 ft,

$$dr = 1 \text{ in} = \dfrac{1}{12} \text{ ft}$$

$$dV = 4\pi r^2 dr = 4\pi(2)^2\left(\dfrac{1}{12}\right) \approx 4.19 \text{ ft}^3$$

87. **(a)** $m = \dfrac{(\Sigma x)(\Sigma y) - n(\Sigma xy)}{(\Sigma x)^2 - n(\Sigma x^2)}$

$$= \dfrac{(1394)(1607) - 8(291,990)}{(1394)^2 - 8(255,214)}$$

$$= \dfrac{-95,762}{-98,476} \approx .97$$

$$b = \dfrac{\Sigma y - m\Sigma x}{n}$$

$$= \dfrac{1607 - .97(1394)}{8}$$

$$\approx 31.5$$

$$y' = .97x + 31.5$$

(b) When x = 190,

$$y' = .97(190) + 31.5$$

$$\approx 216.$$

(c) If x = 130,

$$y' = .97(130) + 31.5$$

$$= 157.6 \approx 158.$$

If x = 142,

$$y' = .97(142) + 31.5$$

$$= 169.24 \approx 169.$$

If x = 200,

$$y' = .97(200) + 31.5$$

$$= 225.5 \approx 226.$$

Blood Sugar Level	Predicted Cholesterol Level	Actual Cholesterol Level
130	158	170
142	169	173
200	226	192

The predicted values are in the vicinity of the actual values, but not "close."

89. Assume blood vessel is cylindrical.

$V = \pi r^2 h$, $r = .7$, $h = 2.7$,
$dr = dh = \pm .1$

$dV = 2\pi rh\,dr + \pi r^2\,dh$
$= 2\pi(.7)(2.7)(\pm .1) + \pi(.7)^2(\pm .1)$
$\approx \pm 1.3$

The possible error is 1.3 cm³.

91. $P(x, y)$

$= .01(-x^2 + 3xy + 160x - 5y^2$
$+ 200y + 2600)$

with $x + y = 280$.

(a) $y = 280 - x$

$P(x) = .01[-x^2 + 3x(280 - x) + 160x$
$- 5(280 - x)^2 + 200(280 - x)$
$+ 2600]$
$= .01(-x^2 + 840x - 3x^2 + 160x$
$- 392,000 + 2800x - 5x^2$
$+ 56,000 - 200x + 2600)$

$P(x) = .01(-9x^2 + 3600x - 333,400)$

$P'(x) = .01(-18x + 3600)$
$.01(-18x + 3600) = 0$
$-18x = -3600$
$x = 200$

If $x < 200$, $P'(x) > 0$, and if
$x > 200$, $P'(x) < 0$. Therefore,
P is maximum when $x = 200$. If
$x = 200$, $y = 80$.

$P(200, 80)$
$= .01[-200^2 + 3(200)(80) + 160(200)$
$- 5(80)^2 + 200(80) + 2600]$
$= .01(26,600)$
$= 266$

Thus, $200 spent on fertilizer and
$80 spent on seed will produce a
maximum profit of $266 per acre.

(b) $P(x, y)$

$= .01(-x^2 + 3xy + 160x - 5y^2$
$+ 200y + 2600)$

$P_x = .01(-2x + 3y + 160)$
$P_y = .01(3x - 10y + 200)$
$.01(-2x + 3y + 160) = 0$
$.01(3x - 10y + 200) = 0$

These equations simplify to

$-2x + 3y = -160$
$3x - 10y = -200$.

Solve the system.

$-6x + 9y = -480$
$\underline{6x - 20y = -400}$
$-11y = -880$
$y = 80$

If $y = 80$,

$3x - 10(80) = -200$
$3x = 600$
$x = 200$.

$P_{xx} = .01(-2) = -.02$
$P_{yy} = .01(-10) = -.1$
$P_{xy} = 0$

For $(200, 80)$, $D = (-.02)(-.1) - 0^2$
$= .002 > 0$ since $P_{xx} < 0$, there is a
relative maximum at $(200, 80)$.

$P(200, 80) = 266$ as in part (a).
Thus, $200 spent on fertilizer and
$80 spent on seed will produce a
maximum profit of $266 per acre.

(c) Maximize P(x, y)

$$= .01(-x^2 + 3xy + 160x - 5y^2$$
$$+ 200y + 2600)$$

subject to x + y = 280.

1. g(x, y) = x + y - 280

2. F(x, y, λ)

$$= .01(-x^2 + 3xy + 160x - 5y^2$$
$$+ 200y + 2600)$$
$$+ \lambda(x + y - 280)$$

3. $F_x = .01(-2x + 3y + 160) + \lambda$
   $F_y = .01(3x - 10y + 200) + \lambda$
   $F_\lambda = x + y - 280$

4. $.01(-2x + 3y + 160) + \lambda = 0$ *(1)*
   $.01(3x - 10y + 200) + \lambda = 0$ *(2)*
   $x + y - 280 = 0$ *(3)*

5. Equations (1) and (2) give

$.01(-2x + 3y + 160) = .01(3x - 10y + 200)$

$-2x + 3y + 160 = 3x - 10y + 200$

$-5x + 13y = 40$

Multiplying equation (3) by 5 gives

$5x + 5y - 1400 = 0$

$$-5x + 13y = 40$$
$$\underline{5x + 5y = 1400}$$
$$18y = 1440$$
$$y = 80$$

If y = 80,

$$5x + 5(80) = 1400$$
$$5x = 1000$$
$$x = 200.$$

Thus, P(200, 80) is a maximum. As before, P(200, 80) = 266.

Thus, $200 spent on fertilizer and $80 spent on seed will produce a maximum profit of $266 per acre.

Extended Application

1. $\dfrac{\partial F}{\partial x_1} = 2(.4 - .7x_1 - .4x_2)(-.7)$

$\qquad + 2(.5)(.15 - .4x_1 - .1x_2)(-.4)$
$\qquad + 2(.4)(.25 - .3x_1 - .3x_2)(.3) + \lambda$

$\qquad = -1.4(.4 - .7x_1 - .4x_2)$
$\qquad - .4(.15 - .4x_1 - .1x_2)$
$\qquad - .24(.25 - .3x_1 - .3x_2) + \lambda$

$\qquad = -.56 + .98x_1 + .56x_2 - .06 + .16x_1$
$\qquad + .04x_2 - .06 + .072x_1 + .072x_2 + \lambda$

$\qquad = -.68 + 1.212x_1 + .672x_2 + \lambda$

$\dfrac{\partial F}{\partial x_2} = 2(.4 - .7x_1 - .4x_2)(-.4)$

$\qquad + 2(.5)(.15 - .4x_1 - .1x_2)(-.1)$
$\qquad + 2(.4)(.25 - .3x_1 - .3x_2)(-.3)$
$\qquad + \lambda$

$\qquad = -.8(.4 - .7x_1 - .4x_2)$
$\qquad - .1(.15 - .4x_1 - .1x_2)$
$\qquad - .24(.25 - .3x_1 - .3x_2) + \lambda$

$\qquad = -.32 + .56x_1 + .32x_2 - .015$
$\qquad + .04x_1 + .01x_2 - .06 + .072x_1$
$\qquad + .072x_2 + \lambda$

$\qquad = -.395 + .672x_1 + .402x_2 + \lambda$

2. $-.68 + 1.212x_1 + .672x_2 + \lambda = 0$
   $-.395 + .672x_1 + .402x_2 + \lambda = 0$
   $-.06 + .06x_3 + \lambda = 0$
   $x_1 + x_2 + x_3 - 1 = 0$

We will solve this system of linear equations using Cramer's Rule.

$1.212x_1 + .672x_2 + 0x_3 + 1\lambda = .68$

$.672x_1 + .402x_2 + 0x_3 + 1\lambda = .395$

$0x_1 + 0x_2 + .06x_3 + 1\lambda = .06$

$1x_1 + 1x_2 + 1x_3 + 0\lambda = 1$

$$x_1 = \frac{\begin{vmatrix} .68 & .672 & 0 & 1 \\ .395 & .402 & 0 & 1 \\ .06 & 0 & .06 & 1 \\ 1 & 1 & 1 & 0 \end{vmatrix}}{\begin{vmatrix} 1.212 & .672 & 0 & 1 \\ .672 & .402 & 0 & 1 \\ 0 & 0 & .06 & 1 \\ 1 & 1 & 1 & 0 \end{vmatrix}}$$

$$= \frac{-.02502}{-.05184} \approx .48$$

$$x_2 = \frac{\begin{vmatrix} 1.212 & .68 & 0 & 1 \\ .672 & .395 & 0 & 1 \\ 0 & .06 & .06 & 1 \\ 1 & 1 & 1 & 0 \end{vmatrix}}{-.05184}$$

$$= \frac{-.00468}{-.05184} \approx .09$$

$$x_3 = \frac{\begin{vmatrix} 1.212 & .672 & .68 & 1 \\ .672 & .402 & .395 & 1 \\ 0 & 0 & .06 & 1 \\ 1 & 1 & 1 & 0 \end{vmatrix}}{-.05184}$$

$$= \frac{-.02214}{-.05184} \approx .43$$

3. Suppose $w_1 = .4$ and $w_2 = .5$ and all other values remain the same. In this case,

$$F = [.4 - .7x_1 - .4x_2]^2$$
$$+ .4[.15 - .4x_1 - .1x_2]^2$$
$$+ .5[.25 - .3x_1 - .3x_2]^2$$
$$+ .03(1 - x_3)^2 + \lambda(x_1 + x_2 + x_3 - 1).$$

Then,

$$\frac{\partial F}{\partial x_1} = 2(.4 - .7x_1 - .4x_2)(-.7)$$
$$+ 2(.4)(.15 - .4x_1 - .1x_2)(-.4)$$
$$+ 2(.5)(.25 - .3x_1 - .3x_2)(-.3) + \lambda$$
$$= -1.4(.4 - .7x_1 - .4x_2)$$
$$- .32(.15 - .4x_1 - .1x_2)$$
$$- .3(.25 - .3x_1 - .3x_2) + \lambda$$

$$= -.56 + .98x_1 + .56x_2 - .048$$
$$+ .128x_1 + .032x_2 - .075$$
$$+ .09x_1 + .09x_2 + \lambda$$
$$= -.683 + 1.198x_1 + .682x_2 + \lambda.$$

$$\frac{\partial F}{\partial x_2} = 2(.4 - .7x_1 - .4x_2)(-.4)$$
$$+ 2(.4)(.15 - .4x_1 - .1x_2)(-.1)$$
$$+ 2(.5)(.25 - .3x_1 - .3x_2)(-.3) + \lambda$$
$$= -.8(.4 - .7x_1 - .4x_2)$$
$$- .08(.15 - .4x_1 - .1x_2)$$
$$- .3(.25 - .3x_1 - .3x_2) + \lambda$$
$$= -.32 + .56x_1 + .32x_2 - .012$$
$$+ .032x_1 + .008x_2 - .075$$
$$+ .09x_1 + .09x_2 + \lambda$$
$$= -.407 + .682x_1 + .418x_2 + \lambda$$

$$\frac{\partial F}{\partial x_3} = 2(.03)(1 - x_3)(-1) + \lambda$$
$$= -.06 + .06x_3 + \lambda$$

$$\frac{\partial F}{\partial \lambda} = x_1 + x_2 + x_3 - 1$$

We set these equal to zero.

$$-.683 + 1.198x_1 + .682x_2 + \lambda = 0$$
$$-.407 + .682x_1 + .418x_2 + \lambda = 0$$
$$-.06 + .06x_3 + \lambda = 0$$
$$x_1 + x_2 + x_3 - 1 = 0$$

We will solve this system of linear equations using Cramer's Rule.

$$1.198x_1 + .682x_2 + 0x_3 + 1\lambda = .683$$
$$.682x_1 + .418x_2 + 0x_3 + 1\lambda = .407$$
$$0x_1 + 0x_2 + .06x_3 + 1\lambda = .06$$
$$1x_1 + 1x_2 + 1x_3 + 0\lambda = 1$$

$$x_1 = \frac{\begin{vmatrix} .683 & .682 & 0 & 1 \\ .407 & .418 & 0 & 1 \\ .06 & 0 & .06 & 1 \\ 1 & 1 & 1 & 0 \end{vmatrix}}{\begin{vmatrix} 1.198 & .682 & 0 & 1 \\ .682 & .418 & 0 & 1 \\ 0 & 0 & .06 & 1 \\ 1 & 1 & 1 & 0 \end{vmatrix}}$$

$$= \frac{-.02448}{-.05076} \approx .48$$

$$x_2 = \frac{\begin{vmatrix} 1.198 & .683 & 0 & 1 \\ .682 & .407 & 0 & 1 \\ 0 & .06 & .06 & 1 \\ 1 & 1 & 1 & 0 \end{vmatrix}}{-.05076}$$

$$= \frac{-.00522}{-.05076} \approx .10$$

$$x_3 = \frac{\begin{vmatrix} 1.198 & .682 & .683 & 1 \\ .682 & .418 & .407 & 1 \\ 0 & 0 & .06 & 1 \\ 1 & 1 & 1 & 0 \end{vmatrix}}{-.05076}$$

$$= \frac{-.02106}{-.05076} \approx .41$$

The predator should spend .48 of its time feed in location 1, .10 of its time in location 2, and .41 of its time on nonfeeding activities.

CHAPTER 16   TEST

1. Find $f(-2, 1)$ for $f(x, y) = 2x^2 - 4xy + 7y$.

2. Find $g(-1, 3)$ for $g(x, y) = \sqrt{2x^2 - xy^2}$.

3. Complete the ordered triples $(0, 0, \quad)$, $(0, \quad , 0)$ and $(\quad , 0, 0)$ for the plane $2x - 4y + 8z = 8$.

4. Graph the first octant portion of the plane $4x + 2y + 8z = 8$.

5. Let $f(x, y) = x^2 + 2y^2$. Find $\dfrac{f(x + h, y) - f(x, y)}{h}$.

6. Let $z = f(x, y) = 3x^3 - 5x^2y + 4y^2$. Find

   (a) $\dfrac{\partial z}{\partial x}$ **(b)** $\dfrac{\partial z}{\partial y}(1, 1)$ **(c)** f_{xx}.

7. Let $f(x, y) = \dfrac{x}{2x + y^2}$. Find

   (a) $f_x(1, -1)$ **(b)** $f_y(2, 1)$.

8. Let $f(x, y) = \sqrt{2x^2 - y^2}$. Find

   (a) f_x **(b)** f_y **(c)** f_{xy}.

9. Let $f(x, y) = xe^{y^2}$. Find

   (a) f_x **(b)** f_y **(c)** f_{yx}.

10. Let $f(x, y) = \ln(2x^2y^2 + 1)$. Find

    (a) f_{xx} **(b)** f_{xy} **(c)** f_{yy}.

11. The production function for a certain country is

$$z = 2x^5y^4$$

where x represents the amount of labor and y the
amount of capital. Find the marginal productivity
of

(a) labor   (b) capital.

**Find all points where the functions defined below have any relative
extrema.   Find any saddle points.**

12. $z = 4x^2 + 2y^2 - 8x$ 13. $z = 2x^2 - 4xy + y^3$

14. $z = 1 - 2y - x^2 - 2xy - 2y^2$

15. A company manufactures two calculator models. The total revenue
from x thousand solar calculators and y thousand battery-operated
calculators is

$$R(x, y) = 4000 - 5y^2 - 8x^2 - 2xy + 42y + 102x.$$

Find x and y so that revenue is maximized.

16. Use Lagrange multipliers to find extrema of $f(x, y) = x^2 - 6xy$
subject to the constraint $x + y = 7$.

17. Find two numbers whose sum is 30 such that x^2y is maximized.

18. A closed box with square ends must have a volume of 64 cubic
inches. Find the dimensions of such a box that has minimum
surface area.

19. Find an equation for the least squares line for the following data.

x	1	2	3	5	9
y	18	26	36	50	82

Estimate y when x = 6.

20. (a) Find dz for $z = e^{x+y} \ln xy$.

(b) Evaluate dz when x = 1, y = 1, dx = .02 and dy = .01.

21. A sphere of radius 3 ft is to receive an insulating coating 1/2 inch thick. Use differentials to approximate the volume of the coating needed.

22. Evaluate $\displaystyle\int_0^4 \int_1^4 \sqrt{x + 3y} \; dx \; dy$.

23. Find the volume under the surface $z = 2xy$ and above the rectangle $0 \le x \le 1$, $0 \le y \le 4$.

24. Evaluate $\displaystyle\int_0^2 \int_{y/2}^3 (x + y) dx \; dy$.

25. Use the region R with boundaries $0 \le y \le 2$ and $0 \le x \le y$ to evaluate

$$\iint_R (6 - x - y) dx \; dy.$$

CHAPTER 16   TEST ANSWERS

1. 23   **2.** 11   **3.** (0, 0, 1), (0, -2, 0), (4, 0, 0)

4.

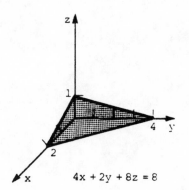

$$4x + 2y + 8z = 8$$

5. $2x + h$ **6.** **(a)** $9x^2 - 10xy$ **(b)** 3

(c) $18x - 10y$

7. **(a)** $\frac{1}{9}$ **(b)** $\frac{-4}{25}$ **8.** **(a)** $\frac{2x}{\sqrt{2x^2 - y^2}}$ **(b)** $\frac{-y}{\sqrt{2x^2 - y^2}}$ **(c)** $\frac{2xy}{(2x^2 - y^2)^{3/2}}$

9. **(a)** e^{y^2} **(b)** $2xye^{y^2}$ **(c)** $2ye^{y^2}$

10. **(a)** $\frac{-8x^2y^4 + 4y^2}{(2x^2y^2 + 1)^2}$ **(b)** $\frac{8xy}{(2x^2y^2 + 1)^2}$ **(c)** $\frac{4x^2 - 8x^4y^2}{(2x^2y^2 + 1)^2}$

11. **(a)** $10x^4y^4$ **(b)** $8x^5y^3$

12. Relative minimum of -4 at (1, 0)

13. Relative minimum of $-\frac{32}{27}$ at $\left(\frac{4}{3}, \frac{4}{3}\right)$, saddle point at (0, 0)

14. Relative maximum of 2 at (1, -1)

15. 6 thousand solar calculators and 3 thousand battery-operated calculators

16. Minimium of -63 at (3, 4)   **17.** $x = 20$, $y = 10$

18. 4 inches by 4 inches by 4 inches   **19.** $y' = 7.96x + 10.56$; 58.32

20. **(a)** $(e^{x+y} \ln (xy) + \frac{e^{x+y}}{x}) dx + (e^{x+y} \ln xy + \frac{e^{x+y}}{y}) dy$ **(b)** $.03e^2 \approx .222$

21. 4.71 cu ft 22. $\frac{4}{45}(993 - 13^{5/2})$ 23. 8 24. $\frac{40}{3}$ 25. 8

CHAPTER 17   PROBABILITY AND CALCULUS

Section 17.1

1. $f(x) = \frac{1}{9}x - \frac{1}{18}$; [2, 5]

Show that condition 1 holds.

$$\int_2^5 \left(\frac{1}{9}x - \frac{1}{18}\right)dx = \frac{1}{9}\int_2^5 \left(x - \frac{1}{2}\right)dx$$

$$= \frac{1}{9}\left(\frac{x^2}{2} - \frac{1}{2}x\right)\Big|_2^5$$

$$= \frac{1}{9}\left(\frac{25}{2} - \frac{5}{2} - \frac{4}{2} + 1\right)$$

$$= \frac{1}{9}(8 + 1)$$

$$= 1$$

Show that condition 2 holds.

Since $2 \le x \le 5$,

$$\frac{2}{9} \le \frac{1}{9}x \le \frac{5}{9}$$

$$\frac{1}{6} \le \frac{1}{9}x - \frac{1}{18} \le \frac{1}{2}.$$

Hence, $f(x) \ge 0$ on [2, 5].
Yes, $f(x)$ is a probability density function.

3. $f(x) = \frac{1}{21}x^2$; [1, 4]

$$\frac{1}{21}\int_1^4 x^2\,dx = \frac{1}{21}\left(\frac{x^3}{3}\right)\Big|_1^4$$

$$= \frac{1}{21}\left(\frac{64}{3} - \frac{1}{3}\right)$$

$$= 1$$

Since $x^2 \ge 0$, $f(x) \ge 0$ on [1, 4].
Yes, $f(x)$ is a probability density function.

5. $f(x) = 4x^3$; [0, 3].

$$4\int_0^3 x^3\,dx = 4\left(\frac{x^4}{4}\right)\Big|_0^3$$

$$= 4\left(\frac{81}{4} - 0\right)$$

$$= 81 \ne 1$$

No, $f(x)$ is not a probability density function.

7. $f(x) = \frac{x^2}{16}$; [-2, 2].

$$\frac{1}{16}\int_{-2}^2 x^2\,dx = \frac{1}{16}\left(\frac{x^3}{3}\right)\Big|_{-2}^2$$

$$= \frac{1}{16}\left(\frac{8}{3} + \frac{8}{3}\right)$$

$$= \frac{1}{3} \ne 1$$

No, $f(x)$ is not a probability density function.

9. $f(x) = kx^{1/2}$; [1, 4]

$$\int_1^4 kx^{1/2}\,dx = \frac{2}{3}kx^{3/2}\Big|_1^4$$

$$= \frac{2}{3}k(8 - 1)$$

$$= \frac{14}{3}k$$

If $\frac{14}{3}k = 1$,

$$k = \frac{3}{14}.$$

Notice that $f(x) = \frac{3}{4}x^{1/2} \ge 0$ for all x in [1, 4].

11. $f(x) = kx^2$; $[0, 5]$

$$\int_0^5 kx^2 \, dx = k\frac{x^3}{3}\Big|_0^5$$

$$= k\left(\frac{125}{3} - 0\right)$$

$$= k\left(\frac{125}{3}\right)$$

If $k\left(\frac{125}{3}\right) = 1$,

$$k = \frac{3}{125}.$$

Notice that $f(x) = \frac{3}{125}x^2 \geq 0$ for all x in $[0, 5]$.

13. $f(x) = kx$; $[0, 3]$

$$\int_0^3 kx \, dx = k\frac{x^2}{2}\Big|_0^3$$

$$= k\left(\frac{9}{2} - 0\right)$$

$$= \frac{9}{2}k$$

If $\frac{9}{2}k = 1$,

$$k = \frac{2}{9}.$$

Notice that $f(x) = \frac{2}{9}x \geq 0$ for all x in $[0, 3]$.

15. $f(x) = kx$; $[1, 5]$

$$\int_1^5 kx \, dx = k\frac{x^2}{2}\Big|_1^5$$

$$= k\left(\frac{25}{2} - \frac{1}{2}\right)$$

$$= 12k$$

If $12k = 1$,

$$k = \frac{1}{12}.$$

17. The total area under the graph of a probability density function always equals 1.

21. $f(x) = \frac{1}{2}(1 + x)^{-3/2}$; $[0, \infty)$

$$\frac{1}{2}\int_0^\infty (1 + x)^{-3/2} \, dx$$

$$= \lim_{a \to \infty} \frac{1}{2}\int_0^a (1 + x)^{-3/2} \, dx$$

$$= \lim_{a \to \infty} \frac{1}{2}(1 + x)^{-1/2}\left(\frac{-2}{1}\right)\Big|_0^a$$

$$= \lim_{a \to \infty} [-(1 + a)^{-1/2} + 1]$$

$$= \lim_{a \to \infty} \left(\frac{-1}{\sqrt{1 + a}} + 1\right)$$

$$= 0 + 1 = 1$$

Since $x \geq 0$, $f(x) \geq 0$.
$f(x)$ is a probability density function.

(a) $P(0 \leq x \leq 2)$

$$= \frac{1}{2}\int_0^2 (1 + x)^{-3/2} \, dx$$

$$= -(1 + x)^{-1/2}\Big|_0^2$$

$$= -3^{-1/2} + 1$$

$$\approx .4226$$

(b) $P(1 \leq x \leq 3)$

$$= \frac{1}{2}\int_1^3 (1 + x)^{-3/2} \, dx$$

$$= -(1 + x)^{-1/2}\Big|_1^3$$

$$= -4^{-1/2} + 2^{-1/2}$$

$$\approx .2071$$

(c) $P(x \leq 5)$

$$= \frac{1}{2} \int_5^\infty (1 + x)^{-3/2} \, dx$$

$$= \lim_{a \to \infty} \frac{1}{2} \int_5^a (1 + x)^{-3/2} \, dx$$

$$= \lim_{a \to \infty} [-(1 + x)^{-1/2}] \Big|_5^a$$

$$= \lim_{a \to \infty} [-(1 + a)^{-1/2} + 6^{-1/2}]$$

$$= \lim_{a \to \infty} \left(\frac{-1}{\sqrt{1 + a}} + 6^{-1/2} \right)$$

$$\approx 0 + .4082$$

$$\approx .4082$$

23. $f(x) = \frac{1}{2} e^{-x/2}$; $[0, \infty)$,

$$\frac{1}{2} \int_0^\infty e^{-x/2} \, dx$$

$$= \lim_{a \to \infty} \frac{1}{2} \int_0^a e^{-x/2} \, dx$$

$$= \lim_{a \to \infty} \frac{1}{2} \left(\frac{-2}{1} e^{-x/2} \right) \Big|_0^a$$

$$= \lim_{a \to \infty} -e^{-x/2} \Big|_0^a$$

$$= \lim_{a \to \infty} \left(\frac{-1}{e^{a/2}} + 1 \right)$$

$$= 0 + 1$$

$$= 1$$

$f(x) > 0$ for all x.
$f(x)$ is a probability density
function.

(a) $P(0 \leq x \leq 1) = \frac{1}{2} \int_0^1 e^{-x/2} \, dx$

$$= -e^{-x/2} \Big|_0^1$$

$$= \frac{-1}{e^{1/2}} + 1$$

$$\approx .3935$$

(b) $P(1 \leq x \leq 3 = \frac{1}{2} \int_1^3 e^{-x/2} \, dx$

$$= -e^{-x/2} \Big|_1^3$$

$$= \frac{-1}{e^{3/2}} + \frac{1}{e^{1/2}}$$

$$\approx .3834$$

(c) $P(x \geq 2) = \frac{1}{2} \int_2^\infty e^{-x/2} \, dx$

$$= \lim_{a \to \infty} \frac{1}{2} \int_2^a e^{-x/2} \, dx$$

$$= \lim_{a \to \infty} (-e^{-x/2}) \Big|_2^a$$

$$= \lim_{a \to \infty} \left(\frac{-1}{e^{a/2}} + \frac{1}{e} \right)$$

$$\approx .3679$$

25. $f(x) = \frac{1}{2} e^{-x/2}$; $[0, \infty)$

(a) $P(0 \leq x \leq 12) = \frac{1}{2} \int_0^{12} e^{-x/2} \, dx$

$$= -e^{-x/2} \Big|_0^{12}$$

$$= \frac{-1}{e^6} + 1$$

$$\approx .9975$$

(b) $P(12 \le x \le 20) = \dfrac{1}{2} \displaystyle\int_{12}^{20} e^{-x/2}\, dx$

$$= -e^{-x/2} \Big|_{12}^{20}$$

$$= \dfrac{-1}{e^{10}} + \dfrac{1}{e^{6}}$$

$$\approx .0024$$

27. $f(x) = \dfrac{1}{2\sqrt{x}};\ [1,\ 4]$

(a) $P(3 \le x \le 4) = \displaystyle\int_{3}^{4} \left(\dfrac{1}{2\sqrt{x}}\right) dx$

$$= \dfrac{1}{2} \int_{3}^{4} x^{-1/2}\, dx$$

$$= \dfrac{1}{2}(2)x^{1/2} \Big|_{3}^{4}$$

$$= (2 - 3^{1/2})$$

$$= .2679$$

(b) $P(1 \le x \le 2) = \displaystyle\int_{1}^{2} \left(\dfrac{1}{2\sqrt{x}}\right) dx$

$$= \dfrac{1}{2}(2)x^{1/2} \Big|_{1}^{2}$$

$$= 2^{1/2} - 1$$

$$= .4142$$

(c) $P(2 \le x \le 3) = \displaystyle\int_{2}^{3} \left(\dfrac{1}{2\sqrt{x}}\right) dx$

$$= \dfrac{1}{2}(2)x^{1/2} \Big|_{2}^{3}$$

$$= 3^{1/2} - 2^{1/2}$$

$$= .3178$$

29. $f(x) = \dfrac{8}{7(x-2)^2}$ for $[3,\ 10]$

(a) $P(3 \le x \le 4) = \dfrac{8}{7} \displaystyle\int_{3}^{4} (x-2)^{-2}\, dx$

$$= -\dfrac{8}{7}(x-2)^{-1} \Big|_{3}^{4}$$

$$= -\dfrac{8}{7}\left(\dfrac{1}{2} - 1\right)$$

$$= -\dfrac{8}{7}\left(-\dfrac{1}{2}\right)$$

$$= \dfrac{4}{7}$$

$$\approx .5714$$

(b) $P(5 \le x \le 10) = \dfrac{8}{7} \displaystyle\int_{5}^{10} (x-2)^{-2}\, dx$

$$= -\dfrac{8}{7}(x-2)^{-1} \Big|_{5}^{10}$$

$$= \dfrac{8}{7}\left(\dfrac{1}{8} - \dfrac{1}{3}\right)$$

$$= -\dfrac{8}{7}\left(-\dfrac{5}{24}\right)$$

$$= \dfrac{5}{21}$$

$$\approx .2381$$

31. $f(x) = \dfrac{105}{4x^2};\ [15,\ 35]$

$$\int_{a}^{b} \dfrac{105}{4x^2}\, dx = \dfrac{105}{4} \int_{a}^{b} x^{-2}\, dx$$

$$= \dfrac{105}{4}\left(-\dfrac{1}{x}\right) \Big|_{a}^{b}$$

$$= \dfrac{105}{4}\left(-\dfrac{1}{b} + \dfrac{1}{a}\right)$$

$$= \dfrac{105(b-a)}{4ab}$$

(a) $P(x \leq 25) = \int_{15}^{25} \frac{105}{4x^2} \, dx$

$$= \frac{105(25 - 15)}{4(15)(25)}$$

$$= .7$$

(b) $P(x \leq 21) = \int_{15}^{21} \frac{105}{4x^2} \, dx$

$$= \frac{105(21 - 15)}{4(15)(21)}$$

$$= .5$$

(c) $P(21 \leq x \leq 25) =$

$$= P(x \leq 25) - P(x \leq 21)$$

$$= .7 - .5$$

$$= .2$$

Section 17.2

1. $f(x) = \frac{1}{4}$; $[3, 7]$

$$E(x) = \mu = \int_{3}^{7} \frac{1}{4}x \, dx = \frac{1}{4}\left(\frac{x^2}{2}\right)\Big|_{3}^{7}$$

$$= \frac{49}{8} - \frac{9}{8}$$

$$= 5$$

$$Var(x) = \int_{3}^{7} (x - 5)^2 \left(\frac{1}{4}\right) dx$$

$$= \frac{1}{4} \cdot \frac{(x - 5)^3}{3}\Big|_{3}^{7}$$

$$= \frac{8}{12} + \frac{8}{12}$$

$$= \frac{4}{3} \approx 1.33$$

$$\sigma \approx \sqrt{Var(x)} \ \sqrt{4/3} \approx 1.15$$

3. $f(x) = \frac{x}{8} - \frac{1}{4}$; $[2, 6]$

$$\mu = \int_{2}^{6} x\left(\frac{x}{8} - \frac{1}{4}\right) dx$$

$$= \int_{2}^{6} \left(\frac{x^2}{8} - \frac{x}{4}\right) dx$$

$$= \left(\frac{x^3}{24} - \frac{x^2}{8}\right)\Big|_{2}^{6}$$

$$= \left(\frac{216}{24} - \frac{36}{8}\right) - \left(\frac{8}{24} - \frac{4}{8}\right)$$

$$= \frac{208}{24} - 4$$

$$= \frac{26}{3} - 4$$

$$= \frac{14}{3} \approx 4.67$$

Use the alternative formula to find

$$Var(x) = \int_{2}^{6} x^2\left(\frac{x}{8} - \frac{1}{4}\right) dx - \left(\frac{14}{3}\right)^2$$

$$= \int_{2}^{6} \left(\frac{x^3}{8} - \frac{x^2}{4}\right) dx - \frac{196}{9}$$

$$= \left(\frac{x^4}{32} - \frac{x^3}{12}\right)\Big|_{2}^{6} - \frac{196}{9}$$

$$= \left(\frac{1296}{32} - \frac{216}{12}\right) - \left(\frac{16}{32} - \frac{8}{12}\right)$$

$$- \frac{196}{9}$$

$$\approx .89.$$

$$\sigma = \sqrt{Var(x)} \approx \sqrt{.89} \approx .94$$

5. $f(x) = 1 - \frac{1}{\sqrt{x}}$; $[1, 4]$

$$\mu = \int_{1}^{4} x(1 - x^{-1/2}) \, dx$$

$$= \int_{1}^{4} (x - x^{1/2}) \, dx$$

$$= \left(\frac{x^2}{2} - \frac{2x^{3/2}}{3}\right)\Big|_1^4$$

$$= \frac{16}{2} - \frac{16}{3} - \frac{1}{2} + \frac{2}{3}$$

$$= \frac{17}{6} \approx 2.83$$

$$\text{Var}(x) = \int_1^4 x^2(1 - x^{-1/2})dx$$

$$- \left(\frac{17}{6}\right)^2$$

$$= \int_1^4 (x^2 - x^{3/2})dx - \frac{289}{36}$$

$$= \left(\frac{x^3}{3} - \frac{2x^{5/2}}{5}\right)\Big|_1^4 - \frac{289}{36}$$

$$= \frac{64}{3} - \frac{64}{5} - \frac{1}{3} + \frac{2}{5} - \frac{289}{36}$$

$$\approx .57$$

$$\sigma \approx \sqrt{\text{Var}(x)} \approx .76$$

7. $f(x) = 4x^{-5}$; for $[1, \infty)$

$$\mu = \int_1^\infty x(4x^{-5})dx$$

$$= \lim_{a \to \infty} \int_1^a 4x^{-4}\, dx$$

$$= \lim_{a \to \infty} \left(\frac{4x^{-3}}{-3}\right)\Big|_1^a$$

$$= \lim_{a \to \infty} \left(\frac{-4}{3a^3} + \frac{4}{3}\right)$$

$$= \frac{4}{3} \approx 1.33$$

$$\text{Var}(x) = \int_1^\infty x^2(4x^{-5})dx - \left(\frac{4}{3}\right)^2$$

$$= \lim_{a \to \infty} \int_1^a 4x^{-3}\, dx - \frac{16}{9}$$

$$= \lim_{a \to \infty} \left(\frac{4x^{-2}}{-2}\right)\Big|_1^a - \frac{16}{9}$$

$$= \lim_{a \to \infty} \left(\frac{-2}{a^2} + 2\right) - \frac{16}{9}$$

$$= 2 - \frac{16}{9} = \frac{2}{9} \approx .22$$

$$\sigma = \sqrt{\text{Var}(x)} = \sqrt{\frac{2}{9}} \approx .47$$

11. $f(x) = \frac{\sqrt{x}}{18}$; $[0, 9]$

(a) $E(x) = \mu = \int_0^9 \frac{x\sqrt{x}}{18}\, dx$

$$= \int_0^9 \frac{x^{3/2}}{18}\, dx$$

$$= \frac{2x^{5/2}}{90}\Big|_0^9 = \frac{x^{5/2}}{45}\Big|_0^9$$

$$= \frac{243}{45} = \frac{27}{5} = 5.40$$

(b) $\text{Var}(x) = \int_0^9 \frac{x^2\sqrt{x}}{18}\, dx - \left(\frac{27}{5}\right)^2$

$$= \int_0^9 \frac{x^{5/2}}{18}\, dx - \left(\frac{27}{5}\right)^2$$

$$= \frac{x^{7/2}}{63}\Big|_0^9 - \left(\frac{27}{5}\right)^2$$

$$= \frac{2187}{63} - \left(\frac{27}{5}\right)^2 \approx 5.55$$

(c) $\sigma = \sqrt{\text{Var}(x)} \approx 2.36$

(d) $P(5.40 < x \leq 9)$

$$= \int_{5.4}^{9} \frac{x^{1/2}}{18} \, dx$$

$$= \frac{x^{3/2}}{27} \Big|_{5.4}^{9}$$

$$= \frac{27}{27} - \frac{(5.4)^{1.5}}{27} \approx .54$$

(e) $P(5.40 - 2.36 \leq x \leq 5.40 + 2.36)$

$$= \int_{3.04}^{7.76} \frac{x^{1/2}}{18} \, dx$$

$$= \frac{x^{3/2}}{27} \Big|_{3.04}^{7.76}$$

$$= \frac{7.76^{3/2}}{27} - \frac{3.04^{3/2}}{27}$$

$$\approx .60$$

13. $f(x) = \frac{1}{2}x$; $[0, 2]$

(a) $E(x) = \mu = \int_{0}^{2} \frac{1}{2}x^2 \, dx = \frac{x^3}{6} \Big|_{0}^{2}$

$$= \frac{8}{6} = \frac{4}{3} \approx 1.33$$

(b) $\text{Var}(x) = \int_{0}^{1} \frac{1}{2}x^3 \, dx - \frac{16}{9}$

$$= \frac{x^4}{8} \Big|_{0}^{2} - \frac{16}{9}$$

$$= 2 - \frac{16}{9} = \frac{2}{9} \approx .22$$

(c) $\sigma = \sqrt{\text{Var}(x)} = \sqrt{\frac{2}{9}} \approx .47$

(d) $P(\frac{4}{3} < x \leq 2) = \int_{4/3}^{2} \frac{x}{2} \, dx$

$$= \frac{x^2}{4} \Big|_{4/3}^{2}$$

$$= 1 - \frac{16}{36} \approx .56$$

(e) $P(\frac{4}{3} - .47 \leq x \leq \frac{4}{3} + .47)$

$$= \int_{.86}^{1.8} \frac{x}{2} \, dx = \frac{x^2}{4} \Big|_{.86}^{1.8}$$

$$= \frac{1.8^2}{4} - \frac{.86^2}{4} \approx .63$$

15. $f(x) = \frac{1}{4}$; $[3, 7]$

(a) m = median: $\int_{3}^{m} \frac{1}{4} \, dx = \frac{1}{2}$

$$\frac{1}{4}x \Big|_{3}^{m} = \frac{1}{2}$$

$$\frac{m}{4} - \frac{3}{4} = \frac{1}{2}$$

$$m - 3 = 2$$

$$m = 5$$

(b) $E(x) = \mu = 5$ (from Exercise 1)

$$P(x = 5) = \int_{5}^{5} \frac{1}{4} \, dx = 0$$

17. $f(x) = \dfrac{x}{8} - \dfrac{1}{4}$; $[2, 6]$

(a) m = median:

$$\int_2^m \left(\dfrac{x}{8} - \dfrac{1}{4}\right) dx = \dfrac{1}{2}$$

$$\left(\dfrac{x^2}{16} - \dfrac{x}{4}\right)\Bigg|_2^m = \dfrac{1}{2}$$

$$\dfrac{m^2}{16} - \dfrac{m}{4} - \dfrac{1}{4} + \dfrac{1}{2} = \dfrac{1}{2}$$

$$m^2 - 4m - 4 + 8 = 8$$

$$m^2 - 4m - 4 = 0$$

$$m = \dfrac{4 \pm \sqrt{16 + 16(1)}}{2}$$

Reject $\dfrac{4 - \sqrt{32}}{2}$ since it is not in

$[2, 6]$.

$$m = \dfrac{4 + \sqrt{32}}{2} \approx 4.83$$

(b) $E(x) = \mu = \dfrac{14}{3}$ (from Exercise 3)

$$P\left(\dfrac{14}{3} \leq x \leq 4.83\right) = \int_{4.67}^{4.83} \left(\dfrac{x}{8} - \dfrac{1}{4}\right) dx$$

$$= \left(\dfrac{x^2}{16} - \dfrac{x}{4}\right)\Bigg|_{4.67}^{4.83}$$

$$= \dfrac{4.83^2}{16} - \dfrac{4.83}{4}$$

$$- \dfrac{4.67^2}{16} + \dfrac{4.67}{4}$$

$$\approx .055$$

19. $f(x) = 4x^{-5}$; $[1, \infty)$

(a) m = median:

$$\int_1^m 4x^{-5}\ dx = \dfrac{1}{2}$$

$$\dfrac{4x^{-4}}{-4}\Bigg|_1^m = \dfrac{1}{2}$$

$$-m^{-4} + 1 = \dfrac{1}{2}$$

$$1 - \dfrac{1}{m^4} = \dfrac{1}{2}$$

$$2m^4 - 2 = m^4$$

$$m^4 = 2$$

$$m = \sqrt[4]{2} \approx 1.19$$

(b) $E(x) = \mu = \dfrac{4}{3}$ (From Exercise 7)

$$P\left(1.19 \leq x \leq \dfrac{4}{3}\right) \approx \int_{1.19}^{1.33} 4x^{-5} dx$$

$$\approx -x^{-4}\Bigg|_{1.19}^{1.33}$$

$$\approx -\dfrac{1}{1.33^4} + \dfrac{1}{1.19^4}$$

$$\approx .18$$

21. $f(x) = \dfrac{1}{58\sqrt{x}}$; $[1, 900]$

(a) $E(x) = \int_1^{900} x \cdot \dfrac{1}{58\sqrt{x}}\ dx$

$$= \dfrac{1}{58} \int_1^{900} x^{1/2}\ dx$$

$$= \dfrac{1}{87} x^{3/2}\Bigg|_1^{900}$$

$$= \dfrac{1}{87}(900x^{3/2} - 1^{3/2})$$

$$\approx 310.3 \text{ hr}$$

(b) Var(x)

$$= \int_{1}^{900} x^2 \frac{1}{58\sqrt{x}}\, dx - (310.3)^2$$

$$= \frac{1}{58} \int_{1}^{900} x^{3/2}\, dx - (310.3)^2$$

$$= \frac{1}{145} x^{5/2} \Big|_{1}^{900} - (310.3)^2$$

$$= \frac{1}{145}(900^{5/2} - 1^{5/2}) - (310.3)^2$$

$$= 71,300.11$$

$$\sigma = \sqrt{71,300.11} \approx 267 \text{ hr}$$

(c) P(310.3 + 267 < x ≤ 900)

$$= \int_{577.3}^{9.00} \frac{1}{58\sqrt{x}}\, dx$$

$$= \frac{2}{58} x^{1/2} \Big|_{577.3}^{900}$$

$$= \frac{2}{58}(900^{1/2} - 577.3^{1/2})$$

$$\approx .206$$

23. $f(x) = \frac{1}{2} e^{-x/2}$; [0, ∞]

(a) E(x) = μ

$$= \int_{0}^{\infty} x \cdot \frac{1}{2} e^{-x/2}\, dx$$

$$= \frac{1}{2} \lim_{a \to \infty} \int_{0}^{a} x e^{-x/2}\, dx$$

Use integration by parts.

$$= \frac{1}{2} \lim_{a \to \infty} (-2xe^{-x/2}) \Big|_{0}^{a}$$

$$\quad - \frac{1}{2} \lim_{a \to \infty} \int_{0}^{a} -2e^{-x/2}\, dx$$

$$= \frac{1}{2} \lim_{a \to \infty} (-2xe^{-x/2}) \Big|_{0}^{a}$$

$$\quad + \lim_{a \to \infty} \int_{0}^{a} e^{-x/2}\, dx$$

$$= -\lim_{a \to \infty} (xe^{-x/2}) \Big|_{0}^{a}$$

$$\quad - 2 \lim_{a \to \infty} (e^{-x/2}) \Big|_{0}^{a}$$

$$= -\lim_{a \to \infty} \left(\frac{a}{e^{a/2}} - \frac{0}{e^0} \right)$$

$$\quad - 2 \lim_{a \to \infty} \left(\frac{1}{e^{a/2}} - \frac{1}{e^0} \right)$$

$$= (0 - 0) - 2(0 - 1)$$

$$= -(0) + 2$$

$$E(x) = 2 \text{ mo}$$

(b) $Var(x) = \frac{1}{2} \int_{0}^{\infty} x^2 e^{-x/2}\, dx - 2^2$

Use integration by parts twice.

$$\frac{1}{2} \int x^2 e^{-x/2}\, dx$$

$$= \frac{1}{2}\left(-2x^2 e^{-x/2} + 4 \int x e^{-x/2}\, dx\right)$$

$$= \frac{1}{2}\Big[-2x^2 e^{-x/2} + 4(-2xe^{-x/2}$$

$$\quad + 2\int e^{-x/2}\, dx)\Big]$$

$$= \frac{1}{2}(-2x^2 e^{-x/2} - 8xe^{-x/2} - 16e^{-x/2})$$

$$= -x^2 e^{-x/2} - 4xe^{-x/2} - 8e^{-x/2}$$

Var(x)

$$= \lim_{a \to \infty} (-x^2 e^{-x/2} - 4xe^{-x/2} - 8e^{-x/2}) \Big|_0^a - 4$$

$$= \lim_{a \to \infty} (-a^2 e^{-a/2} - 4ae^{-a/2} - 8e^{-a/2} + 0$$

$$+ 0 + 8e^0) - 4$$

$$= -0 - 0 - 0 + 0 + 0 + 8 - 4$$

Var(x) = 4 mo

$$\sigma = \sqrt{4} = 2 \text{ mo}$$

(c) $P(0 < x < 2) = \dfrac{1}{2} \displaystyle\int_0^2 e^{-x/2}\,dx$

$$= -e^{-x/2} \Big|_0^2$$

$$= -e^{-1} + 1$$

$$\approx .632$$

25. $f(x) = \dfrac{3}{32}(4x - x^2); \quad [0, 4]$

(a) $E(x) = \displaystyle\int_0^4 \dfrac{3x}{32}(4x - x^2)\,dx$

$$= \int_0^4 \left(\dfrac{3}{8}x^2 - \dfrac{3x^3}{32}\right)dx$$

$$= \left(\dfrac{x^3}{8} - \dfrac{3x^4}{128}\right)\Big|_0^4$$

$$= 8 - \dfrac{768}{128}$$

$$= 2$$

(b) $Var(x) = \displaystyle\int_0^4 \dfrac{3x^2}{32}(4x - x^2)\,dx - 4$

$$= \int_0^4 \left(\dfrac{3x^3}{8} - \dfrac{3x^4}{32}\right)dx - 4$$

$$= \left(\dfrac{3x^4}{32} - \dfrac{3x^5}{160}\right)\Big|_0^4$$

$$= \dfrac{3(4^4)}{32} - \dfrac{3(4^5)}{160} - 4$$

$$= 24 - 19.2 - 4$$

$$= .8$$

$$\sigma = \sqrt{Var(x)} \approx .89$$

(c) $P(2 - .89 \le x \le 2 + .89)$

$$\approx \int_{1.11}^{2.89} \dfrac{3}{32}(4x - x^2)\,dx$$

$$= \left(\dfrac{3}{16}x^2 - \dfrac{x^3}{32}\right)\Big|_{1.11}^{2.89}$$

$$= \dfrac{3(2.89)^2}{16} - \dfrac{2.89^3}{32} - \dfrac{3(1.11)^2}{16}$$

$$+ \dfrac{1.11^3}{32}$$

$$\approx .62$$

27. $f(x) = \dfrac{5.5 - x}{15}$ for $[0, 5]$

(a) $\mu = \displaystyle\int_0^5 x\left(\dfrac{5.5 - x}{15}\right)dx$

$$= \int_0^5 \left(\dfrac{5.5}{15}x - \dfrac{1}{15}x^2\right)dx$$

$$= \left(\dfrac{5.5}{30}x^2 - \dfrac{1}{45}x^3\right)\Big|_0^5$$

$$= \left(\dfrac{5.5}{30} \cdot 25 - \dfrac{1}{45} \cdot 125\right) - 0$$

$$\approx 1.806$$

(b) Var(x)

$$= \int_0^5 x^2 \left(\frac{5.5 - x}{15}\right) dx - \mu^2$$

$$= \int_0^5 \left(\frac{5.5}{15}x^2 - \frac{1}{15}x^3\right) dx - \mu^2$$

$$= \left(\frac{5.5}{45}x^3 - \frac{1}{60}x^4\right)\Big|_0^5 - \mu^2$$

$$= \frac{5.5}{45} \cdot 125 - \frac{1}{60} \cdot 625 - 0 - \mu^2$$

$$\approx 1.60108$$

$$\sigma = \sqrt{\text{Var}(x)} \approx 1.265$$

(c) $P(x \le \mu - \sigma)$

$$= P(x \le 1.806 - 1.265)$$

$$= P(x \le .541)$$

$$= \int_0^{.541} \frac{5.5 - x}{15} \, dx$$

$$= \left(\frac{5.5}{15}x - \frac{1}{30}x^2\right)\Big|_0^{.541}$$

$$= \left(\frac{5.5}{15}(.541) - \frac{1}{30}(.541)^2 - 0\right)$$

$$\approx .1886$$

Section 17.3

1. $f(x) = \frac{5}{4}$ for [4, 4.8]

    This is a uniform distribution:
    a = 4, b = 4.8.

    (a) $\mu = \frac{1}{2}(4.8 + 4) = \frac{1}{2}(8.8)$

    $$= 4.4 \text{ cm}$$

(b) $\sigma = \frac{1}{\sqrt{12}}(4.8 - 4)$

$$= \frac{1}{\sqrt{12}}(.8)$$

$$\approx .23 \text{ cm}$$

(c) $P(4.4 < x < 4.4 + .23)$

$$= \int_{4.4}^{4.63} \frac{5}{4} \, dx$$

$$= \frac{5}{4}x \Big|_{4.4}^{4.63} = \frac{5}{4}(4.63 - 4.4)$$

$$\approx .29$$

3. $f(t) = .03e^{-.03t}$ for [0, ∞)

    This is an exponential distribution:
    a = .03.

    (a) $\mu = \frac{1}{.03} \approx 33.33 \text{ yr}$

    (b) $\sigma = \frac{1}{.03} \approx 33.33 \text{ yr}$

    (c) $P(33.33 < x < 33.33 + 33.33)$

    $$= \int_{33.33}^{66.66} .03e^{-.03t} \, dt$$

    $$= -e^{-.03t}\Big|_{33.33}^{66.66}$$

    $$= \frac{-1}{e^{.03(66.66)}} + \frac{-1}{e^{.03(33.33)}}$$

    $$\approx .23$$

5. $f(x) = e^{-t}$ for [0, ∞)

    This is an exponential distribution:
    a = 1.

    (a) $\mu = \frac{1}{1} = 1 \text{ day}$

    (b) $\sigma = \frac{1}{1} = 1 \text{ day}$

(c) $P(1 < x < 1 + 1)$

$$= \int_1^2 e^{-t} dt$$

$$= -e^{-t} \Big|_1^2$$

$$= -\frac{1}{e^2} + \frac{1}{e}$$

$$\approx .23$$

In Exercises 7–13, use the table in the appendix for areas under the normal curve.

7. $z = 3.50$

Area to the left of $z = 3.50$ is .9998. Given mean $\mu = z = 0$, so area to left of μ is .5. Area between μ and z is

$$.9998 - .5 = .4998.$$

Therefore, this area represents 49.98% of total area under normal curve.

9. Between $z = 1.28$ and $z = 2.05$
Area to left of $z = 2.05$ is .9798 and area to left of $z = 1.28$ is .8997.

$$.9798 - .8997 = .0801$$

Percent of total area = 8.01%.

11. Since 10% = .10, the z-score that corresponds to the area of .10 to the left of z is -1.28.

13. 18% of the total area to the right of z means $1 - .18$ of the total area is to the left of z.

$$1 - .18 = .82$$

The closest z-score that corresponds to the area of .82 is .92.

19. Let m be the median of the exponential distribution $f(x) = ae^{-ax}$ for $[0, \infty)$.

$$\int_0^m ae^{-ax} dx = .5$$

$$-e^{-ax} \Big|_0^m = .5$$

$$-e^{-am} + 1 = .5$$

$$.5 = e^{-am}$$

$$-am = \ln .5$$

$$m = -\frac{\ln .5}{a}$$

or

$$-am = \ln \frac{1}{2}$$

$$-am = -\ln 2$$

$$m = \frac{\ln 2}{a}$$

21. For a uniform distribution,

$$f(x) = \frac{1}{b - a} \text{ for } [a, b].$$

Thus, we have

$$f(x) = \frac{1}{85 - 10} = \frac{1}{75} \text{ for } [10, 85].$$

(a) $\mu = \frac{1}{2}(10 + 85) = \frac{1}{2}(95)$

$$= 47.5 \text{ thousands}$$

Therefore, the agent sells $47,500 in insurance.

(b) $P(50 < x < 85) = \int_{50}^{85} \frac{1}{75} \, dx$

$$= \frac{x}{75} \Big|_{50}^{85}$$

$$= \frac{87}{75} - \frac{50}{75}$$

$$= \frac{35}{75}$$

$$= .47$$

23. (a) Since we have an exponential distribution with $\mu = 4.25$,

$$\mu = \frac{1}{a} = 4.25.$$

$$a = .235.$$

Therefore,

$f(x) = .235e^{-.235x}$ on $[0, \infty)$.

(b) $P(x > 10)$

$$= \int_{10}^{\infty} .235e^{-.235x}$$

$$= \lim_{a \to \infty} \int_{10}^{a} .235e^{-.235x}$$

$$= \lim_{a \to \infty} (-e^{-.235x}) \Big]_{10}^{a}$$

$$= \lim_{a \to \infty} \left(-\frac{1}{e^{.235a}} + \frac{1}{e^{2.35}}\right)$$

$$= \frac{1}{e^{2.35}} = .095$$

25. (a) $\mu = 2.5$, $\sigma = .2$, $x = 2.7$

$$z = \frac{2.7 - 2.5}{.2} = 1$$

Area to the right of $z = 1$ is

$$1 - .8413 = .1587.$$

Probability $= .1587$

(b) Within 1.2 standard deviations of the mean is the area between $z = -1.2$ and $z = 1.2$.

Area to left of $1.2 = .8849$

Area to the left of $-1.2 = .1151$

$$.8849 - .1151 = .7698$$

Probability $= .7698$

27. For a uniform distribution,

$$f(x) = \frac{1}{b - a} \text{ for } [a, b].$$

$$f(x) = \frac{1}{36 - 20} = \frac{1}{16} \text{ for } [20, 36]$$

(a) $\mu = \frac{1}{2}(20 + 36) = \frac{1}{2}(56)$

$$= 28 \text{ days}$$

(b) $P(30 < x \le 36)$

$$= \int_{30}^{36} \frac{1}{16} \, dx = \frac{1}{16}x \Big|_{30}^{36}$$

$$= \frac{1}{16}(36 - 30)$$

$$= .375$$

29. We have an exponential distribution, with $a = 1$.

$$f(t) = e^{-t}, \ [0, \infty)$$

(a) $\mu = \frac{1}{1} = 1 \text{ hr}$

(b) P(t < 30 min)

$$= P(t < .5 \text{ hr})$$

$$= \int_0^{.5} e^{-t} \, dt$$

$$= -e^{-t} \Big|_0^{.5}$$

$$= 1 - e^{-.5} \approx .39$$

31. $f(x) = ae^{-ax}$ for $[0, \infty)$

Since $\mu = 25$ and $\mu = \dfrac{1}{a}$,

$$a = \frac{1}{25} = .04.$$

Thus, $f(x) = .04e^{-.04x}$.

(a) We must find t such that P(x ≤ t) = .90.

$$\int_0^t .04e^{-.04x} \, dx = .90$$

$$-e^{-.04x} \Big|_0^t = .90$$

$$-e^{-.04t} + 1 = .90$$

$$.10 = -e^{-.04t}$$

$$-.04t = \ln .10$$

$$t = \frac{\ln .10}{-.04}$$

$$t \approx 57.56$$

The longest time within which the predator will be 90% certain of finding a prey is approximately 58 min.

(b) P(x ≥ 60)

$$= \int_{60}^{\infty} .04e^{-.04x} \, dx$$

$$= \lim_{b \to \infty} \int_{60}^b .04e^{-.04x} \, dx$$

$$= \lim_{b \to \infty} (-e^{-.04x}) \Big|_{60}^b$$

$$= \lim_{b \to \infty} [-e^{-.04b} + e^{-(.04)(60)}]$$

$$= 0 + e^{-2.4}$$

$$\approx .0907$$

The probability that the predator will have to spend more than one hour looking for a prey is approximately .09.

Exercises 33–37 are to be solved using a computer. The answers may vary depending on the computer and the software that is used.

33. $\displaystyle\int_0^{50} .5e^{-.5x} \, dx \approx 1.00002$

35. $\displaystyle\int_0^{50} .5x^2 e^{-.5x} \, dx = 8.000506$

37. $\displaystyle\int_{-\infty}^{\infty} \frac{1}{\sqrt{2\pi}} e^{-x^2/2} \, dx$

(a) $\mu = 2.7416 \times 10^8 \approx 0$

(b) $\sigma = .999433 \approx 1$

Chapter 17   Review Exercises

1. In a probability function, the y-values (or function values) represent probabilities.

3. A probability density function f for [a, b] must satisfy the following two conditions:

    (1) $\displaystyle\int_a^b f(x)\ dx = 1$

    (2) $f(x) \geq 0$ for all x in the interval [a, b].

5. $f(x) = \sqrt{x} \geq 0$; [4, 9]

    $$\int_4^9 x^{1/2}\ dx = \frac{2}{3}x^{3/2}\ \Big|_4^9$$

    $$= \frac{2}{3}(27 - 8)$$

    $$= \frac{38}{3} \neq 1$$

    f(x) is not a probability density function.

7. $f(x) = e^{-x}$; [0, ∞]

    $$\int_0^\infty e^{-x}\ dx = \lim_{b\to\infty} \int_0^b e^{-x}\ dx$$

    $$= \lim_{b\to\infty} -e^{-x}\ \Big|_0^b$$

    $$= \lim_{b\to\infty} \left(1 - \frac{1}{e^b}\right) = 1$$

    f(x) > 0 for all x.
    f(x) is a probability density function.

9. $f(x) = kx^2$; [0, 3]

    $$\int_0^3 kx^2\ dx = \frac{kx^3}{3}\ \Big|_0^3$$

    $$= 9k$$

    Since f(x) is a probability density function,

    $$9k = 1$$

    $$k = \frac{1}{9}.$$

11. $f(x) = \frac{1}{10}$ for [10, 20]

     (a) $P(x \leq 12)$

     $$= \int_{10}^{12} \frac{1}{10}\ dx$$

     $$= \frac{x}{10}\Big|_{10}^{12}$$

     $$= \frac{1}{5} = .2$$

     (b) $P\left(x \geq \frac{31}{2}\right)$

     $$= \int_{31/2}^{20} \frac{1}{10}\ dx$$

     $$= \frac{x}{20}\Big|_{31/2}^{20}$$

     $$= \frac{9}{20} = .45$$

     (c) $P(10.8 \leq x \leq 16.2)$

     $$= \int_{10.8}^{16.2} \frac{1}{10}\ dx$$

     $$= \frac{x}{10}\Big|_{10.8}^{16.2} = .54$$

13. The distribution that is tallest or most peaked has the smallest standard deviation. This is the distribution pictured in graph (b).

15. $f(x) = \frac{2}{9}(x - 2); [2, 5]$

$$\mu = \int_2^5 \frac{2x}{9}(x - 2)\,dx$$

$$= \int_2^5 \frac{2}{9}(x^2 - 2x)\,dx$$

$$= \frac{2}{9}\left(\frac{x^3}{3} - x^2\right)\Big|_2^5$$

$$= \frac{2}{9}\left(\frac{125}{3} - 25 - \frac{8}{3} + 4\right) = 4$$

$$\text{Var}(x) = \int_2^5 \frac{2x^2}{9}(x - 2)\,dx - (4)^2$$

$$= \int_2^5 \frac{2}{9}(x^3 - 2x^2)\,dx - 16$$

$$= \frac{2}{9}\left(\frac{x^4}{4} - \frac{2x^3}{3}\right)\Big|_2^5 - 16$$

$$= \frac{2}{9}\left(\frac{625}{4} - \frac{250}{3} - 4 + \frac{8}{3}\right) - 16$$

$$= .5$$

$$\sigma = \sqrt{.5} \approx .71$$

17. $f(x) = 5x^{-6}; [1, \infty)$

$$\mu = \int_1^\infty x \cdot 5x^{-6}\,dx$$

$$= \int_1^\infty 5x^{-5}\,dx$$

$$= \lim_{b \to \infty} \int_1^b 5x^{-5}\,dx$$

$$= \lim_{b \to \infty} \frac{5x^{-4}}{-4}\Big|_1^b$$

$$= \lim_{b \to \infty} \frac{5}{4}\left(1 - \frac{1}{b^4}\right)$$

$$= \frac{5}{4}$$

$$\text{Var}(x) = \int_1^\infty x^2 \cdot 5x^{-6}\,dx - \left(\frac{5}{4}\right)^2$$

$$= \lim_{b \to \infty} \int_1^b 5x^{-4}\,dx - \frac{25}{16}$$

$$= \lim_{b \to \infty} \frac{5x^{-3}}{-3}\Big|_1^b - \frac{25}{16}$$

$$= \lim_{b \to \infty} \frac{5}{3}\left(1 - \frac{1}{b^3}\right) - \frac{25}{16}$$

$$= \frac{5}{3} - \frac{25}{16} = \frac{5}{48} \approx .10$$

$$\sigma \approx \sqrt{\text{Var}(x)} \approx .32$$

19. $f(x) = 4x - 3x^2; [0, 1]$

$$\mu = \int_0^1 x(4x - 3x^2)\,dx$$

$$= \int_0^1 (4x^2 - 3x^3)\,dx$$

$$= \left(\frac{4x^3}{3} - \frac{3x^4}{4}\right)\Big|_0^1$$

$$= \frac{4}{3} - \frac{3}{4} = \frac{7}{12} \approx .58$$

Find m such that

$$\int_0^m (4x - 3x^2)\,dx = \frac{1}{2}.$$

$$\int_0^m (4x - 3x^2)\,dx = (2x^2 - x^3)\Big|_0^m$$

$$= 2m^2 - m^3 = \frac{1}{2}$$

Therefore,

$$2m^3 - 4m^2 + 1 = 0.$$

This equation has no rational roots, but trial and error used with synthetic division reveals that $m \approx -.44$, .60, and 1.87. The only one of these in [0, 1] is .60.

$$P\left(\frac{7}{12} < x < .60\right)$$

$$= \int_{7/12}^{.60} (4x - 3x^2)\,dx$$

$$= 2x^2 - x^3 \Big|_{7/12}^{.60}$$

$$= 2(.60)^2 - (.60)^3 - 2\left(\frac{7}{12}\right)^2$$

$$+ \left(\frac{7}{12}\right)^3 \approx .02$$

21. $f(x) = .01e^{-.01x}$ for $[0, \infty)$ is an exponential distribution.

(a) $\mu = \dfrac{1}{.01} = 100$

(b) $\sigma = \dfrac{1}{.01} = 100$

(c) $P(100 - 100 < x < 100 + 100)$

$$= P(0 < x < 200)$$

$$= \int_0^{200} .01e^{-.01x}\,dx$$

$$= -e^{-.01x} \Big|_0^{200}$$

$$= 1 - e^{-2} \approx .86$$

For Exercises 23–29, use the table in the Appendix for area under the normal curve.

23. Area to the left of $z = -.49$ is .3121.
Percent of area is 31.21%.

25. Area between $z = -.98$ and $z = -.15$ is

$$.4404 - .1635 = .2769.$$

Percent of area is 27.69%.

27. Region up to 1.2 standard deviations below the mean is region to the left of $z = -1.2$. Area is .1151 so percent of area is 11.51%.

29. 52% of area is to the right implies that 48% is to the left.

$$P(z < a) = .48 \text{ for } a = -.05$$

Thus, 52% of the area lies to the right of $z = -.05$.

31. $f(t) = \frac{5}{112}(1 - t^{-3/2})$; $[, 25]$

P(No repairs in years 1–3)

= P(First repair needed in
   years 4–25)

$= \int_{4}^{25} \frac{5}{112}(1 - t^{-3/2})dt$

$= \frac{5}{112}(t + 2t^{-1/2})\Big|_{4}^{25}$

$= \frac{5}{112}[25 + \frac{2}{5} - 4 - 1]$

$= \frac{51}{56} \approx .911$

33. $f(x) = \frac{1}{6}e^{-x/6}$ for $[0, \infty)$ is an

exponential distribution.

(a) $\mu = \frac{1}{1/6} = 6$

(b) $\sigma = \frac{1}{1/6} = 6$

(c) $P(x > 6) = \int_{6}^{\infty} \frac{1}{6}e^{-x/6} \, dx$

$= \lim_{b \to \infty} \int_{6}^{b} \frac{1}{6}e^{-x/6} \, dx$

$= \lim_{b \to \infty} -e^{-x/6} \Big|_{6}^{b}$

$= \lim_{b \to \infty} (e^{-1} - \frac{1}{e^{b/6}})$

$= \frac{1}{e} \approx .37$

35. $f(x) = .01e^{-.01x}$ for $[0, \infty)$ is an
exponential distribution.

$P(0 \le x \le 100)$

$= \int_{0}^{100} .01e^{-.01x} \, dx$

$= -e^{-.01x} \Big|_{0}^{100}$

$= 1 - \frac{1}{e} \approx .63$

37. $f(x) = \frac{6}{15,925}(x^2 + x)$ for $[20, 25]$

(a)

$\mu = \int_{20}^{25} \frac{6}{15,925}(x^2 + x)dx$

$= \frac{6}{15,925}\int_{20}^{25} (x^3 + x^2)dx$

$= \frac{6}{15,925}[\frac{x^4}{4} + \frac{x^3}{3}]\Big|_{20}^{25}$

$= \frac{6}{15,925}$

$\cdot [\frac{(25)^4}{4} + \frac{(25)^3}{3} - \frac{(20)^4}{4} - \frac{(20)^3}{3}]$

$\approx 22.68°C$

(b)

$P(x < \mu)$

$= \int_{20}^{22.68} \frac{6}{15,925}(x^2 + x)dx$

$= \frac{6}{15,925}[\frac{x^3}{3} + \frac{x^2}{2}]\Big|_{20}^{22.68}$

$= \frac{6}{15,925}$

$\cdot [\frac{(22.68)^3}{3} + \frac{(22.68)^2}{2} - \frac{(20)^3}{3} - \frac{(20)^2}{2}]$

$\approx .48$

39. Normal distribution, $\mu = 2.4$ g,

$\sigma = .4$ g, x = tension

$$P(x < 1.9) = P\left(\frac{x - 2.4}{.4} < \frac{1.9 - 2.4}{.4}\right)$$

$$= P(z < -1.25)$$

$$= .1056$$

41. Normal distribution, $\mu = 40$,

$\sigma = 13$, x = "take"

$$P(x > 50) = P\left(\frac{x - 40}{13} > \frac{50 - 40}{13}\right)$$

$$= P(z > .77)$$

$$= 1 - P(z \le .77)$$

$$= 1 - .7794$$

$$= .2206$$

43. $f(x) = e^{-x}$ for $[0, \infty)$

$f(x) = 1e^{-1x}$

(a) This is an exponential distribution with a = 1.

(b) The domain of f is $[0, \infty)$.

The range of f is $(0, 1]$.

(c) See the graph in the answer section of your textbook.

(d) For an exponential distribution, $\mu = \frac{1}{a}$ and $\sigma = \frac{1}{a}$.

Thus,

$$\mu = \frac{1}{1} = 1 \quad \text{and} \quad \sigma = \frac{1}{1} = 1.$$

(e) $P(\mu - \sigma \le x \le \mu + \sigma) = P(1 - 1 \le x \le 1 + 1)$

$$= P(0 \le x \le 2)$$

$$= \int_0^2 e^{-x}dx$$

$$= -e^{-x}\Big|_0^2$$

$$= -e^{-2} + 1$$

$$\approx .86$$

Extended Application

1. Expected profit

$$= (S - C_p)\int_a^T Df(D)\,dD + (S - C_p)(T)\int_T^b f(D)\,dD - C_C \int_a^T (T - D)f(D)\,dD - C_A A$$

$$= (10,000 - 5000)\int_{500}^T D\frac{1}{1000}\,dD + (10,000 - 5000)T\int_T^{1500} \frac{1}{1000}\,dD$$

$$- 3000\int_{500}^T (T - D)\frac{1}{1000}\,dD - 100A$$

$$= 5 \cdot \frac{D^2}{2}\bigg|_{500}^T + 5T \cdot D \bigg|_T^{1500} - 3\left(TD - \frac{D^2}{2}\right)\bigg|_{500}^T - 100A$$

$$= 2.5T^2 - 625,000 + 7500T - 5T^2 - 3\left(T^2 - \frac{T^2}{2}\right) + 3(500T - 125,000) - 100A$$

$$= 2.5T^2 - 625,000 + 7500T - 5T^2 - 1.5T^2 + 1500T - 375,000 - 100A$$

Expected profit $= -4.0T^2 + 9000T - 1,000,000 - 100A$

Since $T = .1A + 200$,

Expected profit $= -4.0(.1A + 200)^2 + 9000(.1A + 200) - 1,000,000 - 100A$

$$= -4(.01A^2 + 40A + 40,000) + 900A + 1,800,000 - 1,000,000 - 100A$$

$$= -.04A^2 - 160A - 160,000 + 900A + 1,800,000 - 1,000,000 - 100A$$

$$= -.04A^2 + 640A + 640,000.$$

2. Let $P(A) = -.04A^2 + 640A + 640,000$. We want to maximize this profit function.

$$P'(A) = -.08A + 640$$

If $-.08A + 640 = 0$, then $A = 8000$.

The graph of this function is a parabola opening downward. It has a maximum value when $A = 8000$. Thus, planting 8000 acres will maximize profit.

CHAPTER 17   TEST

In Exercises 1–4, which of the functions are probability density functions on the given intervals? If a function is not a probability density function, tell why.

1. $f(x) = 4$; $[3, 7]$

2. $f(x) = \frac{1}{5}$; $[5, 10]$

3. $f(x) = \frac{1}{2}(x - 4)$; $[4, 6]$

4. $f(x) = e^{-3x}$; $[0, \infty)$

5. Find a value of k that will make $f(x) = kx^{1/3}$ on the interval $[1, 8]$ a probability density function.

6. The probability density function for a random variable x is defined by

$$f(x) = .2 \quad \text{for } [12, 17].$$

Find the following probabilities.

(a) $P(x \leq 14)$ (b) $P(13 < x < 16)$ (c) $P(x < 15)$

7. The probability density function of a random variable x is defined by

$$f(x) = 3x^2 \quad \text{for } [0, 1].$$

Find the following probabilities.

(a) $P(x \leq 1)$ (b) $P(.2 \leq x \leq .7)$ (c) $P(x \geq .3)$

8. The time in years until a particular radioactive particle decays is a random variable with probability density function

$$f(t) = .04e^{-.04t} \quad \text{for } [0, \infty).$$

Find the probability that a certain such particle decays in less than 60 yr.

Find the expected value and the standard deviation for each probability density function defined as follows.

9. $f(x) = \frac{1}{4}$; $[1, 5]$

10. $f(x) = 2x$; $[0, 1]$

11. $f(x) = \frac{3}{2}(x - 1)^2$; $[0, 2]$

12. $f(x) = \frac{1}{4\sqrt{x}}$; $[1, 9]$